Practice Problems *for the*

Chemical Engineering PE Exam

A Companion to the
Chemical Engineering Reference Manual

Sixth Edition

Michael R. Lindeburg, PE

The Power to Pass
www.ppi2pass.com

Professional Publications, Inc. • Belmont, California

PRACTICE PROBLEMS FOR THE CHEMICAL ENGINEERING PE EXAM
Sixth Edition

Current printing of this edition: 2

Printing History

edition number	printing number	update
5	4	Minor corrections. Copyright update.
6	1	Title change. New edition. Copyright update.
6	2	Minor corrections.

Printed in the United States of America

PPI
1250 Fifth Avenue, Belmont, CA 94002
(650) 593-9119
www.ppi2pass.com

ISBN 978-1-59126-008-0

Topics

Mathematical Support

Law and Ethics

Fluids

Thermodynamics

Heat Transfer

Environmental

Mass Transfer

Kinetics

Plant Design

Where do I find help
solving these Practice Problems?

Practice Problems for the Chemical Engineering PE Exam presents complete, step-by-step solutions for 450 problems like those found on the Chemical PE exam. You can find all the background information, including charts and tables of data, that you need to solve these problems in the *Chemical Engineering Reference Manual*.

The *Chemical Engineering Reference Manual* may be purchased from PPI (800-426-1178 or www.ppi2pass.com) or from your favorite bookstore.

Table of Contents

Preface and Acknowledgments

Putting out a practice problems manual for one of my reference manuals has always been a major undertaking. For one thing, it's a lot of work. For another, "good enough" just won't do. Over the years, I have learned the amount of detail that will help you learn from a solution, as opposed to just presenting the numbers. (I have also learned that if I leave out certain steps or items, then I get a lot of inquisitive letters, phone calls, and email!)

The style standards set by PPI's Production Department are as strict as my content standards. There is a proper way to edit, typeset, illustrate, and proofread a book. The Production Department won't do it any other way.

Most textbook authors see their problem sets or solutions manuals as a necessary evil: something required by their contract, something to bang out as quickly as possible, an afterthought stuck between two editions. The finished product seems to say, "Here are the numbers and the answer. Sure, we've been a little sloppy with units, and maybe we've omitted a few steps. But a little struggling is good for you. You can figure it all out. Somehow."

But not PPI. All that struggling with vague content and sloppy production wastes your time. And before an exam, time is one thing that an examinee doesn't have much of. There was no way we were going to cut corners on this book.

PPI's Production Department took on this project while working on at least a zillion other books, labored valiantly without complaint, and managed to pull off a winner. My admiration and gratitude go to John Boykin who proofread and edited, to Kate Hayes who typeset, to Yvonne Sartain who illustrated this book, and to Cathy Schrott, Production Department Manager, who oversaw the printing and kept me on my toes.

For me, it's a little scary to retreat from the security of solutions whose kinks had already been ironed out in previous editions and now pass into the critical, hot spotlight of a new edition. You would think that after all these years of writing problems and solutions I would know virtually all of the ways you and I make mistakes. But even knowing the ways mistakes are made doesn't mean we can avoid them all.

If you think you've found something questionable in a solution, or if you think there is a better way to solve a problem, please use our Errata Report Form at **www.ppi2pass.com/errata** to let us know about it. I like to learn new things, too.

Michael R. Lindeburg, PE

How to Use This Book

This book is a companion to *Chemical Engineering Reference Manual*. Since it is a practice problems book, there are a few, but not many, ways to use it.

Since most of the problems in the book can be solved in either customary U.S. or SI units, your first decision will be which set of units you will work in. Don't get me wrong: The exam doesn't give you such a choice. Exam problems are either in customary U.S. or SI units, not both. So, you have to be proficient with both. I recommend that you solve half of the problems in customary U.S. units and half in SI. Then, if you have time, go back to solve all of the problems a second time, using the alternate units.

The big decision you have to make is whether you really work the practice problems or not. Some people think they can read a problem statement, think about it for about ten seconds, read the solution, and then say "Yes, that's what I was thinking of, and that's what I would have done." Sadly, these people find out too late that the human brain doesn't learn very efficiently that way. Under pressure, they find they know and remember little. For real learning, you have to spend some time with a stubby pencil.

There are so many places where you can get messed up solving a problem. Maybe it's in the use of your calculator, like pushing log instead of ln, or forgetting to set the angle to radians instead of degrees, and so on. Maybe it's rusty math. What is $\ln(e^x)$, anyway?

How do you factor a polynomial? Maybe it's in finding the data needed or the proper unit conversion. Maybe it's just trying to find out if that funky code equation expects L to be in feet or inches. These things take time. And you have to make the mistakes once so that you don't make them again.

If you do decide to get your hands dirty and actually work these problems, you'll have to decide how much reliance to place on the published solutions while solving the problems. It's tempting to turn to a solution when you get slowed down by details or stumped by the subject material. You'll probably want to maximize the number of problems you solve by spending as little time as possible with each problem. I want you to struggle a little bit more than that.

Studying a new subject is analogous to using a machete to cut a path through a dense jungle. By doing the work, you develop pathways that weren't there before. It's a lot different than just looking at the route on a map. You actually get nowhere by looking at a map. But cut that path once, and you're in business until the jungle overgrowth closes in again.

So, do the problems. All of them. Do them in both sets of units. Don't look at the answers until you've sweated a little. And, let's not have any whining. Please.

1 Systems of Units

PRACTICE PROBLEMS

1. Convert 250°F to degrees Celsius.
- (A) 115°C
- (B) 121°C
- (C) 124°C
- (D) 420°C

2. Convert the Stefan-Boltzmann constant (0.1713×10^{-8} Btu/ft^2-hr-°R^4) from English to SI units.
- (A) 5.14×10^{-10} W/m^2·K^4
- (B) 0.95×10^{-8} W/m^2·K^4
- (C) 5.67×10^{-8} W/m^2·K^4
- (D) 7.33×10^{-6} W/m^2·K^4

3. How many U.S. tons (2000 lbm per ton) of coal with a heating value of 13,000 Btu/lbm must be burned to provide as much energy as a complete nuclear conversion of 1 g of its mass? (Hint: Use Einstein's equation: $E = mc^2$.)
- (A) 1.7 tons
- (B) 14 tons
- (C) 779 tons
- (D) 3300 tons

SOLUTIONS

1. The conversion to degrees Celsius is given by

$$°C = \left(\tfrac{5}{9}\right)(°F - 32°F)$$
$$= \left(\tfrac{5}{9}\right)(250°F - 32°F)$$
$$= \left(\tfrac{5}{9}\right)(218°F)$$
$$= \boxed{121.1°C \ (121°C)}$$

The answer is (B).

2. In U.S. customary units, the Stefan-Boltzmann constant, σ, is 0.1713×10^{-8} Btu/hr-ft^2-°R^4.

Use the following conversion factors.

$$1 \text{ Btu/hr} = 0.2931 \text{ W}$$
$$1 \text{ ft} = 0.3048 \text{ m}$$
$$T_{°R} = \frac{9}{5}T_K$$

Performing the conversion gives

$$\sigma = \left(0.1713 \times 10^{-8}\frac{\text{Btu}}{\text{hr-ft}^2\text{-°R}^4}\right)\left(0.2931\frac{\text{W}}{\frac{\text{Btu}}{\text{hr}}}\right)$$
$$\times \left(\frac{1 \text{ ft}}{0.3048 \text{ m}}\right)^2\left(\frac{1°R}{\frac{5}{9}K}\right)^4$$
$$= \boxed{5.67 \times 10^{-8} \text{ W/m}^2\text{·K}^4}$$

The answer is (C).

3. The energy produced from the nuclear conversion of any quantity of mass is given as

$$E = mc^2$$

The speed of light, c, is 3×10^8 m/s.

For a mass of 1 g (0.001 kg),

$$E = mc^2$$
$$= (0.001 \text{ kg})\left(3 \times 10^8 \frac{\text{m}}{\text{s}}\right)^2$$
$$= 9 \times 10^{13} \text{ J}$$

Convert to U.S. customary units with the conversion 1 Btu = 1055 J.

$$E = (9 \times 10^{13} \text{ J}) \left(\frac{1 \text{ Btu}}{1055 \text{ J}} \right)$$

$$= 8.53 \times 10^{10} \text{ Btu}$$

The number of tons of 13,000 Btu/lbm coal is

$$\frac{8.53 \times 10^{10} \text{ Btu}}{\left(13,000 \, \frac{\text{Btu}}{\text{lbm}} \right) \left(2000 \, \frac{\text{lbm}}{\text{ton}} \right)} = \boxed{3281 \text{ tons } (3300 \text{ tons})}$$

The answer is (D).

2 Engineering Drawing Practice

PRACTICE PROBLEMS

1. Two views of an object are shown. Prepare a free-hand drawing of the missing third view.

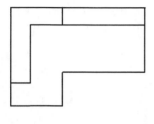

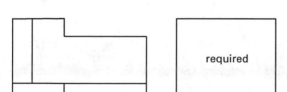

required

2. Two views of an object are shown. Prepare a free-hand drawing of the missing third view.

required

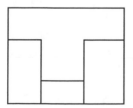

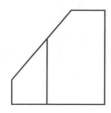

3. Two views of an object are shown. Prepare a free-hand drawing of the missing third view.

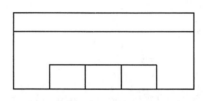

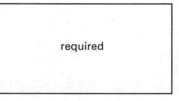

required

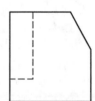

4. A pictorial sketch of an object is shown. Prepare a freehand three-view orthographic drawing set.

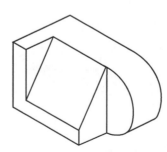

5. Plan and elevation views of an object are shown.

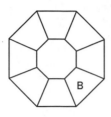

plan view

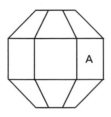

elevation view

(a) Prepare a freehand drawing of the view that shows surface A in true shape.

(b) Referring to part (a), what is this view of surface A called?

(c) Prepare a freehand drawing of the view that shows surface B in true shape.

(d) Referring to part (c), what is the view of surface B called?

6. Prepare a freehand isometric drawing of the object shown. Hint: Lay off horizontal and vertical (i.e., isometric) lines along the isometric axes shown.

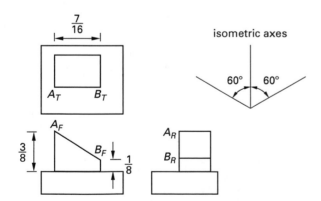

7. Two lines in an oblique position are shown. Prepare a freehand right auxiliary normal view.

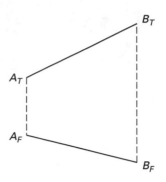

8. Two lines in an oblique position are shown. Prepare a freehand read auxiliary normal view.

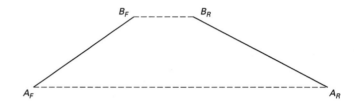

9. A parallel-scale nomograph has been prepared to solve the equation $D = 1.075\sqrt{WH}$. Use the nomograph to estimate the value of D when $H = 10$ and $W = 40$.

$$D = 1.075\sqrt{WH}$$

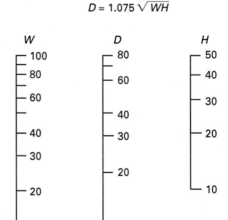

Mathematical Support

SOLUTIONS

1.

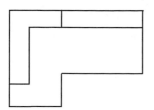

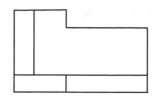

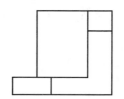

2.

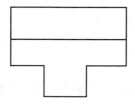

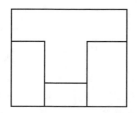

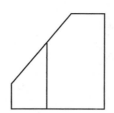

3.

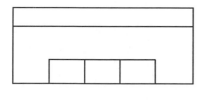

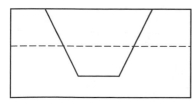

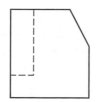

4.

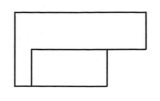

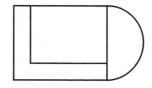

5. (a)

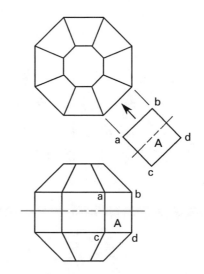

(b) An ┃elevation auxiliary┃ view shows surface A in true shape.

(c)

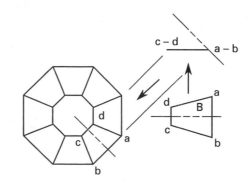

(d) An $\boxed{\text{oblique}}$ view shows surface B in true shape.

6.

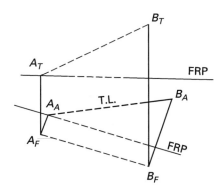

isometric axes

7.

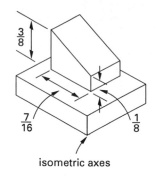

8.

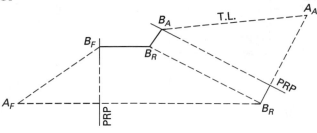

9. A straight line through the points $H = 10$ and $W = 40$ intersects the D-scale at approximately $\boxed{21.5.}$

3 Algebra

PRACTICE PROBLEMS

Series

1. Calculate the following sum.

$$\sum_{j=1}^{5}[(j+1)^2 - 1]$$

(A) 15
(B) 24
(C) 35
(D) 85

Logarithms

2. If every 0.1 sec a quantity increases by 0.1% of its current value, calculate the doubling time.

(A) 14 sec
(B) 70 sec
(C) 690 sec
(D) 69,000 sec

SOLUTIONS

1. Let $S_n = (j+1)^2 - 1$.

For $j = 1$,
$$S_1 = (1+1)^2 - 1 = 3$$

For $j = 2$,
$$S_2 = (2+1)^2 - 1 = 8$$

For $j = 3$,
$$S_3 = (3+1)^2 - 1 = 15$$

For $j = 4$,
$$S_4 = (4+1)^2 - 1 = 24$$

For $j = 5$,
$$S_5 = (5+1)^2 - 1 = 35$$

Substituting the above expressions gives

$$\sum_{j=1}^{5}\left((j+1)^2 - 1\right) = \sum_{j=1}^{5} S_j$$
$$= S_1 + S_2 + S_3 + S_4 + S_5$$
$$= 3 + 8 + 15 + 24 + 35$$
$$= \boxed{85}$$

The answer is (D).

2. Let n represent the number of elapsed periods of 0.1 sec, and let y_n represent the amount present after n periods.

y_0 represents the initial quantity.

$$y_1 = 1.001y_0$$
$$y_2 = 1.001y_1 = (1.001)\left((1.001)y_0\right) = (1.001)^2 y_0$$

Therefore, by deduction,

$$y_n = (1.001)^n y_0$$

The expression for a doubling of the original quantity is

$$2y_0 = y_n$$

Substitute for y_n.

$$2y_0 = (1.001)^n y_0$$
$$2 = (1.001)^n$$

Take the logarithm of both sides.

$$\log(2) = \log(1.001)^n$$
$$= n \log(1.001)$$

Solve for n.

$$n = \frac{\log(2)}{\log(1.001)} = 693.5$$

Since each period is 0.1 sec, the time is given by

$$t = n(0.1 \text{ sec})$$
$$= (693.5)(0.1 \text{ sec}) = \boxed{69.35 \text{ sec}}$$

The answer is (B).

4 Linear Algebra

PRACTICE PROBLEMS

Determinants

1. What is the determinant of matrix $\mathbf{A}$?

$$\mathbf{A} = \begin{bmatrix} 8 & 2 & 0 & 0 \\ 2 & 8 & 2 & 0 \\ 0 & 2 & 8 & 2 \\ 0 & 0 & 2 & 4 \end{bmatrix}$$

   (A) 459
   (B) 832
   (C) 1552
   (D) 1776

Simultaneous Linear Equations

2. Use Cramer's rule to solve for the values of x, y, and z that simultaneously satisfy the following equations.

$$x + y = -4$$
$$x + z - 1 = 0$$
$$2z - y + 3x = 4$$

   (A) $(x, y, z) = (3, 2, 1)$
   (B) $(x, y, z) = (-3, -1, 2)$
   (C) $(x, y, z) = (3, -1, -3)$
   (D) $(x, y, z) = (-1, -3, 2)$

SOLUTIONS

1. Expand by cofactors of the first row since there are two zeros in that row.

$$D = 8 \begin{vmatrix} 8 & 2 & 0 \\ 2 & 8 & 2 \\ 0 & 2 & 4 \end{vmatrix} - 2 \begin{vmatrix} 2 & 0 & 0 \\ 2 & 8 & 2 \\ 0 & 2 & 4 \end{vmatrix} + 0 - 0$$

by first row:

$$\begin{vmatrix} 8 & 2 & 0 \\ 2 & 8 & 2 \\ 0 & 2 & 4 \end{vmatrix} = (8)[(8)(4) - (2)(2)] - (2)[(2)(4) - (2)(0)]$$
$$= (8)(28) - (2)(8) = 208$$

by first row:

$$\begin{vmatrix} 2 & 0 & 0 \\ 2 & 8 & 2 \\ 0 & 2 & 4 \end{vmatrix} = (2)[(8)(4) - (2)(2)]$$
$$= 56$$

$$D = (8)(208) - (2)(56) = \boxed{1552}$$

The answer is (C).

2. Rearrange the equations.

$$\begin{array}{rrrcr} x & + \; y & & = & -4 \\ x & & + \; z & = & 1 \\ 3x & - \; y & + \; 2z & = & 4 \end{array}$$

Write the set of equations in matrix form: $\mathbf{AX} = \mathbf{B}$.

$$\begin{bmatrix} 1 & 1 & 0 \\ 1 & 0 & 1 \\ 3 & -1 & 2 \end{bmatrix} \begin{bmatrix} x \\ y \\ z \end{bmatrix} = \begin{bmatrix} -4 \\ 1 \\ 4 \end{bmatrix}$$

Find the determinant of the matrix $\mathbf{A}$.

$$|\mathbf{A}| = \begin{vmatrix} 1 & 1 & 0 \\ 1 & 0 & 1 \\ 3 & -1 & 2 \end{vmatrix}$$

$$= 1 \begin{vmatrix} 0 & 1 \\ -1 & 2 \end{vmatrix} - 1 \begin{vmatrix} 1 & 0 \\ -1 & 2 \end{vmatrix} + 3 \begin{vmatrix} 1 & 0 \\ 0 & 1 \end{vmatrix}$$

$$= (1)\big((0)(2) - (1)(-1)\big)$$
$$\quad - (1)\big((1)(2) - (-1)(0)\big)$$
$$\quad + (3)\big((1)(1) - (0)(0)\big)$$

$$= (1)(1) - (1)(2) + (3)(1)$$
$$= 1 - 2 + 3$$
$$= 2$$

Find the determinant of the substitutional matrix $\mathbf{A_1}$.

$$|\mathbf{A_1}| = \begin{vmatrix} -4 & 1 & 0 \\ 1 & 0 & 1 \\ 4 & -1 & 2 \end{vmatrix}$$

$$= -4 \begin{vmatrix} 0 & 1 \\ -1 & 2 \end{vmatrix} - 1 \begin{vmatrix} 1 & 0 \\ -1 & 2 \end{vmatrix} + 4 \begin{vmatrix} 1 & 0 \\ 0 & 1 \end{vmatrix}$$

$$= (-4)\big((0)(2) - (1)(-1)\big)$$
$$\quad - (1)\big((1)(2) - (-1)(0)\big)$$
$$\quad + (4)\big((1)(1) - (0)(0)\big)$$
$$= (-4)(1) - (1)(2) + (4)(1)$$
$$= -4 - 2 + 4$$
$$= -2$$

Find the determinant of the substitutional matrix $\mathbf{A_2}$.

$$|\mathbf{A_2}| = \begin{vmatrix} 1 & -4 & 0 \\ 1 & 1 & 1 \\ 3 & 4 & 2 \end{vmatrix}$$

$$= 1 \begin{vmatrix} 1 & 1 \\ 4 & 2 \end{vmatrix} - 1 \begin{vmatrix} -4 & 0 \\ 4 & 2 \end{vmatrix} + 3 \begin{vmatrix} -4 & 0 \\ 1 & 1 \end{vmatrix}$$

$$= (1)\big((1)(2) - (4)(1)\big)$$
$$\quad - (1)\big((-4)(2) - (4)(0)\big)$$
$$\quad + (3)\big((-4)(1) - (1)(0)\big)$$
$$= (1)(-2) - (1)(-8) + (3)(-4)$$
$$= -2 + 8 - 12$$
$$= -6$$

Find the determinant of the substitutional matrix $\mathbf{A_3}$.

$$|\mathbf{A_3}| = \begin{vmatrix} 1 & 1 & -4 \\ 1 & 0 & 1 \\ 3 & -1 & 4 \end{vmatrix}$$

$$= 1 \begin{vmatrix} 0 & 1 \\ -1 & 4 \end{vmatrix} - 1 \begin{vmatrix} 1 & -4 \\ -1 & 4 \end{vmatrix} + 3 \begin{vmatrix} 1 & -4 \\ 0 & 1 \end{vmatrix}$$

$$= (1)\big((0)(4) - (-1)(1)\big)$$
$$\quad - (1)\big((1)(4) - (-1)(-4)\big)$$
$$\quad + (3)\big((1)(1) - (0)(-4)\big)$$
$$= (1)(1) - (1)(0) + (3)(1)$$
$$= 1 - 0 + 3$$
$$= 4$$

Use Cramer's rule.

$$x = \frac{|\mathbf{A_1}|}{|\mathbf{A}|} = \frac{-2}{2} = \boxed{-1}$$

$$y = \frac{|\mathbf{A_2}|}{|\mathbf{A}|} = \frac{-6}{2} = \boxed{-3}$$

$$z = \frac{|\mathbf{A_3}|}{|\mathbf{A}|} = \frac{4}{2} = \boxed{2}$$

The answer is (D).

5 Vectors

PRACTICE PROBLEMS

Dot Products

1. Calculate the dot products for the following vector pairs.

(a) $\mathbf{V}_1 = 2\mathbf{i} + 3\mathbf{j}; \mathbf{V}_2 = 5\mathbf{i} - 2\mathbf{j}$

(b) $\mathbf{V}_1 = 1\mathbf{i} + 4\mathbf{j}; \mathbf{V}_2 = 9\mathbf{i} - 3\mathbf{j}$

(c) $\mathbf{V}_1 = 7\mathbf{i} - 3\mathbf{j}; \mathbf{V}_2 = 3\mathbf{i} + 4\mathbf{j}$

(d) $\mathbf{V}_1 = 2\mathbf{i} - 3\mathbf{j} + 6\mathbf{k}; \mathbf{V}_2 = 8\mathbf{i} + 2\mathbf{j} - 3\mathbf{k}$

(e) $\mathbf{V}_1 = 6\mathbf{i} + 2\mathbf{j} + 3\mathbf{k}; \mathbf{V}_2 = \mathbf{i} + \mathbf{k}$

2. What is the angle between the vectors in Probs. 1(a), 1(b), and 1(c)?

Cross Products

3. Calculate the cross products for each of the five vector pairs in Prob. 1.

SOLUTIONS

1. (a) $\mathbf{V}_1 \cdot \mathbf{V}_2 = \mathbf{V}_{1x}\mathbf{V}_{2x} + \mathbf{V}_{1y}\mathbf{V}_{2y}$
$$= (2)(5) + (3)(-2)$$
$$= \boxed{4}$$

(b) $\quad \mathbf{V}_1 \cdot \mathbf{V}_2 = (1)(9) + (4)(-3)$
$$= \boxed{-3}$$

(c) $\quad \mathbf{V}_1 \cdot \mathbf{V}_2 = (7)(3) + (-3)(4)$
$$= \boxed{9}$$

(d) $\quad \mathbf{V}_1 \cdot \mathbf{V}_2 = \mathbf{V}_{1x}\mathbf{V}_{2x} + \mathbf{V}_{1y}\mathbf{V}_{2y} + \mathbf{V}_{1z}\mathbf{V}_{2z}$
$$= (2)(8) + (-3)(2) + (6)(-3)$$
$$= \boxed{-8}$$

(e) $\quad \mathbf{V}_1 \cdot \mathbf{V}_2 = (6)(1) + (2)(0) + (3)(1)$
$$= \boxed{9}$$

2. (a) $\cos\phi = \dfrac{\mathbf{V}_1 \cdot \mathbf{V}_2}{|\mathbf{V}_1||\mathbf{V}_2|}$
$$= \frac{4}{\left(\sqrt{2^2 + 3^2}\right)\left(\sqrt{5^2 + 2^2}\right)} = 0.206$$
$$\phi = \cos^{-1}(0.206) = \boxed{78.1°}$$

(b) $\quad \cos\phi = \dfrac{\mathbf{V}_1 \cdot \mathbf{V}_2}{|\mathbf{V}_1||\mathbf{V}_2|}$
$$= \frac{-3}{\left(\sqrt{1^2 + 4^2}\right)\left(\sqrt{9^2 + (-3)^2}\right)}$$
$$= -0.077$$
$$\phi = \boxed{94.4°}$$

(c) $\quad \cos\phi = \dfrac{\mathbf{V}_1 \cdot \mathbf{V}_2}{|\mathbf{V}_1||\mathbf{V}_2|}$
$$= \frac{9}{(\sqrt{7^2 + (-3)^2})(\sqrt{3^2 + 4^2})} = 0.236$$
$$\phi = \boxed{76.3°}$$

3. (a) $\quad \mathbf{V}_1 \times \mathbf{V}_2 = \begin{vmatrix} \mathbf{i} & \mathbf{V}_{1x} & \mathbf{V}_{2x} \\ \mathbf{j} & \mathbf{V}_{1y} & \mathbf{V}_{2y} \\ \mathbf{k} & \mathbf{V}_{1z} & \mathbf{V}_{2z} \end{vmatrix}$

$\quad\quad\quad = \begin{vmatrix} \mathbf{i} & 2 & 5 \\ \mathbf{j} & 3 & -2 \\ \mathbf{k} & 0 & 0 \end{vmatrix}$

Expand by the third row.

$$= \mathbf{k} \begin{vmatrix} 2 & 5 \\ 3 & -2 \end{vmatrix} = \boxed{-19\mathbf{k}}$$

(b) $\quad \mathbf{V}_1 \times \mathbf{V}_2 = \begin{vmatrix} \mathbf{i} & 1 & 9 \\ \mathbf{j} & 4 & -3 \\ \mathbf{k} & 0 & 0 \end{vmatrix}$

Expand by the third row.

$$= \mathbf{k} \begin{vmatrix} 1 & 9 \\ 4 & -3 \end{vmatrix} = \boxed{-39\mathbf{k}}$$

(c) $\quad \mathbf{V}_1 \times \mathbf{V}_2 = \begin{vmatrix} \mathbf{i} & 7 & 3 \\ \mathbf{j} & -3 & 4 \\ \mathbf{k} & 0 & 0 \end{vmatrix}$

Expand by the third row.

$$\mathbf{k} \begin{vmatrix} 7 & 3 \\ -3 & 4 \end{vmatrix} = \boxed{37\mathbf{k}}$$

(d) $\quad \mathbf{V}_1 \times \mathbf{V}_2 = \begin{vmatrix} \mathbf{i} & 2 & 8 \\ \mathbf{j} & -3 & 2 \\ \mathbf{k} & 6 & -3 \end{vmatrix}$

Expand by the first column.

$$= \mathbf{i} \begin{vmatrix} -3 & 2 \\ 6 & -3 \end{vmatrix} - \mathbf{j} \begin{vmatrix} 2 & 8 \\ 6 & -3 \end{vmatrix}$$

$$+ \mathbf{k} \begin{vmatrix} 2 & 8 \\ -3 & 2 \end{vmatrix}$$

$$= \boxed{-3\mathbf{i} + 54\mathbf{j} + 28\mathbf{k}}$$

(e) $\quad \mathbf{V}_1 \times \mathbf{V}_2 = \begin{vmatrix} \mathbf{i} & 6 & 1 \\ \mathbf{j} & 2 & 0 \\ \mathbf{k} & 3 & 1 \end{vmatrix}$

Expand by the second row.

$$= -\mathbf{j} \begin{vmatrix} 6 & 1 \\ 3 & 1 \end{vmatrix} + (2) \begin{vmatrix} \mathbf{i} & 1 \\ \mathbf{k} & 1 \end{vmatrix}$$

$$= \boxed{2\mathbf{i} - 3\mathbf{j} - 2\mathbf{k}}$$

6

Trigonometry

PRACTICE PROBLEMS

1. A 5 lbm (5 kg) block sits on a 20° incline without slipping. (a) Draw the freebody with respect to axes parallel and perpendicular to the surface of the incline. (b) Determine the magnitude of the frictional force on the block.

    (A) 1.71 lbf (16.8 N)

    (B) 3.35 lbf (32.9 N)

    (C) 4.70 lbf (46.1 N)

    (D) 5.00 lbf (49.1 N)

SOLUTIONS

1. (a)

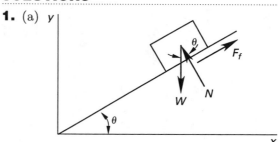

Customary U.S. Solution

(b) The mass of the block is $m = 5$ lbm.

The angle of inclination is $\theta = 20°$. The weight is

$$W = \frac{mg}{g_c}$$

$$= \frac{(5 \text{ lbm}) \left(32.2 \, \dfrac{\text{ft}}{\text{sec}^2} \right)}{32.2 \, \dfrac{\text{lbm-ft}}{\text{lbf-sec}^2}}$$

$$= 5 \text{ lbf}$$

The frictional force is

$$F_f = W \sin \theta$$

$$= (5 \text{ lbf})(\sin 20°)$$

$$= \boxed{1.71 \text{ lbf}}$$

The answer is (A).

SI Solution

(b) The mass of the block is $m = 5$ kg.

The angle of inclination is $\theta = 20°$. The gravitational force is

$$W = mg$$

$$= (5 \text{ kg}) \left(9.81 \, \dfrac{\text{m}}{\text{s}^2} \right)$$

$$= 49.1 \text{ N}$$

The frictional force is

$$F_f = W \sin \theta$$

$$= (49.1 \text{ N})(\sin 20°)$$

$$= \boxed{16.8 \text{ N}}$$

The answer is (A).

7 Analytic Geometry

Mathematical Support

PRACTICE PROBLEMS

1. The diameter of a sphere and the base of a cone are equal. What percentage of that diameter must the cone's height be so that both volumes are equal?

(A) 133%
(B) 150%
(C) 166%
(D) 200%

SOLUTIONS

1. Let d be the diameter of the sphere and the base of the cone.

The volume of the sphere is given by

$$V_{sphere} = \left(\tfrac{4}{3}\right)\pi r^3 = \left(\tfrac{4}{3}\right)\pi\left(\frac{d}{2}\right)^3$$
$$= \left(\frac{\pi}{6}\right)d^3$$

The volume of the circular cone is given by

$$V_{cone} = \left(\tfrac{1}{3}\right)\pi r^2 h = \left(\tfrac{1}{3}\right)\pi\left(\frac{d}{2}\right)^2 h$$
$$= \left(\frac{\pi}{12}\right)d^2 h$$

Since the volume of the sphere and cone are equal,

$$V_{cone} = V_{sphere}$$
$$\left(\frac{\pi}{12}\right)d^2 h = \left(\frac{\pi}{6}\right)d^3$$
$$h = 2d$$

The height of the cone must be 200% of the diameter.

The answer is (D).

8 Differential Calculus

PRACTICE PROBLEMS

1. What are the values of a, b, and c in the following expression such that $n(\infty) = 100$, $n(0) = 10$, and $dn(0)/dt = 0.5$?

$$n(t) = \frac{a}{1 + be^{ct}}$$

(A) $a = 10$, $b = 9$, $c = 1$
(B) $a = 100$, $b = 10$, $c = 1.5$
(C) $a = 100$, $b = 9$, $c = -0.056$
(D) $a = 1000$, $b = 10$, $c = 0.056$

2. Find all minima, maxima, and inflection points for

$$y = x^3 - 9x^2 - 3$$

(A) maximum at $x = 0$
 inflection at $x = 3$
 minimum at $x = 6$
(B) maximum at $x = 0$
 inflection at $x = -3$
 minimum at $x = -6$
(C) minimum at $x = 0$
 inflection at $x = 3$
 maximum at $x = 6$
(D) minimum at $x = 3$
 inflection at $x = 0$
 maximum at $x = -3$

SOLUTIONS

1. If c is positive, then $n(\infty) = \infty$, which is contrary to the data given. Therefore, $c \leq 0$. If $c = 0$, then $n(\infty) = a/(1 + b) = 100$, which is possible depending on a and b. However, $n(0) = a/(1 + b)$ would also equal 100, which is contrary to the given data. Therefore, $c \neq 0$.

c must be less than 0.

Since $c < 0$, then $n(\infty) = a$, so $\boxed{a = 100.}$

Applying the condition $t = 0$ gives

$$n(0) = \frac{a}{1 + b} = 10$$

Since $a = 100$,

$$n(0) = \frac{100}{1 + b}$$
$$100 = (10)(1 + b)$$
$$10 = 1 + b$$
$$\boxed{b = 9}$$

Substitute the results for a and b into the expression.

$$n(t) = \frac{100}{1 + 9e^{ct}}$$

Take the first derivative.

$$\frac{d}{dt}n(t) = \left(\frac{100}{(1 + 9e^{ct})^2}\right)(-9ce^{ct})$$

Apply the initial condition.

$$\frac{d}{dt}n(0) = \left(\frac{100}{(1 + 9e^{c(0)})^2}\right)\left(-9ce^{c(0)}\right) = 0.5$$
$$\left(\frac{100}{(1 + 9)^2}\right)(-9c) = 0.5$$
$$(1)(-9c) = 0.5$$
$$c = \frac{-0.5}{9}$$
$$= \boxed{-0.0556}$$

The answer is (C).

Substitute the terms a, b, and c into the expression.

$$n(t) = \frac{100}{1 + 9e^{-0.0556t}}$$

2. Determine the critical points by taking the first derivative of the function and setting it equal to zero.

$$\frac{dy}{dx} = 3x^2 - 18x = 3x(x - 6)$$
$$3x(x - 6) = 0$$
$$x(x - 6) = 0$$

The critical points are located at $x = 0$ and $x = 6$.

Determine the inflection points by setting the second derivative equal to zero. Take the second derivative.

$$\frac{d^2y}{dx^2} = \left(\frac{d}{dx}\right)\left(\frac{dy}{dx}\right) = \frac{d}{dx}(3x^2 - 18x)$$
$$= 6x - 18$$

Set the second derivative equal to zero.

$$\frac{d^2y}{dx^2} = 0 = 6x - 18 = (6)(x - 3)$$
$$(6)(x - 3) = 0$$
$$x - 3 = 0$$
$$x = 3$$

This inflection point is at $x = 3$.

Determine the local maximum and minimum by substituting the critical points into the expression for the second derivative.

At the critical point $x = 0$,

$$\left.\frac{d^2y}{dx^2}\right|_{x=0} = (6)(x - 3) = (6)(0 - 3)$$
$$= -18$$

Since $-18 < 0$, $x = 0$ is a local maximum.

At the critical point $x = 6$,

$$\left.\frac{d^2y}{dx^2}\right|_{x=6} = (6)(x - 3) = (6)(6 - 3)$$
$$= 18$$

Since $18 > 0$, $x = 6$ is a local minimum.

The answer is (A).

 **Integral Calculus**

PRACTICE PROBLEMS

Elementary Operations

1. Find the integrals.

(a) $$\int \sqrt{1-x}\,dx$$

(b) $$\int \frac{x}{x^2+1}\,dx$$

(c) $$\int \frac{x^2}{x^2+x-6}\,dx$$

Definite Integrals

2. Calculate the definite integrals.

(a) $$\int_1^3 (x^2+4x)\,dx$$

(b) $$\int_{-2}^2 (x^3+1)\,dx$$

(c) $$\int_1^2 (4x^3-3x^2)\,dx$$

Areas by Integration

3. Find the area bounded by $x=1$, $x=3$, $y+x+1=0$, and $y=6x-x^2$.

Fourier Series

4. Find a_0 for the two waveforms shown.

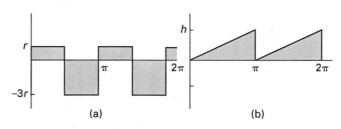

(a) (b)

5. For the two waveforms shown, determine if their Fourier series is of type A, B, or C.

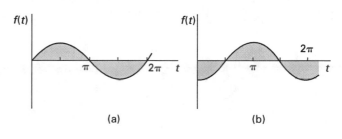

(a) (b)

type A: $f(t) = a_0 + a_2 \cos 2t + b_2 \sin 2t$
$+ a_4 \cos 4t + b_4 \sin 4t + \cdots$

type B: $f(t) = a_0 + b_1 \sin t + b_2 \sin 2t + b_3 \sin 3t + \cdots$

type C: $f(t) = a_0 + a_1 \cos t + a_2 \cos 2t + a_3 \cos 3t + \cdots$

SOLUTIONS

1. (a) $\displaystyle\int \sqrt{1-x}\,dx = \int (1-x)^{\frac{1}{2}}\,dx$

$$= \boxed{\left(-\frac{2}{3}\right)(1-x)^{\frac{3}{2}} + C}$$

(b) $\displaystyle\int \frac{x}{x^2+1}\,dx = \frac{1}{2}\int \frac{2x}{x^2+1}\,dx$

$$= \boxed{\frac{1}{2}\ln\left|(x^2+1)\right| + C}$$

(c) $\displaystyle\frac{x^2}{x^2+x-6} = 1 - \frac{x-6}{x^2+x-6}$

$$= 1 - \frac{x-6}{(x+3)(x-2)}$$

$$= 1 - \frac{\dfrac{9}{5}}{x+3} + \frac{\dfrac{4}{5}}{x-2}$$

$$\int \frac{x^2}{x^2+x-6}\,dx = \int \left(1 - \frac{\dfrac{9}{5}}{x+3} + \frac{\dfrac{4}{5}}{x-2}\right)dx$$

$$= \int dx - \int \frac{\dfrac{9}{5}}{x+3}\,dx + \int \frac{\dfrac{4}{5}}{x-2}\,dx$$

$$= \boxed{x - \frac{9}{5}\ln\left|(x+3)\right| + \frac{4}{5}\ln\left|(x-2)\right| + C}$$

2. (a) $\displaystyle\int_1^3 (x^2+4x)\,dx = \left[\frac{x^3}{3} + 2x^2\right]_1^3 = \boxed{24\tfrac{2}{3}}$

(b) $\displaystyle\int_{-2}^2 (x^3+1)\,dx = \left[\frac{x^4}{4} + x\right]_{-2}^2 = \boxed{4}$

(c) $\displaystyle\int_1^2 (4x^3 - 3x^2)\,dx = \left[x^4 - x^3\right]_1^2 = \boxed{8}$

3.

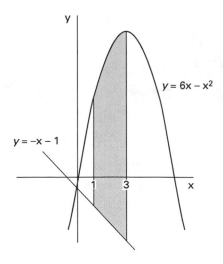

$$\text{area} = \int_1^3 \left((6x - x^2) - (-x-1)\right)dx$$

$$= \int_1^3 (-x^2 + 7x + 1)\,dx$$

$$= \left[-\frac{x^3}{3} + \frac{7}{2}x^2 + x\right]_1^3 = \boxed{21\tfrac{1}{3}}$$

4. For waveform (a):

$$a_0 = \frac{1}{2\pi}\int_0^{2\pi} f(t)\,dt$$

$$= \frac{1}{\pi}\int_0^{\pi} f(t)\,dt$$

$$= \frac{1}{\pi}\left((r)\left(\frac{\pi}{2}\right) + (-3r)\left(\frac{\pi}{2}\right)\right) = \boxed{-r}$$

For waveform (b):

$$a_0 = \frac{1}{2\pi}\int_0^{2\pi} f(t)\,dt$$

$$= \frac{1}{\pi}\int_0^{\pi} f(t)\,dt$$

$$= \left(\frac{1}{\pi}\right)\left(\frac{1}{2}\pi h\right)$$

$$= \boxed{h/2}$$

5. For waveform (a): Since $f(t) = -f(-t)$,
$\boxed{\text{it is type B}}$

For waveform (b): Since $f(t) = f(-t)$, $\boxed{\text{it is type C}}$

10 Differential Equations

PRACTICE PROBLEMS

1. Solve the following differential equation for y.

$$y'' - 4y' - 12y = 0$$

(A) $A_1 e^{6x} + A_2 e^{-2x}$
(B) $A_1 e^{-6x} + A_2 e^{2x}$
(C) $A_1 e^{6x} + A_2 e^{2x}$
(D) $A_1 e^{-6x} + A_2 e^{-2x}$

2. Solve the following differential equation for y.

$$y' - y = 2xe^{2x} \qquad y(0) = 1$$

(A) $y = 2e^{-2x}(x-1) + 3e^{-x}$
(B) $y = 2e^{2x}(x-1) + 3e^{x}$
(C) $y = -2e^{-2x}(x-1) + 3e^{-x}$
(D) $y = 2e^{2x}(x-1) + 3e^{-x}$

3. The oscillation exhibited by the top story of a certain building in free motion is given by the following differential equation.

$$x'' + 2x' + 2x = 0 \qquad x(0) = 0 \qquad x'(0) = 1$$

(a) What is x as a function of time?
(A) $e^{-2t} \sin t$
(B) $e^{t} \sin t$
(C) $e^{-t} \sin t$
(D) $e^{-t} \sin t + e^{-t} \cos t$

(b) What is the building's fundamental natural frequency of vibration?
(A) $^1/_2$
(B) 1
(C) $\sqrt{2}$
(D) 2

(c) What is the amplitude of oscillation?
(A) 0.32
(B) 0.54
(C) 1.7
(D) 6.6

(d) What is x as a function of time if a lateral wind load is applied with a form of $\sin t$?

(A) $\frac{6}{5}e^{-t}\sin t + \frac{2}{5}e^{-t}\cos t$
(B) $\frac{6}{5}e^{t}\sin t + \frac{2}{5}e^{t}\cos t$
(C) $\frac{2}{5}e^{-t}\sin t + \frac{6}{5}e^{-t}\cos t + \frac{2}{5}\sin t$
$\qquad - \frac{1}{5}\cos t$
(D) $\frac{6}{5}e^{-t}\sin t + \frac{2}{5}e^{-t}\cos t + \frac{1}{5}\sin t$
$\qquad - \frac{2}{5}\cos t$

4. (*Time limit: one hour*) A 90 lbm (40 kg) bag of a chemical is accidentally dropped in an aerating lagoon. The chemical is water soluble and nonreacting. The lagoon is 120 ft (35 m) in diameter and filled to a depth of 10 ft (3 m). The aerators circulate and distribute the chemical evenly throughout the lagoon.

Water enters the lagoon at a rate of 30 gal/min (115 L/min). Fully mixed water is pumped into a reservoir at a rate of 30 gal/min (115 L/min).

The established safe concentration of this chemical is 1 ppb (part per billion). How many days will it take for the concentration of the discharge water to reach this level?

(A) 25 days
(B) 50 days
(C) 100 days
(D) 200 days

5. A tank contains 100 gal (100 L) of brine made by dissolving 60 lbm (60 kg) of salt in pure water. Salt water with a concentration of 1 lbm/gal (1 kg/L) enters the tank at a rate of 2 gal/min (2 L/min). A well-stirred mixture is drawn from the tank at a rate of 3 gal/min (3 L/min). Find the mass of salt in the tank after 1 hr.

(A) 13 lbm (13 kg)
(B) 37 lbm (37 kg)
(C) 43 lbm (43 kg)
(D) 51 lbm (51 kg)

SOLUTIONS

1. Obtain the characteristic equation by replacing each derivative with a polynomial term of equal degree.

$$r^2 - 4r - 12 = 0$$

Factor the characteristic equation.

$$(r - 6)(r + 2) = 0$$

The roots are $r_1 = 6$ and $r_2 = -2$.

Since the roots are real and distinct, the solution is

$$y = A_1 e^{r_1 x} + A_2 e^{r_2 x}$$

$$\boxed{= A_1 e^{6x} + A_2 e^{-2x}}$$

The answer is (A).

2. The equation is a first-order linear differential equation of the form

$$y' + p(x)y = g(x)$$
$$p(x) = -1$$
$$g(x) = 2xe^{2x}$$

The integration factor $u(x)$ is given by

$$u(x) = \exp\left(\int p(x)dx\right)$$
$$= \exp\left(\int (-1)dx\right)$$
$$= e^{-x}$$

The closed form of the solution is given by

$$y = \left(\frac{1}{u(x)}\right)\left(\int u(x)g(x)dx + C\right)$$
$$= \left(\frac{1}{e^{-x}}\right)\left(\int (e^{-x})(2xe^{2x})dx + C\right)$$
$$= e^x\left(2(xe^x - e^x) + C\right)$$
$$= e^x\left(2e^x(x - 1) + C\right)$$
$$= 2e^{2x}(x - 1) + Ce^x$$

Apply the initial condition $y(0) = 1$ to obtain the integration constant C.

$$y(0) = 2e^{(2)(0)}(0 - 1) + Ce^0 = 1$$
$$(2)(1)(-1) + C(1) = 1$$
$$-2 + C = 1$$
$$C = 3$$

Substituting in the value for the integration constant C, the solution is

$$y = \boxed{2e^{2x}(x - 1) + 3e^x}$$

The answer is (B).

3. (a) The differential equation is a homogeneous second-order linear differential equation with constant coefficients. Write the characteristic equation.

$$r^2 + 2r + 2 = 0$$

This is a quadratic equation of the form $ar^2 + br + c = 0$ where $a = 1$, $b = 2$, and $c = 2$.

Solve for r.

$$r = \frac{-b \pm \sqrt{b^2 - 4ac}}{2a}$$
$$= \frac{-2 \pm \sqrt{(2)^2 - (4)(1)(2)}}{(2)(1)}$$
$$= \frac{-2 \pm \sqrt{4 - 8}}{2}$$
$$= \frac{-2 \pm \sqrt{-4}}{2}$$
$$= \frac{-2 \pm 2\sqrt{-1}}{2}$$
$$= -1 \pm \sqrt{-1}$$
$$= -1 \pm i$$
$$r_1 = -1 + i \text{ and } r_2 = -1 - i$$

Since the roots are imaginary and of the form $\alpha + i\omega$ and $\alpha - i\omega$ where $\alpha = -1$ and $\omega = 1$, the general form of the solution is given by

$$x(t) = A_1 e^{\alpha t} \cos \omega t + A_2 e^{\alpha t} \sin \omega t$$
$$= A_1 e^{-1t} \cos(1t) + A_2 e^{-1t} \sin(1t)$$
$$= A_1 e^{-t} \cos t + A_2 e^{-t} \sin t$$

Apply the initial conditions $x(0) = 0$ and $x'(0) = 1$ to solve for A_1 and A_2.

First, apply the initial condition $x(0) = 0$.

$$x(t) = A_1 e^0 \cos 0 + A_2 e^0 \sin 0 = 0$$
$$A_1(1)(1) + A_2(1)(0) = 0$$
$$A_1 = 0$$

Substituting, the solution of the differential equation becomes

$$x(t) = A_2 e^{-t} \sin t$$

To apply the second initial condition, take the first derivative.

$$x'(t) = \frac{d}{dt}\left(A_2 e^{-t}\sin t\right)$$
$$= A_2\frac{d}{dt}\left(e^{-t}\sin t\right)$$
$$= A_2\left(\sin t\frac{d}{dt}\left(e^{-t}\right) + e^{-t}\frac{d}{dt}\sin t\right)$$
$$= A_2\left(\sin t\left(-e^{-t}\right) + e^{-t}(\cos t)\right)$$
$$= A_2\left(e^{-t}\right)(-\sin t + \cos t)$$

Apply the initial condition, $x'(0) = 1$.

$$x(0) = A_2\left(e^0\right)(-\sin 0 + \cos 0) = 1$$
$$A_2(1)(0+1) = 1$$
$$A_2 = 1$$

The solution is

$$x(t) = A_2 e^{-t}\sin t$$
$$= (1)e^{-t}\sin t$$
$$= \boxed{e^{-t}\sin t}$$

The answer is (C).

(b) To determine the natural frequency, set the damping term to zero. The equation has the form

$$x'' + 2x = 0$$

This equation has a general solution of the form

$$x(t) = x_0 \cos \omega t + \left(\frac{v_0}{\omega}\right)\sin \omega t$$

ω is the natural frequency. Given the equation $x''+2x = 0$, the characteristic equation is

$$r^2 + 2 = 0$$
$$r = \sqrt{-2}$$
$$= \pm\sqrt{2}i$$

Since the roots are imaginary and of the form $\alpha + i\omega$ and $\alpha - i\omega$ where $\alpha = 0$ and $\omega = \sqrt{2}$, the general form of the solution is given by

$$x(t) = A_1 e^{\alpha t}\cos \omega t + A_2 e^{\alpha t}\sin \omega t$$
$$= A_1 e^{0t}\cos \sqrt{2}t + A_2 e^{0t}\sin \sqrt{2}t$$
$$= A_1(1)\cos \sqrt{2}t + A_2(1)\sin \sqrt{2}t$$
$$= A_1 \cos \sqrt{2}t + A_2 \sin \sqrt{2}t$$

Apply the initial conditions, $x(0) = 0$ and $x'(0) = 1$ to solve for A_1 and A_2. Applying the initial condition $x(0) = 0$ gives

$$x(0) = A_1 \cos \left((\sqrt{2})(0)\right) + A_2 \sin(\sqrt{2})(0) = 0$$
$$A_1 \cos 0 + A_2 \sin 0 = 0$$
$$A_1(1) + A_2(0) = 0$$
$$A_1 = 0$$

Substituting, the solution of the different equation becomes

$$x(t) = A_2 \sin \sqrt{2}t$$

To apply the second initial condition, take the first derivative.

$$x'(t) = \frac{d}{dt}(A_2 \sin \sqrt{2}t)$$
$$= A_2\sqrt{2} \cos \sqrt{2}t$$

Apply the second initial condition, $x'(0) = 1$.

$$x'(0) = A_2\sqrt{2} \cos(\sqrt{2})(0) = 1$$
$$A_2\sqrt{2} \cos(0) = 1$$
$$A_2(\sqrt{2})(1) = 1$$
$$A_2\sqrt{2} = 1$$
$$A_2 = \frac{1}{\sqrt{2}} = \frac{\sqrt{2}}{2}$$

Substituting, the undamped solution becomes

$$x(t) = \left(\frac{\sqrt{2}}{2}\right)\sin \sqrt{2}t$$

Therefore, the undamped natural frequency is $\boxed{\omega = \sqrt{2}.}$

The answer is (C).

(c) The amplitude of the oscillation is the maximum displacement.

Take the derivative of the solution, $x(t) = e^{-t}\sin t$.

$$x'(t) = \frac{d}{dt}(e^{-t}\sin t)$$
$$= \sin t\frac{d}{dt}\left(e^{-t}\right) + e^{-t}\frac{d}{dt}\sin t$$
$$= \sin(t)(-e^{-t}) + e^{-t}\cos t$$
$$= e^{-t}\cos t - \sin t$$

The maximum displacement occurs at $x'(t) = 0$.

Since $e^{-t} \neq 0$ except as t approaches infinity,

$$\cos t - \sin t = 0$$
$$\tan t = 1$$
$$t = \tan^{-1}(1)$$
$$= 0.785 \text{ rad}$$

At $t = 0.785$ rad, the displacement is maximum. Substitute into the orginal solution to obtain a value for the maximum displacement.

$$x(0.785) = e^{-0.785}\sin(0.785)$$
$$= 0.322$$

The amplitude is $\boxed{0.322.}$

The answer is (A).

(d) (An alternative solution using Laplace transforms follows this solution.) The application of a lateral wind load with the form $\sin t$ revises the differential equation to the form
$$x'' + 2x' + 2x = \sin t$$

Express the solution as the sum of the complementary x_c and particular x_p solutions.

$$x(t) = x_c(t) + x_p(t)$$

From part (a),
$$x_c(t) = A_1 e^{-t}\cos t + A_2 e^{-t}\sin t$$

The general form of the particular solution is given by
$$x_p(t) = x^s(A_3\cos t + A_4\sin t)$$

Determine the value of s; check to see if the terms of the particular solution solve the homogeneous equation. Examine the term $A_3\cos(t)$.

Take the first derivative.
$$\frac{d}{dx}(A_3\cos t) = -A_3\sin t$$

Take the second derivative.
$$\frac{d}{dx}\left(\frac{d}{dx}(A_3\cos t)\right) = \frac{d}{dx}(-A_3\sin t)$$
$$= -A_3\cos t$$

Substitute the terms into the homogeneous equation.
$$x'' + 2x' + 2x = -A_3\cos t + (2)(-A_3\sin t)$$
$$+ (2)(-A_3\cos t)$$
$$= A_3\cos t - 2A_3\sin t$$
$$\neq 0$$

Except for the trival solution $A_3 = 0$, the term $A_3\cos t$ does not solve the homogeneous equation.

Examine the second term $A_4\sin t$.

Take the first derivative.
$$\frac{d}{dx}(A_4\sin t) = A_4\cos t$$

Take the second derivative.
$$\frac{d}{dx}\left(\frac{d}{dx}(A_4\sin t)\right) = \frac{d}{dx}(A_4\cos t)$$
$$= -A_4\sin t$$

Substitute the terms into the homogeneous equation.
$$x'' + 2x' + 2x = -A_4\sin t + (2)(A_4\cos t)$$
$$+ (2)(A_4\sin t)$$
$$= A_4\sin t + 2A_4\cos t$$
$$\neq 0$$

Except for the trival solution $A_4 = 0$, the term $A_4\sin t$ does not solve the homogeneous equation.

Neither of the terms satisfies the homogeneous equation $s = 0$; therefore, the particular solution is of the form
$$x_p(t) = A_3\cos t + A_4\sin t$$

Use the method of undetermined coefficients to solve for A_3 and A_4. Take the first derivative.
$$x_p'(t) = \frac{d}{dx}(A_3\cos t + A_4\sin t)$$
$$= -A_3\sin t + A_4\cos t$$

Take the second derivative.
$$x_p''(t) = \frac{d}{dx}\left(\frac{d}{dx}(A_3\cos t + A_4\sin t)\right)$$
$$= \frac{d}{dx}(-A_3\sin t + A_4\cos t)$$
$$= -A_3\cos t - A_4\sin t$$

Substitute the expressions for the derivatives into the differential equation.
$$x'' + 2x' + 2x = (-A_3\cos t - A_4\sin t)$$
$$+ (2)(-A_3\sin t + A_4\cos t)$$
$$+ (2)(A_3\cos t + A_4\sin t)$$
$$= \sin t$$

Rearranging terms gives
$$(-A_3 + 2A_4 + 2A_3)\cos t$$
$$+ (-A_4 - 2A_3 + 2A_4)\sin t = \sin t$$
$$(A_3 + 2A_4)\cos t + (-2A_3 + A_4)\sin t = \sin t$$

Equating coefficients gives
$$A_3 + 2A_4 = 0$$
$$-2A_3 + A_4 = 1$$

Multiplying the first equation by 2 and adding equations gives

$$A_3 + 2A_4 = 0$$
$$+(-2A_3 + A_4) = 1$$

$$5A_4 = 1 \text{ or } A_4 = \tfrac{1}{5}$$

From the first equation for $A_4 = {}^1\!/_5$, $A_3 + (2)({}^1\!/_5) = 0$ and $A_3 = -{}^2\!/_5$.

Substituting for the coefficients, the particular solution becomes

$$x_p(t) = -\tfrac{2}{5}\cos t + \tfrac{1}{5}\sin t$$

Combining the complementary and particular solutions gives

$$x(t) = x_c(t) + x_p(t)$$
$$= A_1 e^{-t}\cos t + A_2 e^{-t}\sin t - \tfrac{2}{5}\cos t + \tfrac{1}{5}\sin t$$

Apply the initial conditions to solve for the coefficients A_1 and A_2; then apply the first initial condition, $x(0) = 0$.

$$x(t) = A_1 e^0 \cos 0 + A_2 e^0 \sin 0$$
$$- \tfrac{2}{5}\cos 0 + \tfrac{1}{5}\sin 0 = 0$$
$$A_1(1)(1) + A_2(1)(0) + \left(-\tfrac{2}{5}\right)(1) + \left(\tfrac{1}{5}\right)(0) = 0$$
$$A_1 - \tfrac{2}{5} = 0$$
$$A_1 = \tfrac{2}{5}$$

Substituting for A_1, the solution becomes

$$x(t) = \tfrac{2}{5}e^{-t}\cos t + A_2 e^{-t}\sin t - \tfrac{2}{5}\cos t + \tfrac{1}{5}\sin t$$

Take the first derivative.

$$x'(t) = \frac{d}{dx}\left(\tfrac{2}{5}e^{-t}\cos t + A_2 e^{-t}\sin t\right)$$
$$+ \left(\left(-\tfrac{2}{5}\right)\cos t + \tfrac{1}{5}\sin t\right)$$
$$= \left(\tfrac{2}{5}\right)\left(-e^{-t}\cos t - e^{-t}\sin t\right)$$
$$+ A_2\left(-e^{-t}\sin t + e^{-t}\cos t\right)$$
$$+ \left(-\tfrac{2}{5}\right)(-\sin t) + \tfrac{1}{5}\cos t$$

Apply the second initial condition, $x'(0) = 1$.

$$x'(0) = \left(\tfrac{2}{5}\right)\left(-e^0\cos 0 - e^0\sin 0\right)$$
$$+ A_2\left(-e^0\sin 0 + e^0\cos 0\right)$$
$$+ \left(-\tfrac{2}{5}\right)(-\sin 0) + \tfrac{1}{5}\cos 0$$
$$= 1$$

$$\left(\tfrac{2}{5}\right)\left(-(1)(1) - (1)(0)\right) + A_2\left(-(1)(0)\right.$$
$$\left. +(1)(1)\right) + \left(-\tfrac{2}{5}\right)(0) + \left(\tfrac{1}{5}\right)(1) = 1$$
$$\left(\tfrac{2}{5}\right)(-1) + A_2(1) + \left(\tfrac{1}{5}\right) = 1$$
$$A_2 = \tfrac{6}{5}$$

Substituting for A_2, the solution becomes

$$x(t) = \boxed{\begin{array}{l} \tfrac{2}{5}e^{-t}\cos t + \tfrac{6}{5}e^{-t}\sin t \\ -\tfrac{2}{5}\cos t + \tfrac{1}{5}\sin t \end{array}}$$

The answer is (D).

(d) *Alternate solution:*

Use the Laplace transform method.

$$x'' + 2x' + 2x = \sin t$$
$$\mathcal{L}(x'') + 2\mathcal{L}(x') + 2\mathcal{L}(x) = \mathcal{L}(\sin t)$$
$$s^2\mathcal{L}(x) - 1 + 2s\mathcal{L}(x) + 2\mathcal{L}(x) = \frac{1}{s^2 + 1}$$
$$\mathcal{L}(x)(s^2 + 2s + 2) - 1 = \frac{1}{s^2 + 1}$$

$$\mathcal{L}(x) = \frac{1}{s^2 + 2s + 2} + \frac{1}{(s^2 + 1)(s^2 + 2s + 2)}$$
$$= \frac{1}{(s + 1)^2 + 1} + \frac{1}{(s^2 + 1)(s^2 + 2s + 2)}$$

Use partial fractions to expand the second term.

$$\frac{1}{(s^2 + 1)(s^2 + 2s + 2)} = \frac{A_1 + B_1 s}{s^2 + 1} + \frac{A_2 + B_2 s}{s^2 + 2s + 2}$$

Cross multiply.

$$= \frac{\begin{array}{l} A_1 s^2 + 2A_1 s + 2A_1 + B_1 s^3 + 2B_1 s^2 \\ + 2B_1 s + A_2 s^2 + A_2 + B_2 s^3 + B_2 s \end{array}}{(s^2 + 1)(s^2 + 2s + 2)}$$

$$= \frac{\begin{array}{l} s^3(B_1 + B_2) + s^2(A_1 + A_2 + 2B_1) \\ + s(2A_1 + 2B_1 + B_2) + 2A_1 + A_2 \end{array}}{(s^2 + 1)(s^2 + 2s + 2)}$$

Compare numerators to obtain the following four simultaneous equations.

$$B_1 + B_2 = 0$$
$$A_1 + A_2 + 2B_1 = 0$$
$$2A_1 + 2B_1 + B_2 = 0$$
$$2A_1 + A_2 = 1$$

Use Cramer's rule to find A_1.

$$A_1 = \frac{\begin{vmatrix} 0 & 0 & 1 & 1 \\ 0 & 1 & 2 & 0 \\ 0 & 0 & 2 & 1 \\ 1 & 1 & 0 & 0 \end{vmatrix}}{\begin{vmatrix} 0 & 0 & 1 & 1 \\ 1 & 1 & 2 & 0 \\ 2 & 0 & 2 & 1 \\ 2 & 1 & 0 & 0 \end{vmatrix}} = \frac{-1}{-5} = \frac{1}{5}$$

The rest of the coefficients are found similarly.

$$A_1 = \tfrac{1}{5}$$
$$A_2 = \tfrac{3}{5}$$
$$B_1 = -\tfrac{2}{5}$$
$$B_2 = \tfrac{2}{5}$$

Then,

$$\mathcal{L}(x) = \frac{1}{(s+1)^2 + 1} + \frac{\tfrac{1}{5}}{s^2 + 1} + \frac{-\tfrac{2}{5}s}{s^2 + 1}$$
$$+ \frac{\tfrac{3}{5}}{s^2 + 2s + 2} + \frac{\tfrac{2}{5}s}{s^2 + 2s + 2}$$

Take the inverse transform.

$$x(t) = \mathcal{L}^{-1}\{\mathcal{L}(x)\}$$
$$= e^{-t}\sin t + \tfrac{1}{5}\sin t - \tfrac{2}{5}\cos t + \tfrac{3}{5}e^{-t}\sin t$$
$$+ \tfrac{2}{5}(e^{-t}\cos t - e^{-t}\sin t)$$
$$\boxed{= \tfrac{6}{5}e^{-t}\sin t + \tfrac{2}{5}e^{-t}\cos t + \tfrac{1}{5}\sin t - \tfrac{2}{5}\cos t}$$

The answer is (D).

4. *Customary U.S. Solution*

The differential equation is given as

$$m'(t) = a(t) - \frac{m(t)o(t)}{V(t)}$$

$a(t)$ = rate of addition of chemical
$m(t)$ = mass of chemical at time t
$o(t)$ = volumetric flow out of the lagoon
     (= 30 gal/min)
$V(t)$ = volume in the lagoon at time t

Water flows into the lagoon at a rate of 30 gal/min, and a water-chemical mix flows out of the lagoon at rate of 30 gal/min. Therefore, the volume of the lagoon at time t is equal to the initial volume.

$$V(t) = \left(\frac{\pi}{4}\right)(\text{diameter of lagoon})^2(\text{depth of lagoon})$$
$$= \left(\frac{\pi}{4}\right)(120\text{ ft})^2(10\text{ ft})$$
$$= 113{,}097\text{ ft}^3$$

Use a conversion factor of 7.48 gal/ft^3.

$$o(t) = \frac{30\ \dfrac{\text{gal}}{\text{min}}}{7.48\ \dfrac{\text{gal}}{\text{ft}^3}}$$
$$= 4.01\text{ ft}^3/\text{min}$$

Substituting into the general form of the differential equation gives

$$m'(t) = a(t) - \frac{m(t)o(t)}{V(t)}$$
$$= (0) - m(t)\left(\frac{4.01\ \dfrac{\text{ft}^3}{\text{min}}}{113{,}097\text{ ft}^3}\right)$$
$$= -\left(\frac{3.55\times10^{-5}}{\text{min}}\right)m(t)$$

$$m'(t) + \left(\frac{3.55\times10^{-5}}{\text{min}}\right)m(t) = 0$$

The differential equation of the problem has a characteristic equation.

$$r + \frac{3.55\times10^{-5}}{\text{min}} = 0$$
$$r = -3.55\times10^{-5}/\text{min}$$

The general form of the solution is given by

$$m(t) = Ae^{rt}$$

Substituting for the root, r, gives

$$m(t) = Ae^{\left(\frac{-3.55\times10^{-5}}{\text{min}}\right)t}$$

Apply the initial condition $m(0) = 90$ lbm at time $t = 0$.

$$m(0) = Ae^{\left(\frac{-3.55\times10^{-5}}{\text{min}}\right)(0)} = 90\text{ lbm}$$
$$Ae^0 = 90\text{ lbm}$$
$$A = 90\text{ lbm}$$

Therefore,

$$m(t) = (90\text{ lbm})\,e^{\left(\frac{-3.55\times10^{-5}}{\text{min}}\right)t}$$

Solve for t.

$$\frac{m(t)}{90\text{ lbm}} = e^{\left(\frac{-3.55\times10^{-5}}{\text{min}}\right)t}$$
$$\ln\left(\frac{m(t)}{90\text{ lbm}}\right) = \ln\left(e^{\left(\frac{-3.55\times10^{-5}}{\text{min}}\right)t}\right)$$
$$= \left(\frac{-3.55\times10^{-5}}{\text{min}}\right)t$$
$$t = \frac{\ln\left(\dfrac{m(t)}{90\text{ lbm}}\right)}{\dfrac{-3.55\times10^{-5}}{\text{min}}}$$

The initial mass of the water in the lagoon is given by

$$m_i = V\rho$$

$$= (113{,}097 \text{ ft}^3)\left(62.4 \; \frac{\text{lbm}}{\text{ft}^3}\right)$$

$$= 7.05 \times 10^6 \text{ lbm}$$

The final mass of chemicals is achieved at a concentration of 1 ppb or

$$m_f = \frac{7.06 \times 10^6 \text{ lbm}}{1 \times 10^9}$$

$$= 7.06 \times 10^{-3} \text{ lbm}$$

Find the time required to achieve a mass of 7.06×10^{-3} lbm.

$$t = \left(\frac{\ln\left(\dfrac{m(t)}{90 \text{ lbm}}\right)}{\dfrac{-3.55 \times 10^{-5}}{\text{min}}}\right)\left(\frac{1 \text{ hr}}{60 \text{ min}}\right)\left(\frac{1 \text{ day}}{24 \text{ hr}}\right)$$

$$= \left(\frac{\ln\left(\dfrac{7.06 \times 10^{-3} \text{ lbm}}{90 \text{ lbm}}\right)}{\dfrac{-3.55 \times 10^{-5}}{\text{min}}}\right)\left(\frac{1 \text{ hr}}{60 \text{ min}}\right)\left(\frac{1 \text{ day}}{24 \text{ hr}}\right)$$

$$= \boxed{185 \text{ days}}$$

The answer is (D).

SI Solution

The differential equation is given as

$$m'(t) = a(t) - \frac{m(t)o(t)}{V(t)}$$

$a(t)$ = rate of addition of chemical
$m(t)$ = mass of chemical at time t
$o(t)$ = volumetric flow out of the lagoon
$\qquad (= 115 \text{ L/min})$
$V(t)$ = volume in the lagoon at time t

Water flows into the lagoon at a rate of 115 L/min, and a water-chemical mix flows out of the lagoon at a rate of 115 L/min. Therefore, the volume of the lagoon at time t is equal to the initial volume.

$$V(t) = \left(\frac{\pi}{4}\right)(\text{diameter of lagoon})^2(\text{depth of lagoon})$$

$$= \left(\frac{\pi}{4}\right)(35 \text{ m})^2(3 \text{ m})$$

$$= 2886 \text{ m}^3$$

Using a conversion factor of $1 \text{ m}^3/1000 \text{ L}$ gives

$$o(t) = \left(115 \; \frac{\text{L}}{\text{min}}\right)\left(\frac{1 \text{ m}^3}{1000 \text{ L}}\right)$$

$$= 0.115 \text{ m}^3/\text{min}$$

Substitute into the general form of the differential equation.

$$m'(t) = a(t) - \frac{m(t)o(t)}{V(t)}$$

$$= 0 - m(t)\left(\frac{0.115 \; \dfrac{\text{m}^3}{\text{min}}}{2886 \text{ m}^3}\right)$$

$$= -\left(\frac{3.985 \times 10^{-5}}{\text{min}}\right)m(t)$$

$$m'(t) + \left(\frac{3.985 \times 10^{-5}}{\text{min}}\right)m(t) = 0$$

The differential equation of the problem has the following characteristic equation.

$$r + \frac{3.985 \times 10^{-5}}{\text{min}} = 0$$

$$r = -3.985 \times 10^{-5}/\text{min}$$

The general form of the solution is given by

$$m(t) = Ae^{rt}$$

Substituting in for the root, r, gives

$$m(t) = Ae^{\left(\frac{-3.985 \times 10^{-5}}{\text{min}}\right)t}$$

Apply the initial condition $m(0) = 40$ kg at time $t = 0$.

$$m(0) = Ae^{\left(\frac{-3.985 \times 10^{-5}}{\text{min}}\right)(0)} = 40 \text{ kg}$$

$$Ae^0 = 40 \text{ kg}$$

$$A = 40 \text{ kg}$$

Therefore,

$$m(t) = (40 \text{ kg})e^{\left(\frac{-3.985 \times 10^{-5}}{\text{min}}\right)t}$$

Solve for t.

$$\frac{m(t)}{40 \text{ kg}} = e^{\left(\frac{-3.985 \times 10^{-5}}{\text{min}}\right)t}$$

$$\ln\left(\frac{m(t)}{40 \text{ kg}}\right) = \ln\left(e^{\left(\frac{-3.985 \times 10^{-5}}{\text{min}}\right)t}\right)$$

$$= \left(\frac{-3.985 \times 10^{-5}}{\text{min}}\right)t$$

$$t = \ln\left(\frac{\dfrac{m(t)}{40 \text{ kg}}}{\dfrac{-3.985 \times 10^{-5}}{\text{min}}}\right)$$

The initial mass of water in the lagoon is given by

$$m_i = V\rho$$

$$= (2886 \text{ m}^3)\left(1000 \ \frac{\text{kg}}{\text{m}^3}\right)$$

$$= 2.886 \times 10^6 \text{ kg}$$

The final mass of chemicals is achieved at a concentration of 1 ppb or

$$m_f = \frac{2.886 \times 10^6 \text{ kg}}{1 \times 10^9}$$

$$= 2.886 \times 10^{-3} \text{ kg}$$

Find the time required to achieve a mass of 2.886×10^{-3} kg.

$$t = \left(\frac{\ln \frac{m(t)}{40 \text{ kg}}}{\frac{-3.985 \times 10^{-5}}{\text{min}}}\right)\left(\frac{1 \text{ h}}{60 \text{ min}}\right)\left(\frac{1 \text{ day}}{24 \text{ h}}\right)$$

$$= \left(\frac{\ln\left(\frac{2.886 \times 10^{-3} \text{ kg}}{40 \text{ kg}}\right)}{\frac{-3.985 \times 10^{-5}}{\text{min}}}\right)\left(\frac{1 \text{ h}}{60 \text{ min}}\right)\left(\frac{1 \text{ day}}{24 \text{ h}}\right)$$

$$= \boxed{166 \text{ days}}$$

The answer is (D).

5. Let

$$m(t) = \text{mass of salt in tank at time } t$$
$$m_0 = 60 \text{ mass units}$$
$$m'(t) = \text{rate at which salt content is changing}$$

2 mass units of salt enter each minute, and 3 volumes leave each minute. The amount of salt leaving each minute is

$$\left(3 \ \frac{\text{vol}}{\text{min}}\right)\left(\text{concentration in } \frac{\text{mass}}{\text{vol}}\right)$$

$$= \left(3 \ \frac{\text{vol}}{\text{min}}\right)\left(\frac{\text{salt content}}{\text{volume}}\right)$$

$$= \left(3 \ \frac{\text{vol}}{\text{min}}\right)\left(\frac{m(t)}{100 - t}\right)$$

$$m'(t) = 2 - (3)\left(\frac{m(t)}{100 - t}\right) \text{ or } m'(t) + \frac{3m(t)}{100 - t}$$

$$= 2 \text{ mass/min}$$

This is a first-order linear differential equation. The integrating factor is

$$m = \exp\left[3 \int \frac{dt}{100 - t}\right]$$

$$= \exp\left[(3)\left(-\ln(100 - t)\right)\right]$$

$$= (100 - t)^{-3}$$

$$m(t) = (100 - t)^3\left[2 \int \frac{dt}{(100 - t)^3} + k\right]$$

$$= 100 - t + (k)(100 - t)^3$$

But $m = 60$ mass units at $t = 0$, so $k = -0.00004$.

$$m(t) = 100 - t - (0.00004)(100 - t)^3$$

At $t = 60$ min,

$$m = 100 - 60 \text{ min} - (0.00004)(100 - 60 \text{ min})^3$$

$$= \boxed{37.44 \text{ mass units}}$$

The answer is (B).

11 Probability and Statistical Analysis of Data

PRACTICE PROBLEMS

Probability

1. Four military recruits whose respective shoe sizes are 7, 8, 9, and 10 report to the supply clerk to be issued boots. The supply clerk selects one pair of boots in each of the four required sizes and hands them at random to the recruits.

(a) What is the probability that all recruits will receive boots of an incorrect size?

- (A) 0.25
- (B) 0.38
- (C) 0.45
- (D) 0.61

(b) What is the probability that exactly three recruits will receive boots of the correct size?

- (A) 0
- (B) 0.063
- (C) 0.17
- (D) 0.25

Probability Distributions

2. The time taken by a toll taker to collect the toll from vehicles crossing a bridge is an exponential distribution with a mean of 23 sec. What is the probability that a random vehicle will be processed in 25 sec or more (i.e., will take longer than 25 sec)?

- (A) 0.17
- (B) 0.25
- (C) 0.34
- (D) 0.52

3. The number of cars entering a toll plaza on a bridge during the hour after midnight follows a Poisson distribution with a mean of 20.

(a) What is the probability that 17 cars will pass through the toll plaza during that hour on any given night?

- (A) 0.076
- (B) 0.12
- (C) 0.16
- (D) 0.23

(b) What is the probability that three or fewer cars will pass through the toll plaza at that hour on any given night?

- (A) 0.0000032
- (B) 0.0019
- (C) 0.079
- (D) 0.11

4. A mechanical component exhibits a negative exponential failure distribution with a mean time to failure of 1000 hr. What is the maximum operating time such that the reliability remains above 99%?

- (A) 3.3 hr
- (B) 5.6 hr
- (C) 8.1 hr
- (D) 10 hr

5. (*Time limit: one hour*) A survey field crew measures one leg of a traverse four times. The following results are obtained.

repetition	measurement	direction
1	1249.529	forward
2	1249.494	backward
3	1249.384	forward
4	1249.348	backward

The crew chief is under orders to obtain readings with confidence limits of 90%.

(a) Which readings are acceptable?
- (A) No readings are acceptable.
- (B) Two readings are acceptable.
- (C) Three readings are acceptable.
- (D) All four readings are acceptable.

(b) Which readings are not acceptable?
- (A) No readings are unacceptable.
- (B) One reading is unacceptable.
- (C) Two readings are unacceptable.
- (D) All four readings are unacceptable.

(c) Explain how to determine which readings are not acceptable.
- (A) Readings inside the 90% confidence limits are unacceptable.
- (B) Readings outside the 90% confidence limits are unacceptable.
- (C) Readings outside the upper 90% confidence limit are unacceptable.
- (D) Readings outside the lower 90% confidence limit are unacceptable.

(d) What is the most probable value of the distance?
- (A) 1249.399
- (B) 1249.410
- (C) 1249.439
- (D) 1249.452

(e) What is the error in the most probable value (at 90% confidence)?
- (A) 0.08
- (B) 0.11
- (C) 0.14
- (D) 0.19

(f) If the distance is one side of a square traverse whose sides are all equal, what is the most probable closure error?
- (A) 0.14
- (B) 0.20
- (C) 0.28
- (D) 0.35

(g) What is the probable error of part (f) expressed as a fraction?
- (A) 1:17,600
- (B) 1:14,200
- (C) 1:12,500
- (D) 1:10,900

(h) What is the order of accuracy of the closure?
- (A) first order
- (B) second order
- (C) third order
- (D) fourth order

(i) Define accuracy and distinguish it from precision.
- (A) If an experiment can be repeated with identical results, the results are considered accurate.
- (B) If an experiment has a small bias, the results are considered precise.
- (C) If an experiment is precise, it cannot also be accurate.
- (D) If an experiment is unaffected by experimental error, the results are accurate.

(j) Give an example of systematic error.
- (A) measuring river depth as a motorized ski boat passes by
- (B) using a steel tape that is too short to measure consecutive distances
- (C) locating magnetic north near a large iron ore deposit along an overland route
- (D) determining local wastewater BOD after a toxic spill

Fault-Tree Analysis

6. A storage tank holds a large quantity of hazardous, flammable liquid. The liquid is pressurized with a 100% nitrogen blanket at a pressure greater than the liquid's vapor pressure. Should the nitrogen blanket fail, a flare vent fitted with a canister of activated carbon is intended to remove toxic fumes from the vented gas. An event time equal to a repair time of one week is selected for this problem.

The following events and their annual frequencies of occurrence are known.

Nitrogen blanket failure (repaired within 1 week)	0.5/yr
Flare failure (repaired within 1 week)	0.5/yr
Adsorbent canister failure (repaired within 1 week)	0.5/yr
Tank overfill failure	0.5/yr
Rupture from corrosion failure	0.05/yr
Rupture from vehicular collision failure	0.2/yr
Miscellaneous tank failure	0.4/yr

(a) Perform a fault-tree analysis to estimate the overall frequency of occurrence of fugitive emissions emanating from the storage tank for any reason. (b) Identify which failure modes have the biggest impact on the overall failure rate.

Statistical Analysis

7. *(Time limit: one hour)* California law requires a statistical analysis of the average speed driven by motorists on a road prior to the use of radar speed control. The following speeds (all in mi/hr) were observed in a random sample of 40 cars.

44, 48, 26, 25, 20, 43, 40, 42, 29, 39, 23, 26, 24, 47, 45, 28, 29, 41, 38, 36, 27, 44, 42, 43, 29, 37, 34, 31, 33, 30, 42, 43, 28, 41, 29, 36, 35, 30, 32, 31

(a) Tabulate the frequency distribution of the data.

(b) Draw the frequency histogram.

(c) Draw the frequency polygon.

(d) Tabulate the cumulative frequency distribution.

(e) Draw the cumulative frequency graph.

(f) What is the upper quartile speed?
- (A) 30 mph
- (B) 35 mph
- (C) 40 mph
- (D) 45 mph

(g) What is the median speed?
- (A) 31 mph
- (B) 33 mph
- (C) 35 mph
- (D) 37 mph

(h) What is the standard deviation of the sample data?
- (A) 2.1 mph
- (B) 6.1 mph
- (C) 6.8 mph
- (D) 7.4 mph

(i) What is the sample standard deviation?
- (A) 7.5 mph
- (B) 18 mph
- (C) 35 mph
- (D) 56 mph

(j) What is the sample variance?
- (A) 56 mi^2/hr^2
- (B) 324 mi^2/hr^2
- (C) 1225 mi^2/hr^2
- (D) 3136 mi^2/hr^2

8. A spot speed study is conducted for a stretch of roadway. During a normal day, the speeds were found to be normally distributed with a mean of 46 and a standard deviation of 3.

(a) What is the 50th percentile speed?
- (A) 39
- (B) 43
- (C) 46
- (D) 49

(b) What is the 85th percentile speed?
- (A) 47.1
- (B) 48.3
- (C) 49.1
- (D) 52.7

(c) What is the upper two standard deviation speed?
- (A) 47.2
- (B) 49.3
- (C) 51.1
- (D) 52.0

(d) The daily average speeds for the same stretch of roadway on consecutive normal days were determined by sampling 25 vehicles each day. What is the upper two-standard deviation average speed?
- (A) 46.6
- (B) 47.2
- (C) 52.0
- (D) 54.7

9. The diameters of bolt holes drilled in structural steel members are normally distributed with a mean of 0.502 in and a standard deviation of 0.005 in. Holes are out of specification if their diameters are less than 0.497 in or more than 0.507 in.

(a) What is the probability that a hole chosen at random will be out of specification?
- (A) 0.16
- (B) 0.22
- (C) 0.32
- (D) 0.68

(b) What is the probability that 2 holes out of a sample of 15 will be out of specification?
- (A) 0.074
- (B) 0.12
- (C) 0.15
- (D) 0.32

Hypothesis Testing

10. 100 bearings were tested to failure. The average life was 1520 hr, and the standard deviation was 120 hr. The manufacturer claims a 1600 hr life. Evaluate using confidence limits of 95% and 99%.
- (A) The claim is accurate at both 95% and 99% confidence.
- (B) The claim is inaccurate only at 95%.
- (C) The claim is inaccurate only at 99%.
- (D) The claim is inaccurate at both 95% and 99% confidence.

Curve Fitting

11. (a) Find the best equation for a line passing through the points given.

(b) Find the correlation coefficient.

x	y
400	370
800	780
1250	1210
1600	1560
2000	1980
2500	2450
4000	3950

12. Find the best equation for a line passing through the points given.

s	t
20	43
18	141
16	385
14	1099

13. The number of vehicles lining up behind a flashing railroad crossing has been observed for five trains of different lengths, as given. What is the mathematical formula that relates the two variables?

no. of cars in train	no. of vehicles
2	14.8
5	18.0
8	20.4
12	23.0
27	29.9

14. The following yield data are obtained from five identical treatment plants.

(a) Develop a mathematical equation to correlate the yield and average temperature.

(b) What is the correlation coefficient?

treatment plant	average temperature (T)	average yield (Y)
1	207.1	92.30
2	210.3	92.58
3	200.4	91.56
4	201.1	91.63
5	203.4	91.83

15. The following data are obtained from a soil compaction test. What is the mathematical formula that relates the two variables?

x	y
−1	0
0	1
1	1.4
2	1.7
3	2
4	2.2
5	2.4
6	2.6
7	2.8
8	3

16. Two resistances, the meter resistance and a shunt resistor, are connected in parallel in an ammeter. Most of the current passing through the meter goes through the shunt resistor. In order to determine the accuracy of the resistance of shunt resistors being manufactured for a line of ammeters, a manufacturer tests a sample of 100 shunt resistors. The numbers of shunt resistors with the resistance indicated (to the nearest hundredth of an ohm) are as follows.

0.200 Ω, 1; 0.210 Ω, 3; 0.220 Ω, 5; 0.230 Ω, 10; 0.240 Ω, 17; 0.250 Ω, 40; 0.260 Ω, 13; 0.270 Ω, 6; 0.280 Ω, 3; 0.290 Ω, 2

(a) What is the mean resistance?
   (A) 0.235 Ω
   (B) 0.247 Ω
   (C) 0.251 Ω
   (D) 0.259 Ω

(b) What is the sample standard deviation?
   (A) 0.0003
   (B) 0.010
   (C) 0.016
   (D) 0.24

(c) What is the median resistance?
   (A) 0.221 Ω
   (B) 0.244 Ω
   (C) 0.252 Ω
   (D) 0.259 Ω

(d) What is the sample variance?
   (A) 0.00027
   (B) 0.0083
   (C) 0.0114
   (D) 0.0163

SOLUTIONS

1. (a) There are $4! = 24$ different possible outcomes. By enumeration, there are 9 completely wrong combinations.

$$p\{\text{all wrong}\} = \frac{9}{24} = \boxed{0.375}$$

correct →	7	8	9	10	all wrong
sizes issued	7	8	9	10	
	7	8	10	9	
	7	9	8	10	
	7	9	10	8	
	7	10	8	9	
	7	10	9	8	
	8	9	10	7	X
	8	9	7	10	
	8	10	9	7	
	8	10	7	9	X
	8	7	9	10	
	8	7	10	9	X
	9	10	7	8	X
	9	10	8	7	X
	9	7	10	8	X
	9	7	8	10	
	9	8	7	10	
	9	8	10	7	
	10	7	8	9	X
	10	7	9	8	
	10	8	7	9	
	10	8	9	7	
	10	9	8	7	X
	10	9	7	8	X

(labels above the table: "sizes" over columns 7, 8, 9, 10)

The answer is (B).

(b) If three recruits get the correct size, the fourth recruit will also since there will be only one pair remaining.

$$p\{\text{exactly } 3\} = \boxed{0}$$

The answer is (A).

2. For an exponential distribution function, the mean is given as

$$\mu = \frac{1}{\lambda}$$

For a mean of 23,

$$\mu = 23 = \frac{1}{\lambda}$$

$$\lambda = 0.0435$$

For an exponential distribution function,

$$p\{X < x\} = F(x) = 1 - e^{-\lambda x}$$
$$p\{X > x\} = 1 - p\{X < x\}$$
$$= 1 - F(x)$$
$$= 1 - \left(1 - e^{-\lambda x}\right)$$
$$= e^{-\lambda x}$$

The probability of a random vehicle being processed in 25 sec or more is given by

$$p\{x > 25\} = e^{-(0.0435)(25)}$$
$$= e^{-1.0875}$$
$$= \boxed{0.337}$$

The answer is (C).

3. (a) The distribution is a Poisson distribution with an average of $\lambda = 20$.

The probability for a Poisson distribution is given by

$$p\{x\} = f(x) = \frac{e^{-\lambda}\lambda^x}{x!}$$

The probability of 17 cars is

$$p\{x = 17\} = f(17) = \frac{e^{-20} \times 20^{17}}{17!}$$
$$= \boxed{0.076 \ (7.6\%)}$$

The answer is (A).

(b) The probability of three or fewer cars is given by

$$p\{x \leq 3\} = p\{x = 0\} + p\{x = 1\} + p\{x = 2\} + p\{x = 3\}$$
$$= f(0) + f(1) + f(2) + f(3)$$
$$= \frac{e^{-20} \times 20^0}{0!} + \frac{e^{-20} \times 20^1}{1!} + \frac{e^{-20} \times 20^2}{2!} + \frac{e^{-20} \times 20^3}{3!}$$
$$= 2 \times 10^{-9} + 4.1 \times 10^{-8} + 4.12 \times 10^{-7} + 2.75 \times 10^{-6}$$
$$= 3.2 \times 10^{-6} \ (3.2 \times 10^{-4}\%)$$
$$= \boxed{0.0000032}$$

The answer is (A).

4. $\lambda = \dfrac{1}{\text{MTTF}}$

$= \dfrac{1}{1000} = 0.001$

The reliability function is

$$R\{t\} = e^{-\lambda t} = e^{-0.001t}$$

Since the reliability is greater than 99%,

$$e^{-0.001t} > 0.99$$

$$\ln(e^{-0.001t}) > \ln(0.99)$$

$$-0.001t > \ln(0.99)$$

$$t < -1000 \ln(0.99)$$

$$\boxed{t < 10.05}$$

The maximum operating time such that the reliability remains above 99% is 10.05 hr.

The answer is (D).

5. Find the average.

$\overline{x} = \dfrac{\sum x_i}{n}$

$= \dfrac{1249.529 + 1249.494 + 1249.384 + 1249.348}{4}$

$= 1249.439$

Since the sample population is small, use the sample standard deviation.

$s = \sqrt{\dfrac{\sum(x_i - \overline{x})^2}{n-1}}$

$= \sqrt{\dfrac{\begin{array}{c}(1249.529-1249.439)^2 + (1249.494-1249.439)^2 \\ +(1249.384-1249.439)^2 + (1249.348-1249.439)^2\end{array}}{4-1}}$

$= 0.08647$

From the standard deviation table, a 90% confidence limit falls within $1.645s$ of $\overline{x}$.

$$1249.439 \pm (1.645)(0.08647) = 1249.439 \pm 0.142$$

Therefore, (1249.297, 1249.581) is the 90% confidence range.

(a) By observation, all the readings fall within the 90% confidence range.

The answer is (D).

(b) No readings are unacceptable.

The answer is (A).

(c) Readings outside the 90% confidence limits are unacceptable.

The answer is (B).

(d) The unbiased estimate of the most probable distance is 1249.439.

The answer is (C).

(e) The error for the 90% confidence range is 0.142.

The answer is (C).

(f) If the surveying crew places a marker, measures a distance x, places a second marker, and then measures the same distance x back to the original marker, the ending point should coincide with the original marker. If, due to measurement errors, the ending and starting points do not coincide, the difference is the closure error.

In this example, the survey crew moves around the four sides of a square, so there are two measurements in the x-direction and two measurements in the y-direction. If the errors E_1 and E_2 are known for two measurements, x_1 and x_2, the error associated with the sum or difference $x_1 \pm x_2$ is

$$E\{x_1 \pm x_2\} = \sqrt{E_1^2 + E_2^2}$$

In this case, the error in the x-direction is

$$E_x = \sqrt{(0.1422)^2 + (0.1422)^2}$$

$$= 0.2011$$

The error in the y-direction is calculated the same way and is also 0.2011. E_x and E_y are combined by the Pythagorean theorem to yield

$$E_{\text{closure}} = \sqrt{(0.2011)^2 + (0.2011)^2}$$

$$= \boxed{0.2844}$$

The answer is (C).

(g) In surveying, error may be expressed as a fraction of one or more legs of the traverse. Assume that the total of all four legs is to be used as the basis.

$$\dfrac{0.2844}{(4)(1249)} = \boxed{\dfrac{1}{17{,}567}}$$

The answer is (A).

(h) In surveying, a class 1 third-order error is smaller than 1/10,000. The error of 1/17,567 is smaller than the third-order error; therefore, the error is within the third-order accuracy.

The answer is (C).

(i) An experiment is accurate if it is unchanged by experimental error. Precision is concerned with the repeatability of the experimental results. If an experiment is repeated with identical results, the experiment is said to be precise. However, it is possible to have a highly precise experiment with a large bias.

The answer is (D).

(j) A systematic error is one that is always present and is unchanged from sample to sample. For example, a steel tape that is 0.02 ft short introduces a systematic error.

The answer is (B).

6. (a) The fault tree is as follows.

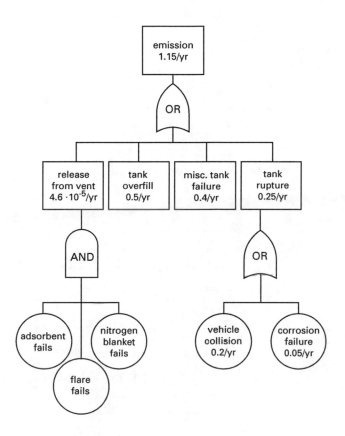

Each of the continuous safety systems is expected to occur 0.5 times per year (i.e., once every two years). Since the repair time is one week or less, the maximum fraction of down time for any one of these three systems is

$$1 \text{ week}/104 \text{ weeks} = 0.0096$$

The probability that all three systems will simultaneously be inoperable is

$$(0.0096)^3 = 8.85 \times 10^{-7}$$

A failure event has been defined as having a one-week duration. There are 52 one-week periods per year. The expected number of one-week periods per year with fugitive emissions is

$$\left(8.85 \times 10^{-7}\right) (52) = 4.6 \times 10^{-5}$$

Failure probabilities are multiplied at AND gates and added at OR gates, from the bottom of the fault tree, working up. The number of tank ruptures per year is $0.2 + 0.05$. In total, fugitive emissions from any cause are likely to occur $4.610 \times 10^{-5} + 0.25 + 0.5 + 0.4 = 1.15$ times per year.

(b) Fugitive emissions from the vent are not a significant factor. Process improvements should be focused on tank overfill prevention and on defining and resolving miscellaneous causes of emissions.

7. (a) and (d) Tabulate the frequency distribution data.

(Note that the lowest speed is 20 mi/hr and the highest speed is 48 mi/hr; therefore, the range is 28 mi/hr. Choose 10 cells with a width of 3 mi/hr.)

midpoint	interval (mi/hr)	frequency	cumulative frequency	cumulative percent
21	20–22	1	1	3
24	23–25	3	4	10
27	26–28	5	9	23
30	29–31	8	17	43
33	32–34	3	20	50
36	35–37	4	24	60
39	38–40	3	27	68
42	41–43	8	35	88
45	44–46	3	38	95
48	47–49	2	40	100

(b)

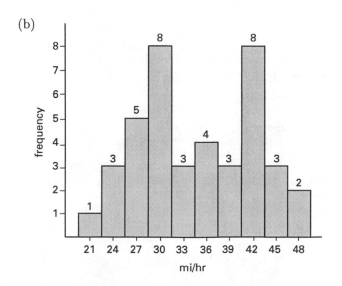

(c)

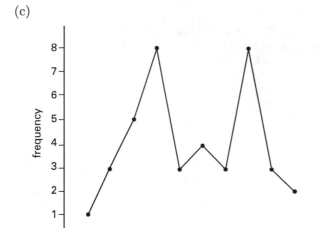

The mean is computed as

$$\overline{x} = \frac{\sum x_i}{n}$$

$$= \frac{1390 \ \dfrac{\text{mi}}{\text{hr}}}{40}$$

$$= \boxed{34.75 \ \text{mi/hr}}$$

The answer is (B).

(h) The standard deviation of the sample data is given as

$$\sigma = \sqrt{\frac{\sum x^2}{n} - \mu^2}$$

$$\sum x^2 = 50{,}496 \ \text{mi}^2/\text{hr}^2$$

Use the sample mean as an unbiased estimator of the population mean, μ.

$$\sigma = \sqrt{\frac{\sum x^2}{n} - \mu^2}$$

$$= \sqrt{\frac{50{,}496 \ \dfrac{\text{mi}^2}{\text{hr}^2}}{40} - \left(34.75 \ \frac{\text{mi}}{\text{hr}}\right)^2}$$

$$= \boxed{7.405 \ \text{mi/hr}}$$

(e)

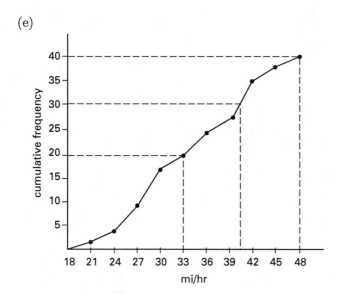

The answer is (D).

(i) The sample standard deviation is given by

$$s = \sqrt{\frac{\sum x^2 - \dfrac{\left(\sum x\right)^2}{n}}{n-1}}$$

$$= \sqrt{\frac{50{,}496 \ \dfrac{\text{mi}^2}{\text{hr}^2} - \dfrac{\left(1390 \ \dfrac{\text{mi}}{\text{hr}}\right)^2}{40}}{40-1}}$$

$$= \boxed{7.500 \ \text{mi/hr}}$$

The answer is (A).

(f) From the cumulative frequency graph in Sol. 6(e), the upper quartile speed occurs at 30 cars or 75%, which corresponds to approximately 40 mi/hr.

The answer is (C).

(g) The mode occurs at two speeds (the frequency at each of the two speeds is 8), which are 30 mi/hr and 42 mi/hr.

From the cumulative frequency chart, the median occurs at 50% or 20 cars and corresponds to 33 mi/hr.

$$\sum x_i = 1390 \ \text{mi/hr}$$

$$n = 40$$

(j) The sample variance is given by the square of the sample standard deviation.

$$s^2 = \left(7.500 \ \frac{\text{mi}}{\text{hr}}\right)^2$$

$$= \boxed{56.25 \ \text{mi}^2/\text{hr}^2}$$

The answer is (A).

8. (a) The 50th percentile speed is the mean speed,

$$46.$$

The answer is (C).

(b) The 85th percentile speed is the speed that is exceeded by only 15% of the measurements. Since this is a normal distribution, App. 11.A can be used. 15% in the upper tail corresponds to 35% between the mean and the 85th percentile. This occurs at approximately $= 1.04\sigma$. The 85th percentile speed is

$$x_{85\%} = \mu + 1.04\sigma$$

$$= 46 + (1.04)(3) = \boxed{49.12}$$

The answer is (C).

(c) The upper 2σ speed is

$$x_{2\sigma} = \mu + 2\sigma$$

$$= 46 + (2)(3) = \boxed{52}$$

The answer is (D).

(d) According to the central limit theorem, the mean of the average speeds is the same as the distribution mean, and the standard deviation of sample means (Eq. 11.68) is

$$s_{\overline{x}} = \frac{\sigma_x}{\sqrt{n}}$$

$$= \frac{3}{\sqrt{25}} = 0.6$$

$$\overline{x}_{2\sigma} = \mu + 2\sigma_{\overline{x}}$$

$$= 46 + (2)(0.6) = \boxed{47.2}$$

The answer is (B).

9. (a) From Eq. 11.43,

$$z_{\text{upper}} = \frac{0.507 \text{ in} - 0.502 \text{ in}}{0.005 \text{ in}} = +1$$

From App. 11.A, the area outside $z = +1$ is

$$0.5 - 0.3413 = 0.1587$$

Since these are symmetrical limits, $z_{\text{lower}} = -1$.

$$\text{total fraction defective} = (2)(0.1587) = \boxed{0.3174}$$

The answer is (C).

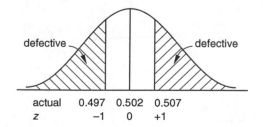

(b) This is a binomial problem.

$$p = p\{\text{defective}\} = 0.3174$$

$$q = 1 - p = 0.6826$$

From Eq. 11.28,

$$f(2) = \binom{15}{2}(0.3174)^2(0.6826)^{13}$$

$$= \left(\frac{15!}{13!2!}\right)(0.3174)^2(0.6826)^{13} = \boxed{0.0739}$$

The answer is (A).

10. This is a typical hypothesis test of two sample population means. The two populations are the original population the manufacturer used to determine the 1600 hr average life value and the new population the sample was taken from. The mean ($\overline{x} = 1520$ hr) of the sample and its standard deviation ($s = 120$ hr) are known, but the mean and standard deviation of a population of average lifetimes are unknown.

Assume that the average lifetime population mean and the sample mean are identical.

$$\overline{x} = \mu = 1520 \text{ hr}$$

The standard deviation of the average lifetime population is

$$\sigma_{\overline{x}} = \frac{s}{\sqrt{n}} = \frac{120 \text{ hr}}{\sqrt{100}} = 12 \text{ hr}$$

The manufacturer can be reasonably sure that the claim of a 1600 hr average life is justified if the average test life is near 1600 hr. "Reasonably sure" must be evaluated based on acceptable probability of being incorrect. If the manufacturer is willing to be wrong with a 5% probability, then a 95% confidence level is required.

Since the direction of bias is known, a one-tailed test is required. To determine if the mean has shifted downward, test the hypothesis that 1600 hr is within the 95% limit of a distribution with a mean of 1520 hr and a standard deviation of 12 hr. From a standard normal table, 5% of a standard normal distribution is outside of $z = 1.645$. Therefore, the 95% confidence limit is

$$1520 \text{ hr} + (1.645)(12 \text{ hr}) = 1540 \text{ hr}$$

The manufacturer can be 95% certain that the average lifetime of the bearings is less than 1600 hr.

If the manufacturer is willing to be wrong with a probability of only 1%, then a 99% confidence limit is required. From the normal table, $z = 2.33$ and the 99% confidence limit is

$$1520 \text{ hr} + (2.33)(12 \text{ hr}) = 1548 \text{ hr}$$

The manufacturer can be 99% certain that the average bearing life is less than 1600 hr.

The answer is (D).

11. (a) Plot the data points to determine if the relationship is linear.

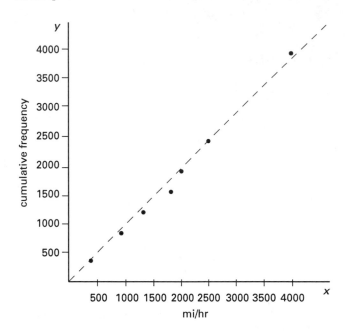

The data appear to be essentially linear. The slope, m, and the y-intercept, b, can be determined using linear regression.

The individual terms are

$$n = 7$$

$$\sum x_i = 400 + 800 + 1250 + 1600 + 2000 + 2500$$
$$+ 4000$$
$$= 12{,}550$$

$$\left(\sum x_i\right)^2 = (12{,}550)^2$$
$$= 1.575 \times 10^8$$

$$\overline{x} = \frac{\sum x_i}{n}$$
$$= \frac{12{,}550}{7}$$
$$= 1792.9$$

$$\sum x_i^2 = (400)^2 + (800)^2 + (1250)^2 + (1600)^2$$
$$+ (2000)^2 + (2500)^2 + (4000)^2$$
$$= 3.117 \times 10^7$$

Similarly,

$$\sum y_i = 370 + 780 + 1210 + 1560 + 1980$$
$$+ 2450 + 3950$$
$$= 12{,}300$$

$$\left(\sum y_i\right)^2 = (12{,}300)^2 = 1.513 \times 10^8$$

$$\overline{y} = \frac{\sum y_i}{n} = \frac{12{,}300}{7} = 1757.1$$

$$\sum y_i^2 = (370)^2 + (780)^2 + (1210)^2 + (1560)^2$$
$$+ (1980)^2 + (2450)^2 + (3950)^2$$
$$= 3.017 \times 10^7$$

Also,

$$\sum x_i y_i = (400)(370) + (800)(780) + (1250)(1210)$$
$$+ (1600)(1560) + (2000)(1980)$$
$$+ (2500)(2450) + (4000)(3950)$$
$$= 3.067 \times 10^7$$

The slope is

$$m = \frac{n \sum x_i y_i - \sum x_i \sum y_i}{n \sum x_i^2 - \left(\sum x_i\right)^2}$$
$$= \frac{(7)(3.067 \times 10^7) - (12{,}550)(12{,}300)}{(7)(3.117 \times 10^7) - (12{,}550)^2}$$
$$= 0.994$$

The y-intercept is

$$b = \overline{y} - m\overline{x}$$
$$= 1757.1 - (0.994)(1792.9)$$
$$= -25.0$$

The least squares equation of the line is

$$y = mx + b$$
$$= \boxed{0.994x - 25.0}$$

(b) The correlation coefficient is

$$r = \frac{n \sum (x_i y_i) - \left(\sum x_i\right)\left(\sum y_i\right)}{\sqrt{\left(n \sum x_i^2 - \left(\sum x_i\right)^2\right)\left(n \sum y_i^2 - \left(\sum y_i\right)^2\right)}}$$
$$= \frac{(7)(3.067 \times 10^7) - (12{,}500)(12{,}300)}{\sqrt{\begin{array}{c}\left((7)(3.117 \times 10^7) - (12{,}500)^2\right)\\ \times \left((7)(3.017 \times 10^7) - (12{,}300)^2\right)\end{array}}}$$
$$\approx \boxed{1.00}$$

12. Plotting the data shows that the relationship is nonlinear.

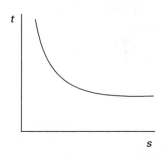

This appears to be an exponential with the form

$$t = ae^{bs}$$

Take the natural log of both sides.

$$\ln t = \ln(ae^{bs})$$
$$= \ln a + \ln(e^{bs})$$
$$= \ln a + bs$$

But, $\ln a$ is just a constant, c.

$$\ln t = c + bs$$

Make the transformation $R = \ln t$.

$$R = c + bs$$

s	R
20	3.76
18	4.95
16	5.95
14	7.00

This is linear.

$$n = 4$$
$$\sum s_i = 20 + 18 + 16 + 14 = 68$$
$$\bar{s} = \frac{\sum s}{n} = \frac{68}{4} = 17$$
$$\sum s_i^2 = (20)^2 + (18)^2 + (16)^2 + (14)^2 = 1176$$
$$\left(\sum s_i\right)^2 = (68)^2 = 4624$$
$$\sum R_i = 3.76 + 4.95 + 5.95 + 7.00 = 21.66$$
$$\bar{R} = \frac{\sum R_i}{n} = \frac{21.66}{4} = 5.415$$
$$\sum R_i^2 = (3.76)^2 + (4.95)^2 + (5.95)^2 + (7.00)^2$$
$$= 123.04$$
$$\left(\sum R_i\right)^2 = (21.66)^2 = 469.16$$
$$\sum s_i R_i = (20)(3.76) + (18)(4.95) + (16)(5.95)$$
$$+ (14)(7.00)$$
$$= 357.5$$

The slope, b, of the transformed line is

$$b = \frac{n\sum s_i R_i - \sum s_i \sum R_i}{n\sum s_i^2 - \left(\sum s_i\right)^2}$$
$$= \frac{(4)(357.5) - (68)(21.66)}{(4)(1176) - (68)^2} = -0.536$$

The intercept is

$$c = \bar{R} - b\bar{s} = 5.415 - (-0.536)(17)$$
$$= 14.527$$

The transformed equation is

$$R = c + bs$$
$$= 14.527 - 0.536s$$

$$\boxed{\ln t = 14.527 - 0.536s}$$

13. The first step is to graph the data.

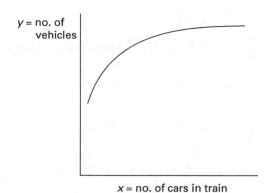

It is assumed that the relationship between the variables has the form $y = a + b \log x$. Therefore, the variable change $z = \log x$ is made, resulting in the following set of data.

z	y
0.301	14.8
0.699	18.0
0.903	20.4
1.079	23.0
1.431	29.9

$$\sum z_i = 4.413$$
$$\sum y_i = 106.1$$
$$\sum z_i^2 = 4.6082$$
$$\sum y_i^2 = 2382.2$$

$$\left(\sum z_i\right)^2 = 19.475$$
$$\left(\sum y_i\right)^2 = 11{,}257.2$$
$$\bar{z} = 0.8826$$
$$\bar{y} = 21.22$$
$$\sum z_i y_i = 103.06$$
$$n = 5$$

The slope is

$$m = \frac{n\sum z_i y_i - \sum z_i \sum y_i}{n\sum z_i^2 - \left(\sum z_i\right)^2}$$
$$= \frac{(5)(103.06) - (4.413)(106.1)}{(5)(4.6082) - 19.475}$$
$$= 13.20$$

The y-intercept is

$$b = \bar{y} - m\bar{z}$$
$$= 21.22 - (13.20)(0.8826)$$
$$= 9.570$$

The resulting equation is

$$y = 9.570 + 13.20z$$

The relationship between x and y is approximately

$$\boxed{y = 9.570 + 13.20 \log x}$$

This is not an optimal correlation, as better correlation coefficients can be obtained if other assumptions about the form of the equation are made. For example, $y = 9.1 + 4\sqrt{x}$ has a better correlation coefficient.

14. (a) Plot the data to verify that they are linear.

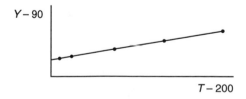

x $T-200$	y $Y-90$
7.1	2.30
10.3	2.58
0.4	1.56
1.1	1.63
3.4	1.83

step 1:

$$\sum x_i = 22.3 \qquad \sum y_i = 9.9$$
$$\sum x_i^2 = 169.43 \qquad \sum y_i^2 = 20.39$$
$$\left(\sum x_i\right)^2 = 497.29 \qquad \left(\sum y_i\right)^2 = 98.01$$
$$\bar{x} = \frac{22.3}{5} = 4.46 \qquad \bar{y} = 1.98$$
$$\sum x_i y_i = 51.54$$

step 2: From Eq. 11.72, the slope is

$$m = \frac{(5)(51.54) - (22.3)(9.9)}{(5)(169.43) - 497.29} = 0.1055$$

step 3: From Eq. 11.73, the y-intercept is

$$b = 1.98 - (0.1055)(4.46) = 1.509$$

The equation of the line is

$$y = 0.1055x + 1.509$$
$$Y - 90 = (0.1055)(T - 200) + 1.509$$
$$Y = \boxed{0.1055T + 70.409}$$

(b) *step 4:* Use Eq. 11.74 to get the correlation coefficient.

$$r = \frac{(5)(51.54) - (22.3)(9.9)}{\sqrt{\left((5)(169.43) - 497.29\right)\left((5)(20.39) - 98.01\right)}}$$
$$= \boxed{0.995}$$

15. Plot the data to see if they are linear.

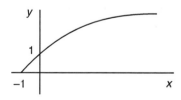

This looks like it could be of the form

$$y = a + b\sqrt{x}$$

However, when x is negative (as in the first point), the function is imaginary. Try shifting the curve to the right, replacing x with $x + 1$.

$$y = a + bz$$
$$z = \sqrt{x + 1}$$

z	y
0	0
1	1
1.414	1.4
1.732	1.7
2	2
2.236	2.2
2.45	2.4
2.65	2.6
2.83	2.8
3	3

Since $y \approx z$, the relationship is

$$y = \sqrt{x + 1}$$

In this problem, the answer was found accidentally. Usually, regression would be necessary.

16. (a)

R	f	fR	fR^2
0.200	1	0.200	0.0400
0.210	3	0.630	0.1323
0.220	5	1.100	0.2420
0.230	10	2.300	0.5290
0.240	17	4.080	0.9792
0.250	40	10.000	2.5000
0.260	13	3.380	0.8788
0.270	6	1.620	0.4374
0.280	3	0.840	0.2352
0.290	2	0.580	0.1682
	100	24.730	6.1421

$$\overline{R} = \frac{\sum fR}{\sum f} = \frac{24.730 \ \Omega}{100} = \boxed{0.2473 \ \Omega}$$

The answer is (B).

(b) The sample standard deviation is given by Eq. 11.63.

$$s = \sqrt{\frac{\sum fR^2 - \dfrac{(\sum fR)^2}{n}}{n - 1}}$$

$$= \sqrt{\frac{6.1421 \ \Omega - \dfrac{(24.73 \ \Omega)^2}{100}}{99}}$$

$$= \boxed{0.0163 \ \Omega}$$

The answer is (C).

(c) 50% of the observations are below the median. 36 observations are below 0.240. 14 more are needed.

$$0.240 \ \Omega + \left(\frac{14}{40}\right)(0.250 \ \Omega - 0.240 \ \Omega) = \boxed{0.2435 \ \Omega}$$

The answer is (B).

(d) $\quad s^2 = (0.0163 \ \Omega)^2 = \boxed{0.0002656 \ \Omega^2}$

The answer is (A).

Mathematical Support

12 Numerical Analysis

PRACTICE PROBLEMS

1. A function is given as $y = 3x^{0.93} + 4.2$. What is the percent error if the value of y at $x = 2.7$ is found by using straight-line interpolation between $x = 2$ and $x = 3$?

    (A) 0.06%
    (B) 0.18%
    (C) 2.5%
    (D) 5.4%

2. Given the following data points, find y by straight-line interpolation for $x = 2.75$.

x	y
1	4
2	6
3	2
4	−14

    (A) 2.1
    (B) 2.4
    (C) 2.7
    (D) 3.0

3. Using the bisection method, find all of the roots of $f(x) = 0$ to the nearest 0.000005.

$$f(x) = x^3 + 2x^2 + 8x - 2$$

SOLUTIONS

1. The actual value at $x = 2.7$ is given by

$$y(x) = 3x^{0.93} + 4.2$$
$$y(2.7) = (3)(2.7)^{0.93} + 4.2$$
$$= 11.756$$

At $x = 3$,
$$y(3) = (3)(3)^{0.93} + 4.2$$
$$= 12.534$$

At $x = 2$,
$$y(2) = (3)(2)^{0.93} + 4.2$$
$$= 9.916$$

Use straight-line interpolation.

$$\frac{x_2 - x}{x_2 - x_1} = \frac{y_2 - y}{y_2 - y_1}$$
$$\frac{3 - 2.7}{3 - 2} = \frac{12.534 - y}{12.534 - 9.916}$$
$$y = 11.749$$

The relative error is given by

$$\frac{\text{actual value} - \text{predicted value}}{\text{actual value}} = \frac{11.756 - 11.749}{11.756}$$

$$= \boxed{0.0006 \ (0.06\%)}$$

The answer is (A).

2. Let $x_1 = 2$; therefore, from the table of data points, $y_1 = 6$. Let $x_2 = 3$; therefore, from the table of data points, $y_2 = 2$.

Let $x = 2.75$. By straight-line interpolation,

$$\frac{x_2 - x}{x_2 - x_1} = \frac{y_2 - y}{y_2 - y_1}$$
$$\frac{3 - 2.75}{3 - 2} = \frac{2 - y}{2 - 6}$$

$$\boxed{y = 3}$$

The answer is (D).

3. $f(x) = x^3 + 2x^2 + 8x - 2$

Try to find an interval in which there is a root.

x	$f(x)$
0	-2
1	9

A root exists in the interval $[0,1]$.

Try $x = \left(\frac{1}{2}\right)(0+1) = 0.5$.

$$f(0.5) = (0.5)^3 + (2)(0.5)^2 + (8)(0.5) - 2 = 2.625$$

A root exists in $[0,0.5]$.

Try $x = 0.25$.

$$f(0.25) = (0.25)^3 + (2)(0.25)^2 + (8)(0.25) - 2 = 0.1406$$

A root exists in $[0,0.25]$.

Try $x = 0.125$.

$$f(0.125) = (0.125)^3 + (2)(0.125)^2 + (8)(0.125) - 2$$
$$= -0.967$$

A root exists in $[0.125,0.25]$.

Try $x = \left(\frac{1}{2}\right)(0.125 + 0.25) = 0.1875$.

Continuing,

$$f(0.1875) = -0.42 \quad [0.1875,0.25]$$
$$f(0.21875) = -0.144 \quad [0.21875,0.25]$$
$$f(0.234375) = -0.002 \quad \text{[This is close enough.]}$$

One root is $x_1 \approx \boxed{0.234375.}$

Try to find the other two roots. Use long division to factor the polynomial.

$$
\begin{array}{r}
x^2 + 2.234375x + 8.52368 \\
x-0.234375 \overline{)\; x^3 + \quad 2x^2 + \quad 8x - 2} \\
-(x^3 - 0.234375x^2) \\
\hline
2.234375x^2 + \quad 8x \\
-(2.234375x^2 - 0.52368x) \\
\hline
8.52368x - 2 \\
-(8.52368x - 1.9977) \\
\hline
\approx 0
\end{array}
$$

Use the quadratic equation to find the roots of $x^2 + 2.234375x + 8.52368$.

$$x_2, x_3 = \frac{-2.234375 \pm \sqrt{(2.234375)^2 - (4)(1)(8.52368)}}{(2)(1)}$$

$$= \boxed{-1.117189 \pm i2.697327} \quad \text{[both imaginary]}$$

13 Properties of Areas

PRACTICE PROBLEMS

1. Locate the centroid of the area.

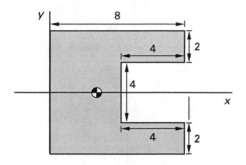

(A) 2.7
(B) 2.9
(C) 3.1
(D) 3.3

2. Replace the distributed load with three concentrated loads, and indicate the points of application.

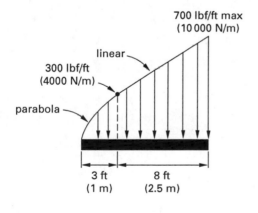

3. Find the centroidal moment of inertia about an axis parallel to the x-axis.

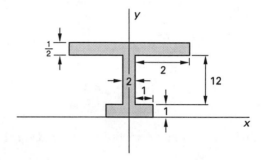

(A) 160 units^4
(B) 290 units^4
(C) 570 units^4
(D) 740 units^4

SOLUTIONS

1. The area is divided into three basic shapes.

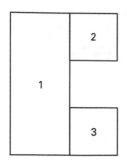

First, calculate the areas of the basic shapes.

$$A_1 = (4)(8) = 32 \text{ units}^2$$
$$A_2 = (4)(2) = 8 \text{ units}^2$$
$$A_3 = (4)(2) = 8 \text{ units}^2$$

Next, find the x-components of the centroids of the basic shapes.

$$x_{c1} = 2 \text{ units}$$
$$x_{c2} = 6 \text{ units}$$
$$x_{c3} = 6 \text{ units}$$

Finally, use Eq. 13.5.

$$x_c = \frac{\sum A_i x_{ci}}{\sum A_i} = \frac{(32)(2) + (8)(6) + (8)(6)}{32 + 8 + 8}$$

$$= \boxed{3.33 \text{ units}}$$

The answer is (D).

2. *Customary U.S. Solution*

The parabolic shape is

$$f(x) = 300\sqrt{\frac{x}{3}}$$

First, use Eq. 13.3 to find the concentrated load given by the area.

$$A = \int f(x)dx = \int_0^3 300\sqrt{\frac{x}{3}}\,dx = \left(\frac{300}{\sqrt{3}}\right)\left[\frac{x^{\frac{3}{2}}}{\frac{3}{2}}\right]_0^3$$

$$= \left(\frac{300}{\left(\frac{3}{2}\right)\sqrt{3}}\right)\left(3^{\frac{3}{2}} - 0^{\frac{3}{2}}\right) = \boxed{600 \text{ lbf}}$$

From Eq. 13.4,

$$dA = f(x)dx = 300\sqrt{\frac{x}{3}}$$

Finally, use Eq. 13.1 to find the location x_c of the concentrated load from the left end.

$$x_c = \frac{\int x\,dx}{A} = \frac{1}{600 \text{ lbf}} \int_0^3 300x\sqrt{\frac{x}{3}}\,dx$$

$$= \frac{300}{600\sqrt{3}} \int_0^3 x^{\frac{3}{2}}\,dx = \left(\frac{300}{600\sqrt{3}}\right)\left[\frac{x^{\frac{5}{2}}}{\frac{5}{2}}\right]_0^3$$

$$= \left(\frac{300}{600\sqrt{3}\left(\frac{5}{2}\right)}\right)\left(3^{\frac{5}{2}} - 0^{\frac{5}{2}}\right) = \boxed{1.8 \text{ ft}}$$

Alternative solution for the parabola:

Use App. 13.A.

$$A = \frac{2bh}{3} = \frac{(2)(300 \text{ lbf})(3 \text{ ft})}{3}$$

$$= \boxed{600 \text{ lbf}}$$

The centroid is located at a distance from the left end of

$$\frac{3h}{5} = \frac{(3)(3 \text{ ft})}{5} = \boxed{1.8 \text{ ft}}$$

The concentrated load for the triangular shape is the area from App. 13.A.

$$A = \frac{bh}{2} = \frac{\left(700 \frac{\text{lbf}}{\text{ft}} - 300 \frac{\text{lbf}}{\text{ft}}\right)(8 \text{ ft})}{2}$$

$$= \boxed{1600 \text{ lbf}}$$

From App. 13.A, the location of the concentrated load from the right end is

$$\frac{h}{3} = \frac{8 \text{ ft}}{3} = \boxed{2.67 \text{ ft}}$$

The concentrated load for the rectangular shape is the area from App. 13.A.

$$A = bh = \left(300 \frac{\text{lbf}}{\text{ft}}\right)(8 \text{ ft})$$

$$= \boxed{2400 \text{ lbf}}$$

From App. 13.A, the location of the concentrated load from the right end is

$$\frac{h}{2} = \frac{8 \text{ ft}}{2} = \boxed{4 \text{ ft}}$$

SI Solution

The parabolic shape is

$$f(x) = 4000\sqrt{x}$$

First, use Eq. 13.3 to find the concentrated load given by the area.

$$A = \int f(x)\,dx = \int_0^1 4000\sqrt{x}\,dx = \left[\frac{4000\,x^{\frac{3}{2}}}{\frac{3}{2}}\right]_0^1$$

$$= \left(\frac{4000}{\frac{3}{2}}\right)\left(1^{\frac{3}{2}} - 0^{\frac{3}{2}}\right) = \boxed{2666.7\ \text{N}}$$

[first concentrated load]

From Eq. 13.4,

$$dA = f(x)\,dx = 4000\sqrt{x}\,dx$$

Finally, use Eq. 13.1 to find the location, x_c, of the concentrated load from the left end.

$$x_c = \frac{\int x\,dA}{A} = \frac{1}{2666.7\ \text{N}}\int_0^1 4000x\sqrt{x}\,dx$$

$$= \frac{4000}{2666.7}\int_0^1 x^{\frac{3}{2}}\,dx = \left(\frac{4000}{2666.7}\right)\left[\frac{x^{\frac{5}{2}}}{\frac{5}{2}}\right]_0^1$$

$$= \left(\frac{4000}{2666.7}\right)\left(\frac{1^{\frac{5}{2}} - 0^{\frac{5}{2}}}{\frac{5}{2}}\right) = \boxed{0.60\ \text{m}} \quad \text{[location]}$$

Alternative solution for the parabola:

Use App. 13.A.

$$A = \frac{2bh}{3} = \frac{(2)\left(4000\ \dfrac{\text{N}}{\text{m}}\right)(1\ \text{m})}{3}$$

$$= \boxed{2666.7\ \text{N}}$$

The centroid is located at

$$\frac{3h}{5} = \frac{(3)(1\ \text{m})}{5} = \boxed{0.6\ \text{m}}$$

The concentrated load for the triangular shape is the area from App. 13.A.

$$A = \frac{bh}{2} = \frac{\left(10\,000\ \dfrac{\text{N}}{\text{m}} - 4000\ \dfrac{\text{N}}{\text{m}}\right)(2.5\ \text{m})}{2}$$

$$= \boxed{7500\ \text{N}} \quad \text{[second concentrated load]}$$

From App. 13.A, the location of the concentrated load from the right end is

$$\frac{h}{3} = \frac{2.5\ \text{m}}{3} = \boxed{0.83\ \text{m}}$$

The concentrated load for the rectangular shape is the area from App. 13.A.

$$A = bh = \left(4000\ \dfrac{\text{N}}{\text{m}}\right)(2.5\ \text{m})$$

$$= \boxed{10\,000\ \text{N}} \quad \text{[third concentrated load]}$$

From App. 13.A, the location of the concentrated load from the right end is

$$\frac{h}{2} = \frac{2.5\ \text{m}}{2} = \boxed{1.25\ \text{m}}$$

3. The area is divided into three basic shapes.

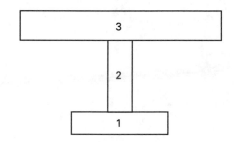

First, calculate the areas of the basic shapes.

$$A_1 = (4)(1) = 4\ \text{units}^2$$
$$A_2 = (2)(12) = 24\ \text{units}^2$$
$$A_3 = (6)(0.5) = 3\ \text{units}^2$$

Next, find the y-components of the centroids of the basic shapes.

$$y_{c1} = 0.5\ \text{units}$$
$$y_{c2} = 7\ \text{units}$$
$$y_{c3} = 13.25\ \text{units}$$

From Eq. 13.6, the centroid of the area is

$$y_c = \frac{\sum A_i y_{ci}}{\sum A_i} = \frac{(4)(0.5) + (24)(7) + (3)(13.25)}{4 + 24 + 3}$$
$$= 6.77\ \text{units}$$

From App. 13.A, the moment of inertia of basic shape 1 about its own centroid is

$$I_{cx1} = \frac{bh^3}{12} = \frac{(4)(1)^3}{12} = 0.33\ \text{units}^4$$

The moment of inertia of basic shape 2 about its own centroid is

$$I_{cx2} = \frac{bh^3}{12} = \frac{(2)(12)^3}{12} = 288 \text{ units}^4$$

The moment of inertia of basic shape 3 about its own centroid is

$$I_{cx3} = \frac{bh^3}{12} = \frac{(6)(0.5)^3}{12} = 0.063 \text{ units}^4$$

From the parallel axis theorem, Eq. 13.20, the moment of inertia of basic shape 1 about the centroidal axis of the section is

$$I_{x1} = I_{cx1} + A_1 d_1^2 = 0.33 + (4)(6.77 - 0.5)^2$$
$$= 157.6 \text{ units}^4$$

The moment of inertia of basic shape 2 about the centroidal axis of the section is

$$I_{x2} = I_{cx2} + A_2 d_2^2 = 288 + (24)(7.0 - 6.77)^2$$
$$= 289.3 \text{ units}^4$$

The moment of inertia of basic shape 3 about the centroidal axis of the section is

$$I_{x3} = I_{cx3} + A_3 d_3^2 = 0.063 + (3)(13.25 - 6.77)^2$$
$$= 126.0 \text{ units}^4$$

The total moment of inertia about the centroidal axis of the section is

$$I_x = I_{x1} + I_{x2} + I_{x3}$$
$$= 157.6 \text{ units}^4 + 289.3 \text{ units}^4 + 126.0 \text{ units}^4$$
$$= \boxed{572.9 \text{ units}^4}$$

The answer is (C).

14 Fluid Properties

PRACTICE PROBLEMS

(Use $g = 32.2$ ft/sec^2 or 9.81 m/s^2 unless told to do otherwise in the problem.)

Pressure

1. What is the absolute pressure if a gauge reads 8.7 psi (60 kPa) vacuum?

    (A) 4 psi (27 kPa)

    (B) 6 psi (41 kPa)

    (C) 8 psi (55 kPa)

    (D) 10 psi (68 kPa)

Viscosity

2. Calculate the kinematic viscosity of air at 80°F (27°C) and 70 psia (480 kPa).

    (A) 3.54×10^{-5} ft^2/sec (3.30×10^{-6} m^2/s)

    (B) 4.25×10^{-5} ft^2/sec (3.96×10^{-6} m^2/s)

    (C) 4.96×10^{-5} ft^2/sec (4.62×10^{-6} m^2/s)

    (D) 6.37×10^{-5} ft^2/sec (5.94×10^{-6} m^2/s)

Solutions

3. Volumes of an 8% solution, a 10% solution, and a 20% solution of nitric acid are to be mixed in order to get 100 mL of a 12% solution. If the 8% solution contributes half of the volume of nitric acid contributed by the 10% and 20% solutions, what volume of 10% acid solution is required?

    (A) 20 mL

    (B) 30 mL

    (C) 50 mL

    (D) 80 mL

SOLUTIONS

1. *Customary U.S. Solution*

$$p_{gage} = -8.7 \text{ lbf/in}^2$$
$$p_{atmospheric} = 14.7 \text{ lbf/in}^2$$

The relationship between absolute, gage, and atmospheric pressure is given by

$$p_{absolute} = p_{gage} + p_{atmospheric}$$
$$= -8.7 \; \frac{\text{lbf}}{\text{in}^2} + 14.7 \; \frac{\text{lbf}}{\text{in}^2}$$
$$= \boxed{6 \text{ lbf/in}^2 \quad (6 \text{ psi})}$$

The answer is (B).

SI Solution

$$p_{gage} = -60 \text{ kPa}$$
$$p_{atmospheric} = 101.3 \text{ kPa}$$

The relationship between absolute, gage, and atmospheric pressure is given by

$$p_{absolute} = p_{gage} + p_{atmospheric}$$
$$= -60 \text{ kPa} + 101.3 \text{ kPa}$$
$$= \boxed{41.3 \text{ kPa}}$$

The answer is (B).

2. *Customary U.S. Solution*

For air at 14.7 psia and 80°F, the absolute viscosity independent of pressure is $\mu = 3.85 \times 10^{-7}$ lbf-sec/ft^2.

Determine the density of air at 70 psia and 80°F. (Assume an ideal gas.)

$$\rho = \frac{p}{RT}$$

For air, $R = 53.3$ lbf-ft/lbm-°R.

Substituting gives

$$\rho = \frac{\left(70 \; \frac{lbf}{in^2}\right)\left(144 \; \frac{in^2}{ft^2}\right)}{\left(53.3 \; \frac{lbf\text{-}ft}{lbm\text{-}°R}\right)(80°F + 460)}$$

$$= 0.350 \; lbm/ft^3$$

The kinematic viscosity, ν, is related to the absolute viscosity by

$$\nu = \frac{\mu g_c}{\rho}$$

$$= \frac{\left(3.85 \times 10^{-7} \; \frac{lbf\text{-}sec}{ft^2}\right)\left(32.2 \; \frac{lbm\text{-}ft}{lbf\text{-}sec^2}\right)}{0.350 \; \frac{lbm}{ft^3}}$$

$$= \boxed{3.54 \times 10^{-5} \; ft^2/sec}$$

The answer is (A).

SI Solution

For air at 480 kPa and 27°C, the absolute viscosity independent of pressure is $\mu = 1.84 \times 10^{-5}$ Pa·s.

Determine the density of air at 480 kPa and 27°C. (Assume an ideal gas.)

$$\rho = \frac{p}{RT}$$

For air, $R = 287$ J/kg·K.

Substituting gives

$$\rho = \frac{(480 \; kPa)\left(1000 \; \frac{Pa}{kPa}\right)}{\left(287 \; \frac{J}{kg\text{-}K}\right)(27°C + 273)}$$

$$= 5.575 \; kg/m^3$$

The kinematic viscosity, ν, is related to the absolute viscosity by

$$\nu = \frac{\mu}{\rho}$$

$$= \frac{1.84 \times 10^{-5} \; Pa\text{-}s}{5.575 \; \frac{kg}{m^3}}$$

$$= \boxed{3.30 \times 10^{-6} \; m^2/s}$$

The answer is (A).

3. Let

$$x = \text{volume of 8\% solution}$$
$$y = \text{volume of 10\% solution}$$
$$z = \text{volume of 20\% solution}$$

The three conditions that must be satisfied are

$$x + y + z = 100 \; mL$$
$$0.08x + 0.10y + 0.20z = (0.12)(100 \; mL) = 12 \; mL$$
$$0.08x = \left(\tfrac{1}{2}\right)(0.10y + 0.20z)$$

Simplifying these equations,

$$\begin{aligned} x \; + \; y \; + \quad z \; &= \; 100 \\ 4x \; + \; 5y \; + \; 10z \; &= \; 600 \\ 8x \; - \; 5y \; - \; 10z \; &= \quad 0 \end{aligned}$$

Adding the second and third equations gives

$$12x = 600$$

$$x = \boxed{50 \; mL}$$

Work with the first two equations to get

$$y + \quad z = 100 - 50 = 50$$
$$5y + 10z = 600 - (4)(50) = 400$$

Multiplying the top equation by -5 and adding to the bottom equation,

$$5z = 150$$
$$z = 30 \; mL$$

From the first equation,

$$y = 20 \; mL$$

The answer is (A).

15 Fluid Statics

PRACTICE PROBLEMS

(Use $g = 32.2$ ft/sec^2 or 9.81 m/s^2 unless told to do otherwise in the problem.)

Buoyancy

1. A blimp contains 10,000 lbm (4500 kg) of hydrogen (specific gas constant = 766.5 ft-lbf/lbm-°R (4124 J/kg·K)) at 56°F (13°C) and 30.2 in Hg (770 mm Hg). What is its lift if the hydrogen and air are in thermal and pressure equilibrium?

  (A) 7.6×10^3 lbf (3.4×10^4 N)
  (B) 1.2×10^4 lbf (5.3×10^4 N)
  (C) 1.3×10^5 lbf (5.9×10^5 N)
  (D) 1.7×10^5 lbf (7.7×10^5 N)

2. A hollow 6 ft (1.8 m) diameter sphere floats half-submerged in seawater. What mass of concrete is required as an external anchor to just submerge the sphere completely?

  (A) 2700 lbm (1200 kg)
  (B) 4200 lbm (1900 kg)
  (C) 5500 lbm (2500 kg)
  (D) 6300 lbm (2700 kg)

SOLUTIONS

1. *Customary U.S. Solution*

Assume the weight of the blimp structure is small (negligible) compared with the weight of the hydrogen.

The lift of the hydrogen-filled blimp (F_{lift}) is equal to the difference between the buoyant force (F_b) and the weight of the hydrogen contained in the blimp (W_H).

$$F_{lift} = F_b - W_H$$

The weight of the hydrogen is calculated from the mass of hydrogen by

$$W_H = \frac{mg}{g_c}$$

$$= \frac{(10{,}000 \text{ lbm})\left(32.2 \, \frac{\text{ft}}{\text{sec}^2}\right)}{32.2 \, \frac{\text{lbm-ft}}{\text{lbf-sec}^2}}$$

$$= 10{,}000 \text{ lbf}$$

The buoyant force is equal to the weight of the displaced air. The weight of the displaced air is calculated by knowing that the volume of the air displaced is equal to the volume of hydrogen enclosed in the blimp. Compute the volume of the hydrogen contained in the blimp by assuming the hydrogen behaves like an ideal gas.

$$V_H = \frac{mRT}{p}$$

For hydrogen, $R = 766.5$ ft-lbf/lbm-°R.

The temperature of the hydrogen is given as 56°F. Convert to absolute temperature (°R).

$$T = 56°F + 460 = 516°R$$

The pressure of the hydrogen is given as 30.2 in Hg. Convert the pressure to units of pounds per square foot.

$$p = (30.2 \text{ in Hg})\left(\frac{1 \, \frac{\text{lbf}}{\text{in}^2}}{2.036 \text{ in Hg}}\right)\left(144 \, \frac{\text{in}^2}{\text{ft}^2}\right)$$

$$= 2136 \text{ lbf/ft}^2$$

Compute the volume of hydrogen.

$$V_H = \frac{mRT}{p}$$

$$= \frac{(10{,}000 \text{ lbm})\left(766.5\ \dfrac{\text{ft-lbf}}{\text{lbm-}^\circ\text{R}}\right)(516^\circ\text{R})}{2136\ \dfrac{\text{lbf}}{\text{ft}^2}}$$

$$= 1.85 \times 10^6 \text{ ft}^3$$

Since the volume of the hydrogen contained in the blimp is equal to the air displaced, the air displaced can be computed from the ideal gas equation by assuming the air behaves like an ideal gas.

$$m = \frac{pV_H}{RT}$$

Since the air and hydrogen are assumed to be in thermal and pressure equilibrium, the temperature and pressure are equal to the value given for the hydrogen.

For air, $R = 53.35$ ft-lbf/lbm-$^\circ$R.

Substituting gives

$$m_{\text{air}} = \frac{pV_H}{RT}$$

$$= \frac{\left(2136\ \dfrac{\text{lbf}}{\text{ft}^2}\right)(1.85 \times 10^6 \text{ ft}^3)}{\left(53.35\ \dfrac{\text{ft-lbf}}{\text{lbm-}^\circ\text{R}}\right)(516^\circ\text{R})}$$

$$= 1.435 \times 10^5 \text{ lbm}$$

Recall that the buoyant force is equal to the weight of the air.

$$F_b = W_{\text{air}} = \frac{mg}{g_c}$$

$$= \frac{(1.435 \times 10^5 \text{ lbm})\left(32.2\ \dfrac{\text{ft}}{\text{sec}^2}\right)}{32.2\ \dfrac{\text{lbm-ft}}{\text{lbf-sec}^2}}$$

$$= 1.435 \times 10^5 \text{ lbf}$$

Therefore, the lift can be calculated as

$$F_{\text{lift}} = F_b - W_H$$

$$= 1.435 \times 10^5 \text{ lbf} - 10{,}000 \text{ lbf}$$

$$= \boxed{1.335 \times 10^5 \text{ lbf}}$$

The answer is (C).

SI Solution

Assume the mass of the blimp structure is small (negligible) compared with the mass of the hydrogen.

The lift of the hydrogen-filled blimp (F_{lift}) is equal to the difference between the buoyant force (F_b) and the weight of the hydrogen contained in the blimp (W_H).

$$F_{\text{lift}} = F_b - W_H$$

The weight of the hydrogen is calculated from the mass of hydrogen by

$$W_H = mg$$

$$= (4500 \text{ kg})\left(9.81\ \frac{\text{m}}{\text{s}^2}\right)$$

$$= 44\,145 \text{ N}$$

The buoyant force is equal to the weight of the displaced air. The weight of the displaced air is calculated by knowing that the volume of the air displaced is equal to the volume of hydrogen enclosed in the blimp. Compute the volume of the hydrogen contained in the blimp by assuming the hydrogen behaves like an ideal gas.

$$V_H = \frac{mRT}{p}$$

For hydrogen, $R = 4124$ J/kg·K.

The temperature of the hydrogen is given as 13°C. Convert to absolute temperature (K).

$$T = 13^\circ\text{C} + 273 = 286\text{K}$$

The pressure of the hydrogen is given as 770 mm Hg. Convert the pressure to units of pascals.

$$p = \frac{(770 \text{ mm Hg})\left(133.4\ \dfrac{\text{kPa}}{\text{m}}\right)}{1000\ \dfrac{\text{mm}}{\text{m}}}$$

$$= 102.7 \text{ kPa}$$

The volume of hydrogen is

$$V_H = \frac{mRT}{p}$$

$$= \frac{(4500 \text{ kg})\left(4124\ \dfrac{\text{J}}{\text{kg·K}}\right)(286\text{K})}{(102.7 \text{ kPa})\left(1000\ \dfrac{\text{Pa}}{\text{kPa}}\right)}$$

$$= 5.168 \times 10^4 \text{ m}^3$$

Since the volume of the hydrogen contained in the blimp is equal to the air displaced, the air displaced can be computed from the ideal gas equation assuming the air behaves like an ideal gas.

$$m = \frac{pV_H}{RT}$$

Since the air and hydrogen are assumed to be in thermal and pressure equilibrium, the temperature and pressure are equal to the value given for the hydrogen.

For air, $R = 287$ J/kg·K.

Substituting gives

$$m_{\text{air}} = \frac{pV_{\text{H}}}{RT}$$

$$= \frac{(102.7 \text{ kPa}) \left(1000 \dfrac{\text{Pa}}{\text{kPa}}\right) (5.168 \times 10^4 \text{ m}^3)}{\left(287 \dfrac{\text{J}}{\text{kg·K}}\right) (286\text{K})}$$

$$= 6.466 \times 10^4 \text{ kg}$$

The buoyant force is equal to the weight of the air, so

$$F_b = W_{\text{air}} = mg$$

$$= (6.466 \times 10^4 \text{ kg}) \left(9.81 \dfrac{\text{m}}{\text{s}^2}\right)$$

$$= 6.34 \times 10^5 \text{ N}$$

Therefore, the lift can be calculated as

$$F_{\text{lift}} = F_b - w_{\text{H}}$$

$$= 6.34 \times 10^5 \text{ N} - 44\,145 \text{ N}$$

$$= \boxed{5.90 \times 10^5 \text{ N}}$$

The answer is (C).

2. *Customary U.S. Solution*

The weight of the sphere is equal to the weight of the displaced volume of water when floating.

The buoyant force is given by

$$F_b = \frac{\rho g V_{\text{displaced}}}{g_c}$$

Since the sphere is half submerged,

$$W_{\text{sphere}} = \left(\tfrac{1}{2}\right) \left(\frac{\rho g V_{\text{sphere}}}{g_c}\right)$$

For seawater, $\rho = 64.0$ lbm/ft^3.

The volume of the sphere is given by

$$V_{\text{sphere}} = \left(\frac{\pi}{6}\right) d^3$$

$$= \left(\frac{\pi}{6}\right) (6 \text{ ft})^3$$

$$= 113.1 \text{ ft}^3$$

The weight of the sphere is

$$W_{\text{sphere}} = \left(\tfrac{1}{2}\right) \left(\frac{\rho g V_{\text{sphere}}}{g_c}\right)$$

$$= \left(\tfrac{1}{2}\right) \left(\frac{\left(64.0 \dfrac{\text{lbm}}{\text{ft}^3}\right) \left(32.2 \dfrac{\text{ft}}{\text{sec}^2}\right) \times (113.1 \text{ ft}^3)}{32.2 \dfrac{\text{lbm-ft}}{\text{lbf-sec}^2}}\right)$$

$$= 3619 \text{ lbf}$$

The buoyant force equation for a fully submerged sphere and anchor can be solved for the concrete volume.

$$W_{\text{sphere}} + W_{\text{concrete}} = (V_{\text{sphere}} + V_{\text{concrete}})\rho_{\text{water}}$$

$$W_{\text{sphere}} + \rho_{\text{concrete}} V_{\text{concrete}} \left(\frac{g}{g_c}\right)$$
$$= (V_{\text{sphere}} + V_{\text{concrete}})\rho_{\text{water}}$$
$$\times \left(\frac{g}{g_c}\right)$$

$$3619 \text{ lbf} + \left(150 \dfrac{\text{lbm}}{\text{ft}^3}\right) (V_{\text{concrete}}) \left(\frac{32.2 \dfrac{\text{ft}}{\text{sec}^2}}{32.2 \dfrac{\text{ft-lbm}}{\text{lbf-sec}^2}}\right)$$

$$= (113.1 \text{ ft}^3 + V_{\text{concrete}})$$

$$\times \left(64.0 \dfrac{\text{lbm}}{\text{ft}^3}\right)$$

$$\times \left(\frac{32.2 \dfrac{\text{ft}}{\text{sec}^2}}{32.2 \dfrac{\text{ft-lbm}}{\text{lbf-sec}^2}}\right)$$

$$V_{\text{concrete}} = 42.09 \text{ ft}^3$$

$$m_{\text{concrete}} = \rho_{\text{concrete}} V_{\text{concrete}}$$

$$= \left(150 \dfrac{\text{lbm}}{\text{ft}^3}\right) (42.09 \text{ ft}^3)$$

$$= \boxed{6314 \text{ lbm}}$$

The answer is (D).

SI Solution

The weight of the sphere is equal to the weight of the displaced volume of water when floating.

The buoyant force is given by

$$F_b = \rho g V_{\text{displaced}}$$

Since the sphere is half submerged,

$$W_{\text{sphere}} = \tfrac{1}{2}\rho g V_{\text{sphere}}$$

For seawater, $\rho = 1024$ kg/m^3.

The volume of the sphere is given by

$$V_{\text{sphere}} = \left(\frac{\pi}{6}\right) d^3$$

$$= \left(\frac{\pi}{6}\right) (1.8 \text{ m})^3$$

$$= 3.054 \text{ m}^3$$

The weight of the sphere required is

$$W_{\text{sphere}} = \tfrac{1}{2}\rho g V_{\text{sphere}}$$

$$= \left(\tfrac{1}{2}\right) \left(1024 \dfrac{\text{kg}}{\text{m}^3}\right) \left(9.81 \dfrac{\text{m}}{\text{s}^2}\right) (3.054 \text{ m}^3)$$

$$= 15\,339 \text{ N}$$

Fluids

The buoyant force equation for a fully submerged sphere and anchor can be solved for the concrete volume.

$$W_{\text{sphere}} + W_{\text{concrete}} = (V_{\text{sphere}} + V_{\text{concrete}})\rho_{\text{water}}$$

$$W_{\text{sphere}} + \rho_{\text{concrete}} g V_{\text{concrete}} = g(V_{\text{sphere}} + V_{\text{concrete}})\rho_{\text{water}}$$

$$15\,339 \text{ N} + \left(2400 \ \frac{\text{kg}}{\text{m}^3}\right)\left(9.81 \ \frac{\text{m}}{\text{s}^2}\right)(V_{\text{concrete}})$$

$$= (3.054 \text{ m}^3 + V_{\text{concrete}})$$

$$\times \left(1024 \ \frac{\text{kg}}{\text{m}^3}\right)\left(9.81 \ \frac{\text{m}}{\text{s}^2}\right)$$

$$V_{\text{concrete}} = 1.136 \text{ m}^3$$

$$m_{\text{concrete}} = \rho_{\text{concrete}} V_{\text{concrete}}$$

$$= \left(2400 \ \frac{\text{kg}}{\text{m}^3}\right)(1.136 \text{ m}^3)$$

$$= \boxed{2726 \text{ kg}}$$

The answer is (D).

16

Fluid Flow Parameters

PRACTICE PROBLEMS

Use the following values unless told to do otherwise in the problem:

$$g = 32.2 \text{ ft/sec}^2 \ (9.81 \text{ m/s}^2)$$

$$\rho_{\text{water}} = 62.4 \text{ lbm/ft}^3 \ (1000 \text{ kg/m}^3)$$

$$p_{\text{atmospheric}} = 14.7 \text{ psia} \ (101.3 \text{ kPa})$$

Hydraulic Radius

1. A 10 in (25 cm) composition pipe is compressed by a tree root until its inside height is only 7.2 in (18 cm). What is its approximate hydraulic radius when flowing half full? State your assumptions.
- (A) 2.2 in (5.5 cm)
- (B) 2.7 in (6.9 cm)
- (C) 3.2 in (8.1 cm)
- (D) 4.5 in (11.4 cm)

2. A pipe with an inside diameter of 18.812 in contains water to a depth of 15.7 in. What is the hydraulic radius? (Work in customary U.S. units only.)
- (A) 4.39 in
- (B) 5.08 in
- (C) 5.72 in
- (D) 6.51 in

SOLUTIONS

1. *Customary U.S. Solution*

The perimeter of the pipe is

$$p = \pi d = \pi(10 \text{ in})$$
$$= 31.42 \text{ in}$$

If the pipe is flowing half-full, the wetted perimeter becomes

$$\text{wetted perimeter} = \tfrac{1}{2}p = \left(\tfrac{1}{2}\right)(31.42 \text{ in})$$
$$= 15.71 \text{ in}$$

Assume the compressed pipe is an elliptical cross section. The ellipse will have a minor axis, b, equal to one-half the height of the compressed pipe or

$$b = \frac{7.2 \text{ in}}{2} = 3.6 \text{ in}$$

When the pipe is compressed, the perimeter of the pipe will remain constant. The perimeter of an ellipse is given by

$$p = 2\pi\sqrt{\tfrac{1}{2}(a^2 + b^2)}$$

Solve for the major axis.

$$a = \sqrt{2\left(\frac{p}{2\pi}\right)^2 - b^2}$$

$$= \sqrt{(2)\left(\frac{31.42 \text{ in}}{2\pi}\right)^2 - (3.6 \text{ in})^2}$$

$$= 6.09 \text{ in}$$

The flow area or area of the ellipse is given by

$$\text{flow area} = \tfrac{1}{2}\pi a b$$
$$= \tfrac{1}{2}\pi(6.09 \text{ in})(3.6 \text{ in})$$
$$= 34.4 \text{ in}^2$$

The hydraulic radius is

$$r_h = \frac{\text{area in flow}}{\text{wetted perimeter}} = \frac{34.4 \text{ in}^2}{15.7 \text{ in}}$$

$$= \boxed{2.19 \text{ in}}$$

The answer is (A).

SI Solution

The perimeter of the pipe is

$$p = \pi d = \pi (25 \text{ cm})$$
$$= 78.54 \text{ cm}$$

If the pipe is flowing half-full, the wetted perimeter becomes

$$\text{wetted perimeter} = \tfrac{1}{2}p = \left(\tfrac{1}{2}\right)(78.54 \text{ cm})$$
$$= 39.27 \text{ cm}$$

Assume the compressed pipe is an elliptical cross section. The ellipse will have a minor axis, b, equal to one-half the height of the compressed pipe or

$$b = \frac{18 \text{ cm}}{2} = 9 \text{ cm}$$

When the pipe is compressed, the perimeter of the pipe will remain constant. The perimeter of an ellipse is given by

$$p = 2\pi\sqrt{\tfrac{1}{2}(a^2 + b^2)}$$

Solve for the major axis.

$$a = \sqrt{2\left(\frac{p}{2\pi}\right)^2 - b^2}$$
$$= \sqrt{(2)\left(\frac{78.54 \text{ cm}}{2\pi}\right)^2 - (9 \text{ cm})^2}$$
$$= 15.2 \text{ cm}$$

The flow area or area of the ellipse is given by

$$\text{flow area} = \tfrac{1}{2}\pi ab = \tfrac{1}{2}\pi(15.2 \text{ cm})(9 \text{ cm})$$
$$= 214.9 \text{ cm}^2$$

The hydraulic radius is

$$r_h = \frac{\text{area in flow}}{\text{wetted perimeter}}$$
$$= \frac{214.9 \text{ cm}^2}{39.27 \text{ cm}}$$
$$= \boxed{5.47 \text{ cm}}$$

The answer is (A).

2. *method 1:* Use App. 7.A for a circular segment.

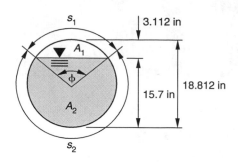

$$r = \frac{d}{2} = \frac{18.812 \text{ in}}{2} = 9.406 \text{ in}$$

$$\phi = (2)\left(\arccos \frac{r-d}{r}\right)$$
$$= (2)\left(\arccos \frac{9.406 \text{ in} - 3.112 \text{ in}}{9.406 \text{ in}}\right) = 1.675 \text{ rad}$$

$$\sin \phi = 0.9946$$

$$A_1 = \tfrac{1}{2}r^2(\phi - \sin \phi)$$
$$= \left(\tfrac{1}{2}\right)(9.406 \text{ in})^2(1.675 - 0.9946) = 30.1 \text{ in}^2$$

$$A_{\text{total}} = A_1 + A_2 = \frac{\pi}{4}D^2 = \left(\frac{\pi}{4}\right)(18.812 \text{ in})^2$$
$$= 277.95 \text{ in}^2$$

$$A_2 = A_{\text{total}} - A_1 = 277.95 \text{ in}^2 - 30.1 \text{ in}^2$$
$$= 247.85 \text{ in}^2$$

$$s_1 = r\phi = (9.406 \text{ in})(1.675) = 15.76 \text{ in}$$

$$s_{\text{total}} = s_1 + s_2 = \pi D = \pi(18.812 \text{ in}) = 59.1 \text{ in}$$

$$s_2 = s_{\text{total}} - s_1 = 59.1 \text{ in} - 15.76 \text{ in} = 43.34 \text{ in}$$

$$r_h = \frac{A_2}{s_2} = \frac{247.85 \text{ in}^2}{43.34 \text{ in}} = \boxed{5.719 \text{ in}}$$

method 2: Use App. 16.A.

$$\frac{d}{D} = \frac{15.7 \text{ in}}{18.812 \text{ in}} = 0.83$$

From App. 16.A, $r_h/D = 0.3041$.

$$r_h = (0.3041)(18.812 \text{ in}) = \boxed{5.72 \text{ in}}$$

The answer is (C).

17 Fluid Dynamics

PRACTICE PROBLEMS

(Use $g = 32.2$ ft/sec^2 (9.81 m/s^2) and 60°F (16°C) water unless told to do otherwise.)

Conservation of Energy

1. 5.0 ft^3/sec (130 L/s) of water flows through a schedule 40 steel pipe that changes gradually in diameter from 6 in (154 mm) at point A to 18 in (429 mm) at point B. Point B is 15 ft (4.6 m) higher than point A. The respective pressures at points A and B are 10 psia (70 kPa) and 7 psia (48.3 kPa). All minor losses are insignificant. What are the direction of flow and velocity at point A?

    (A) 3.2 ft/sec (1 m/s); from A to B
    (B) 25 ft/sec (7.0 m/s); from A to B
    (C) 3.2 ft/sec (1 m/s); from B to A
    (D) 25 ft/sec (7.5 m/s); from B to A

2. Points A and B are separated by 3000 ft of new 6 in schedule-40 steel pipe. 750 gal/min of 60°F water flow from point A to point B. Point B is 60 ft above point A. What must be the pressure at point A if the pressure at B must be 50 psig?

    (A) 87 psig
    (B) 103 psig
    (C) 125 psig
    (D) 167 psig

3. *(Time limit: one hour)* A pipe network connects junctions A, B, C, and D as shown. All pipe sections have a C-value of 150. Water can be added and removed at any of the junctions to achieve the flows listed. Water flows from point A to point D. No flows are backward. The minimum allowable pressure anywhere in the system is 20 psig. All minor losses are insignificant. For simplification, use the nominal pipe sizes.

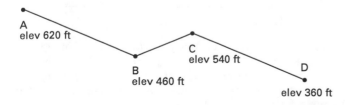

pipe section	length	diameter (in)	flow
A to B	20,000 ft	6	120 gal/min
B to C	10,000 ft	6	160 gal/min
C to D	30,000 ft	4	120 gal/min

(a) What is the pressure at point A?

    (A) 14 psig
    (B) 23 psig
    (C) 31 psig
    (D) 47 psig

(b) What is the elevation of the hydraulic grade line at point A referenced to point D?

    (A) 290 ft
    (B) 330 ft
    (C) 470 ft
    (D) 610 ft

Friction Loss

4. 1.5 ft^3/sec (40 L/s) of 70°F (20°C) water flows through 1200 ft (355 m) of 6 in (nominal) diameter new schedule-40 pipe. What is the friction loss?

    (A) 4 ft (1.2 m)
    (B) 18 ft (5.2 m)
    (C) 36 ft (9.5 m)
    (D) 70 ft (2.1 m)

5. 500 gal/min (30 L/s) of 100°F (40°C) water flows through 300 ft (90 m) of 6 in schedule-40 pipe. The pipe contains two 6 in flanged steel elbows, two full-open gate valves, a full-open 90° angle valve, and a swing check valve. The discharge is located 20 ft (6 m) higher than the entrance. What is the pressure difference between the two ends of the pipe?

    (A) 12 psi (78 kPa)
    (B) 21 psi (140 kPa)
    (C) 45 psi (310 kPa)
    (D) 87 psi (600 kPa)

6. 70°F (20°C) air is flowing at 60 ft/sec (18 m/s) through 300 ft (90 m) of 6 in schedule-40 pipe. The pipe contains two 6 in flanged steel elbows, two full-open gate valves, a full-open 90° angle valve, and a swing check valve. The discharge is located 20 ft (6 m) higher than the entrance. What is the pressure difference between the two ends of the pipe?

- (A) 0.26 psi (1.8 kPa)
- (B) 0.49 psi (3.2 kPa)
- (C) 1.5 psi (10 kPa)
- (D) 13 psi (90 kPa)

Reservoirs

7. Three reservoirs (A, B, and C) are interconnected with a common junction (point D) at elevation 25 ft above an arbitrary reference point. The water levels for reservoirs A, B, and C are at elevations of 50, 40, and 22 ft, respectively. The pipe from reservoir A to the junction is 800 ft of 3 in (nominal) steel pipe. The pipe from reservoir B to the junction is 500 ft of 10 in (nominal) steel pipe. The pipe from reservoir C to the junction is 1000 ft of 4 in (nominal) steel pipe. All pipes are schedule 40 with a friction factor of 0.02. All minor losses and velocity heads can be neglected. What are the direction of flow and pressure at point D?

- (A) out of reservoir B; 500 psf
- (B) out of reservoir B; 930 psf
- (C) into reservoir B; 1100 psf
- (D) into reservoir B; 1260 psf

Water Hammer

8. (*Time limit: one hour*) A cast-iron pipe has an inside diameter of 24 in (600 mm) and a wall thickness of 0.75 in (20 mm). The pipe's modulus of elasticity is 20×10^6 psi (140 GPa). The pipeline is 500 ft (150 m) long. 70°F (20°C) water is flowing at 6 ft/sec (2 m/s).

(a) If a valve is closed instantaneously, what will be the pressure increase experienced in the pipe?

- (A) 48 psi (330 kPa)
- (B) 140 psi (970 kPa)
- (C) 320 psi (2.5 MPa)
- (D) 470 psi (3.2 MPa)

(b) If the pipe is 500 ft (150 m) long, over what length of time must the valve be closed to create a pressure increase equivalent to instantaneous closure?

- (A) 0.25 sec
- (B) 0.68 sec
- (C) 1.6 sec
- (D) 2.1 sec

Parallel Pipe Systems

9. 8 MGD (millions of gallons per day) (350 L/s) of 70°F (20°C) water flows into the new schedule-40 steel pipe network shown. Minor losses are insignificant.

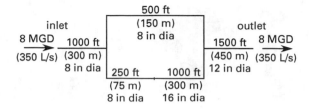

(a) What quantity of water is flowing in the upper branch?

- (A) 1.2 ft^3/sec (0.034 m^3/s)
- (B) 2.9 ft^3/sec (0.081 m^3/s)
- (C) 4.1 ft^3/sec (0.11 m^3/s)
- (D) 5.3 ft^3/sec (0.15 m^3/s)

(b) What is the energy loss per unit mass between the inlet and the outlet?

- (A) 120 ft-lbf/lbm (0.37 kJ/kg)
- (B) 300 ft-lbf/lbm (0.90 kJ/kg)
- (C) 480 ft-lbf/lbm (1.4 kJ/kg)
- (D) 570 ft-lbf/lbm (1.7 kJ/kg)

Pipe Networks

10. A single-loop pipe network is shown. The distance between each junction is 1000 ft. All junctions are on the same elevation. All pipes have a C-value of 100. The volumetric flow rates are to be determined to within 2 gal/min.

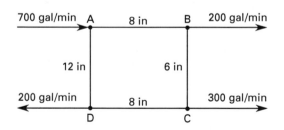

What is the flow rate between junctions B and C?

- (A) 36 gal/min from B to C
- (B) 58 gal/min from B to C
- (C) 84 gal/min from C to B
- (D) 112 gal/min from C to B

11. (*Time limit: one hour*) A double-loop pipe network is shown. The distance between each junction is 1000 ft. The water temperature is 60°F. Elevations and some pressure are known for the junctions. All pipes have a C-value of 100.

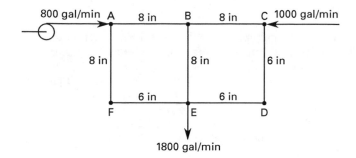

point	pressure	elevation
A		200 ft
B		150 ft
C	40 psig	300 ft
D		150 ft
E		200 ft
F		150 ft

(a) What is the flow rate between junctions B and E?
- (A) 540 gal/min
- (B) 620 gal/min
- (C) 810 gal/min
- (D) 980 gal/min

(b) What is the pressure at point D?
- (A) 51 psig
- (B) 73 psig
- (C) 96 psig
- (D) 120 psig

(c) If the pump receives water at 20 psig (140 kPa), what hydraulic power is required?
- (A) 14 hp
- (B) 28 hp
- (C) 35 hp
- (D) 67 hp

12. *(Time limit: one hour)* The water distribution network shown consists of class A cast-iron pipe installed 1.5 years ago. 1.5 MGD of 50°F water enters at junction A and leaves at junction B. The minimum acceptable pressure at point D is 40 psig (280 kPa).

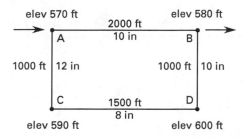

(a) What is the flow rate between junctions A and B?
- (A) 320 gal/min
- (B) 480 gal/min
- (C) 590 gal/min
- (D) 660 gal/min

(b) What is the pressure at point B?
- (A) 32 psig
- (B) 48 psig
- (C) 57 psig
- (D) 88 psig

Tank Discharge

13. The velocity of discharge from a fire hose is 50 ft/sec (15 m/s). The hose is oriented 45° from the horizontal. Disregarding air friction, what is the maximum range of the discharge?
- (A) 45 ft (14 m)
- (B) 78 ft (23 m)
- (C) 91 ft (24 m)
- (D) 110 ft (33 m)

14. A full cylindrical tank 40 ft (12 m) high has a constant diameter of 20 ft. The tank has a 4 in (100 mm) diameter hole in its bottom. The coefficient of discharge for the hole is 0.98. How long will it take for the water level to drop from 40 ft to 20 ft (12 m to 6 m)?
- (A) 950 sec
- (B) 1200 sec
- (C) 1450 sec
- (D) 1700 sec

Venturi Meters

15. A venturi meter with an 8 in diameter throat is installed in a 12 in diameter water line. The venturi is perfectly smooth, so that the discharge coefficient is 1.00. An attached mercury manometer registers a 4 in differential. What is the volumetric flow rate?
- (A) 1.7 ft^3/sec
- (B) 5.2 ft^3/sec
- (C) 6.4 ft^3/sec
- (D) 18 ft^3/sec

16. 60°F (15°C) benzene (specific gravity at 60°F (15°C) of 0.885) flows through an 8 in/3^1/$_2$ in (200 mm/90 mm) venturi meter whose coefficient of discharge is 0.99. A mercury manometer indicates a 4 in difference in the heights of the mercury columns. What is the volumetric flow rate of the benzene?
- (A) 1.2 ft^3/sec (34 L/s)
- (B) 9.1 ft^3/sec (250 L/s)
- (C) 13 ft^3/sec (360 L/s)
- (D) 27 ft^3/sec (760 L/s)

Orifice Meters

17. A sharp-edged orifice meter with a 0.2 ft diameter opening is installed in a 1.0 ft diameter pipe. 70°F water approaches the orifice at 2 ft/sec. What is the indicated pressure drop across the orifice meter?

    (A) 5.9 psi

    (B) 13 psi

    (C) 22 psi

    (D) 47 psi

18. A mercury manometer is used to measure a pressure difference across an orifice meter in a water line. The difference in mercury levels is 7 in (17.8 cm). What is the pressure differential?

    (A) 1.7 psi (12 kPa)

    (B) 3.2 psi (22 kPa)

    (C) 7.9 psi (55 kPa)

    (D) 23 psi (160 kPa)

19. A sharp-edged ISA orifice is used in a schedule 40 steel 12 in (300 mm) water line. (Figure 17.28 is applicable.) The water temperature is 70°F (20°C), and the flow rate is 10 ft^3/sec (250 L/s). The head loss across the orifice must not exceed 25 ft (7.5 m). What is the smallest orifice that can be used?

    (A) 5.5 in (14 cm)

    (B) 7.3 in (19 cm)

    (C) 8.1 in (20 cm)

    (D) 8.9 in (23 cm)

Impulse-Momentum

20. A pipe necks down from 24 in at point A to 12 in at point B. 8 ft^3/sec of 60°F water flow from point A to point B. The pressure head at point A is 20 ft. Friction is insignificant over the distance between points A and B. What are the magnitude and direction of the resultant force on the water?

    (A) 2900 lbf; toward A

    (B) 3500 lbf; toward A

    (C) 2900 lbf; toward B

    (D) 3500 lbf; toward B

21. A 2 in (50 mm) diameter horizontal water jet has an absolute velocity (with respect to a stationary point) of 40 ft/sec (12 m/s) as it strikes a curved blade. The blade is moving horizontally away with an absolute velocity of 15 ft/sec (4.5 m/s). Water is deflected 60° from the horizontal. What is the force on the blade?

    (A) 18 lbf (80 N)

    (B) 26 lbf (110 N)

    (C) 35 lbf (160 N)

    (D) 47 lbf (210 N)

Pumps

22. 2000 gal/min (125 L/s) of brine with a specific gravity of 1.2 pass through an 85% efficient pump. The centerlines of the pump's 12 in (300 mm) inlet and 8 in (200 mm) outlet are at the same elevation. The inlet suction gauge indicates 6 in (150 mm) of mercury below atmospheric. The discharge pressure gauge is located 4 ft (1.2 m) above the centerline of the pump's outlet and indicates 20 psig (138 kPa). What is the input power to the pump?

    (A) 12 hp (8.9 kW)

    (B) 36 hp (26 kW)

    (C) 52 hp (39 kW)

    (D) 87 hp (65 kW)

Turbines

23. 100 ft^3/sec (2.6 kL/s) of water pass through a horizontal turbine. The water's pressure is reduced from 30 psig (210 kPa) to 5 psi (35 kPa) vacuum. Disregarding friction, velocity, and other factors, what power is generated?

    (A) 110 hp (82 kW)

    (B) 380 hp (280 kW)

    (C) 730 hp (540 kW)

    (D) 920 hp (640 kW)

Drag

24. (*Time limit: one hour*) A car traveling through 70°F (20°C) air has the following characteristics.

frontal area	28 ft^2 (2.6 m^2)
mass	3300 lbm (1500 kg)
drag coefficient	0.42
rolling resistance	1% of weight
engine thermal efficiency	28%
fuel heating value	115,000 Btu/gal (460 MJ/L)

(a) Considering only the drag and rolling resistance, what is the fuel consumption when the car is traveling at 55 mi/hr (90 km/h)?

    (A) 0.026 gal/mi (0.0043 L/km)

    (B) 0.038 gal/mi (0.0065 L/km)

    (C) 0.051 gal/mi (0.0097 L/km)

    (D) 0.13 gal/mi (0.022 L/km)

(b) What is the percentage increase in fuel consumption at 65 mi/hr (105 km/h) compared to at 55 mi/hr (90 km/h)?

    (A) 10%

    (B) 20%

    (C) 30%

    (D) 40%

SOLUTIONS

1. *Customary U.S. Solution*

Assume schedule-40 pipe.

$$D_A = 0.5054 \text{ ft}$$
$$D_B = 1.4063 \text{ ft}$$

Let point A be at zero elevation.

The total energy at point A from Bernoulli's equation is

$$E_{tA} = E_p + E_v + E_z$$
$$= \frac{p_A}{\rho} + \frac{v_A^2}{2g_c} + \frac{z_A g}{g_c}$$

At point A, the diameter is 6 in. The velocity at point A is

$$\dot{V} = v_A A_A$$
$$= v_A \left(\frac{\pi}{4}\right) D_A^2$$
$$v_A = \left(\frac{4}{\pi}\right)\left(\frac{\dot{V}}{D_A^2}\right)$$
$$= \left(\frac{4}{\pi}\right)\left(\frac{5.0 \frac{\text{ft}^3}{\text{sec}}}{(0.5054 \text{ ft})^2}\right)$$
$$= \boxed{24.9 \text{ ft/sec}}$$

$$p_A = \left(10 \frac{\text{lbf}}{\text{in}^2}\right)\left(\frac{144 \text{ in}^2}{1 \text{ ft}^2}\right) = 1440 \text{ lbf/ft}^2$$
$$z_A = 0$$

For water, $\rho \approx 62.4 \text{ lbm/ft}^3$.

$$E_{tA} = \frac{p_A}{\rho} + \frac{v_A^2}{2g_c} + \frac{z_A g}{g_c}$$
$$= \frac{1440 \frac{\text{lbf}}{\text{ft}^2}}{62.4 \frac{\text{lbm}}{\text{ft}^3}} + \frac{\left(24.9 \frac{\text{ft}}{\text{sec}}\right)^2}{(2)\left(32.2 \frac{\text{lbm-ft}}{\text{lbf-sec}^2}\right)} + 0$$
$$= 32.7 \text{ ft-lbf/lbm}$$

Similarly, the total energy at point B is

$$v_B = \left(\frac{4}{\pi}\right)\left(\frac{\dot{V}}{D_B^2}\right)$$
$$= \left(\frac{4}{\pi}\right)\left(\frac{5.0 \frac{\text{ft}^3}{\text{sec}}}{(1.4063 \text{ ft})^2}\right)$$
$$= 3.22 \text{ ft/sec}$$

$$p_B = \left(7 \frac{\text{lbf}}{\text{in}^2}\right)\left(\frac{144 \text{ in}^2}{1 \text{ ft}^2}\right) = 1008 \text{ lbf/ft}^2$$
$$z_B = 15 \text{ ft}$$
$$E_{tB} = \frac{p_B}{\rho} + \frac{v_B^2}{2g_c} + \frac{z_B g}{g_c}$$
$$= \frac{1008 \frac{\text{lbf}}{\text{ft}^2}}{62.4 \frac{\text{lbm}}{\text{ft}^3}} + \frac{\left(3.22 \frac{\text{ft}}{\text{sec}}\right)^2}{(2)\left(32.2 \frac{\text{lbm-ft}}{\text{lbf-sec}^2}\right)}$$
$$+ \frac{(15 \text{ ft})\left(32.2 \frac{\text{ft}}{\text{sec}^2}\right)}{32.2 \frac{\text{lbm-ft}}{\text{lbf-sec}^2}}$$
$$= 31.3 \text{ ft-lbf/lbm}$$

Since $E_{tA} > E_{tB}$, the flow is from point A to point B.

The answer is (B).

SI Solution

Let point A be at zero elevation.

The total energy at point A from Bernoulli's equation is

$$E_{tA} = E_p + E_v + E_z$$
$$= \frac{p_A}{\rho} + \frac{v_A^2}{2} + z_A g$$

At point A, the diameter is 154 mm (0.154 m). The velocity at point A is

$$\dot{V} = v_A A_A$$
$$= v_A \left(\frac{\pi}{4}\right) D_A^2$$
$$v_A = \left(\frac{4}{\pi}\right)\left(\frac{\dot{V}}{D_A^2}\right)$$
$$= \left(\frac{4}{\pi}\right)\left(\frac{\left(130 \frac{\text{L}}{\text{s}}\right)\left(\frac{1 \text{ m}^3}{1000 \text{ L}}\right)}{(0.154 \text{ m})^2}\right)$$
$$= \boxed{6.98 \text{ m/s}}$$

$$p_A = 69 \text{ kPa } (69\,000 \text{ Pa})$$
$$z_A = 0$$

For water, $\rho = 1000$ kg/m^3.

$$E_{tA} = \frac{p_A}{\rho} + \frac{v_A^2}{2} + z_A g$$

$$= \frac{70\,000 \text{ Pa}}{1000 \dfrac{\text{kg}}{\text{m}^3}} + \frac{\left(6.98 \dfrac{\text{m}}{\text{s}}\right)^2}{2} + 0$$

$$= 94.36 \text{ J/kg}$$

Similarly, the total energy at point B is

$$v_B = \left(\frac{4}{\pi}\right)\left(\frac{\dot{V}}{D_B^2}\right)$$

$$= \left(\frac{4}{\pi}\right)\left(\frac{\left(130 \dfrac{\text{L}}{\text{s}}\right)\left(\dfrac{1 \text{ m}^3}{1000 \text{ L}}\right)}{(0.429 \text{ m})^2}\right)$$

$$= 0.90 \text{ m/s}$$
$$p_B = 48.3 \text{ kPa (48\,300 Pa)}$$
$$z_B = 4.6 \text{ m}$$

$$E_{tB} = \frac{p_B}{\rho} + \frac{v_B^2}{2} + z_B g$$

$$= \frac{48\,300 \text{ Pa}}{1000 \dfrac{\text{kg}}{\text{m}^3}} + \frac{\left(0.90 \dfrac{\text{m}}{\text{s}}\right)^2}{2}$$

$$+ (4.6 \text{ m})\left(9.81 \dfrac{\text{m}}{\text{s}^2}\right)$$

$$= 93.8 \text{ J/kg}$$

Since $E_{tA} > E_{tB}$, the flow is from point A to point B.

The answer is (B).

2. $\dot{V} = \left(750 \dfrac{\text{gal}}{\text{min}}\right)\left(0.002228 \dfrac{\text{ft}^3\text{-min}}{\text{sec-gal}}\right)$

$$= 1.671 \text{ ft}^3/\text{sec}$$

$D = 0.5054$ ft and $A = 0.2006$ ft^2.

$$v = \frac{\dot{V}}{A} = \frac{1.671 \dfrac{\text{ft}^3}{\text{sec}}}{0.2006 \text{ ft}^2} = 8.33 \text{ ft/sec}$$

For 60°F water,

$$\nu = 1.217 \times 10^{-5} \text{ ft}^2/\text{sec} \quad \text{[App. 14.A]}$$

$$\text{Re} = \frac{vD}{\nu} = \frac{\left(8.33 \dfrac{\text{ft}}{\text{sec}}\right)(0.5054 \text{ ft})}{1.217 \times 10^{-5} \dfrac{\text{ft}^2}{\text{sec}}} = 3.46 \times 10^5$$

For steel,

$$\epsilon = 0.0002 \quad \text{[App. 17.A]}$$

$$\frac{\epsilon}{D} = \frac{0.0002 \text{ ft}}{0.5054 \text{ ft}} \approx 0.0004$$

$$f = 0.0175 \quad \text{[App. 17.B]}$$

From Eq. 17.22,

$$h_f = \frac{(0.0175)(3000 \text{ ft})\left(8.33 \dfrac{\text{ft}}{\text{sec}}\right)^2}{(2)(0.5054 \text{ ft})\left(32.2 \dfrac{\text{ft}}{\text{sec}^2}\right)} = 111.9 \text{ ft}$$

Use the Bernoulli equation. Since velocity is the same at points A and B, it may be omitted.

$$\frac{\left(144 \dfrac{\text{in}^2}{\text{ft}^2}\right)p_1}{62.37 \dfrac{\text{lbf}}{\text{ft}^3}} = \frac{\left(50 \dfrac{\text{lbf}}{\text{in}^2}\right)\left(144 \dfrac{\text{in}^2}{\text{ft}^2}\right)}{62.37 \dfrac{\text{lbf}}{\text{ft}^3}}$$

$$+ 60 \text{ ft} + 111.9 \text{ ft}$$

$$p_1 = \boxed{124.5 \text{ lbf/in}^2 \text{ (psig)}}$$

The answer is (C).

3. From Eq. 17.31, the friction loss from A to B is

$$h_{f,\text{A-B}} = \frac{(10.44)(20,000 \text{ ft})\left(120 \dfrac{\text{gal}}{\text{min}}\right)^{1.85}}{(150)^{1.85}(6 \text{ in})^{4.8655}}$$

$$= 22.6 \text{ ft}$$

Calculate the velocity head.

$$v = \frac{\dot{V}}{A}$$

$$= \frac{\left(120 \dfrac{\text{gal}}{\text{min}}\right)\left(0.002228 \dfrac{\text{ft}^3\text{-min}}{\text{sec-gal}}\right)}{\left(\dfrac{\pi}{4}\right)\left(\dfrac{6 \text{ in}}{12 \dfrac{\text{in}}{\text{ft}}}\right)^2} = 1.36 \text{ ft/sec}$$

$$h_v = \frac{v^2}{2g}$$

$$= \frac{\left(1.36 \dfrac{\text{ft}}{\text{sec}}\right)^2}{(2)\left(32.2 \dfrac{\text{ft}}{\text{sec}^2}\right)} = 0.029 \text{ ft}$$

Velocity heads are low and can be disregarded.

$$h_{f,\text{B-C}} = \frac{(10.44)(10{,}000\text{ ft})\left(160\ \frac{\text{gal}}{\text{min}}\right)^{1.85}}{(150)^{1.85}(6\text{ in})^{4.8655}}$$
$$= 19.25\text{ ft}$$

$$h_{f,\text{C-D}} = \frac{(10.44)(30{,}000\text{ ft})\left(120\ \frac{\text{gal}}{\text{min}}\right)^{1.85}}{(150)^{1.85}(4\text{ in})^{4.8655}}$$
$$= 243.9\text{ ft}$$

Assume a pressure of 20 psig at point A.

$$h_{p,\text{A}} = \frac{\left(20\ \frac{\text{lbf}}{\text{in}^2}\right)\left(144\ \frac{\text{in}^2}{\text{ft}^2}\right)}{62.4\ \frac{\text{lbf}}{\text{ft}^3}} = 46.2\text{ ft}$$

From the Bernoulli equation, ignoring velocity head,

$$h_{p,\text{A}} + z_\text{A} = h_{p,\text{B}} + z_\text{B} + h_{f,\text{A-B}}$$
$$46.2\text{ ft} + 620\text{ ft} = h_{p,\text{B}} + 460\text{ ft} + 22.6\text{ ft}$$
$$h_{p,\text{B}} = 183.6\text{ ft}$$

$$p_\text{B} = \gamma h = \frac{\left(62.4\ \frac{\text{lbf}}{\text{ft}^3}\right)(183.6\text{ ft})}{144\ \frac{\text{in}^2}{\text{ft}^2}}$$
$$= 79.6\text{ lbf/in}^2\text{ (psig)}$$

For B to C,

$$183.6\text{ ft} + 460\text{ ft} = h_{p,\text{C}} + 540\text{ ft} + 19.25\text{ ft}$$
$$h_{p,\text{C}} = 84.35\text{ ft}$$

$$p_\text{C} = \frac{(84.35\text{ ft})\left(62.4\ \frac{\text{lbf}}{\text{ft}^3}\right)}{144\ \frac{\text{in}^2}{\text{ft}^2}}$$
$$= 36.6\text{ lbf/in}^2\text{ (psig)}$$

For C to D,

$$84.35\text{ ft} + 540\text{ ft} = h_{p,\text{D}} + 360\text{ ft} + 243.9\text{ ft}$$
$$h_{p,\text{D}} = 20.45\text{ ft}$$

$$p_\text{D} = \frac{(20.45\text{ ft})\left(62.4\ \frac{\text{lbf}}{\text{ft}^3}\right)}{144\ \frac{\text{in}^2}{\text{ft}^2}}$$
$$= 8.9\text{ lbf/in}^2\text{ (psig)}\quad[\text{too low}]$$

Since p_D is too low, add $20 - 8.9 = 11.1$ lbf/in^2 (psig) to each point.

(a) $\quad p_\text{A} = 20.0\ \frac{\text{lbf}}{\text{in}^2} + 11.1\ \frac{\text{lbf}}{\text{in}^2} = \boxed{31.1\text{ lbf/in}^2\text{ (psig)}}$

$$p_\text{B} = 79.6\ \frac{\text{lbf}}{\text{in}^2} + 11.1\ \frac{\text{lbf}}{\text{in}^2} = 90.7\text{ lbf/in}^2\text{ (psig)}$$

$$p_\text{C} = 36.6\ \frac{\text{lbf}}{\text{in}^2} + 11.1\ \frac{\text{lbf}}{\text{in}^2} = 47.7\text{ lbf/in}^2\text{ (psig)}$$

$$p_\text{D} = 8.9\ \frac{\text{lbf}}{\text{in}^2} + 11.1\ \frac{\text{lbf}}{\text{in}^2} = 20.0\text{ lbf/in}^2\text{ (psig)}$$

The answer is (C).

(b) The elevation of the hydraulic grade line above point D is the sum of the potential and static heads.

$$\Delta h_{\text{A-D}} = z_\text{A} - z_\text{D} + \frac{p_\text{A} - p_\text{D}}{\gamma}$$
$$= 620\text{ ft} - 360\text{ ft}$$
$$+ \frac{\left(31.1\ \frac{\text{lbf}}{\text{in}^2} - 20\ \frac{\text{lbf}}{\text{in}^2}\right)\left(144\ \frac{\text{in}^2}{\text{ft}^2}\right)}{62.4\ \frac{\text{lbf}}{\text{ft}^3}}$$
$$= \boxed{285.6\text{ ft}}$$

The answer is (A).

4. *Customary U.S. Solution*

For 6 in schedule-40 pipe, the internal diameter, D_i, is 0.5054 ft. The internal area is 0.2006 ft^2.

The velocity, v, is calculated from the volumetric flow, $\dot{V}$, and the flow area, A_i, by

$$\text{v} = \frac{\dot{V}}{A_i}$$
$$= \frac{1.5\ \frac{\text{ft}^3}{\text{sec}}}{0.2006\text{ ft}^2}$$
$$= 7.48\text{ ft/sec}$$

For water at 70°F, the kinematic viscosity, ν, is 1.059×10^{-5} ft^2/sec [App. 14.A].

Calculate the Reynolds number.

$$\text{Re} = \frac{D_i \text{v}}{\nu}$$
$$= \frac{(0.5054\text{ ft})\left(7.48\ \frac{\text{ft}}{\text{sec}}\right)}{1.059 \times 10^{-5}\ \frac{\text{ft}^2}{\text{sec}}}$$
$$= 3.57 \times 10^5$$

Since Re > 2100, the flow is turbulent. The friction loss coefficient can be determined from the Moody diagram.

For new steel pipe, the specific roughness, ϵ, is 0.0002 ft.

The relative roughness is

$$\frac{\epsilon}{D_i} = \frac{0.0002 \text{ ft}}{0.5054 \text{ ft}}$$
$$= 0.0004$$

From the Moody diagram with $\text{Re} = 3.57 \times 10^5$ and $e/D_i = 0.0004$, the friction factor, f, can be determined as 0.0174.

Use Darcy's equation to compute the frictional loss.

$$h_f = \frac{fLv^2}{2D_i g}$$

$$= \frac{(0.0174)(1200 \text{ ft}) \left(7.48 \frac{\text{ft}}{\text{sec}}\right)^2}{(2)(0.5054 \text{ ft}) \left(32.2 \frac{\text{ft}}{\text{sec}^2}\right)}$$

$$= \boxed{35.9 \text{ ft}}$$

The answer is (C).

SI Solution

For 6 in pipe, the internal diameter is 154.1 mm and the internal area is $186.5 \times 10^{-4} \text{ m}^2$.

The velocity, v, is calculated from the volumetric flow, $\dot{V}$, and the flow area, A_i, by

$$\text{v} = \frac{\dot{V}}{A_i}$$

$$= \frac{\left(40 \frac{\text{L}}{\text{s}}\right) \left(0.001 \frac{\text{m}^3}{\text{L}}\right)}{186.5 \times 10^{-4} \text{ m}^2}$$

$$= 2.145 \text{ m/s}$$

For water at 20°C, the kinematic viscosity is

$$\nu = \frac{\mu}{\rho} = \frac{1.0050 \times 10^{-3} \text{ Pa·s}}{998.23 \frac{\text{kg}}{\text{m}^3}}$$

$$= 1.007 \times 10^{-6} \text{ m}^2/\text{s}$$

Calculate the Reynolds number.

$$\text{Re} = \frac{D_i \text{v}}{\nu}$$

$$= \frac{(154.1 \text{ mm}) \left(0.001 \frac{\text{m}}{\text{mm}}\right) \left(2.145 \frac{\text{m}}{\text{s}}\right)}{1.007 \times 10^{-6} \frac{\text{m}^2}{\text{s}}}$$

$$= 3.282 \times 10^5$$

Since Re > 2100, the flow is turbulent. The friction loss coefficient can be determined from the Moody diagram.

For new steel pipe, the specific roughness, ϵ, is 6.0×10^{-5} m.

The relative roughness is

$$\frac{\epsilon}{D_i} = \frac{6.0 \times 10^{-5} \text{ m}}{0.1541 \text{ m}}$$
$$= 0.0004$$

From the Moody diagram with $\text{Re} = 3.28 \times 10^5$ and $e/D_i = 0.0004$, the friction factor, f, can be determined as 0.0175.

Use Darcy's equation to compute the frictional loss.

$$h_f = \frac{fLv^2}{2D_i g}$$

$$= \frac{(0.0175)(355 \text{ m}) \left(2.145 \frac{\text{m}}{\text{s}}\right)^2}{(2)(0.1541 \text{ m}) \left(9.81 \frac{\text{m}}{\text{s}^2}\right)}$$

$$= \boxed{9.45 \text{ m}}$$

The answer is (C).

5. *Customary U.S. Solution*

For 6 in schedule-40 pipe, the internal diameter, D_i, is 0.5054 ft. The internal area is 0.2006 ft^2.

The velocity, v, is calculated from the volumetric flow, $\dot{V}$, and the flow area, A_i.

Convert the volumetric flow rate from gal/min to ft^3/sec.

$$\dot{V} = \left(500 \frac{\text{gal}}{\text{min}}\right) \left(\frac{1 \text{ ft}^3}{7.48 \text{ gal}}\right) \left(\frac{1 \text{ min}}{60 \text{ sec}}\right)$$

$$= 1.114 \text{ ft}^3/\text{sec}$$

The velocity is

$$\text{v} = \frac{\dot{V}}{A_i}$$

$$= \frac{1.114 \frac{\text{ft}^3}{\text{sec}}}{0.2006 \text{ ft}^2}$$

$$= 5.55 \text{ ft/sec}$$

Use App. 14.A. For water at 100°F, the kinematic viscosity, ν, is $0.739 \times 10^{-5} \text{ ft}^2/\text{sec}$ and the density is 62.00 lbm/ft^2.

Calculate the Reynolds number.

$$Re = \frac{D_i v}{\nu}$$

$$= \frac{(0.5054 \text{ ft})\left(5.55 \dfrac{\text{ft}}{\text{sec}}\right)}{0.739 \times 10^{-5} \dfrac{\text{ft}^2}{\text{sec}}}$$

$$= 3.80 \times 10^5$$

Since Re > 2100, the flow is turbulent. The friction loss coefficient can be determined from the Moody diagram.

For new steel pipe, the specific roughness, ϵ, is 0.0002 ft.

The relative roughness is

$$\frac{\epsilon}{D_i} = \frac{0.0002 \text{ ft}}{0.5054 \text{ ft}}$$

$$= 0.0004$$

From the Moody diagram with Re $= 3.80 \times 10^5$ and $e/D_i = 0.0004$, the friction factor, f, can be determined as 0.0173.

Use App. 17.D. The equivalent lengths of the valves and fittings are

standard radius elbow	2×8.9 ft =	17.8 ft
gate valve (fully open)	2×3.2 ft =	6.4 ft
90° angle valve (fully open)	1×63.0 ft =	63.0 ft
swing check valve	1×63.0 ft =	63.0 ft
		150.2 ft

The equivalent pipe length is the sum of the straight run of pipe and the equivalent length of pipe for the valves and fittings.

$$L_e = L + L_{\text{fittings}}$$

$$= 300 \text{ ft} + 150.2 \text{ ft}$$

$$= 450.2 \text{ ft}$$

Use Darcy's equation to compute the frictional loss.

$$h_f = \frac{f L_e v^2}{2 D_i g}$$

$$= \frac{(0.0173)(450.2 \text{ ft})\left(5.55 \dfrac{\text{ft}}{\text{sec}}\right)^2}{(2)(0.5054 \text{ ft})\left(32.2 \dfrac{\text{ft}}{\text{sec}^2}\right)}$$

$$= 7.37 \text{ ft}$$

The total difference in head is the sum of the head loss through the pipe, valves, and fittings and the change in elevation.

$$\Delta h = h_f + \Delta z$$

$$= 7.37 \text{ ft} + 20 \text{ ft}$$

$$= 27.37 \text{ ft}$$

The pressure difference between the entrance and discharge can be determined from

$$\Delta p = \gamma \Delta h = \rho h \times \left(\frac{g}{g_c}\right)$$

$$= \frac{\left(62.0 \dfrac{\text{lbm}}{\text{ft}^3}\right)(27.37 \text{ ft})\left(32.2 \dfrac{\text{ft}}{\text{sec}^2}\right)\left(\dfrac{1 \text{ ft}^2}{144 \text{ in}^2}\right)}{32.2 \dfrac{\text{lbm-ft}}{\text{lbf-sec}^2}}$$

$$= \boxed{11.8 \text{ lbf/in}^2 \ (11.8 \text{ psi})}$$

The answer is (A).

SI Solution

For 6 in pipe, the internal diameter is 154.1 mm (0.1541 m).

The internal area is 186.5×10^{-4} m^2.

The velocity, v, is calculated from the volumetric flow, $\dot{V}$, and the flow area, A_i.

$$v = \frac{\dot{V}}{A_i}$$

$$= \frac{\left(30 \dfrac{\text{L}}{\text{s}}\right)\left(0.001 \dfrac{\text{m}^3}{\text{L}}\right)}{186.5 \times 10^{-4} \text{ m}^2}$$

$$= 1.61 \text{ m/s}$$

Use App. 14.B. For water at 40°C, the kinematic viscosity is 6.611×10^{-7} m^2/s and the density is 992.25 kg/m^3.

Calculate the Reynolds number.

$$Re = \frac{D_i v}{\nu}$$

$$= \frac{(0.1541 \text{ m})\left(1.61 \dfrac{\text{m}}{\text{s}}\right)}{6.611 \times 10^{-7} \dfrac{\text{m}^2}{\text{s}}}$$

$$= 3.75 \times 10^5$$

Since Re > 2100, the flow is turbulent. The friction loss coefficient can be determined from the Moody diagram.

For new steel pipe, the specific roughness is 6.0×10^{-5} m.

The relative roughness is

$$\frac{\epsilon}{D_i} = \frac{6.0 \times 10^{-5} \text{ m}}{0.1541 \text{ m}}$$

$$= 0.0004$$

From the Moody diagram with Re $= 3.75 \times 10^5$ and $e/D_i = 0.0004$, the friction factor, f, can be determined as 0.0173.

Use App. 17.D. The equivalent lengths of the valves and fittings are

standard radius elbow	2×2.7 m $=$	5.4 m
gate valve (fully open)	2×1.0 m $=$	2.0 m
90° angle valve (fully open)	1×18.9 m $=$	18.9 m
swing check valve	1×18.9 m $=$	18.9 m
		45.2 m

The equivalent pipe length is the sum of the straight run of pipe and the equivalent length of pipe for the valves and fittings.

$$L_e = L + L_{\text{fittings}}$$
$$= 90 \text{ m} + 45.2 \text{ m}$$
$$= 135.2 \text{ m}$$

Use Darcy's equation to compute the frictional loss.

$$h_f = \frac{fLv^2}{2D_i g}$$
$$= \frac{(0.0173)(135.2 \text{ m})\left(1.61 \frac{\text{m}}{\text{s}^2}\right)^2}{(2)(0.1541 \text{ m})\left(9.81 \frac{\text{m}}{\text{s}^2}\right)}$$
$$= 2.01 \text{ m}$$

The total difference in head is the sum of the head loss through the pipe, valves, and fittings and the change in elevation.

$$\Delta h = h_f + \Delta z$$
$$= 2.01 \text{ m} + 6 \text{ m}$$
$$= 8.01 \text{ m}$$

The pressure difference between the entrance and discharge can be determined from

$$\Delta p = \rho h g$$
$$= \left(992.25 \frac{\text{kg}}{\text{m}^3}\right)(8.01 \text{ m})\left(9.81 \frac{\text{m}}{\text{s}^2}\right)$$
$$= \boxed{77\,969 \text{ Pa} \quad (78 \text{ kPa})}$$

The answer is (A).

6. *Customary U.S. Solution*

For 6 in schedule-40 pipe, the internal diameter, D_i, is 0.5054 ft. The internal area is 0.2006 ft^2.

For air at 70°F, the kinematic viscosity is 16.15×10^{-5} ft^2/sec.

Calculate the Reynolds number.

$$\text{Re} = \frac{D_i v}{\nu}$$
$$= \frac{(0.5054 \text{ ft})\left(60 \frac{\text{ft}}{\text{sec}}\right)}{16.15 \times 10^{-5} \frac{\text{ft}^2}{\text{sec}}}$$
$$= 1.88 \times 10^5$$

Since Re > 2100, the flow is turbulent. The friction loss coefficient can be determined from the Moody diagram.

For new steel pipe, the specific roughness, ϵ, is 0.0002 ft.

The relative roughness is

$$\frac{\epsilon}{D_i} = \frac{0.0002 \text{ ft}}{0.5054 \text{ ft}}$$
$$= 0.0004$$

From the Moody diagram with Re $= 1.88 \times 10^5$ and $e/D_i = 0.0004$, the friction factor, f, can be determined as 0.0184.

From App. 17.D, the equivalent length of the valves and fittings is

standard radius elbow	2×8.9 ft $=$	17.8 ft
gate valve (fully open)	2×3.2 ft $=$	6.4 ft
90° angle valve (fully open)	1×63.0 ft $=$	63.0 ft
swing check valve	1×63.0 ft $=$	63.0 ft
		150.2 ft

The equivalent pipe length is the sum of the straight run of pipe and the equivalent length of pipe for the valves and fittings.

$$L_e = L + L_{\text{fittings}}$$
$$= 300 \text{ ft} + 150.2 \text{ ft}$$
$$= 450.2 \text{ ft}$$

Use Darcy's equation to compute the frictional loss.

$$h_f = \frac{fL_e v^2}{2D_i g}$$
$$= \frac{(0.0184)(450.2 \text{ ft})\left(60 \frac{\text{ft}}{\text{sec}}\right)^2}{(2)(0.5054 \text{ ft})\left(32.2 \frac{\text{ft}}{\text{sec}^2}\right)}$$
$$= 916.2 \text{ ft}$$

The difference in head is the sum of the head loss through the pipe, valves, and fittings and the change in elevation.

$$h = h_f + \Delta z$$
$$= 916.2 \text{ ft} + 20 \text{ ft}$$
$$= 936.2 \text{ ft}$$

Assume the density of the air, ρ, is approximately 0.075 lbm/ft^3.

The pressure difference between the entrance and discharge can be determined from

$$\Delta p = \gamma h = \rho h \times \left(\frac{g}{g_c} \right)$$

$$= \left(0.075 \ \frac{\text{lbm}}{\text{ft}^3} \right) (936.2 \ \text{ft})$$

$$\times \left(\frac{32.2 \ \frac{\text{ft}}{\text{sec}^2}}{32.2 \ \frac{\text{lbm-ft}}{\text{lbf-sec}^2}} \right) \left(\frac{1 \ \text{ft}^2}{144 \ \text{in}^2} \right)$$

$$= \boxed{0.49 \ \text{lbf/in}^2 \quad (0.49 \ \text{psi})}$$

The answer is (B).

SI Solution

For 6 in pipe, the internal diameter, D_i, is 154.1 mm (0.1541 m) and the internal area is 186.5×10^{-4} m^2 [App. 16.C].

For air at 20°C, the kinematic viscosity, ν, is 1.51×10^{-5} m^2/s [App. 14.E].

Calculate the Reynolds number.

$$\text{Re} = \frac{D_i \text{v}}{\nu}$$

$$= \frac{(0.1541 \ \text{m}) \left(18 \ \frac{\text{m}}{\text{s}} \right)}{1.51 \times 10^{-5} \ \frac{\text{m}^2}{\text{s}}}$$

$$= 1.84 \times 10^5$$

Since Re > 2100, the flow is turbulent. The friction loss coefficient can be determined from the Moody diagram.

For new steel pipe, the specific roughness, ϵ, is 6.0×10^{-5} m.

The relative roughness is

$$\frac{\epsilon}{D_i} = \frac{6.0 \times 10^{-5}}{0.1541 \ \text{m}}$$

$$= 0.0004$$

From the Moody diagram with Re = 1.84×10^5 and $e/D_i = 0.0004$, the friction factor, f, can be determined as 0.0185.

Compute the equivalent lengths of the valves and fittings. (Convert from App. 17.D.)

standard radius elbow	2×2.7 m =	5.4 m
gate valve (fully open)	2×1.0 m =	2.0 m
90° angle valve (fully open)	1×18.9 m =	18.9 m
swing check valve	1×18.9 m =	18.9 m
		45.2 m

The equivalent pipe length is the sum of the straight run of pipe and the equivalent length of pipe for the valves and fittings.

$$L_e = L + L_{\text{fittings}}$$

$$= 90 \ \text{m} + 45.2 \ \text{m}$$

$$= 135.2 \ \text{m}$$

Use Darcy's equation to compute the frictional loss.

$$h_f = \frac{f L \text{v}^2}{2 D_i g}$$

$$= \frac{(0.0185)(135.2 \ \text{m}) \left(18 \ \frac{\text{m}}{\text{s}} \right)^2}{(2)(0.1541 \ \text{m}) \left(9.81 \ \frac{\text{m}}{\text{s}^2} \right)}$$

$$= 268.0 \ \text{m}$$

The difference in head is the sum of the head loss through the pipe, valves, and fittings and the change in elevation.

$$\Delta h = h_f + \Delta z$$

$$= 268.0 \ \text{m} + 6 \ \text{m}$$

$$= 274.0 \ \text{m}$$

Assume the density of the air, ρ, is approximately 1.20 kg/m^3.

The pressure difference between the entrance and discharge can be determined from

$$\Delta p = \rho h g$$

$$= \left(1.20 \ \frac{\text{kg}}{\text{m}^3} \right) (274.0 \ \text{m}) \left(9.81 \ \frac{\text{m}}{\text{s}^2} \right)$$

$$= \boxed{3226 \ \text{Pa} \quad (3.23 \ \text{kPa})}$$

The answer is (B).

7. Assume that flows from reservoirs A and B are toward D and then toward C. Then,

$$\dot{V}_{\text{A-D}} + \dot{V}_{\text{B-D}} = \dot{V}_{\text{D-C}}$$

or

$$A_{\text{A}} \text{v}_{\text{A-D}} + A_{\text{B}} \text{v}_{\text{B-D}} - A_{\text{C}} \text{v}_{\text{D-C}} = 0$$

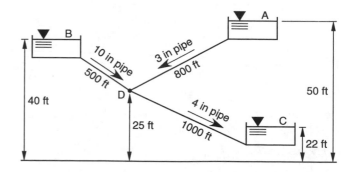

From App. 16.B, for schedule-40 pipe,

$$A_A = 0.05134 \text{ ft}^2 \qquad D_A = 0.2557 \text{ ft}$$
$$A_B = 0.5476 \text{ ft}^2 \qquad D_B = 0.8350 \text{ ft}$$
$$A_C = 0.08841 \text{ ft}^2 \qquad D_C = 0.3355 \text{ ft}$$

$$0.05134v_{A-D} + 0.5476v_{B-D} - 0.08841v_{D-C} = 0 \quad \text{[Eq. 1]}$$

Ignoring the velocity heads, the conservation of energy equation between A and D is

$$z_A = \frac{p_D}{\gamma} + z_D + h_{f,A-D}$$

$$50 \text{ ft} = \frac{p_D}{62.4 \, \frac{\text{lbf}}{\text{ft}^3}} + 25 \text{ ft} + \frac{(0.02)(800 \text{ ft})(v_{A-D})^2}{(2)(0.2557 \text{ ft})\left(32.2 \, \frac{\text{ft}}{\text{sec}^2}\right)}$$

or

$$v_{A-D} = \sqrt{25.73 - 0.0165p_D} \quad \text{[Eq. 2]}$$

Similarly, for B–D,

$$40 \text{ ft} = \frac{p_D}{62.4 \, \frac{\text{lbf}}{\text{ft}^3}} + 25 \text{ ft} + \frac{(0.02)(500 \text{ ft})(v_{B-D})^2}{(2)(0.8350 \text{ ft})\left(32.2 \, \frac{\text{ft}}{\text{sec}^2}\right)}$$

or

$$v_{B-D} = \sqrt{80.66 - 0.0862p_D} \quad \text{[Eq. 3]}$$

For D–C,

$$22 \text{ ft} = \frac{p_D}{62.4 \, \frac{\text{lbf}}{\text{ft}^3}} + 25 \text{ ft} - \frac{(0.02)(1000 \text{ ft})(v_{D-C})^2}{(2)(0.3355 \text{ ft})\left(32.2 \, \frac{\text{ft}}{\text{sec}^2}\right)}$$

or

$$v_{D-C} = \sqrt{3.24 + 0.0173p_D} \quad \text{[Eq. 4]}$$

Equations 1, 2, 3, and 4 must be solved simultaneously. To do this, assume a value for p_D. This value then determines all three velocities in Eqs. 2, 3, and 4. These velocities are substituted into Eq. 1. A trial and error solution yields

$$v_{A-D} = 3.21 \text{ ft/sec}$$
$$v_{B-D} = 0.408 \text{ ft/sec}$$
$$v_{D-C} = 4.40 \text{ ft/sec}$$

$$\boxed{p_D = 933.8 \text{ lbf/ft}^2 \quad \text{(psf)}}$$

$$\boxed{\text{Flow is from B to D.}}$$

The answer is (B).

8. *Customary U.S. Solution*

The composite modulus of elasticity of the pipe and water is given by Eq. 17.205.

For water at 70°F, $E_{\text{water}} = 320 \times 10^3 \text{ lbf/in}^2$.

For cast-iron pipe, $E_{\text{pipe}} = 20 \times 10^6 \text{ lbf/in}^2$.

$$E = \frac{E_{\text{water}} t_{\text{pipe}} E_{\text{pipe}}}{t_{\text{pipe}} E_{\text{pipe}} + d_{\text{pipe}} E_{\text{water}}}$$

$$= \frac{\left(320 \times 10^3 \, \frac{\text{lbf}}{\text{in}^2}\right)(0.75 \text{ in})\left(20 \times 10^6 \, \frac{\text{lbf}}{\text{in}^2}\right)}{(0.75 \text{ in})\left(20 \times 10^6 \, \frac{\text{lbf}}{\text{in}^2}\right)}$$
$$+ (24 \text{ in})\left(320 \times 10^3 \, \frac{\text{lbf}}{\text{in}^2}\right)$$

$$= 2.12 \times 10^5 \text{ lbf/in}^2$$

The speed of sound in the pipe is

$$a = \sqrt{\frac{E g_c}{\rho}}$$

$$= \sqrt{\frac{\left(2.12 \times 10^5 \, \frac{\text{lbf}}{\text{in}^2}\right)\left(\frac{144 \text{ in}^2}{1 \text{ ft}^2}\right)\left(32.2 \, \frac{\text{lbm-ft}}{\text{lbf-sec}^2}\right)}{62.3 \, \frac{\text{lbm}}{\text{ft}^3}}}$$

$$= 3972 \text{ ft/sec}$$

(a) The maximum pressure is given by Eq. 17.204.

$$\Delta p = \frac{\rho a \Delta v}{g_c}$$

$$= \left(\frac{\left(62.3 \, \frac{\text{lbm}}{\text{ft}^3}\right)\left(3972 \, \frac{\text{ft}}{\text{sec}}\right)\left(6 \, \frac{\text{ft}}{\text{sec}}\right)}{32.2 \, \frac{\text{lbm-ft}}{\text{lbf-sec}^2}}\right)$$
$$\times \left(\frac{1 \text{ ft}^2}{144 \text{ in}^2}\right)$$

$$= \boxed{320.2 \text{ lbf/in}^2 \quad (320.2 \text{ psi})}$$

The answer is (C).

(b) The length of time the pressure is constant at the valve is given by

$$t = \frac{2L}{a}$$

$$= \frac{(2)(500 \text{ ft})}{3972 \, \frac{\text{ft}}{\text{sec}}}$$

$$= \boxed{0.252 \text{ sec}}$$

The answer is (A).

SI Solution

The composite modulus of elasticity of the pipe and water is given by Eq. 17.205.

For water at 20°C, $E_{water} = 2.2 \times 10^9$ Pa.

For cast-iron pipe, $E_{pipe} = 1.4 \times 10^{11}$ Pa.

$$E = \frac{E_{water}t_{pipe}E_{pipe}}{t_{pipe}E_{pipe} + d_{pipe}E_{water}}$$

$$= \frac{(2.2 \times 10^9 \text{ Pa})(0.02 \text{ m})(1.4 \times 10^{11} \text{ Pa})}{(0.02 \text{ m})(1.4 \times 10^{11} \text{ Pa}) + (0.6 \text{ m})(2.2 \times 10^9 \text{ Pa})}$$

$$= 1.50 \times 10^9 \text{ Pa}$$

The speed of sound in the pipe is

$$a = \sqrt{\frac{E}{\rho}}$$

$$= \sqrt{\frac{1.50 \times 10^9 \text{ Pa}}{1000 \dfrac{\text{kg}}{\text{m}^3}}}$$

$$= 1225 \text{ m/s}$$

(a) The maximum pressure is given by

$$\Delta p = \rho a \Delta v$$

$$= \left(1000 \frac{\text{kg}}{\text{m}^3}\right)\left(1225 \frac{\text{m}}{\text{s}}\right)\left(2 \frac{\text{m}}{\text{s}}\right)$$

$$= \boxed{2.45 \times 10^6 \text{ Pa} \quad (2450 \text{ kPa})}$$

The answer is (C).

(b) The length of time the pressure is constant at the valve is given by

$$t = \frac{2L}{a}$$

$$= \frac{(2)(150 \text{ m})}{1225 \dfrac{\text{m}}{\text{s}}}$$

$$= \boxed{0.245 \text{ s}}$$

The answer is (A).

9. *Customary U.S. Solution*

First it is necessary to collect data on schedule-40 pipe and water. The fluid viscosity, pipe dimensions, and other parameters can be found in various appendices in Chs. 14 and 16. At 70°F water, $\nu = 1.059 \times 10^{-5}$ ft^2/sec.

From Table 17.2, $\epsilon = 0.0002$ ft.

8 in pipe	$D = 0.6651$ ft	$A = 0.3474$ ft^2
12 in pipe	$D = 0.9948$ ft	$A = 0.7773$ ft^2
16 in pipe	$D = 1.25$ ft	$A = 1.2272$ ft^2

The flow quantity is converted from gallons per minute to cubic feet per second.

$$\dot{V} = \frac{(8 \text{ MGD})\left(10^6 \dfrac{\frac{\text{gal}}{\text{day}}}{\text{MGD}}\right)\left(0.002228 \dfrac{\frac{\text{ft}^3}{\text{sec}}}{\frac{\text{gal}}{\text{min}}}\right)}{\left(24 \dfrac{\text{hr}}{\text{day}}\right)\left(60 \dfrac{\text{min}}{\text{hr}}\right)}$$

$$= 12.378 \text{ ft}^3/\text{sec}$$

For the inlet pipe, the velocity is

$$v = \frac{\dot{V}}{A} = \frac{12.378 \dfrac{\text{ft}^3}{\text{sec}}}{0.3474 \text{ ft}^2} = 35.63 \text{ ft/sec}$$

The Reynolds number is

$$Re = \frac{Dv}{\nu} = \frac{(0.6651 \text{ ft})\left(35.63 \dfrac{\text{ft}}{\text{sec}}\right)}{1.059 \times 10^{-5} \dfrac{\text{ft}^2}{\text{sec}}}$$

$$= 2.24 \times 10^6$$

The relative roughness is

$$\frac{\epsilon}{D} = \frac{0.0002 \text{ ft}}{0.6651 \text{ ft}} = 0.0003$$

From the Moody diagram, $f = 0.015$.

Equation 17.23(b) is used to calculate the frictional energy loss.

$$E_{f,1} = h_f \times \left(\frac{g}{g_c}\right) = \frac{fLv^2}{2Dg_c}$$

$$= \frac{(0.015)(1000 \text{ ft})\left(35.63 \dfrac{\text{ft}}{\text{sec}}\right)^2}{(2)(0.6651 \text{ ft})\left(32.2 \dfrac{\text{lbm-ft}}{\text{lbf-sec}^2}\right)}$$

$$= 444.6 \text{ ft-lbf/lbm}$$

For the outlet pipe, the velocity is

$$v = \frac{\dot{V}}{A} = \frac{12.378 \dfrac{\text{ft}^3}{\text{sec}}}{0.7773 \text{ ft}^2} = 15.92 \text{ ft/sec}$$

The Reynolds number is

$$Re = \frac{Dv}{\nu} = \frac{(0.9948 \text{ ft})\left(15.92 \dfrac{\text{ft}}{\text{sec}}\right)}{1.059 \times 10^{-5} \dfrac{\text{ft}^2}{\text{sec}}}$$

$$= 1.5 \times 10^6$$

The relative roughness is

$$\frac{\epsilon}{D} = \frac{0.0002 \text{ ft}}{0.9948 \text{ ft}} = 0.0002$$

From the Moody diagram, $f = 0.014$.

Equation 17.23(b) is used to calculate the frictional energy loss.

$$E_{f,2} = h_f \times \left(\frac{g}{g_c}\right) = \frac{fLv^2}{2Dg_c}$$

$$= \frac{(0.014)(1500 \text{ ft})\left(15.92 \dfrac{\text{ft}}{\text{sec}}\right)^2}{(2)(0.9948 \text{ ft})\left(32.2 \dfrac{\text{lbm-ft}}{\text{lbf-sec}^2}\right)}$$

$$= 83.1 \text{ ft-lbf/lbm}$$

Assume a 50% split through the two branches. In the upper branch, the velocity is

$$v = \frac{\dot{V}}{A} = \frac{\left(\frac{1}{2}\right)\left(12.378 \dfrac{\text{ft}^3}{\text{sec}}\right)}{0.3474 \text{ ft}^2} = 17.81 \text{ ft/sec}$$

The Reynolds number is

$$\text{Re} = \frac{Dv}{\nu} = \frac{(0.6651 \text{ ft})\left(17.81 \dfrac{\text{ft}}{\text{sec}}\right)}{1.059 \times 10^{-5} \dfrac{\text{ft}^2}{\text{sec}}}$$

$$= 1.1 \times 10^6$$

The relative roughness is

$$\frac{\epsilon}{D} = \frac{0.0002 \text{ ft}}{0.6651 \text{ ft}} = 0.0003$$

From the Moody diagram, $f = 0.015$.

For the 16 in pipe in the lower branch, the velocity is

$$v = \frac{\dot{V}}{A} = \frac{\left(\frac{1}{2}\right)\left(12.378 \dfrac{\text{ft}^3}{\text{sec}}\right)}{1.2272 \text{ ft}^2} = 5.04 \text{ ft/sec}$$

The Reynolds number is

$$\text{Re} = \frac{Dv}{\nu} = \frac{(1.25 \text{ ft})\left(5.04 \dfrac{\text{ft}}{\text{sec}}\right)}{1.059 \times 10^{-5} \dfrac{\text{ft}^2}{\text{sec}}}$$

$$= 5.95 \times 10^5$$

The relative roughness is

$$\frac{\epsilon}{D} = \frac{0.0002 \text{ ft}}{1.25 \text{ ft}} = 0.00016$$

From the Moody diagram, $f = 0.015$.

These values of f for the two branches are fairly insensitive to changes in $\dot{V}$, so they will be used for the rest of the problem in the upper branch.

Eq. 17.23(b) is used to calculate the frictional energy loss in the upper branch.

$$E_{f,\text{upper}} = h_f \times \left(\frac{g}{g_c}\right) = \frac{fLv^2}{2Dg_c}$$

$$= \frac{(0.015)(500 \text{ ft})\left(17.81 \dfrac{\text{ft}}{\text{sec}}\right)^2}{(2)(0.6651 \text{ ft})\left(32.2 \dfrac{\text{lbm-ft}}{\text{lbf-sec}^2}\right)}$$

$$= 55.5 \text{ ft-lbf/lbm}$$

To calculate a loss for any other flow in the upper branch,

$$E_{f,\text{upper 2}} = E_{f,\text{upper}} \left(\frac{\dot{V}}{\left(\frac{1}{2}\right)\left(12.378 \dfrac{\text{ft}^3}{\text{sec}}\right)}\right)^2$$

$$= \left(55.5 \dfrac{\text{ft-lbf}}{\text{lbm}}\right)\left(\frac{\dot{V}}{6.189 \dfrac{\text{ft}^3}{\text{sec}}}\right)^2$$

$$= 1.45\dot{V}^2$$

Similarly, for the lower branch, in the 8 in section,

$$E_{f,\text{lower, 8 in}} = \frac{(0.015)(250 \text{ ft})\left(17.81 \dfrac{\text{ft}}{\text{sec}}\right)^2}{(2)(0.6651 \text{ ft})\left(32.2 \dfrac{\text{lbm-ft}}{\text{lbf-sec}^2}\right)}$$

$$= 27.8 \text{ ft-lbf/lbm}$$

For the lower branch, in the 16 in section,

$$E_{f,\text{lower,16 in}} = \frac{(0.015)(1000 \text{ ft})\left(5.04 \dfrac{\text{ft}}{\text{sec}}\right)^2}{(2)(1.25 \text{ ft})\left(32.2 \dfrac{\text{lbm-ft}}{\text{lbf-sec}^2}\right)}$$

$$= 4.7 \text{ ft-lbf/lbm}$$

The total loss in the lower branch is

$$E_{f,\text{lower}} = E_{f,\text{lower,8 in}} + E_{f,\text{lower,16 in}}$$

$$= 27.8 \dfrac{\text{ft-lbf}}{\text{lbm}} + 4.7 \dfrac{\text{ft-lbf}}{\text{lbm}}$$

$$= 32.5 \text{ ft-lbf/lbm}$$

To calculate a loss for any other flow in the lower branch,

$$E_{f,\text{lower 2}} = E_{f,\text{lower}} \left(\frac{\dot{V}}{\left(\frac{1}{2}\right)\left(12.378 \dfrac{\text{ft}^3}{\text{sec}}\right)} \right)^2$$

$$= \left(32.5 \dfrac{\text{ft-lbf}}{\text{lbm}}\right) \left(\frac{\dot{V}}{6.189 \dfrac{\text{ft}^3}{\text{sec}}} \right)^2$$

$$= 0.85\dot{V}^2$$

Let x be the fraction flowing in the upper branch. Then, because the friction losses are equal,

$$E_{f,\text{upper 2}} = E_{f,\text{lower 2}}$$
$$1.45x^2 = (0.85)(1-x)^2$$
$$x = 0.432$$

(a) $\qquad \dot{V}_{\text{upper}} = (0.432)\left(12.378 \dfrac{\text{ft}^3}{\text{sec}}\right)$

$$= \boxed{5.347 \text{ ft}^3/\text{sec}}$$

The answer is (D).

(b) $\qquad \dot{V}_{\text{lower}} = (1 - 0.432)\left(12.378 \dfrac{\text{ft}^3}{\text{sec}}\right)$

$$= 7.03 \text{ ft}^3/\text{sec}$$

$$E_{f,\text{total}} = E_{f,1} + E_{f,\text{lower 2}} + E_{f,2}$$
$$E_{f,\text{lower 2}} = 0.85\dot{V}_{\text{lower}}^2$$

$$= (0.85)\left(7.03 \dfrac{\text{ft}^3}{\text{sec}}\right)^2$$

$$= 42.0 \text{ ft}$$

$$E_{f,\text{total}} = 444.6 \dfrac{\text{ft-lbf}}{\text{lbm}} + 42.0 \dfrac{\text{ft-lbf}}{\text{lbm}} + 83.1 \dfrac{\text{ft-lbf}}{\text{lbm}}$$

$$= \boxed{569.7 \text{ ft-lbf/lbm}}$$

The answer is (D).

SI Solution

First it is necessary to collect data on schedule-40 pipe and water. The fluid viscosity, pipe dimensions, and other parameters can be found in various appendices in Chs. 14 and 16. At 20°C water, $\nu = 1.007 \times 10^{-6}$ m²/s.

From Table 17.2, $\epsilon = 6 \times 10^{-5}$ m.

8 in pipe	$D = 202.7$ mm $A = 322.7 \times 10^{-4}$ m²
12 in pipe	$D = 303.2$ mm $A = 721.9 \times 10^{-4}$ m²
16 in pipe	$D = 381$ mm $A = 1104 \times 10^{-4}$ m²

For the inlet pipe, the velocity is

$$\text{v} = \frac{\dot{V}}{A} = \frac{\left(350 \dfrac{\text{L}}{\text{s}}\right)\left(\dfrac{1 \text{ m}^3}{1000 \text{ L}}\right)}{322.7 \times 10^{-4} \text{ m}^2} = 10.85 \text{ m/s}$$

The Reynolds number is

$$\text{Re} = \frac{D\text{v}}{\nu} = \frac{(0.2027 \text{ m})\left(10.85 \dfrac{\text{m}}{\text{s}}\right)}{1.007 \times 10^{-6} \dfrac{\text{m}^2}{\text{s}}}$$

$$= 2.18 \times 10^6$$

The relative roughness is

$$\frac{\epsilon}{D} = \frac{6 \times 10^{-5} \text{ m}}{0.2027 \text{ m}} = 0.0003$$

From the Moody diagram, $f = 0.015$.

Equation 17.23(a) is used to calculate the frictional energy loss.

$$E_{f,1} = h_f g = \frac{fL\text{v}^2}{2D}$$

$$= \frac{(0.015)(300 \text{ m})\left(10.85 \dfrac{\text{m}}{\text{s}}\right)^2}{(2)(0.2027 \text{ m})}$$

$$= 1307 \text{ J/kg}$$

For the outlet pipe, the velocity is

$$\text{v} = \frac{\dot{V}}{A} = \frac{\left(350 \dfrac{\text{L}}{\text{s}}\right)\left(\dfrac{1 \text{ m}^3}{1000 \text{ L}}\right)}{721.9 \times 10^{-4} \text{ m}^2} = 4.848 \text{ m/s}$$

The Reynolds number is

$$\text{Re} = \frac{D\text{v}}{\nu} = \frac{(0.3032 \text{ m})\left(4.848 \dfrac{\text{m}}{\text{s}}\right)}{1.007 \times 10^{-6} \dfrac{\text{m}^2}{\text{s}}}$$

$$= 1.46 \times 10^6$$

The relative roughness is

$$\frac{\epsilon}{D} = \frac{6 \times 10^{-5} \text{ m}}{0.3032 \text{ m}} = 0.0002$$

Fluids

From the Moody diagram, $f = 0.014$.

Equation 17.23(a) is used to calculate the frictional energy loss.

$$E_{f,2} = h_f g = \frac{fLv^2}{2D}$$

$$= \frac{(0.014)(450 \text{ m}) \left(4.848 \, \frac{\text{m}}{\text{s}}\right)^2}{(2)(0.3032 \text{ m})}$$

$$= 244.2 \text{ J/kg}$$

Assume a 50% split through the two branches. In the upper branch, the velocity is

$$v = \frac{\dot{V}}{A} = \frac{\left(\frac{1}{2}\right) \left(350 \, \frac{\text{L}}{\text{s}}\right) \left(\frac{1 \text{ m}^3}{1000 \text{ L}}\right)}{322.7 \times 10^{-4} \text{ m}^2} = 5.423 \text{ m/s}$$

The Reynolds number is

$$Re = \frac{Dv}{\nu} = \frac{(0.2027 \text{ m}) \left(5.423 \, \frac{\text{m}}{\text{s}}\right)}{1.007 \times 10^{-6} \, \frac{\text{m}^2}{\text{s}}}$$

$$= 1.1 \times 10^6$$

The relative roughness is

$$\frac{\epsilon}{D} = \frac{6 \times 10^{-5} \text{ m}}{0.2027 \text{ m}} = 0.0003$$

From the Moody diagram, $f = 0.015$.

For the 16 in pipe in the lower branch, the velocity is

$$v = \frac{\dot{V}}{A} = \frac{\left(\frac{1}{2}\right) \left(350 \, \frac{\text{L}}{\text{s}}\right) \left(\frac{1 \text{ m}^3}{1000 \text{ L}}\right)}{1104 \times 10^{-4} \text{ m}^2} = 1.585 \text{ m/s}$$

The Reynolds number is

$$Re = \frac{Dv}{\nu} = \frac{(0.381 \text{ m}) \left(1.585 \, \frac{\text{m}}{\text{s}}\right)}{1.007 \times 10^{-6} \, \frac{\text{m}^2}{\text{s}}}$$

$$= 6.00 \times 10^5$$

The relative roughness is

$$\frac{\epsilon}{D} = \frac{6 \times 10^{-5} \text{ m}}{0.381 \text{ m}} = 0.00016$$

From the Moody diagram, $f = 0.015$.

These values of f for the two branches are fairly insensitive to changes in $\dot{V}$, so they will be used for the rest of the problem in the upper branch.

Eq. 17.23(a) is used to calculate the frictional energy loss in the upper branch.

$$E_{f,\text{upper}} = h_f g = \frac{fLv^2}{2D}$$

$$= \frac{(0.015)(150 \text{ m}) \left(5.423 \, \frac{\text{m}}{\text{s}}\right)^2}{(2)(0.2027 \text{ m})}$$

$$= 163.2 \text{ J/kg}$$

To calculate a loss for any other flow in the upper branch,

$$E_{f,\text{upper 2}} = E_{f,\text{upper}} \left(\frac{\dot{V}}{\left(\frac{1}{2}\right) \left(0.350 \, \frac{\text{m}^3}{\text{s}}\right)}\right)^2$$

$$= \left(163.2 \, \frac{\text{J}}{\text{kg}}\right) \left(\frac{\dot{V}}{0.175 \, \frac{\text{m}^3}{\text{s}}}\right)^2$$

$$= 5329 \dot{V}^2$$

Similarly, for the lower branch, in the 8 in section,

$$E_{f,\text{lower, 8 in}} = \frac{(0.015)(75 \text{ m}) \left(5.423 \, \frac{\text{m}}{\text{s}}\right)^2}{(2)(0.2027 \text{ m})}$$

$$= 81.61 \text{ J/kg}$$

For the lower branch, in the 16 in section,

$$E_{f,\text{lower,16 in}} = \frac{(0.015)(300 \text{ m}) \left(1.585 \, \frac{\text{m}}{\text{s}}\right)^2}{(2)(0.381 \text{ m})}$$

$$= 14.84 \text{ J/kg}$$

The total loss in the lower branch is

$$E_{f,\text{lower}} = E_{f,\text{lower,8 in}} + E_{f,\text{lower,16 in}}$$

$$= 81.61 \, \frac{\text{J}}{\text{kg}} + 14.84 \, \frac{\text{J}}{\text{kg}}$$

$$= 96.45 \text{ J/kg}$$

To calculate a loss for any other flow in the lower branch,

$$E_{f,\text{lower 2}} = E_{f,\text{lower}} \left(\frac{\dot{V}}{\left(\frac{1}{2}\right) \left(0.350 \, \frac{\text{m}^3}{\text{s}}\right)}\right)^2$$

$$= \left(96.45 \, \frac{\text{J}}{\text{kg}}\right) \left(\frac{\dot{V}}{0.175 \, \frac{\text{m}^3}{\text{s}}}\right)^2$$

$$= 3149 \dot{V}^2$$

Let x be the fraction flowing in the upper branch. Then, because the friction losses are equal,

$$E_{f,\text{upper 2}} = E_{f,\text{lower 2}}$$
$$5329x^2 = (3149)(1-x)^2$$
$$x = 0.435$$

(a)
$$\dot{V}_{\text{upper}} = (0.435)\left(0.350\ \frac{m^3}{s}\right)$$

$$= \boxed{0.152\ m^3/s}$$

The answer is (D).

(b)
$$\dot{V}_{\text{lower}} = (1-0.435)\left(0.350\ \frac{m^3}{s}\right)$$

$$= 0.198\ m^3/s$$

$$E_{f,\text{total}} = E_{f,1} + E_{f,\text{lower 2}} + E_{f,2}$$

$$E_{f,\text{lower 2}} = 3149\dot{V}_{\text{lower}}^2$$

$$= (3149)\left(0.198\ \frac{m^3}{s}\right)^2$$

$$= 123.5\ J/kg$$

$$E_{f,\text{total}} = 1307\ \frac{J}{kg} + 123.5\ \frac{J}{kg} + 244.2\ \frac{J}{kg}$$

$$= \boxed{1675\ J/kg\ (1.7\ kJ/kg)}$$

The answer is (D).

10. *steps 1, 2, and 3:*

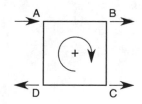

step 4: There is only one loop: ABCD.

step 5: pipe AB: $K' = \dfrac{(10.44)(1000\ \text{ft})}{(100)^{1.85}(8\ \text{in})^{4.87}}$

$$= 8.33 \times 10^{-5}$$

pipe BC: $K' = \dfrac{(10.44)(1000\ \text{ft})}{(100)^{1.85}(6\ \text{in})^{4.87}}$

$$= 3.38 \times 10^{-4}$$

pipe CD: $K' = 8.33 \times 10^{-5}$ [same as AB]

pipe DA: $K' = \dfrac{(10.44)(1000\ \text{ft})}{(100)^{1.85}(12\ \text{in})^{4.87}}$

$$= 1.16 \times 10^{-5}$$

step 6: Assume the flows are as shown in the figure.

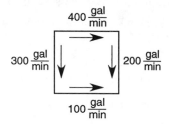

step 7:

$$\delta = \left(\frac{-1}{1.85}\right)$$

$$\times \left(\frac{\begin{array}{c}(8.33 \times 10^{-5})(400)^{1.85} + (3.38 \times 10^{-4})(200)^{1.85} \\ - (8.33 \times 10^{-5})(100)^{1.85} \\ -(1.16 \times 10^{-5})(300)^{1.85}\end{array}}{\begin{array}{c}(8.33 \times 10^{-5})(400)^{0.85} + (3.38 \times 10^{-4})(200)^{0.85} \\ + (8.33 \times 10^{-5})(100)^{0.85} \\ +(1.16 \times 10^{-5})(300)^{0.85}\end{array}} \right)$$

$$= \left(\frac{-1}{1.85}\right)\left(\frac{10.67}{4.98 \times 10^{-2}}\right) = -116\ \text{gal/min}$$

step 8: The adjusted flows are shown.

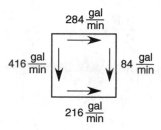

step 7: $\delta = -24$ gal/min

step 8: The adjusted flows are shown.

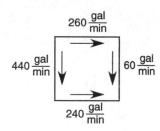

step 7: $\delta = -2$ gal/min [small enough]

step 8: The final adjusted flows are shown.

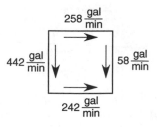

The answer is (B).

11. (a) This is a Hardy Cross problem. The pressure at point C does not change the solution procedure.

step 1: The Hazen-Williams roughness coefficient is given.

step 2: Choose clockwise as positive.

step 3: Nodes are already numbered.

step 4: Choose the loops as shown.

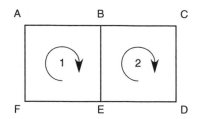

step 5: $\dot{V}$ is in gal/min.
d is in inches.
L is in feet.

Each pipe has the same length.

$$K'_{8 \text{ in}} = \frac{(10.44)(1000 \text{ ft})}{(100)^{1.85}(8 \text{ in})^{4.87}} = 8.33 \times 10^{-5}$$

$$K'_{6 \text{ in}} = \frac{(10.44)(1000 \text{ ft})}{(100)^{1.85}(6 \text{ in})^{4.87}} = 3.38 \times 10^{-4}$$

$$K'_{CE} = 2K'_{6 \text{ in}} = 6.76 \times 10^{-4}$$

$$K'_{EA} = K'_{6 \text{ in}} + K'_{8 \text{ in}} = 4.21 \times 10^{-4}$$

step 6: Assume the flows shown.

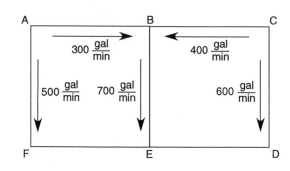

$$AB = 300 \text{ gal/min}$$
$$BE = 700 \text{ gal/min}$$
$$AE = 500 \text{ gal/min}$$
$$CE = 600 \text{ gal/min}$$
$$CB = 400 \text{ gal/min}$$

step 7: If the elevations are included as part of the head loss,

$$\Sigma h = \Sigma K' \dot{V}_a^n + \delta \Sigma n K' \dot{V}_a^{n-1} + z_2 - z_1 = 0$$

However, since the loop closes on itself, $z_2 = z_1$, and the elevations can be omitted.

- First iteration:

Loop 1:

$$\delta_1 = \frac{\begin{array}{l} -\big((8.33 \times 10^{-5})(300)^{1.85} \\ \quad +(8.33 \times 10^{-5})(700)^{1.85} \\ \quad -(4.21 \times 10^{-4})(500)^{1.85}\big) \end{array}}{\begin{array}{l} (1.85)\big((8.33 \times 10^{-5})(300)^{0.85} \\ \quad +(8.33 \times 10^{-5})(700)^{0.85} \\ \quad +(4.21 \times 10^{-4})(500)^{0.85}\big) \end{array}}$$

$$= \frac{-(-22.97)}{0.213} = +108 \text{ gal/min}$$

Loop 2:

$$\delta_2 = \frac{\begin{array}{l} -\big((6.76 \times 10^{-4})(600)^{1.85} \\ \quad -(8.33 \times 10^{-5})(700)^{1.85} \\ \quad -(8.33 \times 10^{-5})(400)^{1.85}\big) \end{array}}{\begin{array}{l} (1.85)\big((6.76 \times 10^{-4})(600)^{0.85} \\ \quad +(8.33 \times 10^{-5})(700)^{0.85} \\ \quad +8.33 \times 10^{-5})(400)^{0.85}\big) \end{array}}$$

$$= \frac{-(72.5)}{0.353} = -205 \text{ gal/min}$$

- Second iteration:

$$
\begin{array}{llll}
AB: & 300 + & 108 & = 408 \text{ gal/min} \\
BE: & 700 + & 108 - (-205) & = 1013 \text{ gal/min} \\
AE: & 500 - & 108 & = 392 \text{ gal/min} \\
CE: & 600 + & (-205) & = 395 \text{ gal/min} \\
CB: & 400 - & (-205) & = 605 \text{ gal/min}
\end{array}
$$

Loop 1:

$$\delta_1 = \frac{-(9.48)}{0.205} = -46 \text{ gal/min}$$

Loop 2:

$$\delta_2 = \frac{-(1.08)}{0.292} = -3.7 \text{ gal/min} \quad [\text{round to } -4]$$

- Third iteration:

$$
\begin{array}{llll}
AB: & 408 + (-46) & = 362 \text{ gal/min} \\
BE: & 1013 + (-46) - (-4) & = 971 \text{ gal/min} \\
AE: & 392 - (-46) & = 438 \text{ gal/min} \\
CE: & 395 + (-4) & = 391 \text{ gal/min} \\
CB: & 605 - (-4) & = 609 \text{ gal/min}
\end{array}
$$

Loop 1:

$$\delta_1 = \frac{-(0.066)}{0.213} = -0.31 \text{ gal/min} \quad [\text{round to } 0]$$

Loop 2:

$$\delta_2 = \frac{-2.4}{0.29} = -8.3 \text{ gal/min} \quad [\text{round to } -8]$$

Use the following flows.

$$
\begin{array}{lll}
\text{AB: } 362 + & 0 & = 362 \text{ gal/min} \\
\text{BE: } 971 + & 0 - (-8) & = \boxed{979 \text{ gal/min}} \\
\text{AE: } 438 - & 0 & = 438 \text{ gal/min} \\
\text{CE: } 391 + (-8) & & = 383 \text{ gal/min} \\
\text{CB: } 609 - (-8) & & = 617 \text{ gal/min}
\end{array}
$$

The answer is (D).

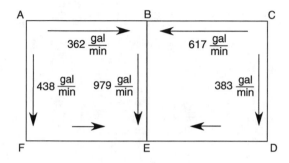

(b) The friction loss in each section is

$$h_{f,\text{AB}} = (8.33 \times 10^{-5})(362)^{1.85} = 4.5 \text{ ft}$$
$$h_{f,\text{BE}} = (8.33 \times 10^{-5})(979)^{1.85} = 28.4 \text{ ft}$$
$$h_{f,\text{AF}} = (8.33 \times 10^{-5})(438)^{1.85} = 6.4 \text{ ft}$$
$$h_{f,\text{FE}} = (3.38 \times 10^{-4})(438)^{1.85} = 26.0 \text{ ft}$$
$$h_{f,\text{CD}} = h_{f,\text{DE}} = (3.38 \times 10^{-4})(383)^{1.85} = 20.3 \text{ ft}$$
$$h_{f,\text{CB}} = (8.33 \times 10^{-5})(617)^{1.85} = 12.1 \text{ ft}$$

$$\gamma = 62.37 \text{ lbf/ft}^3 \quad [\text{at } 60°\text{F}]$$

The pressure at C is 40 psig.

$$h_\text{C} = \frac{\left(40 \ \frac{\text{lbf}}{\text{in}^2}\right)\left(144 \ \frac{\text{in}^2}{\text{ft}^2}\right)}{62.37 \ \frac{\text{lbf}}{\text{ft}^3}} = 92.4 \text{ ft}$$

Next, use the energy continuity equation between adjacent points.

$$h = h_\text{C} + z_\text{C} - z - h_f$$

$$h_\text{D} = 92.4 \text{ ft} + 300 \text{ ft} - 150 \text{ ft} - 20.3 \text{ ft} = 222.1 \text{ ft}$$
$$h_\text{E} = 222.1 \text{ ft} + 150 \text{ ft} - 200 \text{ ft} - 20.3 \text{ ft} = 151.8 \text{ ft}$$
$$h_\text{F} = 151.8 \text{ ft} + 200 \text{ ft} - 150 \text{ ft} + 26 \text{ ft} = 227.8 \text{ ft}$$
$$h_\text{B} = 92.4 \text{ ft} + 300 \text{ ft} - 150 \text{ ft} - 12.1 \text{ ft} = 230.3 \text{ ft}$$
$$h_\text{A} = 230.3 \text{ ft} + 150 \text{ ft} - 200 \text{ ft} + 4.5 \text{ ft} = 184.8 \text{ ft}$$

Using $p = \gamma h$,

$$p_\text{A} = \frac{\left(62.37 \ \frac{\text{lbf}}{\text{ft}^3}\right)(184.8 \text{ ft})}{144 \ \frac{\text{in}^2}{\text{ft}^2}} = 80.0 \text{ lbf/in}^2 \text{ (psig)}$$

$$p_\text{B} = \frac{\left(62.37 \ \frac{\text{lbf}}{\text{ft}^3}\right)(230.3 \text{ ft})}{144 \ \frac{\text{in}^2}{\text{ft}^2}} = 99.7 \text{ psig}$$

$$p_\text{C} = 40 \text{ psig} \quad [\text{given}]$$

$$p_\text{D} = \frac{\left(62.37 \ \frac{\text{lbf}}{\text{ft}^3}\right)(222.1 \text{ ft})}{144 \ \frac{\text{in}^2}{\text{ft}^2}} = \boxed{96.2 \text{ psig}}$$

$$p_\text{E} = \frac{\left(62.37 \ \frac{\text{lbf}}{\text{ft}^3}\right)(151.8 \text{ ft})}{144 \ \frac{\text{in}^2}{\text{ft}^2}} = 65.7 \text{ psig}$$

$$p_\text{F} = \frac{\left(62.37 \ \frac{\text{lbf}}{\text{ft}^3}\right)(227.8 \text{ ft})}{144 \ \frac{\text{in}^2}{\text{ft}^2}} = 98.7 \text{ psig}$$

The answer is (C).

(c) The pressure increase across the pump is

$$\left(80 \ \frac{\text{lbf}}{\text{in}^2} - 20 \ \frac{\text{lbf}}{\text{in}^2}\right)\left(144 \ \frac{\text{in}^2}{\text{ft}^2}\right) = 8640 \text{ lbf/ft}^2$$

Use Table 18.5 to find the hydraulic horsepower.

$$P = \frac{\left(8640 \ \frac{\text{lbf}}{\text{ft}^2}\right)\left(800 \ \frac{\text{gal}}{\text{min}}\right)}{2.468 \times 10^5 \ \frac{\text{lbf-gal}}{\text{min-ft}^2\text{-hp}}} = \boxed{28 \text{ hp}}$$

The answer is (B).

12. This could be solved as a parallel pipe problem or as a pipe network problem.

(a) • Pipe network solution:

step 1: Use the Darcy equation since Hazen-Williams coefficients are not given (or assume C-values based on the age of the pipe).

step 2: Clockwise is positive.

step 3: All nodes are lettered.

step 4: There is only one loop.

step 5: $\epsilon \approx 0.0008$ ft. Assume full turbulence. For class A cast-iron pipe, $D_i = 10.1$ in.

$$\frac{\epsilon}{D} = \frac{0.0008 \text{ ft}}{\dfrac{10.1 \text{ in}}{12 \dfrac{\text{in}}{\text{ft}}}} \approx 0.001$$

$f \approx 0.020$ for full turbulence.

$$K'_{AB} = (1.251 \times 10^{-7})\left(\frac{(0.02)(2000 \text{ ft})}{\left(\dfrac{10.1 \text{ in}}{12 \dfrac{\text{in}}{\text{ft}}}\right)^5} \right) = 11.8 \times 10^{-6}$$

$$K'_{BD} = (1.251 \times 10^{-7})\left(\frac{(0.02)(1000 \text{ ft})}{\left(\dfrac{10.1 \text{ in}}{12 \dfrac{\text{in}}{\text{ft}}}\right)^5} \right) = 5.92 \times 10^{-6}$$

$$K'_{DC} = (1.251 \times 10^{-7})\left(\frac{(0.02)(1500 \text{ ft})}{\left(\dfrac{8.13 \text{ in}}{12 \dfrac{\text{in}}{\text{ft}}}\right)^5} \right) = 26.3 \times 10^{-6}$$

$$K'_{CA} = (1.251 \times 10^{-7})\left(\frac{(0.02)(1000 \text{ ft})}{\left(\dfrac{12.12 \text{ in}}{12 \dfrac{\text{in}}{\text{ft}}}\right)^5} \right) = 2.38 \times 10^{-6}$$

step 6: Assume $\dot{V}_{AB} = 1$ MGD.

$$\dot{V}_{ACDB} = 0.5 \text{ MGD}$$

$$\frac{\dot{V}_{AB}}{\dot{V}_{ACDB}} = \frac{1 \text{ MGD}}{0.5 \text{ MGD}} = 2$$

Convert $\dot{V}$ to gal/min.

$$\dot{V}_{AB} = \frac{(1 \text{ MGD})(1 \times 10^6)}{\left(24 \dfrac{\text{hr}}{\text{day}}\right)\left(60 \dfrac{\text{min}}{\text{hr}}\right)} = 694 \text{ gal/min}$$

$$\dot{V}_{ACDB} = \frac{(0.5 \text{ MGD})(1 \times 10^6)}{\left(24 \dfrac{\text{hr}}{\text{day}}\right)\left(60 \dfrac{\text{min}}{\text{hr}}\right)} = 347 \text{ gal/min}$$

step 7: There is only one loop.

$$694 \frac{\text{gal}}{\text{min}} = (2)\left(347 \frac{\text{gal}}{\text{min}}\right)$$

$$\left(694 \frac{\text{gal}}{\text{min}}\right)^2 = (4)\left(347 \frac{\text{gal}}{\text{min}}\right)^2$$

$$\delta = \frac{(-1)(1 \times 10^{-6})(347)^2\big((11.8)(4) - 5.92 - 26.3 - 2.38\big)}{(2)(1 \times 10^{-6})(347)\big((11.8)(2) + 5.92 + 26.3 + 2.38\big)}$$

$$= -37.6 \text{ gal/min} \quad [\text{use } -38 \text{ gal/min}]$$

step 8: $\dot{V}_{AB} = 694 \dfrac{\text{gal}}{\text{min}} + \left(-38 \dfrac{\text{gal}}{\text{min}}\right)$

$$= 656 \text{ gal/min}$$

$$\dot{V}_{ACDB} = 347 \frac{\text{gal}}{\text{min}} - \left(-38 \frac{\text{gal}}{\text{min}}\right)$$

$$= 385 \text{ gal/min}$$

Repeat step 7.

$$\frac{\dot{V}_{AB}}{\dot{V}_{ACDB}} = \frac{656 \dfrac{\text{gal}}{\text{min}}}{385 \dfrac{\text{gal}}{\text{min}}} = 1.7$$

$$(1.7)^2 = 2.90$$

$$656 \frac{\text{gal}}{\text{min}} = (1.70)\left(385 \frac{\text{gal}}{\text{min}}\right)$$

$$\left(656 \frac{\text{gal}}{\text{min}}\right)^2 = (2.90)\left(385 \frac{\text{gal}}{\text{min}}\right)^2$$

$$\delta = \frac{\begin{array}{c}(-1)(1 \times 10^{-6})(385)^2 \\ \times\big((11.8)(2.90) - 5.92 - 26.3 - 2.38\big)\end{array}}{\begin{array}{c}(2)(1 \times 10^{-6})(385) \\ \times\big((11.8)(1.70) + 5.92 + 26.3 + 2.38\big)\end{array}}$$

$$\approx 0$$

$$\dot{V}_{AB} = \boxed{656 \text{ gal/min}}$$

$$\dot{V}_{ACDB} = 385 \text{ gal/min}$$

Check the Reynolds number in leg AB to verify that $f = 0.02$ was a good choice.

$$A_{10 \text{ in pipe}} = \left(\frac{\pi}{4}\right)\left(\frac{10.1 \text{ in}}{12 \dfrac{\text{in}}{\text{ft}}}\right)^2 = 0.5564 \ [\text{cast-iron pipe}]$$

The flow rate is

$$\left(656 \frac{\text{gal}}{\text{min}}\right)\left(0.002228 \frac{\text{ft}^3\text{-min}}{\text{sec-gal}}\right) = 1.46 \text{ ft}^3/\text{sec}$$

$$v = \frac{\dot{V}}{A} = \frac{1.46 \dfrac{\text{ft}^3}{\text{sec}}}{0.5564 \text{ ft}^2} = 2.62 \text{ ft/sec} \quad [\text{reasonable}]$$

For 50°F water,

$$\nu = 1.410 \times 10^{-5} \ \text{ft}^2/\text{sec}$$

$$\text{Re} = \frac{\left(\dfrac{10.1 \ \text{in}}{12 \ \frac{\text{in}}{\text{ft}}}\right)\left(2.62 \ \dfrac{\text{ft}}{\text{sec}}\right)}{1.410 \times 10^{-5} \ \dfrac{\text{ft}^2}{\text{sec}}} = 1.56 \times 10^5 \ \text{[turbulent]}$$

The assumption of full turbulence made in step 5 is justified.

The answer is (D).

- Alternative closed-form solution:

Use the Darcy equation. Assume $\epsilon = 0.0008$ ft. The relative roughness is

$$\frac{\epsilon}{D} = \frac{0.0008 \ \text{ft}}{\dfrac{10.1 \ \text{in}}{12 \ \frac{\text{in}}{\text{ft}}}} = 0.00095 \quad \text{[use 0.001]}$$

Assume $v_{\max} = 5 \ \text{ft/sec}$ and 50°F temperature. The Reynolds number is

$$\text{Re} = \frac{vD}{\nu} = \frac{\left(5 \ \dfrac{\text{ft}}{\text{sec}}\right)\left(\dfrac{10.1 \ \text{in}}{12 \ \frac{\text{in}}{\text{ft}}}\right)}{1.41 \times 10^{-5} \ \dfrac{\text{ft}^2}{\text{sec}}} = 2.98 \times 10^5$$

From the Moody diagram, $f \approx 0.0205$.

$$h_{f,\text{AB}} = \frac{fLv^2}{2Dg}$$

$$= \frac{(0.0205)(2000 \ \text{ft})\dot{V}_{\text{AB}}^2}{(2)\left(\dfrac{10.1 \ \text{in}}{12 \ \frac{\text{in}}{\text{ft}}}\right)(0.556 \ \text{ft}^2)^2\left(32.2 \ \dfrac{\text{ft}}{\text{sec}^2}\right)}$$

$$= 2.446\dot{V}_{\text{AB}}^2$$

$$h_{f,\text{ACDB}} = \frac{(0.0205)(1000)\dot{V}_{\text{ACDB}}^2}{(2)\left(\dfrac{12.12}{12}\right)(0.801)^2(32.2)}$$

$$+ \frac{(0.0205)(1500)\dot{V}_{\text{ACDB}}^2}{(2)\left(\dfrac{8.13}{12}\right)(0.360)^2(32.2)}$$

$$+ \frac{(0.0205)(1000)\dot{V}_{\text{ACDB}}^2}{(2)\left(\dfrac{10.1}{12}\right)(0.556 \ \text{ft}^2)^2(32.2)}$$

$$= 0.4912\dot{V}_{\text{ACDB}}^2 + 5.438\dot{V}_{\text{ACDB}}^2 + 1.223\dot{V}_{\text{ACDB}}^2$$

$$= 7.152\dot{V}_{\text{ACDB}}^2$$

$$h_{f,\text{AB}} = h_{f,\text{ACDB}}$$

$$2.446\dot{V}_{\text{AB}}^2 = 7.152\dot{V}_{\text{ACDB}}^2$$

$$\dot{V}_{\text{AB}} = \sqrt{\frac{7.152}{2.446}}\dot{V}_{\text{ACDB}} = 1.71\dot{V}_{\text{ACDB}} \quad \text{[Eq. 1]}$$

The total flow rate is

$$\frac{1.5 \ \text{MGD}}{\left(24 \ \dfrac{\text{hr}}{\text{day}}\right)\left(60 \ \dfrac{\text{min}}{\text{hr}}\right)} = 1041.7 \ \text{gal/min}$$

$$\dot{V}_{\text{AB}} + \dot{V}_{\text{ACDB}} = 1041.7 \ \text{gal/min} \quad \text{[Eq. 2]}$$

Solving Eqs. 1 and 2 simultaneously,

$$\dot{V}_{\text{AB}} = \boxed{657.3 \ \text{gal/min}}$$

$$\dot{V}_{\text{ACDB}} = 384.4 \ \text{gal/min}$$

The answer is (D).

This answer is insensitive to the $v_{\max} = 5 \ \text{ft/sec}$ assumption. A second iteration using actual velocities from these flow rates does not change the answer.

- The same technique can be used with the Hazen-Williams equation and an assumed value of C. If $C = 100$ is used, then

$$\dot{V}_{\text{AB}} = \boxed{725.4 \ \text{gal/min}}$$

$$\dot{V}_{\text{ACDB}} = 316.3 \ \text{gal/min}$$

(b) $f \approx 0.021$.

$$h_{f,\text{AB}} = \frac{(0.021)(2000 \ \text{ft})\left(2.62 \ \dfrac{\text{ft}}{\text{sec}}\right)^2}{(2)\left(\dfrac{10.1 \ \text{in}}{12 \ \frac{\text{in}}{\text{ft}}}\right)\left(32.2 \ \dfrac{\text{ft}}{\text{sec}^2}\right)} = 5.32 \ \text{ft}$$

For leg BD, use $f = 0.021$ (assumed).

$$v = \frac{\dot{V}}{A} = \frac{\left(385 \ \dfrac{\text{gal}}{\text{min}}\right)\left(0.002228 \ \dfrac{\text{ft}^3\text{-min}}{\text{sec-gal}}\right)}{0.5564 \ \text{ft}^2}$$

$$= 1.54 \ \text{ft/sec}$$

$$h_{f,\text{DB}} = \frac{(0.021)(1000 \ \text{ft})\left(1.54 \ \dfrac{\text{ft}}{\text{sec}}\right)^2}{(2)\left(\dfrac{10 \ \text{in}}{12 \ \frac{\text{in}}{\text{ft}}}\right)\left(32.2 \ \dfrac{\text{ft}}{\text{sec}^2}\right)} = 0.9 \ \text{ft}$$

At 50°F, $\gamma = 62.4$ lbf/ft^3. From the Bernoulli equation (omitting the velocity term),

$$\frac{p_B}{\gamma} + z_B + h_{f,DB} = \frac{p_D}{\gamma} + z_D$$

$$p_B = \left(\frac{62.4 \, \frac{\text{lbf}}{\text{ft}^3}}{144 \, \frac{\text{in}^2}{\text{ft}^2}}\right)$$

$$\times \left(\frac{\left(40 \, \frac{\text{lbf}}{\text{in}^2}\right)\left(144 \, \frac{\text{in}^2}{\text{ft}^2}\right)}{62.4 \, \frac{\text{lbf}}{\text{ft}^3}} + 600 \text{ ft} - 0.9 \text{ ft} - 580 \text{ ft}\right)$$

$$= \boxed{48.3 \text{ lbf/in}^2 \text{ (psi)}}$$

The answer is (B).

$$\frac{p_A}{\gamma} + z_A = \frac{p_B}{\gamma} + z_B + h_{f,AB}$$

$$p_A = \left(\frac{62.4 \, \frac{\text{lbf}}{\text{ft}^3}}{144 \, \frac{\text{in}^2}{\text{ft}^2}}\right)$$

$$\times \left(\frac{\left(48.3 \, \frac{\text{lbf}}{\text{in}^2}\right)\left(144 \, \frac{\text{in}^2}{\text{ft}^2}\right)}{62.4 \, \frac{\text{lbf}}{\text{ft}^3}} + 580 \text{ ft} + 5.32 \text{ ft} - 540 \text{ ft}\right)$$

$$= 67.9 \text{ lbf/in}^2 \text{ (psi)}$$

13. *Customary U.S. Solution*

Use projectile equations.

The maximum range of the discharge is given by

$$R = v_o^2 \left(\frac{\sin 2\phi}{g}\right)$$

$$= \left(50 \, \frac{\text{ft}}{\text{sec}}\right)^2 \left(\frac{\sin\left((2)(45°)\right)}{32.2 \, \frac{\text{ft}}{\text{sec}^2}}\right)$$

$$= \boxed{77.64 \text{ ft}}$$

The answer is (B).

SI Solution

Use projectile equations.

The maximum range of the discharge is given by

$$R = v_o^2 \left(\frac{\sin 2\phi}{g}\right)$$

$$= \frac{\left(15 \, \frac{\text{m}}{\text{s}}\right)^2 \sin\left((2)(45°)\right)}{9.81 \, \frac{\text{m}}{\text{s}^2}}$$

$$= \boxed{22.94 \text{ m}}$$

The answer is (B).

14.

$$A_o = \left(\frac{\pi}{4}\right)\left(\frac{4 \text{ in}}{12 \, \frac{\text{in}}{\text{ft}}}\right)^2 = 0.08727 \text{ ft}^2$$

$$A_t = \left(\frac{\pi}{4}\right)(20 \text{ ft})^2 = 314.16 \text{ ft}^2$$

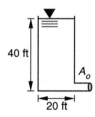

The time to drop from 40 ft to 20 ft is given by Eq. 17.84.

$$t = \frac{(2)(314.16 \text{ ft}^2)\left(\sqrt{40 \text{ ft}} - \sqrt{20 \text{ ft}}\right)}{(0.98)(0.08727 \text{ ft}^2)\sqrt{(2)\left(32.2 \, \frac{\text{ft}}{\text{sec}^2}\right)}}$$

$$= \boxed{1696 \text{ sec}}$$

The answer is (D).

15. $C_d = 1.00$ [given]

$$F_{va} = \frac{1}{\sqrt{1 - \left(\frac{D_2}{D_1}\right)^4}} = \frac{1}{\sqrt{1 - \left(\frac{8 \text{ in}}{12 \text{ in}}\right)^4}}$$

$$= 1.116$$

$$A_2 = \left(\frac{\pi}{4}\right)\left(\frac{8 \text{ in}}{12 \, \frac{\text{in}}{\text{ft}}}\right)^2 = 0.3491 \text{ ft}^2$$

$$p_1 - p_2 = \left(\begin{array}{c} \left(0.491 \dfrac{\text{lbf}}{\text{in}^3} \right)(4 \text{ in}) \\ - \left(0.0361 \dfrac{\text{lbf}}{\text{in}^3} \right)(4 \text{ in}) \end{array} \right) \left(144 \dfrac{\text{in}^2}{\text{ft}^2} \right)$$

$$= 262.0 \text{ lbf/ft}^2$$

$$Q = F_{\text{va}} C_d A_2 \sqrt{\frac{2g(p_1 - p_2)}{\gamma}}$$

$$= (1.116)(1)(0.3491 \text{ ft}^2)$$

$$\times \sqrt{\frac{(2) \left(32.2 \dfrac{\text{ft}}{\text{sec}^2} \right) \left(262 \dfrac{\text{lbf}}{\text{ft}^2} \right)}{62.4 \dfrac{\text{lbf}}{\text{ft}^3}}}$$

$$= \boxed{6.406 \text{ ft}^3/\text{sec (cfs)}}$$

The answer is (C).

16. *Customary U.S. Solution*

The volumetric flow rate of benzene through the venturi meter is given by

$$\dot{V} = C_f A_2 \sqrt{\frac{2g(\rho_m - \rho)h}{\rho}}$$

The density of mercury, ρ_m, at 60°F is approximately 848 lbm/ft^3.

The density of the benzene at 60°F is

$$\rho = (\text{SG})\rho_{\text{water}}$$

$$= (0.885) \left(62.4 \dfrac{\text{lbm}}{\text{ft}^3} \right)$$

$$= 55.22 \text{ lbm/ft}^3$$

The throat area is

$$A_2 = \left(\frac{\pi}{4} \right) D_2^2$$

$$= \left(\frac{\pi}{4} \right) (3.5 \text{ in})^2 \left(\frac{1 \text{ ft}^2}{144 \text{ in}^2} \right) = 0.0668 \text{ ft}^2$$

The flow coefficient is defined as

$$C_f = \frac{C_d}{\sqrt{1 - \beta^4}}$$

β is the ratio of the throat to inlet diameters.

$$\beta = \frac{3.5 \text{ in}}{8 \text{ in}}$$

$$= 0.4375$$

$$C_f = \frac{C_d}{\sqrt{1 - \beta^4}}$$

$$= \frac{0.99}{\sqrt{1 - (0.4375)^4}}$$

$$= 1.00865$$

Find the volumetric flow of benzene.

$$\dot{V} = C_f A_2 \sqrt{\frac{2g(\rho_m - \rho)h}{\rho}}$$

$$= (1.00865)(0.0668 \text{ ft}^2)$$

$$\times \sqrt{\frac{(2) \left(32.2 \dfrac{\text{ft}}{\text{sec}^2} \right) \left(848 \dfrac{\text{lbm}}{\text{ft}^3} - 55.22 \dfrac{\text{lbm}}{\text{ft}^3} \right)(4 \text{ in}) \left(\dfrac{1 \text{ ft}}{12 \text{ in}} \right)}{55.22 \dfrac{\text{lbm}}{\text{ft}^3}}}$$

$$= \boxed{1.183 \text{ ft}^3/\text{sec}}$$

The answer is (A).

SI Solution

The volumetric flow rate of benzene through the venturi meter is given by

$$\dot{V} = C_f A_2 \sqrt{\frac{2g(\rho_m - \rho)h}{\rho}}$$

ρ_m is the density of mercury at 15°C; ρ_m is approximately 13 600 kg/m^3.

The density of the benzene at 15°C is

$$\rho = (\text{SG})\rho_{\text{water}}$$

$$= (0.885) \left(1000 \dfrac{\text{kg}}{\text{m}^3} \right)$$

$$= 885 \text{ kg/m}^3$$

The throat area is

$$A_2 = \left(\frac{\pi}{4} \right) D_2^2$$

$$= \left(\frac{\pi}{4} \right) (0.09 \text{ m})^2$$

$$= 0.0064 \text{ m}^2$$

The flow coefficient is defined as

$$C_f = \frac{C_d}{\sqrt{1 - \beta^4}}$$

β is the ratio of the throat to inlet diameters.

$$\beta = \frac{9 \text{ cm}}{20 \text{ cm}}$$

$$= 0.45$$

$$C_f = \frac{C_d}{\sqrt{1 - \beta^4}}$$

$$= \frac{0.99}{\sqrt{1 - (0.45)^4}}$$

$$= 1.01094$$

Find the volumetric flow of benzene.

$$\dot{V} = C_f A_2 \sqrt{\frac{2g(\rho_m - \rho)h}{\rho}}$$

$$= (1.01094)(0.0064 \ \text{m}^2)$$

$$\times \sqrt{\frac{(2)\left(9.81 \ \dfrac{\text{m}}{\text{s}^2}\right) \times \left(13\,600 \ \dfrac{\text{kg}}{\text{m}^3} - 885 \ \dfrac{\text{kg}}{\text{m}^3}\right)(0.1 \ \text{m})}{885 \ \dfrac{\text{kg}}{\text{m}^3}}}$$

$$= \boxed{0.0344 \ \text{m}^3/\text{s} \quad (34.4 \ \text{L/s})}$$

The answer is (A).

17.   For 70°F water,

$$D_o = 0.2 \ \text{ft}$$

$$v_o = v\left(\frac{D}{D_o}\right)^2 = \left(2 \ \frac{\text{ft}}{\text{sec}}\right)\left(\frac{1 \ \text{ft}}{0.2 \ \text{ft}}\right)^2$$

$$= 50 \ \text{ft/sec}$$

$$\nu = 1.059 \times 10^{-5} \ \text{ft}^2/\text{sec}$$

$$\gamma = 62.3 \ \text{lbf/ft}^3$$

$$\text{Re} = \frac{D_o v_o}{\nu} = \frac{(0.2 \ \text{ft})\left(50 \ \dfrac{\text{ft}}{\text{sec}}\right)}{1.059 \times 10^{-5} \ \dfrac{\text{ft}^2}{\text{sec}}} = 9.44 \times 10^5$$

$$A_o = \left(\frac{\pi}{4}\right)(0.2 \ \text{ft})^2 = 0.0314 \ \text{ft}^2$$

$$A_p = \left(\frac{\pi}{4}\right)(1 \ \text{ft})^2 = 0.7854 \ \text{ft}^2$$

$$\frac{A_o}{A_p} = \frac{0.0314 \ \text{ft}^2}{0.7854 \ \text{ft}^2} = 0.040$$

From Fig. 17.28,

$$C_f \approx 0.60$$

$$\dot{V} = Av = (0.7854 \ \text{ft}^2)\left(2 \ \frac{\text{ft}}{\text{sec}}\right) = 1.571 \ \text{ft}^3/\text{sec}$$

From Eq. 17.160,

$$\Delta p = \left(\frac{\gamma}{2g}\right)\left(\frac{\dot{V}}{C_f A_o}\right)^2$$

$$= \left(\frac{62.3 \ \dfrac{\text{lbf}}{\text{ft}^3}}{(2)\left(32.2 \ \dfrac{\text{ft}}{\text{sec}^2}\right)}\right)\left(\frac{1.571 \ \dfrac{\text{ft}^3}{\text{sec}}}{(0.6)(0.0314 \ \text{ft}^2)}\right)^2$$

$$= 6727 \ \text{lbf/ft}^2 \ (\text{psf})$$

$$\Delta p = \frac{6727 \ \dfrac{\text{lbf}}{\text{ft}^2}}{144 \ \dfrac{\text{in}^2}{\text{ft}^2}} = \boxed{46.7 \ \text{lbf/in}^2 \ (\text{psi})}$$

The answer is (D).

18. *Customary U.S. Solution*

The pressure differential across the orifice meter is given by

$$\Delta p = p_1 - p_2 = (\rho_{\text{mercury}} - \rho_{\text{water}})h \times \left(\frac{g}{g_c}\right)$$

The densities of mercury and water are

$$\rho_{\text{mercury}} = 848 \ \text{lbm/ft}^3$$

$$\rho_{\text{water}} = 62.4 \ \text{lbm/ft}^3$$

Substituting gives

$$\Delta p = (\rho_{\text{mercury}} - \rho_{\text{water}})h \times \left(\frac{g}{g_c}\right)$$

$$= \frac{\left(848 \ \dfrac{\text{lbm}}{\text{ft}^3} - 62.4 \ \dfrac{\text{lbm}}{\text{ft}^3}\right)(7 \ \text{in}) \times \left(\dfrac{1 \ \text{ft}}{12 \ \text{in}}\right)\left(32.2 \ \dfrac{\text{ft}}{\text{sec}^2}\right)}{32.2 \ \dfrac{\text{lbm-ft}}{\text{lbf-sec}^2}}$$

$$= \boxed{458.3 \ \text{lbf/ft}^2 \quad (3.18 \ \text{psi})}$$

The answer is (B).

SI Solution

The pressure differential across the orifice meter is given by

$$\Delta p = p_1 - p_2 = (\rho_{\text{mercury}} - \rho_{\text{water}})hg$$

The densities of mercury and water are

$$\rho_{\text{mercury}} = 13\,600 \ \text{kg/m}^3$$

$$\rho_{\text{water}} = 1000 \ \text{kg/m}^3$$

Substituting gives

$$\Delta p = (\rho_{\text{mercury}} - \rho_{\text{water}})gh$$

$$= \left(13\,600 \ \frac{\text{kg}}{\text{m}^3} - 1000 \ \frac{\text{kg}}{\text{m}^3}\right)(0.178 \ \text{m})\left(9.81 \ \frac{\text{m}}{\text{s}^2}\right)$$

$$= \boxed{22\,002 \ \text{Pa} \quad (22.0 \ \text{kPa})}$$

The answer is (B).

19. *Customary U.S. Solution*

For 12 in pipe (assuming schedule-40),

$$D_i = 0.99483 \ \text{ft} \quad [\text{App. 16.B}]$$

$$A_i = 0.7773 \ \text{ft}^2$$

The velocity is

$$v = \frac{\dot{V}}{A}$$

$$= \frac{10 \frac{\text{ft}^3}{\text{sec}}}{0.7773 \text{ ft}^2}$$

$$= 12.87 \text{ ft/sec}$$

For water at 70°F, $\nu = 1.059 \times 10^{-5} \text{ ft}^2/\text{sec}$.

The Reynolds number in the pipe is

$$\text{Re} = \frac{v D_i}{\nu}$$

$$= \frac{\left(12.87 \frac{\text{ft}}{\text{sec}}\right)(0.99483 \text{ ft})}{1.059 \times 10^{-5} \frac{\text{ft}^2}{\text{sec}}}$$

$$= 1.21 \times 10^6 \quad \text{[fully turbulent in the pipe]}$$

Flow through the orifice will have a higher Reynolds number.

The volumetric flow rate through a sharp-edged orifice is given by

$$\dot{V} = C_f A_o \sqrt{\frac{2g(\rho_m - \rho)h}{\rho}}$$

In terms of pressure,

$$\dot{V} = C_f A_o \sqrt{\frac{2g_c(p_1 - p_2)}{\rho}}$$

Rearranging gives

$$C_f A_o = \frac{\dot{V}}{\sqrt{\frac{2g_c(p_1 - p_2)}{\rho}}}$$

p_1 is the upstream pressure, and p_2 is the downstream pressure.

The maximum head loss must not exceed 25 ft; therefore,

$$\frac{\left(\frac{g_c}{g}\right) \times (p_1 - p_2)}{\rho} = 25 \text{ ft}$$

$$\frac{g_c(p_1 - p_2)}{\rho} = (25 \text{ ft})g$$

Substituting gives

$$C_f A_0 = \frac{10 \frac{\text{ft}^3}{\text{sec}}}{\sqrt{(2)\left(32.2 \frac{\text{ft}}{\text{sec}^2}\right)(25 \text{ ft})}}$$

$$= 0.249 \text{ ft}^2$$

Both C_f and A_o depend on the orifice diameter.

For a 7 in diameter orifice,

$$A_o = \frac{\pi D_o^2}{4} = \frac{\pi \left((7 \text{ in})\left(\frac{1 \text{ ft}}{12 \text{ in}}\right)\right)^2}{4}$$

$$= 0.267 \text{ ft}^2$$

$$\frac{A_o}{A_1} = \frac{0.267 \text{ ft}^2}{0.7773 \text{ ft}^2} = 0.343$$

From a chart of flow coefficients (Fig. 17.28), for $A_o/A_1 = 0.343$ and fully turbulent flow,

$$C_f = 0.645$$

$$C_f A_o = (0.645)(0.267 \text{ ft}^2) = 0.172 \text{ ft}^2 < 0.249 \text{ ft}^2$$

Therefore, a 7 in diameter orifice is too small.

Try a 9 in diameter orifice.

$$A_o = \frac{\pi D_o^2}{4} = \frac{\pi \left((9 \text{ in})\left(\frac{1 \text{ ft}}{12 \text{ in}}\right)\right)^2}{4}$$

$$= 0.442 \text{ ft}^2$$

$$\frac{A_o}{A_1} = \frac{0.442 \text{ ft}^2}{0.7773 \text{ ft}^2} = 0.569$$

From Fig. 17.28, for $A_o/A_1 = 0.569$ and fully turbulent flow,

$$C_f = 0.73$$

$$C_f A_o = (0.73)(0.442 \text{ ft}^2) = 0.323 \text{ ft}^2 > 0.249 \text{ ft}^2$$

Therefore, a 9 in orifice is too large.

Interpolating gives

$$D_o = 7 \text{ in} + \frac{(9 \text{ in} - 7 \text{ in})(0.249 \text{ ft}^2 - 0.172 \text{ ft}^2)}{0.323 \text{ ft}^2 - 0.172 \text{ ft}^2}$$

$$= 8.0 \text{ in}$$

Further iterations yield

$$D_o = \boxed{8.1 \text{ in}}$$

$$C_f A_o = 0.243 \text{ ft}^2$$

The answer is (C).

SI Solution

For 300 mm pipe (assume the nominal diameter is the inner diameter), $D_i = 0.30$ m.

The velocity is

$$v = \frac{\dot{V}}{A} = \frac{\dot{V}}{\frac{\pi D_i^2}{4}}$$

$$= \frac{\left(250 \ \frac{L}{s}\right)\left(\frac{1 \ m^3}{1000 \ L}\right)}{\frac{\pi(0.3 \ m)^2}{4}}$$

$$= 3.54 \ m/s$$

From App. 14.B, for water at 20°C,

$$\nu = 1.007 \times 10^{-6} \ m^2/s$$

The Reynolds number in the pipe is

$$Re = \frac{vD_i}{\nu}$$

$$= \frac{\left(3.54 \ \frac{m}{s}\right)(0.3 \ m)}{1.007 \times 10^{-6} \ \frac{m^2}{s}}$$

$$= 1.05 \times 10^6 \quad \text{[fully turbulent in the pipe]}$$

Flow through the orifice will have a higher Reynolds number.

The volumetric flow rate through a sharp-edged orifice is given by

$$\dot{V} = C_f A_o \sqrt{\frac{2g(\rho_m - \rho)h}{\rho}}$$

In terms of pressure,

$$\dot{V} = C_f A_o \sqrt{\frac{2(p_1 - p_2)}{\rho}}$$

Rearranging gives

$$C_f A_o = \frac{\dot{V}}{\sqrt{\frac{2(p_1 - p_2)}{\rho}}}$$

p_1 is the upstream pressure, and p_2 is the downstream pressure.

The maximum head loss must not exceed 7.5 m; therefore,

$$\frac{p_1 - p_2}{g\rho} = 7.5 \ m$$

$$\frac{p_1 - p_2}{\rho} = (7.5 \ m)g$$

Substituting gives

$$C_f A_o = \frac{0.25 \ \frac{m^3}{s}}{\sqrt{(2)\left(9.81 \ \frac{m}{s^2}\right)(7.5 \ m)}}$$

$$= 0.021 \ m^2$$

Both C_f and A_o depend on the orifice diameter.

For an 18 cm diameter orifice,

$$A_o = \frac{\pi D_o^2}{4} = \frac{\pi(0.18 \ m)^2}{4} = 0.0254 \ m^2$$

$$\frac{A_o}{A_1} = \frac{0.0254 \ m^2}{0.0707 \ m^2} = 0.359$$

From a chart of flow coefficients (Fig. 17.28), for $A_o/A_1 = 0.359$ and fully turbulent flow,

$$C_f = 0.65$$

$$C_f A_o = (0.65)(0.0254 \ m^2) = 0.0165 \ m^2 < 0.021 \ m^2$$

Therefore, an 18 cm diameter orifice is too small.

Try a 23 cm diameter orifice.

$$A_o = \frac{\pi D_o^2}{4} = \frac{\pi(0.23 \ m)^2}{4} = 0.0415 \ m^2$$

$$\frac{A_o}{A_1} = \frac{0.0415 \ m^2}{0.0707 \ m^2} = 0.587$$

From Fig. 17.28, for $A_o/A_1 = 0.587$ and fully turbulent flow,

$$C_f = 0.73$$

$$C_f A_o = (0.73)(0.0415 \ m^2) = 0.0303 \ m^2 > 0.021 \ m^2$$

Therefore, a 23 cm orifice is too large.

Interpolating gives

$$D_o = 18 \ cm$$

$$+ (23 \ cm - 18 \ cm)\left(\frac{0.021 \ m^2 - 0.0165 \ m^2}{0.0303 \ m^2 - 0.0165 \ m^2}\right)$$

$$= 19.6 \ cm$$

Further iteration yields

$$D_o = \boxed{20.0 \ cm}$$

$$C_f = 0.675$$

$$C_f A_o = 0.021 \ m^2$$

The answer is (C).

20. $A_{\text{A}} = \left(\dfrac{\pi}{4}\right)\left(\dfrac{24 \text{ in}}{12 \dfrac{\text{in}}{\text{ft}}}\right)^2 = 3.142 \text{ ft}^2$

$A_{\text{B}} = \left(\dfrac{\pi}{4}\right)\left(\dfrac{12 \text{ in}}{12 \dfrac{\text{in}}{\text{ft}}}\right)^2 = 0.7854 \text{ ft}^2$

$v_{\text{A}} = \dfrac{\dot{V}}{A} = \dfrac{8 \dfrac{\text{ft}^3}{\text{sec}}}{3.142 \text{ ft}^2} = 2.546 \text{ ft/sec}$

$p_{\text{A}} = \gamma h = \left(62.4 \dfrac{\text{lbf}}{\text{ft}^3}\right)(20 \text{ ft}) = 1248 \text{ lbf/ft}^2$

Using the Bernoulli equation to solve for p_{B},

$v_{\text{B}} = \dfrac{\dot{V}}{A} = \dfrac{8 \dfrac{\text{ft}^3}{\text{sec}}}{0.7854 \text{ ft}^2} = 10.19 \text{ ft/sec}$

$p_{\text{B}} = 1248 \dfrac{\text{lbf}}{\text{ft}^2} - \left(\dfrac{\left(10.19 \dfrac{\text{ft}}{\text{sec}}\right)^2 - \left(2.546 \dfrac{\text{ft}}{\text{sec}}\right)^2}{(2)\left(32.2 \dfrac{\text{ft}}{\text{sec}^2}\right)}\right)$

$\times \left(62.4 \dfrac{\text{lbf}}{\text{ft}^3}\right)$

$= 1153.67 \text{ lbf/ft}^2$

With $\theta = 0$, from Eq. 17.200,

$F_x = \left(1153.67 \dfrac{\text{lbf}}{\text{ft}^2}\right)(0.7854 \text{ ft}^2) - \left(1248 \dfrac{\text{lbf}}{\text{ft}^2}\right)(3.142 \text{ ft}^2)$

$\quad + \left(\dfrac{\left(8 \dfrac{\text{ft}^3}{\text{sec}}\right)\left(62.4 \dfrac{\text{lbf}}{\text{ft}^3}\right)}{32.2 \dfrac{\text{ft}}{\text{sec}^2}}\right)\left(10.19 \dfrac{\text{ft}}{\text{sec}} - 2.546 \dfrac{\text{ft}}{\text{sec}}\right)$

$= \boxed{-2897 \text{ lbf on the fluid (toward A)}}$

$F_y = 0$

The answer is (A).

21. *Customary U.S. Solution*

The mass flow rate of the water is

$\dot{m} = \rho\dot{V} = \rho vA = \dfrac{\rho v \pi D^2}{4}$

$= \left(62.4 \dfrac{\text{lbm}}{\text{ft}^3}\right)\left(40 \dfrac{\text{ft}}{\text{sec}}\right)\left(\dfrac{\pi\left((2 \text{ in})\left(\dfrac{1 \text{ ft}}{12 \text{ in}}\right)\right)^2}{4}\right)$

$= 54.45 \text{ lbm/sec}$

The effective mass flow rate of the water is

$\dot{m}_{\text{eff}} = \left(\dfrac{v - v_b}{v}\right)\dot{m}$

$= \left(\dfrac{40 \dfrac{\text{ft}}{\text{sec}} - 15 \dfrac{\text{ft}}{\text{sec}}}{40 \dfrac{\text{ft}}{\text{sec}}}\right)\left(54.45 \dfrac{\text{lbm}}{\text{sec}}\right)$

$= 34.0 \text{ lbm/sec}$

The force in the (horizontal) x-direction is given by

$F_x = \dfrac{\dot{m}_{\text{eff}}(v - v_b)(\cos\theta - 1)}{g_c}$

$= \dfrac{\left(34.0 \dfrac{\text{lbm}}{\text{sec}}\right)\left(40 \dfrac{\text{ft}}{\text{sec}} - 15 \dfrac{\text{ft}}{\text{sec}}\right)(\cos 60° - 1)}{32.2 \dfrac{\text{lbm-ft}}{\text{lbf-sec}^2}}$

$= -13.2 \text{ lbf}$ [the force is acting to the left]

The force in the (vertical) y-direction is given by

$F_y = \dfrac{\dot{m}_{\text{eff}}(v - v_b)\sin\theta}{g_c}$

$= \dfrac{\left(34.0 \dfrac{\text{lbm}}{\text{sec}}\right)\left(40 \dfrac{\text{ft}}{\text{sec}} - 15 \dfrac{\text{ft}}{\text{sec}}\right)(\sin 60°)}{32.2 \dfrac{\text{lbm-ft}}{\text{lbf-sec}^2}}$

$= 22.9 \text{ lbf}$ [the force is acting upward]

The net resultant force is

$F = \sqrt{F_x^2 + F_y^2}$

$= \sqrt{(-13.2 \text{ lbf})^2 + (22.9 \text{ lbf})^2}$

$= \boxed{26.4 \text{ lbf}}$

The answer is (B).

SI Solution

The mass flow rate of the water is

$\dot{m} = \rho\dot{V} = \rho vA = \dfrac{\rho v \pi D^2}{4}$

$= \left(1000 \dfrac{\text{kg}}{\text{m}^3}\right)\left(12 \dfrac{\text{m}}{\text{s}}\right)\left(\dfrac{\pi(0.05 \text{ m})^2}{4}\right)$

$= 23.56 \text{ kg/s}$

The effective mass flow rate of the water is

$\dot{m}_{\text{eff}} = \left(\dfrac{v - v_b}{v}\right)\dot{m}$

$= \left(\dfrac{12 \dfrac{\text{m}}{\text{s}} - 4.5 \dfrac{\text{m}}{\text{s}}}{12 \dfrac{\text{m}}{\text{s}}}\right)\left(23.56 \dfrac{\text{kg}}{\text{s}}\right)$

$= 14.73 \text{ kg/s}$

The force in the (horizontal) x-direction is given by

$$F_x = \dot{m}_{\text{eff}}(\text{v} - \text{v}_b)(\cos\theta - 1)$$

$$= \left(14.73\ \frac{\text{kg}}{\text{s}}\right)\left(12\ \frac{\text{m}}{\text{s}} - 4.5\ \frac{\text{m}}{\text{s}}\right)(\cos 60° - 1)$$

$$= -55.2\ \text{N}\quad [\text{the force is acting to the left}]$$

The force in the (vertical) y-direction is given by

$$F_y = \dot{m}_{\text{eff}}(\text{v} - \text{v}_b)\sin\theta$$

$$= \left(14.73\ \frac{\text{kg}}{\text{s}}\right)\left(12\ \frac{\text{m}}{\text{s}} - 4.5\ \frac{\text{m}}{\text{s}}\right)(\sin 60°)$$

$$= 95.7\ \text{N}\quad [\text{the force is acting upward}]$$

The net resultant force is

$$F = \sqrt{F_x^2 + F_y^2}$$

$$= \sqrt{(-55.2\ \text{N})^2 + (95.7\ \text{N})^2}$$

$$= \boxed{110.5\ \text{N}}$$

The answer is (B).

22. *Customary U.S. Solution*

The power that must be added to the pump is given by

$$P = \frac{\Delta h \dot{m} \times \left(\dfrac{g}{g_c}\right)}{\eta}$$

Assume schedule-40 pipe.

$$D_i = 0.9948\ \text{ft}\quad [\text{App. 16.B}]$$
$$A_i = 0.7773\ \text{ft}^2$$
$$\text{v} = \frac{\dot{V}}{A_i}$$
$$= \frac{\left(2000\ \dfrac{\text{gal}}{\text{min}}\right)\left(\dfrac{0.002228\ \dfrac{\text{ft}^3}{\text{sec}}}{1\ \dfrac{\text{gal}}{\text{min}}}\right)}{0.7773\ \text{ft}^2}$$
$$= 5.73\ \text{ft/sec}$$

(Note that the pressures are in terms of gage pressure, and the density of mercury is 0.491 lbm/in³.)

$$p_i = \left(14.7\ \frac{\text{lbf}}{\text{in}^2} - \frac{(6\ \text{in})\left(0.491\ \dfrac{\text{lbm}}{\text{in}^3}\right)\left(32.2\ \dfrac{\text{ft}}{\text{sec}^2}\right)}{32.2\ \dfrac{\text{lbm-ft}}{\text{lbf-sec}^2}}\right)$$
$$\times\left(\frac{144\ \text{in}^2}{1\ \text{ft}^2}\right)$$
$$= 1692.6\ \text{lbf/ft}^2$$

$$E_{ti} = \frac{p_i}{\rho} + \frac{\text{v}_i^2}{2g_c} + \frac{z_i g}{g_c}$$

Since the pump inlet and outlet are at the same elevation, use $z = 0$ and $\rho = (\text{SG})\rho_{\text{water}}$.

$$E_{ti} = \frac{p_i}{(\text{SG})\rho_{\text{water}}} + \frac{\text{v}_i^2}{2g_c} + 0$$

$$= \frac{1692.6\ \dfrac{\text{lbf}}{\text{ft}^2}}{(1.2)\left(62.4\ \dfrac{\text{lbm}}{\text{ft}^3}\right)} + \frac{\left(5.73\ \dfrac{\text{ft}}{\text{sec}}\right)^2}{(2)\left(32.2\ \dfrac{\text{lbm-ft}}{\text{lbf-sec}^2}\right)}$$

$$= 23.11\ \text{ft-lbf/lbm}$$

Calculate the total head at the inlet.

$$h_{ti} = E_{ti} \times \left(\frac{g_c}{g}\right)$$

$$= \frac{\left(23.11\ \dfrac{\text{ft-lbf}}{\text{lbm}}\right)\left(32.2\ \dfrac{\text{lbm-ft}}{\text{lbf-sec}^2}\right)}{32.2\ \dfrac{\text{ft}}{\text{sec}^2}}$$

$$= 23.11\ \text{ft}$$

At the outlet side of the pump,

$$D_o = 0.6651\ \text{ft}$$
$$A_o = 0.3474\ \text{ft}^2$$
$$Q = \text{v}_o A_o$$
$$\text{v}_o = \frac{Q}{A_o}$$
$$= \frac{\left(2000\ \dfrac{\text{gal}}{\text{min}}\right)\left(\dfrac{0.002228\ \dfrac{\text{ft}^3}{\text{sec}}}{1\ \dfrac{\text{gal}}{\text{min}}}\right)}{0.3474\ \text{ft}^2}$$
$$= 12.83\ \text{ft/sec}$$

(Note that the pressures are in terms of gage pressure and the gauge is located 4 ft above the pump outlet, which adds 4 ft of pressure head at the pump outlet.)

$$p_o = \left(14.7\ \frac{\text{lbf}}{\text{in}^2} + 20\ \frac{\text{lbf}}{\text{in}^2}\right)\left(\frac{144\ \text{in}^2}{1\ \text{ft}^2}\right)$$
$$+ 4\ \text{ft}\left(\frac{(1.2)\left(62.4\ \dfrac{\text{lbm}}{\text{ft}^3}\right)\left(32.2\ \dfrac{\text{ft}}{\text{sec}^2}\right)}{32.2\ \dfrac{\text{lbm-ft}}{\text{lbf-sec}^2}}\right)$$
$$= 5296\ \text{lbf/ft}^2$$

$$E_{to} = \frac{p_o}{\rho} + \frac{\text{v}_o^2}{2g_c} + \frac{z_o g}{g_c}$$

Since the pump inlet and outlet are at the same elevation, use $z = 0$ and $\rho = (SG)\rho_{water}$.

$$E_{to} = \frac{p_o}{(SG)\rho_{water}} + \frac{v_o^2}{2g_c} + 0$$

$$= \frac{5296 \, \frac{lbf}{ft^2}}{(1.2)\left(62.4 \, \frac{lbm}{ft^3}\right)} + \frac{\left(12.83 \, \frac{ft}{sec}\right)^2}{(2)\left(32.2 \, \frac{lbm\text{-}ft}{lbf\text{-}sec^2}\right)}$$

$$= 73.28 \, \text{ft-lbf/lbm}$$

Calculate the total head at the outlet.

$$h_{to} = E_{to} \times \left(\frac{g_c}{g}\right)$$

$$= \left(73.28 \, \frac{ft\text{-}lbf}{lbm}\right)\left(\frac{32.2 \, \frac{lbm\text{-}ft}{lbf\text{-}sec^2}}{32.2 \, \frac{ft}{sec^2}}\right)$$

$$= 73.28 \, \text{ft}$$

Compute the total head required across the pump.

$$\Delta h = h_{to} - h_{ti}$$
$$= 73.28 \, \text{ft} - 23.11 \, \text{ft}$$
$$= 50.17 \, \text{ft}$$

The mass flow rate is

$$\dot{m} = \rho\dot{V}$$

In terms of the specific gravity, the mass flow rate is

$$\dot{m} = (SG)\rho_{water}\dot{V}$$

$$= (1.2)\left(62.4 \, \frac{lbm}{ft^3}\right)\left(2000 \, \frac{gal}{min}\right)\left(\frac{0.002228 \, \frac{ft^3}{sec}}{1 \, \frac{gal}{min}}\right)$$

$$= 333.7 \, \text{lbm/sec}$$

The power that must be added to the pump is

$$P = \frac{\Delta h\dot{m} \times \left(\frac{g}{g_c}\right)}{\eta}$$

$$= \frac{(50.17 \, \text{ft})\left(333.7 \, \frac{lbm}{sec}\right)\left(\frac{32.2 \, \frac{ft}{sec^2}}{32.2 \, \frac{lbm\text{-}ft}{lbf\text{-}sec^2}}\right)}{(0.85)\left(550 \, \frac{ft\text{-}lbf}{hp\text{-}sec}\right)}$$

$$= \boxed{35.8 \, \text{hp}}$$

(Note that it is not necessary to use absolute pressures as has been done in this solution.)

The answer is (B).

SI Solution

The power that must be added to the pump is given by

$$P = \frac{\Delta h\dot{m}g}{\eta}$$

Assume the pipe nominal diameter is equal to the internal diameter.

$$D_i = 0.30 \, \text{m}$$

$$A_i = \frac{\pi D_i^2}{4} = x\frac{\pi(0.30 \, \text{m})^2}{4} = 0.0707 \, \text{m}^2$$

$$v = \frac{\dot{V}}{A_i}$$

$$= \frac{0.125 \, \frac{m^3}{s}}{0.0707 \, \text{m}^2}$$

$$= 1.77 \, \text{m/s}$$

(Note that the pressures are in terms of gage pressure, and the density of mercury is $13\,600 \, \text{kg/m}^3$.)

$$p_i = 1.013 \times 10^5 \, \text{Pa}$$
$$- (0.15 \, \text{m})\left(13\,600 \, \frac{kg}{m^3}\right)\left(9.81 \, \frac{m}{s^2}\right)$$
$$= 8.13 \times 10^4 \, \text{Pa}$$

$$E_{ti} = \frac{p}{\rho} + \frac{v_i^2}{2} + z_i g$$

Since the pump inlet and outlet are at the same elevation, use $z = 0$ and $\rho = (SG)\rho_{water}$.

$$E_{ti} = \frac{p}{(SG)\rho_{water}} + \frac{v_i^2}{2} + 0$$

$$= \frac{8.13 \times 10^4 \, \text{Pa}}{(1.2)\left(1000 \, \frac{kg}{m^3}\right)} + \frac{\left(1.77 \, \frac{m}{s}\right)^2}{2}$$

$$= 69.3 \, \text{J/kg}$$

The total head at the inlet is

$$h_{ti} = \frac{E_{ti}}{g}$$

$$= \frac{69.3 \, \frac{J}{kg}}{9.81 \, \frac{m}{s^2}}$$

$$= 7.06 \, \text{m}$$

Assume the pipe nominal diameter is equal to the internal diameter. On the outlet side of the pump,

$$D_i = 0.20 \text{ m}$$

$$A_o = \frac{\pi D_o^2}{4} = \frac{\pi (0.20 \text{ m})^2}{4} = 0.0314 \text{ m}^2$$

$$v_o = \frac{\dot{V}}{A_o}$$

$$= \frac{0.125 \dfrac{\text{m}^3}{\text{s}}}{0.0314 \text{ m}^2}$$

$$= 3.98 \text{ m/s}$$

(Note that the pressures are in terms of gage pressure and the gauge is located 1.2 m above the pump outlet, which adds 1.2 m of pressure head at the pump outlet.)

$$p_o = 1.013 \times 10^5 \text{ Pa} + 138 \times 10^3 \text{ Pa}$$

$$+ (1.2 \text{ m}) \left((1.2) \left(1000 \frac{\text{kg}}{\text{m}^3} \right) \left(9.81 \frac{\text{m}}{\text{s}^2} \right) \right)$$

$$= 2.53 \times 10^5 \text{ Pa}$$

$$E_{to} = \frac{p_o}{\rho} + \frac{v_o^2}{2} + z_o g$$

Since the pump inlet and outlet are at the same elevation, use $z = 0$ and $\rho = (\text{SG})\rho_{\text{water}}$.

$$E_{to} = \frac{p_o}{(\text{SG})\rho_{\text{water}}} + \frac{v_o^2}{2} + 0$$

$$= \frac{2.53 \times 10^5 \text{ Pa}}{(1.2) \left(1000 \dfrac{\text{kg}}{\text{m}^3} \right)} + \frac{\left(3.98 \dfrac{\text{m}}{\text{s}} \right)^2}{2}$$

$$= 218.8 \text{ J/kg}$$

The total head at the outlet is

$$h_{to} = \frac{E_{to}}{g}$$

$$= \frac{218.8 \dfrac{\text{J}}{\text{kg}}}{22.30 \dfrac{\text{m}}{\text{s}^2}}$$

$$= 11.76 \text{ m}$$

The total head required across the pump is

$$\Delta h = h_{to} - h_{ti}$$

$$= 22.30 \text{ m} - 7.04 \text{ m}$$

$$= 15.26 \text{ m}$$

The mass flow rate is

$$\dot{m} = \rho \dot{V}$$

In terms of the specific gravity, the mass flow rate is

$$\dot{m} = (\text{SG})\rho_{\text{water}} Q$$

$$= (1.2) \left(1000 \frac{\text{kg}}{\text{m}^3} \right) \left(0.125 \frac{\text{m}^3}{\text{s}} \right)$$

$$= 150 \text{ kg/s}$$

The power that must be added to the pump is

$$P = \frac{\Delta h \dot{m} g}{\eta}$$

$$= \frac{(15.26 \text{ m}) \left(150 \dfrac{\text{kg}}{\text{s}} \right) \left(9.81 \dfrac{\text{m}}{\text{s}^2} \right)}{0.85}$$

$$= \boxed{26\,417 \text{ W} \quad (26.4 \text{ kW})}$$

(Note that it is not necessary to use absolute pressures as has been done in this solution.)

The answer is (B).

23. *Customary U.S. Solution*

The power developed by the horizontal turbine is given by

$$P = \dot{m} h_{\text{loss}} \times \left(\frac{g}{g_c} \right)$$

The mass flow rate is

$$\dot{m} = \dot{V} \rho$$

$$= \left(100 \frac{\text{ft}^3}{\text{sec}} \right) \left(62.4 \frac{\text{lbm}}{\text{ft}^3} \right)$$

$$= 6240 \text{ lbm/sec}$$

The head loss across the horizontal turbine is given by

$$h_{\text{loss}} = \left(\frac{\Delta p}{\rho} \right) \times \left(\frac{g_c}{g} \right)$$

$$= \left(\frac{\left(30 \dfrac{\text{lbf}}{\text{in}^2} - \left(-5 \dfrac{\text{lbf}}{\text{in}^2} \right) \right) \left(\dfrac{144 \text{ in}^2}{1 \text{ ft}^2} \right)}{62.4 \dfrac{\text{lbm}}{\text{ft}^3}} \right)$$

$$\times \left(\frac{32.2 \dfrac{\text{lbm-ft}}{\text{lbf-sec}^2}}{32.2 \dfrac{\text{ft}}{\text{sec}^2}} \right)$$

$$= 80.77 \text{ ft}$$

From Table 18.5, the power developed is

$$P = \dot{m}h_{\text{loss}} \times \left(\frac{g}{g_c}\right)$$

$$= \frac{\left(6240 \; \frac{\text{lbm}}{\text{sec}}\right)(80.77 \; \text{ft})\left(32.2 \; \frac{\text{ft}}{\text{sec}^2}\right)}{\left(32.2 \; \frac{\text{lbm-ft}}{\text{lbf-sec}^2}\right)\left(550 \; \frac{\text{ft-lbf}}{\text{hp-sec}}\right)}$$

$$= \boxed{916 \; \text{hp}}$$

The answer is (D).

SI Solution

The power developed by the horizontal turbine is given by

$$P = \dot{m}h_{\text{loss}}g$$

The mass flow rate is

$$\dot{m} = \dot{V}\rho$$

$$= \left(2.6 \; \frac{\text{m}^3}{\text{s}}\right)\left(1000 \; \frac{\text{kg}}{\text{m}^3}\right)$$

$$= 2600 \; \text{kg/s}$$

The head loss across the horizontal turbine is given by

$$h_{\text{loss}} = \frac{\Delta p}{\rho g}$$

$$= \frac{(210 \; \text{kPa} - (-35 \; \text{kPa}))\left(1000 \; \frac{\text{Pa}}{\text{kPa}}\right)}{\left(1000 \; \frac{\text{kg}}{\text{m}^3}\right)\left(9.81 \; \frac{\text{m}}{\text{s}^2}\right)}$$

$$= 25.0 \; \text{m}$$

From Table 18.5, the power developed is

$$P = \dot{m}h_{\text{loss}}g$$

$$= \left(2600 \; \frac{\text{kg}}{\text{s}}\right)(25.0 \; \text{m})\left(9.81 \; \frac{\text{m}}{\text{s}^2}\right)$$

$$= \boxed{637\,650 \; \text{W} \quad (638 \; \text{kW})}$$

The answer is (D).

24. *Customary U.S. Solution*

(a) The drag on the car is given by

$$F_D = \frac{C_D A\rho v^2}{2g_c}$$

For air at 70°F,

$$\rho = \frac{p}{RT} = \frac{\left(14.7 \; \frac{\text{lbf}}{\text{in}^2}\right)\left(144 \; \frac{\text{in}^2}{\text{ft}^2}\right)}{\left(53.35 \; \frac{\text{ft-lbf}}{\text{lbm-°R}}\right)(70°\text{F} + 460)}$$

$$= 0.0749 \; \text{lbm/ft}^3$$

$$v = \left(55 \; \frac{\text{mi}}{\text{hr}}\right)\left(5280 \; \frac{\text{ft}}{\text{mi}}\right)\left(\frac{1 \; \text{hr}}{3600 \; \text{sec}}\right)$$

$$= 80.67 \; \text{ft/sec}$$

Substituting gives

$$F_D = \frac{C_D A\rho v^2}{2g_c}$$

$$= \frac{(0.42)(28 \; \text{ft}^2)\left(0.0749 \; \frac{\text{lbm}}{\text{ft}^3}\right)\left(80.67 \; \frac{\text{ft}}{\text{sec}}\right)^2}{(2)\left(32.2 \; \frac{\text{lbm-ft}}{\text{lbf-sec}^2}\right)}$$

$$= 89.0 \; \text{lbf}$$

The total resisting force is

$$F = F_D + \text{rolling resistance}$$

$$= 89.0 \; \text{lbf} + (0.01)(3300 \; \text{lbm}) \times \left(\frac{g}{g_c}\right)$$

$$= 89.0 \; \text{lbf} + \frac{(0.01)(3300 \; \text{lbm})\left(32.2 \; \frac{\text{ft}}{\text{sec}^2}\right)}{32.2 \; \frac{\text{lbm-ft}}{\text{lbf-sec}^2}}$$

$$= 122.0 \; \text{lbf}$$

The power consumed is

$$P = Fv$$

$$= \frac{(122.0 \; \text{lbf})\left(80.67 \; \frac{\text{ft}}{\text{sec}}\right)}{778 \; \frac{\text{ft-lbf}}{\text{Btu}}}$$

$$= 12.65 \; \text{Btu/sec}$$

The power available from the fuel is

$$P_A = (\text{engine thermal efficiency})(\text{fuel heating value})$$

$$= (0.28)\left(115{,}000 \; \frac{\text{Btu}}{\text{gal}}\right)$$

$$= 32{,}200 \; \text{Btu/gal}$$

The fuel consumption at 55 mi/hr is

$$\frac{P}{P_A} = \frac{12.65 \; \frac{\text{Btu}}{\text{sec}}}{32{,}200 \; \frac{\text{Btu}}{\text{gal}}}$$

$$= 3.93 \times 10^{-4} \; \text{gal/sec}$$

The fuel consumption is

$$\frac{3.93 \times 10^{-4} \frac{\text{gal}}{\text{sec}}}{\left(55 \frac{\text{mi}}{\text{hr}}\right)\left(\frac{1 \text{ hr}}{3600 \text{ sec}}\right)} = \boxed{0.0257 \text{ gal/mi}}$$

The answer is (A).

(b) Similarly, the fuel consumption at 65 mi/hr is

$$v = \left(65 \frac{\text{mi}}{\text{hr}}\right)\left(5280 \frac{\text{ft}}{\text{mi}}\right)\left(\frac{1 \text{ hr}}{3600 \text{ sec}}\right)$$

$$= 95.33 \text{ ft/sec}$$

$$F_D = \frac{C_D A \rho v^2}{2 g_c}$$

$$= \frac{(0.42)(28 \text{ ft}^2)\left(0.0749 \frac{\text{lbm}}{\text{ft}^3}\right)\left(95.33 \frac{\text{ft}}{\text{sec}}\right)^2}{(2)\left(32.2 \frac{\text{lbm-ft}}{\text{lbf-sec}^2}\right)}$$

$$= 124.3 \text{ lbf}$$

The total resisting force is

$$F = F_D + \text{rolling resistance}$$

$$= 124.3 \text{ lbf} + (0.01)(3300 \text{ lbm})\left(\frac{g}{g_c}\right)$$

$$= 124.3 \text{ lbf} + \frac{(0.01)(3300 \text{ lbm})\left(32.2 \frac{\text{ft}}{\text{sec}^2}\right)}{32.2 \frac{\text{lbm-ft}}{\text{lbf-sec}^2}}$$

$$= 157.3 \text{ lbf}$$

The power consumed is

$$P = Fv$$

$$= \frac{(157.3 \text{ lbf})\left(95.33 \frac{\text{ft}}{\text{sec}}\right)}{778 \frac{\text{ft-lbf}}{\text{Btu}}}$$

$$= 19.27 \text{ Btu/sec}$$

The fuel consumption at 65 mi/hr is

$$\frac{P}{P_A} = \frac{19.27 \frac{\text{Btu}}{\text{sec}}}{32{,}200 \frac{\text{Btu}}{\text{gal}}}$$

$$= 5.98 \times 10^{-4} \text{ gal/sec}$$

The fuel consumption is

$$\frac{5.98 \times 10^{-4} \frac{\text{gal}}{\text{sec}}}{\left(65 \frac{\text{mi}}{\text{hr}}\right)\left(\frac{1 \text{ hr}}{3600 \text{ sec}}\right)} = 0.0331 \text{ gal/mi}$$

The relative difference between the fuel consumption at 55 mi/hr and 65 mi/hr is

$$\frac{0.0331 \frac{\text{gal}}{\text{mi}} - 0.0257 \frac{\text{gal}}{\text{mi}}}{0.0257 \frac{\text{gal}}{\text{mi}}} = \boxed{0.288 \;\; (28.8\%)}$$

The answer is (C).

SI Solution

(a) The drag on the car is given by

$$F_D = \frac{C_D A \rho v^2}{2}$$

For air at 20°C,

$$\rho = \frac{p}{RT} = \frac{1.013 \times 10^5 \text{ Pa}}{\left(287 \frac{\text{J}}{\text{kg·K}}\right)(20°\text{C} + 273)}$$

$$= 1.205 \text{ kg/m}^3$$

$$v = \left(90 \frac{\text{km}}{\text{h}}\right)\left(1000 \frac{\text{m}}{\text{km}}\right)\left(\frac{1 \text{ h}}{3600 \text{ s}}\right)$$

$$= 25.0 \text{ m/s}$$

Substituting gives

$$F_D = \frac{C_D A \rho v^2}{2}$$

$$= \left(\tfrac{1}{2}\right)(0.42)(2.6 \text{ m}^2)\left(1.205 \frac{\text{kg}}{\text{m}^3}\right)\left(25.0 \frac{\text{m}}{\text{s}}\right)^2$$

$$= 411.2 \text{ N}$$

The total resisting force is

$$F = F_D + \text{rolling resistance}$$

$$= 411.2 \text{ N} + (0.01)(1500 \text{ kg})g$$

$$= 411.2 \text{ N} + (0.01)(1500 \text{ kg})\left(9.81 \frac{\text{m}}{\text{s}^2}\right)$$

$$= 558.4 \text{ N}$$

The power consumed is

$$P = Fv$$

$$= (558.4 \text{ N})\left(25 \frac{\text{m}}{\text{s}}\right)$$

$$= 13\,960 \text{ W}$$

The power available from the fuel is

$$P_A = (\text{engine thermal efficiency})(\text{fuel heating value})$$

$$= (0.28)\left(4.6 \times 10^8 \frac{\text{J}}{\text{L}}\right)$$

$$= 1.288 \times 10^8 \text{ J/L}$$

The fuel consumption at 90 km/h is

$$\frac{P}{P_A} = \frac{13\,960\text{ W}}{1.288 \times 10^8\ \frac{\text{J}}{\text{L}}}$$

$$= 1.08 \times 10^{-4}\text{ L/s}$$

The fuel consumption is

$$\frac{1.08 \times 10^{-4}\ \frac{\text{L}}{\text{s}}}{\left(90\ \frac{\text{km}}{\text{h}}\right)\left(\frac{1\text{ h}}{3600\text{ s}}\right)} = \boxed{0.00434\text{ L/km}}$$

The answer is (A).

(b) Similarly, the fuel consumption at 105 km/h is

$$v = \left(105\ \frac{\text{km}}{\text{h}}\right)\left(1000\ \frac{\text{m}}{\text{km}}\right)\left(\frac{1\text{ h}}{3600\text{ s}}\right)$$

$$= 29.2\text{ m/s}$$

$$D = \frac{C_D A \rho v^2}{2}$$

$$= \left(\tfrac{1}{2}\right)(0.42)(2.6\text{ m}^2)\left(1.205\ \frac{\text{kg}}{\text{m}^3}\right)\left(29.2\ \frac{\text{m}}{\text{s}}\right)^2$$

$$= 561.0\text{ N}$$

The total resisting force is

$$F = F_D + \text{rolling resistance}$$

$$= 561.0\text{ N} + (0.01)(1500\text{ kg})g$$

$$= 561.0\text{ N} + (0.01)(1500\text{ kg})\left(9.81\ \frac{\text{m}}{\text{s}^2}\right)$$

$$= 708.2\text{ N}$$

The power consumed is

$$P = Fv$$

$$= (708.2\text{ N})\left(29.2\ \frac{\text{m}}{\text{s}}\right)$$

$$= 20\,679\text{ W}$$

The fuel consumption at 105 km/h is

$$\frac{P}{P_A} = \frac{20\,679\text{ W}}{1.288 \times 10^8\ \frac{\text{J}}{\text{L}}}$$

$$= 1.61 \times 10^{-4}\text{ L/s}$$

The fuel consumption is

$$\frac{1.61 \times 10^{-4}\ \frac{\text{L}}{\text{s}}}{\left(105\ \frac{\text{km}}{\text{h}}\right)\left(\frac{1\text{ h}}{3600\text{ s}}\right)} = 0.00552\text{ L/km}$$

The relative difference between the fuel consumption at 90 km/h and 105 km/h is

$$\frac{0.00552\ \frac{\text{L}}{\text{km}} - 0.00434\ \frac{\text{L}}{\text{km}}}{0.00434\ \frac{\text{L}}{\text{km}}} = \boxed{0.272\ \ (27.2\%)}$$

The answer is (C).

18 Hydraulic Machines

PRACTICE PROBLEMS

Pumping Power

1. 2000 gal/min of 60°F thickened sludge with a specific gravity of 1.2 flow through a pump with an inlet diameter of 12 in and an outlet of 8 in. The centerlines of the inlet and outlet are at the same elevation. The inlet pressure is 8 in of mercury (vacuum). A discharge pressure gauge located 4 ft above the pump discharge centerline reads 20 psig. The pump efficiency is 85%. All pipes are schedule-40. What is the input power of the pump?

(A) 26 hp
(B) 31 hp
(C) 37 hp
(D) 53 hp

2. 1.25 ft^3/sec (35 L/s) of 70°F (21°C) water are pumped from the bottom of a tank through 700 ft (230 m) of 4 in (10.2 cm) schedule-40 steel pipe. The line includes a 50 ft (15 m) rise in elevation, two right-angle elbows, a wide-open gate valve, and a swing check valve. All fittings and valves are regular screwed. The inlet pressure is 50 psig (345 kPa), and a working pressure of 20 psig (140 kPa) is needed at the end of the pipe. What is the hydraulic power for this pumping application?

(A) 16 hp (13 kW)
(B) 23 hp (17 kW)
(C) 49 hp (37 kW)
(D) 66 hp (50 kW)

3. 80 gal/min (5 L/s) of 80°F (27°C) water are lifted 12 ft (4 m) vertically by a pump through 50 ft (15 m) of a 2 in (5.1 cm) diameter rubber hose. The discharge end of the hose is submerged in 8 ft (2.5 m) of water as shown. What head is added by the pump?

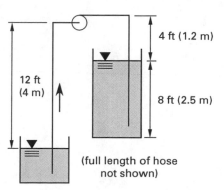

4 ft (1.2 m)

12 ft (4 m)

8 ft (2.5 m)

(full length of hose not shown)

(A) 10 ft (3.0 m)
(B) 13 ft (4.3 m)
(C) 22 ft (6.6 m)
(D) 31 ft (9.3 m)

4. A 20 hp motor drives a centrifugal pump. The pump discharges 60°F (16°C) water at 12 ft/sec (4 m/s) into a 6 in (15.2 cm) steel schedule-40 line. The inlet is 8 in (20.3 cm) schedule-40 steel pipe. The pump suction is 5 psi (35 kPa) below standard atmospheric pressure. The friction head loss in the system is 10 ft (3.3 m). The pump efficiency is 70%. The suction and discharge lines are at the same elevation. What is the maximum height above the pump inlet that water is available at standard atmospheric pressure?

(A) 28 ft (6.9 m)
(B) 37 ft (11 m)
(C) 49 ft (15 m)
(D) 81 ft (25 m)

5. (*Time limit: one hour*) A pump station is used to fill a tank on a hill above from a lake below. The flow rate is 10,000 gal/hr (10.5 L/s) of 60°F (16°C) water. The atmospheric pressure is 14.7 psia (101 kPa). The pump is 12 ft (4 m) above the lake, and the tank surface level is 350 ft (115 m) above the pump. The suction and discharge lines are 4 in (10.2 cm) diameter schedule-40 steel pipe. The equivalent length of the inlet line between the lake and the pump is 300 ft (100 m). The total equivalent length between the lake and the tank is 7000 ft (2300 m), including all fittings, bends, screens, and valves. The cost of electricity is $0.04 per kW-hr. The overall efficiency of the pump and motor set is 70%.

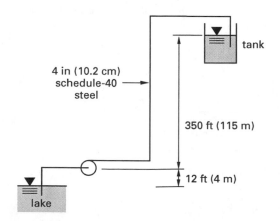

tank

4 in (10.2 cm) schedule-40 steel

350 ft (115 m)

12 ft (4 m)

lake

(a) What does it cost to operate the pump for 1 hr?
- (A) $0.1
- (B) $1
- (C) $3
- (D) $6

(b) What motor power is required?
- (A) 10 hp (7.5 kW)
- (B) 30 hp (25 kW)
- (C) 50 hp (40 kW)
- (D) 75 hp (60 kW)

(c) What is the NPSHA for this application?
- (A) 4 ft (1.2 m)
- (B) 8 ft (2.4 m)
- (C) 12 ft (3.6 m)
- (D) 16 ft (4.5 m)

6. (*Time limit: one hour*) A town with a stable, constant population of 10,000 produces sewage at the average rate of 100 gallons per capita day (gpcd), with peak flows of 250 gpcd. The pipe to the pumping station is 5000 ft in length and has a C-value of 130. The elevation drop along the length is 48 ft. Minor losses in infiltration are insignificant. The pump's maximum suction lift is 10 ft.

(a) If all diameters are available, and if the pipe flows 100% full under gravity flow, what minimum pipe diameter is required?
- (A) 8 in
- (B) 12 in
- (C) 14 in
- (D) 18 in

(b) If constant-speed pumps are used, what is the minimum number of pumps (disregarding spares and backups) that should be used?
- (A) 2
- (B) 3
- (C) 4
- (D) 5

(c) If variable-speed pumps are used, what is the minimum number of pumps (disregarding spares and backups) that should be used?
- (A) 1
- (B) 2
- (C) 3
- (D) 4

(d) If three constant-speed pumps are used with a fourth as backup, what motor power is required?
- (A) 2 hp
- (B) 5 hp
- (C) 8 hp
- (D) 12 hp

(e) If two variable-speed pumps are used with a third as backup, what motor power is required?
- (A) 3 hp
- (B) 8 hp
- (C) 12 hp
- (D) 18 hp

(f) Which of the following are ways of controlling sump pump on-off cycles?

I. detecting sump levels
II. detecting pressure in the sump
III. detecting incoming flow rates
IV. using fixed run times
V. detecting outgoing flow rates
VI. operating manually

- (A) I, II, and III
- (B) I, III, IV, and V
- (C) I, III, and V
- (D) I, II, IV, V, and VI

(g) With intermittent fan operation, approximately how many air changes should the wet well receive per hour?
- (A) 6
- (B) 12
- (C) 20
- (D) 30

(h) With continuous fan operation, approximately how many air changes should the dry well receive per hour?
- (A) 6
- (B) 12
- (C) 20
- (D) 30

7. (*Time limit: one hour*) A pump transfers 3.5 MGD of filtered water from the clear well of a 10 ft by 20 ft (plan) rapid sand filter to a higher elevation. The pump efficiency is 85%, and the motor driving pump has an efficiency of 90%. Minor losses are insignificant. Refer to the following illustration for additional information.

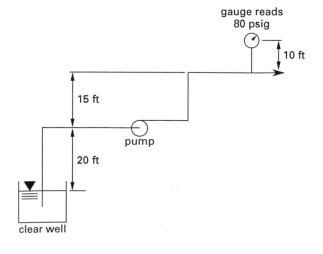

(a) What is the static suction lift?
- (A) 15 ft
- (B) 20 ft
- (C) 35 ft
- (D) 40 ft

(b) What is the static discharge head?
- (A) 15 ft
- (B) 20 ft
- (C) 35 ft
- (D) 40 ft

(c) Based on the information given, what is the approximate total dynamic head?
- (A) 45 ft
- (B) 185 ft
- (C) 210 ft
- (D) 230 ft

(d) What motor power is required?
- (A) 50 hp
- (B) 100 hp
- (C) 150 hp
- (D) 200 hp

(e) If the filter output is increased by 25%, how much surplus will accumulate in 8 hr?
- (A) 60,000 gal
- (B) 190,000 gal
- (C) 250,000 gal
- (D) 370,000 gal

Pumping Other Fluids

8. (*Time limit: one hour*) Gasoline with a specific gravity of 0.7 and viscosity of 6×10^{-6} ft^2/sec (5.6×10^{-7} m^2/s) is transferred from a tanker to a storage tank. The interior of the storage tank is maintained at atmospheric pressure by a vapor-recovery system. The free surface in the storage tank is 60 ft (20 m) above the tanker's free surface. The pipe consists of 500 ft (170 m) of 3 in (7.62 cm) schedule-40 steel pipe with six flanged elbows and two wide-open gate valves. The pump and motor both have individual efficiencies of 88%. Electricity costs $0.045 per kW-hr. The pump's performance data (based on cold, clear water) are known.

flow rate gpm (L/s)	head ft (m)
0 (0)	127 (42)
100 (6.3)	124 (41)
200 (12)	117 (39)
300 (18)	108 (36)
400 (24)	96 (32)
500 (30)	80 (27)
600 (36)	55 (18)

(a) What is the transfer rate?
- (A) 150 gal/min (9.2 L/s)
- (B) 180 gal/min (11 L/s)
- (C) 200 gal/min (12 L/s)
- (D) 230 gal/min (14 L/s)

(b) What is the total cost of operating the pump for 1 hr?
- (A) $0.20
- (B) $0.80
- (C) $1.30
- (D) $2.70

Specific Speed

9. A double-suction water pump moving 300 gal/sec (1.1 kL/s) turns at 900 rpm. The pump adds 20 ft (7 m) of head to the water. What is the specific speed?
- (A) 3000 rpm (52 rpm)
- (B) 6000 rpm (100 rpm)
- (C) 9000 rpm (160 rpm)
- (D) 12,000 rpm (210 rpm)

Cavitation

10. 100 gal/min (6.3 L/s) of pressurized hot water at 281°F and 80 psia (138°C and 550 kPa) is drawn through 30 ft (10 m) of 1.5 in (3.81 cm) schedule-40 steel pipe into a 2 psig (14 kPa) tank. The inlet and outlet are both 20 ft (6 m) below the surface of the water when the tank is full. The inlet line contains two wide-open gate valves and two long-radius elbows. All components are regular screwed. The pump's NPSHR is 10 ft (3 m) for this application. The kinematic viscosity of 281°F (138°C) water is 0.239×10^{-5} ft^2/sec (0.222×10^{-6} m^2/s) and the vapor pressure is 50.02 psia (3.431 bar). Will the pump cavitate?

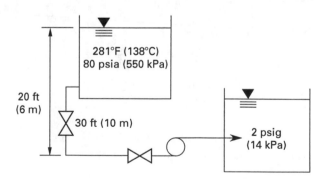

- (A) yes; NPSHA = 4 ft (1.2 m)
- (B) yes; NPSHA = 9 ft (2.7 m)
- (C) no; NPSHA = 24 ft (7.2 m)
- (D) no; NPSHA = 68 ft (21 m)

11. The velocity of the tip of a marine propeller is 4.2 times the boat velocity. The propeller is located 8 ft (3 m) below the surface. The temperature of the seawater is 68°F (20°C). The density is approximately

64.0 lbm/ft^3 (1024 kg/m^3), and the salt content is 2.5% by weight. What is the practical maximum boat velocity, as limited strictly by cavitation?

(A) 9.1 ft/sec (2.7 m/s)

(B) 12 ft/sec (3.8 m/s)

(C) 15 ft/sec (4.5 m/s)

(D) 22 ft/sec (6.6 m/s)

Pump and System Curves

12. The inlet of a centrifugal water pump is 7 ft (2.3 m) above the free surface from which it draws. The suction point is a submerged pipe. The supply line consists of 12 ft (4 m) of 2 in (5.08 cm) schedule-40 steel pipe and contains one long-radius elbow and one check valve. The discharge line is 2 in (5.08 cm) schedule-40 steel pipe and includes two long-radius elbows and an 80 ft (27 m) run. The discharge is 20 ft (6.3 m) above the free surface. All components are regular screwed. The water temperature is 70°F (21°C). The following pump curve data are applicable.

flow rate gpm (L/s)	head ft (m)
0 (0)	110 (37)
10 (0.6)	108 (36)
20 (1.2)	105 (35)
30 (1.8)	102 (34)
40 (2.4)	98 (33)
50 (3.2)	93 (31)
60 (3.6)	87 (29)
70 (4.4)	79 (26)
80 (4.8)	66 (22)
90 (5.7)	50 (17)

(a) What is the flow rate?

(A) 44 gal/min (2.9 L/s)

(B) 69 gal/min (4.5 L/s)

(C) 82 gal/min (5.5 L/s)

(D) 95 gal/min (6.2 L/s)

(b) What can be said about the use of this pump in this installation?

(A) A different pump should be used.

(B) The pump is operating near its most efficient point.

(C) Pressure fluctuations could result from surging.

(D) Overloading will not be a problem.

Affinity Laws

13. A pump was intended to run at 1750 rpm when driven by a 0.5 hp (0.37 kW) motor. What is the required power rating of a motor that will turn the pump at 2000 rpm?

(A) 0.25 hp (0.19 kW)

(B) 0.45 hp (0.34 kW)

(C) 0.65 hp (0.49 kW)

(D) 0.75 hp (0.55 kW)

14. (*Time limit: one hour*) A centrifugal pump running at 1400 rpm has the curve shown. The pump will be installed in an existing pipeline with known head requirements given by the formula $H = 30 + 2Q^2$. H is the system head in feet of water. Q is the flow rate in ft^3/sec.

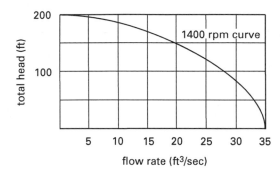

(a) What is the flow rate if the pump is turned at 1400 rpm?

(A) 2000 gal/min

(B) 3500 gal/min

(C) 4000 gal/min

(D) 4500 gal/min

(b) What power is required to drive the pump?

(A) 190 hp

(B) 210 hp

(C) 230 hp

(D) 260 hp

(c) What is the flow rate if the pump is turned at 1200 rpm?

(A) 2000 gal/min

(B) 3500 gal/min

(C) 4000 gal/min

(D) 4500 gal/min

Turbines

15. A horizontal turbine reduces 100 ft^3/sec of water from 30 psia to 5 psia. Friction is negligible. What power is developed?

(A) 350 hp

(B) 500 hp

(C) 650 hp

(D) 800 hp

16. 1000 ft^3/sec of 60°F water flow from a high reservoir through a hydroelectric turbine installation, exiting 625 ft lower. The head loss due to friction is 58 ft. The turbine efficiency is 89%. What power is developed in the turbines?

(A) 40 kW

(B) 18 MW

(C) 43 MW

(D) 71 MW

17. Water at 500 psig and 60°F (3.5 MPa and 16°C) drives a 250 hp (185 kW) turbine at 1750 rpm against a back pressure of 30 psig (210 kPa). The water discharges through a 4 in (100 mm) diameter nozzle at 35 ft/sec (10.5 m/s). The water is deflected 80° by a single blade moving directly away at 10 ft/sec (3 m/s).

(a) What is the specific speed?

(A) 5
(B) 25
(C) 75
(D) 225

(b) What is the total force acting on the blade?

(A) 100 lbf (450 N)
(B) 140 lbf (570 N)
(C) 160 lbf (720 N)
(D) 280 lbf (1300 N)

18. (*Time limit: one hour*) A Francis-design hydraulic reaction turbine with 22 in (560 mm) diameter blades runs at 610 rpm. The turbine develops 250 hp (185 kW) when 25 ft³/sec (700 L/s) of water flow through it. The pressure head at the turbine entrance is 92.5 ft (30.8 m). The elevation of the turbine above the tailwater level is 5.26 ft (1.75 m). The inlet and outlet velocity are both 12 ft/sec (3.6 m/s).

(a) What is the effective head?

(A) 90 ft (30 m)
(B) 95 ft (31 m)
(C) 100 ft (33 m)
(D) 105 ft (35 m)

(b) What is the overall turbine efficiency?

(A) 81%
(B) 88%
(C) 93%
(D) 96%

(c) What will be the turbine speed if the effective head is 225 ft (75 m)?

(A) 600 rpm
(B) 920 rpm
(C) 1100 rpm
(D) 1400 rpm

(d) What horsepower is developed if the effective head is 225 ft (75 m)?

(A) 560 hp (420 kW)
(B) 630 hp (470 kW)
(C) 750 hp (560 kW)
(D) 840 hp (630 kW)

(e) What is the flow rate if the effective head is 225 ft (75 m)?

(A) 25 ft³/sec (700 L/s)
(B) 38 ft³/sec (1100 L/s)
(C) 56 ft³/sec (1600 L/s)
(D) 64 ft³/sec (1800 L/s)

SOLUTIONS

1. $\left(2000 \ \dfrac{\text{gal}}{\text{min}} \right) \left(\dfrac{0.002228 \ \dfrac{\text{ft}^3}{\text{sec}}}{\dfrac{\text{gal}}{\text{min}}} \right) = 4.456 \ \text{ft}^3/\text{sec}$

Assume schedule-40 steel pipe. From App. 16.B,

$12'' : \quad D_1 = 0.9948 \ \text{ft} \quad A_1 = 0.7773 \ \text{ft}^2$
$8'' : \quad D_2 = 0.6651 \ \text{ft} \quad A_2 = 0.3473 \ \text{ft}^2$

$$p_1 = \left(14.7 \ \frac{\text{lbf}}{\text{in}^2} - (8 \ \text{in}) \left(0.491 \ \frac{\text{lbf}}{\text{in}^3} \right) \right) \left(144 \ \frac{\text{in}^2}{\text{ft}^2} \right)$$

$$= 1551.2 \ \text{lbf/ft}^2$$

$$p_2 = \left(14.7 \ \frac{\text{lbf}}{\text{in}^2} + 20 \ \frac{\text{lbf}}{\text{in}^2} \right) \left(144 \ \frac{\text{in}^2}{\text{ft}^2} \right)$$

$$+ (4 \ \text{ft})(1.2) \left(62.4 \ \frac{\text{lbf}}{\text{ft}^3} \right)$$

$$= 5296.3 \ \text{lbf/ft}^2$$

$$v_1 = \frac{4.456 \ \dfrac{\text{ft}^3}{\text{sec}}}{0.7773 \ \text{ft}^2} = 5.73 \ \text{ft/sec}$$

$$v_2 = \frac{4.456 \ \dfrac{\text{ft}^3}{\text{sec}}}{0.3473 \ \text{ft}^2} = 12.83 \ \text{ft/sec}$$

From Eq. 18.9, the total heads at points 1 and 2 are

$$h_{t,1} = \frac{1551.2 \ \dfrac{\text{lbf}}{\text{ft}^2}}{\left(62.4 \ \dfrac{\text{lbf}}{\text{ft}^3} \right)(1.2)} + \frac{\left(5.73 \ \dfrac{\text{ft}}{\text{sec}} \right)^2}{(2) \left(32.2 \ \dfrac{\text{ft}}{\text{sec}^2} \right)} = 21.23 \ \text{ft}$$

$$h_{t,2} = \frac{5296.3 \ \dfrac{\text{lbf}}{\text{ft}^2}}{\left(62.4 \ \dfrac{\text{lbf}}{\text{ft}^3} \right)(1.2)} + \frac{\left(12.83 \ \dfrac{\text{ft}}{\text{sec}} \right)^2}{(2) \left(32.2 \ \dfrac{\text{ft}}{\text{sec}^2} \right)} = 73.29 \ \text{ft}$$

The pump must add $73.29 \ \text{ft} - 21.23 \ \text{ft} = 52.06 \ \text{ft}$ of head.

The power required is given in Table 18.5.

$$P_{\text{ideal}} = \frac{(52.06 \ \text{ft})(1.2) \left(4.456 \ \dfrac{\text{ft}^3}{\text{sec}} \right) \left(62.4 \ \dfrac{\text{lbf}}{\text{ft}^3} \right)}{550 \ \dfrac{\text{ft-lbf}}{\text{hp-sec}}}$$

$$= 31.58 \ \text{hp}$$

The input horsepower is

$$P_{\text{in}} = \frac{P_{\text{ideal}}}{\eta} = \frac{31.58 \ \text{hp}}{0.85} = \boxed{37.15 \ \text{hp}}$$

The answer is (C).

2. *Customary U.S. Solution*

From App. 16.B, data for 4 in schedule-40 steel pipe are

$$D_i = 0.3355 \ \text{ft}$$
$$A_i = 0.08841 \ \text{ft}^2$$

The velocity in the pipe is

$$v = \frac{\dot{V}}{A} = \frac{1.25 \ \dfrac{\text{ft}^3}{\text{sec}}}{0.08841 \ \text{ft}^2} = 14.139 \ \text{ft/sec}$$

From Table 17.3, typical equivalent lengths for schedule-40, screwed steel fittings for 4 in pipes are

90° elbow: 13 ft

gate valve: 2.5 ft

check valve: 38.0 ft

The total equivalent length is

$$(2)(13 \ \text{ft}) + (1)(2.5 \ \text{ft}) + (1)(38 \ \text{ft}) = 66.5 \ \text{ft}$$

At 70°F, from App. 14.A, the density of water is 62.3 lbm/ft^3 and the kinematic viscosity of water, ν, is $1.059 \times 10^{-5} \ \text{ft}^2/\text{sec}$. The Reynolds number is

$$\text{Re} = \frac{Dv}{\nu} = \frac{(0.3355 \ \text{ft}) \left(14.139 \ \dfrac{\text{ft}}{\text{sec}} \right)}{1.059 \times 10^{-5} \ \dfrac{\text{ft}^2}{\text{sec}}}$$

$$= 4.479 \times 10^5$$

From App. 17.A, for steel, $\epsilon = 0.0002 \ \text{ft}$.
So,

$$\frac{\epsilon}{D} = \frac{0.0002 \ \text{ft}}{0.3355 \ \text{ft}} \approx 0.0006$$

From App. 17.B, the friction factor is $f = 0.01835$.
The friction head is given by Eq. 18.6.

$$h_f = \frac{fLv^2}{2Dg}$$

$$= \frac{(0.01835)(700 \ \text{ft} + 66.5 \ \text{ft}) \left(14.139 \ \dfrac{\text{ft}}{\text{sec}} \right)^2}{(2)(0.3355 \ \text{ft}) \left(32.2 \ \dfrac{\text{ft}}{\text{sec}^2} \right)}$$

$$= 130.1 \ \text{ft}$$

The total dynamic head is given by Eq. 18.9. Point s is taken as the bottom of the supply tank.

$$h = \frac{(p_d - p_s)g_c}{\rho g} + \frac{v_d^2 - v_s^2}{2g} + z_d - z_s$$

$$v_s \approx 0$$

$$z_d - z_s = 50 \ \text{ft} \quad [\text{given as rise in elevation}]$$

The discharge and suction pressures are

$$p_d = 20 \text{ psig}$$
$$p_s = 50 \text{ psig}$$

$$h = \frac{\left(20 \frac{\text{lbf}}{\text{in}^2} - 50 \frac{\text{lbf}}{\text{in}^2}\right)}{\left(62.3 \frac{\text{lbm}}{\text{ft}^3}\right)\left(32.2 \frac{\text{ft}}{\text{sec}^2}\right)}$$
$$+ \frac{\left(14.139 \frac{\text{ft}}{\text{sec}}\right)^2}{(2)\left(32.2 \frac{\text{ft}}{\text{sec}^2}\right)} + 50 \text{ ft}$$
$$= -16.2 \text{ ft}$$

The head added is

$$h_A = h + h_f = -16.2 \text{ ft} + 130.1 \text{ ft}$$
$$= 113.9 \text{ ft}$$

The mass flow rate is

$$\dot{m} = \rho \dot{V}$$
$$= \left(62.3 \frac{\text{lbm}}{\text{ft}^3}\right)\left(1.25 \frac{\text{ft}^3}{\text{sec}}\right)$$
$$= 77.875 \text{ lbm/sec}$$

From Table 18.5, the hydraulic horsepower is

$$\text{WHP} = \left(\frac{h_A \dot{m}}{550}\right)\left(\frac{g}{g_c}\right)$$
$$= \left(\frac{(113.9 \text{ ft})\left(77.875 \frac{\text{lbm}}{\text{sec}}\right)}{550 \frac{\text{ft-lbf}}{\text{hp-sec}}}\right)$$
$$\times \left(\frac{32.2 \frac{\text{ft}}{\text{sec}^2}}{32.2 \frac{\text{ft-lbm}}{\text{lbf-sec}^2}}\right)$$
$$= \boxed{16.13 \text{ hp}}$$

The answer is (A).

SI Solution

From App. 16.C, data for 4 in schedule-40 steel pipe are

$$D_i = 102.3 \text{ mm}$$
$$A_i = 82.19 \times 10^{-4} \text{ m}^2$$

The velocity in the pipe is

$$v = \frac{\dot{V}}{A} = \frac{\left(35 \frac{\text{L}}{\text{s}}\right)\left(\frac{1 \text{ m}^3}{1000 \text{ L}}\right)}{82.19 \times 10^{-4} \text{ m}^2} = 4.26 \text{ m/s}$$

From Table 17.3, typical equivalent lengths for schedule-40, screwed steel fittings for 4 in pipes are

    90° elbow: 13 ft

    gate valve: 2.5 ft

    check valve: 38.0 ft

The total equivalent length is

$$(2)(13 \text{ ft}) + (1)(2.5 \text{ ft}) + (1)(38 \text{ ft}) = 66.5 \text{ ft}$$
$$(66.5 \text{ ft})\left(0.3048 \frac{\text{m}}{\text{ft}}\right) = 20.27 \text{ m}$$

At 21°C, from App. 14.B, the water properties are

$$\rho = 998 \text{ kg/m}^3$$
$$\mu = 0.9827 \times 10^{-3} \text{ Pa·s}$$
$$\nu = \frac{\mu}{\rho} = \frac{0.9827 \times 10^{-3} \text{ Pa·s}}{998 \frac{\text{kg}}{\text{m}^3}}$$
$$= 9.85 \times 10^{-7} \text{ m}^2/\text{s}$$

The Reynolds number is

$$\text{Re} = \frac{Dv}{\nu} = \frac{(102.3 \text{ mm})\left(\frac{1 \text{ m}}{1000 \text{ mm}}\right)\left(4.26 \frac{\text{m}}{\text{s}}\right)}{9.85 \times 10^{-7} \frac{\text{m}^2}{\text{s}}}$$
$$= 4.424 \times 10^5$$

From Table 17.2, for steel, $\epsilon = 6 \times 10^{-5}$ m. So,

$$\frac{\epsilon}{D} = \frac{6.0 \times 10^{-5} \text{ m}}{(102.3 \text{ mm})\left(\frac{1 \text{ m}}{1000 \text{ mm}}\right)} = 0.0006$$

From App. 17.B, the friction factor is $f = 0.01836$.

From Eq. 18.6, the friction head is

$$h_f = \frac{fLv^2}{2Dg}$$
$$= \frac{(0.01836)(230 \text{ m} + 20.27 \text{ m})\left(4.26 \frac{\text{m}}{\text{s}}\right)^2}{(2)(102.3 \text{ mm})\left(\frac{1 \text{ m}}{1000 \text{ mm}}\right)\left(9.81 \frac{\text{m}}{\text{s}^2}\right)}$$
$$= 41.5 \text{ m}$$

The total dynamic head is given by Eq. 18.9. Point s is taken as the bottom of the supply tank.

$$h = \frac{p_d - p_s}{\rho g} + \frac{v_d^2 - v_s^2}{2g} + z_d - z_s$$

$$v_s \approx 0$$

$$z_d - z_s = 15 \text{ m} \quad \text{[given as rise in elevation]}$$

The difference between discharge and suction pressure is

$$p_d - p_s = 140 \text{ kPa} - 345 \text{ kPa} = -205 \text{ kPa}$$

$$h = \frac{(-205 \text{ kPa})\left(1000 \dfrac{\text{Pa}}{\text{kPa}}\right)}{\left(998 \dfrac{\text{kg}}{\text{m}^3}\right)\left(9.81 \dfrac{\text{m}}{\text{s}^2}\right)}$$

$$+ \frac{\left(4.26 \dfrac{\text{m}}{\text{s}}\right)^2}{(2)\left(9.81 \dfrac{\text{m}}{\text{s}^2}\right)} + 15 \text{ m}$$

$$= -5.0 \text{ m}$$

The head added by the pump is

$$h_A = h + h_f = -5.0 \text{ m} + 41.5 \text{ m}$$

$$= 36.5 \text{ m}$$

The mass flow rate is

$$\dot{m} = \rho \dot{V}$$

$$= \left(998 \frac{\text{kg}}{\text{m}^3}\right)\left(35 \frac{\text{L}}{\text{s}}\right)\left(\frac{1 \text{ m}^3}{1000 \text{ L}}\right)$$

$$= 34.93 \text{ kg/s}$$

From Table 18.6, the hydraulic power is

$$\text{WkW} = \frac{(9.81)h_A \dot{m}}{1000}$$

$$= \frac{\left(9.81 \dfrac{\text{m}}{\text{s}^2}\right)(36.5 \text{ m})\left(34.93 \dfrac{\text{kg}}{\text{s}}\right)}{1000 \dfrac{\text{W}}{\text{kW}}}$$

$$= \boxed{12.51 \text{ kW}}$$

The answer is (A).

3. *Customary U.S. Solution*

The area of the rubber hose is

$$A = \left(\frac{\pi}{4}\right)D^2 = \left(\frac{\pi}{4}\right)\left(\frac{2 \text{ in}}{12 \dfrac{\text{in}}{\text{ft}}}\right)^2 = 0.0218 \text{ ft}^2$$

The velocity of water in the hose is

$$v = \frac{\dot{V}}{A} = \frac{\left(80 \dfrac{\text{gal}}{\text{min}}\right)\left(0.002228 \dfrac{\dfrac{\text{ft}^3}{\text{sec}}}{\dfrac{\text{gal}}{\text{min}}}\right)}{0.0218 \text{ ft}^2}$$

$$= 8.176 \text{ ft/sec}$$

At 80°F from App. 14.A, the kinematic viscosity of water is $\nu = 0.93 \times 10^{-5} \text{ ft}^2/\text{sec}$.

The Reynolds number is

$$\text{Re} = \frac{vD}{\nu} = \frac{\left(8.176 \dfrac{\text{ft}}{\text{sec}}\right)(2 \text{ in})\left(\dfrac{1 \text{ ft}}{12 \text{ in}}\right)}{0.93 \times 10^{-5} \dfrac{\text{ft}^2}{\text{sec}}}$$

$$= 1.47 \times 10^5$$

Assume that the rubber hose is smooth. From App. 17.B, the friction factor is $f = 0.0166$.

From Eq. 18.6, the friction head is

$$h_f = \frac{fLv^2}{2Dg}$$

$$= \frac{(0.0166)(50 \text{ ft})\left(8.176 \dfrac{\text{ft}}{\text{sec}}\right)^2}{(2)(2 \text{ in})\left(\dfrac{1 \text{ ft}}{12 \text{ in}}\right)\left(32.2 \dfrac{\text{ft}}{\text{sec}^2}\right)}$$

$$= 5.17 \text{ ft}$$

Neglecting entrance and exit losses, the head added by the pump is

$$h_A = h_f + h_z$$

$$= 5.17 \text{ ft} + 12 \text{ ft} - 4 \text{ ft}$$

$$= \boxed{13.17 \text{ ft}}$$

The answer is (B).

SI Solution

The area of the rubber hose is

$$A = \left(\frac{\pi}{4}\right)D^2 = \left(\frac{\pi}{4}\right)(5.1 \text{ cm})^2\left(\frac{1 \text{ m}}{100 \text{ cm}}\right)^2$$

$$= 0.00204 \text{ m}^2$$

The velocity of water in the hose is

$$v = \frac{\dot{V}}{A} = \frac{\left(5 \dfrac{\text{L}}{\text{s}}\right)\left(\dfrac{1 \text{ m}^3}{1000 \text{ L}}\right)}{0.00204 \text{ m}^2} = 2.45 \text{ m/s}$$

At 27°C from App. 14.B, the water data are

$$\rho = 996.5 \text{ kg/m}^3$$
$$\mu = 0.8565 \times 10^{-3} \text{ Pa·s}$$
$$\nu = \frac{\mu}{\rho} = \frac{0.8565 \times 10^{-3} \text{ Pa·s}}{996.5 \frac{\text{kg}}{\text{m}^3}}$$
$$= 8.60 \times 10^{-7} \text{ m}^2/\text{s}$$

The Reynolds number is

$$\text{Re} = \frac{\text{v}D}{\nu} = \frac{\left(2.45 \frac{\text{m}}{\text{s}}\right)(5.1 \text{ cm})\left(\frac{1 \text{ m}}{100 \text{ cm}}\right)}{8.60 \times 10^{-7} \frac{\text{m}^2}{\text{s}}}$$
$$= 1.45 \times 10^5$$

Assume that the rubber hose is smooth. From App. 17.B, the friction factor is $f \approx 0.0166$.

From Eq. 18.6, the friction head is

$$h_f = \frac{fL\text{v}^2}{2Dg}$$
$$= \frac{(0.0166)(15 \text{ m})\left(2.45 \frac{\text{m}}{\text{s}}\right)^2}{(2)(5.1 \text{ cm})\left(\frac{1 \text{ m}}{100 \text{ cm}}\right)\left(9.81 \frac{\text{m}}{\text{s}^2}\right)}$$
$$= 1.49 \text{ m}$$

Neglecting entrance and exit losses, the head added by the pump is

$$h_A = h_f + h_z$$
$$= 1.49 \text{ m} + 4 \text{ m} - 1.2 \text{ m}$$
$$= \boxed{4.29 \text{ m}}$$

The answer is (B).

4. *Customary U.S. Solution*

From App. 16.B, the diameters (inside) for 8 in and 6 in schedule-40 steel pipe are

$$D_1 = 7.981 \text{ in}$$
$$D_2 = 6.065 \text{ in}$$

At 60°F from App. 14.A, the density of water is 62.37 lbm/ft^3.

The mass flow rate through 6 in pipe is

$$\dot{m} = A_2\text{v}_2\rho$$
$$= \left(\frac{\pi}{4}\right)(6.065 \text{ in})^2\left(\frac{1 \text{ ft}^2}{144 \text{ in}^2}\right)\left(12 \frac{\text{ft}}{\text{sec}}\right)\left(62.37 \frac{\text{lbm}}{\text{ft}^3}\right)$$
$$= 150.2 \text{ lbm/sec}$$

The inlet (suction) pressure is

$$14.7 \text{ psia} - 5 \text{ psig} = 9.7 \text{ psia}$$
$$= \left(9.7 \frac{\text{lbf}}{\text{in}^2}\right)\left(144 \frac{\text{in}^2}{\text{ft}^2}\right)$$
$$= 1397 \text{ lbf/ft}^2$$

From Table 18.5, the head added by the pump is

$$h_A = \left(\frac{(550)(\text{BHP})\eta}{\dot{m}}\right)\left(\frac{g_c}{g}\right)$$
$$= \left(\frac{\left(550 \frac{\text{ft-lbf}}{\text{hp-sec}}\right)(20 \text{ hp})(0.70)}{150.2 \frac{\text{lbm}}{\text{sec}}}\right)\left(\frac{32.2 \frac{\text{ft-lbm}}{\text{lbf-sec}^2}}{32.2 \frac{\text{ft}}{\text{sec}^2}}\right)$$
$$= 51.26 \text{ ft}$$

At 1:

$$p_1 = 1397 \text{ lbf/ft}^2$$
$$z_1 = 0$$
$$\text{v}_1 = \frac{\text{v}_2 A_2}{A_1}$$
$$= \text{v}_2\left(\frac{D_2}{D_1}\right)^2$$
$$= \left(12 \frac{\text{ft}}{\text{sec}}\right)\left(\frac{6.065 \text{ in}}{7.981 \text{ in}}\right)^2 = 6.93 \text{ ft/sec}$$

At 2:

$$p_2 = \left(14.7 \frac{\text{lbf}}{\text{in}^2}\right)\left(144 \frac{\text{in}^2}{\text{ft}^2}\right) = 2117 \text{ lbf/ft}^2$$
$$\text{v}_2 = 12 \text{ ft/sec} \quad \text{[given]}$$

Let z_3 be the additional head above atmospheric. From Eq. 18.9(b), the head added by the pump is

$$h_A = \frac{(p_2 - p_1)g_c}{\rho g} + \frac{\text{v}_2^2 - \text{v}_1^2}{2g} + z_2 - z_1 + h_f + z_3$$
$$51.26 \text{ ft} = \left(\frac{2117 \frac{\text{lbf}}{\text{ft}^2} - 1397 \frac{\text{lbf}}{\text{ft}^2}}{62.37 \frac{\text{lbm}}{\text{ft}^3}}\right)\left(\frac{32.2 \frac{\text{ft-lbm}}{\text{lbf-sec}^2}}{32.2 \frac{\text{ft}}{\text{sec}^2}}\right)$$
$$+ \frac{\left(12 \frac{\text{ft}}{\text{sec}}\right)^2 - \left(6.93 \frac{\text{ft}}{\text{sec}}\right)^2}{(2)\left(32.2 \frac{\text{ft}}{\text{sec}^2}\right)}$$
$$+ 0 - 0 + 10 \text{ ft} + z_3$$
$$z_3 = \boxed{28.2 \text{ ft}}$$

The answer is (A).

SI Solution

From App. 16.C, the inside diameters for 8 in and 6 in steel schedule-40 pipe are

$$D_1 = 202.7 \text{ mm}$$
$$D_2 = 154.1 \text{ mm}$$

At 16°C from App. 14.B, the density of water is 998.83 kg/m^3.

The mass flow rate through the 6 in pipe is

$$\dot{m} = A_2 \text{v}_2 \rho$$
$$= \left(\frac{\pi}{4}\right)(154.1 \text{ mm})^2 \left(\frac{1 \text{ m}}{1000 \text{ mm}}\right)^2 \left(4 \frac{\text{m}}{\text{s}}\right)$$
$$\times \left(998.83 \frac{\text{kg}}{\text{m}^3}\right)$$
$$= 74.5 \text{ kg/s}$$

The inlet (suction) pressure is

$$101.3 \text{ kPa} - 35 \text{ kPa} = 66.3 \text{ kPa}$$

From Table 18.6, the head added by the pump is

$$h_A = \frac{(1000)(\text{BkW})\eta}{(9.81)\dot{m}}$$
$$= \frac{\left(1000 \frac{\text{W}}{\text{kW}}\right)(20 \text{ hp})\left(\frac{0.7457 \text{ kW}}{\text{hp}}\right)(0.70)}{\left(9.81 \frac{\text{m}}{\text{s}^2}\right)\left(74.5 \frac{\text{kg}}{\text{s}}\right)}$$
$$= 14.28 \text{ m}$$

At 1:

$$p_1 = 66.3 \text{ kPa}$$
$$z_1 = 0$$
$$\text{v}_1 = \text{v}_2 \left(\frac{A_2}{A_1}\right) = \text{v}_2 \left(\frac{D_2}{D_1}\right)^2$$
$$= \left(4 \frac{\text{m}}{\text{s}}\right)\left(\frac{154.1 \text{ mm}}{202.7 \text{ mm}}\right)^2 = 2.31 \text{ m/s}$$

At 2:
$$p_2 = 101.3 \text{ kPa}$$
$$\text{v}_2 = 4 \text{ m/s} \quad [\text{given}]$$

From Eq. 18.9(a), the head added by the pump is

$$h_A = \frac{p_2 - p_1}{\rho g} + \frac{\text{v}_2^2 - \text{v}_1^2}{2g} + z_2 - z_1 + h_f + z_3$$

$$14.28 \text{ m} = \frac{(101.3 \text{ kPa} - 66.3 \text{ kPa})\left(1000 \frac{\text{Pa}}{\text{kPa}}\right)}{\left(998.83 \frac{\text{kg}}{\text{m}^3}\right)\left(9.81 \frac{\text{m}}{\text{s}^2}\right)}$$
$$+ \frac{\left(4 \frac{\text{m}}{\text{s}}\right)^2 - \left(2.31 \frac{\text{m}}{\text{s}}\right)^2}{(2)\left(9.81 \frac{\text{m}}{\text{s}^2}\right)}$$
$$+ 0 - 0 + 3.3 \text{ m} + z_3$$

$$\boxed{z_3 = 6.86 \text{ m}}$$

The answer is (A).

5. *Customary U.S. Solution*

The flow rate is

$$\dot{V} = \left(10,000 \frac{\text{gal}}{\text{hr}}\right)\left(0.1337 \frac{\text{ft}^3}{\text{gal}}\right) = 1337 \text{ ft}^3/\text{hr}$$

From App. 16.B, data for 4 in schedule-40 steel pipe are

$$D_i = 0.3355 \text{ ft}$$
$$A_i = 0.08841 \text{ ft}^2$$

The velocity in the pipe is

$$\text{v} = \frac{\dot{V}}{A} = \frac{\left(1337 \frac{\text{ft}^3}{\text{hr}}\right)\left(\frac{1 \text{ hr}}{3600 \text{ sec}}\right)}{0.08841 \text{ ft}^2} = 4.20 \text{ ft/sec}$$

At 60°F from App. 14.A, the kinematic viscosity of water is
$$\nu = 1.217 \times 10^{-5} \text{ ft}^2/\text{sec}$$
$$\rho = 62.37 \text{ lbm/ft}^3$$

The Reynolds number is

$$\text{Re} = \frac{D\text{v}}{\nu} = \frac{(0.3355 \text{ ft})\left(4.20 \frac{\text{ft}}{\text{sec}}\right)}{1.217 \times 10^{-5} \frac{\text{ft}^2}{\text{sec}}}$$
$$= 1.16 \times 10^5$$

From App. 17.A, for welded and seamless steel, $\epsilon = 0.0002$ ft.
$$\frac{\epsilon}{D} = \frac{0.0002 \text{ ft}}{0.3355 \text{ ft}} \approx 0.0006$$

From App. 17.B, the friction factor, f, is 0.0205. The friction head is

$$h_f = \frac{fLv^2}{2Dg}$$

$$= \frac{(0.0205)(7000 \text{ ft}) \left(4.2 \frac{\text{ft}}{\text{sec}}\right)^2}{(2)(0.3355 \text{ ft}) \left(32.2 \frac{\text{ft}}{\text{sec}^2}\right)}$$

$$= 117.2 \text{ ft}$$

The head added by the pump is

$$h_A = h_f + h_z$$
$$= 117.2 \text{ ft} + 12 \text{ ft} + 350 \text{ ft}$$
$$= 479.5 \text{ ft}$$

From Table 18.5, the hydraulic horsepower is

$$\text{WHP} = \frac{h_A Q(\text{SG})}{3956}$$

$$= \frac{(479.2 \text{ ft}) \left(10,000 \frac{\text{gal}}{\text{hr}}\right) \left(\frac{1 \text{ hr}}{60 \text{ min}}\right) (1)}{3956 \frac{\text{ft-gal}}{\text{hp-min}}}$$

$$= 20.2 \text{ hp}$$

From Eq. 18.16, the overall efficiency of the pump is

$$\eta = \frac{\text{WHP}}{\text{EHP}}$$

$$\text{EHP} = \frac{20.2 \text{ hp}}{0.7}$$
$$= 28.9 \text{ hp}$$

(a) At \$0.04/kW-hr, power costs for 1 hr are

$$(28.9 \text{ hp}) \left(\frac{0.7457 \text{ kW}}{\text{hp}}\right) (1 \text{ hr}) \left(\frac{\$0.04}{\text{kW-hr}}\right)$$

$$= \boxed{\$0.86 \text{ per hour}}$$

The answer is (B).

(b) The motor horsepower, EHP, is 28.9 hp. Select a

$$\boxed{30 \text{ hp motor.}}$$

The answer is (B).

(c) From Eq. 18.5(b),

$$h_{\text{atm}} = \left(\frac{p_{\text{atm}}}{\rho}\right)\left(\frac{g_c}{g}\right)$$

$$= \left(\frac{\left(14.7 \frac{\text{lbf}}{\text{in}^2}\right)\left(144 \frac{\text{in}^2}{\text{ft}^2}\right)}{62.37 \frac{\text{lbm}}{\text{ft}^3}}\right) \left(\frac{32.2 \frac{\text{ft}}{\text{sec}^2}}{32.2 \frac{\text{ft-lbf}}{\text{lbm-sec}^2}}\right)$$

$$= 33.94 \text{ ft}$$

The friction losses due to 300 ft is

$$h_{f(s)} = \left(\frac{300 \text{ ft}}{7000 \text{ ft}}\right) h_f$$
$$= \left(\frac{300 \text{ ft}}{7000 \text{ ft}}\right)(117.2 \text{ ft})$$
$$= 5.0 \text{ ft}$$

From App. 14.A, the vapor pressure head at 60°F is 0.59 ft.

The NPSHA from Eq. 18.30(a) is

$$\text{NPSHA} = h_{\text{atm}} + h_{z(s)} - h_{f(s)} - h_{\text{vp}}$$
$$= 33.94 \text{ ft} - 12 \text{ ft} - 5.0 \text{ ft} - 0.59 \text{ ft}$$
$$= \boxed{16.35 \text{ ft}}$$

The answer is (D).

SI Solution

From App. 16.C, data for 4 in schedule-40 steel pipe are

$$D_i = 102.3 \text{ mm}$$
$$A_i = 82.19 \times 10^{-4} \text{ m}^2$$

The velocity in the pipe is

$$v = \frac{\dot{V}}{A} = \frac{\left(10.5 \frac{\text{L}}{\text{s}}\right)\left(\frac{1 \text{ m}^3}{1000 \text{ L}}\right)}{82.19 \times 10^{-4} \text{ m}^2} = 1.28 \text{ m/s}$$

From App. 14.B, at 16°C the water data are

$$\rho = 998.83 \text{ kg/m}^3$$
$$\mu = 1.1261 \times 10^{-3} \text{ Pa·s}$$

The Reynolds number is

$$\text{Re} = \frac{\rho v D}{\mu}$$

$$= \frac{\left(998.83 \frac{\text{kg}}{\text{m}^3}\right)\left(1.28 \frac{\text{m}}{\text{s}}\right) \times (102.3 \text{ mm})\left(\frac{1 \text{ m}}{1000 \text{ mm}}\right)}{1.1261 \times 10^{-3} \text{ Pa·s}}$$

$$= 1.16 \times 10^5$$

From Table 17.2, for welded and seamless steel, $\epsilon = 6.0 \times 10^{-5}$ m.

$$\frac{\epsilon}{D} = \frac{6.0 \times 10^{-5} \text{ m}}{(102.3 \text{ mm})\left(\frac{1 \text{ m}}{1000 \text{ mm}}\right)} = 0.0006$$

From App. 17.B, the friction factor is $f = 0.0205$.

From Eq. 18.6, the friction head is

$$h_f = \frac{fLv^2}{2Dg}$$

$$= \frac{(0.0205)(2300 \text{ m})\left(1.28 \ \frac{\text{m}}{\text{s}}\right)^2}{(2)(102.3 \text{ mm})\left(\frac{1 \text{ m}}{1000 \text{ mm}}\right)\left(9.81 \ \frac{\text{m}}{\text{s}^2}\right)}$$

$$= 38.5 \text{ m}$$

The head added by the pump is

$$h_A = h_f + h_z$$

$$= 38.5 \text{ m} + 4 \text{ m} + 115 \text{ m} = 157.5 \text{ m}$$

From Table 18.6, the hydraulic power is

$$\text{WkW} = \frac{(9.81)h_A Q(\text{SG})}{1000}$$

$$= \frac{\left(9.81 \ \frac{\text{m}}{\text{s}^2}\right)(157.5 \text{ m})\left(10.5 \ \frac{\text{L}}{\text{s}}\right)(1)}{1000 \ \frac{\text{W}\cdot\text{L}}{\text{kW}\cdot\text{kg}}}$$

$$= 16.22 \text{ kW}$$

From Eq. 18.16,

$$\text{EHP} = \frac{\text{WHP}}{\eta_{\text{overall}}}$$

$$= \frac{16.22 \text{ kW}}{0.7} = 23.2 \text{ kW}$$

(a) At \$0.04/kW·h, power costs for 1 h are

$$(23.2 \text{ kW})(1 \text{ h})\left(\frac{\$0.04}{\text{kW}\cdot\text{h}}\right) = \boxed{\$0.93 \text{ per hour}}$$

The answer is (B).

(b) The required motor power is 23.2 kW. Select the next higher standard motor size.

The answer is (B).

(c) From Eq. 18.5(a),

$$h_{\text{atm}} = \frac{p}{\rho g}$$

$$= \frac{(101 \text{ kPa})(1000 \text{ Pa})}{\left(998.83 \ \frac{\text{kg}}{\text{m}^3}\right)\left(9.81 \ \frac{\text{m}}{\text{s}^2}\right)} = 10.31 \text{ m}$$

The friction loss due to 100 m is

$$h_{f(s)} = \left(\frac{100 \text{ m}}{2300 \text{ m}}\right)h_f$$

$$= \left(\frac{100 \text{ m}}{2300 \text{ m}}\right)(38.5 \text{ m})$$

$$= 1.67 \text{ m}$$

The vapor pressure at 16°C is 0.01818 bar.

From Eq. 18.5(a),

$$h_{\text{vp}} = \frac{p_{\text{vp}}}{g\rho}$$

$$= \frac{(0.01818 \text{ bar})\left(1 \times 10^5 \ \frac{\text{Pa}}{\text{bar}}\right)}{\left(9.81 \ \frac{\text{m}}{\text{s}^2}\right)\left(998.83 \ \frac{\text{kg}}{\text{m}^3}\right)}$$

$$= 0.19 \text{ m}$$

The NPSHA from Eq. 18.30(a) is

$$\text{NPSHA} = h_{\text{atm}} + h_{z(s)} - h_{f(s)} - h_{\text{vp}}$$

$$= 10.31 \text{ m} - 4 \text{ m} - 1.67 \text{ m} - 0.19 \text{ m}$$

$$= \boxed{4.45 \text{ m}}$$

The answer is (D).

6. (a) $h_f = \Delta z = 48$ ft since $\Delta p = 0$ and $\Delta v = 0$ for open channel flow.

$$Q = \frac{\left(250 \ \frac{\text{gal}}{\text{person}\cdot\text{day}}\right)(10{,}000 \text{ people})}{\left(24 \ \frac{\text{hr}}{\text{day}}\right)\left(60 \ \frac{\text{min}}{\text{hr}}\right)}$$

$$= 1736 \text{ gal/min}$$

Given $C = 130$, solving for d from Eq. 17.31,

$$d_{\text{in}}^{4.8655} = \frac{(10.44)(5000 \text{ ft})\left(1736 \ \frac{\text{gal}}{\text{min}}\right)^{1.85}}{(130)^{1.85}(48 \text{ ft})} = 131{,}462$$

$$d = \boxed{11.27 \text{ in} \quad [\text{round to 12 in minimum}]}$$

The answer is (B).

(b) Without having a specific pump curve, the number of pumps can only be specified based on general rules. Use the *Ten States' Standards*, which states:

- No station will have less than two identical pumps.

- Capacity must be met with one pump out of service.

- Provision must be made to alternate pumps automatically.

$$\boxed{\text{Two pumps are required, plus one spare.}}$$

The answer is (A).

(c) With a variable speed pump, it will be possible to adjust to the wide variations in flow (100 to 250 gpcd). It may be possible to operate with one pump. However, TSS still requires two.

The answer is (B).

(d) With three constant speed pumps,

$$Q = \frac{1736 \ \frac{\text{gal}}{\text{min}}}{3} = 579 \text{ gal/min at maximum capacity}$$

From Table 18.5, assuming specific gravity ≈ 1.00 and using an average pump efficiency of $\eta_{\text{pump}} = 0.80$,

$$\text{rated motor power} = \frac{(10 \text{ ft})\left(579 \ \frac{\text{gal}}{\text{min}}\right)}{\left(3956 \ \frac{\text{gal-ft}}{\text{min-hp}}\right)(0.80)}$$

$$= \boxed{1.83 \text{ hp} \quad [\text{use } 2.0 \text{ hp}]}$$

The answer is (A).

(e) With two variable-speed pumps,

$$Q = \frac{1736 \ \frac{\text{gal}}{\text{min}}}{2} = 868 \text{ gal/min}$$

$$\text{rated motor power} = \frac{(10 \text{ ft})\left(868 \ \frac{\text{gal}}{\text{min}}\right)}{\left(3956 \ \frac{\text{gal-ft}}{\text{min-hp}}\right)(0.80)}$$

$$= \boxed{2.74 \text{ hp} \quad [\text{use } 3.0 \text{ hp}]}$$

The answer is (A).

(f) Incoming flow rate is independent of sump level.

The answer is (D).

(g) In the wet well, the pump (and perhaps motor) is submerged. Forced ventilation air will prevent a concentration of explosive methane. From the *Ten States' Standards*,

$$\boxed{\begin{array}{l} \text{12 air changes per hour if continuous;} \\ \text{30 per hour if intermittent.} \end{array}}$$

The answer is (D).

(h) In the dry well,

$$\boxed{\begin{array}{l} \text{6 air changes per hour if continuous;} \\ \text{30 per hour if intermittent.} \end{array}}$$

The answer is (A).

7. (a) $h_{p(s)} = \boxed{20 \text{ ft}}$

The answer is (B).

(b) $h_{p(d)} = \boxed{15 \text{ ft}}$

The answer is (A).

(c) There is no pipe size specified, so h_v cannot be calculated. Even so, v is typically in the 5 to 10 ft/sec range, and $h_v \approx 0$.

Since pipe lengths are not given, assume $h_f \approx 0$.

$$20 \text{ ft} + 15 \text{ ft} + \frac{\left(80 \ \frac{\text{lbf}}{\text{in}^2}\right)\left(144 \ \frac{\text{in}^2}{\text{ft}^2}\right)}{62.4 \ \frac{\text{lbf}}{\text{ft}^3}} + 10 \text{ ft}$$

$$= \boxed{229.6 \text{ ft of water}}$$

The answer is (D).

(d) The flow rate is

$$\frac{(3.5 \text{ MGD})\left(62.4 \ \frac{\text{lbf}}{\text{ft}^3}\right)}{0.64632 \ \frac{\text{MGD}}{\frac{\text{ft}^3}{\text{sec}}}} = 337.9 \text{ lbf/sec}$$

The rated motor output power does not depend on the motor efficiency.

$$P = \frac{h_A \dot{m}}{550 \eta_{\text{pump}}} = \frac{(229.6 \text{ ft})\left(337.9 \ \frac{\text{lbf}}{\text{sec}}\right)}{\left(550 \ \frac{\text{ft-lbf}}{\text{hp-sec}}\right)(0.85)}$$

$$= 166.0 \text{ hp} \quad [\text{from Table 18.5}]$$

$$\boxed{\text{Use a 200 hp motor.}}$$

The answer is (D).

(e) The chosen motor has more power than is necessary.

$$\dot{m} = \left(\frac{(200 \text{ hp})\left(550 \ \frac{\text{ft-lbf}}{\text{hp-sec}}\right)(0.85)(0.90)}{229.6 \text{ ft}}\right)$$

$$\times \left(\frac{0.64632 \ \frac{\text{MGD}}{\frac{\text{ft}^3}{\text{sec}}}}{62.4 \ \frac{\text{lbf}}{\text{ft}^3}}\right)$$

$$= 3.8 \text{ MGD}$$

The excess required storage is

$$\left(\frac{8\ \text{hr}}{24\ \dfrac{\text{hr}}{\text{day}}}\right)\left((1.25)(3.5\ \text{MGD}) - 3.8\ \text{MGD}\right)$$

$$= \boxed{0.19\ \text{MG}\ (190{,}000\ \text{gal})}$$

The answer is (B).

8. *Customary U.S. Solution*

From App. 16.B, the pipe data for 3 in schedule-40 steel pipe are

$$D_i = 0.2557\ \text{ft}$$
$$A_i = 0.05134\ \text{ft}^2$$

From App. 17.D, the equivalent length for various fittings is

$$\text{flanged elbow, } L_e = 4.4\ \text{ft}$$
$$\text{wide-open gate valve, } L_e = 2.8\ \text{ft}$$

The total equivalent length of pipe and fittings is

$$L_e = 500\ \text{ft} + (6)(4.4\ \text{ft}) + (2)(2.8\ \text{ft})$$
$$= 532\ \text{ft}$$

As a first estimate, assume the flow rate is 100 gal/min.

The velocity in the pipe is

$$\text{v} = \frac{\dot{V}}{A} = \frac{\left(100\ \dfrac{\text{gal}}{\text{min}}\right)\left(0.002228\ \dfrac{\dfrac{\text{ft}^3}{\text{sec}}}{\dfrac{\text{gal}}{\text{min}}}\right)}{0.05134\ \text{ft}^2}$$
$$= 4.34\ \text{ft/sec}$$

The Reynolds number is

$$\text{Re} = \frac{\text{v}D}{\nu} = \frac{\left(4.34\ \dfrac{\text{ft}}{\text{sec}}\right)(0.2557\ \text{ft})}{6 \times 10^{-6}\ \dfrac{\text{ft}^2}{\text{sec}}}$$
$$= 1.85 \times 10^5$$

From App. 17.A, $\epsilon = 0.0002$ ft.

So,

$$\frac{\epsilon}{D} = \frac{0.0002\ \text{ft}}{0.2557\ \text{ft}} \approx 0.0008$$

From the friction factor table, $f \approx 0.0204$.

For higher flow rates, f approaches 0.0186. Since the chosen flow rate was almost the lowest, $f = 0.0186$ should be used.

From Eq. 18.6, the friction head loss is

$$h_f = \frac{fL\text{v}^2}{2Dg}$$

$$= \frac{(0.0186)(532\ \text{ft})\left(4.34\ \dfrac{\text{ft}}{\text{sec}}\right)^2}{(2)(0.2557\ \text{ft})\left(32.2\ \dfrac{\text{ft}}{\text{sec}^2}\right)}$$

$$= 11.3\ \text{ft of gasoline}$$

This neglects the small velocity head. The other system points can be found using

$$\frac{h_{f_1}}{h_{f_2}} = \left(\frac{Q_1}{Q_2}\right)^2$$

$$h_{f_2} = h_{f_1}\left(\frac{Q_2}{100\ \dfrac{\text{gal}}{\text{min}}}\right)^2$$

$$= (11.3\ \text{ft})\left(\frac{Q_2}{100\ \dfrac{\text{gal}}{\text{min}}}\right)^2$$

$$= 0.00113 Q_2^2$$

Q (gal/min)	h_f (ft)	$h_f + 60$ (ft)
100	11.3	71.3
200	45.2	105.2
300	101.7	161.7
400	180.8	240.8
500	282.5	342.5
600	406.8	466.8

(a) Plot the system and pump curves.

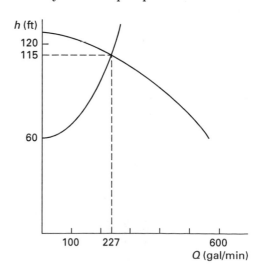

$$h = 115 \text{ ft}$$

$$\boxed{Q = 227 \text{ gal/min}}$$

This value could be used to determine a new friction function.

The answer is (D).

(b) From Table 18.5, the hydraulic horsepower is

$$\text{WHP} = \frac{h_A Q(\text{SG})}{3956} = \frac{(115 \text{ ft}) \left(227 \dfrac{\text{gal}}{\text{min}}\right)(0.7)}{3956 \dfrac{\text{ft-gal}}{\text{hp-min}}}$$

$$= 4.62 \text{ hp}$$

$$\frac{(4.62 \text{ hp}) \left(0.7457 \dfrac{\text{kW}}{\text{hp}}\right)}{(0.88)(0.88)} \times (1 \text{ hr}) \left(0.045 \dfrac{\$}{\text{kW-hr}}\right) = \boxed{\$0.20}$$

The answer is (A).

SI Solution

From App. 16.C, the pipe data for 3 in schedule-40 pipe are

$$D_i = 77.92 \text{ mm}$$

$$A_i = 47.69 \times 10^{-4} \text{ m}^2$$

From App. 17.D, the equivalent length for various fittings is

$$\text{flanged elbow, } L_e = 4.4 \text{ ft}$$

$$\text{wide-open gate valve, } L_e = 2.8 \text{ ft}$$

The total equivalent length of pipe and fittings is

$$L_e = 170 \text{ m} + (6)(4.4 \text{ ft}) \left(0.3048 \dfrac{\text{m}}{\text{ft}}\right)$$

$$+ (2)(2.8 \text{ ft}) \left(0.3048 \dfrac{\text{m}}{\text{ft}}\right)$$

$$= 180 \text{ m}$$

As a first estimate, assume flow rate is 6.3 L/s. The velocity in the pipe is

$$v = \frac{\dot{V}}{A} = \frac{\left(6.3 \dfrac{\text{L}}{\text{s}}\right) \left(\dfrac{1 \text{ m}^3}{1000 \text{ L}}\right)}{47.69 \times 10^{-4} \text{ m}^2} = 1.32 \text{ m/s}$$

The Reynolds number is

$$\text{Re} = \frac{vD}{\nu}$$

$$= \frac{\left(1.32 \dfrac{\text{m}}{\text{s}}\right)(77.92 \text{ mm}) \left(\dfrac{1 \text{ m}}{1000 \text{ mm}}\right)}{5.6 \times 10^{-7} \dfrac{\text{m}^2}{\text{s}}}$$

$$= 1.75 \times 10^5$$

From Table 17.2, $\epsilon = 6.0 \times 10^{-5}$ m.

$$\frac{\epsilon}{D} = \frac{6.0 \times 10^{-5} \text{ m}}{(77.92 \text{ mm}) \left(\dfrac{1 \text{ m}}{1000 \text{ mm}}\right)} \approx 0.0008$$

From the friction factor table (App. 17.B), $f = 0.0205$.

For higher flow rates, f approaches 0.0186. Since the chosen flow rate was almost the lowest, $f = 0.0186$ should be used.

From Eq. 17.22, the friction head loss is

$$h_f = \frac{fLv^2}{2Dg} = \frac{(0.0186)(180 \text{ m}) \left(1.32 \dfrac{\text{m}}{\text{s}}\right)^2}{(2)(77.92 \text{ mm}) \left(\dfrac{1 \text{ m}}{1000 \text{ mm}}\right) \left(9.81 \dfrac{\text{m}}{\text{s}^2}\right)}$$

$$= 3.82 \text{ m of gasoline}$$

This neglects the small velocity head. The other system points can be found using

$$\frac{h_{f_1}}{h_{f_2}} = \left(\frac{Q_1}{Q_2}\right)^2$$

$$h_{f_2} = h_{f_1} \left(\frac{Q_2}{Q_1}\right)^2 = (3.82 \text{ m}) \left(\frac{Q_2}{6.3 \dfrac{\text{L}}{\text{s}}}\right)^2$$

$$= 0.0962 Q_2^2$$

Q (L/s)	h_f (m)	$h_f + 20$ (m)
6.3	3.82	23.82
12	13.85	33.85
18	31.2	51.2
24	55.4	75.4
30	86.6	106.6
36	124.7	144.7

(a) Plot the system and pump curves.

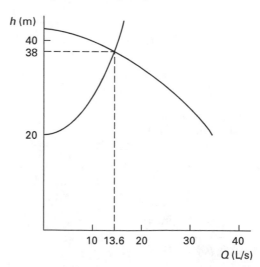

$$h = 38.0 \text{ m}$$

$$\boxed{Q = 13.6 \text{ L/s}}$$

This value could be used to determine a new friction function.

The answer is (D).

(b) From Table 18.6, the hydraulic power is

$$\text{WkW} = \frac{9.81 h_A Q(\text{SG})}{1000}$$

$$= \frac{\left(9.81 \ \frac{\text{m}}{\text{s}^2}\right)(38.0 \text{ m})\left(13.6 \ \frac{\text{L}}{\text{s}}\right)(0.7)}{1000 \ \frac{\text{W}}{\text{kW}}}$$

$$= 3.55 \text{ kW}$$

The cost per hour is

$$= \left(\frac{3.55 \text{ kW}}{(0.88)(0.88)}\right)(1 \text{ h})\left(0.045 \ \frac{\$}{\text{kW·h}}\right)$$

$$= \boxed{\$0.21}$$

The answer is (A).

9. *Customary U.S. Solution*

From Eq. 18.28(b), the specific speed is

$$n_s = \frac{n\sqrt{Q}}{h_A^{0.75}}$$

For a double-suction pump, Q in the preceding equation is half of the full flow rate.

$$n_s = \frac{(900 \text{ rpm})\sqrt{\left(300 \ \frac{\text{gal}}{\text{sec}}\right)\left(60 \ \frac{\text{sec}}{\text{min}}\right)\left(\frac{1}{2}\right)}}{(20 \text{ ft})^{0.75}}$$

$$= \boxed{9028 \text{ rpm}}$$

The answer is (C).

SI Solution

From Eq. 18.28(a), the specific speed is

$$n_s = \frac{n\sqrt{\dot{V}}}{h_A^{0.75}}$$

For a double-suction pump, $\dot{V}$ in the preceding equation is half of the full flow rate.

$$n_s = \frac{(900 \text{ rpm})\sqrt{\left(1.1 \ \frac{\text{kL}}{\text{s}}\right)\left(\frac{1 \text{ m}^3}{1 \text{ kL}}\right)\left(\frac{1}{2}\right)}}{(7 \text{ m})^{0.75}}$$

$$= \boxed{155.1 \text{ rpm}}$$

The answer is (C).

10.

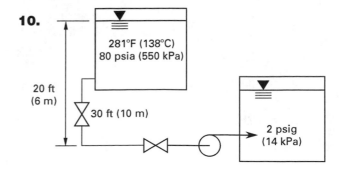

Customary U.S. Solution

From App. 16.B, data for 1.5 in schedule-40 steel pipe are

$$D_i = 0.1342 \text{ ft}$$

$$A_i = 0.01414 \text{ ft}^2$$

The velocity in the pipe is

$$\text{v} = \frac{\dot{V}}{A} = \frac{\left(100 \ \frac{\text{gal}}{\text{min}}\right)\left(0.002228 \ \frac{\text{ft}^3\text{-min}}{\text{sec-gal}}\right)}{0.01414 \text{ ft}^2}$$

$$= 15.76 \text{ ft/sec}$$

From App. 17.D, for screwed steel fittings, the approximate equivalent lengths for fittings are

inlet (square mouth): $L_e = 3.1 \text{ ft}$

long radius 90° ell: $L_e = 3.4 \text{ ft}$

wide-open gate valves: $L_e = 1.2 \text{ ft}$

The total equivalent length is

$$30 \text{ ft} + 3.1 \text{ ft} + (2)(3.4 \text{ ft}) + (2)(1.2 \text{ ft}) = 42.3 \text{ ft}$$

From App. 17.A, for steel, $\epsilon = 0.0002 \text{ ft}$, so

$$\frac{\epsilon}{D} = \frac{0.0002 \text{ ft}}{0.1342 \text{ ft}} = 0.0015$$

At 281°F, $\nu = 0.239 \times 10^{-5} \text{ ft}^2/\text{sec}$. The Reynolds number is

$$\text{Re} = \frac{D\text{v}}{\nu} = \frac{(0.1342 \text{ ft})\left(15.76 \ \frac{\text{ft}}{\text{sec}}\right)}{0.239 \times 10^{-5} \ \frac{\text{ft}^2}{\text{sec}}}$$

$$= 8.85 \times 10^5$$

From App. 17.B, the friction factor is $f = 0.022$.

From Eq. 18.6, the friction head is

$$h_f = \frac{fL\text{v}^2}{2Dg}$$

$$= \frac{(0.022)(42.3 \text{ ft})\left(15.76 \ \frac{\text{ft}}{\text{sec}}\right)^2}{(2)(0.1342 \text{ ft})\left(32.2 \ \frac{\text{ft}}{\text{sec}^2}\right)}$$

$$= 26.74 \text{ ft}$$

HYDRAULIC MACHINES **18-17**

At 281°F, the saturated vapor pressure of pressurized water is

$$p_{\text{vapor}} = 50.02 \text{ psia} \quad [\text{from steam tables}]$$

$$\rho = \frac{1}{v_f} = \frac{1}{0.01727 \dfrac{\text{ft}^3}{\text{lbm}}} = 57.9 \text{ lbm/ft}^3$$

From Eq. 18.5(b),

$$h_{\text{vp}} = \left(\frac{p_{\text{vapor}}}{\rho}\right)\left(\frac{g_c}{g}\right)$$

$$= \left(\frac{\left(50.02 \dfrac{\text{lbf}}{\text{in}^2}\right)\left(144 \dfrac{\text{in}^2}{\text{ft}^2}\right)}{57.9 \dfrac{\text{lbm}}{\text{ft}^3}}\right)\left(\frac{32.2 \dfrac{\text{ft-lbm}}{\text{lbf-sec}^2}}{32.2 \dfrac{\text{ft}}{\text{sec}^2}}\right)$$

$$= 124.4 \text{ ft}$$

From Eq. 18.5(b), the pressure head is

$$h_p = \left(\frac{p}{\rho}\right)\left(\frac{g_c}{g}\right)$$

$$= \left(\frac{\left(80 \dfrac{\text{lbf}}{\text{in}^2}\right)\left(144 \dfrac{\text{in}^2}{\text{ft}^2}\right)}{57.9 \dfrac{\text{lbm}}{\text{ft}^3}}\right)\left(\frac{32.2 \dfrac{\text{ft-lbm}}{\text{lbf-sec}^2}}{32.2 \dfrac{\text{ft}}{\text{sec}^2}}\right)$$

$$= 199.0 \text{ ft}$$

From Eq. 18.30(a), the NPSHA is

$$\text{NPSHA} = h_p + h_{z(s)} - h_{f(s)} - h_{\text{vp}}$$
$$= 199.0 \text{ ft} + 20 \text{ ft} - 26.74 \text{ ft} - 124.4 \text{ ft}$$
$$= 67.9 \text{ ft}$$

Since NPSHR = 10 ft, the pump will not cavitate.

(Note that a pump may not actually be needed in this configuration.)

The answer is (D).

SI Solution

From App. 16.C, data for 1.5 in schedule-40 steel pipe are

$$D_i = 40.89 \text{ mm}$$
$$A_i = 13.13 \times 10^{-4} \text{ m}^2$$

The velocity in the pipe is

$$v = \frac{\dot{V}}{A} = \frac{\left(6.3 \dfrac{\text{L}}{\text{s}}\right)\left(\dfrac{1 \text{ m}^3}{1000 \text{ L}}\right)}{13.13 \times 10^{-4} \text{ m}^2}$$
$$= 4.80 \text{ m/s}$$

From App. 17.D, for screwed steel fittings, the approximate equivalent lengths for fittings are

inlet (square mouth): $L_e = 3.1$ ft

long radius 90° ell: $L_e = 3.4$ ft

wide-open gate valves: $L_e = 1.2$ ft

The total equivalent length is

$$30 \text{ ft} + 3.1 \text{ ft} + (2)(3.4 \text{ ft}) + (2)(1.2 \text{ ft}) = 42.3 \text{ ft}$$

$$(42.3 \text{ ft})\left(0.3048 \frac{\text{m}}{\text{ft}}\right) = 12.89 \text{ m}$$

From Table 17.2, for steel, $\epsilon = 6.0 \times 10^{-5}$ m.

$$\frac{\epsilon}{D} = \frac{6.0 \times 10^{-5} \text{ m}}{(40.89 \text{ mm})\left(\dfrac{1 \text{ m}}{1000 \text{ mm}}\right)} \approx 0.0015$$

At 138°C, $\nu = 0.222 \times 10^{-6}$ m²/s. The Reynolds number is

$$\text{Re} = \frac{Dv}{\nu}$$

$$= \frac{(40.89 \text{ mm})\left(\dfrac{1 \text{ m}}{1000 \text{ mm}}\right)\left(4.80 \dfrac{\text{m}}{\text{s}}\right)}{0.222 \times 10^{-6} \dfrac{\text{m}^2}{\text{s}}}$$

$$= 8.84 \times 10^5$$

From App. 17.B, the friction factor is $f = 0.022$.

From Eq. 18.6, the friction head is

$$h_f = \frac{fLv^2}{2Dg}$$

$$= \frac{(0.022)(12.89 \text{ m})\left(4.8 \dfrac{\text{m}}{\text{s}}\right)^2}{(2)(40.89 \text{ mm})\left(\dfrac{1 \text{ m}}{1000 \text{ mm}}\right)\left(9.81 \dfrac{\text{m}}{\text{s}^2}\right)}$$

$$= 8.14 \text{ m}$$

From Eq. 18.5(a),

$$h_{\text{vp}} = \frac{p_{\text{vapor}}}{\rho g}$$

At 138°C, the saturated vapor pressure of pressurized water is

$$p_{\text{vapor}} = 3.431 \text{ bar} \quad [\text{from steam tables}]$$

$$\rho = \frac{1}{v_f}$$

$$= \frac{1}{\left(1.0777 \dfrac{\text{cm}^3}{\text{g}}\right)\left(1000 \dfrac{\text{g}}{\text{kg}}\right)\left(\dfrac{1 \text{ m}^3}{(100 \text{ cm})^3}\right)}$$

$$= 927.9 \text{ kg/m}^3$$

$$h_{\text{vp}} = \frac{(3.431 \text{ bar})\left(10^5 \dfrac{\text{Pa}}{\text{bar}}\right)}{\left(927.9 \dfrac{\text{kg}}{\text{m}^3}\right)\left(9.81 \dfrac{\text{m}}{\text{s}^2}\right)}$$

$$= 37.69 \text{ m}$$

PROFESSIONAL PUBLICATIONS, INC.

From Eq. 18.5(a), the pressure head is

$$h_p = \frac{p}{\rho g}$$

$$= \frac{(550 \text{ kPa})\left(1000 \, \dfrac{\text{Pa}}{\text{kPa}}\right)}{\left(927.9 \, \dfrac{\text{kg}}{\text{m}^3}\right)\left(9.81 \, \dfrac{\text{m}}{\text{s}^2}\right)}$$

$$= 60.42 \text{ m}$$

From Eq. 18.30(a), the NPSHA is

$$\text{NPSHA} = h_p + h_{z(s)} - h_{f(s)} - h_{\text{vp}}$$

$$= 60.42 \text{ m} + 6 \text{ m} - 8.14 \text{ m} - 37.69 \text{ m}$$

$$= 20.6 \text{ m}$$

Since NPSHR is 3 m, $\boxed{\text{the pump will not cavitate.}}$

(Note that a pump may not actually be needed in this configuration.)

The answer is (D).

11. The solvent is the water (fresh), and the solution is the seawater. Since seawater contains approximately $2\tfrac{1}{2}\%$ salt by weight, 100 lbm of seawater will yield 2.5 lbm salt and 97.5 lbm water. The molecular weight of salt is $23.0 + 35.5 = 58.5$ lbm/lbmol. The number of moles of salt in 100 lbm of seawater is

$$n_{\text{salt}} = \frac{2.5 \text{ lbm}}{58.5 \, \dfrac{\text{lbm}}{\text{lbmol}}} = 0.043 \text{ lbmol}$$

Similarly, the molecular weight of water is 18.016 lbm/lbmol. The number of moles of water is

$$n_{\text{water}} = \frac{97.5 \text{ lbm}}{18.016 \, \dfrac{\text{lbm}}{\text{lbmol}}} = 5.412 \text{ lbmol}$$

The mole fraction of water is

$$\frac{5.412 \text{ lbmol}}{5.412 \text{ lbmol} + 0.043 \text{ lbmol}} = 0.992$$

Customary U.S. Solution

Cavitation will occur when

$$h_{\text{atm}} - h_{\text{v}} < h_{\text{vp}}$$

The density of seawater is 64.0 lbm/ft^3.

From Eq. 18.5(b), the atmospheric head is

$$h_{\text{atm}} = \left(\frac{p}{\rho}\right)\left(\frac{g_c}{g}\right)$$

$$= \left(\frac{\left(14.7 \, \dfrac{\text{lbf}}{\text{in}^2}\right)\left(144 \, \dfrac{\text{in}^2}{\text{ft}^2}\right)}{64.0 \, \dfrac{\text{lbm}}{\text{ft}^3}}\right)\left(\frac{32.2 \, \dfrac{\text{ft-lbm}}{\text{lbf-sec}^2}}{32.2 \, \dfrac{\text{ft}}{\text{sec}^2}}\right)$$

$$= 33.075 \text{ ft} \quad [\text{ft of seawater}]$$

$$h_{\text{depth}} = 8 \text{ ft} \quad [\text{given}]$$

From Eq. 18.7, the velocity head is

$$h_{\text{v}} = \frac{v_{\text{propeller}}^2}{2g} = \frac{(4.2 v_{\text{boat}})^2}{(2)\left(32.2 \, \dfrac{\text{ft}}{\text{sec}^2}\right)}$$

$$= 0.2739 v_{\text{boat}}^2$$

The vapor pressure of 68°F freshwater is $p_{\text{vp}} = 0.3391$ psia.

From App. 14.A, the density of water at 68°F is 62.32 lbm/ft^3. Raoult's law predicts the actual vapor pressure of the solution.

$$p_{\text{vapor,solution}} = (p_{\text{vapor,solvent}})\left(\begin{array}{c}\text{mole fraction} \\ \text{of the solvent}\end{array}\right)$$

$$p_{\text{vapor,seawater}} = (0.992)(0.3391 \text{ psia})$$

$$= 0.3364 \text{ psia}$$

From Eq. 18.5(b), the vapor pressure head is

$$h_{\text{vapor,seawater}} = \left(\frac{p}{\rho}\right)\left(\frac{g_c}{g}\right)$$

$$= \left(\frac{\left(0.3364 \, \dfrac{\text{lbf}}{\text{in}^2}\right)\left(144 \, \dfrac{\text{in}^2}{\text{ft}^2}\right)}{64.0 \, \dfrac{\text{lbm}}{\text{ft}^3}}\right)$$

$$\times \left(\frac{32.2 \, \dfrac{\text{ft-lbm}}{\text{lbf-sec}^2}}{32.2 \, \dfrac{\text{ft}}{\text{sec}^2}}\right)$$

$$= 0.7569 \text{ ft}$$

Then,

$$8 \text{ ft} + 33.075 \text{ ft} - 0.2739 v_{\text{boat}}^2 = 0.7569 \text{ ft}$$

$$v_{\text{boat}} = \boxed{12.13 \text{ ft/sec}}$$

The answer is (B).

SI Solution

Cavitation will occur when

$$h_{\text{atm}} - h_{\text{v}} < h_{\text{vp}}$$

The density of seawater is 1024 kg/m^3.

From Eq. 18.5(a), the atmospheric head is

$$h_{\text{atm}} = \frac{p}{\rho g}$$

$$= \frac{(101.3 \text{ kPa})\left(1000 \dfrac{\text{Pa}}{\text{kPa}}\right)}{\left(1024 \dfrac{\text{kg}}{\text{m}^3}\right)\left(9.81 \dfrac{\text{m}}{\text{s}^2}\right)}$$

$$= 10.08 \text{ m}$$

$$h_{\text{depth}} = 3 \text{ m} \quad [\text{given}]$$

From Eq. 18.7, the velocity head is

$$h_{\text{v}} = \frac{v_{\text{propeller}}^2}{2g} = \frac{(4.2 v_{\text{boat}})^2}{(2)\left(9.81 \dfrac{\text{m}}{\text{s}^2}\right)}$$

$$= 0.899 v_{\text{boat}}^2$$

The vapor pressure of 20°C freshwater is

$$p_{\text{vp}} = (0.02339 \text{ bar})\left(100 \frac{\text{kPa}}{\text{bar}}\right)$$

$$= 2.339 \text{ kPa}$$

From App. 14.B, the density of water at 20°C is 998.23 kg/m^3. Raoult's law predicts the actual vapor pressure of the solution.

$$p_{\text{vapor,solution}} = (p_{\text{vapor,solvent}})\left(\begin{array}{c}\text{mole fraction}\\ \text{of the solvent}\end{array}\right)$$

The solvent is the freshwater and the solution is the seawater.

The mole fraction of water is 0.992.

$$p_{\text{vapor,seawater}} = (0.992)(2.339 \text{ kPa}) = 2.320 \text{ kPa}$$

From Eq. 18.5(a), the vapor pressure head is

$$h_{\text{vapor,seawater}} = \frac{(2.320 \text{ kPa})\left(1000 \dfrac{\text{Pa}}{\text{kPa}}\right)}{\left(9.81 \dfrac{\text{m}}{\text{s}^2}\right)\left(1024 \dfrac{\text{kg}}{\text{m}^3}\right)}$$

$$= 0.231 \text{ m}$$

Then,

$$3 \text{ m} + 10.08 \text{ m} - 0.899 v_{\text{boat}}^2 = 0.231 \text{ m}$$

$$v_{\text{boat}} = \boxed{3.78 \text{ m/s}}$$

The answer is (B).

12.

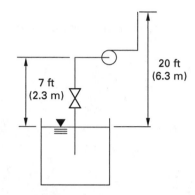

Customary U.S. Solution

From App. 17.D, the approximate equivalent lengths of various screwed steel fittings are

inlet: $L_e = 8.5$ ft [essentially a reentrant inlet]

check valve: $L_e = 19$ ft

long radius elbows: $L_e = 3.6$ ft

The total equivalent length of the 2 in line is

$$L_e = 12 \text{ ft} + 8.5 \text{ ft} + 19.0 \text{ ft} + (3)(3.6 \text{ ft}) + 80 \text{ ft}$$

$$= 130.3 \text{ ft}$$

From App. 16.B, for schedule-40 2 in pipe, the pipe data are

$$D_i = 0.1723 \text{ ft}$$

$$A_i = 0.0233 \text{ ft}^2$$

Since the flow rate is unknown, it must be assumed in order to find velocity. Assume 90 gal/min.

$$\dot{V} = \left(90 \frac{\text{gal}}{\text{min}}\right)\left(0.002228 \frac{\frac{\text{ft}^3}{\text{sec}}}{\frac{\text{gal}}{\text{min}}}\right) = 0.2005 \text{ ft}^3/\text{sec}$$

The velocity is

$$\text{v} = \frac{\dot{V}}{A_i} = \frac{0.2005 \dfrac{\text{ft}^3}{\text{sec}}}{0.0233 \text{ ft}^2} = 8.605 \text{ ft/sec}$$

From App. 14.A, the kinematic viscosity of water at 70°F is $\nu = 1.059 \times 10^{-5}$ ft^2/sec.

The Reynolds number is

$$
\mathrm{Re} = \frac{D\mathrm{v}}{\nu} = \frac{(0.1723\ \mathrm{ft})\left(8.605\ \dfrac{\mathrm{ft}}{\mathrm{sec}}\right)}{1.059 \times 10^{-5}\ \dfrac{\mathrm{ft}^2}{\mathrm{sec}}}
$$

$$
= 1.4 \times 10^5
$$

From App. 17.A, the specific roughness of steel pipe is

$$
\epsilon = 0.0002\ \mathrm{ft}
$$

$$
\frac{\epsilon}{D} = \frac{0.0002\ \mathrm{ft}}{0.1723\ \mathrm{ft}} = 0.0012
$$

From App. 17.B, $f = 0.022$. At 90 gal/min, the friction loss in the line from Eq. 18.6 is

$$
h_f = \frac{fL\mathrm{v}^2}{2Dg}
$$

$$
= \frac{(0.022)(130.3\ \mathrm{ft})\left(8.605\ \dfrac{\mathrm{ft}}{\mathrm{sec}}\right)^2}{(2)(0.1723\ \mathrm{ft})\left(32.2\ \dfrac{\mathrm{ft}}{\mathrm{sec}^2}\right)} = 19.1\ \mathrm{ft}
$$

From Eq. 18.7, the velocity head at 90 gal/min is

$$
h_\mathrm{v} = \frac{\mathrm{v}^2}{2g} = \frac{\left(8.605\ \dfrac{\mathrm{ft}}{\mathrm{sec}}\right)^2}{(2)\left(32.2\ \dfrac{\mathrm{ft}}{\mathrm{sec}^2}\right)} = 1.1\ \mathrm{ft}
$$

In general, the friction head and velocity head are proportional to v^2 and Q^2.

$$
h_f = (19.1\ \mathrm{ft})\left(\frac{Q_2}{90\ \dfrac{\mathrm{gal}}{\mathrm{min}}}\right)^2
$$

$$
h_\mathrm{v} = (1.1\ \mathrm{ft})\left(\frac{Q_2}{90\ \dfrac{\mathrm{gal}}{\mathrm{min}}}\right)^2
$$

(Note that the 7 ft dimension is included in the 20 ft dimension.)

The total system head is

$$
h = h_z + h_\mathrm{v} + h_f
$$

$$
= 20\ \mathrm{ft} + (1.1 + 19.1\ \mathrm{ft})\left(\frac{Q_2}{90\ \dfrac{\mathrm{gal}}{\mathrm{min}}}\right)^2
$$

From this equation, the following table for system head can be generated.

Q_2 (gal/min)	system head, h (ft)
0	20.0
10	20.2
20	21.0
30	22.2
40	24.0
50	26.2
60	29.0
70	32.2
80	36.0
90	40.2
100	44.9
110	50.2

(a) The intersection point of the system curve and the pump curve defines the operating flow rate. The flow rate is 95 gal/min.

The answer is (D).

(b) The intersection point is not in an efficient range for the pump because it is so far down on the system curve that the pumping efficiency will be low.

A different pump should be used.

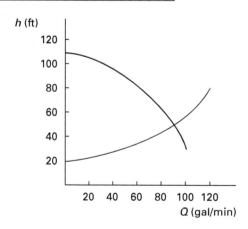

The answer is (A).

SI Solution

Use the approximate quivalent lengths of various screwed steel fittings from the Customary U.S. Solution. The total equivalent length of 5.08 cm schedule-40 pipe is

$$
L_e = 4\ \mathrm{m} + \big(8.5\ \mathrm{ft} + 19.0\ \mathrm{ft} + (3)(3.6\ \mathrm{ft})\big)\left(0.3048\ \frac{\mathrm{m}}{\mathrm{ft}}\right)
$$

$$
+ 27\ \mathrm{m}
$$

$$
= 42.67\ \mathrm{m}
$$

From App. 16.C, for 2 in schedule-40 pipe, the pipe data are

$$D_i = 52.50 \text{ mm}$$
$$A_i = 21.65 \times 10^{-4} \text{ m}^2$$

Since the flow rate is unknown, it must be assumed in order to find velocity. Assume 6 L/s.

$$\dot{V} = \left(6 \ \frac{\text{L}}{\text{s}}\right)\left(\frac{1 \text{ m}^3}{1000 \text{ L}}\right) = 6 \times 10^{-3} \text{ m}^3/\text{s}$$

The velocity is

$$v = \frac{\dot{V}}{A_i} = \frac{6 \times 10^{-3} \ \frac{\text{m}^3}{\text{s}}}{21.65 \times 10^{-4} \text{ m}^2} = 2.77 \text{ m/s}$$

From App. 14.B, the absolute viscosity of water at 21°C is $\mu = 0.9827 \times 10^{-3}$ Pa·s.

The density of water is $\rho = 998 \text{ kg/m}^3$.

The Reynolds number is

$$Re = \frac{\rho v D_i}{\mu} = \frac{\left(998 \ \frac{\text{kg}}{\text{m}^3}\right)\left(2.77 \ \frac{\text{m}}{\text{s}}\right) \times (52.50 \text{ mm})\left(\frac{1 \text{ m}}{1000 \text{ mm}}\right)}{0.9827 \times 10^{-3} \text{ Pa·s}}$$
$$= 1.5 \times 10^5$$

From Table 17.2, the specific roughness of steel pipe is

$$\epsilon = 6.0 \times 10^{-5} \text{ m}$$

$$\frac{\epsilon}{D} = \frac{6.0 \times 10^{-5} \text{ m}}{(52.50 \text{ mm})\left(\frac{1 \text{ m}}{1000 \text{ mm}}\right)} \approx 0.0012$$

From App. 17.B, $f = 0.022$. At 6 L/s, the friction loss in the line from Eq. 18.6 is

$$h_f = \frac{fLv^2}{2Dg}$$
$$= \frac{(0.022)(42.67 \text{ m})\left(2.77 \ \frac{\text{m}}{\text{s}}\right)^2}{(2)(52.50 \text{ mm})\left(\frac{1 \text{ m}}{1000 \text{ mm}}\right)\left(9.81 \ \frac{\text{m}}{\text{s}^2}\right)}$$
$$= 6.99 \text{ m}$$

At 6 L/s, the velocity head from Eq. 18.7 is

$$h_v = \frac{v^2}{2g} = \frac{\left(2.77 \ \frac{\text{m}}{\text{s}}\right)^2}{(2)\left(9.81 \ \frac{\text{m}}{\text{s}^2}\right)} = 0.39 \text{ m}$$

In general, the friction head and velocity head are proportional to v^2 and Q^2.

$$h_f = (6.99 \text{ m})\left(\frac{Q_2}{6 \ \frac{\text{L}}{\text{s}}}\right)^2$$

$$h_v = (0.39 \text{ m})\left(\frac{Q_2}{6 \ \frac{\text{L}}{\text{s}}}\right)^2$$

(Note that the 2.3 m dimension is included in the 6.3 m dimension.)

The total system head is

$$h = h_z + h_v + h_f$$
$$= 6.3 \text{ m} + (0.39 \text{ m} + 6.99 \text{ m})\left(\frac{Q_2}{6 \ \frac{\text{L}}{\text{s}}}\right)^2$$

From this equation the following table for the system head can be generated.

Q_2 (L/s)	h (m)
0	6.3
0.6	6.37
1.2	6.60
1.8	6.96
2.4	7.48
3.2	8.40
3.6	8.96
4.4	10.27
4.8	11.02
5.7	12.96
6.5	14.96
7.0	16.35
7.5	17.83

The intersection point of the system curve and the pump curve will define the operating flow rate.

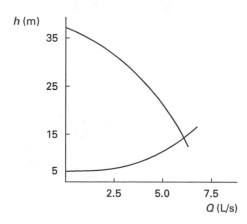

(a) The flow rate is $\boxed{6.2 \text{ L/s}.}$

The answer is (D).

(b) The intersection point is not in an efficient range of the pump because it is so far down on the system curve that the pumping efficiency will be low.

A different pump should be used.

The answer is (A).

13. From Eq. 18.52,

$$P_2 = P_1 \left(\frac{\rho_2 n_2^3 D_2^5}{\rho_1 n_1^3 D_1^5} \right)$$

$$= P_1 \left(\frac{n_2}{n_1} \right)^3 \quad [\rho_2 = \rho_1 \text{ and } D_2 = D_1]$$

Customary U.S. Solution

$$P_2 = (0.5 \text{ hp}) \left(\frac{2000 \text{ rpm}}{1750 \text{ rpm}} \right)^3 = \boxed{0.746 \text{ hp}}$$

The answer is (D).

SI Solution

$$P_2 = P_1 \left(\frac{n_2}{n_1} \right)^3$$

$$= (0.37 \text{ kW}) \left(\frac{2000 \text{ rpm}}{1750 \text{ rpm}} \right)^3$$

$$= \boxed{0.55 \text{ kW}}$$

The answer is (D).

14. (a) Random values of Q are chosen, and the corresponding values of H are determined by the formula $H = 30 + 2Q^2$.

Q (ft^3/sec)	H (ft)
0	30
2.5	42.5
5	80
7.5	142.5
10	230
15	480
20	830
25	1280
30	1830

The intersection of the system curve and the 1400 rpm pump curve defines the operating point at that rpm.

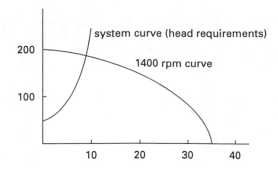

From the intersection of the graphs, at 1400 rpm the flow rate is approximately 9 ft^3/sec and the corresponding head is $30 + (2)(9)^2 \approx 192$ ft.

$$Q = \left(9 \frac{\text{ft}^3}{\text{sec}} \right) \left(448.8 \frac{\frac{\text{gal}}{\text{min}}}{\frac{\text{ft}^3}{\text{sec}}} \right) = \boxed{4039 \text{ gal/min}}$$

The answer is (C).

(b) From Table 18.5, the hydraulic horsepower is

$$\text{WHP} = \frac{h_A \dot{V} (\text{SG})}{8.814}$$

$$= \frac{(192 \text{ ft}) \left(9 \frac{\text{ft}^3}{\text{sec}} \right) (1)}{8.814 \frac{\text{ft}^4}{\text{hp-sec}}}$$

$$= 196 \text{ hp}$$

From Eq. 18.28(b), the specific speed is

$$n_s = \frac{n \sqrt{Q}}{h_A^{0.75}}$$

$$= \frac{(1400 \text{ rpm}) \sqrt{4039 \frac{\text{gal}}{\text{min}}}}{(192 \text{ ft})^{0.75}}$$

$$= 1725$$

From Fig. 18.8 with curve E, $\eta \approx 86\%$.

The minimum pump motor power should be

$$\frac{196 \text{ hp}}{0.86} = \boxed{228 \text{ hp}}$$

The answer is (C).

(c) From Eq. 18.41,

$$Q_2 = Q_1 \left(\frac{n_2}{n_1} \right) = \left(4039 \frac{\text{gal}}{\text{min}} \right) \left(\frac{1200 \text{ rpm}}{1400 \text{ rpm}} \right)$$

$$= \boxed{3462 \text{ gal/min}}$$

The answer is (B).

15. Since turbines are essentially pumps running backward, use Table 18.5.

$$\Delta p = \left(30\ \frac{\text{lbf}}{\text{in}^2} - 5\ \frac{\text{lbf}}{\text{in}^2}\right)\left(144\ \frac{\text{in}^2}{\text{ft}^2}\right) = 3600\ \text{lbf/ft}^2$$

$$P = \frac{\left(3600\ \frac{\text{lbf}}{\text{ft}^2}\right)\left(100\ \frac{\text{ft}^3}{\text{sec}}\right)}{550\ \frac{\text{ft-lbf}}{\text{hp-sec}}} = \boxed{654.5\ \text{hp}}$$

The answer is (C).

16. The flow rate is

$$\gamma\dot{V} = \left(62.4\ \frac{\text{lbf}}{\text{ft}^3}\right)\left(1000\ \frac{\text{ft}^3}{\text{sec}}\right) = 6.24 \times 10^4\ \text{lbf/sec}$$

The head available for work is

$$\Delta h = 625\ \text{ft} - 58\ \text{ft} = 567\ \text{ft}$$

Use Table 18.5. The power is

$$P = (0.89)\left(6.24 \times 10^4\ \frac{\text{lbf}}{\text{sec}}\right)(567\ \text{ft})$$
$$= 3.149 \times 10^7\ \text{ft-lbf/sec}$$

Convert from hp to kW.

$$\frac{\left(3.149 \times 10^7\ \frac{\text{ft-lbf}}{\text{sec}}\right)\left(0.7457\ \frac{\text{kW}}{\text{hp}}\right)}{550\ \frac{\text{ft-lbf}}{\text{hp-sec}}}$$
$$= \boxed{4.27 \times 10^4\ \text{kW (43 MW)}}$$

The answer is (C).

17. *Customary U.S. Solution*

(a) From App. 14.A, the density of water at 60°F is 62.37 lbm/ft^3. From Eq. 18.5(b), the head dropped is

$$h = \left(\frac{\Delta p}{\rho}\right)\left(\frac{g_c}{g}\right)$$
$$= \left(\frac{\left(500\ \frac{\text{lbf}}{\text{in}^2} - 30\ \frac{\text{lbf}}{\text{in}^2}\right)\left(144\ \frac{\text{in}^2}{\text{ft}^2}\right)}{62.37\ \frac{\text{lbm}}{\text{ft}^3}}\right)$$
$$\times \left(\frac{32.2\ \frac{\text{ft-lbm}}{\text{lbf-sec}^2}}{32.2\ \frac{\text{ft}}{\text{sec}^2}}\right)$$
$$= 1085\ \text{ft}$$

From Eq. 18.62(b), the specific speed of a turbine is

$$n_s = \frac{n\sqrt{P}}{(h_t)^{1.25}} = \frac{(1750\ \text{rpm})\sqrt{250\ \text{hp}}}{(1085\ \text{ft})^{1.25}}$$
$$= \boxed{4.443}$$

The answer is (A).

(b) The flow rate, $\dot{V}$, is

$$\dot{V} = A\text{v} = \left(\frac{\pi}{4}\right)\left(\frac{4\ \text{in}}{12\ \frac{\text{in}}{\text{ft}}}\right)^2\left(35\ \frac{\text{ft}}{\text{sec}}\right)$$
$$= 3.054\ \text{ft}^3/\text{sec}$$

The flow rate, considering the blade's moving away at 10 ft/sec, is

$$\dot{V}' = \frac{\left(35\ \frac{\text{ft}}{\text{sec}} - 10\ \frac{\text{ft}}{\text{sec}}\right)\left(3.054\ \frac{\text{ft}^3}{\text{sec}}\right)}{35\ \frac{\text{ft}}{\text{sec}}}$$
$$= 2.181\ \text{ft}^3/\text{sec}$$

The forces in the x-direction and the y-direction are

$$F_x = \left(\frac{\dot{V}'\rho}{g_c}\right)(\text{v}_j - \text{v}_b)(\cos\theta - 1)$$
$$= \left(\frac{\left(2.181\ \frac{\text{ft}^3}{\text{sec}}\right)\left(62.37\ \frac{\text{lbm}}{\text{ft}^3}\right)}{32.2\ \frac{\text{lbm-ft}}{\text{lbf-sec}^2}}\right)$$
$$\times \left(35\ \frac{\text{ft}}{\text{sec}} - 10\ \frac{\text{ft}}{\text{sec}}\right)(\cos 80° - 1)$$
$$= -87.27\ \text{lbf}$$

$$F_y = \left(\frac{\dot{V}'\rho}{g_c}\right)(\text{v}_j - \text{v}_b)\sin\theta$$
$$= \left(\frac{\left(2.181\ \frac{\text{ft}^3}{\text{sec}}\right)\left(62.37\ \frac{\text{lbm}}{\text{ft}^3}\right)}{32.2\ \frac{\text{lbm-ft}}{\text{lbf-sec}^2}}\right)$$
$$\times \left(35\ \frac{\text{ft}}{\text{sec}} - 10\ \frac{\text{ft}}{\text{sec}}\right)\sin 80°$$
$$= 104.0\ \text{lbf}$$

$$R = \sqrt{F_x^2 + F_y^2}$$

$$= \sqrt{(-87.27 \text{ lbf})^2 + (104.0 \text{ lbf})^2} = \boxed{135.8 \text{ lbf}}$$

The answer is (B).

SI Solution

(a) From App. 14.B, the density of water at 16°C is 998.83 kg/m^3. From Eq. 18.5(a), the head dropped is

$$h = \frac{\Delta p}{\rho g}$$

$$= \frac{(3.5 \text{ MPa})\left(10^6 \dfrac{\text{Pa}}{\text{MPa}}\right) - (210 \text{ kPa})\left(1000 \dfrac{\text{Pa}}{\text{kPa}}\right)}{\left(998.83 \dfrac{\text{kg}}{\text{m}^3}\right)\left(9.81 \dfrac{\text{m}}{\text{s}^2}\right)}$$

$$= 335.8 \text{ m}$$

From Eq. 18.62(a), the specific speed of a turbine is

$$n_s = \frac{n\sqrt{P}}{(h_t)^{1.25}}$$

$$= \frac{(1750 \text{ rpm})\sqrt{185 \text{ kW}}}{(335.8 \text{ m})^{1.25}}$$

$$= 16.56$$

Since the lowest suggested value of n_s for a reaction turbine is 38,

recommend an impulse (Pelton) turbine.

The answer is (A).

(b) The flow rate, $\dot{V}$, is

$$\dot{V} = A\text{v} = \left(\frac{\pi}{4}\right)\left((100 \text{ mm})\left(\frac{1 \text{ m}}{1000 \text{ mm}}\right)\right)^2\left(10.5 \frac{\text{m}}{\text{s}}\right)$$

$$= 0.08247 \text{ m}^3/\text{s}$$

The flow rate, considering the blade's moving away at 3 m/s, is

$$\dot{V}' = \frac{\left(10.5 \dfrac{\text{m}}{\text{s}} - 3 \dfrac{\text{m}}{\text{s}}\right)\left(0.08247 \dfrac{\text{m}^3}{\text{s}}\right)}{10.5 \dfrac{\text{m}}{\text{s}}}$$

$$= 0.05891 \text{ m}^3/\text{s}$$

The forces in the x-direction and the y-direction are

$$F_x = \dot{V}'\rho(\text{v}_j - \text{v}_b)(\cos\theta - 1)$$

$$= \left(0.05891 \frac{\text{m}^3}{\text{s}}\right)\left(998.83 \frac{\text{kg}}{\text{m}^3}\right)$$

$$\times \left(10.5 \frac{\text{m}}{\text{s}} - 3 \frac{\text{m}}{\text{s}}\right)(\cos 80° - 1)$$

$$= -364.7 \text{ N}$$

$$F_y = \dot{V}'\rho(\text{v}_j - \text{v}_b)\sin\theta$$

$$= \left(0.05891 \frac{\text{m}^3}{\text{s}}\right)\left(998.83 \frac{\text{kg}}{\text{m}^3}\right)$$

$$\times \left(10.5 \frac{\text{m}}{\text{s}} - 3 \frac{\text{m}}{\text{s}}\right)\sin 80°$$

$$= 434.6 \text{ N}$$

$$R = \sqrt{F_x^2 + F_y^2}$$

$$= \sqrt{(-364.7 \text{ N})^2 + (434.6 \text{ N})^2} = \boxed{567.3 \text{ N}}$$

The answer is (B).

18. *Customary U.S. Solution*

(a) The total effective head is due to the pressure head, velocity head, and tailwater head.

$$h_{\text{eff}} = h_p + h_v - h_{z\text{(tailwater)}}$$

$$= 92.5 \text{ ft} + \frac{\left(12 \dfrac{\text{ft}}{\text{sec}}\right)^2}{(2)\left(32.2 \dfrac{\text{ft}}{\text{sec}^2}\right)} - (-5.26 \text{ ft})$$

$$= \boxed{100 \text{ ft}}$$

The answer is (C).

(b) From Table 18.5, the theoretical hydraulic horsepower is

$$P_{\text{th}} = \frac{h_A \dot{V}(\text{SG})}{8.814} = \frac{(100 \text{ ft})\left(25 \dfrac{\text{ft}^3}{\text{sec}}\right)(1)}{8.814 \dfrac{\text{ft}^4}{\text{hp-sec}}}$$

$$= 283.6 \text{ hp}$$

The overall turbine efficiency is

$$\eta = \frac{P_{\text{brake}}}{P_{\text{th}}} = \frac{250 \text{ hp}}{283.6 \text{ hp}} = \boxed{0.882 \ (88.2\%)}$$

The answer is (B).

(c) From Eq. 18.42 (the affinity laws),

$$n_2 = n_1\sqrt{\frac{h_2}{h_1}} = (610 \text{ rpm})\sqrt{\frac{225 \text{ ft}}{100 \text{ ft}}}$$

$$= \boxed{915 \text{ rpm}}$$

The answer is (B).

(d) Combine Eqs. 18.42 and 18.43.

$$P_2 = P_1 \left(\frac{h_2}{h_1}\right)^{1.5} = (250 \text{ hp}) \left(\frac{225 \text{ ft}}{100 \text{ ft}}\right)^{1.5}$$

$$= \boxed{843.8 \text{ hp}}$$

The answer is (D).

(e) Combine Eqs. 18.41 and 18.42.

$$Q_2 = Q_1 \sqrt{\frac{h_2}{h_1}} = \left(25 \ \frac{\text{ft}^3}{\text{sec}}\right) \sqrt{\frac{225 \text{ ft}}{100 \text{ ft}}}$$

$$= \boxed{37.5 \text{ ft}^3/\text{sec}}$$

The answer is (B).

SI Solution

(a) The total effective head is due to the pressure head, velocity head, and tailwater head.

$$h_{\text{eff}} = h_p + h_v - h_{z(\text{tailwater})}$$

$$= 30.8 \text{ m} + \frac{\left(3.6 \ \frac{\text{m}}{\text{s}}\right)^2}{(2)\left(9.81 \ \frac{\text{m}}{\text{s}^2}\right)} - (-1.75 \text{ m})$$

$$= \boxed{33.21 \text{ m}}$$

The answer is (C).

(b) From Table 18.6, the theoretical hydraulic kilowatt is

$$P_{\text{th}} = \frac{9.81 h_A Q (\text{SG})}{1000}$$

$$= \frac{\left(9.81 \ \frac{\text{m}}{\text{s}^2}\right)(33.21 \text{ m})\left(700 \ \frac{\text{L}}{\text{s}}\right)(1)}{1000 \ \frac{\text{W}}{\text{kW}}}$$

$$= 228.1 \text{ kW}$$

The overall turbine efficiency is

$$\eta = \frac{P_{\text{brake}}}{P_{\text{th}}} = \frac{185 \text{ kW}}{228.1 \text{ kW}}$$

$$= \boxed{0.811 \ (81.1\%)}$$

The answer is (A).

(c) From Eq. 18.42 (the affinity laws),

$$h_2 = n_1 \sqrt{\frac{h_2}{h_1}} = (610 \text{ rpm}) \sqrt{\frac{75 \text{ m}}{33.21 \text{ m}}}$$

$$= \boxed{917 \text{ rpm}}$$

The answer is (B).

(d) Combine Eqs. 18.42 and 18.43.

$$P_2 = P_1 \left(\frac{h_2}{h_1}\right)^{1.5} = (185 \text{ kW}) \left(\frac{75 \text{ m}}{33.21 \text{ m}}\right)^{1.5}$$

$$= \boxed{627.3 \text{ kW}}$$

The answer is (D).

(e) Combine Eqs. 18.41 and 18.42.

$$Q_2 = Q_1 \sqrt{\frac{h_2}{h_1}} = \left(700 \ \frac{\text{L}}{\text{s}}\right) \sqrt{\frac{75 \text{ m}}{33.21 \text{ m}}}$$

$$= \boxed{1051.9 \text{ L/s}}$$

The answer is (B).

19 Special Fluid Topics

PRACTICE PROBLEMS

Pumping Power

1. A 3 ft diameter column is packed with 1 in Intalox saddles. 5000 lbm/hr of air at an average pressure of 1 atm and an average temperature of 70°F flows through the tower. The viscosity and density of air are 0.044 lbm/ft-hr and 0.075 lbm/ft^3, respectively. (a) Use the Ergun equation to estimate the pressure drop through the dry packing. (b) Estimate the pressure drop if 10,000 lbm/hr of 70°F water (density 62.3 lbm/ft^3) flows countercurrent to the air.

2. A packed bed of 0.5 in Berl saddles has a diameter of 2 ft and a packing height of 2 ft. The original 2 ft packing height expands to 5 ft at the onset of fluidization. The fluid stream has a viscosity of 0.120 lbm/ft-hr and a density of 0.025 lbm/ft^3. The packing has an initial packed density of 54 lbm/ft^3. Estimate the flow rate required to fluidize the bed.

SOLUTIONS

1. (a) From Table 19.1,

$$\varepsilon = 0.78$$
$$a_p = 78 \text{ ft}^2/\text{ft}^3$$
$$D_p = \frac{6(1 - \varepsilon)}{a_p} = \frac{(6)(1 - 0.78)}{78 \ \dfrac{\text{ft}^2}{\text{ft}^3}} = 0.0169 \text{ ft}$$

The superficial gas velocity is

$$G_0 = \frac{\dot{m}_g}{A} = \frac{5000 \ \dfrac{\text{lbm}}{\text{hr}}}{\left(\dfrac{\pi}{4}\right)(3 \text{ ft})^2} = 707 \text{ lbm/hr-ft}^2$$

Equation 19.1(b) is used to estimate the pressure drop in dry packing.

$$\frac{\Delta p}{L} = \left(\frac{1 - \varepsilon}{\varepsilon^3}\right) \left(\frac{G_0^2}{D_p g_c \rho_g}\right) \left(\frac{150(1 - \varepsilon)\mu_g'}{D_p G_0} + 1.75\right)$$

$$= \left(\frac{1 - 0.78}{(0.78)^3}\right) \left(\frac{\left(707 \ \dfrac{\text{lbm}}{\text{hr-ft}^2}\right)^2}{(0.0169 \text{ ft})\left(32.2 \ \dfrac{\text{ft-lbm}}{\text{lbf-sec}^2}\right)} \right. $$
$$\left. \times \left(0.075 \ \dfrac{\text{lbm}}{\text{ft}^3}\right)\left(3600 \ \dfrac{\text{sec}}{\text{hr}}\right)^2 \right)$$
$$\times \left(\frac{(150)(1 - 0.78)\left(0.044 \ \dfrac{\text{lbm}}{\text{ft-hr}}\right)}{(0.0169 \text{ ft})\left(707 \ \dfrac{\text{lbm}}{\text{hr-ft}^2}\right)} + 1.75 \right)$$

$$= \boxed{0.82 \ \frac{\text{lbf/ft}^2}{\text{ft}} \quad (0.82 \text{ psf/ft})}$$

(b) The irrigated pressure drop is estimated from Eq. 19.4. From Table 19.1,

$$C_2 = 12 \times 10^{-8} \text{ lbf-hr}^2/\text{lbm-ft}^2$$

$$C_3 = 2.8 \times 10^{-3} \text{ hr/ft}$$

$$L_0 = \frac{\dot{m}_l}{A} = \frac{10{,}000 \dfrac{\text{lbm}}{\text{hr}}}{\left(\dfrac{\pi}{4}\right)(3 \text{ ft})^2} = 1415 \text{ lbm/hr-ft}^2$$

$$\frac{\Delta p}{L} = C_2 \left(\frac{G_0^2}{\rho_g}\right)(10)^{C_3 L_0/\rho_l}$$

$$= \left(12 \times 10^{-8} \frac{\text{lbf-hr}^2}{\text{lbm-ft}^2}\right)$$

$$\times \left(\frac{\left(707 \dfrac{\text{lbm}}{\text{hr-ft}^2}\right)^2}{0.075 \dfrac{\text{lbm}}{\text{ft}^3}}\right)$$

$$\times (10)^{\left(0.0028 \frac{\text{hr}}{\text{ft}}\right)\left(1415 \frac{\text{lbm}}{\text{hr-ft}^2}\right)/\left(62.3 \frac{\text{lbm}}{\text{ft}^3}\right)}$$

$$= \boxed{0.93 \frac{\text{lbf/ft}^2}{\text{ft}} \quad (0.93 \text{ psf/ft})}$$

2. The bed void fraction can be determined from Eq. 19.9.

$$\varepsilon_{\text{mf}} = 1 - \frac{m_s}{L_{\text{mf}} A(\rho_s - \rho_f)}$$

$$= 1 - \frac{\rho_s A L}{L_{\text{mf}} A(\rho_s - \rho_f)} = 1 - \frac{\rho_s L}{L_{\text{mf}}(\rho_s - \rho_f)}$$

$$= 1 - \frac{\left(54 \dfrac{\text{lbm}}{\text{ft}^3}\right)(2 \text{ ft})}{(5 \text{ ft})\left(54 \dfrac{\text{lbm}}{\text{ft}^3} - 0.025 \dfrac{\text{lbm}}{\text{ft}^3}\right)}$$

$$= 0.60$$

Use Eq. 19.3 to calculate the effective particle diameter, using data from Table 19.1.

$$D_p = \frac{6(1-\varepsilon)}{a_p} = \frac{(6)(1-0.60)}{155 \dfrac{\text{ft}^2}{\text{ft}^3}} = 0.0155 \text{ ft}$$

The desired flow rate is obtained from Eq. 19.8.

$$G_{\text{mf}} = \frac{D_p^2 g \rho_f (\rho_s - \rho_f)\phi_p^2 \varepsilon_{\text{mf}}^3}{150 \mu_f'(1-\varepsilon_{\text{mf}})}$$

$$\frac{(0.0155 \text{ ft})^2 \left(32.2 \dfrac{\text{ft}}{\text{sec}^2}\right)\left(3600 \dfrac{\text{sec}}{\text{hr}}\right)^2}{}$$

$$\times \left(0.025 \frac{\text{lbm}}{\text{ft}^3}\right)\left(54 \frac{\text{lbm}}{\text{ft}^3} - 0.025 \frac{\text{lbm}}{\text{ft}^3}\right)$$

$$= \frac{\times (0.80)^2 (0.60)^3}{150 \left(0.120 \dfrac{\text{lbm}}{\text{ft-hr}}\right)(1-0.60)}$$

$$= \boxed{2598 \text{ lbm/hr-ft}^2}$$

20 Inorganic Chemistry

PRACTICE PROBLEMS

Empirical Formula Development

1. The gravimetric analysis of a compound is 40% carbon, 6.7% hydrogen, and 53.3% oxygen. What is the simplest formula for the compound?

(A) HCO
(B) HCO_2
(C) CH_2O
(D) CHO_2

2. At a particular set of conditions, the enthalpy of pure H_2SO_4 is 0 Btu/lbm, the enthalpy of water is 38 Btu/lbm, and the enthalpy of a solution of 25% (by weight) H_2SO_4 is −38 Btu/lbm of solution. Calculate the heat of mixing when 50 lbm of 100% sulfuric acid (H_2SO_4) is diluted to a final concentration of 25% (by weight) at the original conditions.

3. A 50% (by weight) solution of sodium hydroxide (NaOH) is to be produced from solid NaOH and water. The enthalpy of solid NaOH is 458 Btu/lbm, the enthalpy of water is 38 Btu/lbm, and the enthalpy of 50% (by weight) NaOH is 120 Btu/lbm of solution. Calculate the enthalpy of solution per pound of solution formed.

SOLUTIONS

1. Calculate the mole ratios of the atoms by assuming there are 100 g of sample.

For 100 g of sample,

substance	mass	$\dfrac{m}{AW}$ = no. moles	mole ratio
C	40 g	$\dfrac{40}{12} = 3.33$	1
H	6.7 g	$\dfrac{6.7}{1} = 6.7$	2
O	53.3 g	$\dfrac{53.3}{16} = 3.33$	1

The empirical formula is $\boxed{CH_2O.}$

The answer is (C).

2. A 25% (by weight) solution has a mass fraction of solute of 0.25.

$$0.25 = \frac{m_{H_2SO_4}}{m_{solution}}$$
$$= \frac{50 \text{ lbm}}{m_{solution}}$$
$$m_{solution} = 200 \text{ lbm}$$

The total mass of the solution is the sum of the mass of its components, H_2SO_4 and H_2O.

$$m_{solution} = m_{H_2SO_4} + m_{H_2O}$$
$$200 \text{ lbm} = 50 \text{ lbm} + m_{H_2O}$$
$$m_{H_2O} = 200 \text{ lbm} - 50 \text{ lbm}$$
$$= 150 \text{ lbm}$$

150 lbm of water are required to form a 25% (by weight) solution of H_2SO_4.

From Eq. 20.27(d),

$$\Delta h_t = mh - \Sigma m_i h_i$$

$$= m_{\text{solution}} h_{\text{solution}} - m_{H_2SO_4} h_{H_2SO_4} - m_{H_2O} h_{H_2O}$$

$$= (200 \text{ lbm}) \left(-38 \, \frac{\text{Btu}}{\text{lbm}} \right)$$

$$- (150 \text{ lbm}) \left(38 \, \frac{\text{Btu}}{\text{lbm}} \right) - (50 \text{ lbm}) \left(0 \, \frac{\text{Btu}}{\text{lbm}} \right)$$

$$= \boxed{-13{,}300 \text{ Btu} \quad \text{[heat evolved, exothermic]}}$$

3. A 50% (by weight) solution has a mass fraction of solute of 0.50.

$$0.5 = \frac{m_{\text{NaOH}}}{m_{\text{solution}}}$$

$$= \frac{m_{\text{NaOH}}}{1 \text{ lbm}}$$

$$m_{\text{NaOH}} = 0.5 \text{ lbm}$$

The total mass of the solution is the sum of the mass of its components, NaOH and H_2O.

$$m_{\text{solution}} = m_{\text{NaOH}} + m_{H_2O}$$

$$1 \text{ lbm} = 0.5 \text{ lbm} + m_{H_2O}$$

$$m_{H_2O} = 1 \text{ lbm} - 0.5 \text{ lbm}$$

$$= 0.5 \text{ lbm } H_2O$$

The solution is formed by combining 0.5 lbm of water and 0.5 lbm of NaOH.

From Eq. 20.27(d),

$$\Delta h_t = mh - \Sigma m_i h_i$$

$$= m_{\text{solution}} h_{\text{solution}} - m_{H_2O} h_{H_2O} - m_{NaOH_4} h_{NaOH}$$

$$= (1 \text{ lbm}) \left(120 \, \frac{\text{Btu}}{\text{lbm}} \right)$$

$$- (0.5 \text{ lbm}) \left(38 \, \frac{\text{Btu}}{\text{lbm}} \right) - (0.5 \text{ lbm}) \left(458 \, \frac{\text{Btu}}{\text{lbm}} \right)$$

$$= \boxed{-128 \text{ Btu/lbm} \quad \text{[heat evolved, exothermic]}}$$

21 Fuels and Combustion

PRACTICE PROBLEMS

1. Methane (MW = 16.043) with a higher heating value of 24,000 Btu/lbm (55.8 MJ/kg) is burned in a furnace with a 50% efficiency based on the higher heating value. How much water can be heated from 60 to 200°F (15 to 95°C) when 7 ft^3 (200 L) at 60°F and 14.73 psia are burned?

(A) 25 lbm (11 kg)
(B) 35 lbm (16 kg)
(C) 50 lbm (23 kg)
(D) 95 lbm (43 kg)

2. 15 lbm/hr (6.8 kg/h) of propane (C_3H_8, MW = 44.097) are burned stoichiometrically in air. What volume of dry carbon dioxide (CO_2) is formed after cooling to 70°F (21°C) and 14.7 psia (101 kPa)?

(A) 180 ft^3/hr (5.0 m^3/h)
(B) 270 ft^3/hr (7.6 m^3/h)
(C) 390 ft^3/hr (11 m^3/h)
(D) 450 ft^3/hr (13 m^3/h)

3. In a particular installation, 30% by volume excess air at 15 psia (103 kPa) and 100°F (40°C) is needed for the combustion of methane. How much nitrogen (MW = 28.016) passes through the furnace if methane at the same conditions is burned at the rate of 4000 ft^3/hr (31 L/s)?

(A) 270 lbm/hr (0.033 kg/s)
(B) 930 lbm/hr (0.11 kg/s)
(C) 1800 lbm/hr (0.22 kg/s)
(D) 2700 lbm/hr (0.34 kg/s)

4. How much air is required to completely burn one unit mass of a fuel that is 84% carbon, 15.3% hydrogen, 0.3% sulfur, and 0.4% nitrogen by weight?

(A) 9 lbm air/lbm fuel (9 kg air/kg fuel)
(B) 12 lbm air/lbm fuel (12 kg air/kg fuel)
(C) 15 lbm air/lbm fuel (15 kg air/kg fuel)
(D) 18 lbm air/lbm fuel (18 kg air/kg fuel)

5. Propane (C_3H_8) is burned with 20% excess air. What is the gravimetric percentage of carbon dioxide in the flue gas?

(A) 8%
(B) 12%
(C) 15%
(D) 22%

6. The ultimate analysis of a coal is 80% carbon, 4% hydrogen, 2% oxygen, and the rest ash. The flue gases are 60°F and 14.7 psia (15.6°C and 101.3 kPa) when sampled, and are 80% nitrogen, 12% carbon dioxide, 7% oxygen, and 1% carbon monoxide by volume. How much air is required to burn 1 lbm (1 kg) of coal under these conditions?

(A) 11 lbm (11 kg)
(B) 15 lbm (15 kg)
(C) 19 lbm (19 kg)
(D) 23 lbm (23 kg)

7. What is the approximate heating value of an oil with a specific gravity of 40° API?

(A) 20,000 Btu/lbm (46 MJ/kg)
(B) 25,000 Btu/lbm (58 MJ/kg)
(C) 30,000 Btu/lbm (69 MJ/kg)
(D) 35,000 Btu/lbm (81 MJ/kg)

8. The ultimate analysis of a coal is 75% carbon, 5% hydrogen, 3% oxygen, 2% nitrogen, and the rest ash. Atmospheric air is 60°F (16°C) and at standard pressure.

(a) What is the theoretical temperature of the combustion products?

(A) 3500°F (1900°C)
(B) 4000°F (2200°C)
(C) 4500°F (2300°C)
(D) 5000°F (2500°C)

(b) Estimate the actual temperature of the combustion products. State your assumptions.

(A) 650°F (340°C)
(B) 880°F (470°C)
(C) 970°F (520°C)
(D) 1300°F (700°C)

9. A fuel oil has the following ultimate analysis: 85.43% carbon, 11.31% hydrogen, 2.7% oxygen, 0.34% sulfur, and 0.22% nitrogen. The oil is burned with 60% excess air. Evaluate flue gas volumes at 600°F (320°C).

(a) What volume of wet flue gases will be produced?

(A) 450 ft^3 (28 m^3)
(B) 500 ft^3 (32 m^3)
(C) 550 ft^3 (35 m^3)
(D) 600 ft^3 (38 m^3)

(b) What volume of dry flue gases will be produced?
- (A) 480 ft³ (30 m³)
- (B) 560 ft³ (35 m³)
- (C) 630 ft³ (40 m³)
- (D) 850 ft³ (79 m³)

(c) What will be the volumetric fraction of carbon dioxide?
- (A) 6%
- (B) 8%
- (C) 10%
- (D) 13%

10. The ultimate analysis of a coal is 51.45% carbon, 16.69% ash, 15.71% moisture, 7.28% oxygen, 4.02% hydrogen, 3.92% sulfur, and 0.93% nitrogen. 15,395 lbm (6923 kg) of the coal are burned, and 2816 lbm (1267 kg) of ash containing 20.9% carbon (by weight) are recovered. 13.3 lbm (kg) of dry gases are produced per pound (kilogram) of fuel burned. How much air was used per unit mass of fuel?
- (A) 13 lbm air/lbm fuel (13 kg air/kg fuel)
- (B) 14 lbm air/lbm fuel (14 kg air/kg fuel)
- (C) 15 lbm air/lbm fuel (15 kg air/kg fuel)
- (D) 17 lbm air/lbm fuel (17 kg air/kg fuel)

11. A coal is 65% carbon by weight. During combustion, 3% of the coal is lost in the ash pit. Combustion uses 9.87 lbm (kg) of air per pound (kg) of fuel. The flue gas analysis is 81.5% nitrogen, 9.5% carbon dioxide, and 9% oxygen. What is the percentage of excess air by mass?
- (A) 10%
- (B) 30%
- (C) 70%
- (D) 140%

12. A coal has an ultimate analysis of 67.34% carbon, 4.91% oxygen, 4.43% hydrogen, 4.28% sulfur, 1.08% nitrogen, and the rest ash. 3% of the carbon is lost during combustion. The flue gases are 81.9% nitrogen, 15.5% carbon dioxide, 1.6% carbon monoxide, and 1% oxygen by volume. What is the heat loss due to the formation of carbon monoxide?
- (A) 600 Btu/lbm (1.4 MJ/kg)
- (B) 800 Btu/lbm (1.9 MJ/kg)
- (C) 1000 Btu/lbm (2.3 MJ/kg)
- (D) 1200 Btu/lbm (2.8 MJ/kg)

13. (*Time limit: one hour*) A natural gas is 93% methane, 3.73% nitrogen, 1.82% hydrogen, 0.45% carbon monoxide, 0.35% oxygen, 0.25% ethylene, 0.22% carbon dioxide, and 0.18% hydrogen sulfide by volume. The gas is burned with 40% excess air. Atmospheric air is 60°F and at standard atmospheric pressure.

(a) What is the gas density?
- (A) 0.017 lbm/ft³
- (B) 0.043 lbm/ft³
- (C) 0.069 lbm/ft³
- (D) 0.110 lbm/ft³

(b) What are the theoretical air requirements?
- (A) 5 ft³ air/ft³ fuel
- (B) 7 ft³ air/ft³ fuel
- (C) 9 ft³ air/ft³ fuel
- (D) 13 ft³ air/ft³ fuel

(c) What is the percentage of CO_2 in the flue gas (wet basis)?
- (A) 6.9%
- (B) 7.7%
- (C) 8.1%
- (D) 11%

(d) What is the percentage of CO_2 in the flue gas (dry basis)?
- (A) 6.9%
- (B) 7.7%
- (C) 8.1%
- (D) 11%

14. Coal enters a steam generator at 73°F (23°C). The coal has an ultimate analysis of 78.42% carbon, 8.25% oxygen, 5.68% ash, 5.56% hydrogen, 1.09% nitrogen, and 1.0% sulfur. The coal's heating value is 14,000 Btu/lbm (32.6 MJ/kg). 7.03% of the coal's weight is lost in the ash pit. The ash contains 31.5% carbon. Air at 67°F (19°C) wet bulb and 73°F (23°C) dry bulb is supplied. The flue gases consist of 80.08% nitrogen, 14.0% carbon dioxide, 5.5% oxygen, and 0.42% carbon monoxide. The flue gases are at a temperature of 575°F (300°C). Saturated water at 212°F (100°C) and 1.0 atm enters the steam generator. 11.12 lbm (kg) of water are evaporated in the boiler per pound (kilogram) of dry coal consumed. What is the complete heat balance, enumerating all combustion losses in the furnace?

15. Coal has a gravimetric analysis of 83% carbon, 5% hydrogen, 5% oxygen, and 7% noncombustible matter. 10% of the fired coal mass, including all of the noncombustible matter, is recovered in the ash pit. 26 lbm (26 kg) of air at 70°F (21°C) and 1 atmosphere are used per lbm (kg) of coal burned. When loaded into the furnace, the coal temperature is 60°F (15.6°C). The stack gas from coal combustion has a temperature of 550°F (290°C). What percentage of the combustion heat is carried away by the stack gases?
- (A) 18%
- (B) 24%
- (C) 37%
- (D) 49%

16. (*Time limit: one hour*) A utility boiler burns coal with an ultimate analysis of 76.56% carbon, 7.7% oxygen, 6.1% silicon, 5.5% hydrogen, 2.44% sulfur, and 1.7% nitrogen. 410 lbm/hr of refuse are removed with a composition of 30% carbon and 0% sulfur. All the sulfur and the remaining carbon is burned. The power plant has the following characteristics.

- coal feed rate: 15,300 lbm/hr

- electric power rating: 17 MW

- generator efficiency: 95%

- steam generator efficiency: 86%

- cooling water rate: 225 ft³/sec

(a) What is the emission rate of solid particulates in lbm/hr?

(A) 23 lbm/hr
(B) 150 lbm/hr
(C) 810 lbm/hr
(D) 1700 lbm/hr

(b) How much sulfur dioxide is produced per hour?

(A) 220 lbm/hr
(B) 340 lbm/hr
(C) 750 lbm/hr
(D) 1100 lbm/hr

(c) What is the temperature rise of the cooling water?

(A) 2.4°F
(B) 6.5°F
(C) 9.8°F
(D) 13°F

(d) What efficiency must the flue gas particulate collectors have in order to meet a limit of 0.1 lbm of particulates per million Btu per hour (0.155 kg/MW)?

(A) 93.1%
(B) 97.4%
(C) 98.8%
(D) 99.1%

17. (*Time limit: one hour*) 250 SCFM (118 L/s) of propane are mixed with an oxidizer consisting of 60% oxygen and 40% nitrogen by volume in a proportion allowing 40% excess oxygen by weight. The maximum velocity for the two reactants when combined is 400 ft/min (2 m/s) at 14.7 psia (101 kPa) and 60°F (17°C). Maximum velocity for the products is 800 ft/min (4 m/s) at 8 psia (55 kPa) and 460°F (240°C).

(a) Size the inlet pipe.

(A) 8 ft² (0.8 m²)
(B) 11 ft² (1.1 m²)
(C) 14 ft² (1.3 m²)
(D) 17 ft² (1.6 m²)

(b) Size the stack.

(A) 9 ft² (0.8 m²)
(B) 11 ft² (1.1 m²)
(C) 14 ft² (1.3 m²)
(D) 17 ft² (1.6 m²)

(c) What is the actual flow of oxygen?

(A) 150 lbm/min (1.1 kg/s)
(B) 180 lbm/min (1.3 kg/s)
(C) 220 lbm/min (1.6 kg/s)
(D) 270 lbm/min (2.0 kg/s)

(d) What is the volume of flue gases?

(A) 4000 ft³/min (1.9 m³/s)
(B) 7000 ft³/min (3.3 m³/s)
(C) 9000 ft³/min (4.3 m³/s)
(D) 11,000 ft³/min (5.3 m³/s)

(e) What is the dew point of the flue gases?

(A) 40°F (4°C)
(B) 100°F (38°C)
(C) 130°F (55°C)
(D) 180°F (80°C)

18. (*Time limit: one hour*) An industrial process uses hot gas at 3600°R (1980°C) and 14.7 psia (101 kPa). It is proposed that propane be burned stoichiometrically in a mixture of nitrogen and oxygen. After passing through the process, gas will be exhausted through a duct, being cooled slowly to 100°F (38°C) and 14.7 psia (101 kPa) before discharge. The following data are available.

- The enthalpies of formation (at the standard reference temperature) are

$$C_3H_8(g) \quad \Delta H_f = +28,800 \text{ Btu/lbmol}$$
$$(+67.0 \text{ GJ/kmol})$$
$$CO_2(g) \quad \Delta H_f = -169,300 \text{ Btu/lbmol}$$
$$(-393.8 \text{ GJ/kmol})$$
$$H_2O(g) \quad \Delta H_f = -104,040 \text{ Btu/lbmol}$$
$$(-242 \text{ GJ/kmol})$$

- The enthalpy increases from the standard reference temperature to 3600°R (1980°C) are

CO_2 :	39,791 Btu/lbmol	(92.6 GJ/kmol)
H_2O :	31,658 Btu/lbmol	(73.6 GJ/kmol)
N_2 :	24,471 Btu/lbmol	(56.9 GJ/kmol)

(a) What are the proportions by weight of the oxygen and nitrogen?

(b) What is the amount of water vapor, if any, present in the stack gas immediately after combustion?

(c) How much water is removed from the stack gas?

19. (*Time limit: one hour*) An internal combustion engine is normally fuel-injected with gasoline (C_8H_{18}; lower heating value of 23,200 Btu/lbm; 54 MJ/kg). It is desired to switch to alcohol (C_2H_5OH; lower heating value of 11,930 Btu/lbm; 27.7 MJ/kg). Minimal changes to the engine are to be made. The indicated and mechanical efficiencies are unchanged. Stoichiometric air/fuel ratios are being used for initial calculations.

(a) If the power output is to be unchanged, what must be the percentage change in specific fuel consumption?

   (A) 45%
   (B) 65%
   (C) 85%
   (D) 95%

(b) If the fuel-injection velocity is to be unchanged, what is the percentage change in jet size?

   (A) 35%
   (B) 55%
   (C) 80%
   (D) 95%

(c) If no changes are made to the engine, what will be the percentage change in power output?

   (A) −25%
   (B) −45%
   (C) −60%
   (D) −75%

SOLUTIONS

1. *Customary U.S. Solution*

$$T = 60°F + 460 = 520°R$$

$$p = 14.73 \text{ psia}$$

$$R = \frac{R^*}{MW} = \frac{1545.33 \ \frac{\text{ft-lbf}}{\text{lbmol-°R}}}{16.043 \ \frac{\text{lbm}}{\text{lbmol}}}$$

$$= 96.32 \text{ ft-lbf/lbm-°R}$$

$$m = \frac{pV}{RT}$$

$$= \frac{\left(14.73 \ \frac{\text{lbf}}{\text{in}^2}\right)\left(144 \ \frac{\text{in}^2}{\text{ft}^2}\right)(7 \text{ ft}^3)}{\left(96.32 \ \frac{\text{ft-lbf}}{\text{lbm-°R}}\right)(520°R)}$$

$$= 0.296 \text{ lbm}$$

The energy available from methane is

$$Q = \eta m(\text{HHV})$$

$$= (0.5)(0.296 \text{ lbm})\left(24,000 \ \frac{\text{Btu}}{\text{lbm}}\right)$$

$$= 3552 \text{ Btu}$$

This energy is used by water to heat from 60°F to 200°F.

$$Q = m_{\text{water}}c_p(T_2 - T_1)$$

$$m_{\text{water}} = \frac{3552 \text{ Btu}}{\left(1 \ \frac{\text{Btu}}{\text{lbm-°F}}\right)(200°F - 60°F)} = \boxed{25.37 \text{ lbm}}$$

The answer is (A).

SI Solution

$$T = (60°F + 460)\left(\frac{1 \text{ K}}{1.8°R}\right) = 288.89\text{K}$$

$$p = (14.73 \text{ psia})\left(\frac{101.325 \text{ kPa}}{14.696 \text{ psia}}\right) = 101.56 \text{ kPa}$$

$$R = \frac{R^*}{MW} = \frac{8314.3 \ \frac{\text{J}}{\text{kmol·K}}}{16.043 \ \frac{\text{kg}}{\text{kmol}}}$$

$$= 518.25 \text{ J/kg·K}$$

$$m = \frac{pV}{RT}$$

$$= \frac{(101.56 \text{ kPa})\left(1000 \ \frac{\text{Pa}}{\text{kPa}}\right)(200 \text{ L})\left(\frac{1 \text{ m}^3}{1000 \text{ L}}\right)}{\left(518.25 \ \frac{\text{J}}{\text{kg·K}}\right)(288.89\text{K})}$$

$$= 0.136 \text{ kg}$$

The energy available from methane is

$$Q = \eta m (\text{HHV})$$

$$= (0.5)(0.136 \text{ kg}) \left(55.8 \, \frac{\text{MJ}}{\text{kg}} \right) \left(1000 \, \frac{\text{kJ}}{\text{MJ}} \right)$$

$$= 3794 \text{ kJ}$$

This energy is used by water to heat it from 15°C to 95°C.

$$Q = m_{\text{water}} c_p (T_2 - T_1)$$

$$m_{\text{water}} = \frac{3794 \text{ kJ}}{\left(4.1868 \, \frac{\text{kJ}}{\text{kg·C}} \right) (95°\text{C} - 15°\text{C})} = \boxed{11.33 \text{ kg}}$$

The answer is (A).

2. *Customary U.S. Solution*

From Table 21.7,

$$\begin{array}{cccccc}
& \text{C}_3\text{H}_8 & + & 5\text{O}_2 & \longrightarrow & 3\text{CO}_2 & + 4\text{H}_2\text{O} \\
\text{MW} & 44.097 & & (5)(32) & & (3)(44.011) \\
& 44.097 & & 160 & & 132.033
\end{array}$$

The amount of carbon dioxide produced is 132.033 lbm/ 44.097 lbm propane. For 15 lbm/hr of propane, the amount of carbon dioxide produced is

$$\left(\frac{132.033 \text{ lbm}}{44.097 \text{ lbm}} \right) \left(15 \, \frac{\text{lbm}}{\text{hr}} \right) = 44.91 \text{ lbm/hr}$$

$$R = \frac{R^*}{\text{MW}} = \frac{1545.33 \, \frac{\text{ft-lbf}}{\text{lbmol-°R}}}{44.097 \, \frac{\text{lbm}}{\text{lbmol}}}$$

$$= 35.04 \text{ ft-lbf/lbm-°R}$$

$$T = 70°\text{F} + 460 = 530°\text{R}$$

$$V = \frac{mRT}{p}$$

$$= \frac{\left(44.91 \, \frac{\text{lbm}}{\text{hr}} \right) \left(35.04 \, \frac{\text{ft-lbf}}{\text{lbm-°R}} \right) (530°\text{R})}{\left(14.7 \, \frac{\text{lbf}}{\text{in}^2} \right) \left(144 \, \frac{\text{in}^2}{\text{ft}^2} \right)}$$

$$= \boxed{394 \text{ ft}^3/\text{hr}}$$

The answer is (C).

SI Solution

From Table 21.7,

$$\begin{array}{cccccc}
& \text{C}_3\text{H}_8 & + & 5\text{O}_2 & \longrightarrow & 3\text{CO}_2 & + 4\text{H}_2\text{O} \\
\text{MW} & 44.097 & & (5)(32) & & (3)(44.011) \\
& 44.097 & & 160 & & 132.033
\end{array}$$

The amount of carbon dioxide produced is 132.033 kg/ 44.097 kg propane. For 6.8 kg/h of propane, the amount of carbon dioxide produced is

$$\left(\frac{132.033 \text{ kg}}{44.097 \text{ kg}} \right) \left(6.8 \, \frac{\text{kg}}{\text{h}} \right) = 20.36 \text{ kg/h}$$

$$R = \frac{R^*}{\text{MW}} = \frac{8314.3 \, \frac{\text{J}}{\text{kmol·K}}}{44.097 \, \frac{\text{kg}}{\text{kmol}}}$$

$$= 188.55 \text{ J/kg·K}$$

$$T = 21°\text{C} + 273 = 294\text{K}$$

$$V = \frac{mRT}{p}$$

$$= \frac{\left(20.36 \, \frac{\text{kg}}{\text{h}} \right) \left(188.55 \, \frac{\text{J}}{\text{kg·K}} \right) (294\text{K})}{(101 \text{ kPa}) \left(1000 \, \frac{\text{Pa}}{\text{kPa}} \right)}$$

$$= \boxed{11.17 \text{ m}^3/\text{h}}$$

The answer is (C).

3. Use the balanced chemical reaction equation from Table 21.7.

$$\text{CH}_4 + 2\text{O}_2 \longrightarrow \text{CO}_2 + 2\text{H}_2\text{O}$$

With 30% excess air and considering there are 3.773 volumes of nitrogen for every volume of oxygen (Table 21.6), the reaction equation is

$$\text{CH}_4 + (1.3)(2)\text{O}_2 + (1.3)(2)(3.773)\text{N}_2$$
$$\longrightarrow \text{CO}_2 + 2\text{H}_2\text{O} + (1.3)(2)(3.773)\text{N}_2 + 0.6\text{O}_2$$

$$\text{CH}_4 + 2.6\text{O}_2 + 9.81\text{N}_2$$
$$\longrightarrow \text{CO}_2 + 2\text{H}_2\text{O} + 9.81\text{N}_2 + 0.6\text{O}_2$$

Customary U.S. Solution

The volume of nitrogen that accompanies 4000 ft³/hr of entering methane is

$$V_{\text{N}_2} = \left(\frac{9.81 \text{ ft}^3 \text{ N}_2}{1 \text{ ft}^3 \text{ CH}_4} \right) \left(4000 \, \frac{\text{ft}^3}{\text{hr}} \text{ CH}_4 \right)$$

$$= 39{,}240 \text{ ft}^3 \text{ N}_2/\text{hr}$$

This is the "partial volume" of nitrogen in the input stream.

$$R = \frac{R^*}{\text{MW}} = \frac{1545.33 \, \frac{\text{ft-lbf}}{\text{lbmol-°R}}}{28.016 \, \frac{\text{lbm}}{\text{lbmol}}}$$

$$= 55.16 \text{ ft-lbf/lbm-°R}$$

Thermodynamics

The absolute temperature is

$$T = 100°F + 460 = 560°R$$

$$m_{N_2} = \frac{p_{N_2} V_{N_2}}{RT}$$

$$= \frac{\left(15 \frac{lbf}{in^2}\right)\left(144 \frac{in^2}{ft^2}\right)\left(39,240 \frac{ft^3}{hr}\right)}{\left(55.16 \frac{ft\text{-}lbf}{lbm\text{-}°R}\right)(560°R)}$$

$$= \boxed{2744 \text{ lbm/hr}}$$

The answer is (D).

SI Solution

The volume of nitrogen that accompanies 31 L/s of entering methane is

$$\left(\frac{9.81 \text{ m}^3 \text{ N}_2}{1 \text{ m}^3 \text{ CH}_4}\right)\left(31 \frac{L}{s}\right)\left(\frac{1 \text{ m}^3}{1000 \text{ L}}\right) = 0.3041 \text{ m}^3/\text{s}$$

This is the "partial volume" of nitrogen in the input stream.

$$R = \frac{R^*}{MW} = \frac{8314.3 \frac{J}{kmol\cdot K}}{28.016 \frac{kg}{kmol}}$$

$$= 296.8 \text{ J/kg·K}$$

The absolute temperature is

$$T = 40°C + 273 = 313K$$

$$m_{N_2} = \frac{p_{N_2} V_{N_2}}{RT}$$

$$= \frac{(103 \text{ kPa})\left(1000 \frac{Pa}{kPa}\right)\left(0.3041 \frac{m^3}{s}\right)}{\left(296.8 \frac{J}{kg\cdot K}\right)(313K)}$$

$$= \boxed{0.337 \text{ kg/s}}$$

The answer is (D).

4. From Table 21.7, combustion reactions are

$$\begin{array}{cccc} & C & + & O_2 & \longrightarrow & CO_2 \\ MW & 12 & & 32 \end{array}$$

The mass of oxygen required per unit mass of carbon is

$$\frac{32}{12} = 2.67$$

$$\begin{array}{cccc} & 2H_2 & + & O_2 & \longrightarrow & 2H_2O \\ MW & (2)(2) & & 32 \end{array}$$

The mass of oxygen required per unit mass of hydrogen is

$$\frac{32}{(2)(2)} = 8.0$$

$$\begin{array}{cccc} & S & + & O_2 & \longrightarrow & SO_2 \\ MW & 32.1 & & 32 \end{array}$$

The mass of oxygen required per unit mass of sulfur is

$$\frac{32}{32.1} = 1.0$$

Nitrogen does not burn.

The mass of oxygen required per unit mass of fuel is

$$(0.84)(2.67) + (0.153)(8) + (0.003)(1)$$
$$= 3.47 \text{ unit of mass of O}_2/\text{unit mass fuel}$$

From Table 21.6, air is 0.2315 O_2/unit mass, so the air required is

$$\frac{3.47}{0.2315} = \boxed{15.0 \text{ lbm air/lbm fuel}}$$

(Customary U.S. Solution)

$$= \boxed{15.0 \text{ kg air/kg fuel}}$$

(SI Solution)

The answer is (C).

5. The balanced chemical reaction equation is

$$C_3H_8 + 5O_2 \longrightarrow 3CO_2 + 4H_2O$$

With 20% excess air, the oxygen volume is $(1.2)(5) = 6$.

$$C_3H_8 + 6O_2 \longrightarrow 3CO_2 + 4H_2O + O_2$$

From Table 21.6, there are 3.773 volumes of nitrogen for every volume of oxygen.

$$(6)(3.773) = 22.6$$
$$C_3H_8 + 6O_2 + 22.6\,N_2 \longrightarrow 3CO_2 + 4H_2O + O_2 + 22.6\,N_2$$

The percentage of carbon dioxide by weight in flue gas is

$$G_{CO_2} = \frac{(3)(44.011)}{(3)(44.011) + (4)(18.016)} \\ + 32 + (22.6)(28.016)$$

$$= \boxed{0.152 \ (15.2\%)}$$

The answer is (C).

6. The actual air/fuel ratio can be estimated from the flue gas analysis and the fraction of carbon in fuel.

From Eq. 21.9,

$$\frac{m_{air}}{m_{fuel}} = \frac{3.04 B_{N_2} G_C}{B_{CO_2} + B_{CO}}$$

$$= \frac{(3.04)(80\%)(0.80)}{12\% + 1\%}$$

$$= 14.97$$

The air required to burn 1 lbm of coal is $\boxed{14.97 \text{ lbm.}}$

(Customary U.S. Solution)

The air required to burn 1 kg of coal is $\boxed{14.97 \text{ kg.}}$

(SI Solution)

The answer is (B).

Alternate Solution

The use of Eq. 21.9 obscures the process of finding the air/fuel ratio. (The SI solution is similar but is not presented here.)

step 1: Find the mass of oxygen in the stack gases.

$$R_{CO_2} = 35.11 \text{ ft-lbf/lbm-°R}$$

$$R_{CO} = 55.17 \text{ ft-lbf/lbm-°R}$$

$$R_{O_2} = 48.29 \text{ ft-lbf/lbm-°R}$$

The partial densities are

$$\rho_{CO_2} = \frac{p}{RT} = \frac{(0.12)\left(14.7 \frac{\text{lbf}}{\text{in}^2}\right)\left(144 \frac{\text{in}^2}{\text{ft}^2}\right)}{\left(35.11 \frac{\text{ft-lbf}}{\text{lbm-°R}}\right)(60°\text{F} + 460)}$$

$$= 1.391 \times 10^{-2} \text{ lbm/ft}^3$$

$$\rho_{CO} = \frac{(0.01)\left(14.7 \frac{\text{lbf}}{\text{in}^2}\right)\left(144 \frac{\text{in}^2}{\text{ft}^2}\right)}{\left(55.11 \frac{\text{ft-lbf}}{\text{lbm-°R}}\right)(60°\text{F} + 460)}$$

$$= 7.387 \times 10^{-4} \text{ lbm/ft}^3$$

$$\rho_{O_2} = \frac{(0.07)\left(14.7 \frac{\text{lbf}}{\text{in}^2}\right)\left(144 \frac{\text{in}^2}{\text{ft}^2}\right)}{\left(48.29 \frac{\text{ft-lbf}}{\text{lbm-°R}}\right)(60°\text{F} + 460)}$$

$$= 5.901 \times 10^{-3} \text{ lbm/ft}^3$$

The fraction of oxygen in the three components is

$$CO_2: \frac{32.0}{44} = 0.7273$$

$$CO: \frac{16}{28} = 0.5714$$

$$O_2: 1.00$$

In 100 ft^3 of stack gases, the total oxygen mass will be

$$(100 \text{ ft}^3)\left((0.7273)\left(1.391 \times 10^{-2} \frac{\text{lbm}}{\text{ft}^3}\right)\right.$$

$$+ (0.5714)\left(7.387 \times 10^{-4} \frac{\text{lbm}}{\text{ft}^3}\right)$$

$$\left.+ (1.00)\left(5.901 \times 10^{-3} \frac{\text{lbm}}{\text{ft}^3}\right)\right) = 1.644 \text{ lbm}$$

step 2: Since air is 23.15% oxygen by weight, the mass of air per 100 ft^3 of stack gases is

$$\frac{1.644 \text{ lbm}}{0.2315} = 7.102 \text{ lbm}$$

step 3: Find the mass of carbon in the stack gases by a similar process.

$$CO_2: \frac{12}{44} = 0.2727$$

$$CO: \frac{12}{28} = 0.4286$$

$$(100 \text{ ft}^3)\left((0.2727)\left(1.391 \times 10^{-2} \frac{\text{lbm}}{\text{ft}^3}\right)\right.$$

$$\left.+ (0.4286)\left(7.387 \times 10^{-4} \frac{\text{lbm}}{\text{ft}^3}\right)\right)$$

$$= 0.4110 \text{ lbm}$$

step 4: The coal is 80% carbon, so the air per lbm of coal for combustion of the carbon is

$$\left(\frac{0.80 \frac{\text{lbm carbon}}{\text{lbm coal}}}{0.4110 \frac{\text{lbm carbon}}{100 \text{ ft}^3}}\right)(7.102 \text{ lbm})$$

$$= 13.824 \text{ lbm air/lbm coal}$$

step 5: This does not include air to burn hydrogen, since Orsat is a dry analysis.

The theoretical air for the hydrogen is given by Eq. 21.6.

$$R_{a/f,H} = \left(34.5 \; \frac{lbm}{lbm} \right) \left(G_H - \frac{G_O}{8} \right)$$

$$= \left(34.05 \; \frac{lbm}{lbm} \right) \left(0.04 - \frac{0.02}{8} \right)$$

$$= 1.277 \; lbm \; air/lbm \; fuel$$

Ignoring any excess air for the hydrogen, the total air per pound of coal is

$$13.824 \; \frac{lbm \; air}{lbm \; coal} + 1.277 \; \frac{lbm \; air}{lbm \; coal}$$

$$= \boxed{15.10 \; lbm \; air/lbm \; coal}$$

The answer is (B).

7. From Eq. 14.11,

$$SG = \frac{141.5}{°API + 131.5}$$

$$= \frac{141.5}{40 + 131.5} = 0.825$$

Customary U.S. Solution

From Eq. 21.18(b),

$$HHV = 22{,}320 - (3780)(SG)^2$$

$$= 22{,}320 - (3780)(0.825)^2$$

$$= \boxed{19{,}747 \; Btu/lbm}$$

The answer is (A).

SI Solution

From Eq. 21.18(a),

$$HHV = 51.92 - (8.792)(SG)^2$$

$$= 51.92 - (8.792)(0.825)^2$$

$$= \boxed{45.94 \; MJ/kg}$$

The answer is (A).

8. *Customary U.S. Solution*

(a) *step 1:* From Eq. 21.16(b), substituting the lower heating value of hydrogen from App. 21.A, the lower heating value of coal is

$$LHV = 14{,}093 G_C + (51{,}623) \left(G_H - \frac{G_O}{8} \right)$$

$$+ 3983 G_S$$

$$= (14{,}093)(0.75)$$

$$+ (51{,}623) \left(0.05 - \frac{0.03}{8} \right) + (3983)(0)$$

$$= 12{,}957 \; Btu/lbm$$

step 2: The gravimetric analysis of 1 lbm of coal is

carbon: 0.75 lbm

free hydrogen: $\left(G_{H,total} - \dfrac{G_O}{8} \right) = 0.05 - \dfrac{0.03}{8}$

$$= 0.0463$$

The ratio of the molecular weight of water (18) to that of a hydrogen molecule (2) is 9.

water: $(9)(0.05 - 0.0463) = 0.0333$

nitrogen: 0.02

step 3: From Table 21.8, the theoretical stack gases per lbm coal for 0.75 lbm of carbon are

$$CO_2 = (0.75)(3.667 \; lbm) = 2.750 \; lbm$$

$$N_2 = (0.75)(8.883 \; lbm) = 6.662 \; lbm$$

All products are calculated similarly (as the following table summarizes).

	CO_2	N_2	H_2O
from C:	2.750 lbm	6.662 lbm	
from H_2:		1.217 lbm	0.414 lbm
from H_2O:			0.0333 lbm
from O_2:	shows up in	CO_2 and H_2	
from N_2:		0.02 lbm	
total:	2.750 lbm	7.899 lbm	0.4473 lbm

step 4: Assume the combustion gases leave at 1000°F.

$$T_{ave} = \left(\tfrac{1}{2} \right) (60°F + 1000°F) = 530°F$$

$$T_{ave} = 530°F + 460 = 990°R$$

The specific heat values are given in Table 21.1.

$$c_{p CO_2} = 0.251 \; Btu/lbm\text{-}°F$$

$$c_{p N_2} = 0.255 \; Btu/lbm\text{-}°F$$

$$c_{p H_2O} = 0.475 \; Btu/lbm\text{-}°F$$

The energy required to raise the combustion products for 1 lbm of coal 1°F is

$$m_{CO_2} c_{p CO_2} + m_{N_2} c_{p N_2} + m_{H_2O} c_{p H_2O}$$

$$= (2.750 \; lbm) \left(0.251 \; \frac{Btu}{lbm\text{-}°F} \right)$$

$$+ (7.899 \; lbm) \left(0.255 \; \frac{Btu}{lbm\text{-}°F} \right)$$

$$+ (0.4473 \; lbm) \left(0.475 \; \frac{Btu}{lbm\text{-}°F} \right)$$

$$= 2.92 \; Btu/°F$$

step 5: Assuming all combustion heat goes into the stack gases, the temperature is given by Eq. 21.19.

$$T_{max} = T_i + \frac{\text{lower heat of combustion}}{\text{energy required}}$$

$$= 60°F + \frac{12{,}957 \dfrac{\text{Btu}}{\text{lbm}}}{2.92 \dfrac{\text{Btu}}{\text{lbm-°F}}} = \boxed{4497°F}$$

[unreasonable]

The answer is (C).

(b) *step 6:* In reality, assuming 40% excess air and 75% of heat absorbed by the boiler, the excess air (based on 76.85% N_2 by weight) is

$$(0.40)\left(\frac{7.899 \text{ lbm}}{0.7685}\right) = 4.111 \text{ lbm}$$

From Table 21.1, $c_{p_{air}} = 0.249$ Btu/lbm-°R. Therefore,

$$T_{max} = 60°F + \frac{\left(12{,}957 \dfrac{\text{Btu}}{\text{lbm}}\right)(1 - 0.75)}{2.92 \dfrac{\text{Btu}}{°F}}$$
$$+ (4.111 \text{ lbm})\left(0.249 \dfrac{\text{Btu}}{\text{lbm-°R}}\right)$$

$$= \boxed{881°F}$$

The answer is (B).

SI Solution

(a) *step 1:* From Eq. 21.16(a), and substituting the lower heating value of hydrogen from App. 21.A, the lower heating value of coal is

$$\text{LHV} = 32.78 G_C + \left(51{,}623 \frac{\text{Btu}}{\text{lbm}}\right)\left(2.326 \frac{\text{kJ-lbm}}{\text{kg-Btu}}\right)$$
$$\times \left(\frac{1 \text{ MJ}}{1000 \text{ kJ}}\right)\left(G_H - \frac{G_O}{8}\right)$$
$$+ 9.264 \, G_S$$
$$= (32.78)(0.75) + (120.1)\left(0.05 - \frac{0.03}{8}\right)$$
$$+ (9.264)(0)$$
$$= 30.14 \text{ MJ/kg}$$

Steps 2 and 3 are the same as for the U.S. solution except that all masses are in kg.

step 4: Assume the combustion gases leave at 550°C.

$$T_{ave} = \left(\tfrac{1}{2}\right)(16°C + 550°C) + 273$$
$$= 283°C + 273 = 556K$$

Specific heat values are given in Table 21.1. Using the footnote for SI units,

$$c_{p_{CO_2}} = 1.051 \text{ kJ/kg·K}$$
$$c_{p_{N_2}} = 1.068 \text{ kJ/kg·K}$$
$$c_{p_{H_2O}} = 1.989 \text{ kJ/kg·K}$$

The energy required to raise the combustion products for 1 kg of coal 1°C is

$$m_{CO_2} c_{p_{CO_2}} + m_{N_2} c_{p_{N_2}} + m_{H_2O} c_{p_{H_2O}}$$
$$= (2.750 \text{ kg})\left(1.051 \frac{\text{kJ}}{\text{kg·K}}\right)$$
$$+ (7.899 \text{ kg})\left(1.068 \frac{\text{kJ}}{\text{kg·K}}\right)$$
$$+ (0.4473 \text{ kg})\left(1.989 \frac{\text{kJ}}{\text{kg·K}}\right)$$
$$= 12.22 \text{ kJ/K}$$

step 5: Assuming all combustion heat goes into the stack gases, the temperature is given by Eq. 21.19.

$$T_{max} = T_i + \frac{\text{lower heat of combustion}}{\text{energy required}}$$

$$= 16°C + \frac{\left(30.14 \dfrac{\text{MJ}}{\text{kg}}\right)\left(1000 \dfrac{\text{kJ}}{\text{MJ}}\right)}{12.22 \dfrac{\text{kJ}}{\text{K}}}$$

$$= \boxed{2482°C}$$ [unreasonable]

The answer is (C).

(b) *step 6:* In reality, assuming 40% excess air and 75% of heat absorbed by the boiler, the excess air (based on 76.85% N_2 by weight) is

$$(0.40)\left(\frac{7.899 \text{ kg}}{0.7685}\right) = 4.111 \text{ kg}$$

From Table 21.1, using the table footnote, $c_{p_{air}} = 1.043$ kJ/kg·K.

Therefore,

$$T_{max} = 16°C + \frac{\left(30.14 \dfrac{\text{MJ}}{\text{kg}}\right)}{12.22 \dfrac{\text{kJ}}{\text{K}} + (4.111 \text{ kg})\left(1.043 \dfrac{\text{kJ}}{\text{kg·K}}\right)}$$
$$\times \left(1000 \frac{\text{kJ}}{\text{MJ}}\right)(1 - 0.75)$$

$$= \boxed{472.5°C}$$

The answer is (B).

Thermodynamics

9. Assume the oxygen is in the form of moisture in the fuel.

The available hydrogen is

$$G_{\text{H,free}} = G_H - \frac{G_O}{8}$$

$$= 0.1131 - \frac{0.027}{8}$$

$$= 0.1097$$

Customary U.S. Solution

step 1: From Table 21.8, find the stoichiometric oxygen required per lbm of fuel oil.

$$C \longrightarrow CO_2: \ O_2 \text{ required} = (0.8543)(2.667 \text{ lbm})$$
$$= 2.2784 \text{ lbm}$$

$$H_2 \longrightarrow H_2O: \ O_2 \text{ required} = (0.1097)(7.936 \text{ lbm})$$
$$= 0.8706 \text{ lbm}$$

$$S \longrightarrow SO_2: \ O_2 \text{ required} = (0.0034)(0.998 \text{ lbm})$$
$$= 0.0034 \text{ lbm}$$

The total amount of oxygen required per lbm of fuel oil is

$$2.2784 \text{ lbm} + 0.8706 \text{ lbm}$$
$$+ \ 0.0034 \text{ lbm} = 3.1524 \text{ lbm}$$

step 2: With 60% excess air, the excess oxygen is

$$(0.6)(3.1524 \text{ lbm}) = 1.8914 \text{ lbm } O_2/\text{lbm fuel}$$

From Eq. 23.47, this oxygen occupies a volume of

$$V = \frac{mRT}{p}$$

At standard conditions,

$$p = 14.7 \text{ psia}$$
$$T = 60°F + 460 = 520°R$$

From Table 23.7, for oxygen, $R = 48.29$ ft-lbf/lbm-°R.

$$V = \frac{(1.8914 \text{ lbm})\left(48.29 \ \dfrac{\text{ft-lbf}}{\text{lbm-}°\text{R}}\right)(520°\text{R})}{\left(14.7 \ \dfrac{\text{lbf}}{\text{in}^2}\right)\left(144 \ \dfrac{\text{in}^2}{\text{ft}^2}\right)}$$

$$= 22.44 \text{ ft}^3$$

step 3: The theoretical nitrogen based on Table 21.6 is

$$\left(\frac{3.1524 \text{ lbm}}{0.2315}\right)(0.7685) = 10.465 \text{ lbm } N_2/\text{lbm fuel}$$

The actual nitrogen with 60% excess air and nitrogen in the fuel is

$$(10.465 \text{ lbm})(1.6) + 0.0022 \text{ lbm}$$
$$= 16.746 \text{ lbm } N_2/\text{lbm fuel}$$

From Eq. 23.47, this nitrogen occupies a volume of

$$V = \frac{mRT}{p}$$

From Table 23.7, R for nitrogen $= 55.16$ ft-lbf/lbm-°R.

$$V = \frac{(16.746 \text{ lbm})\left(55.16 \ \dfrac{\text{ft-lbf}}{\text{lbm-}°\text{R}}\right)(520°\text{R})}{\left(14.7 \ \dfrac{\text{lbf}}{\text{in}^2}\right)\left(144 \ \dfrac{\text{in}^2}{\text{ft}^2}\right)}$$

$$= 226.91 \text{ ft}^3$$

step 4: From Table 21.8, the 60°F combustion product volumes per lbm of fuel will be

CO_2:	$(0.8543)(31.63 \text{ ft}^3)$	$=$	27.02 ft^3
H_2O:	$(0.1131)(188.25 \text{ ft}^3)$	$=$	21.29 ft^3
SO_2:	$(0.0034)(11.84 \text{ ft}^3)$	$=$	0.040 ft^3
N_2:	from step 3	$=$	226.91 ft^3
O_2:	from step 2	$=$	22.44 ft^3
		total $=$	297.7 ft^3

(a) At 60°F, the wet volume will be 297.7 ft^3.

At 600°F, the wet volume is

$$V_{\text{wet},600°\text{F}} = (297.7 \text{ ft}^3)\left(\frac{600°\text{F} + 460}{60°\text{F} + 460}\right)$$

$$= \boxed{606.9 \text{ ft}^3}$$

The answer is (D).

(b) At 60°F, the dry volume will be

$$297.7 \text{ ft}^3 - 21.29 \text{ ft}^3 = 276.4 \text{ ft}^3$$

At 600°F, the dry volume is

$$V_{\text{dry},600°\text{F}} = (276.4 \text{ ft}^3)\left(\frac{600°\text{F} + 460}{60°\text{F} + 460}\right)$$

$$= \boxed{563.4 \text{ ft}^3}$$

The answer is (B).

(c) The volumetric fraction of dry carbon dioxide is

$$\frac{27.02 \text{ ft}^3}{276.41 \text{ ft}^3} = \boxed{0.098 \ (9.8\%)}$$

The answer is (C).

SI Solution

step 1: From Table 21.8, find the stoichiometric oxygen required per lbm of fuel oil.

$$C \longrightarrow CO_2 : \ O_2 \text{ required} = (0.8543)(2.667 \text{ kg})$$
$$= 2.2784 \text{ kg}$$
$$H_2 \longrightarrow H_2O : \ O_2 \text{ required} = (0.1097)(7.936 \text{ kg})$$
$$= 0.8706 \text{ kg}$$
$$S \longrightarrow SO_2 : \ O_2 \text{ required} = (0.0034)(0.998 \text{ kg})$$
$$= 0.0034 \text{ kg}$$

The total amount of oxygen required per kg of fuel oil is

$$2.2784 \text{ kg} + 0.8706 \text{ kg}$$
$$+ 0.0034 \text{ kg} = 3.1524 \text{ kg}$$

step 2: With 60% excess air, the excess oxygen is

$$(0.6)(3.1524 \text{ kg}) = 1.8914 \text{ kg } O_2/\text{kg fuel}$$

From Eq. 23.47, this oxygen occupies a volume of

$$V = \frac{mRT}{p}$$

At standard conditions,

$$p = 101.3 \text{ kPa}$$
$$T = 16°C + 273 = 289K$$

From Table 23.7, for oxygen, $R = 259.82$ J/kg·K.

$$V = \frac{(1.8914 \text{ kg}) \left(259.82 \ \dfrac{J}{kg\cdot K} \right) (289K)}{(101.3 \text{ kPa}) (1000 \text{ kPa})}$$
$$= 1.402 \text{ m}^3$$

step 3: The theoretical nitrogen based on Table 21.6 is

$$\left(\frac{3.1524 \text{ kg}}{0.2315} \right) (0.7685) = 10.465 \text{ kg } N_2/\text{kg fuel}$$

The actual nitrogen with 60% excess air and nitrogen in the fuel is

$$(10.465 \text{ kg})(1.6) + 0.0022 \text{ kg}$$
$$= 16.746 \text{ kg } N_2/\text{kg fuel}$$

From Eq. 23.47, this nitrogen occupies a volume of

$$V = \frac{mRT}{p}$$

From Table 23.7, R for nitrogen = 296.77 J/kg·K.

$$V = \frac{(16.746 \text{ kg}) \left(296.77 \ \dfrac{J}{kg\cdot K} \right) (289K)}{(101.3 \text{ kPa}) \left(1000 \ \dfrac{Pa}{kPa} \right)}$$

$$= 14.178 \text{ m}^3$$

step 4: From Table 21.8, the 16°C combustion product volumes per kg of fuel will be

CO_2:	$(0.8543)(31.63 \text{ ft}^3)(0.06243)$ =	1.687 m^3
H_2O:	$(0.1131)(188.25 \text{ ft}^3)(0.06243)$ =	1.329 m^3
SO_2:	$(0.0034)(11.84 \text{ ft}^3)(0.06243)$ =	0.003 m^3
N_2:	from step 3 =	14.178 m^3
O_2:	from step 2 =	1.402 m^3
	total =	18.599 m^3

(a) At 16°C, the wet volume will be 18.599 m^3. At 320°C, the wet volume is

$$V_{\text{wet},320°C} = (18.599 \text{ m}^3) \left(\frac{320°C + 273}{16°C + 273} \right)$$

$$= \boxed{38.16 \text{ m}^3}$$

The answer is (D).

(b) At 16°C, the dry volume will be

$$18.599 \text{ m}^3 - 1.329 \text{ m}^3 = 17.27 \text{ m}^3$$

At 320°C, the dry volume is

$$V_{\text{dry},320°C} = (17.27 \text{ m}^3) \left(\frac{320°C + 273}{16°C + 273} \right)$$

$$= \boxed{35.44 \text{ m}^3}$$

The answer is (B).

(c) The volumetric fraction of dry carbon dioxide is

$$\frac{1.687 \text{ m}^3}{17.27 \text{ m}^3} = \boxed{0.098 \ (9.8\%)}$$

The answer is (C).

Thermodynamics

10. *Customary U.S. Solution*

step 1: Based on the 15,395 lbm of coal burned producing 2816 lbm of ash containing 20.9% carbon by weight, the usable percentage of carbon per lbm of fuel is

$$0.5145 - \frac{(2816 \text{ lbm})(0.209)}{15,395 \text{ lbm}} = 0.4763$$

step 2: Since moisture is reported separately, assume all of the oxygen and hydrogen are free. (This is not ordinarily the case.) From Table 21.8, find the stoichiometric oxygen required per lbm of fuel.

$$C \longrightarrow CO_2: O_2 \text{ required} = (0.4763)(2.667 \text{ lbm})$$
$$= 1.2703 \text{ lbm}$$
$$H_2 \longrightarrow H_2O: O_2 \text{ required} = (0.0402)(7.936 \text{ lbm})$$
$$= 0.3190 \text{ lbm}$$
$$S \longrightarrow SO_2: O_2 \text{ required} = (0.0392)(0.998 \text{ lbm})$$
$$= 0.0391 \text{ lbm}$$

The total amount of O_2 required per lbm of fuel is

$$1.2703 \text{ lbm} + 0.3190 \text{ lbm}$$
$$+ 0.0391 \text{ lbm} - 0.0728 \text{ lbm} = 1.5556 \text{ lbm}$$

step 3: The theoretical air based on Table 21.6 is

$$\frac{1.5556 \text{ lbm}}{0.2315} = 6.720 \text{ lbm air/lbm fuel}$$

step 4: Ignoring fly ash, the theoretical dry products are given from Table 21.8.

$$CO_2: (0.4763)(3.667 \text{ lbm}) = 1.7466 \text{ lbm}$$
$$SO_2: (0.0392)(1.998 \text{ lbm}) = 0.0783 \text{ lbm}$$
$$N_2: 0.0093 \text{ lbm}$$
$$+ (6.720 \text{ lbm})$$
$$\times (0.7685 \text{ lbm}) = 5.1736 \text{ lbm}$$
$$\text{total} = 6.999 \text{ lbm}$$

step 5: The excess air is

$$13.3 \text{ lbm} - 6.999 \text{ lbm} = 6.301 \text{ lbm}$$

step 6: The total air supplied per pound of fuel is

$$6.301 \text{ lbm} + 6.720 \text{ lbm} = \boxed{13.02 \text{ lbm}}$$

The answer is (A).

SI Solution

step 1: Based on the 6923 kg of coal burned producing 1267 kg of ash containing 20.9% carbon by weight, the usable percentage of carbon per kg of fuel is

$$0.5145 - \frac{(1267 \text{ kg})(0.209)}{6923 \text{ kg}} = 0.4763$$

step 2: From Table 21.8, find the stoichiometric oxygen required per kg of fuel.

$$C \longrightarrow CO_2: O_2 \text{ required} = (0.4763)(2.667 \text{ kg})$$
$$= 1.2703 \text{ kg}$$
$$H_2 \longrightarrow H_2O: O_2 \text{ required} = (0.0402)(7.936 \text{ kg})$$
$$= 0.3190 \text{ kg}$$
$$S \longrightarrow SO_2: O_2 \text{ required} = (0.0392)(0.998 \text{ kg})$$
$$= 0.0391 \text{ kg}$$

The total amount of O_2 required per kg of fuel is

$$1.2703 \text{ kg} + 0.3190 \text{ kg}$$
$$+ 0.0391 \text{ kg} - 0.0728 \text{ kg} = 1.5556 \text{ kg}$$

step 3: The theoretical air based on Table 21.6 is

$$\frac{1.5556 \text{ kg}}{0.2315} = 6.720 \text{ kg air/kg fuel}$$

step 4: Ignoring fly ash, the theoretical dry products are given from Table 21.8.

$$CO_2: (0.4763)(3.667 \text{ kg}) = 1.7466 \text{ kg}$$
$$SO_2: (0.0392)(1.998 \text{ kg}) = 0.0783 \text{ kg}$$
$$N_2: 0.0093 \text{ kg}$$
$$+ (6.720 \text{ kg})$$
$$\times (0.7685 \text{ kg}) = 5.1736 \text{ kg}$$
$$\text{total} = 6.999 \text{ kg}$$

step 5: The excess air is

$$13.3 \text{ kg} - 6.999 \text{ kg} = 6.301 \text{ kg}$$

step 6: The total air supplied is

$$6.301 \text{ kg} + 6.720 \text{ kg} = \boxed{13.02 \text{ kg}}$$

The answer is (A).

11. (The customary U.S. and SI solutions are essentially identical.)

From Eq. 21.9, the actual air-fuel ratio can be estimated as

$$R_{a/f,\text{actual}} = \frac{3.04 B_{N_2} G_C}{B_{CO_2} + B_{CO}}$$

A fraction of carbon is reduced due to the percentage of coal lost in the ash pit.

$$G_C = (1 - 0.03)(0.65) = 0.6305$$

$$R_{a/f,\text{actual}} = \frac{(3.04)(0.815)(0.6305)}{0.095 + 0}$$

$$= 16.44 \text{ lbm air/lbm fuel (kg air/kg fuel)}$$

Combustion uses 9.87 lbm of air/lbm of fuel.

$$= (9.87 \text{ lbm})(1 - 0.03)$$
$$= 9.57 \text{ lbm of air/lbm of fuel}$$

$$\% \text{ of excess air} = \left(\frac{16.44 \text{ lbm} - 9.57 \text{ lbm}}{9.57 \text{ lbm}}\right)(100\%)$$

$$= \boxed{71.8\%}$$

The answer is (C).

12. From Eq. 21.23, the heat loss due to the formation of carbon monoxide is

$$q = \frac{(\text{HHV}_C - \text{HHV}_{CO})G_C B_{CO}}{B_{CO_2} + B_{CO}}$$

From App. 21.A, the difference in the two heating values is

$$\text{HHV}_C - \text{HHV}_{CO} = 14{,}093 \frac{\text{Btu}}{\text{lbm}} - 4347 \frac{\text{Btu}}{\text{lbm}}$$
$$= 9746 \text{ Btu/lbm} \quad (22.67 \text{ MJ/kg})$$
$$G_C = (1 - 0.03)(0.6734) = 0.6532$$

Customary U.S. Solution

$$q = \frac{\left(9746 \frac{\text{Btu}}{\text{lbm}}\right)(0.6532)(1.6\%)}{15.5\% + 1.6\%}$$

$$= \boxed{596 \text{ Btu/lbm}}$$

The answer is (A).

SI Solution

$$q = \frac{\left(22.67 \frac{\text{MJ}}{\text{kg}}\right)(0.6532)(1.6\%)}{15.5\% + 1.6\%}$$

$$= \boxed{1.39 \text{ MJ/kg}}$$

The answer is (A).

13. (a) For methane, $B = 0.93$.

From ideal gas laws (R for methane $= 96.32$ ft-lbf/lbm-°R),

$$\rho = \frac{p}{RT}$$

$$= \frac{\left(14.7 \frac{\text{lbf}}{\text{in}^2}\right)\left(144 \frac{\text{in}^2}{\text{ft}^2}\right)}{\left(96.32 \frac{\text{ft-lbf}}{\text{lbm-°R}}\right)(60°F + 460)}$$

$$= 0.0422 \text{ lbm/ft}^3$$

From Table 21.9, $K = 9.55$ ft³ air/ft³ fuel.

From Table 21.8,

$$\text{products: 1 ft}^3 \text{ CO}_2, \text{ 2 ft}^3 \text{ H}_2\text{O}$$

From App. 21.A, HHV $= 1013$ Btu/lbm.

Reaction Products for Prob. 13

gas	B	$\rho\left(\frac{\text{lbm}}{\text{ft}^3}\right)$	ft³ air	HHV $\left(\frac{\text{Btu}}{\text{lbm}}\right)$	CO₂	H₂O	other
CH₄	0.93	0.0422	9.556	1013	1	2	–
N₂	0.0373	0.0738	–	–	–	–	1 N₂
CO	0.0045	0.0738	2.389	322	1	–	–
H₂	0.0182	0.0053	2.389	325	–	1	–
C₂H₄	0.0025	0.0739	14.33	1614	2	2	–
H₂S	0.0018	0.0900	7.167	647	–	1	1 SO₂
O₂	0.0035	0.0843	–	–	–	–	–
CO₂	0.0022	0.1160	–	–	1	–	–

Similar results for all the other fuel components are tabulated in the table on the previous page.

The composite density is

$$\rho = \sum B_i \rho_i$$

$$= (0.93)\left(0.0422 \frac{lbm}{ft^3}\right) + (0.0373)\left(0.0738 \frac{lbm}{ft^3}\right)$$

$$+ (0.0045)\left(0.0738 \frac{lbm}{ft^3}\right) + (0.0182)\left(0.0053 \frac{lbm}{ft^3}\right)$$

$$+ (0.0025)\left(0.0739 \frac{lbm}{ft^3}\right) + (0.0018)\left(0.0900 \frac{lbm}{ft^3}\right)$$

$$+ (0.0035)\left(0.0843 \frac{lbm}{ft^3}\right) + (0.0022)\left(0.1160 \frac{lbm}{ft^3}\right)$$

$$= \boxed{0.0433 \; lbm/ft^3}$$

The answer is (B).

(b) The air is 20.9% oxygen by volume. The theoretical air requirements are

$$\sum B_i V_{air,i} - \frac{O_2 \; in \; fuel}{0.209}$$

$$= (0.93)(9.556 \; ft^3) + (0.0373)(0)$$

$$+ (0.0045)(2.389 \; ft^3) + (0.0182)(2.389 \; ft^3)$$

$$+ (0.0025)(14.33 \; ft^3) + (0.0018)(7.167 \; ft^3)$$

$$+ (0.0035)(0) + (0.0022)(0) - \frac{0.0035}{0.209}$$

$$= 8.990 \; ft^3 - 0.01675 \; ft^3$$

$$= \boxed{8.9733 \; ft^3 \; air/ft^3 \; fuel}$$

The answer is (C).

((c) and (d)) The theoretical oxygen will be

$$(8.9733 \; ft^3)(0.209) = 1.875 \; ft^3/ft^3$$

The excess oxygen will be

$$(0.4)(1.875 \; ft^3) = 0.75 \; ft^3/ft^3$$

Similarly, the total nitrogen in the stack gases is

$$(1.4)(0.791)(8.9733 \; ft^3) + 0.0373 = 9.974 \; ft^3/ft^3 \; fuel$$

The stack gases per ft^3 of fuel are

excess O_2:	$= 0.7500 \; ft^3$
excess N_2:	$= 9.974 \; ft^3$
excess SO_2:	$= 0.0018 \; ft^3$
excess CO_2: $(0.93)(1) + (0.0045)(1)$ $+ (0.0025)(2) + (0.0022)(1)$	$= 0.942 \; ft^3$
excess H_2O: $(0.93)(2) + (0.0182)(1)$ $+ (0.0025)(2) + (0.0018)(1)$	$= 1.885 \; ft^3$
	total $= 13.57 \; ft^3$

The total wet volume is 13.57 ft^3/ft^3 fuel.

The total dry volume is 11.68 ft^3/ft^3 fuel.

The volumetric analyses are

	O_2	N_2	SO_2	CO_2	H_2O
wet:	$\frac{0.7500 \; ft^3}{13.57 \; ft^3}$	$\frac{9.974 \; ft^3}{13.57 \; ft^3}$	$\frac{0.0018 \; ft^3}{13.57 \; ft^3}$	$\frac{0.942 \; ft^3}{13.57 \; ft^3}$	$\frac{1.885}{13.57}$
	$= 0.0553$	0.735	$-$	0.069	0.139
dry:	$\frac{0.7500 \; ft^3}{11.68 \; ft^3}$	$\frac{9.974 \; ft^3}{11.68 \; ft^3}$	$\frac{0.0018 \; ft^3}{11.68 \; ft^3}$	$\frac{0.942 \; ft^3}{11.68 \; ft^3}$	
	$= 0.0642$	0.854	$-$	0.081	$-$

$$B_{CO_2,wet} = \boxed{6.9\%}$$

The answer is (A) for Prob. 13(c).

$$B_{CO_2,dry} = \boxed{8.1\%}$$

The answer is (C) for Prob. 13(d).

14. *Customary U.S. Solution*

step 1: From App. 23.B, the heat of vaporization at 212°F is $h_{fg} = 970.3$ Btu/lbm.

The heat absorbed in the boiler is

$$m_{H_2O} h_{fg} = (11.12 \; lbm \; H_2O)\left(970.3 \frac{Btu}{lbm}\right)$$

$$= 10,789.7 \; Btu/lbm \; fuel$$

step 2: The losses for heating stack gases can be found as follows.

The burned carbon per lbm of fuel is

$$0.7842 - (0.315)(0.0703) = 0.7621 \; lbm/lbm \; fuel$$

The mass ratio of dry flue gases to solid fuel is given by Eq. 21.12.

$$\frac{\text{mass of flue gas}}{\text{mass of solid fuel}} = \frac{\left(\begin{array}{c}11 B_{CO_2} + 8 B_{O_2} \\ + (7)(B_{CO} + B_{N_2})\end{array}\right)\left(G_C + \dfrac{G_S}{1.833}\right)}{(3)(B_{CO_2} + B_{CO})}$$

$$= \frac{\left(\begin{array}{c}(11)(14.0) + (8)(5.5) \\ + (7)(0.42 + 80.08)\end{array}\right)\left(0.7621 + \dfrac{0.01}{1.833}\right)}{(3)(14.0 + 0.42)}$$

$$= 13.51 \; lbm \; stack \; gases/lbm \; fuel$$

Properties of nitrogen frequently are assumed for dry flue gas. From Table 21.1, for nitrogen at an average temperature of $(575°F + 73°F)/2 = 324°F$ $(784°R)$, $c_p \approx 0.252$ Btu/lbm-°F.

The losses for heating stack gases are given by Eq. 21.20.

$$q_1 = m_{\text{flue gas}} c_p (T_{\text{flue gas}} - T_{\text{incoming air}})$$

$$= (13.51 \text{ lbm}) \left(0.252 \frac{\text{Btu}}{\text{lbm-°F}}\right) (575°F - 67°F)$$

$$= 1729.5 \text{ Btu/lbm fuel}$$

The heat loss in the vapor formed during the combustion of hydrogen is given by Eq. 21.21.

$$q_2 = 8.94 G_H (h_g - h_f)$$

Assume that the partial pressure of the water vapor is below 1 psia (the lowest App. 23.C goes).

h_g at 575°F and 1 psia can be found from the superheat tables, App. 23.C.

$$h_g \approx 1324.2 \text{ Btu/lbm}$$

h_f at 73°F can be found from App. 23.A.

$$h_f = 41.09 \text{ Btu/lbm}$$

$$G_{H,\text{available}} = G_{H,\text{total}} - \frac{G_O}{8}$$

$$= 0.0556 - \frac{0.0825}{8}$$

$$= 0.0453$$

$$q_2 = (8.94)(0.0453) \left(1324.2 \frac{\text{Btu}}{\text{lbm}} - 41.09 \frac{\text{Btu}}{\text{lbm}}\right)$$

$$= 519.6 \text{ Btu/lbm fuel}$$

Heat is also lost when it is absorbed by the moisture originally in the combustion air.

$$q_3 = \omega m_{\text{combustion air}} (h_g - h'_g)$$

Assume that the partial pressure of the water vapor is below 1 psia (the lowest App. 23.C goes).

From App. 23.B, at 73°F and 0.4 psia, $h'_g \approx 1090$ Btu/lbm. From the psychrometric chart,

$$\omega = 90 \text{ grains/lbm air}$$

$$= 0.0129 \text{ lbm water/lbm air}$$

Considering the sulfur content, find the air/fuel ratio.

$$\frac{\text{lbm air}}{\text{lbm fuel}} = \frac{3.04 B_{N_2} \left(G_C + \frac{G_S}{1.833}\right)}{B_{CO_2} + B_{CO}}$$

$$= \frac{(3.04)(80.08\%) \left(0.7621 + \frac{0.01}{1.833}\right)}{14\% + 0.42\%}$$

$$= 12.96 \text{ lbm air/lbm fuel}$$

$$q_3 = \left(0.0129 \frac{\text{lbm water}}{\text{lbm air}}\right) \left(12.96 \frac{\text{lbm air}}{\text{lbm fuel}}\right)$$

$$\times \left(1324.2 \frac{\text{Btu}}{\text{lbm}} - 1090 \frac{\text{Btu}}{\text{lbm}}\right)$$

$$= 39.2 \text{ Btu/lbm fuel}$$

The energy lost in incomplete combustion is given by Eq. 21.23.

$$q_4 = \frac{(\text{HHV}_C - \text{HHV}_{CO}) G_C B_{CO}}{B_{CO_2} + B_{CO}}$$

$$= \frac{\left(9746 \frac{\text{Btu}}{\text{lbm}}\right)(0.7621)(0.42\%)}{14\% + 0.42\%}$$

$$= 216.3 \text{ Btu/lbm fuel}$$

The energy lost in unburned carbon is given by Eq. 21.24.

$$q_5 = \left(14{,}093 \frac{\text{Btu}}{\text{lbm}}\right) m_{\text{ash}} G_{C,\text{ash}}$$

$$= \left(14{,}093 \frac{\text{Btu}}{\text{lbm}}\right)(0.0703)(0.315)$$

$$= 312.1 \text{ Btu/lbm fuel}$$

The energy lost in radiation and unaccounted for is

$$14{,}000 \frac{\text{Btu}}{\text{lbm}} - 10{,}789.7 \frac{\text{Btu}}{\text{lbm}} - 1729.5 \frac{\text{Btu}}{\text{lbm}}$$

$$- 519.6 \frac{\text{Btu}}{\text{lbm}} - 39.2 \frac{\text{Btu}}{\text{lbm}} - 216.3 \frac{\text{Btu}}{\text{lbm}}$$

$$- 312.1 \frac{\text{Btu}}{\text{lbm}}$$

$$= \boxed{393.6 \text{ Btu/lbm fuel}}$$

SI Solution

step 1: From App. 23.N, the heat of vaporization at 100°C is $h_{fg} = 2257.0$ kJ/kg.

The heat absorbed in the boiler is

$$m_{H_2O}h_{fg} = (11.12 \text{ kg})\left(2257.0 \ \frac{kJ}{kg}\right)$$

$$= 25\,098 \text{ kJ/kg fuel}$$

step 2: From step 2 of the U.S. solution,

$$\frac{\text{mass of flue gas}}{\text{mass of solid fuel}} = 13.51 \text{ kg stack gases/kg fuel}$$

Properties of nitrogen frequently are assumed for dry flue gas. For an average nitrogen temperature of $(1/2)(300°C + 23°C) + 273 = 434.5$ K $(782°R)$, $c_p \approx 0.252$ Btu/lbm-°R.

c_p from Table 21.1 can be found (using the table footnote) as

$$c_p = \left(0.252 \ \frac{Btu}{lbm\text{-}°R}\right)\left(4.187 \ \frac{\frac{kJ}{kg\cdot K}}{\frac{Btu}{lbm\text{-}°R}}\right)$$

$$= 1.055 \text{ kJ/kg}\cdot K$$

The losses for heating stack gases are given by Eq. 21.20.

$$q_1 = m_{\text{flue gas}}c_p(T_{\text{flue gas}} - T_{\text{incoming air}})$$

$$= (13.51 \text{ kg})\left(1.055 \ \frac{kJ}{kg\cdot K}\right)(300°C - 23°C)$$

$$= 3948.1 \text{ kJ/kg fuel}$$

The heat loss in the vapor formed during the combustion of hydrogen is given by Eq. 21.21.

$$q_2 = 8.94G_H(h_g - h_f)$$

Assume that the partial pressure of the water vapor is below 10 kPa (the lowest App. 24.P goes).

h_g at 300°C and 10 kPa can be found from the superheat tables, App. 23.P.

$$h_g = 3076.5 \text{ kJ/kg}$$

h_f at 23°C can be found from App. 23.N.

$$h_f = 96.52 \text{ kJ/kg}$$

$$G_{H,\text{available}} = G_{H,\text{total}} - \frac{G_O}{8}$$

$$= 0.0556 - \frac{0.0825}{8}$$

$$= 0.0453$$

$$q_2 = (8.94)(0.0453)\left(3076.5 \ \frac{kJ}{kg} - 96.52 \ \frac{kJ}{kg}\right)$$

$$= 1206.8 \text{ kJ/kg fuel}$$

Heat is also lost when it is absorbed by the moisture originally in the combustion air.

$$q_3 = \omega m_{\text{combustion air}}(h_g - h'_g)$$

Assume that the partial pressure of the water vapor is low. At 23°C, from App. 23.O, $h'_g \approx 2500$ kJ/kg.

From the psychrometric chart for 19°C wet bulb and 23°C dry bulb, $\omega = 12.2$ g/kg dry air.

The air-fuel ratio from the customary U.S. solution is

$$\frac{\text{kg air}}{\text{kg fuel}} = 12.96 \text{ kg air/kg fuel}$$

$$q_3 = \left(12.2 \ \frac{g}{kg}\right)\left(\frac{1 \text{ kg}}{1000 \text{ g}}\right)\left(12.96 \ \frac{kg}{kg}\right)$$

$$\times \left(3076.5 \ \frac{kJ}{kg} - 2500 \ \frac{kJ}{kg}\right)$$

$$= 91.15 \text{ kJ/kg}$$

Energy lost in incomplete combustion is given by Eq. 21.23.

$$q_4 = \frac{(\text{HHV}_C - \text{HHV}_{CO})\,G_C B_{CO}}{B_{CO_2} + B_{CO}}$$

$$= \left(\frac{\left(22.67 \ \frac{MJ}{kg}\right)(0.7621)(0.42)}{14 + 0.42}\right)\left(1000 \ \frac{kJ}{MJ}\right)$$

$$= \left(0.5032 \ \frac{MJ}{kg}\right)\left(1000 \ \frac{kJ}{MJ}\right)$$

$$= 503.2 \text{ kJ/kg fuel}$$

The energy lost in unburned carbon is given by Eq. 21.24.

$$q_5 = \left(32.8 \ \frac{MJ}{kg}\right)m_{\text{ash}}G_{C,\text{ash}}$$

$$= \left(32.8 \ \frac{MJ}{kg}\right)\left(1000 \ \frac{kJ}{MJ}\right)(0.0703)(0.315)$$

$$= 726.3 \text{ kJ/kg fuel}$$

Energy lost in radiation and unaccounted for is

$$\left(32.6 \ \frac{MJ}{kg}\right)\left(1000 \ \frac{kJ}{MJ}\right) - 25\,098 \ \frac{kJ}{kg}$$

$$- 3948.1 \ \frac{kJ}{kg} - 1206.8 \ \frac{kJ}{kg} - 91.15 \ \frac{kJ}{kg}$$

$$- 503.2 \ \frac{kJ}{kg} - 726.3 \ \frac{kJ}{kg} = \boxed{1026.5 \text{ kJ/kg}}$$

15. *Customary U.S. Solution*

step 1: The incoming reactants on a per-pound basis are

> 0.07 lbm ash
> 0.05 lbm hydrogen
> 0.05 lbm oxygen
> 0.83 lbm carbon

This is an ultimate analysis. Assume that only the hydrogen that is not locked up with oxygen in the form of water is combustible. From Eq. 21.15, the available hydrogen is

$$G_{H,available} = G_{H,total} - \frac{G_O}{8}$$
$$= 0.05 \text{ lbm} - \frac{0.05 \text{ lbm}}{8}$$
$$= 0.04375 \text{ lbm}$$

The mass of water produced is the hydrogen mass plus eight times as much oxygen. The locked hydrogen is

$$0.05 \text{ lbm} - 0.04375 \text{ lbm} = 0.00625 \text{ lbm}$$

$$\frac{\text{lbm of}}{\text{moisture}} = G_H + G_O$$
$$= G_H + 8G_H$$
$$= 0.00625 \text{ lbm} + (8)(0.00625 \text{ lbm})$$
$$= 0.05625 \text{ lbm}$$

The air is 23.15% oxygen by weight (Table 21.6), so other reactants for 26 lbm of air are

$$(0.2315)(26 \text{ lbm}) = 6.019 \text{ lbm O}_2$$
$$(0.7685)(26 \text{ lbm}) = 19.981 \text{ lbm N}_2$$

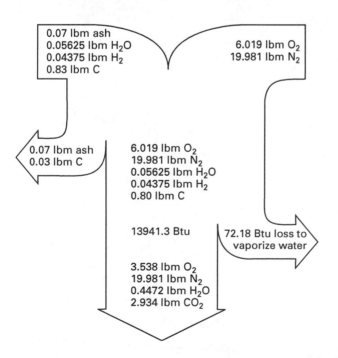

step 2: Ash pit material losses are 10% or 0.1 lbm, which includes all of the ash.

> 0.07 lbm ash (noncombustible matter)
> 0.03 lbm unburned carbon

step 3: Determine what remains.

> 6.019 lbm oxygen
> 19.981 lbm nitrogen
> 0.05625 lbm water
> 0.04375 lbm hydrogen
> 0.80 lbm carbon

step 4: Determine the energy loss in vaporizing the moisture.

$$q = (\text{moisture})(h_g - h_f)$$

From App. 23.C, h_g at 550°F and 14.7 psia is 1311.3 Btu/lbm.

From App. 23.A, h_f at 60°F is 28.08 Btu/lbm.

$$q = (0.05625 \text{ lbm})\left(1311.3 \frac{\text{Btu}}{\text{lbm}} - 28.08 \frac{\text{Btu}}{\text{lbm}}\right)$$
$$= 72.18 \text{ Btu}$$

step 5: Calculate the heating value of the remaining fuel components using App. 21.A.

$$HV_C = (0.80 \text{ lbm})\left(14{,}093 \frac{\text{Btu}}{\text{lbm}}\right)$$
$$= 11274.4 \text{ Btu}$$

$$HV_H = (0.04375 \text{ lbm})\left(60{,}958 \frac{\text{Btu}}{\text{lbm}}\right)$$
$$= 2666.9 \text{ Btu}$$

The heating value after the coal moisture is evaporated is

$$11{,}274.4 \text{ Btu} + 2666.9 \text{ Btu} - 72.18 \text{ Btu}$$
$$= 13{,}869 \text{ Btu}$$

step 6: Using Table 21.8, determine the combustion products.

$$\frac{\text{oxygen required}}{\text{by carbon}} = (0.80)(2.667 \text{ lbm})$$
$$= 2.134 \text{ lbm}$$

$$\frac{\text{oxygen required}}{\text{by hydrogen}} = (0.04375)(7.936 \text{ lbm})$$
$$= 0.3472 \text{ lbm}$$

$$\frac{\text{carbon dioxide}}{\text{produced by carbon}} = (0.8)(3.667 \text{ lbm})$$
$$= 2.934 \text{ lbm}$$

$$\frac{\text{water produced}}{\text{by hydrogen}} = (0.04375)(8.936 \text{ lbm})$$
$$= 0.3910 \text{ lbm}$$

The remaining oxygen is

$$6.019 \text{ lbm} - 2.134 \text{ lbm} - 0.3472 \text{ lbm} = 3.538 \text{ lbm}$$

step 7: The gaseous products must be heated from 70°F to 550°F. The average temperature is

$$\left(\tfrac{1}{2}\right)(70°F + 550°F) = 310°F \ (770°R)$$

From Table 21.1, the specific heat of gaseous products is

gas	c_p $\left(\dfrac{\text{Btu}}{\text{lbm-°R}}\right)$
oxygen	0.228
nitrogen	0.252
water	0.460
carbon dioxide	0.225

$$
\begin{aligned}
Q_{\text{heating}} = &\left((3.538 \text{ lbm})\left(0.228 \frac{\text{Btu}}{\text{lbm-°R}}\right)\right. \\
&+ (19.981 \text{ lbm})\left(0.252 \frac{\text{Btu}}{\text{lbm-°R}}\right) \\
&+ (0.3910 \text{ lbm})\left(0.460 \frac{\text{Btu}}{\text{lbm-°R}}\right) \\
&\left.+ (2.934 \text{ lbm})\left(0.225 \frac{\text{Btu}}{\text{lbm-°R}}\right)\right) \\
&\times (550°F - 70°F) \\
= &\ 3207.3 \text{ Btu}
\end{aligned}
$$

step 8: The percent loss is

$$\frac{3207.3 \text{ Btu} + 72.18 \text{ Btu}}{13,869 \text{ Btu}} = \boxed{0.236 \ \ (23.6\%)}$$

The answer is (B).

SI Solution

Steps 1 through 3 are the same as for the customary U.S. solution except that everything is based on kg.

step 4: Determine the energy loss in the vaporizing moisture.

$$q = (\text{moisture})(h_g - h_f)$$

From App. 23.P, h_g at 290°C and 101.3 kPa is 3054.3 kJ/kg.

h_f at 15.6°C from App. 23.N is 65.51 kJ/kg.

$$q = (0.05625 \text{ kg})\left(3054.3 \frac{\text{kJ}}{\text{kg}} - 65.51 \frac{\text{kJ}}{\text{kg}}\right)$$

$$= 168.1 \text{ kJ}$$

step 5: Calculate the heating value of the remaining fuel components using App. 21.A and the table footnote.

$$\text{HV}_C = (0.80 \text{ kg})\left(14\,093 \frac{\text{Btu}}{\text{lbm}}\right)\left(2.326 \frac{\frac{\text{kJ}}{\text{kg}}}{\frac{\text{Btu}}{\text{lbm}}}\right)$$

$$= 26\,224 \text{ kJ}$$

$$\text{HV}_H = (0.04375 \text{ kg})\left(60\,958 \frac{\text{Btu}}{\text{lbm}}\right)\left(2.326 \frac{\frac{\text{kJ}}{\text{kg}}}{\frac{\text{Btu}}{\text{lbm}}}\right)$$

$$= 6203 \text{ kJ}$$

The heating value after the coal moisture is evaporated is

$$26\,224 \text{ kJ} + 6203 \text{ kJ} - 168.1 \text{ kJ} = 32\,259 \text{ kJ}$$

step 6: This step is the same as for the customary U.S. solution except that all quantities are in kg.

step 7: The gaseous products must be heated from 21°C to 290°C. The average temperature is

$$\left(\tfrac{1}{2}\right)(21°C + 290°C) = 156°C \ (429°K)$$

$$(429\text{K})\left(1.8 \frac{°R}{K}\right) = 771°R$$

From Table 21.1, the specific heat of gaseous products is calculated using the table footnote.

gas	c_p $\left(\dfrac{\text{kJ}}{\text{kg·K}}\right)$
oxygen	0.955
nitrogen	1.055
water	1.926
carbon dioxide	0.942

$$
\begin{aligned}
Q_{\text{heating}} = &\left((3.538 \text{ kg})\left(0.955 \frac{\text{kJ}}{\text{kg·K}}\right)\right. \\
&+ (19.981 \text{ lbm})\left(1.055 \frac{\text{kJ}}{\text{kg·K}}\right) \\
&+ (0.3910 \text{ kg})\left(1.926 \frac{\text{kJ}}{\text{kg·K}}\right) \\
&\left.+ (2.934 \text{ kg})\left(0.942 \frac{\text{kJ}}{\text{kg·K}}\right)\right) \\
&\times (290°C - 21°C) \\
= &\ \boxed{7525.4 \text{ kJ}}
\end{aligned}
$$

step 8: The percentage loss is

$$\frac{7525.4 \text{ kJ} + 168.1 \text{ kJ}}{32\,259 \text{ kJ}} = \boxed{0.238 \quad (23.8\%)}$$

The answer is (B).

16. (a) Silicon in ash is SiO_2 with a molecular weight of

$$28.09 \frac{\text{lbm}}{\text{lbmol}} + (2)\left(16 \frac{\text{lbm}}{\text{lbmol}}\right) = 60.09 \text{ lbm/lbmol}$$

The oxygen used with 6.1% by mass silicon is

$$\left(\frac{(2)(16 \text{ lbm})}{28.09 \text{ lbm}}\right)\left(0.061 \frac{\text{lbm}}{\text{lbm coal}}\right)$$
$$= 0.0695 \text{ lbm/lbm coal}$$

Silicon ash produced per lbm of coal is

$$0.061 \frac{\text{lbm}}{\text{lbm coal}} + 0.0695 \frac{\text{lbm}}{\text{lbm coal}}$$
$$= 0.1305 \text{ lbm/lbm coal}$$

Silicon ash produced per hour is

$$\left(0.1305 \frac{\text{lbm}}{\text{lbm coal}}\right)\left(15,300 \frac{\text{lbm coal}}{\text{hr}}\right)$$
$$= 1996.7 \text{ lbm/hr}$$

The silicon in 410 lbm/hr refuse is

$$\left(410 \frac{\text{lbm}}{\text{hr}}\right)(1 - 0.3) = 287 \text{ lbm/hr}$$

The emission rate is

$$1996.7 \text{ lbm/hr} - 287 \text{ lbm/hr} = \boxed{1709.7 \text{ lbm/hr}}$$

The answer is (D).

(b) From Table 21.7, the stoichiometric reaction for sulfur is

$$\begin{array}{ccccc} & S & + & O_2 & \longrightarrow & SO_2 \\ \text{MW} & 32 & & 32 & & 64 \end{array}$$

Sulfur dioxide produced for 15,300 lbm/hr of coal feed is

$$\left(15,300 \frac{\text{lbm}}{\text{hr}}\right)(0.0244 \text{ lbm S})\left(\frac{64 \text{ lbm } SO_2}{32 \text{ lbm S}}\right)$$
$$= \boxed{746.6 \text{ lbm/hr}}$$

The answer is (C).

(c) From Eq. 21.16(b), the heating value of the fuel is

$$HHV = 14,093 G_C + (60,958)\left(G_H - \frac{G_O}{8}\right) + 3983 G_S$$

$$= \left(14,093 \frac{\text{Btu}}{\text{lbm}}\right)(0.7656)$$

$$+ \left(60,958 \frac{\text{Btu}}{\text{lbm}}\right)\left(0.055 - \frac{0.077}{8}\right)$$

$$+ \left(3983 \frac{\text{Btu}}{\text{lbm}}\right)(0.0244)$$

$$= 13,653 \text{ Btu/lbm}$$

The gross available combustion power is

$$\dot{m}_f(HV) = \left(15,300 \frac{\text{lbm}}{\text{hr}}\right)\left(13,653 \frac{\text{Btu}}{\text{lbm}}\right)$$
$$= 2.089 \times 10^8 \text{ Btu/hr}$$

The carbon in 410 lbm/hr refuse is

$$\left(410 \frac{\text{lbm}}{\text{hr}}\right)(0.3) = 123 \text{ lbm/hr}$$

Power lost in unburned carbon in refuse is $\dot{m}_C(HV)$. From App. 21.A, the gross heat of combustion for carbon is 14,093 Btu/lbm.

$$\left(123 \frac{\text{lbm}}{\text{hr}}\right)\left(14,093 \frac{\text{Btu}}{\text{lbm}}\right) = 1.733 \times 10^6 \text{ Btu/hr}$$

The remaining combustion power is

$$2.089 \times 10^8 \frac{\text{Btu}}{\text{hr}} - 1.733 \times 10^6 \frac{\text{Btu}}{\text{hr}}$$
$$= 2.072 \times 10^8 \text{ Btu/hr}$$

Losses in the steam generator and electrical generator will further reduce this to

$$(0.86)\left(2.072 \times 10^8 \frac{\text{Btu}}{\text{hr}}\right) = 1.782 \times 10^8 \text{ Btu/hr}$$

With an electrical output of 17 MW, thermal energy removed by cooling water is

$$Q = 1.782 \times 10^8 \frac{\text{Btu}}{\text{hr}} - (17 \text{ MW})\left(1000 \frac{\text{kW}}{\text{MW}}\right)$$
$$\times \left(3413 \frac{\frac{\text{Btu}}{\text{hr}}}{\text{kW}}\right)$$
$$= 1.202 \times 10^8 \text{ Btu/hr}$$

Thermodynamics

The temperature rise of the cooling water is

$$\Delta T = \frac{Q}{\dot{m}c_p}$$

At 60°F, the specific heat of water is $c_p = 1$ Btu/lbm-°F.

$$\Delta T = \frac{1.202 \times 10^8 \, \frac{\text{Btu}}{\text{hr}}}{\left(225 \, \frac{\text{ft}^3}{\text{sec}}\right)\left(62.4 \, \frac{\text{lbm}}{\text{ft}^3}\right)\left(3600 \, \frac{\text{sec}}{\text{hr}}\right)\left(1 \, \frac{\text{Btu}}{\text{lbm-°F}}\right)}$$

$$= \boxed{2.38°\text{F}}$$

The answer is (A).

The electrical generation is not cooled by the cooling water. Therefore, it is not correct to include the generation efficiency in the calculation of losses.

(d) Limiting 0.1 lbm of particulates per million Btu per hour, the allowable emission rate is

$$\left(0.1 \, \frac{\text{lbm}}{\text{MBtu}}\right)\left(15{,}300 \, \frac{\text{lbm}}{\text{hr}}\right)$$
$$\times \left(13{,}653 \, \frac{\text{Btu}}{\text{lbm}}\right)\left(\frac{1 \, \text{MBtu}}{10^6 \, \text{Btu}}\right) = 20.89 \, \text{lbm/hr}$$

The efficiency of the flue gas particulate collectors is

$$\eta = \frac{\text{actual emission rate} - \text{allowable emission rate}}{\text{actual emission rate}}$$

$$= \frac{1709.7 \, \frac{\text{lbm}}{\text{hr}} - 20.89 \, \frac{\text{lbm}}{\text{hr}}}{1709.7 \, \frac{\text{lbm}}{\text{hr}}}$$

$$= \boxed{0.988 \; (98.8\%)}$$

The answer is (C).

17. *Customary U.S. Solution*

(c) The stoichiometric reaction for propane is given in Table 21.7.

$$\begin{array}{ccccc} \text{C}_3\text{H}_8 + & 5\text{O}_2 & \longrightarrow & 3\text{CO}_2 & + & 4\text{H}_2\text{O} \\ \text{MW} \;\; 44.097 & (5)(32.000) & & (3)(44.011) & & (4)(18.016) \end{array}$$

With 40% excess O_2 by weight,

$$\begin{array}{cc} \text{C}_3\text{H}_8 & (1.4)(5)\text{O}_2 \\ \text{MW} \;\; 44.097 + & (7)(32.000) \\ 44.097 & 224 \end{array}$$

$$\begin{array}{cccc} & 3\text{CO}_2 & 4\text{H}_2\text{O} & 2\text{O}_2 \\ \longrightarrow & (3)(44.011) + & (4)(18.016) + & (2)(32.000) \\ & 132.033 & 72.064 & 64 \end{array}$$

The excess oxygen is

$$(2)(32) = 64 \, \text{lbm/lbmol C}_3\text{H}_8$$

The mass ratio of nitrogen to oxygen is

$$\frac{G_\text{N}}{G_\text{O}} = \left(\frac{B_\text{N}}{R_\text{N}}\right)\left(\frac{R_\text{O}}{B_\text{O}}\right)$$
$$= \left(\frac{0.40}{55.16 \, \frac{\text{ft-lbf}}{\text{lbm-°R}}}\right)\left(\frac{48.29 \, \frac{\text{ft-lbf}}{\text{lbm-°R}}}{0.60}\right)$$
$$= 0.584$$

(Values of R_N and R_O are taken from Table 23.7.)

The nitrogen accompanying the oxygen is

$$(7)(32)(0.584) = 130.8 \, \text{lbm}$$

The mass balance per mole of propane is

$$\begin{array}{ccccccccc} & \text{C}_3\text{H}_8 + & \text{O}_2 + & \text{N}_2 & \longrightarrow & \text{CO}_2 & + & \text{H}_2\text{O} & + \text{O}_2 + & \text{N}_2 \\ \text{mass} & & & & & & & & \\ \text{per} & 44.097 + & 224 + & 130.8 & \longrightarrow & 132.033 + & & 72.064 + & 64 + & 130.8 \\ \text{mole} & & & & & & & & \end{array}$$

At standard conditions (60°F, 1 atm), the propane density is given by Eq. 23.50.

$$\rho = \frac{p}{RT}$$

The absolute temperature, T, is

$$60°\text{F} + 460 = 520°\text{R}$$

R for propane from Table 23.7 is 35.04 ft-lbf/lbm-°R.

$$\rho = \frac{\left(14.7 \, \frac{\text{lbf}}{\text{in}^2}\right)\left(144 \, \frac{\text{in}^2}{\text{ft}^2}\right)}{\left(35.04 \, \frac{\text{ft-lbf}}{\text{lbm-°R}}\right)(520°\text{R})}$$

$$= 0.1162 \, \text{lbm/ft}^3$$

The mass flow rate of propane based on 250 SCFM of propane is

$$\left(250 \, \frac{\text{ft}^3}{\text{min}}\right)\left(0.1162 \, \frac{\text{lbm}}{\text{ft}^3}\right) = 29.05 \, \text{lbm/min}$$

Scaling the other mass balance factors down by

$$\frac{29.05 \, \frac{\text{lbm}}{\text{min}}}{44.097 \, \frac{\text{lbm}}{\text{lbmol}}} = 0.6588 \, \text{lbmol/min}$$

$$\frac{lbm}{min} \quad \begin{array}{ccc} C_3H_8 + & O_2 & + N_2 \\ 29.05 + & 147.57 + & 86.17 \end{array}$$

$$\longrightarrow \quad \begin{array}{cccc} CO_2 + & H_2O + & O_2 & + N_2 \\ 86.98 + & 47.48 + & 42.16 + & 86.17 \end{array}$$

The oxygen flow rate is $\boxed{147.57 \text{ lbm/min}}$ [part (c)].

The answer is (A).

(a) Using R from Table 23.7, the specific volumes of the reactants are given by Eq. 23.50.

$$v_{C_3H_8} = \frac{RT}{p}$$

$$= \frac{\left(35.04 \; \frac{\text{ft-lbf}}{\text{lbm-}^\circ\text{R}}\right)(80^\circ\text{F} + 460)}{\left(14.7 \; \frac{\text{lbf}}{\text{in}^2}\right)\left(144 \; \frac{\text{in}^2}{\text{ft}^2}\right)}$$

$$= 8.939 \text{ ft}^3/\text{lbm}$$

$$v_{O_2} = \frac{\left(48.29 \; \frac{\text{ft-lbf}}{\text{lbm-}^\circ\text{R}}\right)(80^\circ\text{F} + 460)}{\left(14.7 \; \frac{\text{lbf}}{\text{in}^2}\right)\left(144 \; \frac{\text{in}^2}{\text{ft}^2}\right)}$$

$$= 12.319 \text{ ft}^3/\text{lbm}$$

$$v_{N_2} = \frac{\left(55.16 \; \frac{\text{ft-lbf}}{\text{lbm-}^\circ\text{R}}\right)(80^\circ\text{F} + 460)^\circ\text{R}}{\left(14.7 \; \frac{\text{lbf}}{\text{in}^2}\right)\left(144 \; \frac{\text{in}^2}{\text{ft}^2}\right)}$$

$$= 14.071 \text{ ft}^3/\text{lbm}$$

The total incoming volume is

$$\dot{V} = \left(29.05 \; \frac{\text{lbm}}{\text{min}}\right)\left(8.939 \; \frac{\text{ft}^3}{\text{lbm}}\right)$$

$$+ \left(147.57 \; \frac{\text{lbm}}{\text{min}}\right)\left(12.319 \; \frac{\text{ft}^3}{\text{lbm}}\right)$$

$$+ \left(86.17 \; \frac{\text{lbm}}{\text{min}}\right)\left(14.071 \; \frac{\text{ft}^3}{\text{lbm}}\right)$$

$$= 3290 \text{ ft}^3/\text{min}$$

Since the velocity must be kept below 400 ft/min, the area of inlet pipe is

$$A_{\text{in}} = \frac{\dot{V}}{\text{v}} = \frac{3290 \; \frac{\text{ft}^3}{\text{min}}}{400 \; \frac{\text{ft}}{\text{min}}} = \boxed{8.23 \text{ ft}^2}$$

Use 9 ft^2.

The answer is (A).

(d) Similarly, the specific volumes of the products are

$$v_{CO_2} = \frac{\left(35.11 \; \frac{\text{ft-lbf}}{\text{lbm-}^\circ\text{R}}\right)(460^\circ\text{F} + 460)}{\left(8 \; \frac{\text{lbf}}{\text{in}^2}\right)\left(144 \; \frac{\text{in}^2}{\text{ft}^2}\right)}$$

$$= 28.04 \text{ ft}^3/\text{lbm}$$

$$v_{H_2O} = \frac{\left(85.78 \; \frac{\text{ft-lbf}}{\text{lbm-}^\circ\text{R}}\right)(460^\circ\text{F} + 460)}{\left(8 \; \frac{\text{lbf}}{\text{in}^2}\right)\left(144 \; \frac{\text{in}^2}{\text{ft}^2}\right)}$$

$$= 68.50 \text{ ft}^3/\text{lbm}$$

$$v_{O_2} = \frac{\left(48.29 \; \frac{\text{ft-lbf}}{\text{lbm-}^\circ\text{R}}\right)(460^\circ\text{F} + 460)}{\left(8 \; \frac{\text{lbf}}{\text{in}^2}\right)\left(144 \; \frac{\text{in}^2}{\text{ft}^2}\right)}$$

$$= 38.56 \text{ ft}^3/\text{lbm}$$

$$v_{N_2} = \frac{\left(55.16 \; \frac{\text{ft-lbf}}{\text{lbm-}^\circ\text{R}}\right)(460^\circ\text{F} + 460)}{\left(8 \; \frac{\text{lbf}}{\text{in}^2}\right)\left(144 \; \frac{\text{in}^2}{\text{ft}^2}\right)}$$

$$= 44.05 \text{ ft}^3/\text{lbm}$$

The total exhaust volume is

$$\dot{V} = \left(86.98 \; \frac{\text{lbm}}{\text{min}}\right)\left(28.04 \; \frac{\text{ft}^3}{\text{lbm}}\right)$$

$$+ \left(47.48 \; \frac{\text{lbm}}{\text{min}}\right)\left(68.50 \; \frac{\text{ft}^3}{\text{lbm}}\right)$$

$$+ \left(42.16 \; \frac{\text{lbm}}{\text{min}}\right)\left(38.56 \; \frac{\text{ft}^3}{\text{lbm}}\right)$$

$$+ \left(86.98 \; \frac{\text{lbm}}{\text{min}}\right)\left(44.05 \; \frac{\text{ft}^3}{\text{min}}\right)$$

$$= \boxed{11{,}148 \text{ ft}^3/\text{min}} \quad \text{[part (d)]}$$

The answer is (D).

(b) Since the velocity of the products must be kept below 800 ft/min, the area of the stack is

$$A_{\text{stack}} = \frac{Q}{\text{v}} = \frac{11{,}148 \; \frac{\text{ft}^3}{\text{min}}}{800 \; \frac{\text{ft}}{\text{min}}}$$

$$= \boxed{13.94 \text{ ft}^2} \quad \text{[part (b)]}$$

The answer is (C).

(e) For ideal gases, the partial pressure is volumetrically weighted. The water vapor partial pressure in the stack is

$$(8\text{ psia})\left(\frac{\left(47.48\ \frac{\text{lbm}}{\text{min}}\right)\left(68.50\ \frac{\text{ft}^3}{\text{lbm}}\right)}{11{,}148\ \frac{\text{ft}^3}{\text{lbm}}}\right) = 2.33\text{ psia}$$

The saturation temperature corresponding to 2.33 psia is $T_{\text{dp}} = \boxed{132°\text{F}.}$

The answer is (C).

SI Solution

(c) Following the procedure for the customary U.S. solution, the mass balance per mole of propane is

$$C_3H_8 + O_2 + N_2 \longrightarrow CO_2 + H_2O + O_2 + N_2$$
$$\frac{\text{kg}}{\text{per mole}}\quad 44.097 + 224 + 130.8 \longrightarrow 132.033 + 72.064 + 64 + 130.8$$

At standard conditions (16°C, 101.3 kPa), the propane density is given by Eq. 23.50.

$$\rho = \frac{p}{RT}$$

The absolute temperature is

$$T = 16°\text{C} + 273 = 289\text{K}$$

From Table 23.7, R for propane is 188.55 J/kg·K.

$$\rho = \frac{(101.3\text{ kPa})\left(1000\ \frac{\text{Pa}}{\text{kPa}}\right)}{\left(188.55\ \frac{\text{J}}{\text{kg·K}}\right)(289\text{K})}$$
$$= 1.86\text{ kg/m}^3$$

The mass flow rate of propane based on 118 L/s of propane is

$$\left(118\ \frac{\text{L}}{\text{s}}\right)\left(\frac{1\text{ m}^3}{1000\text{ L}}\right)\left(1.86\ \frac{\text{kg}}{\text{m}^3}\right) = 0.2195\text{ kg/s}$$

Scale the other mass balance factors down.

$$\frac{0.2195\ \frac{\text{kg}}{\text{s}}}{44.097\ \frac{\text{kg}}{\text{kmol}}} = 0.004978\text{ kmol/s}$$

$$C_3H_8 + O_2 + N_2 \longrightarrow CO_2 + H_2O$$
$$\frac{\text{kg}}{\text{s}}\quad 0.2195\quad 1.115\quad 0.6511\quad 0.6572\quad 0.3587$$
$$+ O_2 + N_2$$
$$0.3186\quad 0.6511$$

The oxygen flow rate is $\boxed{1.115\text{ kg/s.}}$

The answer is (A).

(a) Using R from Table 23.7, the specific volumes of the reactants are given by Eq. 23.50.

$$v_{C_3H_8} = \frac{RT}{p} = \frac{\left(188.55\ \frac{\text{J}}{\text{kg·K}}\right)(27°\text{C}+273)}{(101\text{ kPa})\left(1000\ \frac{\text{Pa}}{\text{kPa}}\right)}$$
$$= 0.5600\text{ m}^3/\text{kg}$$

$$v_{O_2} = \frac{\left(259.82\ \frac{\text{J}}{\text{kg·K}}\right)(27°\text{C}+273)}{(101\text{ kPa})\left(1000\ \frac{\text{Pa}}{\text{kPa}}\right)}$$
$$= 0.7717\text{ m}^3/\text{kg}$$

$$v_{N_2} = \frac{\left(296.77\ \frac{\text{J}}{\text{kg·K}}\right)(27°\text{C}+273)}{(101\text{ kPa})\left(1000\ \frac{\text{Pa}}{\text{kPa}}\right)}$$
$$= 0.8815\text{ m}^3/\text{kg}$$

The total incoming volume, $\dot{V}$, is

$$\dot{V} = \left(0.2195\ \frac{\text{kg}}{\text{s}}\right)\left(0.5600\ \frac{\text{m}^3}{\text{kg}}\right)$$
$$+ \left(1.115\ \frac{\text{kg}}{\text{s}}\right)\left(0.7717\ \frac{\text{m}^3}{\text{kg}}\right)$$
$$+ \left(0.6511\ \frac{\text{kg}}{\text{s}}\right)\left(0.8815\ \frac{\text{m}^3}{\text{kg}}\right)$$
$$= 1.557\text{ m}^3/\text{s}$$

Since the velocity for the reactants must be kept below 2 m/s, the area of inlet pipe is

$$A_{\text{in}} = \frac{\dot{V}}{\text{v}} = \frac{1.557\ \frac{\text{m}^3}{\text{s}}}{2\ \frac{\text{m}}{\text{s}}} = \boxed{0.779\text{ m}^2}$$

The answer is (A).

(d) Similarly, the specific volumes of the products are

$$v_{CO_2} = \frac{\left(188.92 \dfrac{J}{kg\cdot}\right)(240°C + 273)}{(55 \text{ kPa})\left(1000 \dfrac{Pa}{kPa}\right)}$$

$$= 1.762 \text{ m}^3/\text{kg}$$

$$v_{H_2O} = \frac{\left(461.5 \dfrac{J}{kg\cdot}\right)(240°C + 273)}{(55 \text{ kPa})\left(1000 \dfrac{Pa}{kPa}\right)}$$

$$= 4.305 \text{ m}^3/\text{kg}$$

$$v_{O_2} = \frac{\left(259.82 \dfrac{J}{kg\cdot K}\right)(240°C + 273)}{(55 \text{ kPa})\left(1000 \dfrac{Pa}{kPa}\right)}$$

$$= 2.423 \text{ m}^3/\text{kg}$$

$$v_{N_2} = \frac{\left(296.77 \dfrac{J}{kg\cdot K}\right)(240°C + 273)}{(55 \text{ kPa})\left(1000 \dfrac{Pa}{kPa}\right)}$$

$$= 2.768 \text{ m}^3/\text{kg}$$

The total exhaust volume is

$$\dot{V} = \left(0.6572 \frac{kg}{s}\right)\left(1.762 \frac{m^3}{kg}\right)$$
$$+ \left(0.3587 \frac{kg}{s}\right)\left(4.305 \frac{m^3}{kg}\right)$$
$$+ \left(0.3186 \frac{kg}{s}\right)\left(2.423 \frac{m^3}{kg}\right)$$
$$+ \left(0.6511 \frac{kg}{s}\right)\left(2.768 \frac{m^3}{kg}\right)$$
$$= \boxed{5.276 \text{ m}^3/\text{s}}$$

The answer is (D).

(b) Since the velocity of products must be kept below 4 m/s, the area of stack is

$$A_{\text{stack}} = \frac{\dot{V}}{v} = \frac{5.276 \dfrac{m^3}{s}}{4 \dfrac{m}{s}}$$

$$= \boxed{1.319 \text{ m}^2}$$

The answer is (C).

(e) For ideal gases, the partial pressure is volumetrically weighted. The water vapor partial pressure in the stack is

$$(55 \text{ kPa})\left(\frac{\left(0.3584 \dfrac{kg}{s}\right)\left(4.305 \dfrac{m^3}{kg}\right)}{5.273 \dfrac{m^3}{s}}\right)$$

$$= 16.093 \text{ kPa}$$

The saturation temperature corresponding to 16.093 kPa is found from App. 23.O to be $T_{\text{dp}} = \boxed{54.5°C.}$

The answer is (C).

18. *Customary U.S. Solution*

(a) Since atmospheric air is not used, the nitrogen and oxygen can be varied independently. Furthermore, since enthalpy increase information is not given for oxygen, a 0% excess oxygen can be assumed.

From Table 21.7,

$$C_3H_8 + 5O_2 \longrightarrow 3CO_2 + 4H_2O$$
$$\text{moles} \quad (1) \qquad (5) \qquad\quad (3) \qquad (4)$$

Subtract the reactant enthalpies from the product enthalpies to calculate the heat of reaction. The enthalpy of formation of oxygen is zero, since it is an element in its natural state.

$$n_{CO_2}(\Delta H_f)_{CO_2} + n_{H_2O}(\Delta H_f)_{H_2O}$$
$$- n_{C_3H_8}(\Delta H_f)_{C_3H_8} - n_{O_2}(\Delta H_f)_{O_2}$$
$$= (3 \text{ lbmol})\left(-169,300 \frac{\text{Btu}}{\text{lbmol}}\right)$$
$$+ (4 \text{ lbmol})\left(-104,040 \frac{\text{Btu}}{\text{lbmol}}\right)$$
$$- (1 \text{ lbmol})\left(28,800 \frac{\text{Btu}}{\text{lbmol}}\right)$$
$$- (5 \text{ lbmol})(0)$$
$$= -952,860 \text{ Btu/lbmol of fuel}$$

The negative sign indicates an exothermal reaction.

Let x be the number of moles of nitrogen per mole of propane. Use the nitrogen to cool the combustion. The above heat of reaction will increase the enthalpy of products from the standard reference temperature to 3600°R. Therefore,

$$952,860 \frac{\text{Btu}}{\text{lbmol}} = (3 \text{ lbmol})\left(39,791 \frac{\text{Btu}}{\text{lbmol}}\right)$$
$$+ (4 \text{ lbmol})\left(31,658 \frac{\text{Btu}}{\text{lbmol}}\right)$$
$$+ x\left(24,471 \frac{\text{Btu}}{\text{lbmol}}\right)$$
$$x = 28.89 \text{ lbmol/lbmol fuel}$$

The mass of nitrogen per lbmole of propane is

$$M_{N_2} = \left(28.89 \, \frac{\text{lbmol}}{\text{lbmol fuel}}\right)\left(28.016 \, \frac{\text{lbm}}{\text{lbmol}}\right)$$

$$= \boxed{809.4 \text{ lbm/lbmol propane}}$$

The mass of oxygen per lbmole of propane is

$$M_{O_2} = (5 \text{ lbmol})\left(32 \, \frac{\text{lbm}}{\text{lbmol}}\right)$$

$$= \boxed{160 \text{ lbm/lbmol fuel}}$$

(b) The partial pressure is volumetrically weighted. This is the same as molar weighting.

product	lbmol	volumetric fraction
CO_2	3	$3/35.89 = 0.0836$
H_2O	4	$4/35.89 = 0.1115$
N_2	28.89	$28.89/35.89 = 0.8049$
O_2	0	$0/35.89 = 0$
	35.89 lbmol	1.000

The partial pressure of water vapor is

$$p_{H_2O} = \left(\frac{n_{H_2O}}{n}\right)p = (0.1115)(14.7 \text{ psia})$$

$$= 1.64 \text{ psia}$$

From App. 23.B, this corresponds to approximately 118°F. Since the stack temperature is 100°F, some of the water will condense. From App. 23.A, the maximum vapor pressure of water is 0.9503 psia. Let n be the number of moles of water in the stack gas.

$$n_{H_2O} = \left(\frac{p_{H_2O}}{p}\right)n = \left(\frac{0.9503 \text{ psia}}{14.7 \text{ psia}}\right)(35.89 \text{ lbmol})$$

$$= \boxed{2.320 \text{ lbmol/lbmol } C_3H_8}$$

(c) The water removed is

$$4 - n_{H_2O} = 4 \text{ lbmol} - 2.320 \text{ lbmol}$$

$$= 1.680 \text{ lbmol of } H_2O/\text{lbmol of } C_3H_8$$

$$m = (1.680 \text{ lbmol})\left(18.016 \, \frac{\text{lbm}}{\text{lbmol}}\right)$$

$$= \boxed{30.27 \text{ lbm } H_2O/\text{lbmol } C_3H_8}$$

SI Solution

(a) From the customary U.S. solution, the heat of reaction is

$$n_{CO_2}(\Delta H_f)_{CO_2} + n_{H_2O}(\Delta H_f)_{H_2O}$$
$$- n_{C_3H_8}(\Delta H_f)_{C_3H_8} - n_{O_2}(\Delta H_f)_{O_2}$$

$$= (3 \text{ kmol})\left(-393.8 \, \frac{\text{GJ}}{\text{kmol}}\right)$$

$$+ (4 \text{ kmol})\left(-242 \, \frac{\text{GJ}}{\text{kmol}}\right)$$

$$- (1 \text{ kmol})\left(67.0 \, \frac{\text{GJ}}{\text{kmol}}\right) - (5 \text{ kmol})(0)$$

$$= -2216.4 \text{ GJ/kmol fuel}$$

The negative sign indicates an exothermal reaction.

Let x be the number of moles of nitrogen per mole of propane. Use the nitrogen to cool the combustion. The above heat of reaction will increase the enthalpy of products from the standard reference temperature to 1980°C. Therefore,

$$2216.4 \, \frac{\text{GJ}}{\text{mol}} = (3 \text{ kmol})\left(92.6 \, \frac{\text{GJ}}{\text{kmol}}\right)$$

$$+ (4 \text{ kmol})\left(73.6 \, \frac{\text{GJ}}{\text{kmol}}\right)$$

$$+ x\left(56.9 \, \frac{\text{GJ}}{\text{kmol}}\right)$$

$$x = 28.90 \text{ kmol/kmol fuel}$$

The mass of nitrogen per mole of propane is

$$M_{N_2} = (28.90 \text{ kmol})\left(28.016 \, \frac{\text{kg}}{\text{kmol}}\right)$$

$$= \boxed{809.7 \text{ kg/kmol fuel}}$$

The mass of oxygen per kmol of propane is

$$M_{O_2} = (5 \text{ kmol})\left(32 \, \frac{\text{kg}}{\text{kmol}}\right)$$

$$= \boxed{160 \text{ kg/kmol propane}}$$

(b) The partial pressure is volumetrically weighted. This is the same as molar weighting.

product	kmol	volumetric fraction
CO_2	3	$3/35.90 = 0.0836$
H_2O	4	$4/35.90 = 0.1114$
N_2	28.90	$28.90/35.90 = 0.8050$
O_2	0	$0/35.90 = 0$
	35.90 kmol	1.000

The partial pressure of water vapor is

$$p_{H_2O} = \left(\frac{n_{H_2O}}{n}\right)p = (0.1114)(101 \text{ kPa})$$
$$= 11.25 \text{ kPa}$$

From App. 23.O, this corresponds to approximately 47.6°C. Since the stack temperature is 38°C, some of the water will condense. From App. 23.N, the maximum vapor pressure of water at the stack temperature is 6.632 kPa. Let n be the number of moles of water in the stack gas.

$$n_{H_2O} = \left(\frac{p_{H_2O}}{p}\right)n = \left(\frac{6.632 \text{ kPa}}{101 \text{ kPa}}\right)(35.90 \text{ kmol})$$
$$= \boxed{2.357 \text{ kmol } H_2O/\text{kmol } C_3H_8}$$

(c) The liquid water removed is

$$4 - n_{H_2O} = 4 \text{ kmol} - 2.357 \text{ kmol}$$
$$= 1.643 \text{ kmol of } H_2O/\text{kmol propane}$$
$$m = (1.643 \text{ lbmol})\left(18.016 \frac{\text{kg}}{\text{kmol}}\right)$$
$$= \boxed{29.6 \text{ kg } H_2O/\text{kmol } C_3H_8}$$

19. *Customary U.S. Solution*

(a) If the power output is to be unchanged,

$$BHP_1 = BHP_2$$
$$(\dot{m}_{F,1})(LHV_1) = (\dot{m}_{F,2})(LHV_2)$$

From Eq. 27.8,

$$\dot{m}_F = (BSFC)(BHP)$$
$$(BSFC_1)(LHV_1) = (BSFC_2)(LHV_2)$$
$$\frac{BSFC_2}{BSFC_1} = \frac{LHV_1}{LHV_2} = \frac{23,200 \frac{\text{Btu}}{\text{lbm}}}{11,930 \frac{\text{Btu}}{\text{lbm}}}$$
$$= 1.945$$
$$\frac{BSFC_2 - BSFC_1}{BSFC_1} = \frac{BSFC_2}{BSFC_1} - 1 = 1.945 - 1$$
$$= \boxed{0.945 \quad [94.5\% \text{ increase}]}$$

The answer is (D).

(b) If the fuel injection velocity is to be unchanged $(v_2 = v_1)$,

$$\dot{m} = \rho A v$$
$$A = \frac{\dot{m}}{\rho v}$$
$$\dot{m}_2 = 1.945\dot{m}_1 \quad [\text{part (a)}]$$
$$\frac{A_2 - A_1}{A_1} = \frac{\dfrac{\dot{m}_2}{\rho_2 v_2} - \dfrac{\dot{m}_1}{\rho_1 v_1}}{\dfrac{\dot{m}_1}{\rho_1 v_1}} = \frac{\dfrac{\dot{m}_2}{\rho_2} - \dfrac{\dot{m}_1}{\rho_1}}{\dfrac{\dot{m}_1}{\rho_1}}$$
$$= \frac{\dfrac{1.945 m_1}{\rho_2} - \dfrac{\dot{m}_1}{\rho_1}}{\dfrac{\dot{m}_1}{\rho_1}} = \frac{\dfrac{1.945}{\rho_2} - \dfrac{1}{\rho_1}}{\dfrac{1}{\rho_1}}$$
$$\frac{A_2 - A_1}{A_1} = (1.945)\left(\frac{\rho_1}{\rho_2}\right) - 1$$

From Table 21.3, for gasoline, $SG_1 = 0.74$. For ethanol, $SG_2 = 0.794$.

$$\frac{\rho_1}{\rho_2} = \frac{SG_1}{SG_2} = \frac{0.74}{0.794}$$
$$\frac{A_2 - A_1}{A_1} = (1.945)\left(\frac{0.74}{0.794}\right) - 1$$
$$= \boxed{0.810 \quad [81\% \text{ increase}]}$$

The answer is (C).

(c) If no changes are made to the engine, power output is proportional to the weight flow and heating value.

$$\frac{P_2 - P_1}{P_1} = \frac{\dot{m}_{F2}(LHV_2) - \dot{m}_{F1}(LHV_1)}{\dot{m}_{F1}(LHV_1)}$$
$$\dot{m}_F = \rho A v$$
$$\frac{P_2 - P_1}{P_1} = \frac{(\rho_2 A_2 v_2)(LHV_2) - (\rho_1 A_1 v_1)(LHV_1)}{(\rho_1 A_1 v_1)(LHV_1)}$$

Since no changes are made to the engine, $A_2 = A_1$ and $v_2 = v_1$.

$$\frac{P_2 - P_1}{P_1} = \frac{\rho_2(LHV_2) - \rho_1(LHV_1)}{\rho_1(LHV_1)}$$
$$= \frac{LHV_2 - \left(\dfrac{\rho_1}{\rho_2}\right)(LHV_1)}{\left(\dfrac{\rho_1}{\rho_2}\right)(LHV_1)}$$

From part (b),

$$\frac{\rho_1}{\rho_2} = \frac{0.74}{0.794}$$

$$\frac{P_2 - P_1}{P_1} = \frac{11{,}930 \, \frac{\text{Btu}}{\text{lbm}} - \left(\frac{0.74}{0.794}\right)\left(23{,}200 \, \frac{\text{Btu}}{\text{lbm}}\right)}{\left(\frac{0.74}{0.794}\right)\left(23{,}200 \, \frac{\text{Btu}}{\text{lbm}}\right)}$$

$$= \boxed{-0.45 \quad [45\% \text{ decrease}]}$$

The answer is (B).

SI Solution

(a) From part (a) of the customary U.S. solution,

$$\frac{\text{BSFC}_2}{\text{BSFC}_1} = \frac{\text{LHV}_1}{\text{LHV}_2} = \frac{54 \, \frac{\text{MJ}}{\text{kg}}}{27.7 \, \frac{\text{MJ}}{\text{kg}}} = 1.949$$

$$\frac{\text{BSFC}_2 - \text{BSFC}_1}{\text{BSFC}_1} = \frac{\text{BSFC}_2}{\text{BSFC}_1} - 1 = 1.949 - 1$$

$$= \boxed{0.949 \quad [94.9\% \text{ increase}]}$$

The answer is (D).

(b) From part (b) of the customary U.S. solution,

$$\frac{A_2 - A_1}{A_1} = \frac{\dfrac{\dot{m}_2}{\rho_2} - \dfrac{\dot{m}_1}{\rho_1}}{\dfrac{\dot{m}_1}{\rho_1}}$$

$$\dot{m}_2 = 1.949 \dot{m}_1 \quad [\text{part (a)}]$$

$$\frac{A_2 - A_1}{A_1} = \frac{\dfrac{1.949 \dot{m}_1}{\rho_2} - \dfrac{\dot{m}_1}{\rho_1}}{\dfrac{\dot{m}_1}{\rho_1}}$$

$$= (1.949)\left(\frac{\rho_1}{\rho_2}\right) - 1$$

$$= (1.949)\left(\frac{0.74}{0.794}\right) - 1$$

$$= \boxed{0.816 \quad [81.6\% \text{ increase}]}$$

The answer is (C).

(c) From part (c) of the customary U.S. solution,

$$\frac{P_2 - P_1}{P_1} = \frac{\text{LHV}_2 - \left(\dfrac{\rho_1}{\rho_2}\right)(\text{LHV}_1)}{\left(\dfrac{\rho_1}{\rho_2}\right)(\text{LHV}_1)}$$

$$= \frac{27.7 \, \frac{\text{MJ}}{\text{kg}} - \left(\dfrac{0.74}{0.794}\right)\left(54 \, \frac{\text{MJ}}{\text{kg}}\right)}{\left(\dfrac{0.74}{0.795}\right)\left(54 \, \frac{\text{MJ}}{\text{kg}}\right)}$$

$$= \boxed{-0.45 \quad [45\% \text{ decrease}]}$$

The answer is (B).

PRACTICE PROBLEMS

Energy

1. A solid cast-iron sphere ($\rho = 0.256\,\text{lbm/in}^3$ (7090 kg/m^3)) of 10 in (25 cm) diameter travels without friction at 30 ft/sec (9 m/s) horizontally. What is its kinetic energy?

(A) 900 ft-lbf (12 kJ)

(B) 1200 ft-lbf (1.6 kJ)

(C) 1600 ft-lbf (2.0 kJ)

(D) 1900 ft-lbf (2.4 kJ)

Work

2. What work is done when a balloon carries a 12 lbm (5.2 kg) load to 40,000 ft (12 000 m) height?

(A) 2.4×10^5 ft-lbf (300 kJ)

(B) 4.8×10^5 ft-lbf (610 kJ)

(C) 7.7×10^5 ft-lbf (980 kJ)

(D) 9.9×10^5 ft-lbf (1.3 MJ)

3. Find the compression of a spring if a 100 lbm (50 kg) weight is dropped from 8 ft (2 m) onto a spring with a constant of 33.33 lbf/in (5.837×10^3 N/m).

(A) 27 in (0.67 m)

(B) 34 in (0.85 m)

(C) 39 in (0.90 m)

(D) 45 in (1.1 m)

4. A punch press flywheel operates at 300 rpm with a moment of inertia of 15 slug-ft^2 (20 kg·m^2). Find the speed in rpm to which the wheel will be reduced after a sudden punching requiring 4500 ft-lbf (6100 J) of work.

(A) 160 rpm

(B) 190 rpm

(C) 220 rpm

(D) 310 rpm

5. A force of 550 lbf (2500 N) making a 40° angle (upward) from the horizontal pushes a box 20 ft (6 m) across the floor. What work is done?

(A) 2200 ft-lbf (3.0 kJ)

(B) 3700 ft-lbf (5.2 kJ)

(C) 4200 ft-lbf (6.0 kJ)

(D) 8400 ft-lbf (12 kJ)

6. A 1000 ft long (300 m long) cable has a mass of 2 lbm per foot (3 kg/m) and is suspended from a winding drum down into a vertical shaft. What work must be done to rewind the cable?

(A) 0.50×10^6 ft-lbf (0.6 MJ)

(B) 0.75×10^6 ft-lbf (0.9 MJ)

(C) 1×10^6 ft-lbf (1.3 MJ)

(D) 2×10^6 ft-lbf (2.6 MJ)

Power

7. What volume in ft^3 (m^3) of water can be pumped to a 130 ft (40 m) height in one hour by a 7 hp (5 kW) pump? Assume 85% efficiency.

(A) 1500 ft^3 (40 m^3)

(B) 1800 ft^3 (49 m^3)

(C) 2000 ft^3 (54 m^3)

(D) 2400 ft^3 (65 m^3)

8. What power in horsepower (kW) is required to lift a 3300 lbm (1500 kg) mass 250 ft (80 m) in 14 seconds?

(A) 40 hp (30 kW)

(B) 70 hp (53 kW)

(C) 90 hp (68 kW)

(D) 110 hp (84 kW)

Thermodynamics

SOLUTIONS

1. *Customary U.S. Solution*

$$E_{kinetic} = \tfrac{1}{2}\left(\frac{m}{g_c}\right)v^2 = \tfrac{1}{2}\left(V\left(\frac{\rho}{g_c}\right)\right)v^2$$

$$= \left(\tfrac{1}{2}\right)\left(\tfrac{4}{3}\pi r^3\right)\left(\frac{\rho}{g_c}\right)v^2 = \left(\tfrac{2}{3}\pi\right)\left(\frac{10}{2}\,\text{in}\right)^3\left(\frac{\rho}{g_c}\right)v^2$$

$$= \left(\tfrac{2}{3}\pi\right)\left(\left(\frac{10}{2}\,\text{in}\right)\left(\frac{\text{ft}}{12\,\text{in}}\right)\right)^3$$

$$\times \left(0.256\,\frac{\text{lbm}}{\text{in}^3}\right)\left(1728\,\frac{\text{in}^3}{\text{ft}^3}\right)$$

$$\times \left(30\,\frac{\text{ft}}{\text{sec}}\right)^2\left(\frac{1}{32.2}\,\frac{\text{lbf-sec}^2}{\text{lbm-ft}}\right)$$

$$= \boxed{1873\ \text{ft-lbf}}$$

The answer is (D).

SI Solution

$$E_{kinetic} = \tfrac{1}{2}mv^2 = \tfrac{1}{2}(\rho V)v^2$$

$$= \left(\tfrac{1}{2}\right)\rho\left(\tfrac{4}{3}\pi r^3\right)v^2 = \left(\tfrac{2}{3}\pi\right)\rho\left(\frac{0.25\,\text{m}}{2}\right)^3 v^2$$

$$= \left(\tfrac{2}{3}\pi\right)\left(\frac{0.25\,\text{m}}{2}\right)^3\left(7.09\times 10^3\,\frac{\text{kg}}{\text{m}^3}\right)\left(9\,\frac{\text{m}}{\text{s}}\right)^2$$

$$= 2.35\times 10^3\ \text{J} = \boxed{2.35\ \text{kJ}}$$

The answer is (D).

2. *Customary U.S. Solution*

$$W = \Delta E_{potential} = m\frac{g}{g_c}\Delta h$$

$$= (12\ \text{lbm})\left(\frac{32.2\,\frac{\text{ft}}{\text{sec}^2}}{32.2\,\frac{\text{lbm-ft}}{\text{sec}^2\text{-lbf}}}\right)(40{,}000\ \text{ft})$$

$$= \boxed{4.8\times 10^5\ \text{ft-lbf}}$$

The answer is (B).

SI Solution

$$W = \Delta E_{potential} = mg\Delta h$$

$$= (5.2\ \text{kg})\left(9.81\,\frac{\text{m}}{\text{s}^2}\right)(12\,000\ \text{m})\left(\frac{\text{kJ}}{1000\ \text{J}}\right)$$

$$= \boxed{612.1\ \text{kJ}}$$

The answer is (B).

3. *Customary U.S. Solution*

$$\Delta E_{potential} = \Delta E_{spring}$$
$$W(\Delta h + \Delta x) = \tfrac{1}{2}k(\Delta x)^2$$

Rearranging,

$$\tfrac{1}{2}k(\Delta x)^2 - W\Delta x - W\Delta h = 0$$

$$\left(\tfrac{1}{2}\right)\left(33.33\,\frac{\text{lbf}}{\text{in}}\right)(\Delta x)^2 - (100\ \text{lbf})\Delta x$$

$$- (100\ \text{lbf})(8\ \text{ft})\left(12\,\frac{\text{in}}{\text{ft}}\right) = 0$$

$$16.665\Delta x^2 - 100\Delta x = 9600$$
$$\Delta x^2 - 6\Delta x = 576$$
$$(\Delta x - 3)^2 = 576 + 9$$
$$\Delta x - 3 = \sqrt{585} = \pm 24.2$$
$$\Delta x = \boxed{27.2\ \text{in}}$$

The answer is (A).

SI Solution

$$\Delta E_{potential} = \Delta E_{spring}$$
$$mg(\Delta h + \Delta x) = \tfrac{1}{2}k(\Delta x)^2$$

Rearranging,

$$\tfrac{1}{2}k(\Delta x)^2 - mg\Delta x - mg\Delta h = 0$$

$$\left(\tfrac{1}{2}\right)\left(5.837\times 10^3\,\frac{\text{N}}{\text{m}}\right)(\Delta x)^2 - (50\ \text{kg})\left(9.81\,\frac{\text{m}}{\text{s}^2}\right)\Delta x$$

$$- (50\ \text{kg})\left(9.81\,\frac{\text{m}}{\text{s}^2}\right)(2\ \text{m}) = 0$$

$$2918.5\Delta x^2 - 490.5\Delta x - 981.0 = 0$$
$$\Delta x^2 - 0.1681\Delta x = 0.3361$$
$$(\Delta x - 0.08403)^2 = 0.3361 + (0.08403)^2$$
$$= 0.3432$$
$$\Delta x - 0.08403 = \sqrt{0.3432} = \pm 0.5858$$
$$\Delta x = \boxed{0.6699\ \text{m}}$$

The answer is (A).

4. *Customary U.S. Solution*

$$W_{done\ by\ wheel} = \Delta E_{rotational}$$
$$= \tfrac{1}{2}I\omega_{initial}^2 - \tfrac{1}{2}I\omega_{final}^2$$

$$\omega_{\text{final}} = \sqrt{(\omega_{\text{initial}})^2 - \frac{2W}{I}} = 2\pi f$$

$$f_{\text{final}} = \left(\frac{1}{2\pi}\right)\left(\frac{60 \text{ rpm}}{\text{rps}}\right)$$

$$\times \sqrt{\left((2\pi)(300 \text{ rpm})\left(\frac{\text{rps}}{60 \text{ rpm}}\right)\right)^2 - \frac{(2)(45 \times 10^2 \text{ ft-lbf})}{15 \text{ slug-ft}^2}}$$

$$= \boxed{187.8 \text{ rpm}}$$

The answer is (B).

SI Solution

$$W_{\text{done by wheel}} = \Delta E_{\text{rotational}}$$
$$= \tfrac{1}{2}I\omega_{\text{initial}}^2 - \tfrac{1}{2}I\omega_{\text{final}}^2$$

$$\omega_{\text{final}} = \sqrt{(\omega_{\text{initial}})^2 - \frac{2W}{I}} = 2\pi f$$

$$f_{\text{final}} = \left(\frac{1}{2\pi}\right)\left(\frac{60 \text{ rpm}}{\text{rps}}\right)$$

$$\times \sqrt{\left((2\pi)\left(\frac{300 \text{ rpm}}{60 \frac{\text{s}}{\text{min}}}\right)\right)^2 - \frac{(2)(6.1 \times 10^3 \text{ J})}{20 \text{ kg·m}^2}}$$

$$= \boxed{185.4 \text{ rpm}}$$

The answer is (B).

5. *Customary U.S. Solution*

$$W_{\text{done on box}} = F_x \Delta x = (F)(\cos\theta)\Delta x$$
$$= (550 \text{ lbf})(\cos 40°)(20 \text{ ft})$$
$$= \boxed{8430 \text{ ft-lbf}}$$

The answer is (D).

SI Solution

$$W_{\text{done on box}} = F_x \Delta x = (F)(\cos\theta)\Delta x$$
$$= (2500 \text{ N})(\cos 40°)(6 \text{ m})\left(\frac{\text{kJ}}{1000 \text{ J}}\right)$$
$$= \boxed{11.5 \text{ kJ}}$$

The answer is (D).

6. *Customary U.S. Solution*

$$W_{\text{to retrieve cable}} = \int_0^l F\,dh$$
$$= \int_0^l ((l-h)w)\,dh$$
$$= \tfrac{1}{2}wl^2 = \left(\tfrac{1}{2}\right)\left(2\,\frac{\text{lbf}}{\text{ft}}\right)(1000 \text{ ft})^2$$
$$= \boxed{10^6 \text{ ft-lbf}}$$

The answer is (C).

SI Solution

$$W_{\text{to retrieve cable}} = \int_0^l F\,dh$$
$$= \int_0^l ((l-h)m_l g)\,dh$$
$$= \tfrac{1}{2}m_l g l^2$$
$$= \left(\tfrac{1}{2}\right)\left(3\,\frac{\text{kg}}{\text{m}}\right)\left(9.81\,\frac{\text{m}}{\text{s}^2}\right)(300 \text{ m})^2$$
$$= 1.32 \times 10^6 \text{ J} = \boxed{1.32 \text{ MJ}}$$

The answer is (C).

7. *Customary U.S. Solution*

$$P_{\text{actual}}\Delta t = W_{\text{done by pump}}$$
$$\eta P_{\text{ideal}}\Delta t = \Delta E_{\text{potential}}$$
$$= m\frac{g}{g_c}\Delta h$$
$$= (\rho V)\frac{g}{g_c}\Delta h$$

$$V = \frac{\eta P_{\text{ideal}}\Delta t}{\rho\frac{g}{g_c}\Delta h}$$

$$= \frac{(0.85)(7 \text{ hp})\left(550\,\frac{\text{ft-lbf}}{\text{hp-sec}}\right)(3600 \text{ sec})}{\left(62.4\,\frac{\text{lbm}}{\text{ft}^3}\right)\left(\frac{32.2\,\frac{\text{ft}}{\text{sec}^2}}{32.2\,\frac{\text{lbm-ft}}{\text{sec}^2\text{-lbf}}}\right)(130 \text{ ft})}$$

$$= \boxed{1450 \text{ ft}^3}$$

The answer is (A).

SI Solution

$$P_{\text{actual}}\Delta t = W_{\text{done by pump}}$$
$$\eta P_{\text{ideal}}\Delta t = \Delta E_{\text{potential}}$$
$$= mg\Delta h$$
$$= (\rho V)g\Delta h$$

$$V = \frac{\eta P_{\text{ideal}} \Delta t}{\rho g \Delta h}$$

$$= \frac{(0.85)(5 \times 10^3 \text{ W})(3600 \text{ s})}{\left(1000 \frac{\text{kg}}{\text{m}^3}\right)\left(9.81 \frac{\text{m}}{\text{s}^2}\right)(40 \text{ m})}$$

$$= \boxed{39.0 \text{ m}^3}$$

The answer is (A).

8. *Customary U.S. Solution*

$$P\Delta t = W = m\frac{g}{g_c}\Delta h$$

$$P = \frac{mg\Delta h}{g_c \Delta t}$$

$$= \frac{(3300 \text{ lbm})\left(32.2 \frac{\text{ft}}{\text{sec}^2}\right)(250 \text{ ft})}{\left(32.2 \frac{\text{lbm-ft}}{\text{sec}^2\text{-lbf}}\right)(14 \text{ sec})\left(550 \frac{\text{ft-lbf}}{\text{hp-sec}}\right)}$$

$$= \boxed{107 \text{ hp}}$$

The answer is (D).

SI Solution

$$P\Delta t = W = mg\Delta h$$

$$P = \frac{mg\Delta h}{\Delta t}$$

$$= \left(\frac{(1500 \text{ kg})\left(9.81 \frac{\text{m}}{\text{s}^2}\right)(80 \text{ m})}{14 \text{ s}}\right)\left(\frac{\text{kW}}{1000 \text{ W}}\right)$$

$$= \boxed{84.1 \text{ kW}}$$

The answer is (D).

PRACTICE PROBLEMS

1. What is the molar enthalpy of 250°F (120°C) steam with a quality of 92%?

(A) 16,000 Btu/lbmole (37 MJ/kmol)
(B) 18,000 Btu/lbmole (41 MJ/kmol)
(C) 20,000 Btu/lbmole (46 MJ/kmol)
(D) 22,000 Btu/lbmole (51 MJ/kmol)

2. What is the ratio of specific heats for air at 600°F (300°C)?

(A) 1.33
(B) 1.38
(C) 1.41
(D) 1.67

3. What is the density of helium at 600°F (300°C) and one standard atmosphere?

(A) 0.0052 lbm/ft^3 (0.085 kg/m^3)
(B) 0.0061 lbm/ft^3 (0.098 kg/m^3)
(C) 0.0076 lbm/ft^3 (0.12 kg/m^3)
(D) 0.0095 lbm/ft^3 (0.15 kg/m^3)

4. The following table shows equilibrium data for a partially miscible water (compound 1)-ether (compound 2) system at five different conditions of temperature and pressure. All three phases (i.e., two liquids and a vapor) coexist in equilibrium at the conditions indicated.

(a) Identify the mixture properties (i.e., the temperatures and the mole fractions of ether) that correspond to points A, B, C, D, and E on the following figure for 1.000 atm. (b) Determine Henry's law constant at 60°C.

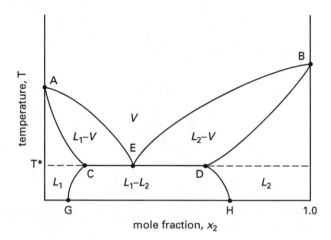

temperature (°C)	total pressure (atm)	x_2 in saturated water phase	x_2 in saturated ether phase	$p_{1,sat}$ (atm)	$p_{2,sat}$ (atm)
34	1.000	0.0123	0.9456	0.053	0.983
40	1.250	0.0116	0.9416	0.073	1.212
50	1.744	0.0103	0.9348	0.121	1.679
60	2.381	0.0093	0.9271	0.196	2.271
70	3.195	0.0075	0.9212	0.306	3.018
80	4.229	0.0069	0.9158	0.467	3.935
90	5.514	0.0058	0.9107	0.691	5.040
100	7.040	–	–	1.000	6.390

SOLUTIONS

1. *Customary U.S. Solution*

From App. 23.A, for 250°F steam, the enthalpy of saturated liquid, h_f, is 218.6 Btu/lbm. The heat of vaporization, h_{fg}, is 945.6 Btu/lbm. The enthalpy is given by Eq. 23.44.

$$h = h_f + xh_{fg}$$
$$= 218.6 \ \frac{\text{Btu}}{\text{lbm}} + (0.92)\left(945.6 \ \frac{\text{Btu}}{\text{lbm}}\right)$$
$$= 1088.6 \ \text{Btu/lbm}$$

The molecular weight of water is 18 lbm/lbmol. The molar enthalpy is given by Eq. 23.18.

$$H = (\text{MW})h$$
$$= \left(18 \ \frac{\text{lbm}}{\text{lbmol}}\right)\left(1088.6 \ \frac{\text{Btu}}{\text{lbm}}\right)$$
$$= \boxed{19{,}595 \ \text{Btu/lbmol}}$$

The answer is (C).

SI Solution

From App. 23.N, for 120°C steam, the enthalpy of saturated liquid, h_f, is 503.71 kJ/kg. The heat of vaporization, h_{fg}, is 2202.6 kJ/kg. The enthalpy is given by Eq. 23.44.

$$h = h_f + xh_{fg}$$
$$= 503.71 \ \frac{\text{kJ}}{\text{kg}} + (0.92)\left(2202.6 \ \frac{\text{kJ}}{\text{kg}}\right)$$
$$= 2530.1 \ \text{kJ/kg}$$

The molecular weight of water is 18 kg/kmol. Molar enthalpy is given by Eq. 23.18.

$$H = (\text{MW})h$$
$$= \left(18 \ \frac{\text{kg}}{\text{kmol}}\right)\left(2530.1 \ \frac{\text{kJ}}{\text{kg}}\right)$$
$$= \boxed{45\,542 \ \text{kJ/kmol}}$$

The answer is (C).

2. *Customary U.S. Solution*

The absolute temperature is

$$600°\text{F} + 460 = 1060°\text{R}$$

From Table 21.1, the specific heat at constant pressure for air at 1060°R is $c_p = 0.250$ Btu/lbm-°R.

From Eq. 23.52, the specific gas constant is

$$R = \frac{R^*}{\text{MW}} = \frac{1545.33 \ \frac{\text{ft-lbf}}{\text{lbmol-°R}}}{28.967 \ \frac{\text{lbm}}{\text{lbmol}}}$$
$$= 53.35 \ \text{ft-lbf/lbm-°R}$$

From Eq. 23.99(b),

$$c_\text{v} = c_p - \frac{R}{J}$$
$$= 0.250 \ \frac{\text{Btu}}{\text{lbm-°R}} - \frac{53.35 \ \frac{\text{ft-lbf}}{\text{lbm-°R}}}{778 \ \frac{\text{ft-lbf}}{\text{Btu}}}$$
$$= 0.1814 \ \text{Btu/lbm-°R}$$

The ratio of specific heats is given by Eq. 23.32.

$$k = \frac{c_p}{c_\text{v}} = \frac{0.250 \ \frac{\text{Btu}}{\text{lbm-°R}}}{0.1814 \ \frac{\text{Btu}}{\text{lbm-°R}}}$$
$$= \boxed{1.378}$$

The answer is (B).

SI Solution

From Table 21.1, the specific heat at constant pressure for air is 0.250 Btu/lbm-°R. From the table footnote, the SI specific heat at constant pressure for air is

$$c_p = \left(0.250 \ \frac{\text{Btu}}{\text{lbm·°R}}\right)\left(4.187 \ \frac{\frac{\text{kJ}}{\text{kg·K}}}{\frac{\text{Btu}}{\text{lbm-°R}}}\right)$$
$$= 1.047 \ \text{kJ/kg·K}$$

From Eq. 23.52, the specific gas constant is

$$R = \frac{R^*}{\text{MW}} = \frac{8314.3 \ \frac{\text{J}}{\text{kmol·K}}}{28.967 \ \frac{\text{kg}}{\text{kmol}}}$$
$$= 287.0 \ \text{J/kg·K}$$

From Eq. 23.99(a),

$$c_v = c_p - R$$
$$= \left(1.047 \; \frac{\text{kJ}}{\text{kg·K}}\right)\left(1000 \; \frac{\text{J}}{\text{kJ}}\right) - 287.0 \; \frac{\text{J}}{\text{kg·K}}$$
$$= 760 \; \text{J/kg·K}$$

The ratio of specific heats is given by Eq. 23.32.

$$k = \frac{c_p}{c_v} = \frac{\left(1.047 \; \frac{\text{kJ}}{\text{kg·K}}\right)\left(1000 \; \frac{\text{J}}{\text{kJ}}\right)}{760 \; \frac{\text{J}}{\text{kg·K}}}$$

$$= \boxed{1.377}$$

The answer is (B).

3. *Customary U.S. Solution*

From Eq. 23.52, the specific gas constant is

$$R = \frac{R^*}{MW} = \frac{1545.33 \; \frac{\text{ft-lbf}}{\text{lbmol-°R}}}{4 \; \frac{\text{lbm}}{\text{lbmol}}}$$
$$= 386.3 \; \text{ft-lbf/lbm-°R}$$

The absolute temperature is

$$600°F + 460 = 1060°R$$

From Eq. 23.54, the density of helium is

$$\rho = \frac{p}{RT}$$
$$= \frac{\left(14.7 \; \frac{\text{lbf}}{\text{in}^2}\right)\left(144 \; \frac{\text{in}^2}{\text{ft}^2}\right)}{\left(386.3 \; \frac{\text{ft-lbf}}{\text{lbm-°R}}\right)(1060°R)}$$

$$= \boxed{0.00517 \; \text{lbm/ft}^3}$$

The answer is (A).

SI Solution

From Eq. 23.52, the specific gas constant is

$$R = \frac{R^*}{MW} = \frac{8314.3 \; \frac{\text{J}}{\text{kmol·K}}}{4 \; \frac{\text{kg}}{\text{kmol}}}$$
$$= 2079 \; \text{J/kg·K}$$

The absolute temperature is

$$300°C + 273 = 573K$$

From Eq. 23.54, the density of helium is

$$\rho = \frac{p}{RT}$$
$$= \frac{1.013 \times 10^5 \; \text{Pa}}{2079 \; \frac{\text{J}}{\text{kg·K}}(573K)}$$

$$= \boxed{0.0850 \; \text{kg/m}^3}$$

The answer is (A).

4. (a) The pressure for this problem (1.000 atm) corresponds to the first line ($T = 34°C$) of the given data, so all three phases co-exist at this temperature. Points C and D can be read directly from the table.

$$\boxed{\begin{array}{l} \text{at C, } T = 34°C \text{ and } x_2 = 0.0123 \\ \text{at D, } T = 34°C \text{ and } x_2 = 0.9456 \end{array}}$$

Point E is identified by equating the vapor mole fraction to the liquid mole fraction. Use Raoult's law for the high-concentration component.

$$x_2 = y_2 = 1 - y_1 = 1 - \frac{p_{1,\text{sat}}(1 - x_2)}{p}$$
$$= 1 - \frac{(0.053 \; \text{atm})(1 - 0.0123)}{1.000 \; \text{atm}}$$
$$= 0.948$$

$$\boxed{\text{at E, } T = 34°C \text{ and } x_2 = 0.948}$$

Points A and B occur at temperatures where the saturation pressure of the components, 1 and 2 respectively, equals the total pressure (1.000 atm).

$$\boxed{\begin{array}{l} \text{at A, } T = 100°C \text{ and } x_2 = 0.000 \\ \text{at B, } T = 34.5°C \text{ and } x_2 = 1.000 \quad \text{[by interpolation]} \end{array}}$$

(b) The Henry's law constant, k, for ether can be determined from the data at 60°C and Eq. 23.4.

$$x_1 + x_2 = 1$$
$$x_1 = \frac{p - k_2}{p_{1,\text{sat}} - k_2}$$
$$x_1 = 1 - x_2 = \frac{p - k_2}{p_{1,\text{sat}} - k_2}$$
$$1 - 0.0093 = \frac{2.381 \; \text{atm} - k_2}{0.196 \; \text{atm} - k_2}$$

$$k_2 = \boxed{235 \; \text{atm}}$$

Thermodynamics

24 Changes in Thermodynamic Properties

PRACTICE PROBLEMS

1. Cast iron is heated from 80°F to 780°F (27°C to 416°C). What heat is required per unit mass?
- (A) 70 Btu/lbm (160 kJ/kg)
- (B) 120 Btu/lbm (280 kJ/kg)
- (C) 170 Btu/lbm (390 kJ/kg)
- (D) 320 Btu/lbm (740 kJ/kg)

2. The ventilation rate in a building is 3×10^5 ft^3/hr (2.4 m^3/s). The air is heated from 35°F to 75°F (2°C to 24°C) by water whose temperature decreases from 180°F to 150°F (82°C to 66°C). What is the water flow rate in gal/min (L/s)?
- (A) 9 gal/min (0.58 L/s)
- (B) 13 gal/min (0.83 L/s)
- (C) 15 gal/min (0.96 L/s)
- (D) 22 gal/min (1.4 L/s)

3. 8.0 ft^3 (0.25 m^3) of 180°F, 14.7 psia (82°C, 101.3 kPa) air are cooled to 100°F (38°C) in a constant-pressure process. What work is done?
- (A) −900 ft-lbf (−1.3 kJ)
- (B) −1100 ft-lbf (−1.5 kJ)
- (C) −1500 ft-lbf (−2.3 kJ)
- (D) −2100 ft-lbf (−3.1 kJ)

4. What is the availability of an isentropic process using steam with an initial quality of 95% and operating between 300 psia and 50 psia (2 MPa and 0.35 MPa)?
- (A) 100 Btu/lbm (230 kJ/kg)
- (B) 130 Btu/lbm (300 kJ/kg)
- (C) 210 Btu/lbm (480 kJ/kg)
- (D) 340 Btu/lbm (780 kJ/kg)

5. (*Time limit: one hour*) A closed air heater receives 540°F, 100 psia (280°C, 700 kPa) air and heats it to 1540°F (840°C). The outside temperature is 100°F (40°C). The pressure of the air drops 20 psi (150 kPa) as it passes through the heater. What is the percentage loss in available energy due to the pressure drop?
- (A) 5%
- (B) 12%
- (C) 18%
- (D) 34%

6. (*Time limit: one hour*) Xenon gas at 20 psia and 70°F (150 kPa and 21°C) is compressed to 3800 psia and 70°F (25 MPa and 21°C) by a compressor/heat exchanger combination. The compressed gas is stored in a 100 ft^3 (3 m^3) rigid tank initially charged with xenon gas at 20 psia (150 kPa).

(a) What is the mass of the xenon gas initially in the tank?
- (A) 35 lbm (18 kg)
- (B) 42 lbm (22 kg)
- (C) 46 lbm (24 kg)
- (D) 51 lbm (27 kg)

(b) What is the average mass flow rate of xenon gas into the tank if the compressor fills the tank in exactly one hour?
- (A) 6300 lbm/hr (0.88 kg/s)
- (B) 9700 lbm/hr (1.3 kg/s)
- (C) 12,000 lbm/hr (1.6 kg/s)
- (D) 14,000 lbm/hr (1.9 kg/s)

(c) If filling takes exactly one hour and electricity costs $0.045 per kW-hr, what is the cost of filling the tank?
- (A) $8
- (B) $14
- (C) $27
- (D) $35

7. (*Time limit: one hour*) The mass of an insulated 20 ft^3 (0.6 m^3) steel tank is 40 lbm (20 kg). The steel has a specific heat of 0.11 Btu/lbm-°R (0.46 kJ/kg·K). The tank is placed in a room where the surrounding air is 70°F and 14.7 psia (21°C and 101.3 kPa). After the tank is evacuated to 1 psia and 70°F (7 kPa and 21°C), a valve is suddenly opened, allowing the tank to fill with room air. The air enters the tank in a well-mixed, turbulent condition. Find the air temperature inside the tank after filling and after the gas and tank have reached thermal equilibrium, but before any heat loss from the tank to the room occurs.
- (A) 73°F (20°C)
- (B) 80°F (30°C)
- (C) 103°F (40°C)
- (D) 190°F (90°C)

SOLUTIONS

1. *Customary U.S. Solution*

From Table 23.2, the approximate value of specific heat for cast iron is $c_p = 0.10$ Btu/lbm-°F.

The heat required per unit mass is

$$q = c_p(T_2 - T_1)$$
$$= \left(0.10\ \frac{\text{Btu}}{\text{lbm-°F}}\right)(780°F - 80°F)$$
$$= \boxed{70\ \text{Btu/lbm}}$$

The answer is (A).

SI Solution

From Table 23.2, the approximate value of specific heat of cast iron is $c_p = 0.42$ kJ/kg·K.

The heat required per unit mass is

$$q = c_p(T_2 - T_1)$$
$$= \left(0.42\ \frac{\text{kJ}}{\text{kg·K}}\right)(416°C - 27°C)$$
$$= \boxed{163.4\ \text{kJ/kg}}$$

The answer is (A).

2. *Customary U.S. Solution*

First calculate the mass flow rate of air to be heated by using the ideal gas law. (Usually the air mass would be evaluated at the entering conditions. This problem is ambiguous. The building conditions are used because a building ventilation rate was specified.)

$$\dot{m}_{\text{air}} = \frac{p\dot{V}}{RT}$$

The absolute temperature is

$$T = 75°F + 460 = 535°R$$
$$\dot{m}_{\text{air}} = \frac{\left(14.7\ \frac{\text{lbf}}{\text{in}^2}\right)\left(144\ \frac{\text{in}^2}{\text{ft}^2}\right)\left(3 \times 10^5\ \frac{\text{ft}^3}{\text{hr}}\right)}{\left(53.3\ \frac{\text{ft-lbf}}{\text{lbm-°R}}\right)(535°R)}$$
$$= 2.227 \times 10^4\ \text{lbm/hr}$$

The heat lost by the water is equal to the heat gained by the air.

$$\dot{m}_w c_{p,w}(T_{1,w} - T_{2,w}) = \dot{m}_{\text{air}} c_{p,\text{air}}(T_{2,\text{air}} - T_{1,\text{air}})$$

$$\dot{m}_w\left(1\ \frac{\text{Btu}}{\text{lbm-°F}}\right)(180°F - 150°F)$$
$$= \left(2.227 \times 10^4\ \frac{\text{lbm}}{\text{hr}}\right)\left(0.241\ \frac{\text{Btu}}{\text{lbm-°F}}\right)(75°F - 35°F)$$

$$\dot{m}_w = 7156.1\ \text{lbm/hr}$$

From App. 32.A, the density of water at 165°F is approximately 61 lbm/ft³. The water volume flow rate is

$$\dot{V}_w = \frac{\dot{m}}{\rho} = \frac{\left(7156.1\ \frac{\text{lbm}}{\text{hr}}\right)\left(\frac{1\ \text{hr}}{3600\ \text{sec}}\right)}{\left(61\ \frac{\text{lbm}}{\text{ft}^3}\right)\left(0.002228\ \frac{\text{ft}^3\text{-min}}{\text{sec-gal}}\right)}$$
$$= \boxed{14.63\ \text{gal/min}}$$

The answer is (C).

SI Solution

First calculate the mass flow rate of air to be heated by using the ideal gas law. (Usually the air mass would be evaluated at the entering conditions. This problem is ambiguous. The building conditions are used because a building ventilation rate was specified.)

$$\dot{m}_{\text{air}} = \frac{p\dot{V}}{RT}$$

The absolute temperature is

$$T = 24°C + 273 = 297\text{K}$$
$$\dot{m}_{\text{air}} = \frac{(1.013 \times 10^5\ \text{Pa})\left(2.4\ \frac{\text{m}^3}{\text{s}}\right)}{\left(287\ \frac{\text{J}}{\text{kg·K}}\right)(297\text{K})}$$
$$= 2.85\ \text{kg/s}$$

The heat lost by the water is equal to the heat gained by the air.

$$\dot{m}_w c_{p,w}(T_{1,w} - T_{2,w}) = \dot{m}_{\text{air}} c_{p,\text{air}}(T_{2,\text{air}} - T_{1,\text{air}})$$

$$\dot{m}_w\left(4.190\ \frac{\text{kJ}}{\text{kg·K}}\right)(82°C - 66°C)$$
$$= \left(2.85\ \frac{\text{kg}}{\text{s}}\right)\left(1.005\ \frac{\text{kJ}}{\text{kg·K}}\right)(24°C - 2°C)$$

$$\dot{m}_w = 0.940\ \text{kg/s}$$

From App. 32.B, the density of water at 74°C is approximately 976 kg/m³. The water volume flow rate is

$$\dot{V}_w = \left(\frac{\dot{m}}{\rho}\right) = \left(0.940 \frac{kg}{s}\right)\left(976 \frac{kg}{m^3}\right)\left(1000 \frac{L}{m^3}\right)$$

$$= \boxed{0.963 \text{ L/s}}$$

The answer is (C).

3. *Customary U.S. Solution*

The mass of air is

$$m = \frac{p_1 V_1}{R T_1}$$

$$= \frac{\left(14.7 \frac{lbf}{in^2}\right)\left(144 \frac{in^2}{ft^2}\right)(8.0 \text{ ft}^3)}{\left(53.3 \frac{ft\text{-}lbf}{lbm\text{-}°R}\right)(180°F + 460)}$$

$$= 0.4964 \text{ lbm}$$

For a constant pressure process from Eq. 24.51, on a per unit mass basis,

$$W = R(T_2 - T_1)$$

The total work for m in lbm is

$$W = mR(T_2 - T_1)$$

$$= (0.4964 \text{ lbm})\left(53.3 \frac{ft\text{-}lbf}{lbm\text{-}°R}\right)(100°F - 180°F)$$

$$= \boxed{-2116.6 \text{ ft-lbf}}$$

This is negative because work is done on the system.

The answer is (D).

SI Solution

The mass of air is

$$m = \frac{p_1 V_1}{R T_1}$$

$$= \frac{(101.3 \text{ kPa})\left(1000 \frac{Pa}{kPa}\right)(0.25 \text{ m}^3)}{\left(287 \frac{J}{kg\cdot K}\right)(82°C + 273)}$$

$$= 0.2486 \text{ kg}$$

For a constant pressure process from Eq. 24.51, on a per unit mass basis,

$$W = R(T_2 - T_1)$$

The total work for m in kg is

$$W = mR(T_2 - T_1)$$

$$= (0.2486 \text{ kg})\left(287 \frac{J}{kg\cdot K}\right)(38°C - 82°C)$$

$$= \boxed{-3139.3 \text{ J}}$$

The answer is (D).

4. *Customary U.S. Solution*

From App. 23.B, for 300 psia, the enthalpy of saturated liquid, h_f, is 394.1 Btu/lbm. The heat of vaporization, h_{fg}, is 809.8 Btu/lbm. The enthalpy is given by Eq. 23.44.

$$h_1 = h_f + x h_{fg}$$

$$= 394.1 \frac{Btu}{lbm} + (0.95)\left(809.8 \frac{Btu}{lbm}\right)$$

$$= 1163.4 \text{ Btu/lbm}$$

From the Mollier diagram, for an isentropic process from 300 psia to 50 psia, $h_2 = 1031$ Btu/lbm.

The availability is calculated from Eq. 24.164 using an isentropic process ($s_1 = s_2$) for unit mass.

$$\text{availability} = h_1 - h_2$$

$$= 1163.4 \frac{Btu}{lbm} - 1031 \frac{Btu}{lbm}$$

$$= \boxed{132.4 \text{ Btu/lbm}}$$

The answer is (B).

SI Solution

From App. 23.0, for 2 MPa, the enthalpy of saturated liquid, h_f, is 908.79 kJ/kg. The heat of vaporization, h_{fg}, is 1890.7 kJ/kg. The enthalpy is given by Eq. 23.44.

$$h_1 = h_f + x h_{fg}$$

$$= 908.79 \frac{kJ}{kg} + (0.95)\left(1890.7 \frac{kJ}{kg}\right)$$

$$= 2705.0 \text{ kJ/kg}$$

From the Mollier diagram, for an isentropic process from 2 MPa to 0.35 MPa, $h_2 = 2405$ kJ/kg.

The availability is calculated from Eq. 24.164 using an isentropic process ($s_1 = s_2$) for unit mass.

$$\text{availability} = h_1 - h_2$$

$$= 2705 \frac{kJ}{kg} - 2405 \frac{kJ}{kg}$$

$$= \boxed{300 \text{ kJ/kg}}$$

The answer is (B).

5. *Customary U.S. Solution*

The absolute temperature at the inlet of the air heater is

$$T_1 = 540°F + 460 = 1000°R$$

The absolute temperature at the outlet of the air heater is

$$T_2 = 1540°F + 460 = 2000°R$$

Since pressures are low and temperatures are high, use an air table.

From App. 23.F at 1000°R,

$$h_1 = 240.98 \text{ Btu/lbm}$$

$$\phi_1 = 0.75042 \text{ Btu/lbm-°R}$$

From App. 23.F at 2000°R,

$$h_2 = 504.71 \text{ Btu/lbm}$$

$$\phi_2 = 0.93205 \text{ Btu/lbm-°R}$$

The availability per unit mass is calculated from Eq. 24.164 using $T_L = 100°F + 460 = 560°R$.

$$W_{\max} = h_1 - h_2 + T_L(s_2 - s_1)$$

For no pressure drop,

$$s_2 - s_1 = \phi_2 - \phi_1$$

$$W_{\max} = h_1 - h_2 + T_L(\phi_2 - \phi_1)$$

$$= 240.98 \frac{\text{Btu}}{\text{lbm}} - 504.71 \frac{\text{Btu}}{\text{lbm}}$$

$$+ (560°R) \left(0.93205 \frac{\text{Btu}}{\text{lbm-°R}} \right.$$

$$\left. -0.75042 \frac{\text{Btu}}{\text{lbm-°R}} \right)$$

$$= -162.02 \text{ Btu/lbm}$$

With a pressure drop from 100 psia to 80 psia, from Eq. 23.43,

$$s_2 - s_1 = \phi_2 - \phi_1 - \left(\frac{R}{J} \right) \ln \left(\frac{p_2}{p_1} \right)$$

$$W_{\max, p \text{ loss}} = h_1 - h_2 + T_L \left(\phi_2 - \phi_1 \right.$$

$$\left. - \left(\frac{R}{J} \right) \ln \left(\frac{p_2}{p_1} \right) \right)$$

$$= 240.98 \frac{\text{Btu}}{\text{lbm}} - 504.71 \frac{\text{Btu}}{\text{lbm}} + (560°R)$$

$$\times \left(0.93205 \frac{\text{Btu}}{\text{lbm-°R}} - 0.75042 \frac{\text{Btu}}{\text{lbm-°R}} \right.$$

$$\left. - \left(\frac{53.3 \frac{\text{ft-lbf}}{\text{lbm-°R}}}{778 \frac{\text{ft-lbf}}{\text{Btu}}} \right) \ln \left(\frac{80 \text{ psia}}{100 \text{ psia}} \right) \right)$$

$$= -153.46 \text{ Btu/lbm}$$

The percentage loss in available energy is

$$\frac{W_{\max} - W_{\max, p \text{ loss}}}{W_{\max}} \times 100\%$$

$$= \frac{-162.02 \frac{\text{Btu}}{\text{lbm}} - \left(-153.46 \frac{\text{Btu}}{\text{lbm}} \right)}{-162.02 \frac{\text{Btu}}{\text{lbm}}} \times 100\%$$

$$= \boxed{5.28\%}$$

The answer is (A).

SI Solution

The absolute temerature at the inlet of the air heater is

$$T_1 = 280°C + 273 = 553K$$

The absolute temperature at the outlet of the air heater is

$$T_2 = 840°C + 273 = 1113K$$

Since pressures are low and temperatures are high, use an air table.

From App. 23.S at 553K,

$$h_1 = 557.9 \text{ kJ/kg}$$

$$\phi_1 = 2.32372 \text{ kJ/kg·K}$$

From App. 23.S at 1113K,

$$h_2 = 1176.2 \text{ kJ/kg}$$

$$\phi_2 = 3.09092 \text{ kJ/kg·K}$$

The availability per unit mass is calculated from Eq. 24.164 using $T_L = 40°C + 273 = 313K$.

$$W_{\max} = h_1 - h_2 + T_L(s_2 - s_1)$$

For no pressure drop,

$$s_2 - s_1 = \phi_2 - \phi_1$$

$$W_{\max} = h_1 - h_2 + T_L(\phi_2 - \phi_1)$$

$$= 557.9 \frac{\text{kJ}}{\text{kg}} - 1176.2 \frac{\text{kJ}}{\text{kg}}$$

$$+ (313K) \left(3.09092 \frac{\text{kJ}}{\text{kg·K}} - 2.32372 \frac{\text{kJ}}{\text{kg·K}} \right)$$

$$= -378.17 \text{ kJ/kg}$$

With a pressure drop from 700 kPa to 550 kPa, from Eq. 23.43,

$$s_2 - s_1 = \phi_2 - \phi_1 - R \ln\left(\frac{p_2}{p_1}\right)$$

$$W_{\text{max},p \text{ loss}} = h_1 - h_2$$

$$+ T_L\left(\phi_2 - \phi_1 - R \ln\left(\frac{p_2}{p_1}\right)\right)$$

$$= 557.9 \frac{\text{kJ}}{\text{kg}} - 1176.2 \frac{\text{kJ}}{\text{kg}} + 313\text{K}$$

$$\times \left(3.09092 \frac{\text{kJ}}{\text{kg·K}} - 2.32372 \frac{\text{kJ}}{\text{kg·K}} \right.$$

$$\left. - \left(\frac{287 \frac{\text{J}}{\text{kg·K}}}{1000 \frac{\text{J}}{\text{kJ}}}\right) \ln\left(\frac{550 \text{ kPa}}{700 \text{ kPa}}\right) \right)$$

$$= -356.50 \text{ kJ/kg}$$

The percentage loss in available energy is

$$\frac{W_{\text{max}} - W_{\text{max},p \text{ loss}}}{W_{\text{max}}} \times 100\%$$

$$= \frac{-378.17 \frac{\text{kJ}}{\text{kg}} - \left(-356.50 \frac{\text{kJ}}{\text{kg}}\right)}{-378.17 \frac{\text{kJ}}{\text{kg}}} \times 100\%$$

$$= \boxed{5.73\%}$$

The answer is (A).

6. *Customary U.S. Solution*

(a) Assume the tank is originally at 70°F. The absolute temperature is

$$T = 70°\text{F} + 460 = 530°\text{R}$$

From Table 23.7, $R = 11.77$ ft-lbf/lbm-°R. From Eq. 23.51,

$$m = \frac{pV}{RT}$$

$$= \frac{\left(20 \frac{\text{lbf}}{\text{in}^2}\right)\left(144 \frac{\text{in}^2}{\text{ft}^2}\right)(100 \text{ ft}^3)}{\left(11.77 \frac{\text{ft-lbf}}{\text{lbm-°R}}\right)(530°\text{R})}$$

$$= \boxed{46.17 \text{ lbm}}$$

The answer is (C).

(b) From Table 23.4, the critical temperature and pressure of xenon are 521.9°R and 58.2 atm, respectively. The reduced variables are

$$T_r = \frac{T}{T_c} = \frac{530°\text{R}}{521.9°\text{R}} = 1.02$$

$$p_r = \frac{p}{p_c} = \frac{3800 \text{ psia}}{(58.2 \text{ atm})\left(14.7 \frac{\text{psia}}{\text{atm}}\right)} = 4.44$$

From App. 23.Z, z is read as 0.61. Using Eq. 23.97,

$$m = \frac{pV}{zRT} = \frac{\left(3800 \frac{\text{lbf}}{\text{in}^2}\right)\left(144 \frac{\text{in}^2}{\text{ft}^2}\right)(100 \text{ ft}^3)}{(0.61)\left(11.77 \frac{\text{ft-lbf}}{\text{lbm-°R}}\right)(530°\text{R})}$$

$$= 14{,}380 \text{ lbm}$$

The average mass flow rate of xenon is

$$\dot{m} = \frac{14{,}380 \text{ lbm} - 46.17 \text{ lbm}}{1 \text{ hr}}$$

$$= \boxed{14{,}334 \text{ lbm/hr}}$$

The answer is (D).

(c) For isothermal compression, the work per unit mass is calculated from Eq. 24.79.

$$W = mRT \ln\left(\frac{p_1}{p_2}\right)$$

$$= \frac{(14{,}334 \text{ lbm})\left(11.77 \frac{\text{ft-lbf}}{\text{lbm-°R}}\right)}{\times (530°\text{R}) \ln\left(\frac{20 \text{ psia}}{3800 \text{ psia}}\right)}{\left(778 \frac{\text{ft-lbf}}{\text{Btu}}\right)\left(3413 \frac{\text{Btu}}{\text{kW·hr}}\right)}$$

$$= -176.7 \text{ kW-hr} \quad (\text{for 1 hr})$$

The cost of electricity is

$$\left(\frac{\$0.045}{\text{kW-hr}}\right)(176.7 \text{ kW-hr}) = \boxed{\$7.95}$$

The answer is (A).

SI Solution

(a) Assume the tank is originally at 21°C. The absolute temperature is

$$T = 21°\text{C} + 273 = 294\text{K}$$

The answer is (C).

From Table 23.7, $R = 63.32$ J/kg·K. From Eq. 23.51,

$$m = \frac{pV}{RT}$$

$$= \frac{(150 \text{ kPa})\left(1000 \ \dfrac{\text{Pa}}{\text{kPa}}\right)(3 \text{ m}^3)}{\left(63.32 \ \dfrac{\text{J}}{\text{kg·K}}\right)(294\text{K})}$$

$$= \boxed{24.17 \text{ kg}}$$

(b) From Table 23.4, the critical temperature and pressure of xenon are 289.9K and 58.2 atm, respectively. The reduced variables are

$$T_r = \frac{T}{T_c} = \frac{294\text{K}}{289.9\text{K}} = 1.01$$

$$p_r = \frac{p}{p_c} = \frac{25 \text{ MPa}}{(58.2 \text{ atm})\left(0.1013 \ \dfrac{\text{MPa}}{\text{atm}}\right)} = 4.24$$

From App. 23.Z, z is read as 0.59. Using Eq. 23.97,

$$m = \frac{pV}{zRT}$$

$$= \frac{(25 \text{ MPa})\left(10^6 \ \dfrac{\text{Pa}}{\text{MPa}}\right)(3 \text{ m}^3)}{(0.59)\left(63.32 \ \dfrac{\text{J}}{\text{kg·K}}\right)(294\text{K})}$$

$$= 6.828 \text{ kg}$$

The average mass flow rate of xenon is

$$\dot{m} = \frac{6828 \text{ kg} - 24.17 \text{ kg}}{(1 \text{ h})\left(3600 \ \dfrac{\text{s}}{\text{h}}\right)}$$

$$= \boxed{1.89 \text{ kg/s}}$$

The answer is (D).

(c) For isothermal compression, the work per unit mass is calculated from Eq. 24.83.

$$W = mRT \ln\left(\frac{p_1}{p_2}\right)$$

$$= (6828 \text{ kg} - 24.17 \text{ kg})\left(63.32 \ \dfrac{\text{J}}{\text{kg·K}}\right)(294\text{K})$$

$$\times \ln\left(\frac{150 \text{ kPa}}{(25 \text{ MPa})\left(1000 \ \dfrac{\text{kPa}}{\text{MPa}}\right)}\right)$$

$$\times \left(\frac{1 \text{ kJ}}{1000 \text{ J}}\right)\left(\frac{1 \text{ h}}{3600 \text{ s}}\right)$$

$$= -180 \text{ kW·h}$$

The cost of electricity is

$$\left(\frac{\$0.045}{\text{kW·h}}\right)(180 \text{ kW·h}) = \boxed{\$8.10}$$

The answer is (A).

7. Choose the control volume to include the air outside the tank that is pushed into the tank (subscript "e" for "entering"), as well as the tank volume.

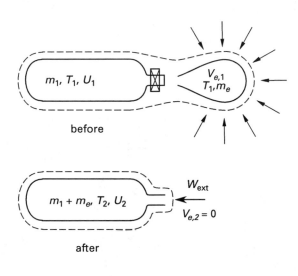

before

after

Customary U.S. Solution

The absolute temperature of the air in the tank when evacuated is

$$T_1 = 70°\text{F} + 460 = 530°\text{R}$$

From Table 23.7, $R = 53.3$ ft-lbf/lbm-°R. From Eq. 23.51,

$$m = \frac{p_1 V_1}{RT_1}$$

$$= \frac{\left(1 \ \dfrac{\text{lbf}}{\text{in}^2}\right)\left(144 \ \dfrac{\text{in}^2}{\text{ft}^2}\right)(20 \text{ ft}^3)}{\left(53.3 \ \dfrac{\text{ft-lbf}}{\text{lbm-°R}}\right)(530°\text{R})}$$

$$= 0.102 \text{ lbm}$$

Assume $T_2 = 300°$F. The absolute temperature is

$$T_2 = 300°\text{F} + 460 = 760°\text{R}$$

From Eq. 23.51,

$$m_2 = m_1 + m_e$$

$$= \frac{p_2 V_2}{R T_2}$$

$$= \frac{\left(14.7 \ \frac{\text{lbf}}{\text{in}^2}\right)\left(144 \ \frac{\text{in}^2}{\text{ft}^2}\right)(20 \ \text{ft}^3)}{\left(53.3 \ \frac{\text{ft-lbf}}{\text{lbm-}^\circ\text{R}}\right)(760^\circ\text{R})}$$

$$m_1 + m_e \approx 1.045 \ \text{lbm}$$

$$m_e \approx 1.045 \ \text{lbm} - m_1$$

$$= 1.045 \ \text{lbm} - 0.102 \ \text{lbm}$$

$$= 0.943 \ \text{lbm}$$

From Eq. 23.51, the initial volume of the external air is

$$V_{e,1} = \frac{mRT}{p}$$

$$= \frac{(0.943 \ \text{lbm})\left(53.3 \ \frac{\text{ft-lbf}}{\text{lbm-}^\circ\text{R}}\right)(530^\circ\text{R})}{\left(14.7 \ \frac{\text{lbf}}{\text{in}^2}\right)\left(144 \ \frac{\text{in}^2}{\text{ft}^2}\right)}$$

$$= 12.58 \ \text{ft}^3$$

From Eq. 24.23, for a closed system,

$$Q = \Delta U + W$$

For an adiabatic system, $Q = 0$.

$$W_{\text{ext}} = \Delta U$$

For a constant pressure, closed system, from Eq. 24.50, the total work is

$$W_{\text{ext}} = p(V_{e,2} - V_{e,1})$$

$$= \frac{\left(14.7 \ \frac{\text{lbf}}{\text{in}^2}\right)\left(144 \ \frac{\text{in}^2}{\text{ft}^2}\right)(0 - 12.58 \ \text{ft}^3)}{778 \ \frac{\text{ft-lbf}}{\text{Btu}}}$$

$$= -34.23 \ \text{Btu} \quad \left[\begin{array}{c}\text{surroundings do work}\\\text{on the system}\end{array}\right]$$

This energy is used to raise the temperature of the air and tank. Consider air as an ideal gas. The process inside the tank is not constant-pressure, as was the compression of the external air mass. However, the process is adiabatic, so $\Delta U = -W$. From Eq. 24.37, $\Delta U = c_v \Delta T$.

$$W_{\text{ext}} = \left((m_1 + m_e)c_v\right)(T_2 - T_1)$$

$$34.23 \ \text{Btu} = \left((1.045 \ \text{lbm})\left(0.171 \ \frac{\text{Btu}}{\text{lbm-}^\circ\text{F}}\right)\right.$$

$$\left. \times (T_2 - 70^\circ\text{F})\right)$$

$$T_2 = 261.6^\circ\text{F}$$

(A second iteration may be required.)

The answer is (B).

Now, allow the 261.6°F air and the 70°F tank to reach thermal equilibrium. The process will be constant-pressure.

$$q_{\text{air}} = q_{\text{tank}}$$

$$c_{p,\text{air}} m_{\text{air}}(T_{\text{air}} - T_{\text{equilibrium}})$$

$$= c_{p,\text{tank}} m_{\text{tank}}(T_{\text{equilibrium}} - T_1)$$

$$\left(0.24 \ \frac{\text{Btu}}{\text{lbm-}^\circ\text{F}}\right)(1.045 \ \text{lbm})(261.6^\circ\text{F} - T_{\text{equilibrium}})$$

$$= \left(0.11 \ \frac{\text{Btu}}{\text{lbm-}^\circ\text{F}}\right)(40 \ \text{lbm})(T_{\text{equilibrium}} - 70^\circ\text{F})$$

$$T_{\text{equilibrium}} = \boxed{80.3^\circ\text{F}}$$

SI Solution

The absolute temperature of the air in the tank when evacuated is

$$T_1 = 21^\circ\text{C} + 273 = 294\text{K}$$

From Table 23.7, $R = 287 \ \text{J/kg·K}$. From Eq. 23.51,

$$m = \frac{p_1 V_1}{R T_1}$$

$$= \frac{(7 \ \text{kPa})\left(1000 \ \frac{\text{Pa}}{\text{kPa}}\right)(0.6 \ \text{m}^3)}{\left(287 \ \frac{\text{J}}{\text{kg·K}}\right)(294\text{K})}$$

$$= 0.0498 \ \text{kg}$$

Assume $T_2 = 127^\circ\text{C}$. The absolute temperature is

$$T_2 = 127^\circ\text{C} + 273 = 400\text{K}$$

From Eq. 23.51,

$$m_2 = m_1 + m_e = \frac{p_2 V_2}{R T_2}$$

$$= \frac{(101.3 \ \text{kPa})\left(1000 \ \frac{\text{Pa}}{\text{kPa}}\right)(0.6 \ \text{m}^3)}{\left(287 \ \frac{\text{J}}{\text{kg·K}}\right)(400\text{K})}$$

$$= 0.5294 \ \text{kg}$$

$$m_e \approx 0.5294 \ \text{kg} - m_1$$

$$= 0.5294 \ \text{kg} - 0.0498 \ \text{kg}$$

$$= 0.4796 \ \text{kg}$$

From Eq. 23.51, the initial volume of the external air is

$$V_{e,1} = \frac{mRT}{p}$$

$$= \frac{(0.4796 \ \text{kg})\left(287 \ \frac{\text{J}}{\text{kg·K}}\right)(294\text{K})}{(101.3 \ \text{kPa})\left(1000 \ \frac{\text{Pa}}{\text{kPa}}\right)}$$

$$= 0.3995 \ \text{m}^3$$

From Eq. 24.23, for a closed system,

$$Q = \Delta U + W$$

For an adiabatic system, $Q = 0$.

$$W_{\text{ext}} = \Delta U$$

For a constant pressure, closed system, from Eq. 24.50, the total work is

$$W_{\text{ext}} = p(V_{e,2} - V_{e,1})$$

$$= (101.3 \text{ kPa}) \left(1000 \, \frac{\text{Pa}}{\text{kPa}} \right) (0 - 0.3995 \text{ m}^3)$$

$$= -40\,469 \text{ J} \quad \begin{bmatrix} \text{surroundings do work} \\ \text{on the system} \end{bmatrix}$$

This energy is used to raise the temperature of the air and tank. Consider air as an ideal gas. The process inside the tank is not constant-pressure, as was the compression of the external air mass. However, the process is adiabatic, so $\Delta U = -W$. From Eq. 24.37, $\Delta U = c_v \Delta T$.

$$W_{\text{ext}} = \big((m_1 + m_e)c_v\big)(T_2 - T_1)$$

$$40\,469 \text{ J} = \left((0.5294 \text{ kg}) \left(718 \, \frac{\text{J}}{\text{kg·K}} \right) (T_2 - 21°\text{C}) \right)$$

$$T_2 = 127.5°\text{C}$$

This is close enough to the assumed value of T_2 that a second iteration is not necessary.

Now, allow the 127.5°C air and the 21°C tank to reach thermal equilibrium. The process will be constant-pressure.

$$q_{\text{air}} = q_{\text{tank}}$$

$$c_{p,\text{air}} m_{\text{air}} (T_{\text{air}} - T_{\text{equilibrium}})$$

$$= c_{p,\text{tank}} m_{\text{tank}} (T_{\text{equilibrium}} - T_1)$$

$$\left(1005 \, \frac{\text{J}}{\text{kg·K}} \right) (0.5294 \text{ kg})(127.5°\text{C} - T_{\text{equilibrium}})$$

$$= \left(460 \, \frac{\text{J}}{\text{kg·K}} \right) (20 \text{ kg})(T_{\text{equilibrium}} - 21°\text{C})$$

$$T_{\text{equilibrium}} = \boxed{26.8°\text{C}}$$

The answer is (B).

25 Vapor Power Cycle Equipment

PRACTICE PROBLEMS

1. What is the isentropic efficiency of a process that expands dry steam from 100 psia to 3 psia (700 kPa to 20 kPa) and 90% quality?

- (A) 40%
- (B) 50%
- (C) 60%
- (D) 70%

2. A 5000 kW steam turbine uses 200 psia (1.5 MPa) steam with 100°F (50°C) of superheat. The condenser is at 1 in Hg (3.4 kPa) absolute.

(a) What is the water rate at full load?

- (A) 4.7 lbm/kW-hr (0.0006 kg/kW·s)
- (B) 8.8 lbm/kW-hr (0.0011 kg/kW·s)
- (C) 25 lbm/kW-hr (0.0031 kg/kW·s)
- (D) 37 lbm/kW-hr (0.0046 kg/kW·s)

(b) If the actual load is only 2500 kW and the steam is throttled to reduce the availability, what is the loss in available energy per unit mass of steam?

- (A) 80 Btu/lbm (190 kJ/kg)
- (B) 130 Btu/lbm (310 kJ/kg)
- (C) 190 Btu/lbm (460 kJ/kg)
- (D) 370 Btu/lbm (890 kJ/kg)

3. A 10,000 kW steam turbine operates on 400 psia (3.0 MPa), 750°F (420°C) dry steam, expanding to 2 in Hg (6.8 kPa) absolute. What is the adiabatic heat drop available for power production?

- (A) 450 Btu/lbm (1100 kJ/kg)
- (B) 600 Btu/lbm (1400 kJ/kg)
- (C) 750 Btu/lbm (1800 kJ/kg)
- (D) 900 Btu/lbm (2200 kJ/kg)

4. A 750 kW steam turbine has a water rate of 20 lbm/kW-hr (2.5×10^{-3} kg/kW·s). Steam with 50°F (30°C) of superheat is expanded from 165 psia (1.0 MPa absolute) to 26 in Hg (90 kPa) absolute. 65°F (18°C) cooling water is available. The terminal temperature difference is zero. Find the quantity of cooling water required.

- (A) 45,000 lbm/hr (5.9 kg/s)
- (B) 60,000 lbm/hr (7.8 kg/s)
- (C) 80,000 lbm/hr (10 kg/s)
- (D) 95,000 lbm/hr (12 kg/s)

5. 332,000 lbm/hr (41.8 kg/s) of 81°F (27°C) water enters a two-pass, counterflow, closed feedwater heater. The feedwater heater is constructed of 1850 ft² (170 m²) of ⅝ in (15.8 mm) copper tubing. Saturated steam is bled from a turbine at 4.45 psia (30 kPa) and condenses to saturated liquid. The heated water leaves at 150°F (65°C) with an enthalpy of 1100 Btu/lbm (2.56 MJ/kg).

(a) What is the overall heat transfer coefficient?

- (A) 1900 Btu/hr-ft²-°F (16 kW/m²·°C)
- (B) 4400 Btu/hr-ft²-°F (37 kW/m²·°C)
- (C) 6500 Btu/hr-ft²-°F (55 kW/m²·°C)
- (D) 7700 Btu/hr-ft²-°F (65 kW/m²·°C)

(b) What is the steam extraction rate?

- (A) 1.7×10^8 Btu/hr (48 MW)
- (B) 3.9×10^8 Btu/hr (110 MW)
- (C) 9.9×10^8 Btu/hr (270 MW)
- (D) 2.5×10^9 Btu/hr (700 MW)

6. A two-pass surface condenser constructed of 1 in (25.4 mm) BWG tubing receives 82,000 lbm/hr (10.3 kg/s) of steam from a turbine. Steam enters the condenser with an enthalpy of 980 Btu/lbm (2.280 MJ/kg). The condenser operates at a pressure of 1 in Hg (3.4 kPa) absolute. Water is circulated at 8 fps (2.4 m/s) through an equivalent length of 120 ft (36 m) of extra strong 30 in (76.2 cm) steel pipe. An additional head loss of 6 in wg (1.5 kPa) is incurred in the intake screens.

(a) What is the head added by the circulating water pump?

- (A) 2.1 ft of water (6.1 kPa)
- (B) 3.2 ft of water (9.3 kPa)
- (C) 6.6 ft of water (19 kPa)
- (D) 9.0 ft of water (26 kPa)

(b) If the water temperature increases 10°F (5.6°C), what is the circulation rate of cooling water?

- (A) 9000 gal/min (34 kL/min)
- (B) 11,000 gal/min (42 kL/min)
- (C) 13,000 gal/min (49 kL/min)
- (D) 15,000 gal/min (57 kL/min)

7. 100 lbm/hr (0.013 kg/s) of 60°F (16°C) water is turned into 14.7 psia (101.3 kPa) saturated steam in an electric boiler. Radiation losses are 35% of the supplied energy. What is the cost if electricity is $0.04 per kW-hr?

    (A) $2/hr
    (B) $3/hr
    (C) $4/hr
    (D) $5/hr

8. A gas burner produces 250 lbm/hr (0.032 kg/s) of 98% dry steam at 40 psia (300 kPa) from 60°F (16°C) feedwater. The fuel gas enters at 80°F (26°C) and 4 in Hg (13.6 kPa) and has a heating value of 550 Btu/ft^3 (20.5 MJ/m^3) at standard industrial conditions. The barometric pressure is 30.2 in Hg (102.4 kPa). 13.5 ft^3/min (6.4 L/s) of fuel gas is consumed. What is the efficiency of the boiler?

    (A) 37%
    (B) 43%
    (C) 57%
    (D) 66%

9. A boiler evaporates 8.23 lbm (8.23 kg) of 120°F (50°C) water per pound (per kilogram) of coal fired, producing 100 psia (700 kPa) saturated steam. The coal is 2% moisture by weight as fired, and dry coal is 5% ash. 12% of the coal is removed from the ash pit. (Ash has the same composition as unfired, dry coal.) Coal is initially at 60°F (16°C), and combustion occurs at 14.7 psia (101.3 kPa). The combustion products leave at 600°F (315°C). The heating value of the coal is 12,800 Btu/lbm (29.80 MJ/kg). What is the efficiency of the boiler?

    (A) 53%
    (B) 68%
    (C) 74%
    (D) 82%

10. 500 psia (3.5 MPa) steam is superheated to 1000°F (500°C) before expanding through a 75% efficient turbine to 5 psia (30 kPa). No subcooling occurs. The pump work is negligible compared to the 200 MW generated.

(a) What quantity of steam is required?

    (A) 8.6×10^5 lbm/hr (120 kg/s)
    (B) 1.4×10^6 lbm/hr (190 kg/s)
    (C) 2.0×10^6 lbm/hr (270 kg/s)
    (D) 4.2×10^6 lbm/hr (310 kg/s)

(b) What heat is removed by the condenser?

    (A) 3.2×10^8 Btu/hr (99 MW)
    (B) 8.7×10^8 Btu/hr (270 MW)
    (C) 2.1×10^9 Btu/hr (650 MW)
    (D) 8.7×10^9 Btu/hr (2.7 GW)

11. 191,000 lbm/hr (24 kg/s) of 635°F (335°C) combustion gases flow through a 20 ft (6 m) wide boiler stack whose front and back plates are 5 ft-10 in (1.78 m) apart. An integral crossflow economizer is being designed to heat water from 212°F to 285°F (100°C to 140°C) by dropping the stack gas temperature to 470°F (240°C). Layers of 24 tubes with dimensions 0.957 in (24.3 mm) I.D., 1.315 in (33.4 mm) O.D., and 20 ft (6 m) length will be placed on a 2.315 in (58.8 mm) pitch in horizontal banks. The overall coefficient of heat transfer for the tubes is 10 Btu/hr-ft^2-°F (57 W/m^2·°C). How many 24-tube layers are required?

    (A) 5
    (B) 9
    (C) 12
    (D) 17

12. (*Time limit: one hour*) Water is used in an adiabatic steam desuperheater. 1000 lbm/hr (0.13 kg/s) of 200 psia (1.5 MPa), 600°F (300°C) steam enters with negligible velocity. 50 lbm/hr (0.0063 kg/s) of 82°F (28°C) water enters with negligible velocity. 100 psia (700 kPa) steam leaves the desuperheater at 2000 ft/sec (600 m/s).

(a) What is the temperature of the leaving steam?

    (A) 330°F (165°C)
    (B) 360°F (180°C)
    (C) 400°F (200°C)
    (D) 470°F (240°C)

(b) What is the quality of the leaving steam?

    (A) 85%
    (B) 88%
    (C) 93%
    (D) 99%

13. (*Time limit: one hour*) Waste steam at 400°F (200°C) and 100 psia (700 kPa) was originally used only for heating cold water. Cold water entered at 70°F (21°C) and 60 psia (400 kPa). 2000 lbm/hr (0.25 kg/s) of hot water at 180°F (80°C) and 20 psia (150 kPa) were produced. Now, the same quantity of steam is to be expanded through a low-pressure turbine. The low-pressure turbine has an isentropic efficiency of 60% and a mechanical efficiency of 96%. The steam will then flow through a mixing heater (see diagram). A pressure drop of 5 psi (30 kPa) occurs through the heater. The heater output must remain at 180°F (80°C) and 20 psia (350 kPa), but the output at point F may decrease.

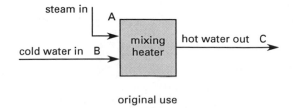

original use

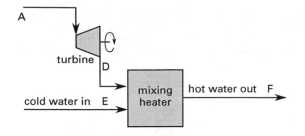

proposed use of waste steam

(a) What power is developed in the turbine?
- (A) 4.7 hp (3.8 kW)
- (B) 8.2 hp (6.3 kW)
- (C) 14 hp (11 kW)
- (D) 27 hp (21 kW)

(b) What is the flow rate at point F?
- (A) 870 lbm/hr (0.10 kg/s)
- (B) 1200 lbm/hr (0.14 kg/s)
- (C) 1900 lbm/hr (0.23 kg/s)
- (D) 3400 lbm/hr (0.41 kg/s)

SOLUTIONS

1. *Customary U.S. Solution*

From App. 23.B, for 100 psia, the enthalpy of dry steam is $h_i = 1187.8$ Btu/lbm.

From the Mollier diagram for an isentropic process from 100 psia to 3 psia, $h_2 = 950$ Btu/lbm.

From App. 23.B, for 3 psia,

$$h_f = 109.39 \text{ Btu/lbm}$$
$$h_{fg} = 1013.1 \text{ Btu/lbm}$$
$$h_2' = h_f + xh_{fg}$$
$$= 109.39 \ \frac{\text{Btu}}{\text{lbm}} + (0.9)\left(1013.1 \ \frac{\text{Btu}}{\text{lbm}}\right)$$
$$= 1021.2 \text{ Btu/lbm}$$

From Eq. 25.17, the isentropic efficiency is

$$\eta_s = \frac{h_1 - h_2'}{h_1 - h_2} = \frac{1187.8 \ \dfrac{\text{Btu}}{\text{lbm}} - 1021.2 \ \dfrac{\text{Btu}}{\text{lbm}}}{1187.8 \ \dfrac{\text{Btu}}{\text{lbm}} - 950 \ \dfrac{\text{Btu}}{\text{lbm}}}$$

$$= \boxed{0.701 \quad (70.1\%)}$$

The answer is (D).

SI Solution

From App. 23.O, for 700 kPa, the enthalpy of dry steam is $h_1 = 2763.5$ kJ/kg.

From the Mollier diagram for an isentropic process from 700 kPa to 20 kPa, $h_2 = 2245$ kJ/kg.

From App. 23.O, for 20 kPa,

$$h_f = 251.4 \text{ kJ/kg}$$
$$h_{fg} = 2358.3 \text{ kJ/kg}$$
$$h_2' = h_f + xh_{fg}$$
$$= 251.4 \ \frac{\text{kJ}}{\text{kg}} + (0.9)\left(2358.3 \ \frac{\text{kJ}}{\text{kg}}\right)$$
$$= 2373.9 \text{ kJ/kg}$$

From Eq. 21.17, the isentropic efficiency is

$$\eta_s = \frac{h_1 - h_2'}{h_1 - h_2}$$

$$= \frac{2763.5 \ \dfrac{\text{kJ}}{\text{kg}} - 2373.9 \ \dfrac{\text{kJ}}{\text{kg}}}{2763.5 \ \dfrac{\text{kJ}}{\text{kg}} - 2245 \ \dfrac{\text{kJ}}{\text{kg}}}$$

$$= \boxed{0.751 \quad (75.1\%)}$$

The answer is (D).

Thermodynamics

2. *Customary U.S. Solution*

(a) From App. 23.B, T_{sat} for 200 psia is 381.86°F.

The steam temperature is

$$381.86°F + 100°F = 481.86°F$$

From App. 23.C, $h_1 = 1258.2$ Btu/lbm.

1 in Hg is approximately 0.5 psia. From the Mollier diagram, assuming isentropic expansion, dropping straight down to the 0.5 psia line, $h_2 \approx 870$ Btu/lbm.

For isentropic expansion, $\eta_{turbine} = 1$, and the steam mass flow rate through the turbine is given by Eq. 25.22.

$$\dot{m} = \frac{P_{turbine}}{h_1 - h_2}$$

$$= \frac{(5000 \text{ kW})\left(3413 \dfrac{\text{Btu}}{\text{hr-kW}}\right)}{1258.2 \dfrac{\text{Btu}}{\text{lbm}} - 870 \dfrac{\text{Btu}}{\text{lbm}}}$$

$$= 4.396 \times 10^4 \text{ lbm/hr}$$

The water rate is

$$\text{WR} = \frac{\dot{m}}{P_{turbine}} = \frac{4.396 \times 10^4 \dfrac{\text{lbm}}{\text{hr}}}{5000 \text{ kW}}$$

$$= \boxed{8.792 \text{ lbm/kW-hr}}$$

The answer is (B).

(b) The loss in available energy per unit mass is

$$\text{loss} = \left(\tfrac{1}{2}\right)(h_1 - h_2)$$

$$= \left(\tfrac{1}{2}\right)\left(1258.2 \dfrac{\text{Btu}}{\text{lbm}} - 870 \dfrac{\text{Btu}}{\text{lbm}}\right)$$

$$= \boxed{194.1 \text{ Btu/lbm}}$$

The answer is (C).

SI Solution

(a) From App. 23.0, T_{sat} for 1.5 MPa is 198.3°C.

The steam temperature is

$$198.3°C + 50°C = 248.3°C$$

From App. 23.P, $h_1 = 2919$ kJ/kg and $s_1 = 6.7000$ kJ/kg·K.

From App. 23.N, for 3.4 kPa, the entropy of saturated liquid, s_f, the entropy of saturated vapor, s_g, the enthalpy of saturated liquid, h_f, and the enthalpy of vaporization, h_{fg}, are

$$s_f = 0.3839 \text{ kJ/kg·K}$$

$$s_g = 8.5329 \text{ kJ/kg·K}$$

$$h_f = 109.82 \text{ kJ/kg}$$

$$h_{fg} = 2439.5 \text{ kJ/kg}$$

For isentropic expansion, $s_1 = s_2 = 6.7000$ kJ/kg·K.

Since $s_2 < s_g$, the expanded steam is in the liquid-vapor region. The quality of the mixture is given by Eq. 23.45.

$$x = \frac{s - s_f}{s_{fg}} = \frac{s - s_f}{s_g - s_f}$$

$$= \frac{6.7000 \dfrac{\text{kJ}}{\text{kg·K}} - 0.3839 \dfrac{\text{kJ}}{\text{kg·K}}}{8.5329 \dfrac{\text{kJ}}{\text{kg·K}} - 0.3839 \dfrac{\text{kJ}}{\text{kg·K}}}$$

$$= 0.7751$$

The final enthalpy is given by Eq. 23.44.

$$h = h_f + x h_{fg}$$

$$= 109.82 \dfrac{\text{kJ}}{\text{kg}} + (0.7751)\left(2439.5 \dfrac{\text{kJ}}{\text{kg}}\right)$$

$$= 2000.7 \text{ kJ/kg}$$

For isentropic expansion, $\eta_{turbine} = 1$, and the steam mass flow rate through a turbine is given by Eq. 25.22.

$$\dot{m} = \frac{P_{turbine}}{h_1 - h_2}$$

$$= \frac{5000 \text{ kW}}{2919 \dfrac{\text{kJ}}{\text{kg}} - 2000.7 \dfrac{\text{kJ}}{\text{kg}}}$$

$$= 5.445 \text{ kg/s}$$

The water rate is

$$\text{WR} = \frac{\dot{m}}{P_{turbine}} = \frac{5.445 \dfrac{\text{kg}}{\text{s}}}{5000 \text{ kW}}$$

$$= \boxed{1.09 \times 10^{-3} \text{ kg/kW·s}}$$

The answer is (B).

(b) Loss in availability per unit mass is

$$\text{loss} = \left(\tfrac{1}{2}\right)(h_1 - h_2)$$

$$= \left(\tfrac{1}{2}\right)\left(2919 \dfrac{\text{kJ}}{\text{kg}} - 2000.7 \dfrac{\text{kJ}}{\text{kg}}\right)$$

$$= \boxed{459.2 \text{ kJ/kg}}$$

The answer is (C).

3. *Customary U.S. Solution*

From App. 23.C, the enthalpy, h_1, of dry steam at 400 psia and 750°F is $h_1 = 1389.6$ Btu/lbm.

From the Mollier diagram, assuming isentropic expansion to 2 in Hg (about 1 psia), $h_2 = 935$ Btu/lbm.

The adiabatic heat drop is

$$h_1 - h_2 = 1389.6 \ \frac{\text{Btu}}{\text{lbm}} - 935 \ \frac{\text{Btu}}{\text{lbm}}$$

$$= \boxed{454.6 \ \text{Btu/lbm}}$$

The answer is (A).

SI Solution

From App. 23.P, the enthalpy, h_1, of dry steam at 3.0 MPa and 420°C is $h_1 = 3276.3$ kJ/kg.

From the Mollier diagram, assuming isentropic expansion, $h_2 \approx 2210$ kJ/kg.

The adiabatic heat drop is

$$h_1 - h_2 = 3276.3 \ \frac{\text{kJ}}{\text{kg}} - 2210 \ \frac{\text{kJ}}{\text{kg}}$$

$$= \boxed{1066.3 \ \text{kJ/kg}}$$

The answer is (A).

4. *Customary U.S. Solution*

The water rate is

$$\text{WR} = \frac{\dot{m}}{P_{\text{turbine}}}$$

The steam flow rate is

$$\dot{m}_{\text{steam}} = (\text{WR})(P_{\text{turbine}})$$

$$= \left(20 \ \frac{\text{lbm}}{\text{kW-hr}}\right)(750 \ \text{kW})$$

$$= 15{,}000 \ \text{lbm/hr}$$

From App. 23.B, for 165 psia, $T_{\text{sat}} = 365.9°\text{F}$.

The steam temperature is

$$365.9°\text{F} + 50°\text{F} = 415.9°\text{F}$$

From App. 23.C, $h_1 = 1225.7$ Btu/lbm.

From Eq. 25.22,

$$P_{\text{turbine}} = \dot{m}(h_1 - h_2)$$

$$h_1 - h_2 = \frac{P_{\text{turbine}}}{\dot{m}} = \frac{(750 \ \text{kW})\left(3413 \ \frac{\text{Btu}}{\text{kW-hr}}\right)}{15{,}000 \ \frac{\text{lbm}}{\text{hr}}}$$

$$= 170.6 \ \text{Btu/lbm}$$

$$h_2 = h_1 - 170.6 \ \frac{\text{Btu}}{\text{lbm}}$$

$$= 1225.7 \ \frac{\text{Btu}}{\text{lbm}} - 170.6 \ \frac{\text{Btu}}{\text{lbm}}$$

$$= 1055.1 \ \text{Btu/lbm}$$

$$p_2 = (26 \ \text{in Hg})\left(0.491 \ \frac{\text{lbf}}{\text{in}^3}\right)$$

$$= 12.77 \ \text{psia}$$

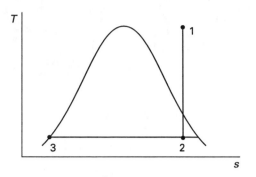

From App. 23.B, for $p_2 = 12.77$ psia,

$$T_{\text{sat}} = T_2 = T_3 \approx 204°\text{F}$$

$$h_{f,3} \approx 171.6 \ \text{Btu/lbm}$$

From App. 32.A, the specific heat of water at 65°F is $c_{p_{\text{water}}} = 0.999$ Btu.

Assuming the water and steam leave in thermal equilibrium, the heat lost by steam is equal to the heat gained by water.

$$\dot{m}_{\text{water}} c_{p_{\text{water}}} (T_{\text{water,out}} - T_{\text{water,in}}) = \dot{m}_{\text{steam}} (h_2 - h_{f,3})$$

Since the terminal temperature difference is zero, $T_{\text{water,out}} = T_3$.

$$\dot{m}_{\text{water}} \left(0.999 \ \frac{\text{Btu}}{\text{lbm-°F}}\right)(204°\text{F} - 65°\text{F})$$

$$= \left(15{,}000 \ \frac{\text{lbm}}{\text{hr}}\right)\left(1055.1 \ \frac{\text{Btu}}{\text{lbm}} - 171.6 \ \frac{\text{Btu}}{\text{lbm}}\right)$$

$$\dot{m}_{\text{water}} = \boxed{9.54 \times 10^4 \ \text{lbm/hr}}$$

The answer is (D).

SI Solution

The steam flow rate is

$$\dot{m}_{\text{steam}} = (\text{WR})(P_{\text{turbine}})$$

$$= \left(2.5 \times 10^{-3} \ \frac{\text{kg}}{\text{kW·s}}\right)(750 \ \text{kW})$$

$$= 1.875 \ \text{kg/s}$$

From App. 23.O for 1 MPa, $T_{\text{sat}} = 179.9°\text{C}$.

The steam temperature is

$$179.9°\text{C} + 30°\text{C} = 209.9°\text{C}$$

From App. 23.P, $h_1 = 2850.6$ kJ/kg.

From Eq. 25.22,

$$P_{\text{turbine}} = \dot{m}(h_1 - h_2)$$

$$h_1 - h_2 = \frac{P_{\text{turbine}}}{\dot{m}} = \frac{750 \text{ kW}}{1.875 \dfrac{\text{kg}}{\text{s}}}$$

$$= 400 \text{ kJ/kg}$$

$$h_2 = h_1 - 400 \frac{\text{kJ}}{\text{kg}}$$

$$= 2850.6 \frac{\text{kJ}}{\text{kg}} - 400 \frac{\text{kJ}}{\text{kg}}$$

$$= 2450.6 \text{ kJ/kg}$$

From App. 23.O for $p_2 = 90$ kPa,

$$T_{\text{sat}} = T_2 = T_3 = 96.71°\text{C}$$
$$h_{f,3} = 405.15 \text{ kJ/kg}$$

From App. 32.B, the specific heat of water at 18°C is $c_p = 4.186$ kJ/kg·K.

Assuming water and steam leave in thermal equilibrium, the heat lost by steam is equal to the heat gained by water.

$$\dot{m}_{\text{water}} c_{p,\text{water}} (T_{\text{water,out}} - T_{\text{water,in}}) = \dot{m}_{\text{steam}} (h_2 - h_{f,3})$$

Since the terminal difference is zero, $T_{\text{water,out}} = T_3$.

$$\dot{m}_{\text{water}} \left(4.186 \frac{\text{kJ}}{\text{kg·K}}\right)(96.71°\text{C} - 18°\text{C})$$

$$= \left(1.875 \frac{\text{kg}}{\text{s}}\right)\left(2450.6 \frac{\text{kJ}}{\text{kg}} - 405.15 \frac{\text{kJ}}{\text{kg}}\right)$$

$$\dot{m}_{\text{water}} = \boxed{11.64 \text{ kg/s}}$$

The answer is (D).

5. *Customary U.S. Solution*

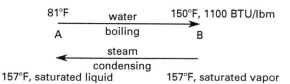

(a) From App. 23.B, for 4.45 psia, $T_{\text{sat}} = 157°\text{F}$.

From App. 23.A, $h_{\text{water},1} = 49.08$ Btu/lbm.

The heat transferred to the water is

$$Q = \dot{m}_{\text{water}}(h_{\text{water},2} - h_{\text{water},1})$$

$$= \left(332{,}000 \frac{\text{lbm}}{\text{hr}}\right)\left(1100 \frac{\text{Btu}}{\text{lbm}} - 49.08 \frac{\text{Btu}}{\text{lbm}}\right)$$

$$= 3.489 \times 10^8 \text{ Btu/hr}$$

The two end temperature differences are

$$\Delta T_A = 157°\text{F} - 81°\text{F} = 76°\text{F}$$
$$\Delta T_B = 157°\text{F} - 150°\text{F} = 7°\text{F}$$

The logarithmic mean temperature difference from Eq. 33.69 is

$$\Delta T_{lm} = \frac{\Delta T_A - \Delta T_B}{\ln\left(\dfrac{\Delta T_A}{\Delta T_B}\right)}$$

$$= \frac{76°\text{F} - 7°\text{F}}{\ln\left(\dfrac{76°\text{F}}{7°\text{F}}\right)} = 28.93°\text{F}$$

Since $T_{\text{steam},A} = T_{\text{steam},B}$, the correction factor (F_c) for ΔT_{lm} is 1.

The overall heat transfer coefficient is calculated from Eq. 33.70.

$$Q = U A F_c \Delta T_{lm}$$

$$U = \frac{Q}{A F_c \Delta T_{lm}}$$

$$= \frac{3.489 \times 10^8 \dfrac{\text{Btu}}{\text{hr}}}{(1850 \text{ ft}^2)(1)(28.93°\text{F})}$$

$$= \boxed{6519 \text{ Btu/hr-ft}^2\text{-}°\text{F}}$$

The answer is (B).

(b) From App. 23.B, the enthalpy of saturated steam at 4.45 psia is $h_1 = 1129$ Btu/lbm.

From App. 23.B, the enthalpy of saturated water at 4.45 psia is $h_2 = 125.1$ Btu/lbm.

The enthalpy change for steam during condensation is

$$\Delta h = h_1 - h_2$$

$$= 1129 \frac{\text{Btu}}{\text{lbm}} - 125.1 \frac{\text{Btu}}{\text{lbm}}$$

$$= 1003.9 \text{ Btu/lbm}$$

The heat transferred (Q) from the steam is equal to the heat gained by water.

$$Q = \dot{m}_{\text{steam}} \Delta h$$

$$\dot{m}_{\text{steam}} = \frac{Q}{\Delta h} = \frac{3.489 \times 10^8 \dfrac{\text{Btu}}{\text{hr}}}{1003.9 \dfrac{\text{Btu}}{\text{lbm}}}$$

$$= 3.475 \times 10^5 \text{ lbm/hr}$$

The energy extraction rate is

$$\dot{m}_{\text{steam}} h_1 = \left(3.475 \times 10^5 \; \frac{\text{lbm}}{\text{hr}} \right) \left(1129 \; \frac{\text{Btu}}{\text{lbm}} \right)$$

$$= \boxed{3.92 \times 10^8 \; \text{Btu/hr}}$$

The answer is (B).

SI Solution

$$
\begin{array}{cc}
27°\text{C} & \text{water} \quad 65°\text{C, 2.56 MJ/kg} \\
\quad\;\; \xrightarrow{\quad\qquad\qquad} & \\
\text{A} \quad & \text{boiling} \qquad\quad \text{B} \\[4pt]
& \text{steam} \\
69.1°\text{C} \xleftarrow{\qquad\qquad\qquad} & 69.1°\text{C} \\
& \text{condensing}
\end{array}
$$

(a) From App. 23.O, for 30 kPa, $T_{\text{sat}} = 69.10°\text{C}$.

From App. 23.N, $h_{\text{water},1} = 113.25$ kJ/kg.

The heat transferred to the water is

$$Q = \dot{m}_{\text{water}} (h_{\text{water},2} - h_{\text{water},1})$$

$$= \left(41.8 \; \frac{\text{kg}}{\text{s}} \right) \left(\left(2.56 \; \frac{\text{MJ}}{\text{kg}} \right) \left(1000 \; \frac{\text{kJ}}{\text{MJ}} \right) \right.$$

$$\left. - 113.25 \; \frac{\text{kJ}}{\text{kg}} \right)$$

$$= 102\,274 \; \text{kJ/s}$$

The two end temperature differences are

$$\Delta T_A = 69.10°\text{C} - 27°\text{C} = 42.1°\text{C}$$
$$\Delta T_B = 69.10°\text{C} - 65°\text{C} = 4.1°\text{C}$$

The logarithmic mean temperature difference is calculated from Eq. 33.69.

$$\Delta T_{lm} = \frac{\Delta T_A - \Delta T_B}{\ln \left(\dfrac{\Delta T_A}{\Delta T_B} \right)} = \frac{42.1°\text{C} - 4.1°\text{C}}{\ln \left(\dfrac{42.1°\text{C}}{4.1°\text{C}} \right)}$$

$$= 16.32°\text{C}$$

Since $T_{\text{steam},A} = T_{\text{steam},B}$, the correction factor (F_c) for ΔT_{lm} is 1.

The overall heat transfer coefficient is calculated from Eq. 33.70.

$$Q = U A F_c \Delta T_{lm}$$

$$U = \frac{Q}{A F_c \Delta T_{lm}} = \frac{\left(102\,274 \; \dfrac{\text{kJ}}{\text{s}} \right) \left(1000 \; \dfrac{\text{W}}{\text{kW}} \right)}{(170 \; \text{m}^2)(1)(16.32°\text{C})}$$

$$= \boxed{36\,863 \; \text{W/m}^2 \cdot °\text{C}}$$

The answer is (B).

(b) From App. 23.O, the enthalpy of saturated steam at 30 kPa is $h_1 = 2625.3$ kJ/kg.

From App. 23.O, the enthalpy of saturated water at 30 kPa is $h_2 = 289.23$ kJ/kg.

The enthalpy change for steam during condensation is

$$\Delta h = h_1 - h_2$$

$$= 2625.3 \; \frac{\text{kJ}}{\text{kg}} - 289.23 \; \frac{\text{kJ}}{\text{kg}}$$

$$= 2336.1 \; \text{kJ/kg}$$

The heat transfer (Q) from the steam is equal to the heat gained by water.

$$Q = \dot{m}_{\text{steam}} \Delta h$$

$$\dot{m}_{\text{steam}} = \frac{Q}{\Delta h} = \frac{102\,274 \; \dfrac{\text{kJ}}{\text{s}}}{2336.1 \; \dfrac{\text{kJ}}{\text{kg}}} = 43.78 \; \text{kg/s}$$

The energy extraction rate is

$$\dot{m}_{\text{steam}} h_1 = \left(43.78 \; \frac{\text{kg}}{\text{s}} \right) \left(2625.3 \; \frac{\text{kJ}}{\text{kg}} \right)$$

$$= \boxed{114\,936 \; \text{kW} \quad (114.9 \; \text{MW})}$$

The answer is (B).

6.

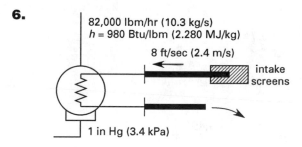

82,000 lbm/hr (10.3 kg/s)
h = 980 Btu/lbm (2.280 MJ/kg)
8 ft/sec (2.4 m/s)
intake screens
1 in Hg (3.4 kPa)

Customary U.S. Solution

(a) From App. 16.B, the dimensions of extra strong 30 in steel pipe are

$$D_i = 2.4167 \; \text{ft}$$
$$A_i = 4.5869 \; \text{ft}^2$$

From Table 17.2, the specific roughness for steel is $\epsilon = 0.0002$ ft.

$$\frac{\epsilon}{D} = \frac{0.0002 \; \text{ft}}{2.4167 \; \text{ft}} = 0.000083$$

Assume fully turbulent flow. (The Reynolds number can also be calculated.)

From the Moody friction factor chart for fully turbulent flow, $f \approx 0.012$.

The head loss from Eq. 17.22 is

$$h_f = \frac{fLv^2}{2Dg}$$

$$= \frac{(0.012)(120 \text{ ft})\left(8 \frac{\text{ft}}{\text{sec}}\right)^2}{(2)(2.4187 \text{ ft})\left(32.2 \frac{\text{ft}}{\text{sec}^2}\right)} = 0.59 \text{ ft}$$

The screen loss is

$$6 \text{ in wg} = (6 \text{ in})\left(\frac{1 \text{ ft}}{12 \text{ in}}\right)$$

$$= 0.5 \text{ ft}$$

$$\text{velocity head} = \frac{v^2}{2g}$$

$$= \frac{\left(8 \frac{\text{ft}}{\text{sec}}\right)^2}{(2)\left(32.2 \frac{\text{ft}}{\text{sec}^2}\right)} = 0.99 \text{ ft}$$

The total head added by a coolant pump (not shown) not including losses inside the condenser is

$$0.59 \text{ ft} + 0.5 \text{ ft} + 0.99 \text{ ft} = \boxed{2.08 \text{ ft}}$$

The answer is (A).

(b) The condenser pressure is

$$(1 \text{ in Hg})\left(0.491 \frac{\text{lbf}}{\text{in}^3}\right) = 0.5 \text{ lbf/in}^2$$

From App. 23.B, the enthalpy of the saturated liquid is $h_f = 47.11$ Btu/lbm.

The specific heat of water is $c_{p,\text{water}} = 1$ Btu/lbm-°F.

The heat lost by the steam is equal to the heat gained by the water.

$$\dot{m}_{\text{water}} c_{p,\text{water}}(\Delta T) = \dot{m}_{\text{steam}}(h_2 - h_f)$$

$$\dot{m}_{\text{water}}\left(1 \frac{\text{Btu}}{\text{lbm-°F}}\right)(10°\text{F})$$

$$= \left(82{,}000 \frac{\text{lbm}}{\text{hr}}\right)\left(980 \frac{\text{Btu}}{\text{lbm}} - 47.11 \frac{\text{Btu}}{\text{lbm}}\right)$$

$$\dot{m}_{\text{water}} = 7.6497 \times 10^6 \text{ lbm/hr}$$

The density of water, ρ, is 62.4 lbm/ft^3.

$$Q = \frac{\dot{m}_{\text{water}}}{\rho}$$

$$= \frac{\left(7.6497 \times 10^6 \frac{\text{lbm}}{\text{hr}}\right)\left(\frac{1 \text{ hr}}{60 \text{ min}}\right)\left(7.48 \frac{\text{gal}}{\text{ft}^3}\right)}{62.4 \frac{\text{lbm}}{\text{ft}^3}}$$

$$= \boxed{1.528 \times 10^4 \text{ gal/min}}$$

The answer is (D).

(The flow rate can also be determined from the velocity and pipe area. However, this does not use the 10° data or perform an energy balance.)

SI Solution

(a) From App. 16.B, the dimensions of extra strong 30 in steel pipe, using the footnote from the table, are

$$D_i = (29.00 \text{ in})\left(25.4 \frac{\text{mm}}{\text{in}}\right)\left(\frac{1 \text{ m}}{1000 \text{ mm}}\right)$$

$$= 0.7366 \text{ m}$$

$$A_i = (660.52 \text{ in}^2)\left(645 \frac{\text{mm}^2}{\text{in}^2}\right)\left(\frac{1 \text{ m}}{1000 \text{ mm}}\right)^2$$

$$= 0.4260 \text{ m}^2$$

From Table 17.2, the specific roughness for steel is $\epsilon = 6.0 \times 10^{-5}$ m.

$$\frac{\epsilon}{D} = \frac{6.0 \times 10^{-5} \text{ m}}{0.7366 \text{ m}} = 0.0000815$$

Assume fully turbulent flow. (The Reynolds number can also be calculated.)

From the Moody friction factor chart for fully turbulent flow, $f \approx 0.012$.

The head loss from Eq. 17.22 is

$$h_f = \frac{fLv^2}{2Dg}$$

$$= \frac{(0.012)(36 \text{ m})\left(2.4 \frac{\text{m}}{\text{s}}\right)^2}{(2)(0.7366 \text{ m})\left(9.81 \frac{\text{m}}{\text{s}^2}\right)} = 0.1722 \text{ m}$$

$$p_f = \rho g h$$

$$= \left(1000 \frac{\text{kg}}{\text{m}^3}\right)\left(9.81 \frac{\text{m}}{\text{s}^2}\right)(0.1722 \text{ m})\left(\frac{1 \text{ kPa}}{1000 \text{ Pa}}\right)$$

$$= 1.69 \text{ kPa}$$

The screen loss is 1.5 kPa.

The velocity head is

$$\frac{v^2}{2g} = \frac{\left(2.4 \, \frac{m}{s}\right)^2}{(2)\left(9.81 \, \frac{m}{s^2}\right)} = 0.2936 \text{ m}$$

$$p_v = \rho g h$$

$$= \left(1000 \, \frac{kg}{m^3}\right)\left(9.81 \, \frac{m}{s^2}\right)(0.2936 \text{ m})\left(\frac{1 \text{ kPa}}{1000 \text{ Pa}}\right)$$

$$= 2.88 \text{ kPa}$$

The total head added by a coolant pump (not shown) not including losses inside the condenser is

$$1.69 \text{ kPa} + 1.5 \text{ kPa} + 2.88 \text{ kPa} = \boxed{6.07 \text{ kPa}}$$

The answer is (A).

(b) The condenser pressure is 3.4 kPa.

From App. 23.N, the enthalpy of saturated liquid is $h_f = 109.8$ kJ/kg.

The specific heat of water is $c_{p,\text{water}} = 4.187$ kJ/kg·K.

The heat lost by the steam is equal to the heat gained by the water.

$$\dot{m}_{\text{water}} c_{p,\text{water}}(\Delta T) = \dot{m}_{\text{steam}}(h_2 - h_f)$$

$$\dot{m}_{\text{water}}\left(4.187 \, \frac{kJ}{kg \cdot K}\right)(5.6°C)$$

$$= \left(10.3 \, \frac{kg}{s}\right)\left(\left(2.280 \, \frac{MJ}{kg}\right)\left(1000 \, \frac{kJ}{MJ}\right) - 109.8 \, \frac{kJ}{kg}\right)$$

$$\dot{m}_{\text{water}} = 953.3 \text{ kg/s}$$

The density of water, ρ, is 1000 kg/m^3.

$$Q = \frac{\dot{m}_{\text{water}}}{\rho}$$

$$= \left(\frac{953.3 \, \frac{kg}{s}}{1000 \, \frac{kg}{m^3}}\right)\left(1000 \, \frac{L}{m^3}\right)\left(60 \, \frac{s}{\text{min}}\right)$$

$$= \boxed{57\,200 \text{ L/min}}$$

The answer is (D).

7. *Customary U.S. Solution*

From App. 23.A, the enthalpy of saturated liquid at 60°F is $h_1 = 28.08$ Btu/lbm.

From App. 23.B, the enthalpy of saturated steam at 14.7 psia is $h_2 = 1150.5$ Btu/lbm.

The heat transfer rate to the water is

$$Q = \dot{m}(h_2 - h_1)$$

$$= \left(100 \, \frac{\text{lbm}}{\text{hr}}\right)\left(1150.5 \, \frac{\text{Btu}}{\text{lbm}} - 28.08 \, \frac{\text{Btu}}{\text{lbm}}\right)$$

$$= 1.122 \times 10^5 \text{ Btu/hr}$$

$$= \left(1.122 \times 10^5 \, \frac{\text{Btu}}{\text{hr}}\right)\left(0.2931 \, \frac{\text{W-hr}}{\text{Btu}}\right)\left(\frac{1 \text{ kW}}{1000 \text{ W}}\right)$$

$$= 32.89 \text{ kW}$$

$$\text{cost} = \frac{(32.89 \text{ kW})\left(\frac{\$0.04}{\text{kW-hr}}\right)}{1 - 0.35}$$

$$= \boxed{\$2.02/\text{hr}}$$

The answer is (A).

SI Solution

From App. 23.N, for saturated liquid at 16°C, $h_1 = 67.19$ kJ/kg.

From App. 23.N, the enthalpy of saturated steam at 101.3 kPa is $h_2 = 2676.1$ kJ/kg.

The heat transfer rate to the water is

$$Q = \dot{m}(h_2 - h_1)$$

$$= \left(0.013 \, \frac{kg}{s}\right)\left(2676.1 \, \frac{kJ}{kg} - 67.19 \, \frac{kJ}{kg}\right)$$

$$= 33.916 \text{ kW}$$

$$\text{cost} = \frac{(33.916 \text{ kW})\left(\frac{\$0.04}{\text{kW·h}}\right)}{1 - 0.35}$$

$$= \boxed{\$2.09/\text{h}}$$

The answer is (A).

8. *Customary U.S. Solution*

From App. 23.A, the enthalpy of saturated liquid at 60°F is $h_1 = 28.08$ Btu/lbm.

From App. 23.B, for 40 psia steam, the enthalpy of saturated liquid, h_f, is 236.16 Btu/lbm. The heat of vaporization, h_{fg}, is 933.8 Btu/lbm. The enthalpy is given by Eq. 23.44.

$$h_2 = h_f + x h_{fg}$$

$$= 236.16 \, \frac{\text{Btu}}{\text{lbm}} + (0.98)\left(933.8 \, \frac{\text{Btu}}{\text{lbm}}\right)$$

$$= 1151.28 \text{ Btu/lbm}$$

Thermodynamics

The heat transfer rate is

$$Q = \dot{m}(h_2 - h_1)$$

$$= \left(250 \; \frac{\text{lbm}}{\text{hr}}\right)\left(1151.28 \; \frac{\text{Btu}}{\text{lbm}} - 28.08 \; \frac{\text{Btu}}{\text{lbm}}\right)$$

$$\times \left(\frac{1 \text{ hr}}{60 \text{ min}}\right)$$

$$= 4680.0 \text{ Btu/min}$$

Find the volume of gas used at standard conditions for a heating gas (60°F).

$$\dot{V}_{\text{std}} = \dot{V}\left(\frac{T_0}{T}\right)\left(\frac{p}{p_0}\right)$$

$$= \left(13.5 \; \frac{\text{ft}^3}{\text{min}}\right)\left(\frac{460 + 60°\text{F}}{460 + 80°\text{F}}\right)$$

$$\times \left(\frac{(4 \text{ in Hg} + 30.2 \text{ in Hg})\left(0.491 \; \frac{\text{lbm}}{\text{in}^3}\right)}{14.7 \; \frac{\text{lbf}}{\text{in}^2}}\right)$$

$$= 14.85 \text{ SCFM}$$

The efficiency of the boiler is

$$\eta = \frac{Q}{\text{heat input}}$$

$$= \frac{4680.0 \; \frac{\text{Btu}}{\text{min}}}{\left(14.85 \; \frac{\text{ft}^3}{\text{min}}\right)\left(550 \; \frac{\text{Btu}}{\text{ft}^3}\right)}$$

$$= \boxed{0.573 \;\; (57.3\%)}$$

The answer is (C).

SI Solution

From App. 23.N, the enthalpy of saturated liquid at 16°C is $h_1 = 67.19$ kJ/kg.

From App. 23.O, for 300 kPa steam, the enthalpy of saturated liquid, h_f, is 561.47 kJ/kg. The enthalpy of vaporization, h_{fg}, is 2163.8 kJ/kg. The enthalpy is given by Eq. 23.44.

$$h_2 = h_f + xh_{fg}$$

$$= 561.47 \; \frac{\text{kJ}}{\text{kg}} + (0.98)\left(2163.8 \; \frac{\text{kJ}}{\text{kg}}\right)$$

$$= 2682.0 \text{ kJ/kg}$$

The heat transfer rate is

$$Q = \dot{m}(h_2 - h_1)$$

$$= \left(0.032 \; \frac{\text{kg}}{\text{s}}\right)\left(2682.0 \; \frac{\text{kJ}}{\text{kg}} - 67.19 \; \frac{\text{kJ}}{\text{kg}}\right)$$

$$= 83.67 \text{ kJ/s}$$

Find the volume of the gas used at standard conditions for a heating gas (16°C).

$$\dot{V}_{\text{std}} = \dot{V}\left(\frac{T_0}{T}\right)\left(\frac{p}{p_0}\right)$$

$$= \left(6.4 \; \frac{\text{L}}{\text{s}}\right)\left(\frac{1 \text{ m}^3}{1000 \text{ L}}\right)\left(\frac{16°\text{C} + 273}{26°\text{C} + 273}\right)$$

$$\times \left(\frac{13.6 \text{ kPa} + 102.4 \text{ kPa}}{101.3 \text{ kPa}}\right)$$

$$= 0.00708 \text{ m}^3/\text{s}$$

The efficiency of the boiler is

$$\eta = \frac{Q}{\text{heat input}}$$

$$= \frac{83.67 \; \frac{\text{kJ}}{\text{s}}}{\left(0.00708 \; \frac{\text{m}^3}{\text{s}}\right)\left(20.5 \; \frac{\text{MJ}}{\text{m}^3}\right)\left(1000 \; \frac{\text{kJ}}{\text{MJ}}\right)}$$

$$= \boxed{0.576 \;\; (57.6\%)}$$

The answer is (C).

9. *Customary U.S. Solution*

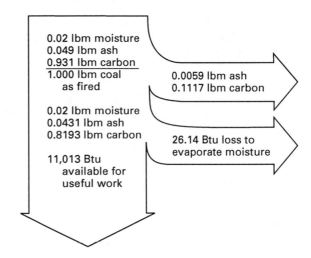

step 1: Determine the actual gravimetric analysis of the coal as fired. Use the successive deletion method on a per-pound basis. 1 lbm of coal contains 0.02 lbm moisture, leaving 0.98 lbm dry coal. Of this, 5% is ash, so the weight of ash is (0.05)(0.98) = 0.049 lbm. The remainder (0.98 lbm − 0.049 lbm = 0.931 lbm) is assumed to be carbon.

step 2: Determine the ash pit material losses. 12% of dry coal goes to the ash pit.

(0.049 lbm)(0.12) = 0.0059 lbm ash

(0.931 lbm)(0.12) = 0.1117 unburned carbon

step 3:   Determine what remains.

0.02 lbm moisture

0.049 lbm − 0.0059 lbm = 0.0431 lbm ash

0.931 lbm − 0.1117 lbm = 0.8193 lbm carbon

step 4:   Determine energy losses. The 0.02 lbm moisture has to be evaporated. The enthalpy of water at 60°F is read from App. 23.A as 28.08 Btu/lbm. For a constant pressure of 14.7 psia, the enthalpy of water vapor at a combustion temperature of 600°F is read from App. 23.C as 1335.2 Btu/lbm. The energy loss in water evaporation is

$$Q = m(h_2 - h_1)$$
$$= (0.02 \text{ lbm}) \left(1335.2 \ \frac{\text{Btu}}{\text{lbm}} - 28.08 \ \frac{\text{Btu}}{\text{lbm}} \right)$$
$$= 26.14 \text{ Btu}$$

step 5:   Calculate the heating value of the remaining coal. The heating value is given per pound of coal, not per pound of carbon.

mass of coal = 0.0431 lbm + 0.8193 lbm
$$= 0.8624 \text{ lbm}$$

The heating value of the remaining coal is

$$(0.8624 \text{ lbm}) \left(12{,}800 \ \frac{\text{Btu}}{\text{lbm}} \right) = 11{,}039 \text{ Btu}$$

step 6:   Subtract the losses.

11,039 Btu − 26.14 Btu = 11,013 Btu

step 7:   Find the energy (Q) required to produce steam. From App. 23.A, the enthalpy of water at 120°F is $h_1 = 88.00$ Btu/lbm.

From App. 23.B, the enthalpy of saturated steam at 100 psia is $h_2 = 1187.8$ Btu/lbm.

$$Q = m(h_2 - h_1)$$
$$= (8.23 \text{ lbm}) \left(1187.8 \ \frac{\text{Btu}}{\text{lbm}} - 88.00 \ \frac{\text{Btu}}{\text{lbm}} \right)$$
$$= 9051.3 \text{ Btu}$$

step 8:   The combustion efficiency is

$$\eta = \frac{Q}{HV} = \frac{9051.3 \text{ Btu}}{11{,}013 \text{ Btu}} = \boxed{0.822 \ \ (82.2\%)}$$

The answer is (D).

SI Solution

Since boiler data are based on 1 unit mass of coal fired, steps 1–3 will be the same as for the customary U.S. solution. Repeat the rest of the steps as follows.

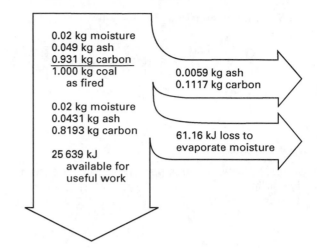

step 4:   Determine the energy losses due to 0.02 kg moisture evaporation. The enthalpy of water at 16°C is taken from App. 23.N as 67.19 kJ/kg. For constant pressure of 101.3 kPa, the enthalpy of 315°C steam is read from App. 23.P as ≈ 312.50 kJ/kg. The energy loss in water evaporation is

$$Q = m(h_2 - h_1)$$
$$= (0.02 \text{ kg}) \left(3125.0 \ \frac{\text{kJ}}{\text{kg}} - 67.19 \ \frac{\text{kJ}}{\text{kg}} \right)$$
$$= 61.16 \text{ kJ}$$

step 5:   Calculate the heating value of the remaining coal. The heating value is given per kilogram of coal, not per kilogram of carbon.

mass of coal = 0.0431 kg + 0.8193 kg
$$= 0.8624 \text{ kg}$$

The heating value of the remaining coal is

$$(0.8624 \text{ kg}) \left(29.80 \ \frac{\text{MJ}}{\text{kg}} \right) \left(1000 \ \frac{\text{kJ}}{\text{MJ}} \right)$$
$$= 25\,700 \text{ kJ}$$

step 6:   Subtract the losses to get a heating value of

$$25\,700 \text{ kJ} - 61.16 \text{ kJ} = 25\,639 \text{ kJ}$$

step 7:   Find the energy (Q) required to produce steam. From App. 23.N, the enthalpy of water at 50°C is $h_1 = 209.33$ kJ/kg.

From App. 23.O, the enthalpy of saturated steam at 700 kPa is $h_2 = 2763.5$ kJ/kg.

$$Q = m(h_2 - h_1)$$
$$= (8.23 \text{ kg}) \left(2763.5 \ \frac{\text{kJ}}{\text{kg}} - 209.33 \ \frac{\text{kJ}}{\text{kg}} \right)$$
$$= 21\,021 \text{ kJ}$$

step 8: The combustion efficiency is

$$\eta = \frac{Q}{HV} = \frac{21\,021 \text{ kJ}}{25\,639 \text{ kJ}} = \boxed{0.820 \ (82.0\%)}$$

The answer is (D).

10. *Customary U.S. Solution*

(a) Refer to Fig. 26.1.

At point D (leaving the superheater and entering the turbine), the enthalpy, h_D, and entropy, s_D, can be obtained from App. 23.C.

$$h_D = 1520.7 \text{ Btu/lbm}$$

$$s_D = 1.7471 \text{ Btu/lbm-°F}$$

For isentropic expansion through the turbine, $s_E = s_D$. From App. 23.B, the enthalpy of saturated liquid, h_f, the enthalpy of evaporation, h_{fg}, the entropy of saturated liquid, s_f, and the entropy of evaporation, s_{fg}, are

$$h_f = 130.17 \text{ Btu/lbm}$$

$$h_{fg} = 1000.9 \text{ Btu/lbm}$$

$$s_f = 0.2349 \text{ Btu/lbm-°F}$$

$$s_{fg} = 1.6093 \text{ Btu/lbm-°F}$$

The quality of the mixture for an isentropic process is given by Eq. 23.45 (at point E).

$$x_E = \frac{s_E - s_f}{s_{fg}}$$

$$= \frac{1.7471 \, \dfrac{\text{Btu}}{\text{lbm-°F}} - 0.2349 \, \dfrac{\text{Btu}}{\text{lbm-°F}}}{1.6093 \, \dfrac{\text{Btu}}{\text{lbm-°F}}}$$

$$= 0.9397$$

The isentropic enthalpy is given by Eq. 23.44 (at point E).

$$h_E = h_f + x_E h_{fg}$$

$$= 130.17 \, \frac{\text{Btu}}{\text{lbm}} + (0.9397)\left(1000.9 \, \frac{\text{Btu}}{\text{lbm}}\right)$$

$$= 1070.7 \text{ Btu/lbm}$$

From Eq. 25.17, the actual enthalpy of steam at point E is

$$h'_E = h_D - \eta_s(h_D - h_E)$$

$$= 1520 \, \frac{\text{Btu}}{\text{lbm}} - (0.75)\left(1520 \, \frac{\text{Btu}}{\text{lbm}} - 1070.7 \, \frac{\text{Btu}}{\text{lbm}}\right)$$

$$= 1183.0 \text{ Btu/lbm}$$

Since the pump work is negligible, the mass flow rate of steam is

$$\dot{m} = \frac{P}{W_{\text{turbine}}} = \frac{P}{h_D - h'_E}$$

$$= \frac{(200 \text{ MW})\left(10^6 \, \dfrac{\text{W}}{\text{MW}}\right)\left(3.4121 \, \dfrac{\text{Btu}}{\text{hr}}\right)}{1520 \, \dfrac{\text{Btu}}{\text{lbm}} - 1183.0 \, \dfrac{\text{Btu}}{\text{lbm}}}$$

$$= \boxed{2.025 \times 10^6 \text{ lbm/hr}}$$

The answer is (C).

(b) The heat removed by the condenser is given by Eq. 25.30.

$$\dot{Q} = \dot{m}(h'_E - h_F)$$

From App. 23.B, at 5 psia, $h_F = 130.17$ Btu/lbm.

$$\dot{Q} = \left(2.025 \times 10^6 \, \frac{\text{lbm}}{\text{hr}}\right)\left(1183.0 \, \frac{\text{Btu}}{\text{lbm}} - 130.17 \, \frac{\text{Btu}}{\text{lbm}}\right)$$

$$= \boxed{2.132 \times 10^9 \text{ Btu/hr}}$$

The answer is (C).

SI Solution

(a) Refer to Fig. 26.1.

At point D (leaving the superheater and entering the turbine), the enthalpy, h_D, and entropy, s_D, can be obtained from the Mollier diagram.

$$h_D \approx 3430 \text{ kJ/kg}$$

$$s_D \approx 7.25 \text{ kJ/kg·K}$$

For isentropic expansion through the turbine, $s_E = s_D$. From App. 23.O, the enthalpy of saturated liquid, h_f, the enthalpy of evaporation, h_{fg}, the entropy of saturated liquid, s_f, and the entropy of saturated vapor, s_g, are

$$h_f = 289.23 \text{ kJ/kg}$$

$$h_{fg} = 2336.1 \text{ kJ/kg}$$

$$s_f = 0.9439 \text{ kJ/kg·K}$$

$$s_g = 7.7686 \text{ kJ/kg·K}$$

The quality of the mixture for an isentropic process is given by Eq. 23.45 (at point ϵ) as

$$x_E = \frac{s_E - s_f}{s_{fg}} = \frac{s_E - s_f}{s_g - s_f}$$

$$= \frac{7.25 \, \dfrac{\text{kJ}}{\text{kg·K}} - 0.9439 \, \dfrac{\text{kJ}}{\text{kg·K}}}{7.7686 \, \dfrac{\text{kJ}}{\text{kg·K}} - 0.9439 \, \dfrac{\text{kJ}}{\text{kg·K}}}$$

$$= 0.9240$$

The isentropic enthalpy is given by Eq. 23.44 (at point E) as

$$h_E = h_f + x_E h_{fg}$$
$$= 289.23 \ \frac{kJ}{kg} + (0.9240)\left(2336.1 \ \frac{kJ}{kg}\right)$$
$$= 2447.8 \ kJ/kg$$

From Eq. 25.17, the actual enthalpy of steam at point E is

$$h'_E = h_D - \eta_s(h_D - h_E)$$
$$= 3430 \ \frac{kJ}{kg} - (0.75)\left(3430 \ \frac{kJ}{kg} - 2447.8 \ \frac{kJ}{kg}\right)$$
$$= 2693.4 \ kJ/kg$$

Since the pump work is negligible, the mass flow rate of steam is

$$\dot{m} = \frac{P}{W_{\text{turbine}}} = \frac{P}{h_D - h'_E}$$
$$= \frac{(200 \ \text{MW})\left(10^3 \ \frac{kW}{MW}\right)}{3430 \ \frac{kJ}{kg} - 2693.4 \ \frac{kJ}{kg}}$$
$$= \boxed{271.5 \ kg/s}$$

The answer is (C).

(b) The heat removed by the condenser is given by Eq. 25.30.
$$\dot{Q} = \dot{m}(h'_E - h_F)$$

From App. 23.O at 30 kPa, $h_F = 289.23 \ kJ/kg$.

$$\dot{Q} = \left(271.5 \ \frac{kg}{s}\right)\left(2693.4 \ \frac{kJ}{kg} - 289.23 \ \frac{kJ}{kg}\right)$$
$$= \boxed{6.527 \times 10^5 \ kW}$$

The answer is (C).

11. *Customary U.S. Solution*

The following illustration shows one of N layers. Each layer consists of 24 tubes, only 3 of which are shown.

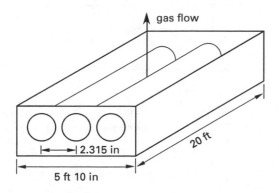

gas flow

20 ft

2.315 in

5 ft 10 in

Assume stack gases consist primarily of nitrogen. The average gas temperature is

$$T_{\text{ave}} = \left(\tfrac{1}{2}\right)(635°F + 470°F) + 460 = 1012.5°R$$

The specific heat of nitrogen is calculated from Eq. 23.98 using constants given in Table 23.9.

$$c_p = A + BT + CT^2 + \frac{D}{\sqrt{T}}$$
$$= 0.2510 + (-1.63 \times 10^{-5})(1012.5°R)$$
$$\quad + (20.4 \times 10^{-9})(1012.5°R)^2 + 0$$
$$= 0.2554 \ \text{Btu/lbm-°R}$$

For the purpose of calculating the logarithmic temperature difference, assume counterflow operation. The two end temperature differences are

$$\Delta T_A = 635°F - 285°F = 350°F$$
$$\Delta T_B = 470°F - 212°F = 258°F$$

635°F ——— gases ———→ 470°F
$\Delta T = 350°F$ $\Delta T = 258°F$
285°F ←——— water ——— 212°F

The logarithmic temperature difference is

$$\Delta T_{lm} = \frac{\Delta T_A - \Delta T_B}{\ln\left(\frac{\Delta T_A}{\Delta T_B}\right)} = \frac{350°F - 258°F}{\ln\left(\frac{350°F}{258°F}\right)}$$
$$= 301.7°F$$

Determine F_c for crossflow.

$$R = \frac{635°F - 470°F}{285°F - 212°F} = 2.26$$
$$x = \frac{285°F - 212°F}{635°F - 212°F} = 0.17$$

F_c for all heat exchanger configurations is close to 1.0 for this set of parameters.

The heat transfer from the temperature gain of water is equal to the heat transfer based on the logarithmic mean temperature difference.

$$Q = \dot{m}c_p\Delta T = U_o A_o F_c \Delta T_{lm}$$
$$\left(191{,}000 \ \frac{lbm}{hr}\right)\left(0.2554 \ \frac{Btu}{lbm\text{-}°R}\right)(635°F - 470°F)$$
$$= \left(10 \ \frac{Btu}{hr\text{-}ft^2\text{-}°F}\right)A_o(1.0)(301.7°F)$$
$$A_o = 2667.9 \ ft^2$$

The tube area per bank is

$$A_{bank} = N\pi D_o L$$

$$= 24\pi(1.315 \text{ in}) \left(\frac{1 \text{ ft}}{12 \text{ in}}\right)(20 \text{ ft})$$

$$= 165.2 \text{ ft}^2$$

$$\text{no. of layers} = \frac{A_o}{A_{bank}}$$

$$= \frac{2667.9 \text{ ft}^2}{165.2 \text{ ft}^2} = \boxed{16.1 \quad [\text{say } 17]}$$

The answer is (D).

SI Solution

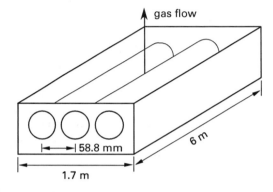

Assume stack gas consists primarily of nitrogen. The average gas temperature is

$$T_{ave} = \left(\tfrac{1}{2}\right)(335°C + 240°C) + 273 = 560.5 \text{K} \quad (1009°\text{R})$$

Use the value of the specific heat of nitrogen that was calculated for the customary U.S. solution since the two temperatures are almost the same.

$$c_p = \left(0.2554 \ \frac{\text{Btu}}{\text{lbm-}°\text{R}}\right)\left(4186.8 \ \frac{\frac{\text{J}}{\text{kg}·°\text{C}}}{\frac{\text{Btu}}{\text{lbm-}°\text{R}}}\right)$$

$$= 1069 \text{ J/kg·}°\text{C}$$

For the purpose of calculating the logarithmic temperature difference, assume counterflow operation. The two end temperature differences are

$$\Delta T_A = 335°C - 140°C = 195°C$$

$$\Delta T_B = 240°C - 100°C = 140°C$$

	gases	
335°C	→	240°C
$\Delta T = 195°C$		$\Delta T = 140°C$
140°C	← water	100°C

The logarithmic mean temperature difference is

$$\Delta T_{lm} = \frac{\Delta T_A - \Delta T_B}{\ln\left(\frac{\Delta T_A}{\Delta T_B}\right)} = \frac{195°C - 140°C}{\ln\left(\frac{195°C}{140°C}\right)}$$

$$= 166.0°C$$

$$R = \frac{335°C - 240°C}{140°C - 100°C} = 2.38$$

$$x = \frac{140°C - 100°C}{335°C - 100°C} = 0.17$$

F_c for all heat exchanger configurations is close to 1.0 for this set of conditions.

The heat transfer from the temperature gain of water is equal to the heat transfer based on the logarithmic mean temperature difference.

$$Q = \dot{m}c_p\Delta T = U_o A_o F_c \Delta T_{lm}$$

$$\times \left(24 \ \frac{\text{kg}}{\text{s}}\right)\left(1069 \ \frac{\text{J}}{\text{kg·}°\text{C}}\right)(335°C - 240°C)$$

$$= \left(57 \ \frac{\text{W}}{\text{m}^2·°\text{C}}\right)A_o(1.0)(166.0°C)$$

$$A_o = 257.6 \text{ m}^2$$

The tube area per bank is

$$A_{bank} = N\pi D_o L$$

$$= 24\pi(33.4 \text{ mm})\left(\frac{1 \text{ m}}{1000 \text{ mm}}\right)(6 \text{ m})$$

$$= 15.1 \text{ m}^2$$

$$\text{no. of layers} = \frac{A_o}{A_{bank}}$$

$$= \frac{257.6 \text{ m}^2}{15.1 \text{ m}^2} = \boxed{17.1 \quad [\text{say } 18]}$$

The answer is (D).

12.

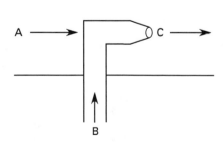

Customary U.S. Solution

(a) The enthalpy, h_A, and entropy, s_A, for steam at 600°F and 200 psia can be obtained from App. 23.C.

$$h_A = 1322.1 \text{ Btu/lbm}$$

$$s_A = 1.6767 \text{ Btu/lbm-}°\text{R}$$

For water at 82°F, enthalpy can be obtained for saturated water from App. 23.A and corrected for compression to 200 psia using App. 23.D. (Correction for compression is optional.)

$$h_B = 50.08 \; \frac{\text{Btu}}{\text{lbm}} + 0.55 \; \frac{\text{Btu}}{\text{lbm}}$$
$$= 50.63 \; \text{Btu/lbm}$$

From App. 23.A, $s_B = 0.09701$ Btu/lbm-°F. From an energy balance equation (with $Q = 0$, $v_1 = 0$, $\Delta z = 0$, and $W = 0$), Eq. 24.30(b) can be written as

$$0 = h_C(m_A + m_B) - (h_A m_A + h_B m_B)$$
$$+ \left(\frac{v_C^2}{2g_c J}\right)(m_A + m_B)$$

$$= h_C \left(1000 \; \frac{\text{lbm}}{\text{hr}} + 50 \; \frac{\text{lbm}}{\text{hr}}\right) - \left(\left(1322.1 \; \frac{\text{Btu}}{\text{lbm}}\right)\right.$$

$$\times \left(1000 \; \frac{\text{lbm}}{\text{hr}}\right) + \left(50.63 \; \frac{\text{Btu}}{\text{lbm}}\right)\left(50 \; \frac{\text{lbm}}{\text{hr}}\right)\bigg)$$

$$+ \frac{\left(2000 \; \frac{\text{ft}}{\text{sec}}\right)^2 \left(1000 \; \frac{\text{lbm}}{\text{hr}} + 50 \; \frac{\text{lbm}}{\text{hr}}\right)}{(2)\left(32.2 \; \frac{\text{lbm-ft}}{\text{lbf-sec}^2}\right)\left(778 \; \frac{\text{ft-lbf}}{\text{Btu}}\right)}$$

$$h_C = 1181.7 \; \text{Btu/lbm}$$
$$p_C = 100 \; \text{psia}$$

Since the enthalpy of saturated steam at 100 psia, h_g, from App. 23.B, is greater than h_C, steam leaving the desuperheater is not 100% saturated. From App. 23.B, the temperature of the saturated steam mix is $\boxed{327.86°F.}$

The answer is (A).

(b) The quality of the steam leaving can be determined using Eq. 23.44.

$$h_C = h_f + x h_{fg}$$

From App. 23.B, for 100 psia, $h_f = 298.6$ Btu/lbm and $h_{fg} = 889.2$ Btu/lbm.

$$x = \frac{1181.7 \; \frac{\text{Btu}}{\text{lbm}} - 298.6 \; \frac{\text{Btu}}{\text{lbm}}}{889.2 \; \frac{\text{Btu}}{\text{lbm}}}$$
$$= \boxed{0.993}$$

The answer is (D).

SI Solution

(a) The enthalpy, h_A, and entropy, s_A, for steam at 300°C and 1.5 MPa can be obtained from App. 23.P.

$$h_A = 3037.6 \; \text{kJ/kg}$$
$$s_A = 6.9179 \; \text{kJ/kg·K}$$

App. 23.Q does not go down to 1.5 MPa, so disregard the effect of compression. For water at 28°C from App. 23.N, $h_B = 117.43$ kJ/kg.

Similarly, from App. 23.N, $s_B = 0.4093$ kJ/kg·K.

From an energy balance equation (with $Q = 0$, $v_1 = 0$, $\Delta z = 0$, and $W = 0$), Eq. 24.30(a) can be written as

$$0 = h_C(m_A + m_B) - (m_A h_A + m_B h_B)$$
$$- \left(\frac{v_C^2}{2}\right)(m_A + m_B)$$

$$= h_C \left(0.13 \; \frac{\text{kg}}{\text{s}} + 0.0063 \; \frac{\text{kg}}{\text{s}}\right) - \left(\left(0.13 \; \frac{\text{kg}}{\text{s}}\right)\right.$$

$$\times \left(3037.6 \; \frac{\text{kJ}}{\text{kg}}\right) + \left(0.0063 \; \frac{\text{kg}}{\text{s}}\right)\left(117.43 \; \frac{\text{kJ}}{\text{kg}}\right)\bigg)$$

$$- \left(\frac{\left(600 \; \frac{\text{m}}{\text{s}}\right)^2}{2}\right)\left(0.13 \; \frac{\text{kg}}{\text{s}} + 0.0063 \; \frac{\text{kg}}{\text{s}}\right)$$

$$\times \left(\frac{1 \; \text{kJ}}{1000 \; \text{J}}\right)$$

$$h_C = 2722.6 \; \text{kJ/kg}$$
$$p_C = 700 \; \text{kPa}$$

Since the enthalpy of saturated steam at 700 kPa, from App. 23.O, is greater than h_C, steam leaving the desuperheater is not 100% saturated. From App. 23.O, the temperature of the steam mix is $\boxed{165.0°C.}$

The answer is (A).

(b) The quality of steam leaving can be determined using Eq. 23.44.

$$h_C = h_f + x h_{fg}$$

From App. 23.O, for 700 kPa, $h_f = 697.22$ kJ/kg and $h_{fg} = 2066.3$ kJ/kg.

$$x = \frac{2722.6 \; \frac{\text{kJ}}{\text{kg}} - 697.22 \; \frac{\text{kJ}}{\text{kg}}}{2066.3 \; \frac{\text{kJ}}{\text{kg}}} = \boxed{0.9802}$$

The answer is (D).

13. *Customary U.S. Solution*

(a) Work with the original system to find the steam flow.

From App. 23.C, at 100 psia and 400°F, $h_A = 1227.5$ Btu/lbm.

From App. 23.A, at 70°F, $h_B = 38.09$ Btu/lbm.

From App. 23.A, at 180°F, $h_C = 147.99$ Btu/lbm.

Let $x =$ fraction of steam in mixture.

From the energy balance equation,

$$m_A h_A + m_B h_B = m_C h_C$$

$$x \left(1227.5 \frac{\text{Btu}}{\text{lbm}} \right)$$
$$+ (1-x) \left(38.09 \frac{\text{Btu}}{\text{lbm}} \right) = (1) \left(147.99 \frac{\text{Btu}}{\text{lbm}} \right)$$
$$x = 0.0924$$

The steam flow is

$$\dot{m} = x \left(2000 \frac{\text{lbm}}{\text{hr}} \right) = (0.0924) \left(2000 \frac{\text{lbm}}{\text{hr}} \right)$$
$$= 184.8 \text{ lbm/hr}$$

Since the pressure drop across the heater is 5 psi,

$$p_D = p_F + 5 \text{ psi}$$
$$= 20 \text{ psia} + 5 \text{ psi} = 25 \text{ psia}$$

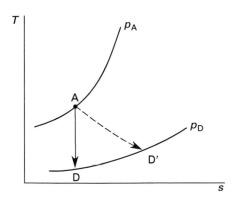

With isentropic expansion through the turbine, using the Mollier diagram, $h_D \approx 1115$ Btu/lbm (liquid-vapor mixture).

From Eq. 25.17,

$$h'_D = h_A - \eta_s(h_A - h_D)$$
$$= 1227.5 \frac{\text{Btu}}{\text{lbm}} - (0.60) \left(1227.5 \frac{\text{Btu}}{\text{lbm}} - 1115 \frac{\text{Btu}}{\text{lbm}} \right)$$
$$= 1160 \text{ Btu/lbm}$$

The power output is

$$P = \eta \dot{m}(h_A - h'_D)$$

$$= \frac{(0.96) \left(184.8 \frac{\text{lbm}}{\text{hr}} \right)}{\times \left(1227.5 \frac{\text{Btu}}{\text{lbm}} - 1160 \frac{\text{Btu}}{\text{lbm}} \right) \left(778 \frac{\text{ft-lbf}}{\text{Btu}} \right)}{\left(3600 \frac{\text{sec}}{\text{hr}} \right) \left(550 \frac{\text{ft-lbf}}{\text{hp-sec}} \right)}$$

$$= \boxed{4.71 \text{ hp}}$$

The answer is (A).

(b) Let $x =$ fraction of steam entering the heater.

From the energy balance equation,

$$m_A h'_D + m_B h_B = m_C h_C$$

$$x \left(1160 \frac{\text{Btu}}{\text{lbm}} \right)$$
$$+ (1-x) \left(38.09 \frac{\text{Btu}}{\text{lbm}} \right) = (1) \left(147.99 \frac{\text{Btu}}{\text{lbm}} \right)$$
$$x = 0.098$$

$$\dot{m}_F = \frac{184.8 \frac{\text{lbm}}{\text{hr}}}{0.098}$$

$$= \boxed{1886 \text{ lbm/hr}}$$

The answer is (C).

SI Solution

(a) Work with the original system to find the steam flow.

From App. 23.N, at 700 kPa and 200°C, $h_A = 2884.8$ kJ/kg.

From App. 23.N, at 21°C, $h_B = 88.14$ kJ/kg.

From App. 23.N, at 80°C, $h_C = 334.91$ kJ/kg.

Let $x =$ fraction of steam in mixture.

From the energy balance equation,

$$m_A h_A + m_B h_B = m_C h_C$$

$$x \left(2884.8 \frac{\text{kJ}}{\text{kg}} \right)$$
$$+ (1-x) \left(88.14 \frac{\text{kJ}}{\text{kg}} \right) = (1) \left(334.91 \frac{\text{kJ}}{\text{kg}} \right)$$
$$x = 0.0882$$

The steam flow is

$$\dot{m} = x \left(0.25 \frac{\text{kg}}{\text{s}} \right) = (0.0882) \left(0.25 \frac{\text{kg}}{\text{s}} \right)$$
$$= 0.0221 \text{ kg/s}$$

Since the pressure drop across the heater is 30 kPa,

$$p_D = p_F + 30 \text{ kPa}$$
$$= 150 \text{ kPa} + 30 \text{ kPa}$$
$$= 180 \text{ kPa}$$

For isentropic expansion through the turbine from the Mollier diagram, $h_D \approx 2583$ kJ/kg (liquid-vapor mixture).

From Eq. 25.17,

$$h_D' = h_A - \eta_s(h_A - h_D)$$
$$= 2884.8 \frac{\text{kJ}}{\text{kg}} - (0.60)\left(2884.8 \frac{\text{kJ}}{\text{kg}} - 2583 \frac{\text{kJ}}{\text{kg}}\right)$$
$$= 2703.7 \text{ kJ/kg}$$

The power output is

$$P = \eta \dot{m}(h_A - h_D')$$
$$= (0.96)\left(0.0221 \frac{\text{kg}}{\text{s}}\right)\left(2884.8 \frac{\text{kJ}}{\text{kg}} - 2703.7 \frac{\text{kJ}}{\text{kg}}\right)$$
$$= \boxed{3.84 \text{ kW}}$$

The answer is (A).

(b) Let x = fraction of steam entering the heater.

From the energy balance equation,

$$m_A h_D' + m_B h_B = m_C h_C$$
$$x\left(2703.7 \frac{\text{kJ}}{\text{kg}}\right)$$
$$+ (1-x)\left(88.14 \frac{\text{kJ}}{\text{kg}}\right) = (1)\left(334.91 \frac{\text{kJ}}{\text{kg}}\right)$$
$$x = 0.0943$$

$$\dot{m}_F = \frac{0.0221 \frac{\text{kg}}{\text{s}}}{0.0943}$$
$$= \boxed{0.234 \text{ kg/s}}$$

The answer is (C).

Thermodynamics

PRACTICE PROBLEMS

1. A steam cycle operates between 650°F and 100°F (340°C and 38°C). What is the maximum possible thermal efficiency?

   (A) 42%
   (B) 49%
   (C) 54%
   (D) 58%

2. A steam Carnot cycle operates between 650°F and 100°F (340°C and 38°C). The turbine and compressor (pump) isentropic efficiencies are 90% and 80%, respectively. What is the thermal efficiency?

   (A) 31%
   (B) 36%
   (C) 41%
   (D) 47%

3. A steam turbine cycle produces 600 MWe of energy. The condenser load is 3.07×10^9 Btu/hr (900 MW). What is the thermal efficiency?

   (A) 32%
   (B) 37%
   (C) 40%
   (D) 44%

4. A steam Rankine cycle operates with 100 psia (700 kPa) saturated steam that is reduced to 1 atm through expansion in a turbine with an isentropic efficiency of 80%. Water is at 80°F (27°C) and 1 atm when it enters the boilerfeed pump. The pump's isentropic efficiency is 60%. What is the cycle's thermal efficiency?

   (A) 10%
   (B) 14%
   (C) 21%
   (D) 26%

5. A turbine and the boilerfeed pumps in a reheat cycle have isentropic efficiencies of 88% and 96%, respectively. The cycle starts with water at 60°F (16°C) at the inlet to the boilerfeed pump and produces 600°F (300°C), 600 psia (4 MPa) steam. The steam is reheated when its pressure drops during the first expansion to 20 psia (150 kPa). What is the thermal efficiency of the cycle?

   (A) 28%
   (B) 34%
   (C) 39%
   (D) 42%

6. The cycle in Prob. 5 is modified to include a bleed of 270°F (130°C) steam (within the second expansion) for feedwater heating in a closed feedwater heater. The condenser pressure is unchanged. Steam leaves the feedwater heater as saturated liquid. The terminal temperature difference in the heater is 6°F (3°C). Condensate and drip pump work are both negligible. What is the thermal efficiency of the cycle?

   (A) 23%
   (B) 28%
   (C) 36%
   (D) 44%

7. (*Time limit: one hour*) A reheat steam cycle operates as shown. The pump work between points F and G is 0.15 Btu/lbm (0.3 kJ/kg).

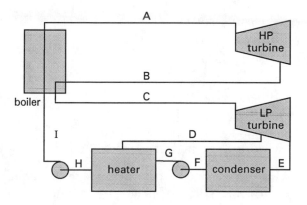

At A: 900 psia (6.2 MPa)
      800°F (420°C)

At B: 200 psia (1.5 MPa)
      1270 Btu/lbm (2960 kJ/kg)

At C: 190 psia (1.4 MPa)
      800°F (420°C)

At D: 50 psia (350 kPa)
      1280 Btu/lbm (2980 kJ/kg)

At E: 2 in Hg absolute
      1075 Btu/lbm (2500 kJ/kg)

At F: 69.73 Btu/lbm (162.5 kJ/kg)

At H: 250.2 Btu/lbm (583.0 kJ/kg)
    0.0173 ft³/lbm

At I: 253.1 Btu/lbm (589.7 kJ/kg)

(a) What is the isentropic efficiency of the high-pressure turbine?

(A) 54%
(B) 61%
(C) 69%
(D) 75%

(b) What is the thermal efficiency of the cycle?

(A) 22%
(B) 29%
(C) 34%
(D) 41%

8. (*Time limit: one hour*) A precision air turbine is used to drive a small dentist's drill. 140°F (60°C) air enters the turbine at the rate of 15 lbm/hr (1.9 g/s). The output of the turbine is 0.25 hp (0.19 kW). The turbine exhausts to 15 psia (103.5 kPa). The flow is steady and the process is adiabatic. The isentropic efficiency of the expansion process is 60%.

(a) What is the exhaust temperature?

(A) 370°R (210K)
(B) 420°R (230K)
(C) 450°R (240K)
(D) 610°R (340K)

(b) What is the inlet air pressure?

(A) 160 psia (1.2 MPa)
(B) 210 psia (1.5 MPa)
(C) 240 psia (1.7 MPa)
(D) 290 psia (2.0 MPa)

(c) What is the change in entropy through the turbine?

(A) 0.078 Btu/lbm-°R (0.36 kJ/kg·K)
(B) 0.10 Btu/lbm-°R (0.46 kJ/kg·K)
(C) 0.18 Btu/lbm-°R (0.82 kJ/kg·K)
(D) 0.32 Btu/lbm-°R (1.5 kJ/kg·K)

SOLUTIONS

1. *Customary U.S. Solution*

The maximum possible thermal efficiency is given by the Carnot cycle. From Eq. 26.8,

$$\eta_{th} = \frac{T_{high} - T_{low}}{T_{high}}$$

The absolute temperatures are

$$T_{high} = 650°F + 460 = 1110°R$$
$$T_{low} = 100°F + 460 = 560°F$$
$$\eta_{th} = \frac{1110°R - 560°R}{1110°R} = \boxed{0.495 \quad (49.5\%)}$$

The answer is (B).

SI Solution

The maximum possible thermal efficiency is given by the Carnot cycle. From Eq. 26.8,

$$\eta_{th} = \frac{T_{high} - T_{low}}{T_{high}}$$

The absolute temperatures are

$$T_{high} = 340°C + 273 = 613K$$
$$T_{low} = 38°C + 273 = 311K$$
$$\eta_{th} = \frac{613K - 311K}{613K} = \boxed{0.493 \quad (49.3\%)}$$

The answer is (B).

2. *Customary U.S. Solution*

Refer to the Carnot cycle (Fig. 26.2).

At point A, from App. 23.A, for saturated liquid at $T_A = 650°F$,

$$h_A = 696.5 \text{ Btu/lbm}$$
$$s_A = 0.8836 \text{ Btu/lbm-°F}$$

At point B, from App. 23.A, for saturated vapor at $T_B = 650°F$,

$$h_B = 1118.7 \text{ Btu/lbm}$$
$$s_B = 1.2643 \text{ Btu/lbm}$$

At point C,

$$T_C = 100°F$$
$$s_C = s_B = 1.2643 \text{ Btu/lbm-°F}$$

From App. 23.A,

$$s_f = 0.1296 \text{ Btu/lbm-°F}$$
$$s_g = 1.9822 \text{ Btu/lbm-°F}$$
$$h_f = 68.05 \text{ Btu/lbm}$$
$$h_{fg} = 1037 \text{ Btu/lbm}$$

$$x_C = \frac{s_C - s_f}{s_g - s_f} = \frac{1.2643 \frac{\text{Btu}}{\text{lbm-°F}} - 0.1296 \frac{\text{Btu}}{\text{lbm-°F}}}{1.9822 \frac{\text{Btu}}{\text{lbm-°F}} - 0.1296 \frac{\text{Btu}}{\text{lbm-°F}}}$$

$$= 0.612$$

$$h_C = h_f + x_C h_{fg}$$

$$= 68.05 \frac{\text{Btu}}{\text{lbm}} + (0.612)\left(1037 \frac{\text{Btu}}{\text{lbm}}\right)$$

$$= 702.7 \text{ Btu/lbm}$$

At point D,

$$T_D = 100°\text{F}$$
$$s_D = s_A = 0.8836 \text{ Btu/lbm-°F}$$

$$x_D = \frac{s_D - s_f}{s_g - s_f} = \frac{0.8836 \frac{\text{Btu}}{\text{lbm-°F}} - 0.1296 \frac{\text{Btu}}{\text{lbm-°F}}}{1.9822 \frac{\text{Btu}}{\text{lbm-°F}} - 0.1296 \frac{\text{Btu}}{\text{lbm-°F}}}$$

$$= 0.407$$

$$h_D = h_f + x_D h_{fg}$$

$$= 68.05 \frac{\text{Btu}}{\text{lbm}} + (0.407)\left(1037 \frac{\text{Btu}}{\text{lbm}}\right)$$

$$= 490.1 \text{ Btu/lbm}$$

From Eq. 26.9, due to the inefficiency of the turbine,

$$h'_C = h_B - \eta_{s,\text{turbine}}(h_B - h_C)$$

$$= 1118.7 \frac{\text{Btu}}{\text{lbm}} - (0.9)\left(1118.7 \frac{\text{Btu}}{\text{lbm}} - 702.7 \frac{\text{Btu}}{\text{lbm}}\right)$$

$$= 744.3 \text{ Btu/lbm}$$

From Eq. 26.10, due to the inefficiency of the pump,

$$h'_A = h_D + \frac{h_A - h_D}{\eta_{s,\text{pump}}}$$

$$= 490.1 \frac{\text{Btu}}{\text{lbm}} + \frac{696.5 \frac{\text{Btu}}{\text{lbm}} - 490.1 \frac{\text{Btu}}{\text{lbm}}}{0.8}$$

$$= 748.1 \text{ Btu/lbm}$$

From Eq. 26.8, the thermal efficiency of the entire cycle is

$$\eta_{\text{th}} = \frac{(h_B - h'_C) - (h'_A - h_D)}{h_B - h'_A}$$

$$= \frac{\left(1118.7 \frac{\text{Btu}}{\text{lbm}} - 744.3 \frac{\text{Btu}}{\text{lbm}}\right) - \left(748.1 \frac{\text{Btu}}{\text{lbm}} - 490.1 \frac{\text{Btu}}{\text{lbm}}\right)}{1118.7 \frac{\text{Btu}}{\text{lbm}} - 748.1 \frac{\text{Btu}}{\text{lbm}}}$$

$$= \boxed{0.314 \ (31.4\%)}$$

The answer is (A).

SI Solution

At point A, $T_A = 340°\text{C}$.

From App. 23.N, for saturated liquid,

$$h_A = 1594.2 \text{ kJ/kg}$$
$$s_A = 3.6594 \text{ kJ/kg·K}$$

At point B, $T_B = 340°\text{C}$.

From App. 23.N, for saturated vapor,

$$h_B = 2622.0 \text{ kJ/kg}$$
$$s_B = 5.3357 \text{ kJ/kg·K}$$

At point C,

$$T_C = 38°\text{C}$$
$$s_C = s_B = 5.3357 \text{ kJ/kg·K}$$

From App. 23.N,

$$s_f = 0.5458 \text{ kJ/kg·K}$$
$$s_g = 8.2950 \text{ kJ/kg·K}$$
$$h_f = 159.21 \text{ kJ/kg}$$
$$h_{fg} = 2411.5 \text{ kJ/kg}$$

$$x_C = \frac{s_C - s_f}{s_g - s_f} = \frac{5.3357 \frac{\text{kJ}}{\text{kg·K}} - 0.5458 \frac{\text{kJ}}{\text{kg·K}}}{8.2950 \frac{\text{kJ}}{\text{kg·K}} - 0.5458 \frac{\text{kJ}}{\text{kg·K}}}$$

$$= 0.618$$

$$h_C = h_f + x_C h_{fg}$$

$$= 159.21 \frac{\text{kJ}}{\text{kg}} + (0.618)\left(2411.5 \frac{\text{kJ}}{\text{kg}}\right)$$

$$= 1649.5 \text{ kJ/kg}$$

At point D,

$$T_D = 38°C$$

$$s_D = s_A = 3.6594 \text{ kJ/kg·K}$$

$$x_D = \frac{s_D - s_f}{s_g - s_f}$$

$$= \frac{3.6594 \, \dfrac{\text{kJ}}{\text{kg·K}} - 0.5458 \, \dfrac{\text{kJ}}{\text{kg·K}}}{8.2950 \, \dfrac{\text{kJ}}{\text{kg·K}} - 0.5458 \, \dfrac{\text{kJ}}{\text{kg·K}}}$$

$$= 0.402$$

$$h_D = h_f + x_D h_{fg}$$

$$= 159.21 \, \frac{\text{kJ}}{\text{kg}} + (0.402)\left(2411.5 \, \frac{\text{kJ}}{\text{kg}}\right)$$

$$= 1128.6 \text{ kJ/kg}$$

From Eq. 26.9, due to the inefficiency of the turbine,

$$h'_C = h_B - \eta_{s,\text{turbine}}(h_B - h_C)$$

$$= 2622.0 \, \frac{\text{kJ}}{\text{kg}} - (0.9)\left(2622.0 \, \frac{\text{kJ}}{\text{kg}} - 1649.5 \, \frac{\text{kJ}}{\text{kg}}\right)$$

$$= 1746.7 \text{ kJ/kg}$$

From Eq. 26.10, due to the inefficiency of the pump,

$$h'_A = h_D + \frac{h_A - h_D}{\eta_{s,\text{pump}}}$$

$$= 1128.6 \, \frac{\text{kJ}}{\text{kg}} + \frac{1594.2 \, \dfrac{\text{kJ}}{\text{kg}} - 1128.6 \, \dfrac{\text{kJ}}{\text{kg}}}{0.8}$$

$$= 1710.6 \text{ kJ/kg}$$

From Eq. 26.8, the thermal efficiency of the entire cycle is

$$\eta_{th} = \frac{(h_B - h'_C) - (h'_A - h_D)}{h_B - h'_A}$$

$$= \frac{\left(2622 \, \dfrac{\text{kJ}}{\text{kg}} - 1746.7 \, \dfrac{\text{kJ}}{\text{kg}}\right) - \left(1710.6 \, \dfrac{\text{kJ}}{\text{kg}} - 1128.6 \, \dfrac{\text{kJ}}{\text{kg}}\right)}{2622 \, \dfrac{\text{kJ}}{\text{kg}} - 1710.6 \, \dfrac{\text{kJ}}{\text{kg}}}$$

$$= \boxed{0.322 \ (32.2\%)}$$

The answer is (A).

3. *Customary U.S. Solution*

From Eq. 26.1, the thermal efficiency is

$$\eta_{th} = \frac{Q_{in} - Q_{out}}{Q_{in}}$$

The condenser load is $Q_{out} = 3.07 \times 10^9$ Btu/hr.

The net work is

$$W_{net} = Q_{in} - Q_{out}$$

The boiler load is

$$Q_{in} = W_{net} + Q_{out}$$

$$= (600 \text{ MW})\left(1000 \, \frac{\text{kW}}{\text{MW}}\right)\left(3412 \, \frac{\frac{\text{Btu}}{\text{hr}}}{\text{kW}}\right)$$

$$+ 3.07 \times 10^9 \, \frac{\text{Btu}}{\text{hr}}$$

$$= 5.12 \times 10^9 \text{ Btu/hr}$$

$$\eta_{th} = \frac{5.12 \times 10^9 \, \dfrac{\text{Btu}}{\text{hr}} - 3.07 \times 10^9 \, \dfrac{\text{Btu}}{\text{hr}}}{5.12 \times 10^9 \, \dfrac{\text{Btu}}{\text{hr}}}$$

$$= \boxed{0.400 \ (40\%)}$$

The answer is (C).

SI Solution

From Eq. 26.1, the thermal efficiency is

$$\eta_{th} = \frac{Q_{in} - Q_{out}}{Q_{in}}$$

The condenser load is $Q_{out} = 900$ MW.

The boiler load is

$$Q_{in} = W_{net} + Q_{out}$$

$$= 600 \text{ MW} + 900 \text{ MW}$$

$$= 1500 \text{ MW}$$

$$\eta_{th} = \frac{1500 \text{ MW} - 900 \text{ MW}}{1500 \text{ MW}}$$

$$= \boxed{0.400 \ (40\%)}$$

The answer is (C).

4. *Customary U.S. Solution*

At point A, $p_A = 100$ psia.

From App. 23.B, the enthalpy of saturated liquid is $h_A = 298.6$ Btu/lbm.

At point B, $p_B = 100$ psia.

From App. 23.B, the enthalpy and entropy of saturated vapor are

$$h_B = 1187.8 \text{ Btu/lbm}$$

$$s_B = 1.6034 \text{ Btu/lbm-°R}$$

At point C,

$$p_C = 1 \text{ atm}$$
$$s_C = s_B = 1.6034 \text{ Btu/lbm-}°\text{R}$$

From App. 23.B, the entropy and enthalpy of saturated liquid and entropy and enthalpy of vaporization are

$$s_f = 0.3121 \text{ Btu/lbm-}°\text{R}$$
$$h_f = 180.15 \text{ Btu/lbm}$$
$$s_{fg} = 1.4446 \text{ Btu/lbm-}°\text{R}$$
$$h_{fg} = 970.4 \text{ Btu/lbm}$$

$$x_C = \frac{s_C - s_f}{s_{fg}} = \frac{1.6034 \dfrac{\text{Btu}}{\text{lbm-}°\text{R}} - 0.3121 \dfrac{\text{Btu}}{\text{lbm-}°\text{R}}}{1.4446 \dfrac{\text{Btu}}{\text{lbm-}°\text{R}}}$$

$$= 0.894$$

$$h_C = h_f + x_C h_{fg}$$
$$= 180.15 \frac{\text{Btu}}{\text{lbm}} + (0.894)\left(970.4 \frac{\text{Btu}}{\text{lbm}}\right)$$
$$= 1047.7 \text{ Btu/lbm}$$

At point D, $T = 80°\text{F}$ and $p_D = 1$ atm (subcooled). h and v are essentially independent of pressure.

From App. 23.A, the enthalpy and specific volume of saturated liquid are

$$h_D = 48.09 \text{ Btu/lbm}$$
$$v_D = 0.01607 \text{ ft}^3/\text{lbm}$$

At point E, $p_E = p_A = 100$ psia.

From Eq. 26.14,

$$h_E = h_D + v_D(p_E - p_D)$$
$$= 48.09 \frac{\text{Btu}}{\text{lbm}} + \left(0.01607 \frac{\text{ft}^3}{\text{lbm}}\right)$$
$$\times \frac{\left(100 \dfrac{\text{lbf}}{\text{in}^2} - 14.7 \dfrac{\text{lbf}}{\text{in}^2}\right)\left(144 \dfrac{\text{in}^2}{\text{ft}^2}\right)}{778 \dfrac{\text{ft-lbf}}{\text{Btu}}}$$
$$= 48.34 \text{ Btu/lbm}$$

From Eq. 26.18, due to the inefficiency of the turbine,

$$h'_C = h_B - \eta_{s,\text{turbine}}(h_B - h_C)$$
$$= 1187.8 \frac{\text{Btu}}{\text{lbm}} - (0.80)\left(1187.8 \frac{\text{Btu}}{\text{lbm}}\right.$$
$$\left. - 1047.7 \frac{\text{Btu}}{\text{lbm}}\right)$$
$$= 1075.7 \text{ Btu/lbm}$$

From Eq. 26.19, due to the inefficiency of the pump,

$$h'_E = h_D + \frac{h_E - h_D}{\eta_{s,\text{pump}}}$$
$$= 48.09 \frac{\text{Btu}}{\text{lbm}} + \frac{48.34 \dfrac{\text{Btu}}{\text{lbm}} - 48.09 \dfrac{\text{Btu}}{\text{lbm}}}{0.6}$$
$$= 48.51 \text{ Btu/lbm}$$

From Eq. 26.17, the thermal efficiency of the cycle is

$$\eta_{th} = \frac{(h_B - h'_C) - (h'_E - h_D)}{h_B - h'_E}$$

$$= \frac{\left(1187.8 \dfrac{\text{Btu}}{\text{lbm}} - 1075.7 \dfrac{\text{Btu}}{\text{lbm}}\right)}{1187.8 \dfrac{\text{Btu}}{\text{lbm}} - 48.51 \dfrac{\text{Btu}}{\text{lbm}}} \frac{-48.51 \dfrac{\text{Btu}}{\text{lbm}} - 48.09 \dfrac{\text{Btu}}{\text{lbm}}}{}$$

$$= \boxed{0.098 \ (9.8\%)}$$

The answer is (A).

SI Solution

At point A, $p_A = 700$ kPa.

From App. 23.O, the enthalpy of saturated liquid is $h_A = 670.56$ kJ/kg.

At point B, $p_B = 700$ kPa.

From App. 23.O, the enthalpy and entropy of saturated vapor are

$$h_B = 2763.5 \text{ kJ/kg}$$
$$s_B = 6.7080 \text{ kJ/kg·K}$$

At point C,

$$p_C = 1 \text{ atm}$$
$$s_C = s_B = 6.7080 \text{ kJ/kg·K}$$

From App. 23.O, the entropy and enthalpy of saturated liquid, the entropy of saturated vapor, and the enthalpy of vaporization are

$$s_f = 1.3026 \text{ kJ/kg·K}$$
$$h_f = 417.46 \text{ kJ/kg}$$
$$s_g = 7.3594 \text{ kJ/kg·K}$$
$$h_{fg} = 2258.0 \text{ kJ/kg}$$

Thermodynamics

$$x_C = \frac{s_C - s_f}{s_g - s_f}$$

$$= \frac{6.7080 \, \frac{\text{kJ}}{\text{kg·K}} - 1.3026 \, \frac{\text{kJ}}{\text{kg·K}}}{7.3594 \, \frac{\text{kJ}}{\text{kg·K}} - 1.3026 \, \frac{\text{kJ}}{\text{kg·K}}}$$

$$= 0.892$$

$$h_C = h_f + x_C h_{fg}$$

$$= 417.46 \, \frac{\text{kJ}}{\text{kg}} + (0.892)\left(2258.0 \, \frac{\text{kJ}}{\text{kg}}\right)$$

$$= 2431.6 \text{ kJ/kg}$$

At point D, $T = 27°C$ and $p_D = 1$ atm (subcooled). h and v are essentially independent of pressure.

From App. 23.N, the enthalpy and specific volume of saturated liquid are

$$h_D = 113.25 \text{ kJ/kg}$$
$$v_D = 1.0035 \text{ cm}^3/\text{g}$$

At point E, $p_E = p_A = 700$ kPa.

From Eq. 26.14,

$$h_E = h_D + v_D(p_E - p_D)$$

$$= 113.25 \, \frac{\text{kJ}}{\text{kg}} + \left(1.0035 \, \frac{\text{cm}^3}{\text{g}}\right)\left(\frac{1 \text{ m}^3}{10^6 \text{ cm}^3}\right)$$

$$\times \left(1000 \, \frac{\text{g}}{\text{kg}}\right)(700 \text{ kPa} - 101 \text{ kPa})$$

$$= 113.85 \text{ kJ/kg}$$

From Eq. 26.18, due to the inefficiency of the turbine,

$$h'_C = h_B - \eta_{s,\text{turbine}}(h_B - h_C)$$

$$= 2763.5 \, \frac{\text{kJ}}{\text{kg}} - (0.80)\left(2763.5 \, \frac{\text{kJ}}{\text{kg}} - 2431.6 \, \frac{\text{kJ}}{\text{kg}}\right)$$

$$= 2498.0 \text{ kJ/kg}$$

From Eq. 26.19, due to the inefficiency of the pump,

$$h'_E = h_D + \frac{h_E - h_D}{\eta_{s,\text{pump}}}$$

$$= 113.25 \, \frac{\text{kJ}}{\text{kg}} + \frac{113.85 \, \frac{\text{kJ}}{\text{kg}} - 113.25 \, \frac{\text{kJ}}{\text{kg}}}{0.6}$$

$$= 114.25 \text{ kJ/kg}$$

From Eq. 26.17, the thermal efficiency of the cycle is

$$\eta_{th} = \frac{(h_B - h'_C) - (h'_E - h_D)}{h_B - h'_E}$$

$$= \frac{\left(2763.5 \, \frac{\text{kJ}}{\text{kg}} - 2498.0 \, \frac{\text{kJ}}{\text{kg}}\right)}{2763.5 \, \frac{\text{kJ}}{\text{kg}} - 114.25 \, \frac{\text{kJ}}{\text{kg}}}$$

$$= \boxed{0.10 \quad (10\%)}$$

The answer is (A).

5. *Customary U.S. Solution*

Refer to the reheat cycle (Fig. 26.8).

At point B, $p_B = 600$ psia.

From App. 23.B, the enthalpy of the saturated liquid is $h_B = 471.7$ Btu/lbm.

At point C, $p_C = 600$ psia.

From App. 23.B, the enthalpy of the saturated vapor is $h_C = 1204.1$ Btu/lbm.

At point D,

$$T_D = 600°F$$
$$p_D = 600 \text{ psia}$$

From App. 23.C, the enthalpy and entropy of superheated vapor are

$$h_D = 1289.5 \text{ Btu/lbm}$$
$$s_D = 1.5320 \text{ Btu/lbm-°R}$$

At point E,

$$p_E = 20 \text{ psia}$$
$$s_E = s_D = 1.5320 \text{ Btu/lbm-°R}$$

From App. 23.B, the various saturation properties are

$$s_f = 0.3358 \text{ Btu/lbm-°R}$$
$$s_{fg} = 1.3962 \text{ Btu/lbm-°R}$$
$$h_f = 196.26 \text{ Btu/lbm}$$
$$h_{fg} = 960.1 \text{ Btu/lbm}$$

$$x_E = \frac{s_E - s_f}{s_g - s_f} = \frac{1.5320 \, \frac{\text{Btu}}{\text{lbm-°R}} - 0.3358 \, \frac{\text{Btu}}{\text{lbm-°R}}}{1.3962 \, \frac{\text{Btu}}{\text{lbm-°R}}}$$

$$= 0.857$$

$$h_E = h_f + x_E h_{fg}$$

$$= 196.26 \, \frac{\text{Btu}}{\text{lbm}} + (0.857)\left(960.1 \, \frac{\text{Btu}}{\text{lbm}}\right)$$

$$= 1019.1 \text{ Btu/lbm}$$

From Eq. 26.38, due to the inefficiency of the turbine,

$$h'_E = h_D - \eta_{s,\text{turbine}}(h_D - h_E)$$

$$= 1289.5\,\frac{\text{Btu}}{\text{lbm}} - (0.88)\left(1289.5\,\frac{\text{Btu}}{\text{lbm}}\right.$$

$$\left. - 1019.1\,\frac{\text{Btu}}{\text{lbm}}\right)$$

$$= 1051.5\ \text{Btu/lbm}$$

At point F, the temperature has been returned to 600°F, but the pressure stays at the expansion pressure, p_E.

$$p_F = 20\ \text{psia}$$
$$T_F = 600°F$$

From App. 23.C, the enthalpy and entropy of the superheated vapor is

$$h_F = 1334.8\ \text{Btu/lbm}$$
$$s_F = 1.9395\ \text{Btu/lbm-°R}$$

At point G,

$$T_G = 60°F$$
$$s_G = s_F = 1.9395\ \text{Btu/lbm-°R}$$

From App. 23.A, various saturation properties are

$$s_f = 0.05555\ \text{Btu/lbm-°R}$$
$$s_g = 2.0943\ \text{Btu/lbm-°R}$$
$$h_f = 28.08\ \text{Btu/lbm}$$
$$h_{fg} = 1059.6\ \text{Btu/lbm}$$

$$x_G = \frac{s_G - s_f}{s_g - s_f} = \frac{1.9395\,\frac{\text{Btu}}{\text{lbm-°R}} - 0.05555\,\frac{\text{Btu}}{\text{lbm-°R}}}{2.0943\,\frac{\text{Btu}}{\text{lbm-°R}} - 0.05555\,\frac{\text{Btu}}{\text{lbm-°R}}}$$

$$= 0.924$$

$$h_G = h_f + x_G h_{fg}$$

$$= 28.08\,\frac{\text{Btu}}{\text{lbm}} + (0.924)\left(1059.6\,\frac{\text{Btu}}{\text{lbm}}\right)$$

$$= 1007.2\ \text{Btu/lbm}$$

From Eq. 26.39, due to the inefficiency of the turbine,

$$h'_G = h_F - \eta_{s,\text{turbine}}(h_F - h_G)$$

$$= 1334.8\,\frac{\text{Btu}}{\text{lbm}} - (0.88)\left(1334.8\,\frac{\text{Btu}}{\text{lbm}}\right.$$

$$\left. - 1007.2\,\frac{\text{Btu}}{\text{lbm}}\right)$$

$$= 1046.5\ \text{Btu/lbm}$$

At point H, $T_H = 60°F$.

From App. 23.A, the saturation pressure, enthalpy, and specific volume of the saturated liquid are

$$p_H = 0.2563\ \text{psia}$$
$$h_H = 28.08\ \text{Btu/lbm}$$
$$v_H = 0.01604\ \text{ft}^3/\text{lbm}$$

At point A, $p_A = 600$ psia.

From Eq. 26.14,

$$h_A = h_H + v_H(p_A - p_H)$$

$$= 28.08\,\frac{\text{Btu}}{\text{lbm}}$$

$$+ \frac{\left(0.01604\,\frac{\text{ft}^3}{\text{lbm}}\right)\times\left(600\,\frac{\text{lbf}}{\text{in}^2} - 0.2563\,\frac{\text{lbf}}{\text{in}^2}\right)\left(144\,\frac{\text{in}^2}{\text{ft}^2}\right)}{778\,\frac{\text{lbf-ft}}{\text{Btu}}}$$

$$= 29.86\ \text{Btu/lbm}$$

From Eq. 26.40, due to the inefficiency of the pump,

$$h'_A = h_H + \frac{h_A - h_H}{\eta_{s,\text{pump}}}$$

$$= 28.08\,\frac{\text{Btu}}{\text{lbm}} + \frac{29.86\,\frac{\text{Btu}}{\text{lbm}} - 28.08\,\frac{\text{Btu}}{\text{lbm}}}{0.96}$$

$$= 29.93\ \text{Btu/lbm}$$

From Eq. 26.37, the thermal efficiency of the cycle for a non-isentropic process for the turbine and the pump is

$$\eta_{th} = \frac{(h_D - h'_A) + (h_F - h'_E) - (h'_G - h_H)}{(h_D - h'_A) + (h_F - h'_E)}$$

$$= \frac{\begin{aligned}&\left(1289.5\,\tfrac{\text{Btu}}{\text{lbm}} - 29.93\,\tfrac{\text{Btu}}{\text{lbm}}\right)\\&+\left(1334.8\,\tfrac{\text{Btu}}{\text{lbm}} - 1051.5\,\tfrac{\text{Btu}}{\text{lbm}}\right)\\&-\left(1046.5\,\tfrac{\text{Btu}}{\text{lbm}} - 28.08\,\tfrac{\text{Btu}}{\text{lbm}}\right)\end{aligned}}{\begin{aligned}&\left(1289.5\,\tfrac{\text{Btu}}{\text{lbm}} - 29.93\,\tfrac{\text{Btu}}{\text{lbm}}\right)\\&+\left(1334.8\,\tfrac{\text{Btu}}{\text{lbm}} - 1051.5\,\tfrac{\text{Btu}}{\text{lbm}}\right)\end{aligned}}$$

$$= \boxed{0.340\ \ (34.0\%)}$$

The answer is (B).

SI Solution

Refer to the reheat cycle (Fig. 26.8).

At point B, $p_B = 4$ MPa.

From App. 23.O, the enthalpy of saturated liquid is $h_B = 1087.3$ kJ/kg.

At point C, $p_C = 4$ MPa.

From App. 23.O, the enthalpy of saturated vapor is $h_C = 2801.4$ kJ/kg.

At point D, $p_D = 4$ MPa and $T_D = 300°C$.

From the Mollier diagram, the enthalpy of superheated vapor is $h_D = 2980$ kJ/kg.

At point E, from the Mollier diagram, assuming isentropic expansion, $h_E = 2395$ kJ/kg.

From Eq. 26.38, due to the inefficiency of the turbine,

$$h'_E = h_D - \eta_{s,\text{turbine}}(h_D - h_E)$$
$$= 2980 \frac{\text{kJ}}{\text{kg}} - (0.88)\left(2980 \frac{\text{kJ}}{\text{kg}} - 2395 \frac{\text{kJ}}{\text{kg}}\right)$$
$$= 2465.2 \text{ kJ/kg}$$

At point F,
$$p_F = 150 \text{ kPa}$$
$$T_F = 300°C$$

From App. 23.P, the enthalpy and entropy of the superheated vapor is

$$h_F = 3073.1 \text{ kJ/kg}$$
$$s_F = 8.0720 \text{ kJ/kg·K}$$

At point G,

$$T_G = 16°C$$
$$s_G = s_F = 8.0720 \text{ kJ/kg·K}$$

From App. 23.N, the various saturation properties are

$$s_f = 0.2390 \text{ kJ/kg·K}$$
$$s_g = 8.7582 \text{ kJ/kg·K}$$
$$h_f = 67.19 \text{ kJ/kg}$$
$$h_{fg} = 2463.6 \text{ kJ/kg}$$

$$x_G = \frac{s_G - s_f}{s_g - s_f} = \frac{8.0720 \frac{\text{kJ}}{\text{kg·K}} - 0.2390 \frac{\text{kJ}}{\text{kg·K}}}{8.7582 \frac{\text{kJ}}{\text{kg·K}} - 0.2390 \frac{\text{kJ}}{\text{kg·K}}}$$
$$= 0.919$$

$$h_G = h_f + x_G h_{fg}$$
$$= 67.19 \frac{\text{kJ}}{\text{kg}} + (0.919)\left(2463.6 \frac{\text{kJ}}{\text{kg}}\right)$$
$$= 2331.2 \text{ kJ/kg}$$

From Eq. 26.39, due to the inefficiency of the turbine,

$$h'_G = h_F - \eta_{s,\text{turbine}}(h_F - h_G)$$
$$= 3073.1 \frac{\text{kJ}}{\text{kg}} - (0.88)\left(3073.1 \frac{\text{kJ}}{\text{kg}} - 2331.2 \frac{\text{kJ}}{\text{kg}}\right)$$
$$= 2420.2 \text{ kJ/kg}$$

At point H, $T_H = 16°C$.

From App. 23.N, the saturation pressure, enthalpy, and specific volume of the saturated liquid are

$$p_H = (0.01818 \text{ bar})\left(100 \frac{\text{kPa}}{\text{bar}}\right) = 1.818 \text{ kPa}$$
$$h_H = 67.19 \text{ kJ/kg}$$
$$v_H = \left(1.0011 \frac{\text{cm}}{\text{g}^3}\right)\left(1000 \frac{\text{g}}{\text{kg}}\right)\left(\frac{1 \text{ m}}{100 \text{ cm}}\right)^3$$
$$= 1.0011 \times 10^{-3} \text{ m}^3/\text{kg}$$

At point A,

$$p_A = (4 \text{ MPa})\left(1000 \frac{\text{kPa}}{\text{MPa}}\right)$$
$$= 4000 \text{ kPa}$$

From Eq. 26.14,

$$h_A = h_H + v_H(p_A - p_H)$$
$$= 67.19 \frac{\text{kJ}}{\text{kg}} + \left(1.0011 \times 10^{-3} \frac{\text{m}^3}{\text{kg}}\right)$$
$$\times (4000 \text{ kPa} - 1.818 \text{ kPa})$$
$$= 71.19 \text{ kJ/kg}$$

From Eq. 26.40, due to the inefficiency of the pump,

$$h'_A = h_H + \frac{h_A - h_H}{\eta_{s,\text{pump}}}$$
$$= 67.19 \frac{\text{kJ}}{\text{kg}} + \frac{71.19 \frac{\text{kJ}}{\text{kg}} - 67.19 \frac{\text{kJ}}{\text{kg}}}{0.96}$$
$$= 71.36 \text{ kJ/kg}$$

From Eq. 26.37, the thermal efficiency of the cycle for a non-isentropic process for the turbine and the pump is

$$\eta_{th} = \frac{(h_D - h_A') + (h_F - h_E') - (h_G' - h_H)}{(h_D - h_A') + (h_F - h_E')}$$

$$= \frac{\left(2980\,\dfrac{kJ}{kg} - 71.36\,\dfrac{kJ}{kg}\right)}{} $$

$$= \frac{\begin{aligned} &\left(2980\,\tfrac{kJ}{kg} - 71.36\,\tfrac{kJ}{kg}\right) \\ &+ \left(3073.1\,\tfrac{kJ}{kg} - 2465.2\,\tfrac{kJ}{kg}\right) \\ &- \left(2420.2\,\tfrac{kJ}{kg} - 67.19\,\tfrac{kJ}{kg}\right) \end{aligned}}{\begin{aligned} &\left(2980\,\tfrac{kJ}{kg} - 71.36\,\tfrac{kJ}{kg}\right) \\ &+ \left(3073.1\,\tfrac{kJ}{kg} - 2465.2\,\tfrac{kJ}{kg}\right) \end{aligned}}$$

$$= \boxed{0.331 \quad (33.1\%)}$$

The answer is (B).

6. *Customary U.S. Solution*

Refer to the following diagram.

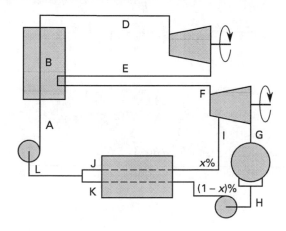

From Prob. 5,

$$h_B = 471.7 \text{ Btu/lbm}$$
$$h_D = 1289.5 \text{ Btu/lbm}$$
$$h_E' = 1051.5 \text{ Btu/lbm}$$
$$h_F = 1334.8 \text{ Btu/lbm}$$
$$s_F = 1.9395 \text{ Btu/lbm-}°\text{R}$$
$$h_G' = 1046.5 \text{ Btu/lbm}$$
$$h_H = 28.08 \text{ Btu/lbm}$$
$$p_H = 0.2563 \text{ psia}$$
$$T_H = 60°\text{F}$$

At point I,

$$T_I = 270°\text{F}$$
$$s_I = s_F = 1.9395 \text{ Btu/lbm-}°\text{R}$$

These conditions are for superheated steam. From App. 23.C,

$$p_I = 4.5 \text{ psia}$$
$$h_I = 1181.0 \text{ Btu/lbm}$$

From Eq. 26.39, due to the inefficiency of the turbine,

$$\begin{aligned} h_I' &= h_F - \eta_{s,\text{turbine}}(h_F - h_I) \\ &= 1334.8\,\frac{\text{Btu}}{\text{lbm}} - (0.88)\left(1334.8\,\frac{\text{Btu}}{\text{lbm}}\right. \\ &\quad \left. -1181.0\,\frac{\text{Btu}}{\text{lbm}}\right) \\ &= 1199.5 \text{ Btu/lbm} \end{aligned}$$

At point J, from App. 23.B, the saturated liquid enthalpy and temperature at 4.5 psia are approximately $h_J = 125.5$ Btu/lbm and 157.6°F, respectively.

At point K, the temperature is

$$157.6°\text{F} \ - 6°\text{F} \approx 152°\text{F}$$

Since the water is subcooled, enthalpy is a function of temperature only.

$$h_K = 119.96 \text{ Btu/lbm}$$

The condensate pump will match the pressure at J: $p_K = 4.5$ psia.

From the energy balance in the heater,

$$(1-x)(h_K - h_H) = x(h_I' - h_J)$$
$$\begin{aligned} (1-x)&\left(119.96\,\frac{\text{Btu}}{\text{lbm}}\right. \\ &\left. - 28.08\,\frac{\text{Btu}}{\text{lbm}}\right) = x\left(1199.5\,\frac{\text{Btu}}{\text{lbm}} - 125.5\,\frac{\text{Btu}}{\text{lbm}}\right) \end{aligned}$$
$$1165.9x = 91.52$$
$$x = 0.0785$$

At point L, $p_L = 4.5$ psia.

$$\begin{aligned} h_L &= xh_J + (1-x)h_K \\ &= (0.0785)\left(125.5\,\frac{\text{Btu}}{\text{lbm}}\right) \\ &\quad + (1 - 0.0785)\left(119.96\,\frac{\text{Btu}}{\text{lbm}}\right) \\ &= 120.4 \text{ Btu/lbm} \end{aligned}$$

Since this is a subcooled liquid, enthalpy is a function of temperature only, and from App. 23.A,

$$T_L = 152.4°F$$

$$v_L = 0.01635 \text{ ft}^3/\text{lbm}$$

At point A, $p_A = 600$ psia.

From Eq. 26.14,

$$h_A = h_L + v_L(p_A - p_L)$$

$$= 120.4 \frac{\text{Btu}}{\text{lbm}}$$

$$+ \frac{\left(0.01635 \frac{\text{ft}^3}{\text{lbm}}\right) \times \left(600 \frac{\text{lbf}}{\text{in}^2} - 4.5 \frac{\text{lbf}}{\text{in}^2}\right) \left(144 \frac{\text{in}^2}{\text{ft}^2}\right)}{778 \frac{\text{lbf-ft}}{\text{lbm}}}$$

$$= 122.2 \text{ Btu/lbm}$$

From Eq. 26.40, due to the inefficiency of the pump,

$$h'_A = h_L + \frac{h_A - h_L}{\eta_{s,\text{pump}}}$$

$$= 120.4 \frac{\text{Btu}}{\text{lbm}} + \frac{122.2 \frac{\text{Btu}}{\text{lbm}} - 120.4 \frac{\text{Btu}}{\text{lbm}}}{0.96}$$

$$= 122.3 \text{ Btu/lbm}$$

From Eq. 26.37, the thermal efficiency of the entire cycle neglecting condensation and drip pump, is

$$\eta_{\text{th}} = \frac{W_{\text{turbines}} - W_{\text{pump}}}{Q_{\text{in}}}$$

$$= \frac{\begin{pmatrix} (h_D - h'_E) + (h_F - h'_I) \\ +(1 - x)(h'_I - h'_G) - (h'_A - h_L) \end{pmatrix}}{(h_D - h'_A) + (h_F - h'_E)}$$

$$= \frac{\begin{pmatrix} \left(1289.5 \frac{\text{Btu}}{\text{lbm}} - 1051.5 \frac{\text{Btu}}{\text{lbm}}\right) \\ + \left(1334.8 \frac{\text{Btu}}{\text{lbm}} - 1199.5 \frac{\text{Btu}}{\text{lbm}}\right) \\ + (1 - 0.0785)\left(1199.5 \frac{\text{Btu}}{\text{lbm}} - 1046.5 \frac{\text{Btu}}{\text{lbm}}\right) \\ - \left(122.3 \frac{\text{Btu}}{\text{lbm}} - 120.4 \frac{\text{Btu}}{\text{lbm}}\right) \end{pmatrix}}{\begin{pmatrix} \left(1289.5 \frac{\text{Btu}}{\text{lbm}} - 122.3 \frac{\text{Btu}}{\text{lbm}}\right) \\ + \left(1334.8 \frac{\text{Btu}}{\text{lbm}} - 1051.5 \frac{\text{Btu}}{\text{lbm}}\right) \end{pmatrix}}$$

$$= \boxed{0.353 \ (35.3\%)}$$

The answer is (C).

SI Solution

From Prob. 5,

$$h_B = 1087.3 \text{ kJ/kg}$$

$$h_D = 2980 \text{ kJ/kg}$$

$$h'_E = 2465.2 \text{ kJ/kg}$$

$$h_F = 3073.1 \text{ kJ/kg}$$

$$s_F = 8.0720 \text{ kJ/kg·K}$$

$$h'_G = 2420.2 \text{ kJ/kg}$$

$$h_H = 67.19 \text{ kJ/kg}$$

At point I,

$$T_I = 130°C$$

$$s_I = s_F$$

$$= 8.0720 \text{ kJ/kg·K}$$

These conditions are for superheated steam. From App. 23.P,

$$p_I = 37.73 \text{ kPa} \quad [0.3773 \text{ bars}]$$

$$h_I = 2742.2 \text{ kJ/kg}$$

From Eq. 26.39, due to inefficiency of the turbine,

$$h'_I = h_F - \eta_{s,\text{turbine}}(h_F - h_I)$$

$$= 3073.1 \frac{\text{kJ}}{\text{kg}} - (0.88)\left(3073.1 \frac{\text{kJ}}{\text{kg}} - 2742.2 \frac{\text{kJ}}{\text{kg}}\right)$$

$$= 2781.9 \text{ kJ/kg}$$

At point J, from App. 23.O, the saturated liquid enthalpy and temperature at 37.73 kPa are $h_J = 311.1$ kJ/kg and 74.3°C, respectively.

At point K, the temperature is

$$74.3°C - 3°C = 71.3°C$$

Since the water is subcooled, enthalpy is a function of temperature only.

$$h_K = 298.4 \text{ kJ/kg}$$

The condensate pump will match the pressure at J: $p_k = 37.73$ kPa.

From an energy balance in the heater,

$$(1 - x)(h_K - h_H) = x(h'_I - h_J)$$

$$(1 - x)\left(298.4 \frac{\text{kJ}}{\text{kg}} - 67.19 \frac{\text{kJ}}{\text{kg}}\right) = x\left(2781.9 \frac{\text{kJ}}{\text{kg}} - 311.1 \frac{\text{kJ}}{\text{kg}}\right)$$

$$2702.0x = 231.2$$

$$x = 0.0856$$

At point L, $p_L = 37.73$ kPa.

$$h_L = xh_J + (1-x)h_K$$
$$= (0.0856)\left(311.1\ \frac{kJ}{kg}\right) + (1 - 0.0856)\left(298.4\ \frac{kJ}{kg}\right)$$
$$= 299.5\ kJ/kg$$

Since this is a subcooled liquid, enthalpy is a function of temperature only, and from App. 23.N,

$$T_L = 71.6°C$$
$$v_L = 1.0238\ cm^3/g$$

At point A, $p_A = 4$ MPa.

From Eq. 26.14,

$$h_A = h_L + v_L(p_A - p_L)$$
$$h_A = 299.5\ \frac{kJ}{kg} + \left(1.0238\ \frac{cm^3}{g}\right)\left(1000\ \frac{g}{kg}\right)$$
$$\times \left(\frac{1\ m}{100\ cm}\right)^3 \left((4\ MPa)\left(1000\ \frac{kPa}{MPa}\right)\right.$$
$$\left. -(0.3773\ bar)\left(100\ \frac{kPa}{bar}\right)\right)$$
$$= 303.6\ kJ/kg$$

From Eq. 26.40, due to the inefficiency in the pump,

$$h'_A = h_L + \frac{h_A - h_L}{\eta_{s,pump}}$$
$$= 299.5\ \frac{kJ}{kg} + \frac{303.6\ \frac{kJ}{kg} - 299.5\ \frac{kJ}{kg}}{0.96}$$
$$= 303.8\ kJ/kg$$

From Eq. 26.37, the thermal efficiency of the entire cycle, neglecting condensation and drip pump, is

$$\eta_{th} = \frac{W_{turbine} - W_{pump}}{Q_{in}}$$

$$= \frac{\begin{aligned}&((h_D - h'_E) + (h_F - h'_I)\\&+ (1-x)(h'_I - h'_G)) - (h'_A - h_L)\end{aligned}}{(h_D - h'_A) + (h_F - h'_E)}$$

$$= \frac{\begin{aligned}&\left(\left(2980\ \frac{kJ}{kg} - 2465.2\ \frac{kJ}{kg}\right)\right.\\&+ \left(3073.1\ \frac{kJ}{kg} - 2781.9\ \frac{kJ}{kg}\right)\\&\left.+ (1 - 0.0856)\left(2781.9\ \frac{kJ}{kg} - 2420.2\ \frac{kJ}{kg}\right)\right)\\&- \left(303.8\ \frac{kJ}{kg} - 299.5\ \frac{kJ}{kg}\right)\end{aligned}}{\begin{aligned}&\left(2980\ \frac{kJ}{kg} - 303.8\ \frac{kJ}{kg}\right)\\&+ \left(3073.1\ \frac{kJ}{kg} - 2465.2\ \frac{kJ}{kg}\right)\end{aligned}}$$

$$= \boxed{0.345\ \ (34.5\%)}$$

The answer is (C).

7. *Customary U.S. Solution*

(a) Refer to the given illustration for Prob. 7 and to the following diagram.

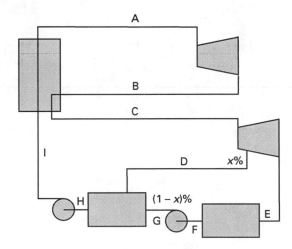

At point A,

$$p_A = 900\ psia$$
$$T_A = 800°F$$

Using the superheated steam table, App. 23.C,

$$h_A = 1393.4\ Btu/lbm$$
$$s_A = 1.5810\ Btu/lbm\text{-}°R$$

At point B, $h'_B = 1270$ Btu/lbm.

From the Mollier diagram, assuming isentropic expansion to 200 psia, $h_B = 1230$ Btu/lbm.

Isentropic efficiency of the high pressure turbine is

$$\eta_{s,\text{turbine}} = \frac{h_A - h'_B}{h_A - h_B}$$

$$= \left(\frac{1393.4\,\frac{\text{Btu}}{\text{lbm}} - 1270\,\frac{\text{Btu}}{\text{lbm}}}{1393.4\,\frac{\text{Btu}}{\text{lbm}} - 1230\,\frac{\text{Btu}}{\text{lbm}}}\right)(100\%)$$

$$= \boxed{75.52\%}$$

The answer is (D).

(b) At point C,

$$p_C = 190 \text{ psia}$$
$$T_C = 800°\text{F}$$

Using the superheated steam table from App. 23.C, $h_C = 1425.7$ Btu/lbm.

At point D: $h'_D = 1280$ Btu/lbm
At point E: $h'_E = 1075$ Btu/lbm
At point F: $h_F = 69.73$ Btu/lbm
At point G: $W_{\text{pump}} = 0.15$ Btu/lbm
$\quad W_{\text{pump}} = h'_G - h_F$
$\quad h'_G = W_{\text{pump}} + h_F$
$$= 0.15\,\frac{\text{Btu}}{\text{lbm}} + 69.73\,\frac{\text{Btu}}{\text{lbm}}$$
$$= 69.88 \text{ Btu/lbm}$$
At point H: $h_H = 250.2$ Btu/lbm
At point I: $h'_I = 253.1$ Btu/lbm

From an energy balance in the heater,

$$xh'_D + (1-x)h'_G = h_H$$
$$x(h'_D - h'_G) = h_H - h'_G$$
$$x = \frac{h_H - h'_G}{h'_D - h'_G}$$
$$= \frac{250.2\,\frac{\text{Btu}}{\text{lbm}} - 69.88\,\frac{\text{Btu}}{\text{lbm}}}{1280\,\frac{\text{Btu}}{\text{lbm}} - 69.88\,\frac{\text{Btu}}{\text{lbm}}}$$
$$= 0.149$$

The thermal efficiency of the cycle is

$$\eta_{th} = \frac{W_{\text{out}} - W_{\text{in}}}{Q_{\text{in}}}$$

$$= \frac{\begin{array}{l}(h_A - h'_B) + (h_C - h'_D) + (1-x)\\ \times (h'_D - h'_E) - (h'_I - h_H) - (1-x)(h'_G - h_F)\end{array}}{(h_A - h'_I) + (h_C - h'_B)}$$

$$= \frac{\begin{array}{l}\left(1393.4\,\frac{\text{Btu}}{\text{lbm}} - 1270\,\frac{\text{Btu}}{\text{lbm}}\right)\\ + \left(1425.7\,\frac{\text{Btu}}{\text{lbm}} - 1280\,\frac{\text{Btu}}{\text{lbm}}\right)\\ + (1-0.149)\left(1280\,\frac{\text{Btu}}{\text{lbm}} - 1075\,\frac{\text{Btu}}{\text{lbm}}\right)\\ - \left(253.1\,\frac{\text{Btu}}{\text{lbm}} - 250.2\,\frac{\text{Btu}}{\text{lbm}}\right)\\ - (1-0.149)\left(69.88\,\frac{\text{Btu}}{\text{lbm}} - 69.73\,\frac{\text{Btu}}{\text{lbm}}\right)\end{array}}{\begin{array}{l}\left(1393.4\,\frac{\text{Btu}}{\text{lbm}} - 253.1\,\frac{\text{Btu}}{\text{lbm}}\right)\\ + \left(1425.7\,\frac{\text{Btu}}{\text{lbm}} - 1270\,\frac{\text{Btu}}{\text{lbm}}\right)\end{array}}$$

$$= \boxed{0.340 \ (34.0\%)}$$

The answer is (C).

SI Solution

(a) Refer to the illustration for Prob. 7 and to the diagram for the customary U.S. solution.

At point A,

$$p_A = 6.2 \text{ MPa}$$
$$T_A = 420°\text{C}$$

Using the Mollier diagram for steam, App. 23.R,

$$h_A = 3235.0 \text{ kJ/kg}$$
$$s_A = 6.65 \text{ kJ/kg·K}$$

At point B, $h'_B = 2960$ kJ/kg.

From the Mollier diagram, assuming isentropic expansion to 1.5 MPa, $h_B = 2860$ kJ/kg.

Isentropic efficiency of the high pressure turbine is

$$\eta_{s,\text{turbine}} = \frac{h_A - h'_B}{h_A - h_B}$$

$$= \frac{3235.0 \; \frac{\text{kJ}}{\text{kg}} - 2960 \; \frac{\text{kJ}}{\text{kg}}}{3235.0 \; \frac{\text{kJ}}{\text{kg}} - 2860 \; \frac{\text{kJ}}{\text{kg}}}$$

$$= \boxed{0.733 \; (73.3\%)}$$

The answer is (D).

(b) At point C,

$$p_C = 1.4 \text{ MPa}$$
$$T_C = 420°\text{C}$$

Using the superheated steam table from App. 23.P, h_C = 3300.6 kJ/kg.

At point D, h'_D = 2980 kJ/kg.

At point E, h'_E = 2500 kJ/kg.

At point F, h_F = 162.5 kJ/kg.

At point G,

$$W_{\text{pump}} = 0.3 \text{ kJ/kg}$$
$$= h'_G - h_F$$
$$h'_G = W_{\text{pump}} + h_F$$
$$= 0.3 \; \frac{\text{kJ}}{\text{kg}} + 162.5 \; \frac{\text{kJ}}{\text{kg}}$$
$$= 162.8 \text{ kJ/kg}$$

At point H, h_H = 583.0 kJ/kg.

At point I, h'_I = 589.7 kJ/kg.

From an energy balance in the heater,

$$xh'_D + (1-x)h'_G = h_H$$
$$x(h'_D - h'_G) = h_H - h'_G$$
$$x = \frac{h_H - h_G}{h'_D - h'_G}$$

$$= \frac{583.0 \; \frac{\text{kJ}}{\text{kg}} - 162.8 \; \frac{\text{kJ}}{\text{kg}}}{2980 \; \frac{\text{kJ}}{\text{kg}} - 162.8 \; \frac{\text{kJ}}{\text{kg}}}$$

$$= 0.149$$

The thermal efficiency of the cycle is

$$\eta_{\text{th}} = \frac{W_{\text{out}} - W_{\text{in}}}{Q_{\text{in}}}$$

$$= \frac{\begin{array}{l}(h_A - h'_B) + (h_C - h'_D) + (1-x) \\ \times (h'_D - h'_E) - (h'_I - h_H) - (1-x)(h'_G - h_F)\end{array}}{(h_A - h'_I) + (h_C - h'_B)}$$

$$= \frac{\begin{array}{l}\left(3235.0 \; \frac{\text{kJ}}{\text{kg}} - 2960 \; \frac{\text{kJ}}{\text{kg}}\right) \\ + \left(3300.6 \; \frac{\text{kJ}}{\text{kg}} - 2980 \; \frac{\text{kJ}}{\text{kg}}\right) \\ + (1-0.149)\left(2980 \; \frac{\text{kJ}}{\text{kg}} - 2500 \; \frac{\text{kJ}}{\text{kg}}\right) \\ - \left(598.7 \; \frac{\text{kJ}}{\text{kg}} - 583.0 \; \frac{\text{kJ}}{\text{kg}}\right) \\ - (1-0.149)\left(162.8 \; \frac{\text{kJ}}{\text{kg}} - 162.5 \; \frac{\text{kJ}}{\text{kg}}\right)\end{array}}{\begin{array}{l}\left(3235.0 \; \frac{\text{kJ}}{\text{kg}} - 598.7 \; \frac{\text{kJ}}{\text{kg}}\right) \\ + \left(3300.6 \; \frac{\text{kJ}}{\text{kg}} - 2960 \; \frac{\text{kJ}}{\text{kg}}\right)\end{array}}$$

$$= \boxed{0.332 \; (33.2\%)}$$

The answer is (C).

8.

15 lbm/hr (1.9 g/s)
140°F (60°C)

0.25 hp (0.19 kW)

15 psia (103.5 kPa)

Customary U.S. Solution

The drill power is

$$P = 0.25 \text{ hp} \quad \text{[given]}$$

$$= (0.25 \text{ hp})\left(2545 \; \frac{\text{Btu}}{\text{hp-hr}}\right) = 636.25 \text{ Btu/hr}$$

The absolute inlet temperature is

$$T_I = 140°\text{F} + 460 = 600°\text{R}$$

From App. 23.F, the properties of air entering the turbine are

$$h_I = 143.47 \text{ Btu/lbm}$$
$$p_{r,1} = 2.005$$
$$\phi_I = 0.62607 \text{ Btu/lbm-}°\text{R}$$

Thermodynamics

From Eq. 25.18,

$$P = \dot{m}(h_1 - h_2')$$

$$\eta_{s,\text{turbine}} = \frac{h_1 - h_2'}{h_1 - h_2}$$

So,

$$P = \dot{m}\eta_{s,\text{turbine}}(h_1 - h_2)$$

$$h_2 = h_1 - \frac{P}{\dot{m}\eta_{s,\text{turbine}}}$$

$$= 143.47 \ \frac{\text{Btu}}{\text{lbm}} - \frac{636.25 \ \dfrac{\text{Btu}}{\text{hr}}}{\left(15 \ \dfrac{\text{lbm}}{\text{hr}}\right)(0.60)}$$

$$= 72.776 \ \text{Btu/lbm}$$

Appendix 23.F doesn't go low enough. From the Keenan and Kayes *Gas Tables*, for $h = 72.776$ Btu/lbm,

$$T_2 = 305°\text{R}$$

$$p_{r,2} = 0.18851$$

(a) Due to the irreversibility of the expansion from Eq. 25.19,

$$h_2' = h_1 - \eta_s(h_1 - h_2)$$

$$= 143.47 \ \frac{\text{Btu}}{\text{lbm}}$$

$$- (0.60)\left(143.47 \ \frac{\text{Btu}}{\text{lbm}} - 72.776 \ \frac{\text{Btu}}{\text{lbm}}\right)$$

$$= 101.05 \ \text{Btu/lbm}$$

From App. 23.F for $h = 101.05$ Btu/lbm,

$$T_2' = \boxed{423°\text{R}}$$

$$\phi_2 = 0.54225 \ \text{Btu/lbm-°R}$$

The answer is (B).

(b) Since $p_1/p_2 = p_{r,1}/p_{r,2}$,

$$p_1 = p_2\left(\frac{p_{r,1}}{p_{r,2}}\right) = (15 \ \text{psia})\left(\frac{2.005}{0.18851}\right)$$

$$= \boxed{159.5 \ \text{psia}}$$

The answer is (A).

(c) From Eq. 23.43, the entropy change is

$$s_2 - s_1 = \phi_2 - \phi_1 - R\ln\left(\frac{p_2}{p_1}\right)$$

From Table 23.7, $R = 53.35$ ft-lbf/lbm-°R.

$$s_2 - s_1 = 0.54225 \ \frac{\text{Btu}}{\text{lbm-°R}} - 0.62607 \ \frac{\text{Btu}}{\text{lbm-°R}}$$

$$- \left(\frac{53.35 \ \dfrac{\text{ft-lbf}}{\text{lbm-°R}}}{778 \ \dfrac{\text{ft-lbf}}{\text{Btu}}}\right)\ln\left(\frac{15 \ \text{psia}}{159.5 \ \text{psia}}\right)$$

$$= \boxed{0.07829 \ \text{Btu/lbm-°R}}$$

The answer is (A).

SI Solution

The absolute inlet temperature is

$$T_{\text{I}} = 60°\text{C} + 273 = 333\text{K}$$

From App. 23.S, the properties of air entering the turbine are

$$h_1 = 333.70 \ \text{kJ/kg}$$

$$p_{r,1} = 2.0064$$

$$\phi_1 = 1.80784 \ \text{kJ/kg·K}$$

From Eq. 25.18,

$$P = \dot{m}(h_1 - h_2')$$

$$\eta_{s,\text{turbine}} = \frac{h_1 - h_2'}{h_1 - h_2}$$

$$P = \dot{m}\eta_{s,\text{turbine}}(h_1 - h_2)$$

$$h_2 = h_1 - \frac{P}{\dot{m}\eta_{s,\text{turbine}}}$$

$$= 333.70 \ \frac{\text{kJ}}{\text{kg}} - \frac{0.19 \ \text{kW}}{\left(1.9 \ \dfrac{\text{g}}{\text{s}}\right)\left(\dfrac{1 \ \text{kg}}{1000 \ \text{g}}\right)(0.60)}$$

$$= 167.03 \ \text{kJ/kg}$$

Appendix 24.S doesn't go low enough. From gas tables, for $h = 167.03$ kJ/kg,

$$T_2 = 164.4\text{K}$$

$$p_{r,2} = 0.16980$$

(a) Due to the irreversibility of the expansion from Eq. 25.19,

$$h_2' = h_1 - \eta_s(h_1 - h_2)$$

$$= 333.70 \ \frac{\text{kJ}}{\text{kg}} - (0.60)\left(330.70 \ \frac{\text{kJ}}{\text{kg}} - 167.03 \ \frac{\text{kJ}}{\text{kg}}\right)$$

$$= 235.50 \ \text{kJ/kg}$$

From App. 23.S, for $h = 235.50$ kJ/kg,

$$T_2' = \boxed{235.3\text{K}}$$

$$\phi_2 = 1.45819 \text{ kJ/kg·K}$$

The answer is (B).

(b) Since $p_1/p_2 = p_{r,1}/p_{r,2}$,

$$p_1 = (103.5 \text{ kPa}) \left(\frac{2.0064}{0.16980} \right)$$

$$= \boxed{1223 \text{ kPa}}$$

The answer is (A).

(c) From Table 23.7, $R = 287.03$ kJ/kg·K. From Eq. 23.43, the entropy change is

$$s_2 - s_1 = \phi_2 - \phi_1 - R \ln \left(\frac{p_2}{p_1} \right)$$

$$= 1.45819 \frac{\text{kJ}}{\text{kg·K}} - 1.80784 \frac{\text{kJ}}{\text{kg·K}}$$

$$- \left(287.03 \frac{\text{kJ}}{\text{kg·K}} \right) \left(\frac{1 \text{ kJ}}{1000 \text{ J}} \right)$$

$$\times \ln \left(\frac{103.5 \text{ kPa}}{1223 \text{ kPa}} \right)$$

$$= \boxed{0.3592 \text{ kJ/kg·K}}$$

The answer is (A).

27 Combustion Power Cycles

PRACTICE PROBLEMS

1. Air expands isentropically at the rate of 10 ft^3/sec (280 L/s) from 200 psia and 1500°F (1.4 MPa and 820°C) to 50 psia (350 kPa).

(a) What is the air's final temperature?
- (A) 1275°R (710K)
- (B) 1325°R (740K)
- (C) 1375°R (770K)
- (D) 1425°R (790K)

(b) What is the air's final volumetric flow rate?
- (A) 28 ft^3/sec (780 L/s)
- (B) 31 ft^3/sec (860 L/s)
- (C) 39 ft^3/sec (1100 L/s)
- (D) 45 ft^3/sec (1300 L/s)

(c) What is the air's enthalpy change?
- (A) −110 Btu/lbm (−250 kJ/kg)
- (B) −160 Btu/lbm (−370 kJ/kg)
- (C) −230 Btu/lbm (−530 kJ/kg)
- (D) −350 Btu/lbm (−770 kJ/kg)

2. A 10 in (250 mm) bore, 18 in (460 mm) stroke, two-cylinder, four-stroke internal combustion engine operates with an indicated mean effective pressure of 95 psig (660 kPa) at 200 rpm. The actual torque developed is 600 ft-lbf (820 N·m). What is the friction horsepower?
- (A) 17 hp (13 kW)
- (B) 22 hp (17 kW)
- (C) 45 hp (33 kW)
- (D) 78 hp (60 kW)

3. A four-cycle internal combustion engine has a bore of 3.1 in (80 mm) and a stroke of 3.8 in (97 mm). When the engine is running at 4000 rpm, the air-fuel mixture enters the cylinders during the intake stroke with a velocity of 100 ft/sec (30 m/s). The intake valve opens at TDC and closes at 40° past BDC. The volumetric efficiency is 65%. What is the effective area of the intake valve?
- (A) 0.012 ft^2 (0.0012 m^2)
- (B) 0.023 ft^2 (0.0023 m^2)
- (C) 0.034 ft^2 (0.0034 m^2)
- (D) 0.058 ft^2 (0.0058 m^2)

4. An engine runs on the Otto cycle. The compression ratio is 10:1. The total intake volume of the cylinders is 11 ft^3 (0.3 m^3). Air enters at 14.2 psia and 80°F (98 kPa and 27°C). After the compression stroke, 160 Btu (179 kJ) of heat are added.

(a) What is the temperature after the heat addition?
- (A) 1290°R (720K)
- (B) 1550°R (860K)
- (C) 1750°R (970K)
- (D) 2320°R (1340K)

(b) What is the thermal efficiency of the cycle?
- (A) 45%
- (B) 52%
- (C) 57%
- (D) 64%

5. A 4.25 in × 6 in (110 mm × 150 mm), six-cylinder diesel engine runs at 1200 rpm while consuming 28 lbm of fuel per hour (0.0035 kg/s). The actual volumetric fraction of carbon dioxide in the exhaust is 9% (dry basis). At another throttle setting, when the air-fuel ratio is 15:1, there is 13.7% (dry basis) carbon dioxide in the exhaust. The atmospheric conditions are 14.7 psia and 70°F (101.3 kPa and 21°C). What is the volumetric efficiency at 1200 rpm?
- (A) 59%
- (B) 65%
- (C) 72%
- (D) 80%

6. At standard atmospheric conditions, a fully loaded diesel engine runs at 2000 rpm and develops 1000 bhp (750 kW). Metered fuel injection is used. The brake specific fuel consumption is 0.45 lbm/bhp-hr (76 kg/GJ). The mechanical efficiency is 80%, independent of the altitude.

(a) At an altitude of 5000 ft (1500 m), what will be the brake horsepower?
- (A) 810 hp (630 kW)
- (B) 860 hp (650 kW)
- (C) 890 hp (690 kW)
- (D) 950 hp (740 kW)

(b) At an altitude of 5000 ft (1500 m), what will be the brake specific fuel consumption?
- (A) 0.26 lbm/hp-hr (44 kg/GJ)
- (B) 0.52 lbm/hp-hr (88 kg/GJ)
- (C) 1.4 lbm/hp-hr (24 kg/GJ)
- (D) 2.2 lbm/hp-hr (37 kg/GJ)

7. In an air-standard gas turbine, air at 14.7 psia and 60°F (101.3 kPa and 16°C) enters a compressor and is compressed through a volume ratio of 5:1. The compressor efficiency is 83%. Air enters the turbine at 1500°F (820°C) and expands to 14.7 psia (101.3 kPa). The turbine efficiency is 92%. What is the thermal efficiency of the cycle?
- (A) 33%
- (B) 39%
- (C) 44%
- (D) 51%

8. A 65% efficient regenerator is added to the gas turbine described in Prob. 7. Assume the specific heat remains constant. What is the new thermal efficiency?
- (A) 24%
- (B) 28%
- (C) 34%
- (D) 41%

9. (*Time limit: one hour*) A gasoline-fueled internal combustion engine runs at 4600 rpm. The engine is four-stroke and V-8 in configuration with a displacement of 265 in^3 (4.3 L). The indicated work required to compress the air-fuel mixture is 1200 ft-lbf (1.6 kJ) per cycle. The indicated work done by the exhaust gases in expansion is 1500 ft-lbf (2.0 kJ) per cycle. The input energy from fuel combustion is 1.27 Btu (1.33 kJ) per cycle. Atmospheric air is at 14.7 psia and 70°F (101.3 kPa and 21°C). The air-fuel ratio is 20:1. The heating value of gasoline is 18,900 Btu/lbm (44 MJ/kg). Neglect the effects of friction.

(a) What is the indicated horsepower?
- (A) 140 hp (110 kW)
- (B) 170 hp (120 kW)
- (C) 200 hp (160 kW)
- (D) 240 hp (190 kW)

(b) What is the thermal efficiency?
- (A) 30%
- (B) 35%
- (C) 39%
- (D) 46%

(c) What is the mass per hour of gasoline consumed?
- (A) 37 lbm/hr (0.0046 kg/s)
- (B) 52 lbm/hr (0.0065 kg/s)
- (C) 74 lbm/hr (0.0093 kg/s)
- (D) 99 lbm/hr (0.012 kg/s)

(d) What is the specific fuel consumption?
- (A) 0.055 lbm/hp-hr (10 kg/GJ)
- (B) 0.11 lbm/hp-hr (19 kg/GJ)
- (C) 0.22 lbm/hp-hr (38 kg/GJ)
- (D) 0.44 lbm/hp-hr (76 kg/GJ)

10. (*Time limit: one hour*) A mixture of carbon dioxide and helium in an engine undergoes the following processes in a cycle.

A to B: compression and heat removal

B to C: constant volume heating

C to A: isentropic expansion

point	temperature	pressure
A	520°R (290K)	14.7 psia (101.3 kPa)
B	1240°R (690K)	unknown
C	1600°R (890K)	568.6 psia (3.920 MPa)

Both gases are ideal gases.

(a) What are the gravimetric fractions of the carbon dioxide and helium in the mixture?
- (A) CO_2, 0.86; He, 0.14
- (B) CO_2, 0.77; He, 0.23
- (C) CO_2, 0.65; He, 0.35
- (D) CO_2, 0.54; He, 0.46

(b) What work is done during the isentropic expansion process?
- (A) 190 Btu/lbm (440 kJ/kg)
- (B) 260 Btu/lbm (610 kJ/kg)
- (C) 330 Btu/lbm (760 kJ/kg)
- (D) 450 Btu/lbm (1000 kJ/kg)

(c) Draw the temperature-entropy and pressure-volume diagrams for the cycle.

11. (*Time limit: one hour*) When the atmospheric conditions are 14.7 psia and 80°F (101.3 kPa and 27°C), a diesel engine with metered fuel injection has the following operating characteristics.

brake horsepower:	200 bhp (150 kW)
brake specific fuel consumption:	0.48 lbm/hp-hr (81 kg/GJ)
air-fuel ratio:	22:1
mechanical efficiency:	86%

The engine is moved to an altitude where the atmospheric conditions are 12.2 psia and 60°F (84 kPa and 16°C). The running speed is unchanged. What are the corresponding operating characteristics?

12. (*Time limit: one hour*) A gas turbine operating on the Brayton cycle with an 8:1 pressure ratio is located at 7000 ft (2100 m) altitude. The conditions at that altitude are 12 psia and 35°F (82 kPa and 2°C). While consuming 0.609 lbm/hp-hr (100 kg/GJ) of fuel and 50,000 cfm (23 500 L/s) of air, the turbine develops 6000 bhp (4.5 MW). The turbine efficiency is 80%, and the compressor efficiency is 85%. The fuel has a lower heating value of 19,000 Btu/lbm (44 MJ/kg). The fuel mass is small compared to the air mass. The turbine receives combustor gases at 1800°F (980°C). The air inlet filter area is 254 ft² (22.9 m²). The turbine is moved to sea level where the conditions are 14.7 psia and 70°F (101.3 kPa and 21°C). The combustion efficiency and combustor temperature remain the same.

(a) What is the new brake horsepower?
(A) 2800 hp (2.1 MW)
(B) 4400 hp (3.4 MW)
(C) 5100 hp (4.0 MW)
(D) 6300 hp (4.8 MW)

(b) What is the new brake specific fuel consumption?
(A) 0.42 lbm/hp-hr (67 kg/GJ)
(B) 0.51 lbm/hp-hr (82 kg/GJ)
(C) 0.63 lbm/hp-hr (100 kg/GJ)
(D) 0.87 lbm/hp-hr (140 kg/GJ)

SOLUTIONS

1. *Customary U.S. Solution*

(a) The absolute temperature is

$$T_1 = 1500°F + 460 = 1960°R$$

From air tables (App. 23.F) at 1960°R,

$$h_1 = 493.64 \text{ Btu/lbm}$$
$$p_{r,1} = 160.48$$
$$v_{r,1} = 4.53$$

After expansion,

$$p_{r,2} = p_{r,1}\left(\frac{p_2}{p_1}\right)$$
$$= (160.48)\left(\frac{50 \text{ psia}}{200 \text{ psia}}\right)$$
$$= 40.12$$

From air tables (App. 23.F) at $p_{r,2} = 40.12$,

$$T_2 = \boxed{1375°R}$$
$$h_2 = 336.39 \text{ Btu/lbm}$$
$$v_{r,2} = 12.721$$

The answer is (C).

(b)
$$\dot{V}_2 = \dot{V}_1\left(\frac{v_{r,2}}{v_{r,1}}\right)$$
$$= \left(10 \frac{\text{ft}^3}{\text{sec}}\right)\left(\frac{12.721}{4.53}\right)$$
$$= \boxed{28.1 \text{ ft}^3/\text{sec}}$$

The answer is (A).

(c) The enthalpy change is

$$\Delta h = h_2 - h_1$$
$$= 336.39 \frac{\text{Btu}}{\text{lbm}} - 493.64 \frac{\text{Btu}}{\text{lbm}}$$
$$= \boxed{-157.25 \text{ Btu/lbm} \quad [\text{decrease}]}$$

The answer is (B).

SI Solution

(a) The absolute temperature is

$$T_1 = 820°C + 273 = 1093K$$

From air tables (App. 23.S) at 1093K,

$$h_1 = 1153 \text{ kJ/kg}$$

$$p_{r,1} = 162.94$$

$$v_{r,1} = 19.275$$

After expansion,

$$p_{r,2} = p_{r,1} \left(\frac{p_2}{p_1} \right)$$

$$= (162.94) \left(\frac{350 \text{ kPa}}{(1.4 \text{ MPa}) \left(1000 \frac{\text{kPa}}{\text{MPa}} \right)} \right)$$

$$= 40.74$$

From air tables (App. 23.S) at $p_{r,2} = 40.74$,

$$T_2 = \boxed{767.2\text{K}}$$

$$h_2 = 786.05 \text{ kJ/kg}$$

$$v_{r,2} = 53.99$$

The answer is (C).

(b)
$$\dot{V}_2 = \dot{V}_1 \left(\frac{v_{r,2}}{v_{r,1}} \right)$$

$$= \left(280 \frac{\text{L}}{\text{s}} \right) \left(\frac{53.99}{19.275} \right)$$

$$= \boxed{784.3 \text{ L/s}}$$

The answer is (A).

(c) The enthalpy change is

$$\Delta h = h_2 - h_1$$

$$= 786.05 \frac{\text{kJ}}{\text{kg}} - 1153 \frac{\text{kJ}}{\text{kg}}$$

$$= \boxed{-366.95 \text{ kJ/kg} \quad [\text{decrease}]}$$

The answer is (B).

2. *Customary U.S. Solution*

The actual brake horsepower from Eq. 27.10(b) is

$$\text{BHP} = \frac{nT}{5252} = \frac{(200 \text{ rpm})(600 \text{ ft-lbf})}{5252}$$

$$= 22.85 \text{ hp}$$

From Eq. 27.41, the number of power strokes per minute is

$$N = \frac{(2n)(\text{no. cylinders})}{\text{no. strokes per cycle}}$$

$$= \frac{(2)(200 \text{ rpm})(2)}{4} = 200 \text{ power strokes/min}$$

The stroke is

$$L = (18 \text{ in}) \left(\frac{1 \text{ ft}}{12 \text{ in}} \right) = 1.5 \text{ ft}$$

The bore area is

$$\left(\frac{\pi}{4} \right) (10 \text{ in})^2 = 78.54 \text{ in}^2$$

From Eq. 27.40(b), the ideal (indicated) horsepower is

$$\text{hp} = \frac{pLAN}{33,000}$$

$$= \frac{\left(95 \frac{\text{lbf}}{\text{in}^2} \right) (1.5 \text{ ft})(78.54 \text{ in}^2) \left(200 \frac{\text{strokes}}{\text{min}} \right)}{33,000 \frac{\text{ft-lbf}}{\text{hp-min}}}$$

$$= 67.83 \text{ hp}$$

The friction horsepower is

$$\text{FHP} = \text{IHP} - \text{BHP}$$

$$= \text{ideal hp} - \text{actual BHP} = 67.83 \text{ hp} - 22.85 \text{ hp}$$

$$= \boxed{44.98 \text{ hp}}$$

The answer is (C).

SI Solution

The actual brake power from Eq. 27.10(a) is

$$\text{BkW} = \frac{nT}{5252} = \frac{(200 \text{ rpm})(820 \text{ N·m})}{9549}$$

$$= 17.17 \text{ kW}$$

From Eq. 27.41, the number of power strokes per minute is

$$N = \frac{(2n)(\text{no. cylinders})}{\text{no. strokes per cycle}}$$

$$= \frac{(2)(200 \text{ rpm})(2)}{4} = 200 \text{ power strokes/min}$$

The stroke is

$$\frac{460 \text{ mm}}{1000 \frac{\text{mm}}{\text{m}}} = 0.46 \text{ m}$$

The bore area is

$$\left(\frac{\pi}{4} \right) \left(\frac{250 \text{ mm}}{1000 \frac{\text{mm}}{\text{m}}} \right)^2 = 4.909 \times 10^{-2} \text{ m}^2$$

From Eq. 27.40(a), the ideal power is

$$kW = pLAN$$

$$= \frac{(660 \text{ kPa})(0.46 \text{ m})}{60 \frac{s}{\text{min}}} \times (4.909 \times 10^{-2} \text{ m}^2)\left(200 \frac{\text{strokes}}{\text{min}}\right)$$

$$= 49.68 \text{ kW}$$

The friction power is

$$\text{ideal kW} - \text{actual kW} = 49.68 \text{ kW} - 17.17 \text{ kW}$$

$$= \boxed{32.51 \text{ kW}}$$

The answer is (C).

3. *Customary U.S. Solution*

The number of degrees that the valve is open is

$$180° + 40° = 220°$$

The time that the valve is open is

$$\left(\frac{220°}{360°}\right)\left(\frac{\text{time}}{\text{rev}}\right) = \left(\frac{220°}{360°}\right)\left(\frac{60 \frac{\text{sec}}{\text{min}}}{4000 \text{ rpm}}\right)$$

$$= 9.167 \times 10^{-3} \text{ sec}$$

The displacement is

$$\left(\frac{\pi}{4}\right)(\text{bore})^2(\text{stroke}) = \left(\frac{\pi}{4}\right)(3.1 \text{ in})^2\left(\frac{1 \text{ ft}}{12 \text{ in}}\right)^2$$

$$\times (3.8 \text{ in})\left(\frac{1 \text{ ft}}{12 \text{ in}}\right)$$

$$= 0.0166 \text{ ft}^3$$

The actual incoming volume per intake stroke is

$$V = (\text{volumetric efficiency})(\text{displacement})$$

$$= (0.65)(0.0166 \text{ ft}^3) = 0.01079 \text{ ft}^3$$

The area is

$$A = \frac{V}{vt} = \frac{0.01079 \text{ ft}^3}{\left(100 \frac{\text{ft}}{\text{sec}}\right)(9.167 \times 10^{-3} \text{ sec})}$$

$$= \boxed{0.0118 \text{ ft}^2 \quad (1.69 \text{ in}^2)}$$

The answer is (A).

SI Solution

The number of degrees that the valve is open is

$$180° + 40° = 220°$$

The time that the valve is open is

$$\left(\frac{220°}{360°}\right)\left(\frac{\text{time}}{\text{rev}}\right) = \left(\frac{220°}{360°}\right)\left(\frac{60 \frac{s}{\text{min}}}{4000 \text{ rpm}}\right)$$

$$= 9.167 \times 10^{-3} \text{ s}$$

The displacement is

$$\left(\frac{\pi}{4}\right)(\text{bore})^2(\text{stroke}) = \left(\frac{\pi}{4}\right)(0.08 \text{ m})^2(0.097 \text{ m})$$

$$= 4.876 \times 10^{-4} \text{ m}^3$$

The actual incoming volume per intake stroke is

$$V = (\text{volumetric efficiency})(\text{displacement})$$

$$= (0.65)(4.876 \times 10^{-4} \text{ m}^3) = 3.169 \times 10^{-4} \text{ m}^3$$

The area is

$$A = \frac{V}{vt} = \frac{3.169 \times 10^{-4} \text{ m}^3}{\left(30 \frac{\text{m}}{\text{s}}\right)(9.167 \times 10^{-3} \text{ s})}$$

$$= \boxed{1.152 \times 10^{-3} \text{ m}^2}$$

The answer is (A).

4. *Customary U.S. Solution*

Refer to the air-standard Otto cycle diagram (Fig. 27.3).

(a) At A:

$$V_A = 11 \text{ ft}^3$$

The absolute temperature is

$$T_A = 80°F + 460 = 540°R$$

For an ideal gas, the mass of the air in the intake volume is

$$m = \frac{pV}{RT} = \frac{\left(14.2 \frac{\text{lbf}}{\text{in}^2}\right)\left(144 \frac{\text{in}^2}{\text{ft}^2}\right)(11 \text{ ft}^3)}{\left(53.3 \frac{\text{ft-lbf}}{\text{lbm-°R}}\right)(540°R)}$$

$$= 0.781 \text{ lbm}$$

From air tables (App. 23.F) at 540°R,

$$v_{r,A} = 144.32$$

$$u_A = 92.04 \text{ Btu/lbm}$$

At B:

The compression ratio is a ratio of volumes.

$$V_B = \left(\frac{1}{10}\right) V_A = \left(\frac{1}{10}\right)(11 \text{ ft}^3)$$
$$= 1.1 \text{ ft}^3$$

Since the compression from A to B is isentropic,

$$v_{r,B} = \frac{v_{r,A}}{10} = \frac{144.32}{10}$$
$$= 14.432$$

From the air tables (App. 23.F) for $v_r = 14.432$,

$$T_B \approx 1314°R$$
$$u_B \approx 230.5 \text{ Btu/lbm}$$

At C:

$$u_C = u_B + \frac{Q_{in,B\text{-}C}}{m}$$
$$= 230.5 \frac{\text{Btu}}{\text{lbm}} + \frac{160 \text{ Btu}}{0.781 \text{ lbm}}$$
$$= 435.4 \text{ Btu/lbm}$$

From air tables (App. 23.F) at $u = 435.4$ Btu/lbm,

$$T_C = \boxed{2319°R}$$
$$v_{r,C} = 2.694$$

At D:

Since expansion is isentropic and the ratio of volumes is the same,

$$v_{r,D} = 10 v_{r,C} = (10)(2.694)$$
$$= 26.94$$

From air tables (App. 23.F), at $v_r = 26.94$,

$$T_D = 1044°R$$
$$u_D = 180.38 \text{ Btu/lbm}$$

The answer is (D).

(b) The heat input is

$$q_{in} = \frac{Q}{m} = \frac{160 \text{ Btu}}{0.781 \text{ lbm}}$$
$$= 204.9 \text{ Btu/lbm}$$

The heat rejected during a constant volume process is $q_{out} = \Delta u$.

Heat is rejected between D and A. Therefore,

$$q_{out} = u_D - u_A = 180.38 \frac{\text{Btu}}{\text{lbm}} - 92.04 \frac{\text{Btu}}{\text{lbm}}$$
$$= 88.34 \text{ Btu/lbm}$$

From Eq. 27.38, the thermal efficiency is

$$\eta_{th} = \frac{q_{in} - q_{out}}{q_{in}} = \frac{204.9 \frac{\text{Btu}}{\text{lbm}} - 88.34 \frac{\text{Btu}}{\text{lbm}}}{204.9 \frac{\text{Btu}}{\text{lbm}}}$$
$$= \boxed{0.569 \quad (56.9\%)}$$

The answer is (C).

SI Solution

(a) At A:

The absolute temperature is

$$T_A = 27°C + 273 = 300K$$

From the ideal gas law, the mass of air in the intake volume is

$$m = \frac{pV}{RT} = \frac{(98 \text{ kPa})\left(1000 \frac{\text{Pa}}{\text{kPa}}\right)(0.3 \text{ m}^3)}{\left(287 \frac{\text{J}}{\text{kg·K}}\right)(300K)}$$
$$= 0.3415 \text{ kg}$$

From air tables (App. 23.S) at 300K,

$$v_{r,A} = 621.2$$
$$u_A = 214.07 \text{ kJ/kg}$$

At B:

The compression ratio is a ratio of volumes.

$$V_B = \left(\frac{1}{10}\right) V_A = \frac{0.3 \text{ m}^3}{10}$$
$$= 0.03 \text{ m}^3$$

Since the compression from A to B is isentropic,

$$v_{r,B} = \frac{v_{r,A}}{10} = \frac{621.2}{10}$$
$$= 62.12$$

From air tables (App. 23.S) for this value of $v_{r,B}$,

$$T_B = 730K$$
$$u_B = 536.0 \text{ kJ/kg}$$

At C:

$$u_C = u_B + \frac{Q_{in,B\text{-}C}}{m}$$

$$= 536.0 \ \frac{kJ}{kg} + \frac{179 \ kJ}{0.3415 \ kg}$$

$$= 1060.2 \ kJ/kg$$

From air tables (App. 23.S) at $u = 1060.2 \ kJ/kg$,

$$T_C = \boxed{1341.4 \ K}$$

$$v_{r,C} = 10.215$$

At D:

Since expansion is isentropic and the ratio of volumes is the same,

$$v_{r,D} = 10 v_{r,C} = (10)(10.215)$$

$$= 102.15$$

From air tables (App. 23.S), at $v_r = 102.15$,

$$T_D = 607.9 \ K$$

$$u_D = 440.8 \ kJ/kg$$

The answer is (D).

(b) The heat input is

$$q_{in} = \frac{Q}{m} = \frac{179 \ kJ}{0.3415 \ kg}$$

$$= 524.2 \ kJ/kg$$

The heat rejected during a constant volume process is

$$q_{out} = \Delta u$$

Heat is rejected between D and A. Therefore,

$$q_{out} = u_D - u_A = 440.8 \ \frac{kJ}{kg} - 214.07 \ \frac{kJ}{kg}$$

$$= 226.7 \ kJ/kg$$

From Eq. 27.38, the thermal efficiency is

$$\eta_{th} = \frac{q_{in} - q_{out}}{q_{in}} = \frac{524.2 \ \frac{kJ}{kg} - 226.7 \ \frac{kJ}{kg}}{524.2 \ \frac{kJ}{kg}}$$

$$= \boxed{0.568 \ (56.8\%)}$$

The answer is (C).

5. *Customary U.S. Solution*

step 1: Find the ideal mass of air ingested.

From Eq. 27.41, the number of power strokes per second is

$$N = \frac{(2n)(\text{no. cylinders})}{\text{no. strokes per cycle}}$$

$$= \frac{(2)\left(1200 \ \frac{rev}{min}\right)\left(\frac{1 \ min}{60 \ sec}\right)(6 \ \text{cylinders})}{4 \ \text{strokes}}$$

$$= 60/\sec$$

The swept volume is

$$V_s = \left(\frac{\pi}{4}\right)(\text{bore})^2(\text{stroke})$$

$$= \left(\frac{\pi}{4}\right)(4.25 \ in)^2\left(\frac{1 \ ft^2}{144 \ in^2}\right)(6 \ in)\left(\frac{1 \ ft}{12 \ in}\right)$$

$$= 0.04926 \ ft^3$$

The ideal volume of air taken in per second is

$$\dot{V}_i = (\text{swept volume})\left(\frac{\text{intake strokes}}{\sec}\right)$$

$$= V_s N$$

$$= (0.04926 \ ft^3)\left(60 \ \frac{1}{\sec}\right) = 2.956 \ ft^3/\sec$$

The absolute temperature is

$$70°F + 460 = 530°R$$

From the ideal gas law, the ideal mass of air in the swept volume is

$$\dot{m} = \frac{p\dot{V}}{RT}$$

$$= \frac{\left(14.7 \ \frac{lbf}{in^2}\right)\left(144 \ \frac{in^2}{ft^2}\right)\left(2.956 \ \frac{ft^3}{\sec}\right)}{\left(53.35 \ \frac{ft\text{-}lbf}{lbm\text{-}°R}\right)(530°R)}$$

$$= 0.2213 \ lbm/\sec$$

step 2: Find the carbon dioxide volume in the exhaust assuming complete combustion when the air/fuel ratio is 15 ($\%CO_2 = 13.7\%$, dry basis).

Air is 76.85% (by weight) nitrogen, so the nitrogen/fuel ratio is

$$(0.7685)(15) = 11.528 \text{ lbm N}_2/\text{lbm fuel}$$

From the ideal gas law, the nitrogen per pound of fuel burned is

$$V_{N_2} = \frac{mRT}{p}$$

$$= \frac{\left(11.528 \frac{\text{lbm N}_2}{\text{lbm fuel}}\right)\left(55.16 \frac{\text{ft-lbf}}{\text{lbm-}^\circ\text{R}}\right)(530^\circ\text{R})}{\left(14.7 \frac{\text{lbf}}{\text{in}^2}\right)\left(144 \frac{\text{in}^2}{\text{ft}^2}\right)}$$

$$= 159.2 \text{ ft}^3/\text{lbm fuel}$$

Similarly, air is 23.15% oxygen by weight, so the oxygen/fuel ratio is

$$(0.2315)(15) = 3.472 \text{ lbm O}_2/\text{lbm fuel}$$

From the ideal gas law, the oxygen per pound of fuel burned is

$$V_{O_2} = \frac{mRT}{p}$$

$$= \frac{\left(3.472 \frac{\text{lbm O}_2}{\text{lbm fuel}}\right)\left(48.29 \frac{\text{ft-lbf}}{\text{lbm-}^\circ\text{R}}\right)(530^\circ\text{R})}{\left(14.7 \frac{\text{lbf}}{\text{in}^2}\right)\left(144 \frac{\text{in}^2}{\text{ft}^2}\right)}$$

$$= 41.98 \text{ ft}^3/\text{lbm fuel}$$

When oxygen forms carbon dioxide, the chemical equation is $C + O_2 \longrightarrow CO_2$.

It takes one volume of oxygen to form one volume of carbon dioxide. Considering nitrogen and excess oxygen in the exhaust, the percentage of carbon dioxide in the exhaust is found from

$$\%CO_2 = \frac{\text{vol CO}_2}{\text{vol CO}_2 + \text{vol O}_2 + \text{vol N}_2}$$

$$\text{vol CO}_2 = x \text{ [unknown], in ft}^3$$

$$\text{vol O}_2 = 41.98 \text{ ft}^3 - \text{oxygen used to make CO}_2$$

$$= 41.98 \text{ ft}^3 - x$$

$$\%CO_2 = 0.137 \text{ [given]}$$

$$0.137 = \frac{x}{x + (41.98 \text{ ft}^3 - x) + 159.3 \text{ ft}^3}$$

$$x = 27.58 \text{ ft}^3/\text{lbm fuel}$$

Assuming complete combustion, the volume of CO_2 will be constant regardless of the amount of air used.

step 3: Calculate the excess air if the percentage of carbon dioxide in the exhaust is 9%.

$$\%CO_2 = \frac{\text{vol CO}_2}{\begin{array}{c} \text{vol CO}_2 + \text{vol O}_2 \\ + \text{vol N}_2 + \text{vol excess air} \end{array}}$$

$$0.09 = \frac{27.58 \text{ ft}^3}{\begin{array}{c} 27.58 \text{ ft}^3 + (41.99 \text{ ft}^3 - 27.58 \text{ ft}^3) \\ + 159.3 \text{ ft}^3 + \text{vol excess air} \end{array}}$$

$$\begin{array}{c}\text{vol} \\ \text{excess} \\ \text{air}\end{array} = 105.2 \text{ ft}^3/\text{lbm fuel}$$

From the ideal gas law, the mass of excess air is

$$m_{\text{excess}} = \frac{\left(14.7 \frac{\text{lbf}}{\text{in}^2}\right)\left(144 \frac{\text{in}^2}{\text{ft}^2}\right)(105.2 \text{ ft}^3)}{\left(53.35 \frac{\text{lbf-ft}}{\text{lbm-}^\circ\text{R}}\right)(530^\circ\text{R})}$$

$$= 7.876 \text{ lbm/lbm fuel}$$

step 4: The actual air/fuel ratio is

$$15 \frac{\text{lbm air}}{\text{lbm fuel}} + 7.876 \frac{\text{lbm air}}{\text{lbm fuel}}$$

$$= 22.876 \text{ lbm air/lbm fuel}$$

The actual air mass per second is

$$\left(22.876 \frac{\text{lbm air}}{\text{lbm fuel}}\right)\left(28 \frac{\text{lbm fuel}}{\text{hr}}\right)\left(\frac{1 \text{ hr}}{3600 \text{ sec}}\right)$$

$$= 0.178 \text{ lbm/sec}$$

step 5: The volumetric efficiency is

$$\eta_v = \frac{0.178 \frac{\text{lbm}}{\text{sec}}}{0.2213 \frac{\text{lbm}}{\text{sec}}} = \boxed{0.804 \ (80.4\%)}$$

The answer is (D).

SI Solution

step 1: Find the ideal mass of air ingested.

From Eq. 27.41, the number of power strokes per second is

$$N = \frac{(2n)(\text{no. cylinders})}{\text{no. strokes per cycle}}$$

$$= \frac{(2)\left(1200 \frac{\text{rev}}{\text{min}}\right)\left(\frac{1 \text{ min}}{60 \text{ s}}\right)(6 \text{ cylinders})}{4 \text{ strokes}}$$

$$= 60/\text{s}$$

The swept volume is

$$V_s = \left(\frac{\pi}{4}\right)(\text{bore})^2(\text{stroke})$$

$$= \left(\frac{\pi}{4}\right)(0.110 \text{ m})^2(0.150 \text{ m})$$

$$= 1.425 \times 10^{-3} \text{ m}^3$$

The ideal volume of air taken in per second is

$$\dot{V}_i = (\text{swept volume})\left(\frac{\text{intake strokes}}{\text{s}}\right)$$

$$= V_s N$$

$$= (1.425 \times 10^{-3})\left(60 \ \frac{1}{\text{s}}\right) = 0.0855 \text{ m}^3/\text{s}$$

The absolute temperature is

$$21°C + 273 = 294\text{K}$$

From the ideal gas law, the ideal mass of air in the swept volume is

$$\dot{m} = \frac{p\dot{V}}{RT}$$

$$= \frac{(101.3 \text{ kPa})\left(1000 \ \frac{\text{Pa}}{\text{kPa}}\right)\left(0.0855 \ \frac{\text{m}^3}{\text{s}}\right)}{\left(287.03 \ \frac{\text{J}}{\text{kg·K}}\right)(294\text{K})}$$

$$= 0.1026 \text{ kg/s}$$

step 2: Find the carbon dioxide volume in the exhaust assuming complete combustion when the air/fuel ratio is 15 (%CO_2 = 13.7% dry basis).

Air is 76.85% nitrogen by weight, so the nitrogen/fuel ratio is

$$(0.7685)(15) = 11.528 \text{ kg N}_2/\text{kg fuel}$$

From the ideal gas law, the nitrogen per kg of fuel burned is

$$V_{N_2} = \frac{mRT}{p}$$

$$= \frac{\left(11.528 \ \frac{\text{kg N}_2}{\text{kg fuel}}\right)\left(296.77 \ \frac{\text{J}}{\text{kg·K}}\right)(294\text{K})}{(101.3 \text{ kPa})\left(1000 \ \frac{\text{Pa}}{\text{kPa}}\right)}$$

$$= 9.929 \text{ m}^3/\text{kg fuel}$$

Similarly, air is 23.15% oxygen by weight, so the oxygen/fuel ratio is

$$(0.2315)(15) = 3.473 \text{ kg O}_2/\text{kg fuel}$$

From the ideal gas law, the oxygen per kg of fuel burned is

$$V_{O_2} = \frac{mRT}{p}$$

$$= \frac{\left(3.473 \ \frac{\text{kg O}_2}{\text{kg fuel}}\right)\left(259.82 \ \frac{\text{J}}{\text{kg·K}}\right)(294\text{K})}{(101.3 \text{ kPa})\left(1000 \ \frac{\text{Pa}}{\text{kPa}}\right)}$$

$$= 2.619 \text{ m}^3/\text{kg fuel}$$

From the customary U.S. solution,

$$\%CO_2 = \frac{\text{vol } CO_2}{\text{vol } CO_2 + \text{vol } O_2 + \text{vol } N_2}$$

$$\text{vol } CO_2 = x \ [\text{unknown}], \text{ in m}^3$$

$$\text{vol } O_2 = 2.619 \text{ m}^3 - O_2 \text{ used to make } CO_2$$

$$= 2.619 \text{ m}^3 - x$$

$$\%CO_2 = 0.137 \ [\text{given}]$$

$$0.137 = \frac{x}{x + (2.619 \text{ m}^3 - x) + 9.929 \text{ m}^3}$$

$$x = 1.719 \text{ m}^3/\text{kg fuel}$$

Assuming complete combustion, the volume of carbon dioxide will be constant regardless of the amount of air used.

step 3: Calculate the excess air if the percentage of carbon dioxide in the exhaust is 9%. Therefore,

$$\%CO_2 = \frac{\text{vol } CO_2}{\begin{array}{c}\text{vol } CO_2 + \text{vol } O_2 \\ + \text{ vol } N_2 + \text{vol excess air}\end{array}}$$

$$0.09 = \frac{1.719 \text{ m}^3}{\begin{array}{c}1.719 \text{ m}^3 + (2.619 \text{ m}^3 - 1.719 \text{ m}^3) \\ + 9.929 \text{ m}^3 + \text{vol excess air}\end{array}}$$

$$\begin{array}{c}\text{vol}\\\text{excess}\\\text{air}\end{array} = 6.552 \text{ m}^3/\text{kg fuel}$$

From the ideal gas law, the mass of the excess air is

$$m_{\text{excess}} = \frac{(101.3 \text{ kPa})\left(1000 \ \frac{\text{Pa}}{\text{kPa}}\right)(6.552 \text{ m}^3)}{\left(287.03 \ \frac{\text{J}}{\text{kg·K}}\right)(294\text{K})}$$

$$= 7.865 \text{ kg/kg fuel}$$

step 4: The actual air/fuel ratio is

$$15 \ \frac{\text{kg air}}{\text{kg fuel}} + 7.865 \ \frac{\text{kg air}}{\text{kg fuel}} = 22.865 \ \text{kg air/kg fuel}$$

The actual air mass per second is

$$\left(22.865 \ \frac{\text{kg air}}{\text{kg fuel}}\right)\left(0.0035 \ \frac{\text{kg}}{\text{s}}\right) = 0.0800 \ \text{kg/s}$$

step 5: The volumetric efficiency is

$$\eta_v = \frac{0.0800 \ \dfrac{\text{kg}}{\text{s}}}{0.1026 \ \dfrac{\text{kg}}{\text{s}}} = \boxed{0.772 \ \ (77.2\%)}$$

The answer is (D).

6. *Customary U.S. Solution*

(a) *step 1:* From App. 27.A,

Altitude 1: standard atmospheric condition, 14.696 psia, 518.7°R

Altitude 2: $z = 5000$ ft, 12.225 psia, 500.9°R

step 2: Calculate the friction power (Eq. 27.68).

$$\text{IHP}_1 = \frac{\text{BHP}_1}{\eta_{m,1}} = \frac{1000 \ \text{hp}}{0.80} = 1250 \ \text{hp}$$

step 3: Not needed since η_m is constant with altitude.

step 4: From the ideal gas law,

$$\rho_{a1} = \frac{p}{RT} = \frac{\left(14.696 \ \dfrac{\text{lbf}}{\text{in}^2}\right)\left(144 \ \dfrac{\text{in}^2}{\text{ft}^2}\right)}{\left(53.35 \ \dfrac{\text{lbf-ft}}{\text{lbm-°R}}\right)(518.7°\text{R})}$$

$$= 0.0765 \ \text{lbm/ft}^3$$

Similarly,

$$\rho_{a2} = \frac{p}{RT} = \frac{\left(12.225 \ \dfrac{\text{lbf}}{\text{in}^2}\right)\left(144 \ \dfrac{\text{in}^2}{\text{ft}^2}\right)}{\left(53.35 \ \dfrac{\text{ft-lbf}}{\text{lbm-°R}}\right)(500.9°\text{R})}$$

$$= 0.0659 \ \text{lbm/ft}^3$$

step 5: Calculate the new frictionless power (Eq. 27.70).

$$\text{IHP}_2 = \text{IHP}_1\left(\frac{\rho_{a2}}{\rho_{a1}}\right)$$

$$= (1250 \ \text{hp})\left(\frac{0.0659 \ \dfrac{\text{lbm}}{\text{ft}^3}}{0.0765 \ \dfrac{\text{lbm}}{\text{ft}^3}}\right)$$

$$= 1076.8 \ \text{hp}$$

steps 6 and 7: Calculate the new net power using Eq. 27.72.

$$\text{BHP}_2 = \eta_{m,2}(\text{IHP}_2) = (0.80)(1076.8 \ \text{hp})$$

$$= \boxed{861.4 \ \text{hp}}$$

The answer is (B).

(b) *step 8:* Not needed.

step 9: The original fuel rate (Eq. 27.74) is

$$\dot{m}_{f1} = (\text{BSFC}_1)(\text{BHP}_1)$$

$$= \left(0.45 \ \frac{\text{lbm}}{\text{bhp-hr}}\right)(1000 \ \text{bhp})$$

$$= 450 \ \text{lbm/hr}$$

step 10: Not needed.

step 11: $\dot{m}_{f2} = \dot{m}_{f1} = 450 \ \text{lbm/hr}$

step 12: The new fuel consumption (Eq. 27.79) is

$$\text{BSFC}_2 = \frac{\dot{m}_{f2}}{\text{BHP}_2} = \frac{450 \ \dfrac{\text{lbm}}{\text{hr}}}{861.4 \ \text{hp}}$$

$$= \boxed{0.522 \ \text{lbm/bhp-hr}}$$

The answer is (B).

SI Solution

(a) *step 1:* From App. 27.A,

Altitude 1: standard atmospheric condition, 1.01325 bar, 288.15K

Altitude 2: 0.8456 bar, 278.4K

$$z = 1500 \ \text{m}$$

step 2: Calculate the friction power (Eq. 27.68).

$$\text{IHP}_1 = \frac{\text{BHP}_1}{\eta_{m,1}} = \frac{750 \ \text{kW}}{0.80} = 937.5 \ \text{kW}$$

step 3: Not needed since η_m is constant with altitude.

step 4: From the ideal gas law, the air densities are

$$\rho_{a1} = \frac{p}{RI} = \frac{(1.01325 \ \text{bar})\left(10^5 \ \dfrac{\text{Pa}}{\text{bar}}\right)}{\left(287.03 \ \dfrac{\text{J}}{\text{kg·K}}\right)(288.15\text{K})}$$

$$= 1.225 \ \text{kg/m}^3$$

Similarly,

$$\rho_{a2} = \frac{(0.8456 \text{ bar})\left(10^5 \frac{\text{Pa}}{\text{bar}}\right)}{\left(287.03 \frac{\text{J}}{\text{kg·K}}\right)(278.4\text{K})}$$

$$= 1.058 \text{ kg/m}^3$$

step 5: Calculate the new frictionless power (Eq. 27.70).

$$\text{IHP}_2 = \text{IHP}_1 \left(\frac{\rho_{a2}}{\rho_{a1}}\right)$$

$$= (937.5 \text{ kW})\left(\frac{1.058 \frac{\text{kg}}{\text{m}^3}}{1.225 \frac{\text{kg}}{\text{m}^3}}\right)$$

$$= 809.7 \text{ kW}$$

steps 6 and 7: Calculate the new net power (Eq. 27.72).

$$\text{BHP}_2 = \eta_{m,2}(\text{IHP}_2) = (0.80)(809.7 \text{ kW})$$

$$= \boxed{647.8 \text{ kW}}$$

The answer is (A).

(b) step 8: Not needed.

step 9: The original fuel rate (Eq. 27.74) is

$$\dot{m}_{f1} = (\text{BSFC}_1)(\text{BHP}_1)$$

$$= \left(76 \frac{\text{kg}}{\text{GJ}}\right)(750 \text{ kW})\left(\frac{\text{GJ}}{10^6 \text{ kJ}}\right)$$

$$= 0.057 \text{ kg/s}$$

step 10: Not needed.

step 11: $\dot{m}_{f2} = \dot{m}_{f1} = 0.057 \text{ kg/s}$

step 12: The new fuel consumption (Eq. 27.79) is

$$\text{BSFC}_2 = \frac{\dot{m}_{f2}}{\text{BHP}_2}$$

$$= \frac{0.057 \frac{\text{kg}}{\text{s}}}{(647.8 \text{ kW})\left(\frac{\text{GW}}{10^6 \text{ kW}}\right)}$$

$$= \boxed{88.0 \text{ kg/GJ}}$$

The answer is (B).

7. *Customary U.S. Solution*

(Use an air table. The SI solution assumes an ideal gas.) Refer to Fig. 27.11.

At A:

$$T_A = 60°\text{F} + 460 = 520°\text{R} \quad [\text{given}]$$

$$p_A = 14.7 \text{ psia} \quad [\text{given}]$$

From the air table (App. 23.F),

$$v_{r,A} = 158.58$$

$$p_{r,A} = 1.2147$$

$$h_A = 124.27 \text{ Btu/lbm}$$

At B:

The process from A to B is isentropic.

$$v_{r,B} = v_{r,A}\left(\frac{V_B}{V_A}\right) = (158.58)\left(\frac{1}{5}\right) = 31.716$$

Locate this volume ratio in the air table (App. 23.F).

$$T_B \approx 980°\text{R}$$

$$h_B \approx 236.02 \text{ Btu/lbm}$$

$$p_{r,B} = 11.430$$

Since process A-B is isentropic,

$$p_B = \left(\frac{p_{r,B}}{p_{r,A}}\right)p_A = \left(\frac{11.430}{1.2147}\right)(14.7 \text{ psia}) = 138.3 \text{ psia}$$

At C:

$$T_C = 1500°\text{F} + 460 = 1960°\text{R} \quad [\text{given}]$$

$$p_C = p_B = 138.3 \text{ psia}$$

Locate the temperature in the air table.

$$h_C = 493.64 \text{ Btu/lbm}$$

$$p_{r,C} = 160.48$$

At D:

$$p_D = 14.7 \text{ psia}$$

Since process C-D is isentropic,

$$p_{r,D} = p_{r,C}\left(\frac{p_D}{p_C}\right) = (160.48)\left(\frac{14.7 \text{ psia}}{138.3 \text{ psia}}\right)$$

$$= 17.057$$

Locate this pressure ratio in the air table.

$$T_D = 1094°R$$

$$h_D = 264.49 \text{ Btu/lbm}$$

Since the efficiency of compression is 83%, from Eq. 27.98,

$$h'_B = h_A + \frac{h_B - h_A}{\eta_{s,\text{compressor}}}$$

$$= 124.27 \frac{\text{Btu}}{\text{lbm}} + \frac{236.02 \frac{\text{Btu}}{\text{lbm}} - 124.27 \frac{\text{Btu}}{\text{lbm}}}{0.83}$$

$$= 258.9 \text{ Btu/lbm}$$

Since the efficiency of the expansion process is 92%, from Eq. 27.100,

$$h'_D = h_C - \eta_{s,\text{turbine}}(h_C - h_D)$$

$$= 493.64 \frac{\text{Btu}}{\text{lbm}}$$

$$- (0.92) \left(493.64 \frac{\text{Btu}}{\text{lbm}} - 264.49 \frac{\text{Btu}}{\text{lbm}} \right)$$

$$= 282.8 \text{ Btu/lbm}$$

From Eq. 27.96, the thermal efficiency is

$$\eta_{\text{th}} = \frac{(h_C - h'_B) - (h'_D - h_A)}{h_C - h'_B}$$

$$= \frac{\left(493.64 \frac{\text{Btu}}{\text{lbm}} - 258.9 \frac{\text{Btu}}{\text{lbm}} \right) - \left(282.8 \frac{\text{Btu}}{\text{lbm}} - 124.27 \frac{\text{Btu}}{\text{lbm}} \right)}{493.64 \frac{\text{Btu}}{\text{lbm}} - 258.9 \frac{\text{Btu}}{\text{lbm}}}$$

$$= \boxed{0.325 \ (32.5\%)}$$

The answer is (A).

SI Solution

Refer to Fig. 27.11.

At A:
$$T_A = 16°C + 273 = 289K \quad \text{[given]}$$
$$p_A = 101.3 \text{ kPa} \quad \text{[given]}$$

At B:

$$T_B = T_A \left(\frac{v_A}{v_B} \right)^{k-1} = (289K)(5)^{1.4-1} = 550.2K$$

$$p_B = p_A \left(\frac{v_A}{v_B} \right)^k = (101.3 \text{ kPa})(5)^{1.4} = 964.2 \text{ kPa}$$

At C:
$$T_C = 820°C + 273 = 1093K \quad \text{[given]}$$
$$p_C = p_B = 964.2 \text{ kPa}$$

At D:
$$p_D = 101.3 \text{ kPa} \quad \text{[given]}$$

$$T_D = T_C \left(\frac{p_D}{p_C} \right)^{\frac{k-1}{k}} = (1093K) \left(\frac{101.3 \text{ kPa}}{964.2 \text{ kPa}} \right)^{\frac{1.4-1}{1.4}}$$

$$T_D = 574.2K$$

For ideal gases, the specific heats are constant. Therefore, the change in internal energy (and enthalpy, approximately) is proportional to the change in temperature. The actual temperature (Eq. 27.99) is

$$T'_B = T_A + \frac{T_B - T_A}{\eta_{s,\text{compressor}}}$$

$$= 289K + \frac{550.2K - 289K}{0.83}$$

$$= 603.7K$$

From Eq. 27.101,

$$T'_D = T_C - \eta_{s,\text{turbine}}(T_C - T_D)$$

$$= 1093K - (0.92)(1093K - 574.2K)$$

$$= 615.7K$$

From Eq. 27.97, the thermal efficiency is

$$\eta_{\text{th}} = \frac{(T_C - T'_B) - (T'_D - T_A)}{T_C - T'_B}$$

$$= \frac{(1093K - 603.7K) - (615.7K - 289K)}{1093K - 603.7K}$$

$$= \boxed{0.332 \ (33.2\%)}$$

The answer is (A).

8. *Customary U.S. Solution*

Since specific heats are constant, use ideal gas rather than air tables.

Refer to Fig. 27.12.

At A:
$$T_A = 60°F + 460 = 520°R \quad \text{[given]}$$
$$p_A = 14.7 \text{ psia} \quad \text{[given]}$$

At B:

$$T_B = T_A \left(\frac{v_A}{v_B} \right)^{k-1} = (520°R)(5)^{1.4-1} = 989.9°R$$

$$p_B = p_A \left(\frac{v_A}{v_B} \right)^k = (14.7 \text{ psia})(5)^{1.4}$$

$$= 139.9 \text{ psia}$$

At D:

$$T_D = 1500°F + 460 = 1960°R \quad \text{[given]}$$

$$p_D = p_B = 139.9 \text{ psia}$$

At E:

$$p_E = 14.7 \text{ psia} \quad \text{[given]}$$

$$T_E = T_D \left(\frac{p_E}{p_D}\right)^{\frac{k-1}{k}} = (1960°R)\left(\frac{14.7 \text{ psia}}{139.9 \text{ psia}}\right)^{\frac{1.4-1}{1.4}}$$

$$= 1029.6°R$$

From Eq. 27.99,

$$T'_B = T_A + \frac{T_B - T_A}{\eta_{s,\text{compressor}}} = 520°R + \frac{989.9°R - 520°R}{0.83}$$

$$= 1086.1°R$$

From Eq. 27.101,

$$T'_E = T_D - \eta_{s,\text{turbine}}(T_D - T_E)$$

$$= 1960°R - (0.92)(1960°R - 1029.6°R)$$

$$= 1104.0°R$$

From Eq. 27.102,

$$\eta_{\text{regenerator}} = \frac{h_C - h'_B}{h'_E - h'_B}$$

For $c_p \approx$ constant,

$$\eta_{\text{regenerator}} = \frac{T_C - T'_B}{T'_E - T'_B}$$

$$0.65 = \frac{T_C - 1086.1°R}{1104.0°R - 1086.1°R}$$

$$T_C = 1097.7°R$$

From Eq. 27.103 with constant specific heats,

$$\eta_{\text{th}} = \frac{(T_D - T'_E) - (T'_B - T_A)}{T_D - T_C}$$

$$= \frac{(1960°R - 1104.0°R) - (1086.1°R - 520°R)}{1960°R - 1097.7°R}$$

$$= \boxed{0.336 \quad (33.6\%)}$$

The answer is (C).

SI Solution

Refer to Fig. 27.12. From Prob. 7,

$$T_A = 289K$$

$$T'_B = 603.7K$$

$$T_D = 1093K$$

$$T'_E = 615.7K$$

For constant specific heats and from Eq. 27.102,

$$\eta_{\text{regenerator}} = \frac{T_C - T'_B}{T'_E - T'_B}$$

$$0.65 = \frac{T_C - 603.7K}{615.7K - 603.7K}$$

$$T_C = 611.5K$$

From Eq. 27.103 with constant specific heats,

$$\eta_{\text{th}} = \frac{(T_D - T'_E) - (T'_B - T_A)}{T_D - T_C}$$

$$= \frac{(1093K - 615.7K) - (603.7K - 289K)}{1093K - 611.5K}$$

$$= \boxed{0.338 \quad (33.8\%)}$$

The answer is (C).

9. *Customary U.S. Solution*

(a) From Eq. 27.41, the number of power strokes per minute is

$$N = \frac{(2n)(\text{no. cylinders})}{\text{no. strokes per cycle}}$$

$$= \frac{\left(2 \frac{\text{strokes}}{\text{rev}}\right)\left(4600 \frac{\text{rev}}{\text{min}}\right)(8 \text{ cylinders})}{4 \frac{\text{strokes}}{\text{power stroke}}}$$

$$= 18,400 \text{ power strokes/min}$$

The net work per cycle is

$$W_{\text{net}} = W_{\text{out}} - W_{\text{in}} = 1500 \text{ ft-lbf} - 1200 \text{ ft-lbf}$$

$$= 300 \text{ ft-lbf/cycle}$$

The indicated horsepower is

$$\text{IHP} = \frac{\left(18,400 \frac{\text{power strokes}}{\text{min}}\right)(300 \text{ ft-lbf})}{33,000 \frac{\text{ft-lbf}}{\text{hp-min}}}$$

$$= \boxed{167.27 \text{ hp}}$$

The answer is (B).

(b) From Eq. 27.3, the thermal efficiency is

$$\eta_{\text{th}} = \frac{W_{\text{out}} - W_{\text{in}}}{Q_{\text{in}}} = \frac{W_{\text{net}}}{Q_{\text{in}}}$$

$$= \frac{300 \; \frac{\text{ft-lbf}}{\text{cycle}}}{\left(1.27 \; \frac{\text{Btu}}{\text{cycle}}\right)\left(778 \; \frac{\text{ft-lbf}}{\text{Btu}}\right)}$$

$$= \boxed{0.304 \; (30.4\%)}$$

The answer is (A).

(c) The lower heating value of gasoline is LHV = 18,900 Btu/lbm.

The fuel consumption is

$$\dot{m}_F = \frac{\left(1.27 \; \frac{\text{Btu}}{\text{cycle}}\right)\left(18,400 \; \frac{\text{power strokes}}{\text{min}}\right)\left(60 \; \frac{\text{min}}{\text{hr}}\right)}{18,900 \; \frac{\text{Btu}}{\text{lbm}}}$$

$$= \boxed{74.18 \; \text{lbm/hr}}$$

The answer is (C).

(d) Specific fuel consumption is given by Eq. 27.8.

$$\text{SFC} = \frac{\text{fuel usage rate}}{\text{power generated}} = \frac{74.18 \; \frac{\text{lbm}}{\text{hr}}}{167.27 \; \text{hp}}$$

$$= \boxed{0.443 \; \text{lbm/hp-hr}}$$

The answer is (D).

SI Solution

(a) From Eq. 27.41, the number of power strokes per minute is

$$N = \frac{(2n)(\text{no. cylinders})}{\text{no. strokes per cycle}}$$

$$= \frac{\left(2 \; \frac{\text{strokes}}{\text{rev}}\right)\left(4600 \; \frac{\text{rev}}{\text{min}}\right)(8 \; \text{cylinders})}{4 \; \text{strokes per power stroke}}$$

$$= 18\,400 \; \text{power strokes/min}$$

The net work per cycle is

$$W_{\text{net}} = W_{\text{out}} - W_{\text{in}} = 2.0 \; \text{kJ} - 1.6 \; \text{kJ}$$

$$= 0.4 \; \text{kJ/cycle}$$

The indicated power is

$$\text{IkW} = \left(18\,400 \; \frac{\text{power strokes}}{\text{min}}\right)\left(\frac{1 \; \text{min}}{60 \; \text{sec}}\right)(0.4 \; \text{kJ})$$

$$= \boxed{122.7 \; \text{kW}}$$

The answer is (B).

(b) From Eq. 27.3, the thermal efficiency is

$$\eta_{\text{th}} = \frac{W_{\text{out}} - W_{\text{in}}}{Q_{\text{in}}} = \frac{W_{\text{net}}}{Q_{\text{in}}}$$

$$= \frac{0.4 \; \frac{\text{kJ}}{\text{cycle}}}{1.33 \; \frac{\text{kJ}}{\text{cycle}}} = \boxed{0.300 \; (30\%)}$$

The answer is (A).

(c) The heating value of gasoline is LHV = 44 MJ/kg.

The fuel consumption is

$$\dot{m}_F = \frac{\left(1.33 \; \frac{\text{kJ}}{\text{cycle}}\right)\left(18\,400 \; \frac{\text{power strokes}}{\text{min}}\right)\left(\frac{1 \; \text{min}}{60 \; \text{s}}\right)}{\left(44 \; \frac{\text{MJ}}{\text{kg}}\right)\left(1000 \; \frac{\text{kJ}}{\text{MJ}}\right)}$$

$$= \boxed{0.00927 \; \text{kg/s}}$$

The answer is (C).

(d) The specific fuel consumption (Eq. 27.8) is

$$\text{SFC} = \frac{\text{fuel usage rate}}{\text{power generated}} = \frac{0.00927 \; \frac{\text{kg}}{\text{s}}}{122.7 \; \text{kW}}$$

$$= \boxed{7.555 \times 10^{-5} \; \text{kg/kJ}}$$

The answer is (D).

10. *Customary U.S. Solution*

(a) At A:
$$T_A = 520°\text{R}$$
$$p_A = 14.7 \; \text{psia}$$

At C:
$$T_C = 1600°$$
$$p_C = 568.6 \; \text{psia}$$

For isentropic process C-A, from Eq. 24.91,

$$p_A = p_C \left(\frac{T_A}{T_C}\right)^{\frac{k}{k-1}}$$

Therefore,

$$14.7 \text{ psia} = (568.6 \text{ psia}) \left(\frac{520°\text{R}}{1600°\text{R}} \right)^{\frac{k}{k-1}}$$

$$0.02585 = (0.325)^{\frac{k}{k-1}}$$

$$\log(0.02585) = \left(\frac{k}{k-1} \right) \log(0.325)$$

$$\frac{k}{k-1} = 3.252$$

$$k = 1.444$$

From Eq. 23.58, the molar specific heat of the mixture is

$$C_{p,\text{mixture}} = \frac{R^* k}{k-1}$$

$$= \frac{\left(1545 \frac{\text{ft-lbf}}{\text{lbmol-°R}} \right)(1.444)}{\left(778 \frac{\text{ft-lbf}}{\text{Btu}} \right)(1.444 - 1)}$$

$$= 6.459 \text{ Btu/lbmol-°R}$$

From Table 23.7,

$$(c_p)_{\text{He}} = 1.240 \text{ Btu/lbm-°R}$$
$$(\text{MW})_{\text{He}} = 4.003 \text{ lbm/lbmol}$$
$$(c_p)_{\text{CO}_2} = 0.207 \text{ Btu/lbm-°R}$$
$$(\text{MW})_{\text{CO}_2} = 44.011 \text{ lbm/lbmol}$$
$$C_{p,\text{He}} = (\text{MW})_{\text{He}}(c_p)_{\text{He}}$$
$$= \left(4.003 \frac{\text{lbm}}{\text{lbmol}} \right) \left(1.240 \frac{\text{Btu}}{\text{lbm-°R}} \right)$$
$$= 4.96 \text{ Btu/lbmol-°R}$$
$$C_{p,\text{CO}_2} = (\text{MW})_{\text{CO}_2}(c_p)_{\text{CO}_2}$$
$$= \left(44.011 \frac{\text{lbm}}{\text{lbmol}} \right) \left(0.207 \frac{\text{Btu}}{\text{lbm-°R}} \right)$$
$$= 9.11 \text{ Btu/lbmol-°R}$$

Let x be the mole fraction of helium in the mixture.

$$x = \frac{n_{\text{He}}}{n_{\text{He}} + n_{\text{CO}_2}}$$

From Eq. 23.86, on a mole basis,

$$c_{p,\text{mixture}} = x(c_p)_{\text{He}} + (1-x)(c_p)_{\text{CO}_2}$$

$$6.459 \frac{\text{Btu}}{\text{lbmol-°R}} = x \left(4.96 \frac{\text{Btu}}{\text{lbmol-°R}} \right)$$
$$+ (1-x) \left(9.11 \frac{\text{Btu}}{\text{lbmol-°R}} \right)$$

$$x = 0.639$$

On a per mole basis, the mass of helium would be

$$m_{\text{He}} = x(\text{MW})_{\text{He}}$$
$$= (0.639) \left(4.003 \frac{\text{lbm}}{\text{lbmol}} \right)$$
$$= 2.558 \text{ lbm}$$

Similarly, the mass of carbon dioxide on a per mole basis would be

$$m_{\text{CO}_2} = (1-x)(\text{MW})_{\text{CO}_2}$$
$$= (1 \text{ lbmol} - 0.639 \text{ lbmol}) \left(44.011 \frac{\text{lbm}}{\text{lbmol}} \right)$$
$$= 15.888 \text{ lbm}$$

The molecular weight of the mixture is

$$(\text{MW})_{\text{mixture}} = 2.558 \text{ lbm} + 15.888 \text{ lbm}$$
$$= 18.446 \text{ lbm/lbmol}$$

The gravimetric (mass) fraction of the gases is

$$G_{\text{He}} = \frac{m_{\text{He}}}{m_{\text{He}} + m_{\text{CO}_2}}$$

$$= \frac{2.558 \text{ lbm}}{2.558 \text{ lbm} + 15.888 \text{ lbm}} = \boxed{0.139}$$

$$G_{\text{CO}_2} = 1 - G_{\text{He}} = 1 - 0.139 = \boxed{0.861}$$

The answer is (A).

(b) From Eq. 23.56,

$$C_{v,\text{mixture}} = C_{p,\text{mixture}} - R^*$$

$$= 6.459 \frac{\text{Btu}}{\text{lbmol-°R}} - \frac{1545 \frac{\text{ft-lbf}}{\text{lbmol-°R}}}{778 \frac{\text{ft-lbf}}{\text{Btu}}}$$

$$= 4.473 \text{ Btu/lbmol-°R}$$

From Eq. 24.102, the work done during the isentropic expansion process is

$$W = c_v(T_1 - T_2) = \left(\frac{C_v}{\text{MW}} \right)_{\text{mixture}} \times (T_\text{C} - T_\text{A})$$

$$= \left(\frac{4.473 \frac{\text{Btu}}{\text{lbmol-°R}}}{18.446 \frac{\text{lbm}}{\text{lbmol}}} \right) (1600°\text{R} - 520°\text{R})$$

$$= \boxed{261.9 \text{ Btu/lbm}}$$

The answer is (B).

(c)

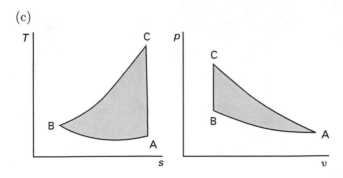

SI Solution

(a) At A:

$$T_A = 290K$$
$$p_A = 101.3 \text{ kPa}$$

At C:

$$T_C = 890K$$
$$p_C = 3.920 \text{ MPa}$$

For isentropic process C-A from Eq. 24.91,

$$p_A = p_C \left(\frac{T_A}{T_C}\right)^{\frac{k}{k-1}}$$

Therefore,

$$101.3 \text{ kPa} = (3.920 \text{ MPa}) \left(1000 \frac{\text{kPa}}{\text{MPa}}\right) \left(\frac{290K}{890K}\right)^{\frac{k}{k-1}}$$

$$0.02584 = (0.3258)^{\frac{k}{k-1}}$$

$$\log(0.02584) = \left(\frac{k}{k-1}\right) \log(0.3258)$$

$$\frac{k}{k-1} = 3.2599$$

$$k = 1.442$$

From Eq. 23.58, the molar specific heat of the mixture is

$$C_{p,\text{mixture}} = \frac{R^* k}{k-1}$$

$$= \frac{\left(8.3143 \frac{\text{kJ}}{\text{kmol·K}}\right)(1.442)}{1.442 - 1}$$

$$= 27.125 \text{ kJ/kmol·K}$$

From Table 23.7,

$$(c_p)_{\text{He}} = \left(5192 \frac{\text{J}}{\text{kg·K}}\right)\left(\frac{1 \text{ kJ}}{1000 \text{ J}}\right)$$

$$= 5.192 \text{ kJ/kg·K}$$

$$(\text{MW})_{\text{He}} = 4.003 \text{ kg/kmol}$$

$$(c_p)_{\text{CO}_2} = \left(867 \frac{\text{J}}{\text{kg·K}}\right)\left(\frac{1 \text{ kJ}}{1000 \text{ J}}\right)$$

$$= 0.867 \text{ kJ/kg·K}$$

$$(\text{MW})_{\text{CO}_2} = 44.011 \text{ kg/kmol}$$

$$C_{p,\text{He}} = (\text{MW})_{\text{He}}(c_p)_{\text{He}}$$

$$= \left(4.003 \frac{\text{kg}}{\text{kmol}}\right)\left(5.192 \frac{\text{kJ}}{\text{kmol}}\right)$$

$$= 20.784 \text{ kJ/kmol·K}$$

$$C_{p,\text{CO}_2} = (\text{MW})_{\text{CO}_2}(c_p)_{\text{CO}_2}$$

$$= \left(44.011 \frac{\text{kg}}{\text{kmol}}\right)\left(0.867 \frac{\text{kJ}}{\text{kg·K}}\right)$$

$$= 38.158 \text{ kJ/kmol·K}$$

Let x be the mole fraction of helium in the mixture.

$$x = \frac{n_{\text{He}}}{n_{\text{He}} + n_{\text{CO}_2}}$$

From Eq. 23.86 on a mole basis,

$$c_{p,\text{mixture}} = x(c_p)_{\text{He}} + (1-x)(c_p)_{\text{CO}_2}$$

$$27.125 \frac{\text{kJ}}{\text{kmol·K}} = x\left(20.784 \frac{\text{kJ}}{\text{kmol·K}}\right)$$
$$+ (1-x)\left(38.158 \frac{\text{kJ}}{\text{kmol·K}}\right)$$

$$x = 0.635 \text{ kmol}$$

On a per mole basis, the mass of helium would be

$$m_{\text{He}} = x(\text{MW})_{\text{He}}$$

$$= (0.635 \text{ kmol})\left(4.003 \frac{\text{kg}}{\text{kmol}}\right)$$

$$= 2.542 \text{ kg}$$

Similarly, the mass of CO_2 on a per mole basis would be

$$m_{\text{CO}_2} = (1-x)(\text{MW})_{\text{CO}_2}$$

$$= (1 \text{ kmol} - 0.635 \text{ kmol})\left(44.011 \frac{\text{kg}}{\text{kmol}}\right)$$

$$= 16.064 \text{ kg}$$

The molecular weight of the mixture is

$$(\text{MW})_{\text{mixture}} = 2.542 \text{ kg} + 16.064 \text{ kg}$$

$$= 18.606 \text{ kg/kmol}$$

The gravimetric (mass) fraction of the gases is

$$G_{He} = \frac{m_{He}}{m_{He} + m_{CO_2}}$$

$$= \frac{2.542 \text{ kg}}{2.542 \text{ kg} + 16.064 \text{ kg}} = \boxed{0.137}$$

$$G_{CO_2} = 1 - G_{He} = 1 - 0.137 = \boxed{0.863}$$

The answer is (A).

(b) From Eq. 23.56,

$$C_{v,mixture} = C_{p,mixture} - R^*$$

$$= 27.125 \ \frac{\text{kJ}}{\text{kmol·K}} - 8.3143 \ \frac{\text{kJ}}{\text{kmol·K}}$$

$$= 18.811 \text{ kJ/kmol·K}$$

From Eq. 24.102, the work done during the isentropic expansion process is

$$W = c_v(T_1 - T_2) = \left(\frac{C_v}{MW}\right)_{mixture} \times (T_C - T_A)$$

$$= \left(\frac{18.811 \ \frac{\text{kJ}}{\text{kmol·K}}}{18.606 \ \frac{\text{kg}}{\text{kmol}}}\right)(890K - 290K)$$

$$= \boxed{606.6 \text{ kJ/kg}}$$

The answer is (B).

(c)

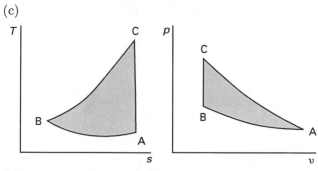

11. *Customary U.S. Solution*

step 1: Not needed.

step 2: From Eq. 27.68, calculate the frictionless power.

$$IHP_1 = \frac{BHP_1}{\eta_{m,1}} = \frac{200 \text{ hp}}{0.86} = 232.6 \text{ hp}$$

step 3: From Eq. 27.69, calculate the friction power, which is assumed to be constant at constant speed.

$$FHP = IHP_1 - BHP_1$$

$$= 232.6 \text{ hp} - 200 \text{ hp} = 32.6 \text{ hp}$$

step 4: Calculate the air densities ρ_{a1} and ρ_{a2} from the ideal gas law. The absolute temperatures are

$$T_1 = 80°F + 460 = 540°R$$

$$T_2 = 60°F + 460 = 520°R$$

$$\rho_{a1} = \frac{p_1}{RT_1} = \frac{\left(14.7 \ \frac{\text{lbf}}{\text{in}^2}\right)\left(144 \ \frac{\text{in}^2}{\text{ft}^2}\right)}{\left(53.3 \ \frac{\text{lbf-ft}}{\text{lbm-°R}}\right)(540°R)}$$

$$= 0.0735 \text{ lbm/ft}^3$$

$$\rho_{a2} = \frac{p_2}{RT_1} = \frac{\left(12.2 \ \frac{\text{lbf}}{\text{in}^2}\right)\left(144 \ \frac{\text{in}^2}{\text{ft}^2}\right)}{\left(53.3 \ \frac{\text{lbf-ft}}{\text{lbm-°R}}\right)(520°R)}$$

$$= 0.0634 \text{ lbm/ft}^3$$

step 5: From Eq. 27.70, calculate the new frictionless power.

$$IHP_2 = IHP_1 \left(\frac{\rho_{a2}}{\rho_{a1}}\right)$$

$$= (232.6 \text{ hp})\left(\frac{0.0634 \ \frac{\text{lbm}}{\text{ft}^3}}{0.0735 \ \frac{\text{lbm}}{\text{ft}^3}}\right)$$

$$= 200.6 \text{ hp}$$

step 6: From Eq. 27.71, calculate the new net power.

$$BHP_2 = IHP_2 - FHP$$

$$= 200.6 \text{ hp} - 32.6 \text{ hp}$$

$$= \boxed{168.0 \text{ hp}}$$

step 7: From Eq. 27.72, calculate the new efficiency.

$$\eta_{m,2} = \frac{BHP_2}{IHP_2}$$

$$= \frac{168.0 \text{ hp}}{200.6 \text{ hp}}$$

$$= \boxed{0.837}$$

step 8: From Eq. 27.73, the volumetric air flow rates are the same since the engine speed is constant.

$$\dot{V}_{a2} = \dot{V}_{a1}$$

step 9: The original air and fuel rates from Eqs. 27.74 through 27.76 are

$$\dot{m}_{f1} = (\text{BSFC}_1)(\text{BHP}_1)$$

$$= \left(0.48 \ \frac{\text{lbm}}{\text{hp-hr}}\right)(200 \ \text{hp}) = 96 \ \text{lbm/hr}$$

$$\dot{m}_{a1} = (\text{AFR})(\dot{m}_{f1})$$

$$= (22)\left(96 \ \frac{\text{lbm}}{\text{hr}}\right) = 2112 \ \text{lbm/hr}$$

$$\dot{V}_{a1} = \frac{\dot{m}_{a1}}{\rho_{a1}}$$

$$= \frac{2112 \ \frac{\text{lbm}}{\text{hr}}}{0.0735 \ \frac{\text{lbm}}{\text{ft}^3}} = 28{,}735 \ \text{ft}^3/\text{hr}$$

$$\dot{V}_{a2} = \dot{V}_{a1} = 28{,}735 \ \text{ft}^3/\text{hr}$$

step 10: From Eq. 27.77, the new air mass flow rate is

$$\dot{m}_{a2} = \dot{V}_{a2}\rho_{a2}$$

$$= \left(28{,}735 \ \frac{\text{ft}^3}{\text{hr}}\right)\left(0.0634 \ \frac{\text{lbm}}{\text{ft}^3}\right)$$

$$= 1821.8 \ \text{lbm/hr}$$

step 11: For engines with metered injection,

$$\dot{m}_{f2} = \dot{m}_{f1} = 96 \ \text{lbm/hr}$$

step 12: From Eq. 27.79, the new fuel consumption is

$$\text{BSFC}_2 = \frac{\dot{m}_{f2}}{\text{BHP}_2} = \frac{96 \ \frac{\text{lbm}}{\text{hr}}}{168.0 \ \text{hp}}$$

$$= \boxed{0.571 \ \text{lbm/hp-hr}}$$

From Eq. 27.75, the new air/fuel ratio is

$$\text{AFR}_2 = \frac{\dot{m}_{a2}}{\dot{m}_{f2}} = \frac{1821.8 \ \frac{\text{lbm}}{\text{hr}}}{96 \ \frac{\text{lbm}}{\text{hr}}} = \boxed{18.98}$$

SI Solution

step 1: Not needed.

step 2: From Eq. 27.68, calculate the frictionless power.

$$\text{IkW}_1 = \frac{\text{BkW}_1}{\eta_{m,1}} = \frac{150 \ \text{kW}}{0.86} = 174.4 \ \text{kW}$$

step 3: From Eq. 27.69, calculate the friction power, which is assumed to be constant at constant speed.

$$\text{FkW} = \text{IkW}_1 - \text{BkW}_1$$

$$= 174.4 \ \text{kW} - 150 \ \text{kW} = 24.4 \ \text{kW}$$

step 4: Calculate the air densities ρ_{a1} and ρ_{a2} from the ideal gas law. The absolute temperatures are

$$T_1 = 27°\text{C} + 273 = 300\text{K}$$

$$T_2 = 16°\text{C} + 273 = 289\text{K}$$

$$\rho_{a1} = \frac{p_1}{RT_1} = \frac{(101.3 \ \text{kPa})\left(1000 \ \frac{\text{Pa}}{\text{kPa}}\right)}{\left(287.03 \ \frac{\text{kJ}}{\text{kg·K}}\right)(300\text{K})}$$

$$= 1.176 \ \text{kg/m}^3$$

$$\rho_{a2} = \frac{p_2}{RT_2} = \frac{(84 \ \text{kPa})\left(1000 \ \frac{\text{Pa}}{\text{kPa}}\right)}{\left(287.03 \ \frac{\text{kJ}}{\text{kg·K}}\right)(289\text{K})}$$

$$= 1.013 \ \text{kg/m}^3$$

step 5: From Eq. 27.70, calculate the new frictionless power.

$$\text{IkW}_2 = \text{IkW}_1\left(\frac{\rho_{a2}}{\rho_{a1}}\right)$$

$$= (174.4 \ \text{kW})\left(\frac{1.013 \ \frac{\text{kg}}{\text{m}^3}}{1.176 \ \frac{\text{kg}}{\text{m}^3}}\right) = 150.2 \ \text{kW}$$

step 6: From Eq. 27.71, calculate the new net power.

$$\text{BkW}_2 = \text{IkW}_2 - \text{FkW} = 150.2 \ \text{kW} - 24.4 \ \text{kW}$$

$$= \boxed{125.8 \ \text{kW}}$$

step 7: From Eq. 27.72, calculate the new efficiency.

$$\eta_{m,2} = \frac{\text{BkW}_2}{\text{IkW}_2} = \frac{125.8\ \text{kW}}{150.2\ \text{kW}} = \boxed{0.838}$$

step 8: From Eq. 27.73, the volumetric air flow rates are the same since the engine speed is constant.

$$\dot{V}_{a2} = \dot{V}_{a1}$$

step 9: The original air and fuel rates from Eqs. 27.74 through 27.76 are

$$\dot{m}_{f1} = (\text{BSFC}_1)(\text{BkW}_1)$$
$$= \left(81\ \frac{\text{kg}}{\text{GJ}}\right)\left(\frac{1\ \text{GJ}}{10^6\ \text{J}}\right)(150\ \text{kW})$$
$$= 0.01215\ \text{kg/s}$$

$$\dot{m}_{a1} = (\text{AFR})\dot{m}_{f1}$$
$$= (22)\left(0.01215\ \frac{\text{kg}}{\text{s}}\right) = 0.2673\ \text{kg/s}$$

$$\dot{V}_{a1} = \frac{\dot{m}_{a1}}{\rho_{a1}} = \frac{0.2673\ \dfrac{\text{kg}}{\text{s}}}{1.176\ \dfrac{\text{kg}}{\text{m}^3}} = 0.2273\ \text{m}^3/\text{s}$$

$$\dot{V}_{a2} = \dot{V}_{a1} = 0.2273\ \text{m}^3/\text{s}$$

step 10: From Eq. 27.77, the new air mass flow rate is

$$\dot{m}_{a2} = \dot{V}_{a2}\rho_{a2}$$
$$= \left(0.2273\ \frac{\text{m}^3}{\text{s}}\right)\left(1.013\ \frac{\text{kg}}{\text{m}^3}\right) = 0.2303\ \text{kg/s}$$

step 11: For engines with metered injection,

$$\dot{m}_{f2} = \dot{m}_{f1} = 0.01215\ \text{kg/s}$$

step 12: From Eq. 27.79, the new fuel consumption is

$$\text{BSFC}_2 = \frac{\dot{m}_{f2}}{\text{BkW}_2}$$
$$= \frac{0.01215\ \dfrac{\text{kg}}{\text{s}}}{(125.8\ \text{kW})\left(\dfrac{1\ \text{GJ}}{10^6\ \text{kJ}}\right)}$$
$$= \boxed{96.58\ \text{kg/GJ}}$$

From Eq. 27.75, the new air/fuel ratio is

$$\text{AFR}_2 = \frac{\dot{m}_{a2}}{\dot{m}_{f2}} = \frac{0.2303\ \dfrac{\text{kg}}{\text{s}}}{0.01215\ \dfrac{\text{kg}}{\text{s}}}$$
$$= \boxed{18.95}$$

12.

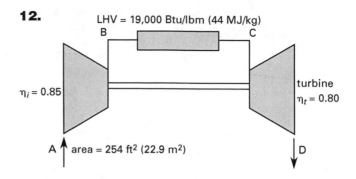

$$\text{LHV} = 19{,}000\ \text{Btu/lbm}\ (44\ \text{MJ/kg})$$

$\eta_i = 0.85$

turbine $\eta_t = 0.80$

A ↑ area = 254 ft² (22.9 m²) D ↓

Customary U.S. Solution

(a) *At 7000 ft altitude:*

$$\dot{V}_{a1} = 50{,}000\ \text{ft}^3/\text{min}$$
$$\text{BHP}_1 = 6000\ \text{hp}$$
$$\text{BSCF}_1 = 0.609\ \text{lbm/hp-hr}$$

At A:

$$p_A = 12\ \text{psia}$$
$$T_A = 35°\text{F}$$

The absolute temperature at A is

$$T_A = 35°\text{F} + 460 = 495°\text{R}$$

Interpolating from App. 23.F (air table),

$$h_A = 118.28\ \text{Btu/lbm}$$
$$p_{r,a} = 1.0238$$

From the ideal gas law, air density is

$$\rho_{a1} = \frac{p_A}{RT_A} = \frac{\left(12\ \dfrac{\text{lbf}}{\text{in}^2}\right)\left(144\ \dfrac{\text{in}^2}{\text{ft}^2}\right)}{\left(53.35\ \dfrac{\text{ft-lbf}}{\text{lbm-°R}}\right)(495°\text{R})}$$
$$= 0.0654\ \text{lbm/ft}^3$$

From Eq. 27.76,

$$\dot{m}_{a1} = \dot{V}_{a1}\rho_{a1}$$
$$= \left(50{,}000\ \frac{\text{ft}^3}{\text{min}}\right)\left(0.0654\ \frac{\text{lbm}}{\text{ft}^3}\right)$$
$$= 3270\ \text{lbm/min}$$

At B:

$$p_B = 8p_A = (8)(12\ \text{psia}) = 96\ \text{psia}$$

Assuming isentropic compression,

$$p_{r,B} = 8p_{r,A} = (8)(1.0238)$$
$$= 8.1904$$

From App. 23.F (air table), this $p_{r,B}$ corresponds to

$$T_B = 893.3°R$$
$$h_B = 214.62 \text{ Btu/lbm}$$

Due to the inefficiency of the compressor, from Eq. 27.99,

$$h'_B = h_A + \frac{h_B - h_A}{\eta_{s,\text{compression}}}$$

$$= 118.28 \frac{\text{Btu}}{\text{lbm}} + \frac{214.62 \frac{\text{Btu}}{\text{lbm}} - 118.28 \frac{\text{Btu}}{\text{lbm}}}{0.85}$$

$$= 231.62 \text{ Btu/lbm}$$

From App. 23.F, this corresponds to $T'_B = 962.3°R$.

$$W_{\text{compression}} = h'_B - h_A$$

$$= 231.62 \frac{\text{Btu}}{\text{lbm}} - 118.28 \frac{\text{Btu}}{\text{lbm}}$$

$$= 113.34 \text{ Btu/lbm}$$

At C:

The absolute temperature is

$$T_C = 1800°F + 460 = 2260°R \quad \text{[no change if moved]}$$

Assuming there is no pressure drop across the combustor, $p_C = 96$ psia.

From App. 23.F (air table),

$$h_C = 577.52 \text{ Btu/lbm}$$
$$p_{r,C} = 286.7$$

The energy requirement from the fuel is

$$\dot{m}\Delta h = \dot{m}_{a1}(h_C - h'_B)$$

$$= \left(3270 \frac{\text{lbm}}{\text{min}}\right)\left(577.52 \frac{\text{Btu}}{\text{lbm}} - 231.62 \frac{\text{Btu}}{\text{lbm}}\right)$$

$$= 1.133 \times 10^6 \text{ Btu/min}$$

The ideal fuel rate is

$$\frac{\dot{m}\Delta h}{\text{LHV}} = \frac{\left(1.133 \times 10^6 \frac{\text{Btu}}{\text{min}}\right)\left(60 \frac{\text{min}}{\text{hr}}\right)}{19{,}000 \frac{\text{Btu}}{\text{lbm}}}$$

$$= 3577.9 \text{ lbm/hr}$$

From Eq. 27.74, the ideal BSFC is

$$(\text{BSFC})_{\text{ideal}} = \frac{3577.9 \frac{\text{lbm}}{\text{hr}}}{6000 \text{ hp}} = 0.596 \text{ lbm/hp-hr}$$

The combustor efficiency is

$$\eta_{\text{combustor}} = \frac{(\text{BSFC})_{\text{ideal}}}{(\text{BSFC})_{\text{actual}}} = \frac{0.596 \frac{\text{lbm}}{\text{hp-hr}}}{0.609 \frac{\text{lbm}}{\text{hp-hr}}}$$

$$= 0.979 \quad (97.9\%)$$

At D:

As determined by atmospheric conditions, $p_D = 12$ psia. If expansion is isentropic,

$$p_{r,D} = \frac{p_{r,C}}{8} = \frac{286.7}{8} = 35.8375$$

From App. 23.F (air table) this $p_{r,D}$ corresponds to

$$T_D = 1334.8°R$$
$$h_D = 325.95 \text{ Btu/lbm}$$

Due to the inefficiency of the turbine, from Eq. 27.101,

$$h'_D = h_C - \eta_{s,\text{turbine}}(h_C - h_D)$$

$$= 577.52 \frac{\text{Btu}}{\text{lbm}} - (0.80)\left(577.52 \frac{\text{Btu}}{\text{lbm}}\right.$$

$$\left. - 325.95 \frac{\text{Btu}}{\text{lbm}}\right)$$

$$= 376.26 \frac{\text{Btu}}{\text{lbm}}$$

$$W_{\text{turbine}} = h_C - h'_D = 577.52 \frac{\text{Btu}}{\text{lbm}} - 376.26 \frac{\text{Btu}}{\text{lbm}}$$

$$= 201.26 \text{ Btu/lbm}$$

The theoretical net horsepower is

$$\text{IHP} = \dot{m}_{a1}(W_{\text{turbine}} - W_{\text{compression}})$$

$$= \frac{\left(3270 \frac{\text{lbm}}{\text{min}}\right)\left(201.26 \frac{\text{Btu}}{\text{lbm}} - 113.34 \frac{\text{Btu}}{\text{lbm}}\right)\left(778 \frac{\text{ft-lbf}}{\text{Btu}}\right)}{33{,}000 \frac{\text{ft-lbf}}{\text{hp-min}}}$$

$$= 6778 \text{ hp}$$

From Eq. 27.69, the friction horsepower is

$$\text{FHP} = \text{IHP} - \text{BHP} = 6778 \text{ hp} - 6000 \text{ hp}$$

$$= 778 \text{ hp}$$

At sea level (zero altitude):

At A:

The absolute temperature is

$$T_A = 70°F + 460 = 530°R$$
$$p_A = 14.7 \text{ psia}$$

From App. 23.F (air table),

$$h_A = 126.86 \text{ Btu/lbm}$$
$$p_{r,A} = 1.2998$$

From the ideal gas law, the air density is

$$\rho_{a2} = \frac{p_A}{RT_A} = \frac{\left(14.7 \dfrac{\text{lbf}}{\text{in}^2}\right)\left(144 \dfrac{\text{in}^2}{\text{ft}^2}\right)}{\left(53.35 \dfrac{\text{lbf-ft}}{\text{lbm-°R}}\right)(530°R)}$$
$$= 0.0749 \text{ lbm/ft}^3$$

From Eq. 27.76, the air mass flow rate is

$$\dot{m}_{a2} = \dot{V}_{a2}\rho_{a2}$$
$$= \left(50{,}000 \frac{\text{ft}^3}{\text{min}}\right)\left(0.0749 \frac{\text{lbm}}{\text{ft}^3}\right)$$
$$= 3745 \text{ lbm/min}$$

At B:

$$p_B = 8p_A = (8)(14.7 \text{ psia}) = 117.6 \text{ psia}$$

Assuming isentropic compression,

$$p_{r,B} = 8p_{r,A} = (8)(1.2998) = 10.3984$$

From App. 23.F (air table), this $p_{r,B}$ corresponds to

$$T_B = 954.5°R$$
$$h_B = 229.70 \text{ Btu/lbm}$$

Due to the inefficiency of the compressor, from Eq. 27.99,

$$h'_B = h_A + \frac{h_B - h_A}{\eta_{s,\text{compression}}}$$
$$= 126.86 \frac{\text{Btu}}{\text{lbm}} + \frac{229.70 \dfrac{\text{Btu}}{\text{lbm}} - 126.86 \dfrac{\text{Btu}}{\text{lbm}}}{0.85}$$
$$= 247.85 \text{ Btu/lbm}$$

From App. 23.F (air table), $h'_B = 247.85$ Btu/lbm corresponds to $T'_B = 1027.6°R$.

$$W_{\text{compression}} = h'_B - h_A$$
$$= 247.85 \frac{\text{Btu}}{\text{lbm}} - 126.86 \frac{\text{Btu}}{\text{lbm}}$$
$$= 120.99 \text{ Btu/lbm}$$

At C:

The absolute temperature is

$$T_C = 2260°R \quad [\text{no change}]$$

Assuming there is no pressure drop across the combustor, $p_C = 117.6$ psia.

From App. 23.F (air table),

$$h_C = 577.52 \text{ Btu/lbm}$$
$$p_{r,C} = 286.7$$

The energy requirement from the fuel is $\dot{m}\Delta h$.

$$\dot{m}_{a2}(h_C - h'_B)$$
$$= \left(3745 \frac{\text{lbm}}{\text{min}}\right)\left(577.52 \frac{\text{Btu}}{\text{lbm}} - 247.85 \frac{\text{Btu}}{\text{lbm}}\right)$$
$$= 1.235 \times 10^6 \text{ Btu/min}$$

Assuming a constant combustion efficiency of 97.9%, the actual fuel rate is

$$\frac{\dot{m}\Delta h}{\text{LHV}} = \frac{\left(1.235 \times 10^6 \dfrac{\text{Btu}}{\text{min}}\right)\left(60 \dfrac{\text{min}}{\text{hr}}\right)}{\left(19{,}000 \dfrac{\text{Btu}}{\text{lbm}}\right)(0.979)}$$
$$= 3983.7 \text{ lbm/hr}$$

At D:

$$p_D = 14.7 \text{ psia}$$

If expansion is isentropic,

$$p_{r,D} = \frac{p_{r,C}}{8} = \frac{286.7}{8} = 35.8375 \quad [\text{no change}]$$

From App. 23.F (air table), this corresponds to

$$T_D = 1334.8°R$$
$$h_D = 325.95 \text{ Btu/lbm}$$
$$h'_D = 376.26 \text{ Btu/lbm} \quad [\text{no change}]$$
$$W_{\text{turbine}} = 201.26 \text{ Btu/lbm} \quad [\text{no change}]$$

The theoretical net horsepower is

$$\text{IHP} = \dot{m}_{a2}(W_{\text{turbine}} - W_{\text{compression}})$$

$$= \frac{\left(3745 \, \frac{\text{lbm}}{\text{min}}\right)\left(201.26 \, \frac{\text{Btu}}{\text{lbm}}\right.}{\left.- 120.99 \, \frac{\text{Btu}}{\text{lbm}}\right)\left(778 \, \frac{\text{ft-lbf}}{\text{Btu}}\right)}{33,000 \, \frac{\text{ft-lbf}}{\text{hp-min}}}$$

$$= 7087 \text{ hp}$$

Assuming the frictional horsepower is constant,

$$\text{BHP} = \text{IHP} - \text{FHP}$$

$$= 7087 \text{ hp} - 778 \text{ hp}$$

$$= \boxed{6309 \text{ hp}}$$

The answer is (D).

(b) From Eq. 27.74, BSFC is

$$\text{BSFC} = \frac{3983.7 \, \frac{\text{lbm}}{\text{hr}}}{6309 \text{ hp}} = \boxed{0.631 \text{ lbm/hp-hr}}$$

The answer is (C).

SI Solution

(a) *At 2100 m altitude:*

At A:

$$p_A = 82 \text{ kPa} \quad \text{[given]}$$

The absolute temperature is

$$T_A = 2°\text{C} + 273 = 275\text{K} \quad \text{[given]}$$

From App. 23.S (air table),

$$h_A = 275.12 \text{ kJ/kg}$$

$$p_{r,A} = 1.0240$$

From the ideal gas law, the air density is

$$\rho_{a1} = \frac{p_A}{RT_A} = \frac{(82 \text{ kPa})\left(1000 \, \frac{\text{Pa}}{\text{kPa}}\right)}{\left(287.03 \, \frac{\text{J}}{\text{kg·K}}\right)(275\text{K})}$$

$$= 1.0389 \text{ kg/m}^3$$

From Eq. 27.76, the air mass flow rate is

$$\dot{m}_{a1} = \dot{V}_{a1}\rho_{a1}$$

$$= \left(23{,}500 \, \frac{\text{L}}{\text{s}}\right)\left(\frac{1 \text{ m}^3}{1000 \text{ L}}\right)\left(1.0389 \, \frac{\text{kg}}{\text{m}^3}\right)$$

$$= 24.41 \text{ kg/s}$$

At B:

$$p_C = 8p_A = (8)(82 \text{ kPa}) = 656 \text{ kPa}$$

Assuming isentropic compression,

$$p_{r,B} = 8p_{r,A} = (8)(1.0240) = 8.192$$

From App. 23.S (air table), this $p_{r,B}$ corresponds to

$$T_B = 496.3\text{K}$$

$$h_B = 499.33 \text{ kJ/kg}$$

Due to the inefficiency of the compressor, from Eq. 27.99,

$$h_B' = h_A + \frac{h_B - h_A}{\eta_{s,\text{compression}}}$$

$$= 275.12 \, \frac{\text{kJ}}{\text{kg}} + \frac{499.33 \, \frac{\text{kJ}}{\text{kg}} - 275.12 \, \frac{\text{kJ}}{\text{kg}}}{0.85}$$

$$= 538.90 \text{ kJ/kg}$$

From App. 23.S (air table), this value of $h = 538.90$ kJ/kg corresponds to $T_B' = 534.7\text{K}$.

$$W_{\text{compression}} = h_B' - h_A$$

$$= 538.90 \, \frac{\text{kJ}}{\text{kg}} - 275.12 \, \frac{\text{kJ}}{\text{kg}} = 263.78 \text{ kJ/kg}$$

At C:

The absolute temperature is

$$T_C = 980°\text{C} + 273 = 1253\text{K}$$

Assuming there is no pressure drop across the combustor, $p_C = 656$ kPa.

From App. 23.S (air table),

$$h_C = 1340.28 \text{ kJ/kg}$$

$$p_{r,C} = 284.3$$

The energy requirement from the fuel is

$$\dot{m}\Delta h = \dot{m}_{a1}(h_C - h_B')$$

$$= \left(24.41 \, \frac{\text{kg}}{\text{s}}\right)\left(1340.28 \, \frac{\text{kJ}}{\text{kg}} - 538.90 \, \frac{\text{kJ}}{\text{kg}}\right)$$

$$= 19\,562 \text{ kJ/s}$$

The ideal fuel rate is

$$\frac{\dot{m}\Delta h}{\text{LHV}} = \frac{19\,562 \, \frac{\text{kJ}}{\text{s}}}{\left(44 \, \frac{\text{MJ}}{\text{kg}}\right)\left(1000 \, \frac{\text{kJ}}{\text{MJ}}\right)} = 0.4446 \text{ kg/s}$$

From Eq. 27.74, the ideal BSFC is

$$(\text{BSFC})_{\text{ideal}} = \left(\frac{0.4446 \, \frac{\text{kg}}{\text{s}}}{4.5 \, \text{MW}} \right) \left(1000 \, \frac{\text{MW}}{\text{GW}} \right)$$

$$= 98.8 \, \text{kg/GJ}$$

The combustor efficiency is

$$\eta_{\text{combustor}} = \frac{(\text{BSFC})_{\text{ideal}}}{(\text{BSFC})_{\text{actual}}} = \frac{98.8 \, \frac{\text{kg}}{\text{GJ}}}{100 \, \frac{\text{kg}}{\text{GJ}}}$$

$$= 0.988 \quad (98.8\%)$$

At D:

Determined by atmospheric conditions, $p_D = 82$ kPa. If expansion is isentropic,

$$p_{r,D} = \frac{p_{r,C}}{8} = \frac{284.3}{8} = 35.54$$

From App. 23.S (air table), this $p_{r,D}$ corresponds to

$$T_D = 740\text{K}$$
$$h_D = 756.44 \, \text{kJ/kg}$$

From Eq. 27.101, due to the inefficiency of the turbine,

$$h'_D = h_C - \eta_{s,\text{turbine}}(h_C - h_D)$$

$$= 1340.28 \, \frac{\text{kJ}}{\text{kg}} - (0.80) \left(1340.28 \, \frac{\text{kJ}}{\text{kg}} \right.$$

$$\left. - 756.44 \, \frac{\text{kJ}}{\text{kg}} \right)$$

$$= 873.21 \, \text{kJ/kg}$$

$$W_{\text{turbine}} = h_C - h'_D = 1340.28 \, \frac{\text{kJ}}{\text{kg}} - 873.21 \, \frac{\text{kJ}}{\text{kg}}$$

$$= 467.07 \, \text{kJ/kg}$$

The theoretical net power is

$$\text{IkW} = \dot{m}_{a1}(W_{\text{turbine}} - W_{\text{compression}})$$

$$= \left(24.41 \, \frac{\text{kg}}{\text{s}} \right) \left(467.07 \, \frac{\text{kJ}}{\text{kg}} - 263.78 \, \frac{\text{kJ}}{\text{kg}} \right)$$

$$\times \left(\frac{1 \, \text{MW}}{1000 \, \text{kW}} \right) = 4.962 \, \text{MW}$$

From Eq. 27.69, the friction horsepower is

$$\text{FkW} = \text{IkW} - \text{BkW} = 4.962 \, \text{MW} - 4.5 \, \text{MW}$$

$$= 0.462 \, \text{MW}$$

At sea level (zero altitude):

At A:

The absolute temperature is

$$T_A = 21°\text{C} + 273 = 294\text{K}$$
$$p_A = 101.3 \, \text{kPa}$$

From App. 23.S,

$$h_A = 294.17 \, \text{kJ/kg}$$
$$p_{r,A} = 1.2917$$

From the ideal gas law, the air density is

$$\rho_{a2} = \frac{p_A}{RT_A} = \frac{(101.3 \, \text{kPa}) \left(1000 \, \frac{\text{Pa}}{\text{kPa}} \right)}{\left(287.03 \, \frac{\text{J}}{\text{kg·K}} \right) (294\text{K})}$$

$$= 1.2004 \, \text{kg/m}^3$$

From Eq. 27.76, the air mass flow rate is

$$\dot{m}_{a2} = \dot{V}_{a2}\rho_{a2}$$

$$= \left(23\,500 \, \frac{\text{L}}{\text{s}} \right) \left(\frac{1 \, \text{m}^3}{1000 \, \text{L}} \right) \left(1.2004 \, \frac{\text{kg}}{\text{m}^3} \right)$$

$$= 28.21 \, \text{kg/s}$$

At B:

$$p_B = 8p_A = (8)(101.3 \, \text{kPa}) = 810.4 \, \text{kPa}$$

Assuming isentropic expansion,

$$p_{r,B} = 8p_{r,A} = (8)(1.2917) = 10.3336$$

From App. 23.S (air table), this $p_{r,B}$ corresponds to

$$T_B = 529.5\text{K}$$
$$h_B = 533.46 \, \text{kJ/kg}$$

Due to the inefficiency of the compressor, from Eq. 27.99,

$$h'_B = h_A + \frac{h_B - h_A}{\eta_{s,\text{compression}}}$$

$$= 294.17 \, \frac{\text{kJ}}{\text{kg}} + \frac{533.46 \, \frac{\text{kJ}}{\text{kg}} - 294.17 \, \frac{\text{kJ}}{\text{kg}}}{0.85}$$

$$= 575.69 \, \text{kJ/kg}$$

Thermodynamics

From App. 23.S (air table), this value of h_B corresponds to

$$T'_B = 570K$$

$$W_{compression} = h'_B - h_A$$

$$= 575.69 \frac{kJ}{kg} - 294.17 \frac{kJ}{kg}$$

$$= 281.52 \; kJ/kg$$

At C:

The absolute temperature is

$$T_C = 1253K \quad \text{[no change]}$$

Assuming there is no pressure drop across the combustor, $p_C = 810.4$ kPa.

From App. 23.S (air table),

$$h_C = 1340.28 \; kJ/kg$$

$$p_{r,C} = 284.3$$

The energy requirement from the fuel is

$$\dot{m}\Delta h = \dot{m}_{a2}(h_C - h'_B)$$

$$= \left(28.21 \frac{kg}{s}\right)\left(1340.28 \frac{kJ}{kg} - 575.69 \frac{kJ}{kg}\right)$$

$$= 21\,569 \; kJ/s$$

Assuming a constant combustion efficiency of 98.8%, the actual fuel rate is

$$\frac{\dot{m}\Delta h}{LHV} = \frac{\left(21\,569 \frac{kJ}{s}\right)\left(\frac{1 \; MJ}{1000 \; kJ}\right)}{\left(44 \frac{MJ}{kg}\right)(0.988)} = 0.496 \; kg/s$$

At D:

$$p_D = 101.3 \; kPa$$

If expansion is isentropic,

$$p_{r,D} = \frac{p_{r,C}}{8} = \frac{284.3}{8} = 35.538 \quad \text{[no change]}$$

From App. 23.S (air table), this $p_{r,D}$ corresponds to

$$T_D = 740K$$

$$h_D = 756.44 \; kJ/kg$$

$$h'_D = 873.21 \; kJ/kg \quad \text{[no change]}$$

$$W_{turbine} = 467.07 \; kJ/kg$$

The theoretical net power is

$$IkW = \dot{m}_{a2}(W_{turbine} - W_{compression})$$

$$= \left(28.21 \frac{kg}{s}\right)\left(467.07 \frac{kJ}{kg} - 281.52 \frac{kJ}{kg}\right)$$

$$\times \left(\frac{1 \; MW}{1000 \; kW}\right) = 5.234 \; MW$$

Assuming the frictional horsepower is constant,

$$BkW = IkW - FkW = 5.234 \; MW - 0.462 \; MW$$

$$= \boxed{4.772 \; MW}$$

The answer is (D).

(b) From Eq. 27.74,

$$BSFC = \frac{0.496 \frac{kg}{s}}{(4.772 \; MW)\left(\frac{1 \; GW}{1000 \; MW}\right)}$$

$$= \boxed{103.9 \; kg/GJ \; (100 \; kg/GJ)}$$

The answer is (C).

28 Nuclear Power Cycles

PRACTICE PROBLEMS

1. Radioactive sodium emits 2.75 MeV gamma rays to initiate the photodisintegration of deuterium.

$$\lambda + {}_1D^2 \longrightarrow {}_1H^1 + {}_0n^1$$

What is the kinetic energy of the neutron?

(A) 0.15 MeV

(B) 0.26 MeV

(C) 0.31 MeV

(D) 0.55 MeV

2. The half-life of Cs-132 is approximately 6.47 days. How long will it take to reduce the activity of a Cs-132 sample to 5% of the original value?

(A) 28 days

(B) 39 days

(C) 64 days

(D) 81 days

3. A source with an activity of 2 curies emits 2 MeV gamma rays. For this intensity, the linear attenuation coefficient and build-up factor for a lead shield are 0.5182 cm^{-1} and 2.78, respectively. How thick must a lead shield be to reduce the activity to 1%?

(A) 5 cm

(B) 8 cm

(C) 11 cm

(D) 25 cm

4. A 10 cm thick lead plate shields a 2 MeV gamma source. The source flux density is 1,000,000 λ/cm^2·s. For this intensity, the linear attenuation coefficient and build-up factor for a lead shield are approximately 0.5182 cm^{-1} and 2.78, respectively. Use 0.0238 cm^2/g as the energy absorption coefficient in air.

(a) What is the approximate total exit flux?

(A) 5.6×10^3 λ/cm^2·s

(B) 9.9×10^3 λ/cm^2·s

(C) 1.1×10^4 λ/cm^2·s

(D) 1.6×10^4 λ/cm^2·s

(b) What is the approximate dose rate for exposure in air?

(A) 20 mR/hr

(B) 30 mR/hr

(C) 40 mR/hr

(D) 50 mR/hr

5. A neutron flux of 1×10^8 neutrons/cm^2·s irradiates a 50°C gold-foil target for 24 hours. At 20°C, gold's absorption cross section is 98 b and its density is 19.32 g/cm^3. The half-life of activated gold, Au-198, is 2.7 days. If the target is a thin disk with a diameter of 25 mm and a volume of 0.4909 cm^3, what is the removal activity?

(A) 1.1×10^{-3} curie

(B) 1.5×10^{-3} curie

(C) 3.2×10^{-3} curie

(D) 8.4×10^{-3} curie

6. A neutron point source surrounded on all sides by 20°C water emits 1×10^7 neutrons/s. The average diffusion coefficient for water is approximately 0.16 cm, and its diffusion length is approximately 2.85 cm. What is the neutron flux 20 cm from the point source?

(A) 92 n/cm^2·s

(B) 150 n/cm^2·s

(C) 220 n/cm^2·s

(D) 350 n/cm^2·s

7. The thermal fission and absorption cross sections for natural uranium are 4.18 b and 7.68 b, respectively. What is the probability that a 20°C neutron will cause fission in natural uranium?

(A) 8%

(B) 35%

(C) 45%

(D) 54%

8. A source of 1 MeV gamma rays has an activity of 1×10^8 λ/s. For this intensity, the linear attenuation coefficient and build-up factor for an iron shield are approximately 0.4683 cm^{-1} and 14.93, respectively. 0.0280 cm^2/g is the energy absorption coefficient in air. What thickness of spherical iron shield will reduce the exposure rate to 1 mR/hr outside the shield?

(A) 8 cm

(B) 15 cm

(C) 37 cm

(D) 64 cm

9. A 1 GW breeder reactor is being designed to process 2000 kg of a 20% Pu-239 and 80% U-238 mixture. The specific fuel usage is 1.23 g/MW-day. For fast neutrons, Pu-239 has an absorption cross section of 1.95 b, has a fission cross section of 1.8 b, and releases 2.95 neutrons per fission. For fast neutrons, U-238 has an absorption cross section of 0.59 b, has a fission cross section of 0.5 b, and releases 2.45 neutrons per fission. The fast fission factor is 1.05. What is the exponential doubling time?

(A) 1400 days

(B) 2100 days

(C) 3600 days

(D) 4500 days

10. The maximum neutron flux in a bare spherical reactor is 4.5×10^{15} n/cm²·s. The radius of the reactor is 40 cm, and the macroscopic fission cross section of the fuel is 0.005 cm⁻¹. It takes 3.1×10^{10} fissions to generate one watt of power. What power is generated in the reactor?

(A) 18 MW

(B) 31 MW

(C) 59 MW

(D) 140 MW

11. A nuclear reactor produces 500,000 kW thermal. The cylindrical core is 10 ft in diameter and 10 ft high. The core contains 100,000 lbm of uranium enriched to 2% U-235. The U-235 fission cross section is 547 barns. What is the average thermal neutron flux?

(A) 1.2×10^{13} n/cm²·s

(B) 7.5×10^{13} n/cm²·s

(C) 3.4×10^{14} n/cm²·s

(D) 6.9×10^{14} n/cm²·s

SOLUTIONS

1. Use carbon-based atomic masses. The mass increase is

$$\Delta m = m_H + m_n - m_D$$
$$= 1.007825 \text{ amu} + 1.008665 \text{ amu} - 2.01410 \text{ amu}$$
$$= 0.00239 \text{ amu}$$

Express the mass increase as energy. Convert to electron volts.

$$\Delta E = (0.00239 \text{ amu}) \left(931.46 \frac{\text{MeV}}{\text{amu}} \right)$$
$$= 2.226 \text{ MeV}$$

Thus, 2.75 MeV − 2.226 MeV = 0.524 MeV are shared by the neutron and the hydrogen atom. Assuming the energy is manifested as kinetic energy, and since the neutron and hydrogen atom have roughly the same mass, the neutron receives half.

$$\Delta E_n = \frac{0.524 \text{ MeV}}{2}$$
$$= \boxed{0.262 \text{ MeV}}$$

The answer is (B).

2. The decay constant is

$$\lambda = \frac{0.693}{6.47 \text{ days}} = 0.1071 \text{ day}^{-1}$$

The activity can be predicted by

$$\frac{A}{A_0} = e^{-\lambda t}$$
$$0.05 = e^{\left(-0.1071 \frac{1}{\text{day}}\right)t}$$

Take the natural logarithm of both sides to determine t.

$$\boxed{t = 27.97 \text{ days}}$$

The answer is (A).

3. The linear attenuation coefficient and build-up factor for 2 MeV gamma radiation are, respectively, approximately

$$\mu_l = 0.5182 \text{ cm}^{-1}; \ \text{B} = 2.78$$

(If necessary, an initial estimate, say 10 cm, of shield thickness can be used with a table of mass attenuation coefficients. In that case, the solution will be iterative.)

$$\frac{I}{I_0} = Be^{-\mu_l x}$$
$$0.01 = 2.78 e^{-\left(0.5182 \frac{1}{\text{cm}}\right)x}$$

Solve for x by taking the natural logarithm of both sides.

$$x = 10.86 \text{ cm}$$

The answer is (C).

4. (a) The uncollided exit flux is

$$\phi = \phi_0 e^{-\mu x} = \left(10^6 \frac{1}{\text{cm}^2 \cdot \text{s}}\right) e^{-\left(0.5182 \frac{1}{\text{cm}}\right)(10 \text{ cm})}$$

$$= 5.62 \times 10^3 \frac{1}{\text{cm}^2 \cdot \text{s}}$$

The build-up flux is

$$\phi_B = BI = (2.78)\left(5.62 \times 10^3 \frac{1}{\text{cm}^2 \cdot \text{s}}\right)$$

$$= 1.56 \times 10^4 \frac{1}{\text{cm}^2 \cdot \text{s}}$$

The answer is (D).

(b) There are many approximate correlations between dose rate and intensity in air. One of them is

$$D' = \left(0.0659 \frac{\text{g} \cdot \text{s} \cdot \text{mR}}{\text{MeV} \cdot \text{hr}}\right) E_0 \phi_B \mu_{a,\text{air}}$$

$$= (0.0659E)(2 \text{ MeV})\left(1.56 \times 10^4 \frac{1}{\text{cm}^2 \cdot \text{s}}\right)$$

$$\times \left(0.0238 \frac{\text{cm}^2}{\text{g}}\right)$$

$$= 48.9 \text{ mR/hr}$$

The answer is (D).

5. The absorption cross section at 50°C is

$$\overline{\sigma} = \frac{\sigma_{a,\text{thermal}}}{1.128} \sqrt{\frac{273 + 20°}{273 + T_{\circ}C}}$$

$$= \frac{98 \text{ b}}{1.128} \sqrt{\frac{273 + 20°C}{273 + 50°C}}$$

$$= 82.75 \text{ b}$$

The number of atoms in the target foil is

$$N = \frac{mN_o}{A} = \frac{\rho V N_o}{A}$$

$$= \frac{\left(19.32 \frac{\text{g}}{\text{cm}^3}\right)(0.4909 \text{ cm}^3)}{197 \frac{\text{g}}{\text{mole}}} \times \left(6.022 \times 10^3 \frac{\text{atoms}}{\text{mole}}\right)$$

$$= 2.9 \times 10^{22} \text{ atoms}$$

The decay constant for the activated gold is

$$\lambda = \frac{0.693}{2.7 \text{ days}} = 0.2567 \text{ day}^{-1}$$

The removal activity is

$$A_R = J\overline{\sigma}_a N(1 - e^{-\lambda t})$$

$$= \frac{\left(10^8 \frac{1}{\text{cm}^2 \cdot \text{s}}\right)(82.75 \text{ b})\left(1 \times 10^{-24} \frac{\text{cm}^2}{\text{b}}\right)}{3.7 \times 10^{10} \frac{1}{\text{curie} \cdot \text{s}}}$$

$$\times (2.9 \times 10^{22} \text{ atoms})\left(1 - e^{\left(-0.2567 \frac{1}{\text{day}}\right)(1 \text{ day})}\right)$$

$$= 1.47 \times 10^{-3} \text{ curie}$$

The answer is (B).

6. For an isotropic point source, the diffused flux is given by

$$\phi = \frac{A_{0,\text{isotropic}} e^{-\frac{r}{L}}}{4\pi \overline{D} r}$$

$$= \frac{\left(1 \times 10^7 \frac{1}{\text{s}}\right) e^{-\frac{20 \text{ cm}}{2.85 \text{ cm}}}}{4\pi (0.16 \text{ cm})(20 \text{ cm})}$$

$$= 223 \frac{1}{\text{cm}^2 \times \text{s}}$$

The answer is (C).

7.
$$p\{f\} = \frac{\sigma_f}{\sigma_a} = \frac{4.18 \text{ b}}{7.68 \text{ b}}$$

$$= 0.544 \quad (54.4\%)$$

The answer is (D).

8. The build-up flux in a spherical shield from an isotropic source is

$$\phi_B = B\frac{A_{0,\text{isotropic}} e^{-\mu_t r}}{4\pi r^2}$$

$$= (14.93)\frac{\left(1 \times 10^8 \frac{1}{\text{s}}\right) e^{-\left(0.4683 \frac{1}{\text{cm}}\right)r}}{4\pi r^2}$$

$$= \frac{\left(1.19 \times 10^8 \frac{1}{\text{s}}\right) e^{-\left(0.4683 \frac{1}{\text{cm}}\right)r}}{r^2}$$

The approximate dose (exposure) rate equation used is

$$D' = \left(0.0659 \; \frac{\text{g} \cdot \text{s} \cdot \text{mR}}{\text{MeV} \cdot \text{hr}}\right) E_0 \phi_B \mu_{a,\text{air}}$$

$$1 \; \frac{\text{mR}}{\text{hr}} = \frac{\left(0.0659 \; \frac{\text{g} \cdot \text{s} \cdot \text{mR}}{\text{MeV} \cdot \text{hr}}\right) (1 \; \text{MeV}) \times \left(1.19 \times 10^8 \; \frac{1}{\text{cm}^2 \cdot \text{s}}\right) e^{-\left(0.4683 \; \frac{1}{\text{cm}}\right) r} \times \left(0.0280 \; \frac{\text{cm}^2}{\text{g}}\right)}{r^2}$$

By trial and error, $r = 14.8$ cm.

The answer is (B).

9. Calculate the average number of neutrons generated per fast neutron released.

$$\eta = G_{\text{Pu-239}} n_{\text{Pu-239}} \left(\frac{\sigma_{f,\text{Pu-239}}}{\sigma_{a,\text{Pu-239}}}\right)$$
$$+ G_{\text{U-238}} n_{\text{U-238}} \left(\frac{\sigma_{f,\text{U-238}}}{\sigma_{a,\text{U-238}}}\right)$$
$$= (0.2)(2.95)\left(\frac{1.8 \; \text{b}}{1.95 \; \text{b}}\right) + (0.8)(2.45)\left(\frac{0.5 \; \text{b}}{0.59 \; \text{b}}\right)$$
$$= 2.206$$

The conversion ratio is

$$\text{CR} = \eta \varepsilon - 1 = (2.206)(1.05) - 1 = 1.316$$

The linear doubling time is

$$t_{d,\text{linear}} = \frac{m}{(\text{CR} - 1)m'P}$$
$$= \frac{(2000 \; \text{kg})\left(1000 \; \frac{\text{g}}{\text{kg}}\right)}{(1.316 - 1)\left(1.23 \; \frac{\text{g}}{\text{MW} \cdot \text{day}}\right)} \times (1 \; \text{GW})\left(1000 \; \frac{\text{MW}}{\text{GW}}\right)$$
$$= 5146 \; \text{days}$$

The exponential doubling time is

$$t_{d,\text{exponential}} = 0.693 t_{d,\text{linear}}$$
$$= (0.693)(5146 \; \text{days})$$
$$= 3566 \; \text{days}$$

The answer is (C).

10. In a spherical reactor, the ratio of maximum to average neutron flux is 3.29.

Reactor power is predicted by

$$P = \frac{\overline{\phi}\Sigma_f V}{3.1 \times 10^{10} \; \frac{1}{\text{W} \cdot \text{s}}}$$
$$= \frac{\left(\dfrac{4.5 \times 10^{15} \; \frac{1}{\text{cm}^2 \cdot \text{s}}}{3.29}\right)\left(0.005 \; \frac{1}{\text{cm}}\right) \times \left(\frac{4\pi}{3}\right)(40 \; \text{cm})^3}{3.1 \times 10^{10} \; \frac{1}{\text{W} \cdot \text{s}}}$$
$$= 5.9 \times 10^7 \; \text{W}$$

The answer is (C).

11. The mass of uranium is

$$m = (100{,}000 \; \text{lbm})\left(0.4536 \; \frac{\text{kg}}{\text{lbm}}\right)\left(1000 \; \frac{\text{g}}{\text{kg}}\right)$$
$$= 4.54 \times 10^7 \; \text{g}$$

The number of fuel molecules per unit volume (cm³) is

$$N_f = \frac{G_f m N_o}{V(MW)}$$
$$= \frac{(0.02)(4.54 \times 10^7 \; \text{g})\left(6.022 \times 10^{23} \; \frac{\text{atoms}}{\text{mol}}\right)}{V\left(235 \; \frac{\text{g}}{\text{mol}}\right)}$$
$$= \frac{2.33 \times 10^{27}}{V}$$

From the thermal power equation,

$$\overline{\phi} = \frac{\left(3.1 \times 10^{10} \; \frac{1}{\text{W} \cdot \text{s}}\right) P_{\text{thermal}}}{\Sigma_f V}$$
$$= \frac{\left(3.1 \times 10^{10} \; \frac{1}{\text{W} \cdot \text{s}}\right) P_{\text{thermal}}}{\sigma_f N_f V}$$
$$= \frac{\left(3.1 \times 10^{10} \; \frac{\text{fissions}}{\text{s}}{\text{W}}\right)(5 \times 10^8 E)}{(547 \; \text{b})\left(1 \times 10^{-24} \; \frac{\text{cm}^2}{\text{b}}\right)(2.33 \times 10^{27})}$$
$$= 1.22 \times 10^{13} \; \frac{\text{fissions}}{\text{cm}^2 \cdot \text{s}}$$

The answer is (A).

29 Gas Compression Cycles

PRACTICE PROBLEMS

1. What is the isentropic efficiency of a compressor that takes in air at 8 psia and $-10°F$ (55 kPa and $-20°C$) and discharges it at 40 psia and 315°F (275 kPa and 160°C)?

(A) 61%
(B) 66%
(C) 74%
(D) 81%

2. A reciprocating air compressor has a 7% clearance and discharges 65 psia (450 kPa) air at the rate of 48 lbm/min (0.36 kg/s). The polytropic exponent is 1.33. The inlet air pressure is 14.7 psia (101 kPa). What mass of air is compressed each minute?

(A) 42 lbm/min (0.32 kg/s)
(B) 51 lbm/min (0.38 kg/s)
(C) 56 lbm/min (0.42 kg/s)
(D) 74 lbm/min (0.55 kg/s)

3. Air at 14.7 psia and 500°F (101.3 kPa and 260°C) is compressed in a centrifugal compressor to 6 atm. The isentropic efficiency of the compression process is 65%.

(a) What is the compression work?
(A) 140 Btu/lbm (320 kJ/kg)
(B) 190 Btu/lbm (440 kJ/kg)
(C) 230 Btu/lbm (540 kJ/kg)
(D) 350 Btu/lbm (810 kJ/kg)

(b) What is the final temperature?
(A) 1550°R (860K)
(B) 1650°R (920K)
(C) 1750°R (970K)
(D) 1850°R (1030K)

(c) What is the increase in entropy?
(A) 0.048 Btu/lbm-°R (0.21 kJ/kg·K)
(B) 0.099 Btu/lbm-°R (0.47 kJ/kg·K)
(C) 0.23 Btu/lbm-°R (1.0 kJ/kg·K)
(D) 0.44 Btu/lbm-°R (2.1 kJ/kg·K)

4. (*Time limit: one hour*) Compressors A and B both discharge into a common tank. The storage tank contains 100 psia, 90°F air (700 kPa, 32°C). Both compressors receive air at 14.7 psia and 80°F (101.3 kPa and 27°C). The flow rate through compressor A is 600 cfm (300 L/s). Process C uses 100 cfm of 80 psia, 85°F (50 L/s of 550 kPa, 29°C) air. Process D uses 120 cfm of 85 psia, 80°F (60 L/s of 590 kPa, 27°C) air. Process E uses 8 lbm/min (0.06 kg/s) of 85°F (29°C) air. Air is an ideal gas, and compressibility effects are to be disregarded. Calculate the flow rate through compressor B.

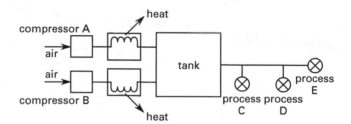

(A) 620 ft³/min (320 L/s)
(B) 740 ft³/min (370 L/s)
(C) 900 ft³/min (450 L/s)
(D) 1100 ft³/min (550 L/s)

5. (*Time limit: one hour*) 300 cfm (150 L/s) of air at 14.7 psia and 90°F (101.3 kPa and 32°C) enter a compressor. The compressor discharges air into a water-cooled heat exchanger. The compressed air is stored at 300 psig and 90°F (2.07 MPa and 32°C) in a 1000 ft³ (27 m³) tank. The tank feeds three air-driven tools with the flow rates and properties listed. The pressures to the air-driven tools are regulated and remain constant at the minimum required operating pressure as the tank pressure changes. Air is an ideal gas, and compressibility effects are to be disregarded. How long can the system run?

	tool 1	tool 2	tool 3
flow rate			
(cfm)	40	15	unknown
(L/s)	19	7	unknown
flow rate			
(lbm/min)	unknown	unknown	6
(kg/s)	unknown	unknown	0.045
minimum pressure			
(psig)	90	50	80
(kPa)	620	350	550
temperature			
(°F)	90	85	80
(°C)	32	29	27

(A) 0.6 hr (1.2 hr)

(B) 1.8 hr (3.6 hr)

(C) 3.4 hr (6.8 hr)

(D) 6.6 hr (13 hr)

SOLUTIONS

1.

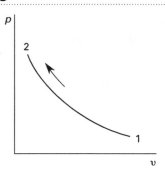

Customary U.S. Solution

Assume air is an ideal gas.

From Eq. 27.18, the isentropic temperature at point 2 is

$$T_2 = T_1 \left(\frac{p_2}{p_1}\right)^{\frac{k-1}{k}}$$

The absolute temperature at point 1 is

$$T_1 = -10°F + 460 = 450°R$$

$$T_2 = (450°R)\left(\frac{40 \text{ psia}}{8 \text{ psia}}\right)^{\frac{1.4-1}{1.4}}$$

$$= 712.7°R$$

$$T_2 = 712.7°R - 460 = 252.7°F$$

The efficiency of the compressor is given by Eq. 27.99.

$$\eta_{s,\text{compressor}} = \frac{T_2 - T_1}{T_2' - T_1}$$

$$= \frac{252.7°F - (-10°F)}{315°F - (-10°F)}$$

$$= \boxed{0.808 \quad (80.8\%)}$$

The answer is (D).

SI Solution

Assume air is an ideal gas.

From Eq. 27.18, the isentropic temperature at point 2 is

$$T_2 = T_1 \left(\frac{p_2}{p}\right)^{\frac{k-1}{k}}$$

The absolute temperature at point 1 is

$$T_1 = -20°C + 273 = 253K$$

$$T_2 = (253K)\left(\frac{275 \text{ kPa}}{55 \text{ kPa}}\right)^{\frac{1.4-1}{1.4}}$$

$$= 400.7K$$

$$T_2 = 400.7K - 273 = 127.7°C$$

The efficiency of the compressor is given by Eq. 27.99.

$$\eta_{s,\text{compressor}} = \frac{T_2 - T_1}{T_2' - T_1}$$
$$= \frac{127.7°C - (-20°C)}{160°C - (-20°C)}$$
$$= \boxed{0.821 \quad (82.1\%)}$$

The answer is (D).

2.

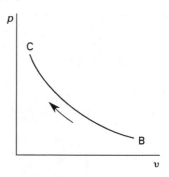

Customary U.S. Solution

The compression ratio for a reciprocating compressor is defined by Eq. 29.4.

$$r_p = \frac{p_C}{p_B}$$
$$= \frac{65 \text{ psia}}{14.7 \text{ psia}} = 4.42$$

The volumetric efficiency is given by Eq. 29.6.

$$\eta_v = 1 - \left(r_p^{\frac{1}{n}} - 1\right)\left(\frac{c}{100}\right)$$
$$= 1 - \left(4.42^{\frac{1}{1.33}} - 1\right)\left(\frac{7\%}{100}\right)$$
$$= 0.856 \quad (85.6\%)$$

The mass of air compressed per minute from Eq. 29.8 is

$$\dot{m} = \frac{\dot{m}_{\text{actual}}}{\eta_v} = \frac{48 \frac{\text{lbm}}{\text{min}}}{0.856}$$
$$= \boxed{56.07 \text{ lbm/min}}$$

The answer is (C).

SI Solution

The compression ratio for the reciprocating compressor is defined by Eq. 29.4.

$$r_p = \frac{p_C}{p_B} = \frac{450 \text{ kPa}}{101 \text{ kPa}} = 4.46$$

The volumetric efficiency is given by Eq. 29.6.

$$\eta_v = 1 - \left(r_p^{\frac{1}{n}} - 1\right)\left(\frac{c}{100}\right)$$
$$= 1 - \left(4.46^{\frac{1}{1.33}} - 1\right)\left(\frac{7\%}{100}\right)$$
$$= 0.855 \quad (86\%)$$

The mass of air compressed per minute from Eq. 29.8 is

$$\dot{m} = \frac{\dot{m}_{\text{actual}}}{\eta_v} = \frac{0.36 \frac{\text{kg}}{\text{s}}}{0.855}$$
$$= \boxed{0.4211 \text{ kg/s}}$$

The answer is (C).

3. *Customary U.S. Solution*

((a) and (b)) Although the ideal gas laws can be used, it is more expedient to use air tables for the low pressures.

The absolute inlet temperature is

$$T_1 = 500°F + 460 = 960°R$$

From App. 23.F,

$$h_1 = 231.06 \text{ Btu/lbm}$$
$$p_{r,1} = 10.61$$
$$\phi_1 = 0.74030 \text{ Btu/lbm-°R}$$

Since $p_1/p_2 = p_{r,1}/p_{r,2}$ and the compression ratio is 6, for isentropic compression,

$$p_{r,2} = 6p_{r,1}$$
$$= (6)(10.61) = 63.66$$

From App. 23.F, at $p_{r,2} = 63.66$,

$$T_2 = 1552°R$$
$$h_2 = 382.95 \text{ Btu/lbm}$$

The actual enthalpy from the definition of isentropic efficiency is

$$h_2' = h_1 + \frac{h_2 - h_1}{\eta_s}$$
$$= 231.06 \frac{\text{Btu}}{\text{lbm}} + \frac{382.95 \frac{\text{Btu}}{\text{lbm}} - 231.06 \frac{\text{Btu}}{\text{lbm}}}{0.65}$$
$$= 464.74 \text{ Btu/lbm}$$

From App. 23.F, this corresponds to $\boxed{T_2' = 1855°R}$ and $\phi_2' = 0.91129$ Btu/lbm-°R.

The answer is (D) for Prob. 3(b).

The compression work is

$$W = h_2' - h_1 = 464.74 \ \frac{\text{Btu}}{\text{lbm}} - 231.06 \ \frac{\text{Btu}}{\text{lbm}}$$

$$\boxed{= 233.68 \ \text{Btu/lbm}}$$

The answer is (C) for Prob. 3(a).

(c) From Eq. 23.39, the increase in entropy is

$$\Delta s = \phi_2' - \phi_1 - R \ln\left(\frac{p_2}{p_1}\right)$$

$$= 0.91129 \ \frac{\text{Btu}}{\text{lbm-°R}} - 0.74030 \ \frac{\text{Btu}}{\text{lbm-°R}}$$

$$- \left(\frac{53.3 \ \frac{\text{ft-lbf}}{\text{lbm-°R}}}{778 \ \frac{\text{ft-lbf}}{\text{Btu}}}\right) \ln\left(\frac{6 \ \text{atm}}{1 \ \text{atm}}\right)$$

$$\boxed{= 0.04824 \ \text{Btu/lbm-°R}}$$

The answer is (A).

SI Solution

((a) and (b)) Although the ideal gas laws can be used, it is more expedient to use air tables for the lower pressures.

The absolute inlet temperature is

$$T_1 = 260°C + 273 = 533K$$

From App. 23.S,

$$h_1 = 537.09 \ \text{kJ/kg}$$

$$p_{r,1} = 10.59$$

$$\phi_1 = 2.28161 \ \text{kJ/kg·K}$$

Since $p_1/p_2 = p_{r,1}/p_{r,2}$ and the compression ratio is 6, for isentropic compression,

$$p_{r,2} = 6p_{r,1}$$

$$= (6)(10.59) = 63.54$$

From App. 23.S, at $p_{r,2} = 63.54$,

$$T_2 = 861.5K$$

$$h_2 = 889.9 \ \text{kJ/kg}$$

The actual enthalpy from the definition of isentropic efficiency is

$$h_2' = h_1 + \frac{h_2 - h_1}{\eta_s}$$

$$= 537.09 \ \frac{\text{kJ}}{\text{kg}} + \frac{889.9 \ \frac{\text{kJ}}{\text{kg}} - 537.09 \ \frac{\text{kJ}}{\text{kg}}}{0.65}$$

$$= 1079.87 \ \text{kJ/kg}$$

From App. 23.S, this corresponds to $\boxed{T_2' = 1029.6K}$ and $\phi_2' = 3.00099$ kJ/kg·K.

The answer is (D) for Prob. 3(b).

The compression work is

$$W = h_2' - h_1 = 1079.87 \ \frac{\text{kJ}}{\text{kg}} - 537.09 \ \frac{\text{kJ}}{\text{kg}}$$

$$\boxed{= 542.78 \ \text{kJ/kg}}$$

The answer is (C) for Prob. 3(a).

(c) From Eq. 23.39, the increase in entropy is

$$\Delta s = \phi_2' - \phi_1 - R \ln\left(\frac{p_2}{p_1}\right)$$

$$= 3.00099 \ \frac{\text{kJ}}{\text{kg·K}} - 2.28161 \ \frac{\text{kJ}}{\text{kg·K}}$$

$$- \left(287.03 \ \frac{\text{J}}{\text{kg·K}}\right)\left(\frac{1 \ \text{kJ}}{1000 \ \text{J}}\right) \ln\left(\frac{6 \ \text{atm}}{1 \ \text{atm}}\right)$$

$$\boxed{= 0.20509 \ \text{kJ/kg·K}}$$

The answer is (A).

4. *Customary U.S. Solution*

The absolute temperature for process C is 85°F + 460 = 545°R. Using ideal gas laws, the mass flow rate for process C is

$$\dot{m}_C = \frac{p_C \dot{V}_C}{R T_C} = \frac{\left(80 \ \frac{\text{lbf}}{\text{in}^2}\right)\left(144 \ \frac{\text{in}^2}{\text{ft}^2}\right)\left(100 \ \frac{\text{ft}^3}{\text{min}}\right)}{\left(53.3 \ \frac{\text{ft-lbf}}{\text{lbm-°R}}\right)(545°R)}$$

$$= 39.66 \ \text{lbm/min}$$

Similarly, the mass flow rate for process D with an absolute temperature of 80°F + 460 = 540°R is

$$\dot{m}_D = \frac{p_D \dot{V}_D}{R T_D} = \frac{\left(85 \ \frac{\text{lbf}}{\text{in}^2}\right)\left(144 \ \frac{\text{in}^2}{\text{ft}^2}\right)\left(120 \ \frac{\text{ft}^3}{\text{min}}\right)}{\left(53.3 \ \frac{\text{ft-lbf}}{\text{lbm-°R}}\right)(540°R)}$$

$$= 51.03 \ \text{lbm/min}$$

For process E, the mass flow rate is given as $\dot{m}_E = 8$ lbm/min.

The total mass flow rate for all three processes is

$$\dot{m}_{total} = \dot{m}_C + \dot{m}_D + \dot{m}_E$$
$$= 39.66 \frac{lbm}{min} + 51.03 \frac{lbm}{min} + 8 \frac{lbm}{min}$$
$$= 98.69 \text{ lbm/min}$$

The mass flow rate into compressor A is with an absolute temperature of $80°F + 460 = 540°R$.

$$\dot{m}_A = \frac{p_A \dot{V}_A}{RT_A} = \frac{\left(14.7 \frac{lbf}{in^2}\right)\left(144 \frac{in^2}{ft^2}\right)\left(600 \frac{ft^3}{min}\right)}{\left(53.3 \frac{ft\text{-}lbf}{lbm\text{-}°R}\right)(540°R)}$$
$$= 44.13 \text{ lbm/min}$$

The required input for compressor B is

$$\dot{m}_B = \dot{m}_{total} - \dot{m}_A$$
$$= 98.69 \frac{lbm}{min} - 44.13 \frac{lbm}{min}$$
$$= 54.56 \text{ lbm/min}$$

The volumetric flow rate for compressor B is

$$\dot{V}_B = \frac{\dot{m}_B RT_B}{p_B}$$
$$= \frac{\left(54.56 \frac{lbm}{min}\right)\left(53.3 \frac{ft\text{-}lbf}{lbm\text{-}°R}\right)(540°R)}{\left(14.7 \frac{lbf}{in^2}\right)\left(144 \frac{in^2}{ft^2}\right)}$$
$$= \boxed{742 \text{ cfm}}$$

The answer is (B).

SI Solution

The absolute temperature for process C is $29°C + 273 = 302K$. Using ideal gas laws, the mass flow rate for process C is

$$\dot{m}_C = \frac{p_C \dot{V}_C}{RT_C}$$
$$= \frac{(550 \text{ kPa})\left(1000 \frac{Pa}{kPa}\right)\left(50 \frac{L}{s}\right)\left(\frac{1 \text{ m}^3}{1000 \text{ L}}\right)}{\left(287 \frac{J}{kg\cdot K}\right)(302K)}$$
$$= 0.3173 \text{ kg/s}$$

The absolute temperature for process D is $27°C + 273 = 300K$. Using ideal gas laws, the mass flow rate for process D is

$$\dot{m}_D = \frac{p_D \dot{V}_D}{RT_D}$$
$$= \frac{(590 \text{ kPa})\left(1000 \frac{Pa}{kPa}\right)\left(60 \frac{L}{s}\right)\left(\frac{1 \text{ m}^3}{1000 \text{ L}}\right)}{\left(287 \frac{J}{kg\cdot K}\right)(300K)}$$
$$= 0.4111 \text{ kg/s}$$

For process E, the mass flow rate is given as $\dot{m}_E = 0.06$ kg/s.

The total mass flow rate for all three processes is

$$\dot{m}_{total} = \dot{m}_C + \dot{m}_D + \dot{m}_E$$
$$= 0.3173 \frac{kg}{s} + 0.4111 \frac{kg}{s} + 0.06 \frac{kg}{s}$$
$$= 0.7884 \text{ kg/s}$$

The absolute temperature for air into compressors A and B is $27°C + 273 = 300K$.

The mass flow rate into compressor A is

$$\dot{m}_A = \frac{p_A \dot{V}_A}{RT_A}$$
$$= \frac{(101.3 \text{ kPa})\left(1000 \frac{Pa}{kPa}\right)\left(300 \frac{L}{s}\right)\left(\frac{1 \text{ m}^3}{1000 \text{ L}}\right)}{\left(287 \frac{J}{kg\cdot K}\right)(300K)}$$
$$= 0.3530 \text{ kg/s}$$

The required input for compressor B is

$$\dot{m}_B = \dot{m}_{total} - \dot{m}_A$$
$$= 0.7884 \frac{kg}{s} - 0.3530 \frac{kg}{s}$$
$$= 0.4354 \text{ kg/s}$$

The volumetric flow rate for compressor B is

$$\dot{V}_B = \frac{\dot{m}_B RT_B}{p_B}$$
$$= \frac{\left(0.4354 \frac{kg}{s}\right)\left(287 \frac{J}{kg\cdot K}\right)(300K)\left(1000 \frac{L}{m^3}\right)}{(101.3 \text{ kPa})\left(1000 \frac{Pa}{kPa}\right)}$$
$$= \boxed{370 \text{ L/s}}$$

The answer is (B).

5.

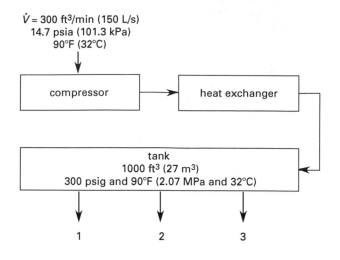

$\dot{V} = 300$ ft³/min (150 L/s)
14.7 psia (101.3 kPa)
90°F (32°C)

Customary U.S. Solution

Assume steady flow and constant properties.

The absolute temperature for compressor air is $90°F + 460 = 550°R$.

Using ideal gas laws, the mass flow rate of air into the compressor is

$$\dot{m} = \frac{p\dot{V}}{RT}$$

$$= \frac{\left(14.7 \; \dfrac{\text{lbf}}{\text{in}^2}\right)\left(144 \; \dfrac{\text{in}^2}{\text{ft}^2}\right)\left(300 \; \dfrac{\text{ft}^3}{\text{min}}\right)}{\left(53.3 \; \dfrac{\text{ft-lbf}}{\text{lbm-°R}}\right)(550°R)}$$

$$= 21.7 \; \text{lbm/min}$$

The absolute temperature and absolute pressure of stored compressed air are

$$T = 90°F + 460$$
$$= 550°R$$
$$p = 300 \text{ psig} + 14.7$$
$$= 314.7 \text{ psia}$$

The mass of stored compressed air in a 1000 ft³ tank is

$$m_{\text{tank}} = \frac{pV}{RT}$$

$$= \frac{\left(314.7 \; \dfrac{\text{lbf}}{\text{in}^2}\right)\left(144 \; \dfrac{\text{in}^2}{\text{ft}^2}\right)(1000 \text{ ft}^3)}{\left(53.3 \; \dfrac{\text{ft-lbf}}{\text{lbm-°R}}\right)(550°R)}$$

$$= 1545.9 \; \text{lbm}$$

Assuming that each tool operates at its minimum pressure, the mass leaving the system can be calculated as follows.

Tool 1:

The absolute pressure is

$$90 \text{ psig} + 14.7 = 104.7 \text{ psia}$$

The absolute temperature is

$$90°F + 460 = 550°R$$

$$\dot{m}_{\text{tool 1}} = \frac{p\dot{V}}{RT}$$

$$= \frac{\left(104.7 \; \dfrac{\text{lbf}}{\text{in}^2}\right)\left(144 \; \dfrac{\text{in}^2}{\text{ft}^2}\right)\left(40 \; \dfrac{\text{ft}^3}{\text{min}}\right)}{\left(53.3 \; \dfrac{\text{ft-lbf}}{\text{lbm-°R}}\right)(550°R)}$$

$$= 20.57 \; \text{lbm/min}$$

Tool 2:

The absolute pressure is

$$50 \text{ psig} + 14.7 = 64.7 \text{ psia}$$

The absolute temperature is

$$85°F + 460 = 545°R$$

$$\dot{m}_{\text{tool 2}} = \frac{p\dot{V}}{RT}$$

$$= \frac{\left(64.7 \; \dfrac{\text{lbf}}{\text{in}^2}\right)\left(144 \; \dfrac{\text{in}^2}{\text{ft}^2}\right)\left(15 \; \dfrac{\text{ft}^3}{\text{min}}\right)}{\left(53.3 \; \dfrac{\text{ft-lbf}}{\text{lbm-°R}}\right)(545°R)}$$

$$= 4.81 \; \text{lbm/min}$$

Tool 3:

$$\dot{m}_{\text{tool 3}} = 6 \text{ lbm/min} \quad \text{[given]}$$

The total mass flow leaving the system is

$$\dot{m}_{\text{total}} = \dot{m}_{\text{tool 1}} + \dot{m}_{\text{tool 2}} + \dot{m}_{\text{tool 3}}$$

$$= 20.57 \; \frac{\text{lbm}}{\text{min}} + 4.81 \; \frac{\text{lbm}}{\text{min}} + 6 \; \frac{\text{lbm}}{\text{min}}$$

$$= 31.38 \; \text{lbm/min}$$

The critical pressure of 104.7 psia is required for tool 1 operation. The mass in the tank when critical pressure is achieved is

$$m_{\text{tank,critical}} = \frac{pV}{RT}$$

$$= \frac{\left(104.7 \; \dfrac{\text{lbf}}{\text{in}^2}\right)\left(144 \; \dfrac{\text{in}^2}{\text{ft}^2}\right)(1000 \text{ ft}^3)}{\left(53.3 \; \dfrac{\text{ft-lbf}}{\text{lbm-°R}}\right)(550°R)}$$

$$= 514.3 \; \text{lbm}$$

The amount in the tank to be depleted is

$$m_{\text{depleted}} = m_{\text{tank}} - m_{\text{tank,critical}}$$
$$= 1545.9 \text{ lbm} - 514.3 \text{ lbm} = 1031.6 \text{ lbm}$$

The net flow rate of air to the tank is

$$\dot{m}_{\text{net}} = \dot{m}_{\text{in}} - \dot{m}_{\text{out}} = 21.7 \frac{\text{lbm}}{\text{min}} - 31.38 \frac{\text{lbm}}{\text{min}}$$
$$= -9.68 \text{ lbm/min}$$

The time the system can run is

$$\frac{m_{\text{depleted}}}{\dot{m}_{\text{net}}} = \frac{1031.6 \text{ lbm}}{9.68 \dfrac{\text{lbm}}{\text{min}}} = \boxed{106.6 \text{ min} \quad (1.78 \text{ hr})}$$

The answer is (B).

SI Solution

The absolute temperature of the compressor air is

$$32°C + 273 = 305K$$

Using ideal gas laws, the mass flow rate of air into the compressor is

$$\dot{m} = \frac{p\dot{V}}{RT}$$
$$= \frac{(101.3 \text{ kPa}) \left(1000 \dfrac{\text{Pa}}{\text{kPa}} \right) \left(150 \dfrac{\text{L}}{\text{s}} \right) \left(\dfrac{1 \text{ m}^3}{1000 \text{ L}} \right)}{\left(287 \dfrac{\text{J}}{\text{kg·K}} \right) (305K)}$$
$$= 0.1736 \text{ kg/s}$$

The absolute temperature and pressure of stored compressed air are

$$T = 32°C + 273 = 305K$$
$$p = (2.07 \text{ MPa}) \left(10^6 \dfrac{\text{Pa}}{\text{MPa}} \right)$$
$$= 2.07 \times 10^6 \text{ Pa}$$

The mass of stored compressed air in a 1000 ft^3 tank is

$$m_{\text{tank}} = \frac{pV}{RT}$$
$$= \frac{(2.07 \times 10^6 \text{ Pa})(27 \text{ m}^3)}{\left(287 \dfrac{\text{J}}{\text{kg·K}} \right) (305K)}$$
$$= 638.5 \text{ kg}$$

Assuming that each tool operates at its minimum pressure, the mass leaving the system can be calculated as follows.

Tool 1:

The absolute temperature is $32°C + 273 = 305K$.

$$\dot{m}_{\text{tool 1}} = \frac{p\dot{V}}{RT}$$
$$= \frac{(620 \text{ kPa}) \left(1000 \dfrac{\text{Pa}}{\text{kPa}} \right) \left(19 \dfrac{\text{L}}{\text{s}} \right) \left(\dfrac{1 \text{ m}^3}{1000 \text{ L}} \right)}{\left(287 \dfrac{\text{J}}{\text{kg·K}} \right) (305K)}$$
$$= 0.1346 \text{ kg/s}$$

Tool 2:

The absolute temperature is $29°C + 273 = 302K$.

$$\dot{m}_{\text{tool 2}} = \frac{p\dot{V}}{RT}$$
$$= \frac{(350 \text{ kPa}) \left(1000 \dfrac{\text{Pa}}{\text{kPa}} \right) \left(7 \dfrac{\text{L}}{\text{s}} \right) \left(\dfrac{1 \text{ m}^3}{1000 \text{ L}} \right)}{\left(287 \dfrac{\text{J}}{\text{kg·K}} \right) (302K)}$$
$$= 0.02827 \text{ kg/s}$$

Tool 3:

$$\dot{m}_{\text{tool 3}} = 0.045 \text{ kg/s} \quad \text{[given]}$$

The total mass flow rate leaving the system is

$$\dot{m}_{\text{total}} = \dot{m}_{\text{tool 1}} + \dot{m}_{\text{tool 2}} + \dot{m}_{\text{tool 3}}$$
$$= 0.1346 \frac{\text{kg}}{\text{s}} + 0.02827 \frac{\text{kg}}{\text{s}} + 0.045 \frac{\text{kg}}{\text{s}}$$
$$= 0.2079 \text{ kg/s}$$

The critical pressure of 620 kPa is required for tool 1 operation. The mass in the tank when critical pressure is achieved is

$$m_{\text{tank,critical}} = \frac{pV}{RT}$$
$$= \frac{(620 \text{ kPa}) \left(1000 \dfrac{\text{Pa}}{\text{kPa}} \right) (27 \text{ m}^3)}{\left(287 \dfrac{\text{J}}{\text{kg·K}} \right) (305K)}$$
$$= 191.2 \text{ kg}$$

The amount in the tank to be depleted is

$$m_{\text{depleted}} = m_{\text{tank}} - m_{\text{tank,critical}}$$
$$= 638.5 \text{ kg} - 191.2 \text{ kg} = 447.3 \text{ kg}$$

The net flow rate of air into the tank is

$$\dot{m}_{\text{net}} = \dot{m}_{\text{in}} - \dot{m}_{\text{out}} = 0.1736 \frac{\text{kg}}{\text{s}} - 0.2079 \frac{\text{kg}}{\text{s}}$$
$$= -0.0343 \text{ kg/s}$$

The time the system can run is

$$\frac{m_{\text{depleted}}}{\dot{m}_{\text{net}}} = \frac{447.3 \text{ kg}}{0.0343 \frac{\text{kg}}{\text{s}}} = \boxed{13\,041 \text{ s} \quad (3.62 \text{ h})}$$

The answer is (B).

30 Refrigeration Cycles

PRACTICE PROBLEMS

Carnot Refrigeration Cycle

1. A heat pump operates on the Carnot cycle between 40°F and 700°F (4°C and 370°C). What is the coefficient of performance?

(A) 1.5
(B) 1.8
(C) 2.2
(D) 2.7

2. A heat pump using refrigerant R-12 operates on the Carnot cycle. The refrigerant evaporates and is compressed at 35.7 psia and 172.4 psia (246 kPa and 1190 kPa), respectively. What is the coefficient of performance?

(A) 3.2
(B) 3.9
(C) 4.3
(D) 5.8

3. A refrigerator uses refrigerant R-12. The input power is 585 W. Heat absorbed from the cooled space is 450 Btu/hr (0.13 kW). What is the coefficient of performance?

(A) 0.2
(B) 0.4
(C) 0.7
(D) 0.9

4. Ammonia is used in a reversed Carnot cycle with reservoirs at 110°F (45°C) and 10°F (−10°C). 1000 BTU/hr (1000 kJ/h) are to be removed.

(a) Find the coefficient of performance.
(A) 2.3
(B) 2.9
(C) 4.7
(D) 5.2

(b) Find the work input.
(A) 210 Btu/hr (210 kJ/h)
(B) 340 Btu/hr (340 kJ/h)
(C) 400 Btu/hr (400 kJ/h)
(D) 630 Btu/hr (630 kJ/h)

(c) Find the rejected heat.
(A) 1000 Btu/hr (1000 kJ/h)
(B) 1200 Btu/hr (1200 kJ/h)
(C) 1400 Btu/hr (1400 kJ/h)
(D) 1600 Btu/hr (1600 kJ/h)

5. A refrigerator cools a continuous aqueous solution ($c_p = 1$ BTU/lbm-°F; 4.19 kJ/kg·°C) flow of 100 gal/min (0.4 m^3/min) from 80°F (25°C) to 20°F (−5°C) in an 80°F (25°C) environment. What is the minimum power requirement?

(A) 82 hp (63 kW)
(B) 100 hp (74 kW)
(C) 130 hp (90 kW)
(D) 150 hp (94 kW)

Vapor Compression Cycle

6. An ammonia compressor is used in a heat pump cycle. Suction pressure is 30 psia (200 kPa). Discharge pressure is 160 psia (1.0 MPa). Saturated liquid ammonia enters the throttle valve. The refrigeration effect is 500 BTU/lbm (1200 kJ/kg) ammonia. Find the coefficient of performance as a heat pump.

(A) 2.1
(B) 4.5
(C) 5.3
(D) 5.9

7. A refrigeration cycle uses Freon-12 (R-12) as a refrigerant between a 70°F (20°C) environment and a −30°F (−30°C) heat source. If the vapor leaving the evaporator and liquid leaving the condenser are both saturated, find the volume flow of refrigerant leaving the evaporator per ton (kW) of refrigeration. Assume isentropic compression.

(A) 9.8 ft^3/min-ton (0.00081 m^3/min-kW)
(B) 12 ft^3/min-ton (0.00099 m^3/min-kW)
(C) 25 ft^3/min-ton (0.0021 m^3/min-kW)
(D) 44 ft^3/min-ton (0.0036 m^3/min-kW)

Air Refrigeration Cycle

8. An air refrigeration cycle compresses air from 70°F (20°C) and 14.7 psia (101 kPa) to 60 psia (400 kPa) in a 70% efficient compressor. The air is cooled to 25°F (−4.0°C) in a constant pressure process before entering a turbine with isentropic efficiency of 0.80. Assume air is an ideal gas.

(a) What is the temperature of the air leaving the compressor?

    (A) 720°R (400K)
    (B) 790°R (440K)
    (C) 860°R (470K)
    (D) 900°R (490K)

(b) What is the temperature of the air leaving the turbine?

    (A) 360°R (200K)
    (B) 430°R (240K)
    (C) 490°R (270K)
    (D) 560°R (310K)

(c) What is the coefficient of performance of the cycle?

    (A) 0.7
    (B) 0.8
    (C) 0.9
    (D) 1.1

SOLUTIONS

1. *Customary U.S. Solution*

The coefficient of performance for a heat pump operating on the Carnot cycle is given by Eq. 30.9.

$$\text{COP}_{\text{heat pump}} = \frac{T_{\text{high}}}{T_{\text{high}} - T_{\text{low}}}$$

The absolute temperatures are

$$T_{\text{high}} = 700°\text{F} + 460 = 1160°\text{R}$$
$$T_{\text{low}} = 40°\text{F} + 460 = 500°\text{R}$$
$$\text{COP}_{\text{heat pump}} = \frac{1160°\text{R}}{1160°\text{R} - 500°\text{R}} = \boxed{1.76}$$

The answer is (B).

SI Solution

The coefficient of performance for a heat pump operating on the Carnot cycle is given by Eq. 30.9.

$$\text{COP}_{\text{heat pump}} = \frac{T_{\text{high}}}{T_{\text{high}} - T_{\text{low}}}$$

The absolute temperatures are

$$T_{\text{high}} = 370°\text{C} + 273 = 643\text{K}$$
$$T_{\text{low}} = 4°\text{C} + 273 = 277\text{K}$$
$$\text{COP}_{\text{heat pump}} = \frac{643\text{K}}{643\text{K} - 277\text{K}} = \boxed{1.76}$$

The answer is (B).

2. *Customary U.S. Solution*

The coefficient of performance for a heat pump operating on the Carnot cycle is given by Eq. 30.9.

$$\text{COP}_{\text{heat pump}} = \frac{T_{\text{high}}}{T_{\text{high}} - T_{\text{low}}}$$

The temperature T_{high} is the saturation temperature at 172.4 psia, and the temperature T_{low} is the saturation temperature at 35.7 psia. From App. 23.H,

$$T_{\text{high}} = 120.2°\text{F} + 460 = 580.2°\text{R}$$
$$T_{\text{low}} = 19.5°\text{F} + 460 = 479.5°\text{R}$$
$$\text{COP}_{\text{heat pump}} = \frac{580.2°\text{R}}{580.2°\text{R} - 479.5°\text{R}} = \boxed{5.76}$$

The answer is (D).

SI Solution

The coefficient of performance for a heat pump operating on the Carnot cycle is given by Eq. 30.9.

$$COP_{\text{heat pump}} = \frac{T_{\text{high}}}{T_{\text{high}} - T_{\text{low}}}$$

The temperature T_{high} is the saturation temperature at 1190 kPa, and the temperature T_{low} is the saturation temperature at 246 kPa. From App. 23.T,

$$T_{\text{high}} = 49°C + 273 = 322K$$

$$T_{\text{low}} = -6.8°C + 273 = 266.2K$$

$$COP_{\text{heat pump}} = \frac{322K}{322K - 266.2K} = \boxed{5.77}$$

The answer is (D).

3. *Customary U.S. Solution*

The coefficient of performance for a refrigerator is given by Eq. 30.1.

$$COP_{\text{refrigerator}} = \frac{Q_{\text{in}}}{W_{\text{in}}}$$

$$= \frac{450 \frac{\text{Btu}}{\text{hr}}}{(585W)\left(3.4121 \frac{\frac{\text{Btu}}{\text{hr}}}{W}\right)}$$

$$= \boxed{0.225}$$

The answer is (A).

SI Solution

From Eq. 30.1, the coefficient of performance for a refrigerator is

$$COP_{\text{refrigerator}} = \frac{Q_{\text{in}}}{W_{\text{in}}}$$

$$= \frac{(0.13 \text{ kW})\left(1000 \frac{W}{kW}\right)}{585W}$$

$$= \boxed{0.222}$$

The answer is (A).

4. *Customary U.S. Solution*

(a) $$COP = \frac{T_{\text{low}}}{T_{\text{high}} - T_{\text{low}}} = \frac{460 + 10°F}{110°F - 10°F}$$

$$= \boxed{4.7}$$

The answer is (C).

(b) $$W_{\text{in}} = \frac{Q_{\text{in}}}{COP} = \frac{1000 \frac{\text{Btu}}{\text{hr}}}{4.7}$$

$$= \boxed{212.77 \text{ Btu/hr}}$$

The answer is (A).

(c) $$Q_{\text{out}} = Q_{\text{in}} + W_{\text{in}} = 1000 \frac{\text{Btu}}{\text{hr}} + 212.77 \frac{\text{Btu}}{\text{hr}}$$

$$= \boxed{1212.77 \text{ Btu/hr}}$$

The answer is (B).

SI Solution

(a) $$COP = \frac{T_{\text{low}}}{T_{\text{high}} - T_{\text{low}}}$$

$$= \frac{-10°C + 273.15}{45°C - (-10°C)}$$

$$= \boxed{4.785}$$

The answer is (C).

(b) $$W_{\text{in}} = \frac{Q_{\text{in}}}{COP} = \frac{1000 \frac{\text{kJ}}{\text{h}}}{4.785}$$

$$= \boxed{208.99 \text{ kJ/h}}$$

The answer is (A).

(c) $$Q_{\text{out}} = Q_{\text{in}} + W_{\text{in}} = 1000 \frac{\text{kJ}}{\text{h}} + 208.99 \frac{\text{kJ}}{\text{h}}$$

$$= \boxed{1208.99 \text{ kJ/h}}$$

The answer is (B).

5. *Customary U.S. Solution*

$$\dot{m} = \dot{V}\rho = \left(100 \frac{\text{gal}}{\text{min}}\right)\left(0.1337 \frac{\text{ft}^3}{\text{gal}}\right)$$

$$\times \left(62.4 \frac{\text{lbm}}{\text{ft}^3}\right)$$

$$= 834.5 \text{ lbm/min}$$

Thermodynamics

$$\dot{Q}_{\text{in}} = \dot{m}c_p \Delta T$$

$$= \left(834.5 \, \frac{\text{lbm}}{\text{min}}\right) \left(1 \, \frac{\text{Btu}}{\text{lbm -°F}}\right) (80°\text{F} - 20°\text{F})$$

$$= 50{,}070 \, \text{Btu/min}$$

$$\text{COP} = \frac{T_{\text{low}}}{T_{\text{high}} - T_{\text{low}}} = \frac{460 + 20°\text{F}}{80°\text{F} - 20°\text{F}}$$

$$= 8$$

$$W_{\text{in,hp}} = \frac{4.715 \, Q_{\text{in,tons}}}{\text{COP}}$$

$$= \frac{(4.715) \left(\dfrac{50{,}070 \, \dfrac{\text{Btu}}{\text{min}}}{200 \, \dfrac{\text{Btu}}{\text{min-ton}}} \right)}{8}$$

$$= \boxed{147.5 \, \text{hp}}$$

The answer is (D).

SI Solution

$$\dot{m} = \dot{V}\rho = \left(0.4 \, \frac{\text{m}^3}{\text{min}}\right) \left(1000 \, \frac{\text{kg}}{\text{m}^3}\right)$$

$$= 400 \, \text{kg/min}$$

$$\dot{Q}_{\text{in}} = \dot{m}c_p \Delta T$$

$$= \left(400 \, \frac{\text{kg}}{\text{min}}\right) \left(4.19 \, \frac{\text{kJ}}{\text{kg·°C}}\right) (25°\text{C} - (-5°\text{C}))$$

$$= 50\,280 \, \text{kJ/min}$$

$$\text{COP} = \frac{T_{\text{low}}}{T_{\text{high}} - T_{\text{low}}} = \frac{-5°\text{C} + 273.15}{25°\text{C} - (-5°\text{C})}$$

$$= 8.938$$

$$\dot{W}_{\text{in}} = \frac{\dot{Q}_{\text{in}}}{\text{COP}} = \left(\frac{50\,280 \, \dfrac{\text{kJ}}{\text{min}}}{8.938} \right) \left(\frac{\text{min}}{60 \, \text{s}} \right)$$

$$= \boxed{93.76 \, \text{kW}}$$

The answer is (D).

6. *Customary U.S. Solution*

At a: $p_a = 160 \, \text{psia}$

Interpolating between 140 psia and 170 psia,

$$\frac{h_a - 126 \, \dfrac{\text{Btu}}{\text{lbm}}}{139.3 \, \dfrac{\text{Btu}}{\text{lbm}} - 126 \, \dfrac{\text{Btu}}{\text{lbm}}} = \frac{160 \, \text{psia} - 140 \, \text{psia}}{170 \, \text{psia} - 140 \, \text{psia}}$$

$$h_a = 134.9 \, \text{Btu/lbm}$$

At b: $h_b = h_a = 134.9 \, \text{Btu/lbm}$

At c: $h_c = h_b + q_{\text{in}} = 134.9 \, \dfrac{\text{Btu}}{\text{lbm}} + 500 \, \dfrac{\text{Btu}}{\text{lbm}}$

$$= 634.9 \, \text{Btu/lbm}$$

From superheated ammonia table at 30 psia,

$$s_c = 1.3845 \, \text{Btu/lbm-°F}$$

At d: $s_d = s_c = 1.3845 \, \text{Btu/lbm-°F}$

$$p_d = p_a = 160 \, \text{psia}$$

Interpolating between 250°F and 300°F,

$$\frac{h_d - 737.6 \, \dfrac{\text{Btu}}{\text{lbm}}}{1.3845 \, \dfrac{\text{Btu}}{\text{lbm-°F}} - 1.3675 \, \dfrac{\text{Btu}}{\text{lbm-°F}}}$$

$$= \frac{767.1 \, \dfrac{\text{Btu}}{\text{lbm}} - 737.6 \, \dfrac{\text{Btu}}{\text{lbm}}}{1.4076 \, \dfrac{\text{Btu}}{\text{lbm-°F}} - 1.3675 \, \dfrac{\text{Btu}}{\text{lbm-°F}}}$$

$$h_d = 750.1 \, \text{Btu/lbm}$$

$$\text{COP} = \frac{q_{\text{out}}}{W_{\text{in}}} = \frac{h_c - h_b}{h_d - h_c} + 1$$

$$= \frac{634.9 \, \dfrac{\text{Btu}}{\text{lbm}} - 134.9 \, \dfrac{\text{Btu}}{\text{lbm}}}{750.1 \, \dfrac{\text{Btu}}{\text{lbm}} - 634.9 \, \dfrac{\text{Btu}}{\text{lbm}}} + 1$$

$$= \boxed{5.34}$$

The answer is (C).

SI Solution

At a: $p_a = 1.0 \, \text{MPa}$

From the saturated ammonia table,

$$h_a = 443.65 \, \text{kJ/kg}$$

At b: $h_b = h_a = 443.65 \, \text{kJ/kg}$

At c: $h_c = h_b + q_{\text{in}} = 443.65 \, \dfrac{\text{kJ}}{\text{kg}} + 1200 \, \dfrac{\text{kJ}}{\text{kg}}$

$$= 1643.65 \, \text{kJ/kg (superheated)}$$

$$p_c = 200 \, \text{kPa}$$

Interpolating from the superheated ammonia table at 200 kPa (0.20 MPa),

$$s_c = 6.4960 \, \frac{\text{kJ}}{\text{kg·K}} + \left(\frac{1643.65 \, \dfrac{\text{kJ}}{\text{kg}} - 1625.18 \, \dfrac{\text{kJ}}{\text{kg}}}{1647.94 \, \dfrac{\text{kJ}}{\text{kg}} - 1625.18 \, \dfrac{\text{kJ}}{\text{kg}}} \right)$$

$$\times \left(6.5759 \, \frac{\text{kJ}}{\text{kg·K}} - 6.4960 \, \frac{\text{kJ}}{\text{kg·K}} \right)$$

$$= 6.5608 \, \text{kJ/kg·K}$$

At d: $s_d = s_c = 6.5608$ kJ/kg·K

$$p_d = p_a = 1.0 \text{ MPa}$$

Interpolating from the superheated ammonia table at 1 MPa,

$$h_d = 1900.02 \frac{\text{kJ}}{\text{kg}} + \left(\frac{6.5608 \frac{\text{kJ}}{\text{kg·K}} - 6.5376 \frac{\text{kJ}}{\text{kg·K}}}{6.5965 \frac{\text{kJ}}{\text{kg·K}} - 6.5376 \frac{\text{kJ}}{\text{kg·K}}} \right)$$
$$\times \left(1924.43 \frac{\text{kJ}}{\text{kg}} - 1900.02 \frac{\text{kJ}}{\text{kg}} \right)$$
$$= 1909.63 \text{ kJ/kg}$$

$$\text{COP} = \frac{q_{\text{out}}}{W_{\text{in}}} = \frac{h_c - h_b}{h_d - h_c} + 1$$

$$= \frac{1643.65 \frac{\text{kJ}}{\text{kg}} - 443.65 \frac{\text{kJ}}{\text{kg}}}{1909.63 \frac{\text{kJ}}{\text{kg}} - 1643.65 \frac{\text{kJ}}{\text{kg}}} + 1$$

$$= \boxed{4.512}$$

The answer is (B).

7. *Customary U.S. Solution*

At a: $T_a = 70°\text{F}$ from saturated Freon-12 table
$\quad h_a = 23.9$ Btu/lbm
At b: $h_b = h_a = 23.9$ Btu/lbm
At c: $T_c = T_b = -30°\text{F}$
$\quad h_c = 74.7$ Btu/lbm
$\quad v_c = 3.088$ ft³/lbm

$$q_{\text{in}} = h_c - h_b = 74.7 \frac{\text{Btu}}{\text{lbm}} - 23.9 \frac{\text{Btu}}{\text{lbm}}$$
$$= 50.8 \text{ Btu/lbm}$$

$$\dot{m} = \frac{200 \frac{\text{Btu}}{\text{min-ton}}}{50.8 \frac{\text{Btu}}{\text{lbm}}}$$
$$= 3.94 \text{ lbm/min-ton}$$

$$\dot{V} = \dot{m} v_c = \left(3.94 \frac{\text{lbm}}{\text{min-ton}} \right) \left(3.088 \frac{\text{ft}^3}{\text{lbm}} \right)$$

$$= \boxed{12.16 \text{ ft}^3/\text{min-ton}}$$

The answer is (B).

SI Solution

At a: $T_a = 20°\text{C}$
$\quad = 293\text{K}$
From the saturated Freon-12 table,

$$h_a = 83.57 \text{ kJ/kg}$$

At b: $h_b = h_a = 83.57$ kJ/kg

At c: $T_c = T_b = -30°\text{C} = 243\text{K}$

$$h_c \approx 243 \text{ kJ/kg}$$
$$v_c \approx 0.1580 \text{ m}^3/\text{kg}$$

$$q_{\text{in}} = h_c - h_b = 243 \frac{\text{kJ}}{\text{kg}} - 83.57 \frac{\text{kJ}}{\text{kg}}$$
$$= 159.43 \text{ kJ/kg}$$

$$\dot{m} = \frac{Q}{q_{\text{in}}} = \frac{1 \text{ kW}}{159.43 \frac{\text{kJ}}{\text{kg}}} = 0.006272 \text{ kg/s}$$

$$\dot{V} = \dot{m} v_c = \left(0.006272 \frac{\text{kg}}{\text{s}} \right) \left(0.1580 \frac{\text{m}^3}{\text{kg}} \right)$$

$$= \boxed{0.000991 \text{ m}^3/\text{s}}$$

The answer is (B).

8. *Customary U.S. Solution*

Assuming ideal gas,

$$\frac{T_d}{T_c} = \left(\frac{p_{\text{high}}}{p_{\text{low}}} \right)^{\frac{k-1}{k}}$$

For air, $k = 1.4$.

$$T_d = T_c \left(\frac{60 \text{ psia}}{14.7 \text{ psia}} \right)^{\frac{1.4-1}{1.4}}$$
$$= (460 + 70°\text{F})(1.495)$$
$$= 792.1°\text{R}$$

(a) Temperature leaving compressor if process is not isentropic:

$$T_d' = T_c + \frac{T_d - T_c}{\eta_{\text{compressor}}}$$

$$= 530°\text{R} + \frac{792.1°\text{R} - 530°\text{R}}{0.7}$$

$$= \boxed{904.5°\text{R} \ (444.5°\text{F})}$$

The answer is (D).

(b)
$$\frac{T_a}{T_b} = \left(\frac{p_{\text{high}}}{p_{\text{low}}}\right)^{\frac{k-1}{k}}$$

$$T_b = \frac{T_a}{\left(\dfrac{60 \text{ psia}}{14.7 \text{ psia}}\right)^{\frac{1.4-1}{1.4}}} = \frac{460 + 25°\text{F}}{1.495}$$

$$= 324.4°\text{R}$$

Temperature leaving the turbine if process is not isentropic:

$$T_b' = T_a - \eta_{\text{turbine}}(T_a - T_b)$$

$$= 485°\text{R} - (0.80)(485°\text{R} - 324.4°\text{R})$$

$$= \boxed{356.5°\text{R} \ (-103.5°\text{F})}$$

The answer is (A).

(c) $\quad$ COP $= \dfrac{T_c - T_b'}{(T_d' - T_a) - (T_c - T_b')}$

$$= \frac{530°\text{R} - 356.5°\text{R}}{(904.5°\text{R} - 485°\text{R}) - (530°\text{R} - 356.5°\text{R})}$$

$$= \boxed{0.705}$$

The answer is (A).

SI Solution

(a) Assuming air is an ideal gas with $k = 1.4$,

$$\frac{T_d}{T_c} = \left(\frac{p_{\text{high}}}{p_{\text{low}}}\right)^{\frac{k-1}{k}}$$

$$T_c = 20\text{K} + 273.15 = 293.15$$

$$T_d = T_c \left(\frac{p_{\text{high}}}{p_{\text{low}}}\right)^{\frac{k-1}{k}} = (293.15\text{K})\left(\frac{400 \text{ kPa}}{101 \text{ kPa}}\right)^{\frac{1.4-1}{1.4}}$$

$$= 434.4\text{K}$$

The temperature leaving the compressor is

$$T_d' = T_c + \frac{T_d - T_c}{\eta_{\text{compressor}}}$$

$$= 293.15\text{K} + \frac{434.4\text{K} - 293.15\text{K}}{0.7}$$

$$= \boxed{494.9\text{K} \ (221.8°\text{C})}$$

The answer is (D).

(b)
$$T_a = -4°\text{C} + 273.15 = 269.15\text{K}$$

$$T_b = \frac{T_a}{\left(\dfrac{p_{\text{high}}}{p_{\text{low}}}\right)^{\frac{k-1}{k}}} = \frac{269.15\text{K}}{\left(\dfrac{400 \text{ kPa}}{101 \text{ kPa}}\right)^{\frac{1.4-1}{1.4}}}$$

$$= 181.6\text{K}$$

The temperature leaving the turbine is

$$T_b' = T_a - \eta_{\text{turbine}}(T_a - T_b)$$

$$= 269.15\text{K} - (0.80)(269.15\text{K} - 181.6\text{K})$$

$$= \boxed{199.1\text{K} \ (-74.0°\text{C})}$$

The answer is (A).

(c) $\quad$ COP $= \dfrac{T_c - T_b'}{(T_d' - T_a) - (T_c - T_b')}$

$$= \frac{293.15\text{K} - 199.1\text{K}}{(494.9\text{K} - 269.15\text{K}) - (293.15\text{K} - 199.1\text{K})}$$

$$= \boxed{0.714}$$

The answer is (A).

31 Heat Transfer: Conduction

PRACTICE PROBLEMS

Thermal Conductivity

1. Experiments have shown that the thermal conductivity, k, of a particular material varies with temperature, T, according to the following relationship.

$$k_T = (0.030)(1 + 0.0015T)$$

What is the value of k that should be used for a transfer of heat through the material if the hot-side temperature is 350° (°F or °C) and the cold-side temperature is 150° (°F or °C)?

- (A) 0.041
- (B) 0.055
- (C) 0.13
- (D) 0.22

Conduction

2. What is the heat flow through insulating brick, 1.0 ft (30 cm) thick, in an oven wall with a thermal conductivity of 0.038 Btu-ft/hr-ft^2-°F (0.066 W/m·K)? The thermal gradient is 350°F (195°C).

- (A) 9.4 Btu/hr-ft^2 (31 W/m^2)
- (B) 13 Btu/hr-ft^2 (43 W/m^2)
- (C) 19 Btu/hr-ft^2 (63 W/m^2)
- (D) 31 Btu/hr-ft^2 (100 W/m^2)

3. A composite wall is made up of 3.0 in (7.6 cm) of material A exposed to 1000°F (540°C), 5.0 in (13 cm) of material B, and 6.0 in (15 cm) of material C exposed to 200°F (90°C). The mean thermal conductivities for materials A, B, and C are 0.06, 0.5, and 0.8 Btu-ft/hr-ft^2-°F (0.1, 0.9, and 1.4 W/m·K), respectively. What are the temperatures at the A-B and B-C material interfaces?

Unsteady Heat Flow

4. The heat supply of a large building is turned off at 5:00 P.M. when the interior temperature is 70°F (21°C). The outdoor temperature is constant at 40°F (4°C). The thermal capacity of the building and its contents is 100,000 Btu/°F (60 MJ/K), and the conductance is 6500 Btu/hr-°F (1.1 kW/K). What is the interior temperature at 1:00 A.M.?

- (A) 45°F (7.2°C)
- (B) 51°F (11°C)
- (C) 58°F (14°C)
- (D) 64°F (18°C)

5. Steel ball bearings varying in diameter from $1/4$ in to $1^1/2$ in (6.35 to 38.1 mm) are quenched from 1800°F (980°C) in an oil bath that remains at 110°F (43°C). The ball bearings are removed when their internal (center) temperature reaches 250°F (120°C). The average film coefficient is 56 Btu/hr-ft^2-°F (320 W/m^2·K).

(a) What is the time constant for the cooling process?

(b) Derive a linear equation for the time the ball bearings remain in the oil bath as a function of diameter.

Internal Heat Generation

6. A 0.4 in (1.0 cm) diameter uranium dioxide fuel rod with a thermal conductivity of 1.1 Btu-ft/hr-ft^2-°F (1.9 W/m·K) is clad with 0.020 in (0.5 mm) of stainless steel. The fuel rod generates 4×10^7 Btu/hr-ft^3 (4.1×10^8 W/m^3) internally. A coolant at 500°F (260°C) circulates around the clad rod. The outside film coefficient is 10,000 Btu/hr-ft^2-°F (57 kW/m^2·K). What is the temperature at the longitudinal centerline of the rod?

- (A) 2100°F (1150°C)
- (B) 2400°F (1320°C)
- (C) 2700°F (1480°C)
- (D) 3100°F (1650°C)

Finned Radiators

7. Two long pieces of 1/16 in (1.6 mm) copper wire are connected end-to-end with a hot soldering iron. The minimum melting temperature of the solder is 450°F (230°C). The surrounding air temperature is 80°F (27°C). The unit film coefficient is 3 Btu/hr-ft^2-°F (17 W/m^2·K). Disregard radiation losses. What is the minimum rate of heat application to keep the solder molten?

- (A) 4.3 Btu/hr (1.3 W)
- (B) 11 Btu/hr (3.3 W)
- (C) 20 Btu/hr (6.0 W)
- (D) 47 Btu/hr (14 W)

Interface Temperatures

(The following scenario applies to Probs. 8–10.)

A composite furnace wall consists of two insulating materials. (Refer to the accompanying illustration for the thermal conductivities of each insulating material, k_1 and k_2, the thickness of each material, and the external film coefficient, h_o.) The inside furnace temperature is 900°C, and the conditions inside the furnace are turbulent. The ambient air temperature is 21°C. There is no resistance to heat flow at the insulation interface. The system is at steady state. The temperature at the interface of the two insulators is designated T_{1-2}, and the outside wall temperature is designated T_o.

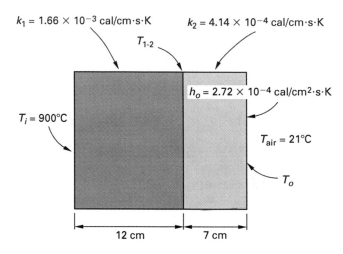

8. What is the heat loss from the outside of the furnace?

   (A) 1.9×10^{-2} cal/cm²·s
   (B) 3.2×10^{-2} cal/cm²·s
   (C) 5.3×10^{-2} cal/cm²·s
   (D) 7.8×10^{-2} cal/cm²·s

9. What is the outside wall temperature, T_o?

   (A) 20°C
   (B) 100°C
   (C) 140°C
   (D) 200°C

10. What is the temperature, T_{1-2}, at the insulation interface?

   (A) 270°C
   (B) 500°C
   (C) 580°C
   (D) 670°C

(The following scenario applies to Probs. 11 and 12.)

A room has an inside wall temperature of 25°C and an outside wall temperature of 0°C. The wall is composed of an outside layer of concrete (thermal conductivity $k_1 = 1.5 \times 10^{-1}$ cal/cm·s·K) and an inside layer of steel (thermal conductivity $k_2 = 2.2 \times 10^{-3}$ cal/cm·s·K). The steel section is 1 cm thick, and the concrete section is 5 cm thick. Conductive heat transfer occurs at steady state.

11. What is the heat flux through the wall?

   (A) 1.1×10^{-3} cal/cm²·s
   (B) 1.1×10^{-2} cal/cm²·s
   (C) 5.1×10^{-2} cal/cm²·s
   (D) 5.9×10^{-2} cal/cm²·s

12. How much would the heat flux change if an additional 1 cm of steel were added to the inside layer?

   (A) 1.0×10^{-4} cal/cm²·s
   (B) 2.4×10^{-2} cal/cm²·s
   (C) 3.2×10^{-2} cal/cm²·s
   (D) 1.9×10^{-1} cal/cm²·s

SOLUTIONS

1. Use the value of k at an average temperature of $(1/2)(T_1 + T_2)$.

$$T = \left(\tfrac{1}{2}\right)(150° + 350°) = 250°$$
$$k = (0.030)(1 + 0.0015T)$$
$$= (0.030)\big(1 + (0.0015)(250°)\big) = \boxed{0.04125}$$

The answer is (A).

2. *Customary U.S. Solution*

From Fourier's law of heat conduction (Eq. 31.16),

$$\frac{Q_{1-2}}{A} = \frac{k\Delta T}{L}$$
$$= \frac{\left(0.038\ \dfrac{\text{Btu-ft}}{\text{hr-ft}^2\text{-°F}}\right)(350°\text{F})}{1.0\ \text{ft}}$$
$$= \boxed{13.3\ \text{Btu/hr-ft}^2}$$

The answer is (B).

SI Solution

From Fourier's law of heat conduction (Eq. 31.16),

$$\frac{Q_{1-2}}{A} = \frac{k\Delta T}{L}$$
$$= \frac{\left(0.066\ \dfrac{\text{W}}{\text{m·K}}\right)(195\text{K})}{(30\ \text{cm})\left(\dfrac{1\ \text{m}}{100\ \text{cm}}\right)}$$
$$= \boxed{42.9\ \text{W/m}^2}$$

The answer is (B).

3.

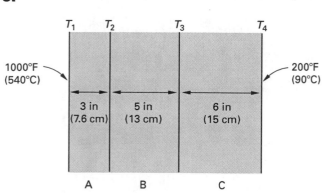

Customary U.S. Solution

Since the wall temperatures are given, it is not necessary to consider films.

From Eq. 31.21, the heat flow through the composite wall is

$$Q = \frac{A(T_1 - T_4)}{\displaystyle\sum_{i=1}^{n} \frac{L_i}{k_i}}$$

On a per unit area basis,

$$\frac{Q}{A} = \frac{1000°\text{F} - 200°\text{F}}{\dfrac{(3\ \text{in})\left(\dfrac{1\ \text{ft}}{12\ \text{in}}\right)}{0.06\ \dfrac{\text{Btu-ft}}{\text{hr-ft}^2\text{-°F}}} + \dfrac{(5\ \text{in})\left(\dfrac{1\ \text{ft}}{12\ \text{in}}\right)}{0.5\ \dfrac{\text{Btu-ft}}{\text{hr-ft}^2\text{-°F}}}}$$
$$+ \dfrac{(6\ \text{in})\left(\dfrac{1\ \text{ft}}{12\ \text{in}}\right)}{0.8\ \dfrac{\text{Btu-ft}}{\text{hr-ft}^2\text{-°F}}}$$

$$= 142.2\ \text{Btu/hr-ft}^2$$

$$= \frac{T_1 - T_4}{\displaystyle\sum_{i=1}^{n} \frac{L_i}{k_i}} = \frac{T_1 - T_2}{\dfrac{L_A}{k_A}} = \frac{T_2 - T_3}{\dfrac{L_B}{k_B}}$$

To find the temperature at the A-B interface (T_2), use

$$\frac{Q}{A} = \frac{T_1 - T_2}{\dfrac{L_A}{k_A}}$$
$$T_2 = T_1 - \left(\frac{Q}{A}\right)\left(\frac{L_A}{k_A}\right)$$
$$= 1000°\text{F} - \frac{\left(142.2\ \dfrac{\text{Btu}}{\text{hr-ft}^2}\right)(3\ \text{in})\left(\dfrac{1\ \text{ft}}{12\ \text{in}}\right)}{0.06\ \dfrac{\text{Btu-ft}}{\text{hr-ft}^2\text{-°F}}}$$
$$= \boxed{407.5°\text{F}}$$

To find the temperature at the B-C interface (T_3), use

$$\frac{Q}{A} = \frac{T_2 - T_3}{\dfrac{L_B}{k_B}}$$
$$T_3 = T_2 - \left(\frac{Q}{A}\right)\left(\frac{L_B}{K_B}\right)$$
$$= 407.5°\text{F} - \left(142.2\ \dfrac{\text{Btu}}{\text{hr-ft}^2}\right)\left(\frac{(5\ \text{in})\left(\dfrac{1\ \text{ft}}{12\ \text{in}}\right)}{0.5\ \dfrac{\text{Btu-ft}}{\text{hr-ft}^2\text{-°F}}}\right)$$
$$= \boxed{289°\text{F}}$$

SI Solution

On a per unit area basis,

$$\frac{Q}{A} = \frac{540°C - 90°C}{\dfrac{(7.6 \text{ cm})\left(\dfrac{1 \text{ m}}{100 \text{ cm}}\right)}{0.1 \dfrac{W}{m \cdot K}} + \dfrac{(13 \text{ cm})\left(\dfrac{1 \text{ m}}{100 \text{ cm}}\right)}{0.9 \dfrac{W}{m \cdot K}} + \dfrac{(15 \text{ cm})\left(\dfrac{1 \text{ m}}{100 \text{ cm}}\right)}{1.4 \dfrac{W}{m \cdot K}}}$$

$$= 444.8 \text{ W/m}^2$$

$$T_2 = T_1 - \left(\frac{Q}{A}\right)\left(\frac{L_A}{k_A}\right)$$

$$= 540°C - \left(444.8 \frac{W}{m^2}\right)\left(\frac{(7.6 \text{ cm})\left(\dfrac{1 \text{ m}}{100 \text{ cm}}\right)}{0.1 \dfrac{W}{m \cdot K}}\right)$$

$$= \boxed{202.0°C}$$

$$T_3 = T_2 - \left(\frac{Q}{A}\right)\left(\frac{L_B}{k_B}\right)$$

$$= 202.0°C - \left(444.8 \frac{W}{m^2}\right)\left(\frac{(13 \text{ cm})\left(\dfrac{1 \text{ m}}{100 \text{ cm}}\right)}{0.9 \dfrac{W}{m \cdot K}}\right)$$

$$= \boxed{137.8°C}$$

4. *Customary U.S. Solution*

This is a transient problem. The total time is from 5 P.M. to 1 A.M., which is 8 hr.

The thermal capacitance (capacity), C_e, is given as 100,000 Btu/°F.

The thermal resistance is

$$R_e = \frac{1}{\text{thermal conductance}}$$

$$= \frac{1}{6500 \dfrac{\text{Btu}}{\text{hr-°F}}} = 0.0001538 \text{ hr-°F/Btu}$$

From Eq. 31.50,

$$T_t = T_\infty + (T_0 - T_\infty)e^{-\frac{t}{R_e C_e}}$$

$$T_{8 \text{ hr}} = 40°F + (70°F - 40°F)$$

$$\times \exp\left(\frac{-8 \text{ hr}}{\left(0.0001538 \dfrac{\text{hr-°F}}{\text{Btu}}\right)\left(100{,}000 \dfrac{\text{Btu}}{°F}\right)}\right)$$

$$= \boxed{57.8°F}$$

The answer is (C).

SI Solution

The thermal capacitance (capacity) is

$$C_e = \left(60 \frac{\text{MJ}}{K}\right)\left(1000 \frac{\text{kJ}}{\text{MJ}}\right)$$

$$= 60{,}000 \text{ kJ/K} \quad \text{[given]}$$

The thermal resistance is

$$R_e = \frac{1}{\text{thermal conductance}}$$

$$= \frac{1}{1.1 \dfrac{\text{kW}}{K}} = 0.909 \text{ K/kW}$$

From Eq. 31.50,

$$T_t = T_\infty + (T_0 - T_\infty)e^{-\frac{t}{R_e C_e}}$$

$$= 4°C + (21°C - 4°C)$$

$$\times \exp\left(\frac{(-8 \text{ h})\left(3600 \dfrac{s}{h}\right)}{\left(0.909 \dfrac{K}{kW}\right)\left(60\,000 \dfrac{\text{kJ}}{K}\right)}\right)$$

$$= \boxed{14.0°C}$$

The answer is (C).

5. *Customary U.S. Solution*

(a) This is a transient problem. Check the Biot number to see if the lumped parameter method can used.

The characteristic site length from Eq. 31.8 is

$$L_c = \frac{V}{A_s} = \frac{\left(\dfrac{\pi}{6}\right)d^3}{\pi d^2} = \frac{d}{6}$$

For the largest ball, $d = 1.5$ in.

$$L_c = \frac{d}{6} = \frac{(1.5 \text{ in})\left(\dfrac{1 \text{ ft}}{12 \text{ in}}\right)}{6} = 0.0208 \text{ ft}$$

Evaluate the thermal conductivity, k, of steel at

$$\left(\tfrac{1}{2}\right)(1800°F + 250°F) = 1025°F$$

From App. 31.B, for steel, $k \approx 22.0$ Btu-ft/hr-ft²-°F.

From Eq. 31.10, the Biot number is

$$Bi = \frac{hL_c}{k} = \frac{\left(56 \dfrac{\text{Btu}}{\text{hr-ft}^2\text{-°F}}\right)(0.0208 \text{ ft})}{22.0 \dfrac{\text{Btu-ft}}{\text{hr-ft}^2\text{-°F}}}$$

$$= 0.053$$

For small balls, Bi will be even smaller.

Since Bi < 0.10, the lumped parameter method can be used.

The assumptions are

- homogeneous body temperature

- minimal radiation losses

- oil bath temperature remains constant

- h remains constant

From Eqs. 31.48 and 31.49, the time constant is

$$C_e R_e = c_p \rho V \left(\frac{1}{hA_s} \right) = \left(\frac{c_p \rho}{h} \right) \left(\frac{V}{A_s} \right)$$
$$= \left(\frac{c_p \rho}{h} \right) L_c = \left(\frac{c_p \rho}{h} \right) \left(\frac{d}{6} \right)$$

From App. 31.B, $\rho = 490$ lbm/ft^3 and $c_p = 0.11$ Btu/lbm-°F, even though those values are for 32°F.

The time constant is (measuring d in inches),

$$C_e R_e = \left(\frac{\left(0.11 \, \frac{\text{Btu}}{\text{lbm-°F}} \right) \left(490 \, \frac{\text{lbm}}{\text{ft}^3} \right)}{56 \, \frac{\text{Btu}}{\text{hr-ft}^2\text{-°F}}} \right)$$
$$\times \left(\frac{d}{6} \right) \left(\frac{1 \, \text{ft}}{12 \, \text{in}} \right)$$
$$= \boxed{0.01337d}$$

(b) Taking the natural log of the transient equation (Eq. 31.51),

$$\ln(T_t - T_\infty) = \ln \left(\Delta T e^{-\frac{t}{R_e C_e}} \right)$$
$$= \ln \Delta T + \ln \left(e^{-\frac{t}{R_e C_e}} \right)$$
$$= \ln \Delta T - \frac{t}{R_e C_e}$$
$$T_t = 250°\text{F}$$
$$T_\infty = 110°\text{F}$$
$$\Delta T = 1800°\text{F} - 110°\text{F} = 1690°\text{F}$$
$$\ln(250°\text{F} - 110°\text{F}) = \ln(1690°\text{F}) - \frac{t}{0.01337d}$$
$$4.942 = 7.432 - \frac{t}{0.01337d}$$
$$t = \boxed{0.0333d}$$

SI Solution

For the largest ball, the characteristic length is

$$L_c = \frac{d}{6} = \frac{(38.1 \, \text{mm}) \left(\frac{1 \, \text{m}}{1000 \, \text{mm}} \right)}{6} = 6.35 \times 10^{-3} \, \text{m}$$

Evaluate the thermal conductivity, k, of steel at

$$\left(\tfrac{1}{2} \right) (980°\text{C} + 120°\text{C}) = 550°\text{C}$$

From App. 31.B and its footnote,

$$k \approx (22.0) \left(\frac{1.7307 \, \text{W·hr·ft}^2\text{·°F}}{\text{m·K·Btu·ft}} \right)$$
$$= 38.08 \, \text{W/m·K}$$

From Eq. 31.10, the Biot number is

$$\text{Bi} = \frac{hL_c}{k} = \frac{\left(320 \, \frac{\text{W}}{\text{m}^2\text{·K}} \right) (6.35 \times 10^{-3} \, \text{m})}{38.08 \, \frac{\text{W}}{\text{m·K}}}$$
$$= 0.053$$

(a) For small balls, Bi will be even smaller.

Since Bi < 0.10, the lumped method can be used. The assumptions are given in the customary U.S. solution. From Eqs. 31.48 and 31.49, the time constant is

$$C_e R_e = c_p \rho V \left(\frac{1}{hA_s} \right) = \left(\frac{c_p \rho}{h} \right) \left(\frac{V}{A_s} \right)$$
$$= \left(\frac{c_p \rho}{h} \right) L_c = \left(\frac{c_p \rho}{h} \right) \left(\frac{d}{6} \right)$$

From App. 31.B and its footnote,

$$\rho = \left(490 \, \frac{\text{lbm}}{\text{ft}^3} \right) \left(16.0185 \, \frac{\text{kg·ft}^3}{\text{m}^3\text{·lbm}} \right)$$
$$= 7849.1 \, \text{kg/m}^3$$
$$c_p = (0.11) \left(4186.8 \, \frac{\text{J·lbm·°F}}{\text{kg·K·Btu}} \right)$$
$$= 460.5 \, \text{J/kg·K}$$
$$C_e R_e = \left(\frac{\left(460.5 \, \frac{\text{J}}{\text{kg·K}} \right) \left(7849.1 \, \frac{\text{kg}}{\text{m}^3} \right)}{\left(320 \, \frac{\text{W}}{\text{m}^2\text{·K}} \right)} \right) \left(\frac{d}{6} \right)$$
$$= \boxed{1882.6d}$$

(b) From the customary U.S. solution,

$$\ln(T_t - T_\infty) = \ln \Delta T - \frac{t}{R_e C_e}$$

$$T_t = 120°C$$

$$T_\infty = 43°C$$

$$\Delta T = 980°C - 43°C = 937°C$$

$$\ln(120°C - 43°C) = \ln(937°C) - \frac{t}{R_e C_e}$$

$$4.344 = 6.843 - \frac{t}{R_e C_e}$$

$$t = (2.499)R_e C_e$$

$$= (2.499)(1882.6 \text{ d})$$

$$= \boxed{4704.6 \text{ d}}$$

6. *Customary U.S. Solution*

The volume of the rod is

$$V = \frac{\pi}{4}d^2 L$$

$$\frac{V}{L} = \frac{\pi}{4}d^2 = \left(\frac{\pi}{4}\right)(0.4 \text{ in})^2 \left(\frac{1 \text{ ft}}{12 \text{ in}}\right)^2$$

$$= 8.727 \times 10^{-4} \text{ ft}^3/\text{ft}$$

The heat output per unit length of rod is

$$\frac{Q}{L} = \left(\frac{V}{L}\right)G$$

$$= \left(8.727 \times 10^{-4} \frac{\text{ft}^3}{\text{ft}}\right)\left(4 \times 10^7 \frac{\text{Btu}}{\text{hr-ft}^3}\right)$$

$$= 3.491 \times 10^4 \text{ Btu/hr-ft}$$

The diameter of the cladding is

$$d_o = 0.4 \text{ in} + (2)(0.020 \text{ in}) = 0.44 \text{ in}$$

The surface area per unit length of cladding is

$$A = \pi d_o = \pi(0.44 \text{ in})\left(\frac{1 \text{ ft}}{12 \text{ in}}\right) = 0.1152 \text{ ft}^2/\text{ft}$$

From Eq. 31.23, the surface temperature of the cladding is

$$T_s = \frac{Q}{hA} + T_\infty$$

$$= \frac{3.491 \times 10^4 \dfrac{\text{Btu}}{\text{hr-ft}}}{\left(10{,}000 \dfrac{\text{Btu}}{\text{hr-ft}^2\text{-°F}}\right)\left(0.1152 \dfrac{\text{ft}^2}{\text{ft}}\right)} + 500°F$$

$$= 530.3°F$$

For the cladding,

$$r_o = \frac{d_o}{2} = \left(\frac{0.44 \text{ in}}{2}\right)\left(\frac{1 \text{ ft}}{12 \text{ in}}\right) = 0.01833 \text{ ft}$$

$$r_i = \frac{d}{2} = \left(\frac{0.4 \text{ in}}{2}\right)\left(\frac{1 \text{ ft}}{12 \text{ in}}\right) = 0.01667 \text{ ft}$$

From App. 31.B, k for stainless steel (at 572°F) is 11 Btu-ft/hr-ft²-°F. This is reasonable because the inside cladding temperature is greater than the surface temperature (530.3°F).

For a cylinder from Eqs. 31.5 and 31.19,

$$T_{\text{inside}} - T_{\text{outside}} = \frac{Q\ln\left(\frac{r_o}{r_i}\right)}{2\pi kL} = \frac{\left(\frac{Q}{L}\right)\ln\left(\frac{r_o}{r_i}\right)}{2\pi k}$$

$$= \frac{\left(3.491 \times 10^4 \dfrac{\text{Btu}}{\text{hr-ft}}\right)\ln\left(\dfrac{0.01833 \text{ ft}}{0.01667 \text{ ft}}\right)}{2\pi\left(11 \dfrac{\text{Btu-ft}}{\text{hr-ft}^2\text{-°F}}\right)}$$

$$= 47.9°F$$

$$T_{\substack{\text{inside} \\ \text{cladding}}} = T_{\substack{\text{outside} \\ \text{fuel rod}}} + 47.9°F$$

$$= 530.3°F + 47.9°F = 578.2°F$$

From Eq. 31.56,

$$T_{\text{center}} = T_o + \frac{Gr_o^2}{4k}$$

$$= 578.2°F + \frac{\left(4 \times 10^7 \dfrac{\text{Btu}}{\text{hr-ft}^3}\right)(0.01667 \text{ ft})^2}{(4)\left(1.1 \dfrac{\text{Btu-ft}}{\text{hr-ft}^2\text{-°F}}\right)}$$

$$= \boxed{3104°F}$$

The answer is (D).

SI Solution

$$\frac{V}{L} = \frac{\pi}{4}d = \left(\frac{\pi}{4}\right)(1.0 \text{ cm})^2 \left(\frac{1 \text{ m}}{100 \text{ cm}}\right)^2$$

$$= 7.854 \times 10^{-5} \text{ m}^3/\text{m}$$

The heat output per unit length of rod is

$$\frac{Q}{L} = \left(\frac{V}{L}\right)G = \left(7.854 \times 10^{-5} \frac{\text{m}^3}{\text{m}}\right)\left(4.1 \times 10^8 \frac{\text{W}}{\text{m}^3}\right)$$

$$= 32\,201.4 \text{ W/m}$$

From Eq. 31.23, the surface temperature of the cladding is

$$T_s = \frac{Q}{hA} + T_\infty$$

The diameter of the cladding is

$$d_o = \left(1.0 \text{ cm} + (2)(0.5 \text{ mm})\left(\frac{1 \text{ cm}}{10 \text{ mm}}\right)\right)\left(\frac{1 \text{ m}}{100 \text{ cm}}\right)$$

$$= 0.011 \text{ m}$$

The surface area per unit length of cladding is

$$A = \pi d_o = \pi(0.011 \text{ m}) = 0.0346 \text{ m}^2/\text{m}$$

$$T_s = \frac{32\,201.4\ \dfrac{\text{W}}{\text{m}}}{\left(57\ \dfrac{\text{kW}}{\text{m}^2\cdot\text{K}}\right)\left(1000\ \dfrac{\text{W}}{\text{kW}}\right)\left(0.0346\ \dfrac{\text{m}^2}{\text{m}}\right)} + 260°\text{C}$$

$$= 276.3°\text{C}$$

For the cladding,

$$r_o = \frac{d_o}{2} = \frac{0.011 \text{ m}}{2} = 0.0055 \text{ m}$$

$$r_i = \frac{d_i}{2} = \frac{0.01 \text{ m}}{2} = 0.0050 \text{ m}$$

From App. 31.B and the table's footnote, k for stainless steel at 300°C is

$$k \approx \left(11\ \frac{\text{Btu-ft}}{\text{hr-ft}^2\text{-}°\text{F}}\right)\left(1.7307\ \frac{\text{W}\cdot\text{hr}\cdot\text{ft}^2\cdot°\text{F}}{\text{m}\cdot\text{K}\cdot\text{Btu}\cdot\text{ft}}\right)$$

$$= 19.038 \text{ W/m}\cdot\text{K}$$

This is reasonable because the inside cladding temperature is greater than the surface temperature (276.3°C).

For a cylinder from Eqs. 31.5 and 31.19,

$$T_{\text{inside}} - T_{\text{outside}} = \frac{Q \ln\left(\dfrac{r_o}{r_i}\right)}{2\pi k L} = \left(\frac{Q}{L}\right)\left(\frac{\ln\left(\dfrac{r_o}{r_i}\right)}{2\pi k}\right)$$

$$= \left(32\,201\ \frac{\text{W}}{\text{m}}\right)\left(\frac{\ln\left(\dfrac{0.0055 \text{ m}}{0.0050 \text{ m}}\right)}{2\pi\left(19.038\ \dfrac{\text{W}}{\text{m}\cdot\text{K}}\right)}\right)$$

$$= 25.7°\text{C}$$

$$T_{\substack{\text{inside}\\\text{cladding}}} = T_{\substack{\text{outside}\\\text{fuel rod}}} + 25.7°\text{C}$$

$$= 276.3°\text{C} + 25.7°\text{C} = 302°\text{C}$$

From Eq. 31.56,

$$T_{\text{center}} = T_o + \frac{G r_o^2}{4k}$$

$$= 302°\text{C} + \frac{\left(4.1 \times 10^8\ \dfrac{\text{W}}{\text{m}^3}\right)(0.0050 \text{ m})^2}{(4)\left(1.9\ \dfrac{\text{W}}{\text{m}\cdot\text{K}}\right)}$$

$$= \boxed{1651°\text{C}}$$

The answer is (D).

7. *Customary U.S. Solution*

Consider this an infinite cylindrical fin with

$$T_b = 450°\text{F}$$
$$T_\infty = 80°\text{F}$$
$$h = 3 \text{ Btu/hr-ft}^2\text{-}°\text{F}$$

From Eq. 31.66, the perimeter length is

$$P = \pi d = \pi\left(\frac{1}{16} \text{ in}\right)\left(\frac{1 \text{ ft}}{12 \text{ in}}\right) = 0.01636 \text{ ft}$$

From App. 31.B, k at 450°F is approximately 215 Btu-ft/hr-ft^2-°F.

From Eq. 31.65, the cross-sectional area of the fin at its base is

$$A_b = \pi r^2 = \pi\left(\frac{d}{2}\right)^2 = \frac{\pi}{4}d^2$$

$$= \left(\frac{\pi}{4}\right)\left(\frac{1}{16} \text{ in}\right)^2\left(\frac{1 \text{ ft}}{12 \text{ in}}\right)^2$$

$$= 2.131 \times 10^{-5} \text{ ft}^2$$

From Eq. 31.64, with two fins joined at the middle,

$$Q = 2\sqrt{hPkA_b}(T_b - T_\infty)$$

$$= (2)\sqrt{\begin{array}{c}\left(3\ \dfrac{\text{Btu}}{\text{hr-ft}^2\text{-}°\text{F}}\right)(0.01636 \text{ ft})\\[2mm] \times\left(215\ \dfrac{\text{Btu-ft}}{\text{hr-ft}^2\text{-}°\text{F}}\right)(2.131 \times 10^{-5} \text{ ft}^2)\end{array}}$$

$$\times (450°\text{F} - 80°\text{F})$$

$$= \boxed{11.1 \text{ Btu/hr}} \quad \text{[This disregards radiation.]}$$

The answer is (B).

SI Solution

Consider this an infinite cylindrical fin with

$$T_b = 230°\text{C}$$
$$T_\infty = 27°\text{C}$$
$$h = 17 \text{ W/m}^2\cdot\text{K}$$

From Eq. 31.66, the perimeter length is

$$P = \pi d = \pi(1.6 \text{ mm})\left(\frac{1 \text{ m}}{1000 \text{ mm}}\right) = 5.027 \times 10^{-3} \text{ m}$$

From App. 31.B and the table footnote, k at 230°C is

$$k \approx \left(215\ \frac{\text{Btu-ft}}{\text{hr-ft}^2\text{-}°\text{F}}\right)\left(1.7307\ \frac{\text{W}\cdot\text{hr}\cdot\text{ft}^2\cdot°\text{F}}{\text{m}\cdot\text{K}\cdot\text{Btu}\cdot\text{ft}}\right)$$

$$= 372.1 \text{ W/m}\cdot\text{K}$$

From Eq. 31.65, the cross-sectional area of the fin at its base is

$$A_b = \pi r^2 = \pi \left(\frac{d}{2}\right)^2 = \frac{\pi}{4} d^2$$

$$= \left(\frac{\pi}{4}\right)(1.6 \text{ mm})^2 \left(\frac{1 \text{ m}}{1000 \text{ mm}}\right)^2$$

$$= 2.011 \times 10^{-6} \text{ m}^2$$

From Eq. 31.64, with two fins joined at the middle,

$$Q = 2\sqrt{hPkA_b}(T_b - T_\infty)$$

$$= (2)\sqrt{\begin{array}{c}\left(17 \dfrac{\text{W}}{\text{m}^2 \cdot \text{K}}\right)(5.027 \times 10^{-3} \text{ m}) \\ \times \left(372.1 \dfrac{\text{W}}{\text{m} \cdot \text{K}}\right)(2.011 \times 10^{-6} \text{ m}^2)\end{array}}$$

$$\times (230°\text{C} - 27°\text{C})$$

$$= \boxed{3.25 \text{ W}} \quad \text{[This disregards radiation.]}$$

The answer is (B).

8. Since the conditions inside the furnace are turbulent, the inside film coefficient is disregarded (i.e., it is very high). Calculate the overall heat transfer coefficient using Eq. 31.25.

$$U = \frac{1}{\sum_i \dfrac{L_i}{k_i} + \sum_j \dfrac{1}{h_j}} = \frac{1}{\dfrac{L_1}{k_1} + \dfrac{L_2}{k_2} + \dfrac{1}{h_o}}$$

$$= \frac{1}{\dfrac{12 \text{ cm}}{1.66 \times 10^{-3} \dfrac{\text{cal}}{\text{cm} \cdot \text{s} \cdot \text{K}}} + \dfrac{7 \text{ cm}}{4.14 \times 10^{-4} \dfrac{\text{cal}}{\text{cm} \cdot \text{s} \cdot \text{K}}}}{} + \frac{1}{2.72 \times 10^{-4} \dfrac{\text{cal}}{\text{cm}^2 \cdot \text{s} \cdot \text{K}}}$$

$$= 3.60 \times 10^{-5} \text{ cal/cm}^2 \cdot \text{s} \cdot \text{K}$$

The heat flux, Q/A, by conduction at steady state is calculated from Eq. 31.26 (recognizing that $\Delta T_{°\text{C}} = \Delta T_K$).

$$\frac{Q}{A} = U \Delta T$$

$$= \left(3.60 \times 10^{-5} \frac{\text{cal}}{\text{cm}^2 \cdot \text{s} \cdot \text{K}}\right)(900°\text{C} - 21°\text{C})$$

$$= \boxed{3.16 \times 10^{-2} \text{ cal/cm}^2 \cdot \text{s}}$$

The answer is (B).

9. The heat flux through the outside film is the same as the heat flux through the wall. The conduction equation can be written in terms of the film resistance and the temperatures on either side of the film.

$$\frac{Q}{A} = h\Delta T = h_o(T_o - T_{\text{air}})$$

$$3.16 \times 10^{-2} \frac{\text{cal}}{\text{cm}^2 \cdot \text{s}} = \left(2.72 \times 10^{-4} \frac{\text{cal}}{\text{cm}^2 \cdot \text{s} \cdot \text{K}}\right)$$

$$\times (T_o - 21°\text{C})$$

$$T_o = \boxed{137°\text{C}}$$

The answer is (C).

10. The heat flux through the inner insulation is the same as the heat flux through the entire wall. The conduction equation can be written in terms of the inner conductive resistance and the temperatures on either side of the insulation.

$$\frac{Q_{1\text{-}2}}{A} = \frac{k_1(T_i - T_{1\text{-}2})}{L_1}$$

$$3.16 \times 10^{-2} \frac{\text{cal}}{\text{cm}^2 \cdot \text{s}} = \frac{\left(1.66 \times 10^{-3} \dfrac{\text{cal}}{\text{cm} \cdot \text{s} \cdot \text{K}}\right)}{\times (900°\text{C} - T_{1\text{-}2})}{12 \text{ cm}}$$

$$T_{1\text{-}2} = \boxed{672°\text{C}}$$

The answer is (D).

11. The heat flux through the wall layers is determined using Eq. 31.17. For the concrete layer, with $T_{1\text{-}2}$ representing the temperature at the concrete-steel interface,

$$\frac{Q}{A} = \frac{k_1(T_{1\text{-}2} - T_o)}{L_1}$$

For the steel layer,

$$\frac{Q}{A} = \frac{k_2(T_i - T_{1\text{-}2})}{L_2}$$

There are two equations with two unknowns, Q/A and $T_{1\text{-}2}$. At steady state, the heat flux through the concrete equals the heat flux through the steel. Solve each equation for the unknown temperature, $T_{1\text{-}2}$.

$$T_{1\text{-}2} = T_o + \left(\frac{Q}{A}\right)\left(\frac{L_1}{k_1}\right) \quad \text{[Eq. I]}$$

$$T_{1\text{-}2} = T_i - \left(\frac{Q}{A}\right)\left(\frac{L_2}{k_2}\right) \quad \text{[Eq. II]}$$

Equate the two terms and solve for heat flux, Q/A (recognizing that $\Delta T_{°C} = \Delta T_K$).

$$T_o + \left(\frac{Q}{A}\right)\left(\frac{L_1}{k_1}\right) = T_i - \left(\frac{Q}{A}\right)\left(\frac{L_2}{k_2}\right)$$

$$\frac{Q}{A} = \frac{T_i - T_o}{\dfrac{L_1}{k_1} + \dfrac{L_2}{k_2}}$$

$$= \frac{25°C - 0°C}{\dfrac{5 \text{ cm}}{1.5 \times 10^{-1}\dfrac{\text{cal}}{\text{cm·s·K}}} + \dfrac{1 \text{ cm}}{2.2 \times 10^{-3}\dfrac{\text{cal}}{\text{cm·s·K}}}}$$

$$= \boxed{5.1 \times 10^{-2} \text{ cal/cm}^2\text{·s}}$$

The answer is (C).

12. The calculation of the heat flux follows the same process as in Prob. 11, except the thickness of the steel wall is increased from 1 cm to 2 cm.

$$\frac{Q}{A} = \frac{T_i - T_o}{\dfrac{L_1}{k_1} + \dfrac{L_2}{k_2}}$$

$$= \frac{25°C - 0°C}{\dfrac{5 \text{ cm}}{1.5 \times 10^{-1}\dfrac{\text{cal}}{\text{cm·s·K}}} + \dfrac{2 \text{ cm}}{2.2 \times 10^{-3}\dfrac{\text{cal}}{\text{cm·s·K}}}}$$

$$= 2.7 \times 10^{-2} \text{ cal/cm}^2\text{·s}$$

The change in heat flux is

$$\Delta\left(\frac{Q}{A}\right) = \left(\frac{Q}{A}\right)_{2 \text{ cm steel}} - \left(\frac{Q}{A}\right)_{1 \text{ cm steel}}$$

$$= 2.7 \times 10^{-2}\frac{\text{cal}}{\text{cm}^2\text{·s}} - 5.1 \times 10^{-2}\frac{\text{cal}}{\text{cm}^2\text{·s}}$$

$$= \boxed{-2.4 \times 10^{-2} \text{ cal/cm}^2\text{·s} \quad [\text{decrease}]}$$

The answer is (B).

Heat Transfer

32 Heat Transfer: Natural Convection

PRACTICE PROBLEMS

1. What is the density of 87% wet steam at 50 psia (350 kPa)?

   (A)   0.75 lbm/ft^3 (12 kg/m^3)
   (B)   0.89 lbm/ft^3 (14 kg/m^3)
   (C)   0.94 lbm/ft^3 (15 kg/m^3)
   (D)   1.07 lbm/ft^3 (17 kg/m^3)

2. What is the viscosity of 100°F (38°C) water in units of lbm/hr-ft (kg/s·m)?

   (A)   1.2 lbm/ft-hr (0.00052 kg/m·s)
   (B)   1.4 lbm/ft-hr (0.00060 kg/m·s)
   (C)   1.6 lbm/ft-hr (0.00068 kg/m·s)
   (D)   1.8 lbm/ft-hr (0.00077 kg/m·s)

3. A fluid in a tank is maintained at 85°F (29°C) by a copper tube carrying hot water. The water decreases in temperature from 190°F (88°C) to 160°F (71°C) as it flows through the tube. At what temperature should the fluid's film coefficient be evaluated?

   (A)   130°F (54°C)
   (B)   160°F (71°C)
   (C)   175°F (79°C)
   (D)   190°F (88°C)

4. A bare, horizontal conductor with circular cross section with an outside diameter of 0.6 in (1.5 cm) dissipates 8.0 W/ft (25 W/m). The conductor is cooled by free convection, and the surrounding air temperature is 60°F (15°C). Assume the film temperature is 100°F (38°C). What is the conductor's surface temperature?

   (A)   85°F (29°C)
   (B)   110°F (43°C)
   (C)   130°F (54°C)
   (D)   160°F (67°C)

SOLUTIONS

1. *Customary U.S. Solution*

If the steam is 87% wet, the quality is $x = 0.13$.

From App. 23.B at 50 psia,

$$v_f = 0.01727 \text{ ft}^3/\text{lbm}$$
$$v_g = 8.52 \text{ ft}^3/\text{lbm}$$

From Eq. 23.47, the specific volume of steam is

$$v = v_f + xv_{fg}$$
$$= 0.01727 \frac{\text{ft}^3}{\text{lbm}} + (0.13)\left(8.52 \frac{\text{ft}^3}{\text{lbm}} - 0.01727 \frac{\text{ft}^3}{\text{lbm}}\right)$$
$$= 1.123 \text{ ft}^3/\text{lbm}$$

The density is

$$\rho = \frac{1}{v} = \frac{1}{1.123 \dfrac{\text{ft}^3}{\text{lbm}}} = \boxed{0.890 \text{ lbm/ft}^3}$$

The answer is (B).

SI Solution

If the steam is 87% wet, the quality is $x = 0.13$.

From App. 23.O at 350 kPa,

$$v_f = 1.0786 \text{ cm}^3/\text{g}$$
$$v_g = 524.3 \text{ cm}^3/\text{g}$$

From Eq. 23.47, the specific volume of steam is

$$v = v_f + xv_{fg}$$
$$= 1.0786 \frac{\text{cm}^3}{\text{g}} + (0.13)\left(524.3 \frac{\text{cm}^3}{\text{g}}\right)$$
$$= 69.24 \text{ cm}^3/\text{g}$$

The density is

$$\rho = \frac{1}{v} = \left(\frac{1}{69.24 \dfrac{\text{cm}^3}{\text{g}}}\right)\left(100 \frac{\text{cm}}{\text{m}}\right)^3 \left(\frac{1 \text{ kg}}{1000 \text{ g}}\right)$$

$$= \boxed{14.44 \text{ kg/m}^3}$$

The answer is (B).

2. *Customary U.S. Solution*

From App. 32.A, the viscosity of water at 100°F is

$$\mu = \left(0.458 \times 10^{-3} \ \frac{\text{lbm}}{\text{ft-sec}}\right)\left(3600 \ \frac{\text{sec}}{\text{hr}}\right)$$

$$= \boxed{1.6488 \ \text{lbm/ft-hr}}$$

The answer is (C).

SI Solution

From App. 32.B, the viscosity of water at 38°C is

$$\mu = \boxed{0.682 \times 10^{-3} \ \text{kg/m·s}}$$

The answer is (C).

3. *Customary U.S. Solution*

The midpoint tube temperature is

$$T_s = \left(\tfrac{1}{2}\right)(190°\text{F} + 160°\text{F}) = 175°\text{F}$$

$$T_\infty = 85°\text{F} \quad [\text{given}]$$

From Eq. 32.11, h should be evaluated at $(1/2)(T_s + T_\infty)$.

The film temperature is

$$T_h = \left(\tfrac{1}{2}\right)(175°\text{F} + 85°\text{F}) = \boxed{130°\text{F}}$$

The answer is (A).

SI Solution

The midpoint tube temperature is

$$T_2 = \left(\tfrac{1}{2}\right)(88°\text{C} + 71°\text{C}) = 79.5°\text{C}$$

$$T_\infty = 29°\text{C} \quad [\text{given}]$$

From Eq. 32.11, h should be evaluated at $(1/2)(T_s + T_\infty)$.

The film temperature is

$$T_h = \left(\tfrac{1}{2}\right)(79.5°\text{C} + 29°\text{C}) = \boxed{54.3°\text{C}}$$

The answer is (A).

4. *Customary U.S. Solution*

The heat loss per unit length is

$$\frac{Q}{L} = \left(8 \ \frac{\text{W}}{\text{ft}}\right)\left(3.413 \ \frac{\text{Btu}}{\text{hr-W}}\right) = 27.3 \ \text{Btu/hr-ft}$$

From App. 32.C, the air properties are at 100°F film.

$$\text{Pr} = 0.72$$

$$\frac{g\beta\rho^2}{\mu^2} = 1.76 \times 10^6 \ \frac{1}{\text{ft}^3\text{-°F}}$$

From Eq. 32.4, the Grashof number is

$$\text{Gr} = L^3 \Delta T \left(\frac{g\beta\rho^2}{\mu^2}\right)$$

The characteristic length is the wire diameter.

$$L = (0.6 \ \text{in})\left(\frac{1 \ \text{ft}}{12 \ \text{in}}\right) = 0.05 \ \text{ft}$$

The temperature gradient is

$$\Delta T = T_s - T_\infty = T_{\text{wire}} - 60°\text{F}$$

T_{wire} is unknown, so assume $T_{\text{wire}} = 150°\text{F}$.

$$\text{Gr} = (0.05 \ \text{ft})^3(150°\text{F} - 60°\text{F})\left(1.76 \times 10^6 \ \frac{1}{\text{ft}^3\text{-°F}}\right)$$

$$= 19{,}800$$

$$\text{Pr}\,\text{Gr} = (0.72)(19{,}800) = 14{,}256$$

From Table 32.3,

$$h \approx (0.27)\left(\frac{T_{\text{wire}} - T_\infty}{d}\right)^{0.25}$$

$$= (0.27)\left(\frac{150°\text{F} - 60°\text{F}}{0.05 \ \text{ft}}\right)^{0.25}$$

$$= 1.76 \ \text{Btu/hr-ft}^2\text{-°F}$$

The heat transfer from the wire is given by Eq. 32.1.

$$Q = Aq = \pi dLh(T_{\text{wire}} - T_\infty)$$

$$\frac{Q}{L} = \pi dh(T_{\text{wire}} - T_\infty)$$

$$T_{\text{wire}} = \frac{\dfrac{Q}{L}}{\pi dh} + T_\infty$$

$$= \frac{27.3 \ \dfrac{\text{Btu}}{\text{hr-ft}}}{\pi(0.05 \ \text{ft})\left(1.76 \ \dfrac{\text{Btu}}{\text{hr-ft}^2\text{-°F}}\right)} + 60°\text{F}$$

$$= 158.7°\text{F}$$

Perform one more iteration with $T_{\text{wire}} = 158°\text{F}$.

$$\text{Gr} = (0.05)^3(158°\text{F} - 60°\text{F})\left(1.76 \times 10^6 \ \frac{1}{\text{ft}^3\text{-°F}}\right)$$

$$= 21{,}560$$

$$\text{Pr}\,\text{Gr} = (0.72)(21{,}560) = 15{,}523$$

From Table 32.3,

$$h = (0.27) \left(\frac{T_{\text{wire}} - T_\infty}{d} \right)^{0.25}$$

$$= (0.27) \left(\frac{158°\text{F} - 60°\text{F}}{0.05 \text{ ft}} \right)^{0.25}$$

$$= 1.80 \text{ Btu/hr-ft}^2\text{-}°\text{F}$$

$$T_{\text{wire}} = \frac{\frac{Q}{L}}{\pi dh} + T_\infty$$

$$= \frac{27.3 \dfrac{\text{Btu}}{\text{hr-ft}}}{\pi (0.05 \text{ ft}) \left(1.80 \dfrac{\text{Btu}}{\text{hr-ft}^2\text{-}°\text{F}} \right)} + 60°\text{F}$$

$$= \boxed{156.6°\text{F}}$$

There is no need to repeat iterations since the assumed temperature and the calculated temperature are about the same.

The answer is (D).

SI Solution

From App. 32.D, for 38°C film, the air properties are

$$\text{Pr} = 0.705$$

$$\frac{g\beta\rho^2}{\mu^2} = 1.12 \times 10^8 \frac{1}{\text{K·m}^3}$$

From Eq. 32.4, the Grashof number is

$$\text{Gr} = L^3 \Delta T \left(\frac{g\beta\rho^2}{\mu^2} \right)$$

The characteristic length is the wire diameter.

$$L = (1.5 \text{ cm}) \left(\frac{1 \text{ m}}{100 \text{ cm}} \right) = 0.015 \text{ m}$$

The temperature gradient is

$$\Delta T = T_s - T_\infty = T_{\text{wire}} - 15°\text{C}$$

$$\text{Gr} = (0.015 \text{ m})^3 (T_{\text{wire}} - 15°\text{C}) \left(1.12 \times 10^8 \frac{1}{\text{K·m}^3} \right)$$

Assume $T_{\text{wire}} = 70°\text{C}$.

$$\text{Gr} = (0.015 \text{ m})^3 (70°\text{C} - 15°\text{C}) \left(1.12 \times 10^8 \frac{1}{\text{K·m}^3} \right)$$

$$= 20\,790$$

$$\text{Pr Gr} = (0.705)(20\,790) = 14\,657$$

From Table 32.3,

$$h \approx (1.32) \left(\frac{T_{\text{wire}} - T_\infty}{d} \right)^{0.25}$$

$$= (1.32) \left(\frac{70°\text{C} - 15°\text{C}}{0.015 \text{ m}} \right)^{0.25}$$

$$= 10.27 \text{ W/m}^2\text{·K}$$

From the customary U.S. solution,

$$T_{\text{wire}} = \frac{\frac{Q}{L}}{\pi dh} + T_\infty$$

$$= \frac{25 \dfrac{\text{W}}{\text{m}}}{\pi (0.015 \text{ m}) \left(10.27 \dfrac{\text{W}}{\text{m}^2\text{·K}} \right)} + 15°\text{C}$$

$$= \boxed{66.7°\text{C}}$$

There is no need to perform another iteration since the assumed temperature and the calculated temperature are about the same.

The answer is (D).

33 Heat Transfer: Forced Convection

PRACTICE PROBLEMS

Logarithmic Mean Temperature Difference

1. Water is heated from 55°F to 87°F (15°C to 30°C) by stack gases that are cooled from 350°F to 270°F (175°C to 130°C). What is the logarithmic mean temperature difference?

- (A) 190°F (105°C)
- (B) 210°F (117°C)
- (C) 235°F (130°C)
- (D) 270°F (150°C)

2. A fluid in a tank is maintained at 85°F (30°C) by an immersed hot water coil. The hot water enters at 190°F (90°C) and leaves at 160°F (70°C). What is the logarithmic mean temperature difference?

- (A) 89°F (49°C)
- (B) 96°F (53°C)
- (C) 111°F (62°C)
- (D) 127°F (71°C)

Film Temperature

3. What film temperature should be used for the heated fluid in Prob. 2?

- (A) 110°F (43°C)
- (B) 130°F (55°C)
- (C) 140°F (60°C)
- (D) 150°F (66°C)

4. If the fluid in Prob. 2 is gradually heated from 85°F (30°C) to 110°F (45°C), what initial film temperature should be used?

- (A) 110°F (43°C)
- (B) 130°F (55°C)
- (C) 140°F (60°C)
- (D) 150°F (66°C)

Reynolds Number

5. Light No. 10 oil is heated from 95°F to 105°F (35°C to 41°C) in a 0.6 in (1.52 cm) inside diameter tube. The average velocity of the oil is 2.0 ft/sec (0.6 m/s). The viscosity of the oil at 100°F (38°C) is 45 centistokes. What is the Reynolds number?

- (A) 200
- (B) 2500
- (C) 4600
- (D) 19,000

Convective Heat Transfer

6. A white, uninsulated, rectangular duct passes through a 50 ft (15 m) wide room. The duct is 18 in (45 cm) wide and 12 in (30 cm) high. The room and its contents are at 70°F (21°C). Air at 100°F (40°C) enters the duct flowing at 800 ft/min (4.0 m/s). The combined convection and radiation film coefficient is 2.0 Btu/hr-ft^2-°F (11 W/m^2·K).

(a) What is the heat transfer to the room?

- (A) 4300 Btu/hr (1.4 kW)
- (B) 5800 Btu/hr (1.9 kW)
- (C) 7400 Btu/hr (2.4 kW)
- (D) 9100 Btu/hr (2.9 kW)

(b) What is the temperature of the air after it has traveled in the duct the full 50 ft (15 m)?

- (A) 85°F (29°C)
- (B) 88°F (31°C)
- (C) 91°F (33°C)
- (D) 94°F (36°C)

(c) What is the pressure drop in in (cm) of water due to friction in the duct?

- (A) 0.020 in wg (4.6 Pa)
- (B) 0.028 in wg (7.5 Pa)
- (C) 0.041 in wg (9.4 Pa)
- (D) 0.055 in wg (13 Pa)

Laminar Flow in Tubes

7. A steel pipe carrying 350°F (175°C) air is 100 ft (30 m) long. The outside and inside diameters are 4.00 in and 3.50 in (10 cm and 9.0 cm), respectively. The pipe is covered with 2.0 in (5.0 cm) of insulation with a thermal conductivity of 0.05 Btu-ft/hr-ft^2-°F (0.086 W/m·K). The pipe passes through a 50°F (10°C) basement. Flow is laminar and fully developed. What is the heat loss?

- (A) 3100 Btu/hr (0.93 kW)
- (B) 3500 Btu/hr (1.1 kW)
- (C) 4100 Btu/hr (1.2 kW)
- (D) 8700 Btu/hr (2.0 kW)

Turbulent Flow in Tubes

8. An uninsulated horizontal pipe with 4.00 in (10 cm) outside diameter carries saturated 300 psia (2.1 MPa) steam through a 70°F (21°C) room. The steam flow rate is 5000 lbm/hr (0.63 kg/s). What decrease in quality will occur in the first 50 ft (15 m) of length?

(A) 1.6%

(B) 2.2%

(C) 2.8%

(D) 4.3%

Heat Exchangers

9. A tubular feedwater heater is being designed to heat 2940 lbm/hr (0.368 kg/s) of water from 70°F (21°C) to 190°F (90°C). Saturated steam at 134 psia (923 kPa) is condensing on the outside of the tubes. The tubes are a copper alloy. Each tube has a 1 in (2.54 cm) outside diameter and a 0.9 in (2.29 cm) inside diameter. The water velocity inside the tubes is 3 ft/sec (0.9 m/s). What outside tube surface area is required?

(A) 2.9 ft^2 (0.28 m^2)

(B) 3.5 ft^2 (0.34 m^2)

(C) 10 ft^2 (0.97 m^2)

(D) 18 ft^2 (1.7 m^2)

10. A surface feedwater heater with one shell pass and two tube passes is being designed to heat 500,000 lbm/hr (60 kg/s) of water from 200°F to 390°F (100°C to 200°C). The water flows at 5 ft/sec (1.5 m/s) through the tubes. Dry, saturated steam at 400°F (205°C) is to be used as the heating medium. The heater is to operate straight condensing (i.e., the condensed steam will not be mixed with the heated water). Saturated water at 400°F (205°C) is removed from the heater. The tubes in the heater are 7/8 in (2.2 cm) outside diameter with 1/16 in (1.6 mm) walls. The overall heat transfer coefficient is estimated as 700 Btu/hr-ft^2-°F.

(a) How many tubes are required?

(A) 80

(B) 110

(C) 140

(D) 170

(b) What should be the tube length?

(A) 30 ft (9 m)

(B) 40 ft (12 m)

(C) 60 ft (17 m)

(D) 80 ft (24 m)

11. A single-pass heat exchanger is tested in a clean condition and is found to heat 100 gal/min (6.3 L/s) of 70°F (21°C) water to 140°F (60°C). The hot side uses 230°F (110°C) steam. The tube's inner surface area is 50 ft^2 (4.7 m^2). After being used in the field for several months, the exchanger heats 100 gal/min (6.3 L/s) of 70°F (21°C) water to 122°F (50°C). What is the fouling factor?

(A) 0.00044 hr-ft^2-°F/Btu (0.000075 m^2·K/W)

(B) 0.00081 hr-ft^2-°F/Btu (0.00014 m^2·K/W)

(C) 0.0011 hr-ft^2-°F/Btu (0.00019 m^2·K/W)

(D) 0.0023 hr-ft^2-°F/Btu (0.00039 m^2·K/W)

Tubes in Crossflow

12. A glass thermometer has an outside diameter of 0.35 in (8.9 mm). Its uniform temperature is 100°F (38°C). The thermometer is inserted perpendicularly into a 100 ft/sec (30 m/s) airflow. The air temperature is 150°F (66°C). What is the film coefficient on the outside of the thermometer?

(A) 36 Btu/hr-ft^2-°F (210 W/m^2·K)

(B) 45 Btu/hr-ft^2-°F (260 W/m^2·K)

(C) 66 Btu/hr-ft^2-°F (380 W/m^2·K)

(D) 91 Btu/hr-ft^2-°F (530 W/m^2·K)

SOLUTIONS

1. *Customary U.S. Solution*

The logarithmic mean temperature difference will be different for different types of flow.

Parallel flow:

55°F (15°C) ────────→ 87°F (30°C)
    A   B
350°F (175°C) ──────→ 270°F (130°C)

$$\Delta T_A = 350°F - 55°F = 295°F$$
$$\Delta T_B = 270°F - 87°F = 183°F$$

From Eq. 33.67, the logarithmic mean temperature difference is

$$\Delta T_{lm} = \frac{\Delta T_A - \Delta T_B}{\ln\left(\frac{\Delta T_A}{\Delta T_B}\right)} = \frac{295°F - 183°F}{\ln\left(\frac{295°F}{183°F}\right)} = \boxed{234.6°F}$$

Counterflow:

55°F (15°C) ────────→ 87°F (30°C)
    A   B
270°F (130°C) ←────── 350°F (175°C)

$$\Delta T_A = 270°F - 55°F = 215°F$$
$$\Delta T_B = 350°F - 87°F = 263°F$$

From Eq. 33.67, the logarithmic mean temperature difference is

$$\Delta T_{lm} = \frac{215°F - 263°F}{\ln\left(\frac{215°F}{263°F}\right)} = \boxed{238.2°F}$$

The answer is (C).

SI Solution

Parallel flow:

$$\Delta T_A = 175°C - 15°C = 160°C$$
$$\Delta T_B = 130°C - 30°C = 100°C$$

From Eq. 33.67, the logarithmic mean temperature difference is

$$\Delta T_{lm} = \frac{160°C - 100°C}{\ln\left(\frac{160°C}{100°C}\right)} = \boxed{127.7°C}$$

Counterflow:

$$\Delta T_A = 130°C - 15°C = 115°C$$
$$\Delta T_B = 175°C - 30°C = 145°C$$

From Eq. 33.67, the logarithmic mean temperature difference is

$$\Delta T_{lm} = \frac{115°C - 145°C}{\ln\left(\frac{115°C}{145°C}\right)} = \boxed{129.4°C}$$

The answer is (C).

2. *Customary U.S. Solution*

$$\Delta T_A = 190°F - 85°F = 105°F$$
$$\Delta T_B = 160°F - 85°F = 75°F$$

From Eq. 33.67, the logarithmic mean temperature difference is

$$\Delta T_{lm} = \frac{\Delta T_A - \Delta T_B}{\ln\left(\frac{\Delta T_A}{\Delta T_B}\right)} = \frac{105°F - 75°F}{\ln\left(\frac{105°F}{75°F}\right)} = \boxed{89.2°F}$$

The answer is (A).

SI Solution

$$\Delta T_A = 90°C - 30°C = 60°C$$
$$\Delta T_B = 70°C - 30°C = 40°C$$

From Eq. 33.67, the logarithmic mean temperature difference is

$$\Delta T_{lm} = \frac{60°C - 40°C}{\ln\left(\frac{60°C}{40°C}\right)} = \boxed{49.3°C}$$

The answer is (A).

3. *Customary U.S. Solution*

$$T_{tank} = 85°F$$
$$T_{coil} = \left(\tfrac{1}{2}\right)(190°F + 160°F) = 175°F$$

The film temperature is

$$T_f = \left(\tfrac{1}{2}\right)(T_{tank} + T_{coil})$$
$$= \left(\tfrac{1}{2}\right)(85°F + 175°F) = \boxed{130°F}$$

The answer is (B).

SI Solution

$$T_{\text{tank}} = 30°C$$
$$T_{\text{coil}} = \left(\tfrac{1}{2}\right)(90°C + 70°C) = 80°C$$

The film temperature is

$$T_f = \left(\tfrac{1}{2}\right)(T_{\text{tank}} + T_{\text{coil}})$$
$$= \left(\tfrac{1}{2}\right)(30°C + 80°C) = \boxed{55°C}$$

The answer is (B).

4. *Customary U.S. Solution*

$$T_{\text{coil,initial}} = \left(\tfrac{1}{2}\right)(190°F + 160°F) = 175°F$$

The initial film temperature is

$$T_{f,\text{initial}} = \left(\tfrac{1}{2}\right)(T_{\text{tank,initial}} + T_{\text{coil,initial}})$$
$$= \left(\tfrac{1}{2}\right)(85°F + 175°F) = \boxed{130°F}$$

The answer is (B).

SI Solution

$$T_{\text{coil,initial}} = \left(\tfrac{1}{2}\right)(90°C + 70°C) = 80°C$$

The initial film temperature is

$$T_{f,\text{initial}} = \left(\tfrac{1}{2}\right)(T_{\text{tank,initial}} + T_{\text{coil,initial}})$$
$$= \left(\tfrac{1}{2}\right)(30°C + 80°C) = \boxed{55°C}$$

The answer is (B).

5. *Customary U.S. Solution*

The Reynolds number is

$$\text{Re}_d = \frac{vD}{\nu}$$

$$D = (0.6 \text{ in})\left(\frac{1 \text{ ft}}{12 \text{ in}}\right) = 0.05 \text{ ft}$$

$$\nu = (45 \text{ cS})\left(\frac{1 \text{ S}}{100 \text{ cS}}\right)\left(\frac{1 \text{ ft}^2}{929 \text{ sec-stoke}}\right)$$
$$= 4.84 \times 10^{-4} \text{ ft}^2/\text{sec}$$

$$v = 2 \text{ ft/sec}$$

$$\text{Re} = \frac{\left(2 \frac{\text{ft}}{\text{sec}}\right)(0.05 \text{ ft})}{4.84 \times 10^{-4} \frac{\text{ft}^2}{\text{sec}}} = \boxed{206.6}$$

The answer is (A).

SI Solution

The Reynolds number is

$$\text{Re}_d = \frac{vD}{\nu}$$

$$D = (1.52 \text{ cm})\left(\frac{1 \text{ m}}{100 \text{ cm}}\right) = 0.0152 \text{ m}$$

$$v = 0.6 \text{ m/s}$$

$$\nu = (45 \text{ cS})\left(1 \frac{\frac{\mu\text{m}^2}{\text{s}}}{\text{cS·s}}\right)\left(\frac{1 \text{ m}^2}{10^6 \text{ }\mu\text{m}^2}\right)$$
$$= 45 \times 10^{-6} \text{ m}^2/\text{s}$$

$$\text{Re}_d = \frac{\left(0.6 \frac{\text{m}}{\text{s}}\right)(0.0152 \text{ m})}{45 \times 10^{-6} \frac{\text{m}^2}{\text{s}}} = \boxed{202.7}$$

The answer is (A).

6. *Customary U.S. Solution*

(a) The exposed duct area is

$$A = (2W + 2H)L$$
$$= ((2)(18 \text{ in}) + (2)(12 \text{ in}))\left(\frac{1 \text{ ft}}{12 \text{ in}}\right)(50 \text{ ft})$$
$$= 250 \text{ ft}^2$$

The duct is noncircular; therefore, the hydraulic diameter of the duct will be used. From Eq. 33.48, the hydraulic diameter is

$$d_H = (4)\left(\frac{\text{area in flow}}{\text{wetted perimeter}}\right) = (4)\left(\frac{WH}{(2)(W+H)}\right)$$
$$= (4)\left(\frac{(18 \text{ in})(12 \text{ in})}{(2)(18 \text{ in} + 12 \text{ in})}\right)\left(\frac{1 \text{ ft}}{12 \text{ in}}\right)$$
$$= 1.2 \text{ ft}$$

From App. 32.C for air at 100°F,

$$\nu = 18.0 \times 10^{-5} \text{ ft}^2/\text{sec}$$
$$\rho = 0.0710 \text{ lbm/ft}^3$$

The Reynolds number is

$$\text{Re} = \frac{vD}{\nu} = \frac{\left(800 \frac{\text{ft}}{\text{min}}\right)(1.2 \text{ ft})\left(\frac{1 \text{ min}}{60 \text{ sec}}\right)}{18.0 \times 10^{-5} \frac{\text{ft}^2}{\text{sec}}}$$
$$= 8.90 \times 10^4$$

This is a turbulent flow. From Eq. 33.39,

$$h_i \approx (0.00351 + 0.000001583 T_{°F}) \left(\frac{(G_{\text{lbm/hr-ft}^2})^{0.8}}{(d_{\text{ft}})^{0.2}} \right)$$

$$G = \rho v = \left(0.0710 \, \frac{\text{lbm}}{\text{ft}^3} \right) \left(800 \, \frac{\text{ft}}{\text{min}} \right) \left(60 \, \frac{\text{min}}{\text{hr}} \right)$$

$$= 3408.0 \, \text{lbm/hr-ft}^2$$

$$h_i = (0.00351 + (0.000001583)(100°F))$$

$$\times \left(\frac{\left(3408.0 \, \frac{\text{lbm}}{\text{hr-ft}^2} \right)^{0.8}}{(1.2 \, \text{ft})^{0.2}} \right)$$

$$= 2.37 \, \text{Btu/hr-ft}^2\text{-}°F$$

Disregarding the duct thermal resistance, the overall heat transfer coefficient (from Eq. 33.72) is

$$\frac{1}{U} \approx \frac{1}{h_i} + \frac{1}{h_o}$$

$$= \frac{1}{2.37 \, \frac{\text{Btu}}{\text{hr-ft}^2\text{-}°F}} + \frac{1}{2.0 \, \frac{\text{Btu}}{\text{hr-ft}^2\text{-}°F}}$$

$$= 0.922 \, \text{hr-ft}^2\text{-}°F/\text{Btu}$$

$$U = 1.08 \, \text{Btu/hr-ft}^2\text{-}°F$$

The heat transfer to the room is

$$Q = UA(T_{\text{ave}} - T_\infty) = UA \left(\left(\tfrac{1}{2} \right) (T_{\text{in}} + T_{\text{out}}) - T_\infty \right)$$

Since T_{out} is unknown, an iteration procedure may be required. Assume $T_{\text{out}} \approx 95°F$.

$$Q = \left(1.08 \, \frac{\text{Btu}}{\text{hr-ft}^2\text{-}°F} \right) (250 \, \text{ft}^2)$$

$$\times \left(\left(\tfrac{1}{2} \right) (100°F + 95°F) - 70°F \right)$$

$$= \boxed{7425 \, \text{Btu/hr}}$$

The answer is (C).

Notice that ΔT (not ΔT_{lm}) is used in accordance with standard conventions in the HVAC industry.

(b) Temperature T_{out} can be verified by using

$$Q = \dot{m} c_p (T_{\text{in}} - T_{\text{out}})$$

$$\dot{m} = GA_{\text{flow}} = GWH$$

$$= \left(3408.0 \, \frac{\text{lbm}}{\text{hr-ft}^2} \right) (18 \, \text{in})(12 \, \text{in})$$

$$\times \left(\frac{1 \, \text{ft}^2}{144 \, \text{in}^2} \right)$$

$$= 5112 \, \text{lbm/hr}$$

$$c_p = 0.240 \, \text{Btu/lbm-}°F$$

[from App. 32.C at 100°F]

$$7425 \, \frac{\text{Btu}}{\text{hr}} = \left(5112 \, \frac{\text{lbm}}{\text{hr}} \right) \left(0.240 \, \frac{\text{Btu}}{\text{lbm-}°F} \right)$$

$$\times (100°F - T_{\text{out}})$$

$$T_{\text{out}} = \boxed{94°F} \quad \text{[close enough]}$$

The answer is (D).

(c) Assume clean galvanized ductwork with $\epsilon = 0.0005$ ft and about 25 joints per 100 ft. The equivalent diameter of a rectangular duct can be estimated from

$$D_e = \frac{(1.3)(\text{short side} \times \text{long side})^{\frac{5}{8}}}{(\text{short side} + \text{long side})^{\frac{1}{4}}}$$

$$= \frac{(1.3)\left((12 \, \text{in})(18 \, \text{in})\right)^{\frac{5}{8}}}{(12 \, \text{in} + 18 \, \text{in})^{\frac{1}{4}}} = 16 \, \text{in}$$

The flow rate is

$$\dot{V} = vA = \frac{\left(800 \, \frac{\text{ft}}{\text{min}} \right) (12 \, \text{in})(18 \, \text{in})}{144 \, \frac{\text{in}^2}{\text{ft}^2}}$$

$$= 1200 \, \text{cfm}$$

From App. 33.E, the friction loss is 0.056 in wg/100 ft. Therefore,

$$\Delta p = (0.056 \, \text{in wg}) \left(\frac{50 \, \text{ft}}{100 \, \text{ft}} \right) = \boxed{0.028 \, \text{in wg}}$$

The answer is (B).

(Notice that the flow rate and not the flow velocity must be used with D_e in App. 33.E.)

SI Solution

(a) The exposed duct area is

$$A = (2W + 2H)L$$

$$= \left((2)(45 \, \text{cm}) + (2)(30 \, \text{cm})\right) \left(\frac{1 \, \text{m}}{100 \, \text{cm}} \right) (15 \, \text{m})$$

$$= 22.5 \, \text{m}^2$$

The duct is noncircular; therefore, the hydraulic diameter of the duct will be used. From Eq. 33.48, the hydraulic diameter is

$$d_H = (4) \left(\frac{\text{area in flow}}{\text{wetted perimeter}} \right)$$

$$= (4) \left(\frac{WH}{(2)(W+H)} \right)$$

$$= (4) \left(\frac{(45 \text{ cm})(30 \text{ cm})}{(2)(45 \text{ cm} + 30 \text{ cm})} \right) \left(\frac{1 \text{ m}}{100 \text{ cm}} \right)$$

$$= 0.36 \text{ m}$$

From App. 32.D, for air at 40°C,

$$\mu = 1.91 \times 10^{-5} \text{ kg/m·s}$$

$$\rho = 1.130 \text{ kg/m}^3$$

$$c_p = 1.0051 \text{ kJ/kg·K}$$

$$k = 0.02718 \text{ W/m·K}$$

The Reynolds number is

$$\text{Re} = \frac{\rho v D}{\mu} = \frac{\left(1.130 \ \frac{\text{kg}}{\text{m}^3}\right) \left(4.0 \ \frac{\text{m}}{\text{s}}\right) (0.36 \text{ m})}{1.91 \times 10^{-5} \ \frac{\text{kg}}{\text{m·s}}}$$

$$= 8.52 \times 10^4$$

This is a turbulent flow. From Eq. 33.34, the Nusselt number is

$$\text{Nu} = (0.023)(\text{Re}^{0.8})$$

The film coefficient is

$$h = (0.023)(\text{Re}^{0.8}) \left(\frac{k}{d} \right)$$

$$= (0.023)(8.52 \times 10^4)^{0.8} \left(\frac{0.02718 \ \frac{\text{W}}{\text{m·K}}}{0.36 \text{ m}} \right)$$

$$= 15.28 \text{ W/m}^2\text{·K}$$

Disregarding the duct thermal resistance, the overall heat transfer coefficient from Eq. 33.72 is

$$\frac{1}{U} = \frac{1}{h_i} + \frac{1}{h_o}$$

$$= \frac{1}{11 \ \frac{\text{W}}{\text{m}^2\text{·K}}} + \frac{1}{15.28 \ \frac{\text{W}}{\text{m}^2\text{·K}}} = 0.1564 \text{ m}^2\text{·K/W}$$

$$U = 6.39 \text{ W/m}^2\text{·K}$$

The heat transfer to the room is

$$Q = UA(T_{\text{ave}} - T_\infty) = UA \left(\left(\tfrac{1}{2}\right)(T_{\text{in}} + T_{\text{out}}) - T_\infty \right)$$

Since T_{out} is unknown, an iterative procedure may be required. Assume $T_{\text{out}} = 36$°C.

$$Q = \left(6.39 \ \frac{\text{W}}{\text{m}^2\text{·K}}\right) (22.5 \text{ m}^2)$$

$$\times \left(\left(\tfrac{1}{2}\right)(40°\text{C} + 36°\text{C}) - 21°\text{C} \right)$$

$$= \boxed{2444.2 \text{ W}}$$

The answer is (C).

Notice that ΔT (not ΔT_{lm}) is used in accordance with standard conventions in the HVAC industry.

(b) Verify temperature T_{out}.

$$Q = \dot{m} c_p (T_{\text{in}} - T_{\text{out}})$$

$$\dot{m} = \rho A_{\text{flow}} v$$

$$= \left(1.130 \ \frac{\text{kg}}{\text{m}^3}\right) (45 \text{ cm})(30 \text{ cm}) \left(\frac{1 \text{ m}}{100 \text{ cm}} \right)^2$$

$$\times \left(4 \ \frac{\text{m}}{\text{s}}\right)$$

$$= 0.6102 \text{ kg/s}$$

$$2444.2 \text{ W} = \left(0.6102 \ \frac{\text{kg}}{\text{s}}\right) \left(1.0051 \ \frac{\text{kJ}}{\text{kg·K}}\right) \left(1000 \ \frac{\text{J}}{\text{kJ}}\right)$$

$$\times (40°\text{C} - T_{\text{out}})$$

$$T_{\text{out}} = \boxed{36°\text{C}} \quad \text{[same as assumed]}$$

The answer is (D).

(c) Assume clean galvanized ductwork with $\epsilon = 0.15$ mm and about 1 joint per meter. The equivalent diameter of a rectangular duct can be estimated from

$$D_e = \frac{(1.3)(\text{short side} \times \text{long side})^{\frac{5}{8}}}{(\text{short side} + \text{long side})^{\frac{1}{4}}}$$

$$= \left(\frac{(1.3)((30 \text{ cm})(45 \text{ cm}))^{\frac{5}{8}}}{(30 \text{ cm} + 45 \text{ cm})^{\frac{1}{4}}} \right) \left(\frac{10 \text{ mm}}{1 \text{ cm}} \right)$$

$$= 400 \text{ mm}$$

The flow rate is

$$\dot{V} = vA = \left(4 \ \frac{\text{m}}{\text{s}}\right)(0.45 \text{ m})(0.30 \text{ m}) \left(1000 \ \frac{\text{L}}{\text{m}^3} \right)$$

$$= 540 \text{ L/s}$$

From App. 33.E, the friction loss is 0.5 Pa/m. For 15 m,

$$\Delta p = \left(0.5 \ \frac{\text{Pa}}{\text{m}}\right)(15 \text{ m}) = \boxed{7.5 \text{ Pa}}$$

The answer is (B).

(Notice that the flow rate and not the flow velocity must be used with D_e in App. 33.F.)

7. *Customary U.S. Solution*

Refer to Fig. 31.3. The corresponding radii are

$$r_a = \frac{d_i}{2} = \frac{(3.5 \text{ in})\left(\dfrac{1 \text{ ft}}{12 \text{ in}}\right)}{2} = 0.1458 \text{ ft}$$

$$r_b = \frac{d_o}{2} = \frac{(4 \text{ in})\left(\dfrac{1 \text{ ft}}{12 \text{ in}}\right)}{2} = 0.1667 \text{ ft}$$

$$r_c = r_b + t_{\text{insulation}} = 0.1667 \text{ ft} + (2 \text{ in})\left(\frac{1 \text{ ft}}{12 \text{ in}}\right)$$

$$= 0.3334 \text{ ft}$$

From App. 31.B, for steel at 350°F, $k_{\text{pipe}} \approx 25.6$ Btu-ft/hr-ft²-°F.

Initially assume a typical value of $h_c = 1.5$ Btu/hr-ft²-°F.

For fully developed laminar flow from Eq. 33.28, $\text{Nu}_d = 3.658$.

From App. 32.C, for air at 350°F,

$$k_{\text{air}} \approx 0.0203 \text{ Btu/hr-ft-°F}$$

$$\text{Nu}_d = \frac{h_a d_i}{k_{\text{air}}} = 3.658$$

$$h_a = \frac{(3.658) k_{\text{air}}}{d_i} = \frac{(3.658)\left(0.0203 \ \dfrac{\text{Btu}}{\text{hr-ft-°F}}\right)}{(3.5 \text{ in})\left(\dfrac{1 \text{ ft}}{12 \text{ in}}\right)}$$

$$= 0.255 \text{ Btu/hr-ft}^2\text{-°F}$$

Neglect thermal resistance between pipe and insulation. From Eq. 31.34, the heat transfer is

$$Q = \frac{2\pi L (T_i - T_\infty)}{\dfrac{1}{r_a h_a} + \dfrac{\ln\left(\dfrac{r_b}{r_a}\right)}{k_{\text{pipe}}} + \dfrac{\ln\left(\dfrac{r_c}{r_b}\right)}{k_{\text{insulation}}} + \dfrac{1}{r_c h_c}}$$

$$= \frac{2\pi(100 \text{ ft})(350°\text{F} - 50°\text{F})}{\dfrac{1}{(0.1458 \text{ ft})\left(0.255 \ \dfrac{\text{Btu}}{\text{hr-ft}^2\text{-°F}}\right)} + \dfrac{\ln\left(\dfrac{0.1667 \text{ ft}}{0.1458 \text{ ft}}\right)}{25.6 \ \dfrac{\text{Btu-ft}}{\text{hr-ft}^2\text{-°F}}}}$$

$$\quad + \dfrac{\ln\left(\dfrac{0.3334 \text{ ft}}{0.1667 \text{ ft}}\right)}{0.05 \ \dfrac{\text{Btu-ft}}{\text{hr-ft}^2\text{-°F}}} + \dfrac{1}{(0.3334 \text{ ft})\left(1.5 \ \dfrac{\text{Btu}}{\text{hr-ft}^2\text{-°F}}\right)}$$

$$= \frac{188{,}496 \text{ ft-°F}}{26.90 \ \dfrac{\text{hr-ft-°F}}{\text{Btu}} + 0.00523 \ \dfrac{\text{hr-ft-°F}}{\text{Btu}}}$$

$$\quad + 13.863 \ \dfrac{\text{hr-ft-°F}}{\text{Btu}} + 2.00 \ \dfrac{\text{hr-ft-°F}}{\text{Btu}}$$

$$= 4407 \text{ Btu/hr}$$

Using Eq. 31.34 to find T_2, use all resistances except the outer $(T_i - T_2)$ resistance.

$$\left(\frac{4407 \ \dfrac{\text{Btu}}{\text{hr}}}{2\pi(100 \text{ ft})}\right)\left(26.9 \ \frac{\text{hr-ft-°F}}{\text{Btu}}\right.$$

$$\left. + 0.00523 \ \frac{\text{hr-ft-°F}}{\text{Btu}} + 13.863 \ \frac{\text{hr-ft-°F}}{\text{Btu}}\right)$$

$$= 285.9°\text{F}$$

$$T_2 = T_i - 285.9°\text{F} = 350°\text{F} - 285.9°\text{F} = 64.1°\text{F}$$

To evaluate h_c, use film temperature.

$$T_{\text{film}} = \left(\tfrac{1}{2}\right)(T_2 + T_\infty) = \left(\tfrac{1}{2}\right)(64.1°\text{F} + 50°\text{F})$$

$$= 57°\text{F}$$

From App. 32.C air at 57°F,

$$\text{Pr} = 0.72$$

$$\frac{g\beta\rho^2}{\mu^2} = 2.645 \times 10^6 \ \frac{1}{\text{ft}^3\text{-°F}}$$

From Eq. 32.4, the Grashof number is

$$\text{Gr} = \frac{L^3 g \beta \rho^2 (T_2 - T_\infty)}{\mu^2}$$

For pipe,

$$L = d_c = 2r_c = (2)(0.3334 \text{ ft}) = 0.6668 \text{ ft}$$

$$\text{Gr} = (0.6668 \text{ ft})^3 \left(2.645 \times 10^6 \, \frac{1}{\text{ft}^3\text{-}^\circ\text{F}}\right)$$
$$\times (64.1^\circ\text{F} - 50^\circ\text{F})$$
$$= 1.1 \times 10^7$$
$$\text{Gr Pr} = (1.1 \times 10^7)(0.72)$$
$$= 7.9 \times 10^6$$

From Table 32.3,

$$h_c \approx (0.27)\left(\frac{T_2 - T_\infty}{d_c}\right)^{\frac{1}{4}}$$
$$= (0.27)\left(\frac{64.1^\circ\text{F} - 50^\circ\text{F}}{0.6668 \text{ ft}}\right)^{\frac{1}{4}}$$
$$= 0.579 \text{ Btu/hr-ft}^2\text{-}^\circ\text{F}$$

At the second iteration, the heat transfer is

$$Q = \frac{188{,}496 \text{ ft-}^\circ\text{F}}{26.90 \, \frac{\text{hr-ft-}^\circ\text{F}}{\text{Btu}} + 0.00523 \, \frac{\text{hr-ft-}^\circ\text{F}}{\text{Btu}}}$$
$$+ 13.863 \, \frac{\text{hr-ft-}^\circ\text{F}}{\text{Btu}}$$
$$+ \frac{1}{(0.3334 \text{ ft})\left(0.579 \, \frac{\text{Btu}}{\text{hr-ft-}^\circ\text{F}}\right)}$$

$$= \boxed{4102 \text{ Btu/hr}}$$

The answer is (C).

Additional iterations will improve the accuracy further.

SI Solution

Refer to Fig. 31.3. The corresponding radii are

$$r_a = \frac{d_i}{2} = \frac{(9.0 \text{ cm})\left(\dfrac{1 \text{ m}}{100 \text{ cm}}\right)}{2} = 0.045 \text{ m}$$

$$r_b = \frac{d_o}{2} = \frac{(10 \text{ cm})\left(\dfrac{1 \text{ m}}{100 \text{ cm}}\right)}{2} = 0.050 \text{ m}$$

$$r_c = r_b + t_{\text{insulation}} = 0.050 \text{ m} + (5.0 \text{ cm})\left(\frac{1 \text{ m}}{100 \text{ cm}}\right)$$
$$= 0.100 \text{ m}$$

From App. 31.B and the table footnote, for steel at 175°C (~347°F),

$$k_{\text{pipe}} \approx \left(25.6 \, \frac{\text{Btu}}{\text{hr-ft-}^\circ\text{F}}\right)\left(1.7307 \, \frac{\text{W-hr-ft-}^\circ\text{F}}{\text{m-K-Btu}}\right)$$
$$= 44.31 \text{ W/m-K}$$

Initially assume a typical value of $h_c = 3.5$ W/m^2·K.

For fully developed laminar flow from Eq. 33.28,

$$\text{Nu}_d = \frac{h_a d_i}{k_{\text{air}}} = 3.658$$

From App. 32.D for air at 175°C,

$$k_{\text{air}} \approx 0.03709 \text{ W/m-K}$$

$$h_a = \frac{3.658 k_{\text{air}}}{d_i}$$
$$= \frac{(3.658)\left(0.03709 \, \dfrac{\text{W}}{\text{m-K}}\right)}{(9.0 \text{ cm})\left(\dfrac{1 \text{ m}}{100 \text{ cm}}\right)}$$
$$= 1.508 \text{ W/m}^2\text{-K}$$

Neglect thermal resistance between pipe and insulation. From Eq. 31.34, the heat transfer is

$$Q = \frac{2\pi L(T_i - T_\infty)}{\dfrac{1}{r_a h_a} + \dfrac{\ln\left(\dfrac{r_b}{r_a}\right)}{k_{\text{pipe}}} + \dfrac{\ln\left(\dfrac{r_c}{r_b}\right)}{k_{\text{insulation}}} + \dfrac{1}{r_c h_c}}$$

$$= \frac{2\pi(30 \text{ m})(175^\circ\text{C} - 10^\circ\text{C})}{\dfrac{1}{(0.045 \text{ m})\left(1.508 \, \dfrac{\text{W}}{\text{m}^2\text{-K}}\right)} + \dfrac{\ln\left(\dfrac{0.050 \text{ m}}{0.045 \text{ m}}\right)}{44.31 \, \dfrac{\text{W}}{\text{m-K}}}}$$

$$+ \dfrac{\ln\left(\dfrac{0.10 \text{ m}}{0.050 \text{ m}}\right)}{0.086 \, \dfrac{\text{W}}{\text{m-K}}} + \dfrac{1}{(0.10 \text{ m})\left(3.5 \, \dfrac{\text{W}}{\text{m}^2\text{-K}}\right)}$$

$$= \frac{31\,101.8 \text{ m-}^\circ\text{C}}{14.74 \, \dfrac{\text{m-K}}{\text{W}} + 0.00238 \, \dfrac{\text{m-K}}{\text{W}}}$$
$$+ 8.06 \, \frac{\text{m-K}}{\text{W}} + 2.86 \, \frac{\text{m-K}}{\text{W}}$$

$$= 1212 \text{ W}$$

Use Eq. 31.34 to find T_2 by using all resistances except the outer resistance.

$$T_i - T_2 = \left(\frac{1212 \text{ W}}{2\pi(30 \text{ m})}\right)$$
$$\times \left(14.74 \, \frac{\text{m-K}}{\text{W}} + 0.00238 \, \frac{\text{m-K}}{\text{W}} + 8.06 \, \frac{\text{m-K}}{\text{W}}\right)$$
$$= 146.6^\circ\text{C}$$
$$T_2 = T_i - 146.6^\circ\text{C} = 175^\circ\text{C} - 146.6^\circ\text{C}$$
$$= 28.4^\circ\text{C}$$

To evaluate h_c, use film temperature.

$$T_{\text{film}} = \left(\tfrac{1}{2}\right)(T_2 - T_\infty) = \left(\tfrac{1}{2}\right)(28.4°C + 10°C)$$
$$= 19.2°C$$

From App. 32.D for air at 19.2°C,

$$Pr = 0.710$$
$$\frac{g\beta\rho^2}{\mu^2} = 1.52 \times 10^8 \; \frac{1}{\text{K·m}^3}$$

From Eq. 32.4, the Grashof number is

$$Gr = \frac{L^3 g\beta\rho^2(T_2 - T_\infty)}{\mu^2}$$

For pipe,

$$L = d_c = 2r_c = (2)(0.10 \text{ m}) = 0.20 \text{ m}$$
$$Gr = (0.20 \text{ m})^3 \left(1.52 \times 10^8 \; \frac{1}{\text{K·m}^3}\right)(28.4°C - 10°C)$$
$$= 2.24 \times 10^7$$
$$Gr\, Pr = (2.24 \times 10^7)(0.710)$$
$$= 1.59 \times 10^7$$

From Table 32.3,

$$h_c \approx (1.37)\left(\frac{T_2 - T_\infty}{d_c}\right)^{\frac{1}{4}}$$
$$= (1.37)\left(\frac{28.4°C - 10°C}{0.20 \text{ m}}\right)^{\frac{1}{4}}$$
$$= 4.24 \text{ W/m}^2\text{·K} \quad \begin{bmatrix} \text{versus assumed value of} \\ 3.5 \text{ W/m}^2\text{·K} \end{bmatrix}$$

Further iteration will improve accuracy.

$$Q = \frac{31\,101.8 \text{ m·°C}}{14.74 \; \dfrac{\text{m·K}}{\text{W}} + 0.00238 \; \dfrac{\text{m·K}}{\text{W}}}$$
$$+ 8.06 \; \dfrac{\text{m·K}}{\text{W}} + \dfrac{1}{(0.10 \text{ m})\left(4.24 \; \dfrac{\text{W}}{\text{m}^2\text{·K}}\right)}$$

$$= \boxed{1236 \text{ W}}$$

The answer is (C).

8. *Customary U.S. Solution*

Neglecting pipe resistance (no information for pipe is given), $T_{\text{pipe}} = T_{\text{sat}}$.

From App. 23.B for 300 lbf/in^2 steam, $T_{\text{sat}} = 417.43°F$. When a vapor condenses, the vapor and condensed liquid are at the same temperature. Therefore, the entire

pipe is assumed to be at 417.43°F. The outside film coefficient should be evaluated from Eq. 33.11.

$$T_{\text{film}} = \left(\tfrac{1}{2}\right)(T_s + T_\infty)$$
$$= \left(\tfrac{1}{2}\right)(417.43°F + 70°F)$$
$$= 243.7°F$$

From App. 32.C for air at 243.7°F,

$$Pr = 0.715$$
$$\frac{g\beta\rho^2}{\mu^2} = 0.673 \times 10^6 \; \frac{1}{\text{ft}^3\text{-°F}}$$

From Eq. 32.4, the Grashof number is

$$Gr = \frac{L^3 g\beta\rho^2(T_s - T_\infty)}{\mu^2}$$
$$L = d_{\text{outside}} = (4 \text{ in})\left(\frac{1 \text{ ft}}{12 \text{ in}}\right) = 0.3333 \text{ ft}$$
$$Gr = (0.3333 \text{ ft})^3 \left(0.673 \times 10^6 \; \frac{1}{\text{ft}^3\text{-°F}}\right)$$
$$\times (417.43°F - 70°F)$$
$$= 8.66 \times 10^6$$
$$Gr\, Pr = (8.66 \times 10^6)(0.715)$$
$$= 6.19 \times 10^6$$

From Table 32.3,

$$h_c \approx (0.27)\left(\frac{T_s - T_\infty}{d_{\text{outside}}}\right)^{\frac{1}{4}}$$
$$= (0.27)\left(\frac{417.43°F - 70°F}{0.3333 \text{ ft}}\right)^{\frac{1}{4}}$$
$$= 1.53 \text{ Btu/hr-ft}^2\text{-°F}$$

From Eq. 32.1, the heat transfer for the first 50 ft due to convection is

$$Q = h_c A_{\text{outside}}(T_s - T_\infty)$$
$$= h_c(\pi d_{\text{outside}} L)(T_s - T_\infty)$$
$$= \left(1.53 \; \frac{\text{Btu}}{\text{hr-ft}^2\text{-°F}}\right)\pi(0.3333 \text{ ft})(50 \text{ ft})$$
$$\times (417.43°F - 70°F)$$
$$Q_{\text{convection}} = 27{,}830 \text{ Btu/hr}$$

To determine heat transfer due to radiation, assume oxidized steel pipe, completely enclosed.

$$F_a = 1$$

The absolute temperatures are

$$T_1 = 417.43°F + 460 = 877.43°R$$
$$T_2 = 70°F + 460 = 530°R$$
$$F_e = \epsilon_{\text{pipe}} = 0.80$$

The radiation heat transfer is

$$E_{net} = \sigma F_a F_e \left(T_1^4 - T_2^4\right)$$
$$= \left(0.1713 \times 10^{-8} \frac{Btu}{hr\text{-}ft^2\text{-}°R^4}\right)$$
$$\times (1)(0.80)\left((877.43°R)^4 - (530°R)^4\right)$$
$$= 704 \text{ Btu/hr-ft}^2$$

$$Q_{radiation} = E_{net}A = E_{net}(\pi d_{outside}L)$$
$$= \left(704 \frac{Btu}{hr\text{-}ft^2}\right)\pi(0.3333 \text{ ft})(50 \text{ ft})$$
$$= 36{,}858 \text{ Btu/hr}$$

The total heat loss is

$$Q_{total} = Q_{convection} + Q_{radiation}$$
$$= 27{,}830 \frac{Btu}{hr} + 36{,}858 \frac{Btu}{hr}$$
$$= 64{,}688 \text{ Btu/hr}$$
$$Q_{total} = \dot{m}\Delta h$$

The enthalpy decrease per pound is

$$\Delta h = \frac{Q_{total}}{\dot{m}_{steam}} = \frac{64{,}688 \dfrac{Btu}{hr}}{5000 \dfrac{lbm}{hr}} = 12.94 \text{ Btu/lbm}$$

This is a quality loss of

$$\Delta x = \frac{\Delta h}{h_{fg}} = \frac{12.94 \dfrac{Btu}{lbm}}{809.8 \dfrac{Btu}{lbm}}$$
$$= \boxed{0.0160 \quad (1.6\%)}$$

The answer is (A).

SI Solution

From the customary U.S. solution, T_{pipe} is the same for the entire length.

From App. 23.O for 2.1 MPa, $T_{sat} = 214.72°C$. The outside film coefficient should be evaluated from Eq. 33.11 as

$$T_{film} = \left(\tfrac{1}{2}\right)(T_s + T_\infty) = \left(\tfrac{1}{2}\right)(214.72°C + 21°C)$$
$$= 117.9°C$$

From App. 32.D for air at 117.9°C,

$$Pr = 0.692$$
$$\frac{g\beta\rho^2}{\mu^2} = 0.403 \times 10^8 \ \frac{1}{K\cdot m^3}$$

From Eq. 32.4, the Grashof number is

$$Gr = \frac{L^3 g\beta\rho^2 (T_s - T_\infty)}{\mu^2}$$
$$L = d_{outside} = (10 \text{ cm})\left(\frac{1 \text{ m}}{100 \text{ cm}}\right) = 0.10 \text{ m}$$
$$Gr = (0.10 \text{ m})^3 \left(0.403 \times 10^8 \ \frac{1}{K\cdot m^3}\right)$$
$$\times (214.72°C - 21°C)$$
$$= 7.807 \times 10^6$$
$$Gr\,Pr = (7.807 \times 10^6)(0.692)$$
$$= 5.40 \times 10^6$$

From Table 32.3,

$$h_c \approx (1.32)\left(\frac{T_s - T_\infty}{d_{outside}}\right)^{\frac{1}{4}}$$
$$= (1.32)\left(\frac{214.72°C - 21°C}{0.10 \text{ m}}\right)^{\frac{1}{4}}$$
$$= 8.76 \text{ W/m}^2\cdot\text{K}$$

From Eq. 32.1, the heat transfer for the first 15 m due to convection is

$$Q_{convection} = h_c(\pi d_{outside}L)(T_s - T_\infty)$$
$$= \left(8.76 \frac{W}{m^2\cdot K}\right)\pi(0.10 \text{ m})(15 \text{ m})$$
$$\times (214.72°C - 21°C)$$
$$= 7997 \text{ W}$$

To determine heat transfer due to radiation, assume oxidized steel pipe, completely enclosed.

$$F_a = 1$$

The absolute temperatures are

$$T_1 = 214.72°C + 273 = 487.72K$$
$$T_2 = 21°C + 273 = 294K$$
$$F_e = \epsilon_{pipe} = 0.80$$

The radiation heat transfer is

$$E_{net} = \sigma F_a F_e \left(T_1^4 - T_2^4\right)$$
$$= \left(5.67 \times 10^{-8} \frac{W}{m^2\cdot K^4}\right)$$
$$\times (1)(0.80)\left((487.72K)^4 - (294K)^4\right)$$
$$= 2228 \text{ W/m}^2$$

$$Q_{radiation} = E_{net}(\pi d_{outside}L)$$
$$= \left(2228 \frac{W}{m^2}\right)\pi(0.10 \text{ m})(15 \text{ m})$$
$$= 10\,499 \text{ W}$$

The total heat loss is

$$Q_{total} = Q_{convection} + Q_{radiation}$$
$$= 7997 \text{ W} + 10\,499 \text{ W}$$
$$= 18\,496 \text{ W}$$
$$Q_{total} = \dot{m}\Delta h$$

The enthalpy decrease per kilogram is

$$\Delta h = \frac{Q_{total}}{\dot{m}_{steam}} = \frac{(18\,496 \text{ W})\left(\frac{1 \text{ kJ}}{1000 \text{ J}}\right)}{0.63 \frac{\text{kg}}{\text{s}}}$$
$$= 29.36 \text{ kJ/kg}$$

This is a quality loss of

$$\Delta x = \frac{\Delta h}{h_{fg}} = \frac{29.36 \frac{\text{kJ}}{\text{kg}}}{1880.8 \frac{\text{kJ}}{\text{kg}}}$$

$$= \boxed{0.0156 \ (1.6\%)}$$

The answer is (A).

9. *Customary U.S. Solution*

The bulk temperature of the water is

$$T_b = \left(\tfrac{1}{2}\right)(T_{in} + T_{out})$$
$$= \left(\tfrac{1}{2}\right)(70°\text{F} + 190°\text{F}) = 130°\text{F}$$

From App. 32.A, the properties of water at 130°F are

$$c_p = 0.999 \text{ Btu/lbm-°F}$$
$$\nu = 0.582 \times 10^{-5} \text{ ft}^2/\text{sec}$$
$$\text{Pr} = 3.45$$
$$k = 0.376 \text{ Btu/hr-ft-°F}$$

The heat transfer is found from the temperature gain of the water.

$$Q = \dot{m}c_p\Delta T$$
$$= \left(2940 \frac{\text{lbm}}{\text{hr}}\right)\left(0.999 \frac{\text{Btu}}{\text{lbm-°F}}\right)(190°\text{F} - 70°\text{F})$$
$$= 352{,}447 \text{ Btu/hr}$$

The Reynolds number is

$$\text{Re} = \frac{\text{v}D}{\nu} = \frac{\left(3 \frac{\text{ft}}{\text{sec}}\right)(0.9 \text{ in})\left(\frac{1 \text{ ft}}{12 \text{ in}}\right)}{0.582 \times 10^{-5} \frac{\text{ft}^2}{\text{sec}}}$$
$$= 3.87 \times 10^4$$

From Eq. 33.33, the film coefficient is

$$h = (0.023)(\text{Re})^{0.8}(\text{Pr})^n \left(\frac{k}{d}\right)$$
$$= (0.023)(3.87 \times 10^4)^{0.8}(3.45)^{0.4}$$
$$\times \left(\frac{0.376 \frac{\text{Btu}}{\text{hr-ft-°F}}}{(0.9 \text{ in})\left(\frac{1 \text{ ft}}{12 \text{ in}}\right)}\right)$$
$$= 885 \text{ Btu/hr-ft}^2\text{-°F}$$

The saturation temperature for 134 psia steam is $\approx$ 350°F. Assume the wall is 20°F lower ($\approx$ 330°F). The film properties are evaluated at the average of the wall and saturation temperatures.

$$T_h = \left(\tfrac{1}{2}\right)(T_{sat,v} + T_s) = \left(\tfrac{1}{2}\right)(350°\text{F} + 330°\text{F})$$
$$= 340°\text{F}$$

Film properties are obtained from App. 32.A for liquid water and App. 23.A for vapor.

$$k_{340°\text{F}} = 0.392 \text{ Btu/hr-ft-°F}$$
$$\mu_{340°\text{F}} = \left(0.109 \times 10^{-3} \frac{\text{lbm}}{\text{sec-ft}}\right)\left(3600 \frac{\text{sec}}{\text{hr}}\right)$$
$$= 0.392 \text{ lbm/ft-hr}$$
$$\rho_{l,340°\text{F}} = \frac{1}{v_{f,340°\text{F}}} = \frac{1}{0.01787 \frac{\text{ft}^3}{\text{lbm}}}$$
$$= 55.96 \text{ lbm/ft}^3$$
$$\rho_{v,340°\text{F}} = \frac{1}{v_{g,340°\text{F}}} = \frac{1}{3.792 \frac{\text{ft}^3}{\text{lbm}}}$$
$$= 0.2637 \text{ lbm/ft}^3$$
$$h_{fg,134 \text{ psia}} = 871 \text{ Btu/lbm}$$
$$d = (0.9 \text{ in})\left(\frac{1 \text{ ft}}{12 \text{ in}}\right) = 0.075 \text{ ft}$$
$$g = \left(32.2 \frac{\text{ft}}{\text{sec}^2}\right)\left(3600 \frac{\text{sec}}{\text{hr}}\right)^2$$
$$= 4.17 \times 10^8 \text{ ft/hr}^2$$

From Eq. 32.31, the film coefficient is

$$h_o = (0.725) \left(\frac{\rho_l(\rho_l - \rho_v)gh_{fg}(k_l)^3}{d\mu_l(T_{\text{sat},v} - T_s)} \right)$$

$$= (0.725)$$

$$\times \left(\frac{\begin{array}{c} \left(55.96 \, \frac{\text{lbm}}{\text{ft}^3}\right)\left(55.96 \, \frac{\text{lbm}}{\text{ft}^3} - 0.2637 \, \frac{\text{lbm}}{\text{ft}^3}\right) \\ \times \left(4.17 \times 10^8 \, \frac{\text{ft}}{\text{hr}^2}\right)\left(871 \, \frac{\text{Btu}}{\text{lbm}}\right) \\ \times \left(0.392 \, \frac{\text{Btu}}{\text{hr-ft-}^\circ\text{F}}\right)^3 \end{array}}{(0.075 \, \text{ft})\left(0.392 \, \frac{\text{lbm}}{\text{ft-hr}}\right)(350^\circ\text{F} - 330^\circ\text{F})} \right)$$

$$= 2379 \, \text{Btu/hr-ft}^2\text{-}^\circ\text{F}$$

From App. 31.B, for copper alloy (70% Cu), at 330°F, $k = 62$ Btu-ft/hr-ft^2-°F.

From Eq. 33.69, the overall heat transfer coefficient based on the outside area is

$$\frac{1}{U_o} = \frac{1}{h_o} + \left(\frac{r_o}{k_{\text{tube}}} \right) \ln \left(\frac{r_o}{r_i} \right) + \frac{r_o}{r_i h_i}$$

$$r_o = \frac{d_o}{2} = \frac{(1 \, \text{in})\left(\frac{1 \, \text{ft}}{12 \, \text{in}} \right)}{2} = 0.0417 \, \text{ft}$$

$$r_i = \frac{d_i}{2} = \frac{(0.9 \, \text{in})\left(\frac{1 \, \text{ft}}{12 \, \text{in}} \right)}{2} = 0.0375 \, \text{ft}$$

$$U_o = \frac{1}{\begin{array}{c} \dfrac{1}{2379 \, \dfrac{\text{Btu}}{\text{hr-ft}^2\text{-}^\circ\text{F}}} \\[2mm] + \left(\dfrac{0.0417 \, \text{ft}}{62 \, \dfrac{\text{Btu-ft}}{\text{hr-ft}^2\text{-}^\circ\text{F}}} \right) \ln \left(\dfrac{0.0417 \, \text{ft}}{0.0375 \, \text{ft}} \right) \\[2mm] + \dfrac{0.0417 \, \text{ft}}{(0.0375 \, \text{ft})\left(885 \, \dfrac{\text{Btu}}{\text{hr-ft}^2\text{-}^\circ\text{F}}\right)} \end{array}}$$

$$= 572 \, \text{Btu/hr-ft}^2\text{-}^\circ\text{F}$$

For crossflow operation (same result for parallel flow),

```
70°F  ───────────────────▶  190°F
          A                      B
350°F ◀───────────────────  350°F
```

$$\Delta T_A = 350^\circ\text{F} - 70^\circ\text{F} = 280^\circ\text{F}$$

$$\Delta T_B = 350^\circ\text{F} - 190^\circ\text{F} = 160^\circ\text{F}$$

From Eq. 33.67, the logarithmic mean temperature difference is

$$\Delta T_{lm} = \frac{\Delta T_A - \Delta T_B}{\ln \left(\dfrac{\Delta T_A}{\Delta T_B} \right)} = \frac{280^\circ\text{F} - 160^\circ\text{F}}{\ln \left(\dfrac{280^\circ\text{F}}{160^\circ\text{F}} \right)} = 214.4^\circ\text{F}$$

The heat transfer is known; therefore, the outside area can be calculated from Eq. 33.68.

$$Q = U_o A_o F_c \Delta T_{lm}$$

For steam condensation, the temperature of steam remains constant. Therefore, $F_c = 1$.

$$A_o = \frac{352{,}447 \, \dfrac{\text{Btu}}{\text{hr}}}{\left(572 \, \dfrac{\text{Btu}}{\text{hr-ft}^2\text{-}^\circ\text{F}}\right)(1)(214.4^\circ\text{F})} = \boxed{2.87 \, \text{ft}^2}$$

At this point the assumption that $T_{\text{sat},v} - T_s = 20^\circ\text{F}$ could be checked using $Q = U_{\text{partial}} A_o \Delta T_{\text{partial}}$, working from the inside (at 130°F) to the outside (at T_s). For the first iteration, $U_{\text{partial}} = 753$ Btu/hr-ft^2 and $T_s = 293^\circ\text{F}$. The solution converges to $T_s = 302^\circ\text{F}$ and $A_o = 2.6 \, \text{ft}^2$.

The answer is (A).

SI Solution

The bulk temperature of the water is

$$T_b = \left(\tfrac{1}{2} \right)(T_{\text{in}} + T_{\text{out}})$$
$$= \left(\tfrac{1}{2} \right)(21^\circ\text{C} + 90^\circ\text{C}) = 55.5^\circ\text{C}$$

From App. 32.B, the properties of water at 55.5°C are

$$c_p = 4.186 \, \text{kJ/kg·K}$$
$$\rho = 986.6 \, \text{kg/m}^3$$
$$\mu = 0.523 \times 10^{-3} \, \text{kg/m·s}$$
$$k = 0.6503 \, \text{W/m·K}$$
$$\text{Pr} = 3.37$$

The heat transfer is found from the temperature gain of the water.

$$Q = \dot{m} c_p \Delta T$$
$$= \left(0.368 \, \frac{\text{kg}}{\text{s}}\right)\left(4.186 \, \frac{\text{kJ}}{\text{kg·K}}\right)\left(1000 \, \frac{\text{J}}{\text{kJ}}\right)$$
$$\times (90^\circ\text{C} - 21^\circ\text{C})$$
$$= 106\,291 \, \text{W}$$

The Reynolds number is

$$\text{Re} = \frac{\rho v D}{\mu}$$

$$= \frac{\left(986.6 \ \frac{\text{kg}}{\text{m}^3}\right)\left(0.9 \ \frac{\text{m}}{\text{s}}\right)(2.29 \ \text{cm})\left(\frac{1 \ \text{m}}{100 \ \text{cm}}\right)}{0.523 \times 10^{-3} \ \frac{\text{kg}}{\text{m} \cdot \text{s}}}$$

$$= 3.89 \times 10^4$$

From Eq. 33.33, the film coefficient is

$$h = (0.023)(\text{Re})^{0.8}(\text{Pr})^n \left(\frac{k}{d}\right)$$

$$= (0.023)(3.89 \times 10^4)^{0.8}(3.37)^{0.4}$$

$$\times \left(\frac{0.6503 \ \frac{\text{W}}{\text{m} \cdot \text{K}}}{(2.29 \ \text{cm})\left(\frac{1 \ \text{m}}{100 \ \text{cm}}\right)}\right)$$

$$= 4989 \ \text{W/m}^2 \cdot \text{K}$$

The saturation temperature for 923 kPa steam is 176.4°C. Assume the wall to be at 10°C lower (or 166.4°C). The film properties are evaluated at the average of the wall and saturation temperatures.

$$T_h = \left(\tfrac{1}{2}\right)\left(T_{\text{sat},v} + T_s\right) = \left(\tfrac{1}{2}\right)(176.4°\text{C} + 166.4°\text{C})$$

$$= 171.4°\text{C}$$

Film properties are obtained from App. 32.B for liquid water and App. 23.N for vapor.

$$k_{171.4°\text{C}} = 0.6745 \ \text{W/m} \cdot \text{K}$$

$$\mu_{171.4°\text{C}} = 0.1712 \times 10^{-3} \ \text{kg/m} \cdot \text{s}$$

$$\rho_{l,171.4°\text{C}} = \frac{1}{v_{f,171.4°\text{C}}}$$

$$= \frac{1}{\left(1.1161 \ \frac{\text{cm}^3}{\text{g}}\right)\left(1000 \ \frac{\text{g}}{\text{kg}}\right)\left(\frac{1 \ \text{m}}{100 \ \text{cm}}\right)^3}$$

$$= 896.0 \ \text{kg/m}^3$$

$$\rho_{v,171.4°\text{C}} = \frac{1}{v_{g,171.4°\text{C}}}$$

$$= \frac{1}{\left(235.3 \ \frac{\text{cm}^3}{\text{g}}\right)\left(1000 \ \frac{\text{g}}{\text{kg}}\right)\left(\frac{1 \ \text{m}}{100 \ \text{cm}}\right)^3}$$

$$= 4.25 \ \text{kg/m}^3$$

$$h_{fg,923 \ \text{kPa}} = \left(2027.5 \ \frac{\text{kJ}}{\text{kg}}\right)\left(1000 \ \frac{\text{J}}{\text{kg}}\right)$$

$$= 2.0275 \times 10^6 \ \text{J/kg}$$

$$d = (2.29 \ \text{cm})\left(\frac{1 \ \text{m}}{100 \ \text{cm}}\right) = 0.0229 \ \text{m}$$

From Eq. 32.31, the film coefficient is

$$h_o = (0.725)\left(\frac{\rho_l(\rho_l - \rho_v)gh_{fg}(k_l)^3}{d\mu_l(T_{\text{sat},v} - T_s)}\right)^{\frac{1}{4}}$$

$$= (0.725)$$

$$\times \left(\frac{\left(896 \ \frac{\text{kg}}{\text{m}^3}\right)\left(896 \ \frac{\text{kg}}{\text{m}^3} - 4.25 \ \frac{\text{kg}}{\text{m}^3}\right)\left(9.81 \ \frac{\text{m}}{\text{s}^2}\right) \times \left(2.0275 \times 10^6 \ \frac{\text{J}}{\text{kg}}\right)\left(0.6745 \ \frac{\text{W}}{\text{m} \cdot \text{K}}\right)^3}{(0.0229 \ \text{m})\left(0.1712 \times 10^{-3} \ \frac{\text{kg}}{\text{m} \cdot \text{s}}\right) \times (176.4°\text{C} - 166.4°\text{C})}\right)^{\frac{1}{4}}$$

$$= 13\,616 \ \text{W/m}^2 \cdot \text{K}$$

From App. 31.B and the table footnote, for copper alloy (70% Cu) at 166.4°C,

$$k \approx \left(62 \ \frac{\text{Btu}}{\text{hr} \cdot \text{ft} \cdot °\text{F}}\right)\left(1.7307 \ \frac{\text{W} \cdot \text{hr} \cdot \text{ft} \cdot °\text{F}}{\text{m} \cdot \text{K} \cdot \text{Btu}}\right)$$

$$= 107.3 \ \text{W/m} \cdot \text{K}$$

From Eq. 33.69, the overall heat transfer coefficient based on outside area is

$$\frac{1}{U_o} = \frac{1}{h_o} + \left(\frac{r_o}{k_{\text{tube}}}\right)\ln\left(\frac{r_o}{r_i}\right) + \frac{r_o}{r_i h_i}$$

$$r_o = \frac{d_o}{2} = \frac{(2.54 \ \text{cm})\left(\frac{1 \ \text{m}}{100 \ \text{cm}}\right)}{2} = 0.0127 \ \text{m}$$

$$r_i = \frac{d_i}{2} = \frac{0.0229 \ \text{m}}{2} = 0.0115 \ \text{m}$$

$$U_o = \frac{1}{\dfrac{1}{13\,616 \ \frac{\text{W}}{\text{m}^2 \cdot \text{K}}} + \left(\dfrac{0.0127 \ \text{m}}{107.3 \ \frac{\text{W}}{\text{m} \cdot \text{K}}}\right)\ln\left(\dfrac{0.0127 \ \text{m}}{0.0115 \ \text{m}}\right) + \dfrac{0.0127 \ \text{m}}{(0.0115 \ \text{m})\left(4989 \ \frac{\text{W}}{\text{m}^2 \cdot \text{K}}\right)}}$$

$$= 3262 \ \text{W/m}^2 \cdot \text{K}$$

```
21°C  ─────────────────────▶  90°C
        A                B
176.4°C ◀─────────────────── 176.4°C
```

$$\Delta T_A = 176.4°\text{C} - 21°\text{C} = 155.4°\text{C}$$

$$\Delta T_B = 176.4°\text{C} - 90°\text{C} = 86.4°\text{C}$$

From Eq. 33.67, the logarithmic mean temperature difference is

$$\Delta T_{lm} = \frac{\Delta T_A - \Delta T_B}{\ln\left(\frac{\Delta T_A}{\Delta T_B}\right)} = \frac{155.4°\text{C} - 86.4°\text{C}}{\ln\left(\frac{155.4°\text{C}}{86.4°\text{C}}\right)}$$

$$= 117.5°\text{C}$$

The heat transfer is known; therefore, the outside area can be calculated from Eq. 33.68.

$$Q = U_o A_o F_c \Delta T_{lm}$$

For steam condensation, the temperature of steam remains constant. Therefore, $F_c = 1$.

$$A_o = \frac{106\,291 \text{ W}}{\left(3262 \ \frac{\text{W}}{\text{m}^2 \cdot \text{K}}\right)(1)(117.5°\text{C})}$$

$$= \boxed{0.277 \text{ m}^2}$$

At this point the assumption that $T_{\text{sub},v} - T_s = 10°\text{C}$ could be checked using $Q = U_{\text{partial}} A_o \Delta T_{\text{partial}}$.

The answer is (A).

10.

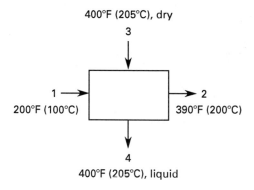

400°F (205°C), dry
3

1 →
200°F (100°C)

→ 2
390°F (200°C)

4
400°F (205°C), liquid

Customary U.S. Solution

(a) From App. 23.A, the enthalpy of each point is

$$h_1 = 168.07 \text{ Btu/lbm}$$

$$h_2 = 364.3 \text{ Btu/lbm}$$

$$h_3 = 1202.0 \text{ Btu/lbm}$$

$$h_4 = 375.1 \text{ Btu/lbm}$$

The heat transfer is due to the temperature gain of water.

$$Q = \dot{m}(h_2 - h_1)$$

$$= \left(500,000 \ \frac{\text{lbm}}{\text{hr}}\right)\left(364.3 \ \frac{\text{Btu}}{\text{lbm}} - 168.07 \ \frac{\text{Btu}}{\text{lbm}}\right)$$

$$= 9.812 \times 10^7 \text{ Btu/hr}$$

The mass flow rate per tube is

$$\dot{m}_{\text{tube}} = \rho A_{\text{tube}} v$$

Select ρ where the volume is greatest (at 390°F). From App. 23.A,

$$v_f = 0.01850 \text{ ft}^3/\text{lbm}$$

$$\rho = \frac{1}{v_f} = \frac{1}{0.01850 \ \frac{\text{ft}^3}{\text{lbm}}}$$

$$= 54.05 \text{ lbm/ft}^3$$

The inside diameter of the tube is

$$d_i = d_o - (2)(\text{wall})$$

$$= \frac{7}{8} \text{ in} - (2)\left(\frac{1}{16} \text{ in}\right)$$

$$= 0.750 \text{ in}$$

The area per tube is

$$A_{\text{tube}} = \left(\frac{\pi}{4}\right)(d_i)^2 = \left(\frac{\pi}{4}\right)(0.750 \text{ in})^2\left(\frac{1 \text{ ft}}{12 \text{ in}}\right)^2$$

$$= 0.003068 \text{ ft}^2$$

$$\dot{m}_{\text{tube}} = \left(54.05 \ \frac{\text{lbm}}{\text{ft}^3}\right)(0.003068 \text{ ft}^2)$$

$$\times \left(5 \ \frac{\text{ft}}{\text{sec}}\right)\left(3600 \ \frac{\text{sec}}{\text{hr}}\right)$$

$$= 2985 \text{ lbm/hr}$$

The required number of tubes is

$$N = \frac{500,000 \ \frac{\text{lbm}}{\text{hr}}}{2985 \ \frac{\text{lbm}}{\text{hr}}} = \boxed{167.5} \quad [\text{say } 168]$$

The answer is (D).

(b) Consider counterflow or parallel flow. Since one fluid temperature remains constant, it will not make a difference. Also, $F_c = 1$.

$$\Delta T_A = 400°\text{F} - 200°\text{F} = 200°\text{F}$$

$$\Delta T_B = 400°\text{F} - 390°\text{F} = 10°\text{F}$$

From Eq. 33.67, the logarithmic mean temperature difference is

$$\Delta T_{lm} = \frac{\Delta T_A - \Delta T_B}{\ln\left(\frac{\Delta T_A}{\Delta T_B}\right)} = \frac{200°\text{F} - 10°\text{F}}{\ln\left(\frac{200°\text{F}}{10°\text{F}}\right)}$$

$$= 63.4°\text{F}$$

The heat transfer is known. So,

$$Q = U A \Delta T_{lm} F_c$$

The surface area is

$$A = (\pi D_o L) N \times 2 \text{ passes}$$
$$Q = 2U(\pi D_o L) N \Delta T_{lm} F_c$$
$$L = \frac{Q}{2U\pi D_o \Delta T_{lm} F_c N}$$

$$= \frac{9.812 \times 10^7 \dfrac{\text{Btu}}{\text{hr}}}{(2)\left(700 \dfrac{\text{Btu}}{\text{hr-ft}^2\text{-}°\text{F}}\right) \pi \left(\dfrac{7}{8} \text{ in}\right)}$$
$$\times \left(\dfrac{1 \text{ ft}}{12 \text{ in}}\right)(63.4°\text{F})(1)(168)$$

$$= \boxed{28.7 \text{ ft}}$$

The answer is (C).

This is the approximate length of the tube bundle and one half of the total straight length of the bent tubes.

SI Solution

(a) From App. 23.N, the enthalpy of each point is

$$h_1 = 419.04 \text{ kJ/kg}$$
$$h_2 = 852.45 \text{ kJ/kg}$$
$$h_3 = 2795.9 \text{ kJ/kg}$$
$$h_4 = 875.11 \text{ kJ/kg}$$

The heat transfer is due to the temperature gain of water.

$$Q = \dot{m}(h_2 - h_1)$$
$$= \left(60 \frac{\text{kg}}{\text{s}}\right)\left(852.45 \frac{\text{kJ}}{\text{kg}} - 419.04 \frac{\text{kJ}}{\text{kg}}\right)$$
$$= 26\,005 \text{ kW}$$

The mass flow rate per tube is

$$\dot{m}_{\text{tube}} = \rho A_{\text{tube}} \text{v}$$

Select ρ where the volume is greatest (at 200°C). From App. 23.N,

$$v_f = \left(1.1565 \frac{\text{cm}^3}{\text{g}}\right)\left(1000 \frac{\text{g}}{\text{kg}}\right)\left(\frac{1 \text{ m}}{100 \text{ cm}}\right)^3$$
$$= 1.1565 \times 10^{-3} \text{ m}^3/\text{kg}$$
$$\rho = \frac{1}{v_f} = \frac{1}{1.1565 \times 10^{-3} \dfrac{\text{m}^3}{\text{kg}}}$$
$$= 864.7 \text{ kg/m}^3$$

The inside diameter of the tube is

$$d_i = d_o - (2)(\text{wall})$$

$$= \left(2.2 \text{ cm} - (2)(1.6 \text{ mm})\left(\frac{1 \text{ cm}}{10 \text{ mm}}\right)\right)\left(\frac{1 \text{ m}}{100 \text{ cm}}\right)$$

$$= 0.0188 \text{ m}$$

The area per tube is

$$A_{\text{tube}} = \left(\frac{\pi}{4}\right)(d_i)^2 = \left(\frac{\pi}{4}\right)(0.0188 \text{ m})^2$$
$$= 2.776 \times 10^{-4} \text{ m}^2$$
$$\dot{m}_{\text{tube}} = \left(864.7 \frac{\text{kg}}{\text{m}^3}\right)(2.776 \times 10^{-4} \text{ m}^2)\left(1.5 \frac{\text{m}}{\text{s}}\right)$$
$$= 0.360 \text{ kg/s}$$

The required number of tubes is

$$N = \frac{60 \dfrac{\text{kg}}{\text{s}}}{0.360 \dfrac{\text{kg}}{\text{s}}} = \boxed{166.7} \quad [\text{say } 167]$$

The answer is (D).

(b) Since one fluid remains at constant temperature, $F_c = 1$. Also, it will not make a difference whether counterflow or parallel flow is considered.

$$\Delta T_A = 205°\text{C} - 100°\text{C} = 105°\text{C}$$
$$\Delta T_B = 205°\text{C} - 200°\text{C} = 5°\text{C}$$

From Eq. 33.67, the logarithmic mean temperature difference is

$$\Delta T_{lm} = \frac{\Delta T_A - \Delta T_B}{\ln\left(\dfrac{\Delta T_A}{\Delta T_B}\right)} = \frac{105°\text{C} - 5°\text{C}}{\ln\left(\dfrac{105°\text{C}}{5°\text{C}}\right)}$$
$$= 32.8°\text{C}$$

The heat transfer is known. So,

$$Q = UA\Delta T_{lm} F_c$$

The surface area is

$$A = (\pi D_o L) N \times 2 \text{ passes}$$
$$Q = 2U(\pi D_o L) N \Delta T_{lm} F_c$$
$$L = \frac{Q}{2U\pi D_o N \Delta T_{lm} F_c}$$

$$= \frac{(26\,005 \text{ kW})\left(1000 \dfrac{\text{W}}{\text{kW}}\right)}{(2)\left(700 \dfrac{\text{Btu}}{\text{hr-ft}^2\text{-}°\text{F}}\right)\left(5.6783 \dfrac{\text{W}}{\text{m}^2\text{·}°\text{C}}\right)\pi}$$
$$\times (2.2 \text{ cm})\left(\frac{1 \text{ m}}{100 \text{ cm}}\right)(167)(32.8°\text{C})(1)$$

$$= \boxed{17.3 \text{ m}}$$

This is the approximate length of the tube bundle and one half of the total straight length of the bent tubes.

The answer is (C).

11. *Customary U.S. Solution*

The water's bulk temperature is

$$T_{b,\text{water}} = \left(\tfrac{1}{2}\right)(70°F + 140°F) = 105°F$$

The fluid properties at 105°F are obtained from App. 32.A.

$$\rho_{105°F} = 61.92 \text{ lbm/ft}^3$$

$$c_{p,105°F} = 0.998 \text{ Btu/lbm-°F}$$

The mass flow rate of water is

$$\dot{m}_{\text{water}} = \dot{V}\rho$$

$$= \frac{\left(100 \dfrac{\text{gal}}{\text{min}}\right)\left(60 \dfrac{\text{min}}{\text{hr}}\right)\left(61.92 \dfrac{\text{lbm}}{\text{ft}^3}\right)}{7.48 \dfrac{\text{gal}}{\text{ft}^3}}$$

$$= 49{,}668 \text{ lbm/hr}$$

The heat transfer is found from the temperature gain of the water.

$$Q_{\text{clean}} = \dot{m}c_p\Delta T$$

$$= \left(49{,}668 \dfrac{\text{lbm}}{\text{hr}}\right)\left(0.998 \dfrac{\text{Btu}}{\text{lbm-°F}}\right)$$

$$\times (140°F - 70°F)$$

$$= 3.470 \times 10^6 \text{ Btu/hr}$$

$$\Delta T_A = 230°F - 70°F = 160°F$$

$$\Delta T_B = 230°F - 140°F = 90°F$$

From Eq. 33.67, the logarithmic mean temperature difference is

$$\Delta T_{lm} = \frac{\Delta T_A - \Delta T_B}{\ln\left(\dfrac{\Delta T_A}{\Delta T_B}\right)} = \frac{160°F - 90°F}{\ln\left(\dfrac{160°F}{90°F}\right)} = 121.66°F$$

The heat transfer is known. Therefore, the overall heat transfer coefficient can be calculated from Eq. 33.68.

$$Q_{\text{clean}} = U_{\text{clean}}AF_c\Delta T_{lm}$$

For steam condensation, the temperature of steam remains constant. Therefore, $F_c = 1$.

$$U_{\text{clean}} = \frac{Q_{\text{clean}}}{AF_c\Delta T_{lm}}$$

$$= \frac{3.470 \times 10^6 \dfrac{\text{Btu}}{\text{hr}}}{(50 \text{ ft}^2)(1)(121.66°F)}$$

$$= 570.4 \text{ Btu/hr-ft}^2\text{-°F}$$

After fouling,

$$Q_{\text{fouled}} = \left(49{,}668 \dfrac{\text{lbm}}{\text{hr}}\right)\left(0.998 \dfrac{\text{Btu}}{\text{lbm-°F}}\right)$$

$$\times (122°F - 70°F)$$

$$= 2.578 \times 10^6 \text{ Btu/hr}$$

$$\Delta T_B = 230°F - 122°F = 108°F$$

$$\Delta T_A = 230°F - 70°F = 160°F$$

From Eq. 33.67, the logarithmic mean temperature difference is

$$\Delta T_{lm} = \frac{160°F - 108°F}{\ln\left(\dfrac{160°F}{108°F}\right)} = 132.3°F$$

$$U_{\text{fouled}} = \frac{2.578 \times 10^6 \dfrac{\text{Btu}}{\text{hr}}}{(50 \text{ ft}^2)(1)(132.3°F)} = 389.7 \text{ Btu/hr-ft}^2\text{-°F}$$

From Eq. 33.74, the fouling factor is

$$R_f = \frac{1}{U_{\text{fouled}}} - \frac{1}{U_{\text{clean}}}$$

$$= \frac{1}{389.7 \dfrac{\text{Btu}}{\text{hr-ft}^2\text{-°F}}} - \frac{1}{570.4 \dfrac{\text{Btu}}{\text{hr-ft}^2\text{-°F}}}$$

$$= \boxed{0.000813 \text{ hr-ft}^2\text{-°F/Btu}}$$

The answer is (B).

SI Solution

The water's bulk temperature is

$$T_{b,\text{water}} = \left(\tfrac{1}{2}\right)(21°C + 60°C) = 40.5°C$$

The fluid properties at 40.5°C are obtained from App. 32.B.

$$\rho_{40.5°C} = 993.5 \text{ kg/m}^3$$

$$c_{p,40.5°C} = 4.183 \text{ kJ/kg·K}$$

The mass flow rate of water is

$$\dot{m}_{\text{water}} = \dot{V}\rho$$

$$= \left(6.3 \dfrac{\text{L}}{\text{s}}\right)\left(\dfrac{1 \text{ m}^3}{1000 \text{ L}}\right)\left(993.5 \dfrac{\text{kg}}{\text{m}^3}\right)$$

$$= 6.26 \text{ kg/s}$$

The heat transfer is found from the temperature gain of the water.

$$Q_{\text{clean}} = \dot{m}c_p\Delta T$$

$$= \left(6.26 \dfrac{\text{kg}}{\text{s}}\right)\left(4.183 \dfrac{\text{kJ}}{\text{kg·K}}\right)\left(1000 \dfrac{\text{J}}{\text{kJ}}\right)$$

$$\times (60°C - 21°C)$$

$$= 1.021 \times 10^6 \text{ W}$$

$$\Delta T_A = 110°C - 21°C = 89°C$$

$$\Delta T_B = 110°C - 60°C = 50°C$$

From Eq. 33.67, the logarithmic mean temperature difference is

$$\Delta T_{lm} = \frac{\Delta T_A - \Delta T_B}{\ln\left(\frac{\Delta T_A}{\Delta T_B}\right)} = \frac{89°C - 50°C}{\ln\left(\frac{89°C}{50°C}\right)} = 67.64°C$$

The heat transfer is known. Therefore, the overall heat transfer coefficient can be calculated from Eq. 33.68.

$$Q_{\text{clean}} = U_{\text{clean}} A F_c \Delta T_{lm}$$

For steam condensation, the temperature of steam remains constant. Therefore, $F_c = 1$.

$$U_{\text{clean}} = \frac{Q_{\text{clean}}}{A F_c \Delta T_{lm}}$$
$$= \frac{1.021 \times 10^6 \text{ W}}{(4.7 \text{ m}^2)(1)(67.64°C)}$$
$$= 3211.6 \text{ W/m}^2\cdot\text{K}$$

After fouling,

$$Q_{\text{fouled}} = \left(6.26 \frac{\text{kg}}{\text{s}}\right)\left(4.183 \frac{\text{kJ}}{\text{kg}\cdot\text{K}}\right)\left(1000 \frac{\text{J}}{\text{kJ}}\right)$$
$$\times (50°C - 21°C)$$
$$= 7.594 \times 10^5 \text{ W}$$
$$\Delta T_A = 110°C - 21°C = 89°C$$
$$\Delta T_B = 110°C - 50°C = 60°C$$

From Eq. 33.67, the logarithmic mean temperature difference is

$$\Delta T_{lm} = \frac{89°C - 60°C}{\ln\left(\frac{89°C}{60°C}\right)} = 73.55°C$$

$$U_{\text{fouled}} = \frac{7.594 \times 10^5 \text{ W}}{(4.7 \text{ m}^2)(1)(73.55°C)} = 2196.8 \text{ W/m}^2\cdot\text{K}$$

From Eq. 33.74, the fouling factor is

$$R_f = \frac{1}{U_{\text{fouled}}} - \frac{1}{U_{\text{clean}}}$$
$$= \frac{1}{2196.8 \frac{\text{W}}{\text{m}^2\cdot\text{K}}} - \frac{1}{3211.6 \frac{\text{W}}{\text{m}^2\cdot\text{K}}}$$
$$= \boxed{0.000144 \text{ m}^2\cdot\text{K/W}}$$

The answer is (B).

12. *Customary U.S. Solution*

The film temperature of the air from Eq. 32.11 is

$$T_h = \left(\tfrac{1}{2}\right)(T_s + T_\infty) = \left(\tfrac{1}{2}\right)(100°F + 150°F) = 125°F$$

From App. 32.C, the air properties at 125°F are

$$\nu = 0.195 \times 10^{-3} \text{ ft}^2/\text{sec}$$
$$k = 0.0159 \text{ Btu/hr-ft-°F}$$
$$\text{Pr} = 0.72$$

The Reynolds number is

$$\text{Re}_d = \frac{\text{v}d}{\nu} = \frac{\left(100 \frac{\text{ft}}{\text{sec}}\right)(0.35 \text{ in})\left(\frac{1 \text{ ft}}{12 \text{ in}}\right)}{0.195 \times 10^{-3} \frac{\text{ft}^2}{\text{sec}}}$$
$$= 1.50 \times 10^4$$

From Eq. 33.50, the film coefficient is

$$h = C_1(\text{Re}_d)^n (\text{Pr})^{\frac{1}{3}} \left(\frac{k}{d}\right)$$

From Table 33.3, $C_1 = 0.193$ and $n = 0.618$.

$$h = (0.193)(1.50 \times 10^4)^{0.618}(0.72)^{\frac{1}{3}}$$
$$\times \left(\frac{0.0159 \frac{\text{Btu}}{\text{hr-ft-°F}}}{(0.35 \text{ in})\left(\frac{1 \text{ ft}}{12 \text{ in}}\right)}\right)$$
$$= \boxed{35.9 \text{ Btu/hr-ft}^2\text{-°F}}$$

The answer is (A).

SI Solution

From Eq. 32.11, the film temperature of the air is

$$T_h = \left(\tfrac{1}{2}\right)(T_s + T_\infty) = \left(\tfrac{1}{2}\right)(38°C + 66°C) = 52°C$$

From App. 32.D, the air properties at 52°C are

$$\rho = 1.09 \text{ kg/m}^3$$
$$\mu = 1.966 \times 10^{-5} \text{ kg/m}\cdot\text{s}$$
$$k = 0.02815 \text{ W/m}\cdot\text{K}$$
$$\text{Pr} = 0.703$$

The Reynolds number is

$$\text{Re}_d = \frac{\rho \text{v}d}{\mu}$$
$$= \frac{\left(1.09 \frac{\text{kg}}{\text{m}^3}\right)\left(30 \frac{\text{m}}{\text{s}}\right)(8.9 \text{ mm})\left(\frac{1 \text{ m}}{1000 \text{ mm}}\right)}{1.966 \times 10^{-5} \frac{\text{kg}}{\text{m}\cdot\text{s}}}$$
$$= 1.48 \times 10^4$$

From Eq. 33.50, the film coefficient is

$$h = C_1 (\mathrm{Re}_d)^h (\mathrm{Pr})^{\frac{1}{3}} \left(\frac{k}{d} \right)$$

From Table 33.3, $C_1 = 0.193$ and $n = 0.618$.

$$h = (0.193)(1.48 \times 10^4)^{0.618}(0.703)^{\frac{1}{3}}$$

$$\times \left(\frac{0.02815 \; \frac{\mathrm{W}}{\mathrm{m \cdot K}}}{(8.9 \; \mathrm{mm}) \left(\dfrac{1 \; \mathrm{m}}{1000 \; \mathrm{mm}} \right)} \right)$$

$$= \boxed{205.0 \; \mathrm{W/m^2 \cdot K}}$$

The answer is (A).

34 Heat Transfer: Radiation

PRACTICE PROBLEMS

Arrangement Factors

1. A 6 in (15 cm) thick furnace wall has a 3 in (8 cm) square peephole. The interior of the furnace is at 2200°F (1200°C). The surrounding air temperature is 70°F (20°C). What is the heat loss due to radiation when the peephole is open?

(A) 450 Btu/hr (150 W)
(B) 1300 Btu/hr (440 W)
(C) 2000 Btu/hr (680 W)
(D) 7900 Btu/hr (2.7 kW)

Combined Heat Transfer

2. A 9 in (23 cm) diameter duct is painted with white lacquer. The surface of the duct is at 200°F (95°C). The duct carries hot air through a room whose walls are 70°F (20°C). The air in the room is at 80°F (27°C). What is the unit heat transfer?

(A) 700 Btu/hr-ft length (700 W/m length)
(B) 900 Btu/hr-ft length (900 W/m length)
(C) 1100 Btu/hr-ft length (1100 W/m length)
(D) 1500 Btu/hr-ft length (1500 W/m length)

3. The walls of a cold storage unit have the cross section shown.

aluminum foil (ϵ = 0.1)
silver paint (ϵ = 0.5)

−60°F (−50°C) metal wall (surface) | high vacuum | high vacuum | | 80°F (27°C) insulation (surface) k=0.025 Btu-ft/hr-ft^2-°F (0.045 W/m^2·K)

2 in (5 cm) — 2 in (5 cm) — 4 in (10 cm)

(a) What is the heat transfer per unit area of wall?

(A) 3 Btu/hr-ft^2 (10 W/m^2)
(B) 7 Btu/hr-ft^2 (23 W/m^2)
(C) 14 Btu/hr-ft^2 (46 W/m^2)
(D) 53 Btu/hr-ft^2 (180 W/m^2)

(b) What is the temperature of the aluminum foil?

(A) 420°R (230K)
(B) 460°R (260K)
(C) 475°R (264K)
(D) 490°R (270K)

4. Dry air at 1 atmospheric pressure flows at 500 ft^3/min (0.25 m^3/s) through 50 ft (15 m) of 12 in (30 cm) diameter uninsulated duct. The emissivity of the duct surface is 0.28. Air enters the duct at 45°F (7°C). The walls, air, and contents of the room through which the duct passes are at 80°F (27°C). An engineer states that the air leaving the duct will be at 50°F (10°C). Consider both convection and radiation to prove or disprove the engineer's statement.

5. A steel pipe is painted on the outside with dull gray (oil-based) paint. The pipe is 35 ft (10 m) long. The pipe is 4.00 in (10.2 cm) inside diameter and 4.25 in (10.8 cm) outside diameter. The pipe carries 200 ft^3/min (0.1 m^3/s) of 500°F (260°C), 25 psig (170 kPa) air through a 70°F (20°C) room. The conditions of the air at the end of the pipe are 350°F (180°C) and 15 psig (100 kPa). The thermal resistance of the pipe material and the inside film may be disregarded.

(a) What is the overall coefficient of heat transfer?

(A) 1.0 Btu/hr-ft^2-°F (6.1 W/m^2·K)
(B) 1.8 Btu/hr-ft^2-°F (11 W/m^2·K)
(C) 3.6 Btu/hr-ft^2-°F (22 W/m^2·K)
(D) 49 Btu/hr-ft^2-°F (300 W/m^2·K)

(b) Using theoretical methods or empirical correlations, what is the calculated overall coefficient of heat transfer?

(A) 1.0 Btu/hr-ft^2-°F (6.1 W/m^2·K)
(B) 1.8 Btu/hr-ft^2-°F (11 W/m^2·K)
(C) 3.8 Btu/hr-ft^2-°F (22 W/m^2·K)
(D) 54 Btu/hr-ft^2-°F (330 W/m^2·K)

(c) Explain the possible reasons for differences between the actual and calculated overall coefficients of heat transfer.

6. A semiconductor device is modeled as an upright circular cylinder 0.75 in (19 mm) in diameter and 1.5 in (38 mm) high. The device emits 5.0 W and is cooled by a combination of natural convection and radiation. The surface emissivity is 0.65. The base is insulated and transmits no heat. The air and surroundings are at 14.7 psia (101 kPa) and 75°F (24°C).

(a) What is the surface temperature of the device?

(b) What percentages of heat are lost through convection and radiation?

7. The temperature of a gas in a duct with 600°F (315°C) walls is evaluated with a 0.5 in (13 mm) diameter thermocouple probe. The emissivity of the probe is 0.8. The gas flow rate is 3480 lbm/hr-ft² (4.7 kg/s·m²). The gas velocity is 400 ft/min (2 m/s). The film coefficient on the probe is given empirically as

$$h = \frac{0.024 G^{0.8}}{D^{0.4}}$$

h in Btu/hr-ft²-°F

G in lbm/hr-ft²

D in ft

$$h = \frac{2.9 G^{0.8}}{D^{0.4}}$$

h in W/m²·K

G in kg/s·m²

D in m

(a) If the actual gas temperature is 300°F (150°C), what is the probe's reading?

   (A) 700°R (390K)

   (B) 740°R (410K)

   (C) 760°R (420K)

   (D) 780°R (480K)

(b) If the probe reading indicates that the gas temperature is 300°F (150°C), what is the actual gas temperature?

   (A) 650°R (360K)

   (B) 740°R (350K)

   (C) 770°R (430K)

   (D) 810°R (450K)

Furnaces

8. A small, oxidized metal tube with a surface temperature of 550K is placed in a very large fire-brick furnace. The tube has a 0.025 m diameter and is 0.65 m long. The air surrounding the tube has a temperature of 1000K. The emissivity of the metal tube is 0.45 at 550K and 0.60 at 1000K. Determine the radiant heat transfer to the tube.

9. Referring to Prob. 8, assume the heat transfer coefficient for natural convection (in W/m²·K) for a tube of diameter D (in m) can be estimated by

$$h_{\text{convection}} = 1.32 \left(\frac{T_1 - T_2}{D} \right)^{0.25}$$

(a) Calculate the radiant heat transfer coefficient. (b) Calculate the combined radiation and convective heat transfer rate to the tube.

10. A refractory furnace flue with a 1 m diameter carries stack gases containing 5% CO_2 at 1200K and 1 atm pressure and with a gas emissivity of 0.075. The convective heat transfer coefficient between the gas and the flue wall is 7.5 W/m²·K. The flue wall is at a temperature of 1100K and has an emissivity of 0.9. (a) Calculate the radiant heat flux. (b) Calculate the convective heat flux.

SOLUTIONS

1. *Customary U.S. Solution*

The absolute temperatures are

$$T_{\text{furnace}} = 2200°F + 460 = 2660°R$$
$$T_\infty = 70°F + 460 = 530°R$$

Assuming that the walls are reradiating, nonconducting, and varying in temperature from 2200°F at the inside to 70°F at the outside, Fig. 34.3, curve 6, can be used to find F_{1-2} using $x = 3\text{ in}/6\text{ in} = 0.5$; $F_{1-2} = 0.38$.

The radiation heat loss is

$$Q = AE_{\text{net}}$$
$$= A\sigma F_{1-2}(T_{\text{furnace}}^4 - T_\infty^4)$$
$$= (3\text{ in})^2 \left(\frac{1\text{ ft}^2}{144\text{ in}^2}\right)\left(0.1713 \times 10^{-8}\ \frac{\text{Btu}}{\text{hr-ft}^2\text{-}°R^4}\right)$$
$$\times (0.38)\left((2660°R)^4 - (530°R)^4\right)$$
$$= \boxed{2033.6\text{ Btu/hr}}$$

The answer is (C).

SI Solution

The absolute temperatures are

$$T_{\text{furnace}} = 1200°C + 273 = 1473\text{K}$$
$$T_\infty = 20°C + 273 = 293\text{K}$$

Making the same assumptions as for the customary U.S. solution, Fig. 34.3, curve 6, can be used to find F_{1-2} using $x = 8\text{ cm}/15\text{ cm} = 0.533$; $F_{1-2} = 0.4$.

The radiation heat loss is

$$Q = AE_{\text{net}}$$
$$= A\sigma F_{1-2}(T_{\text{furnace}}^4 - T_\infty^4)$$
$$= (8\text{ cm})^2 \left(\frac{1\text{ m}}{100\text{ cm}}\right)^2\left(5.67 \times 10^{-8}\ \frac{\text{W}}{\text{m}^2\text{·K}^4}\right)$$
$$\times (0.4)\left((1473\text{K})^4 - (293\text{K})^4\right)$$
$$= \boxed{682.3\text{ W}}$$

The answer is (C).

2. *Customary U.S. Solution*

The absolute temperatures are

$$T_\infty = 80°F + 460 = 540°R$$
$$T_{\text{duct}} = 200°F + 460 = 660°R$$
$$T_{\text{wall}} = 70°F + 460 = 530°R$$

Assume laminar flow. From Table 32.3, the convective film coefficient on the outside of the duct is approximately

$$h_{\text{convective}} = (0.27)\left(\frac{T_{\text{duct}} - T_\infty}{L}\right)^{0.25}$$
$$= (0.27)\left(\frac{660°R - 540°R}{(9\text{ in})\left(\frac{1\text{ ft}}{12\text{ in}}\right)}\right)^{0.25}$$
$$= 0.96\text{ Btu/hr-ft}^2\text{-}°F$$

The duct area per unit length is

$$\frac{A}{L} = \pi D = \pi(9\text{ in})\left(\frac{1\text{ ft}}{12\text{ in}}\right) = 2.356\text{ ft}^2/\text{ft}$$

The convection losses (per unit length) are

$$\frac{Q_{\text{convection}}}{L} = h\left(\frac{A}{L}\right)\Delta T = h\left(\frac{A}{L}\right)(T_{\text{duct}} - T_\infty)$$
$$= \left(0.96\ \frac{\text{Btu}}{\text{hr-ft}^2\text{-}°F}\right)\left(2.356\ \frac{\text{ft}^2}{\text{ft}}\right)$$
$$\times (660°R - 540°R)$$
$$= 271.4\text{ Btu/hr-ft}$$

Assume $\epsilon_{\text{duct}} \approx 0.97$. Then $F_e = \epsilon_{\text{duct}} = 0.97$. $F_a = 1$ since the duct is enclosed. The radiation losses (per unit length) are

$$\frac{E_{\text{net}}}{L} = \left(\frac{A}{L}\right)\sigma F_a F_e \left((T_{\text{duct}})^4 - (T_{\text{wall}})^4\right)$$
$$= (2.356\text{ ft}^2)\left(0.1713 \times 10^{-8}\frac{\text{Btu}}{\text{hr-ft}^2\text{-}°R^4}\right)$$
$$\times (1)(0.97)\left((660°R)^4 - (530°R)^4\right)$$
$$= 433.9\text{ Btu/hr-ft}$$

The total heat transfer per unit length is

$$\frac{Q_{\text{total}}}{L} = \frac{Q_{\text{convection}}}{L} + \frac{E_{\text{net}}}{L}$$
$$= 271.4\ \frac{\text{Btu}}{\text{hr-ft}} + 433.9\ \frac{\text{Btu}}{\text{hr-ft}}$$
$$= \boxed{705.3\text{ Btu/hr-ft length}}$$

The answer is (A).

SI Solution

The absolute temperatures are

$$T_{\text{duct}} = 95°C + 273 = 368\text{K}$$
$$T_{\text{wall}} = 20°C + 273 = 293\text{K}$$
$$T_\infty = 27°C + 273 = 300\text{K}$$

Assume laminar flow. From Table 32.3, the convective film coefficient on the outside of the duct is approximately

$$h_{\text{convective}} = (1.32)\left(\frac{T_{\text{duct}} - T_{\infty}}{L}\right)^{0.25}$$

$$= (1.32)\left(\frac{368\text{K} - 300\text{K}}{(23\text{ cm})\left(\frac{1\text{ m}}{100\text{ cm}}\right)}\right)^{0.25}$$

$$= 5.47\text{ W/m}^2\cdot\text{K}$$

The duct area per unit length is

$$\frac{A}{L} = \pi D = \pi(23\text{ cm})\left(\frac{1\text{ m}}{100\text{ cm}}\right) = 0.723\text{ m}^2/\text{m}$$

The convective losses per unit length are

$$\frac{Q_{\text{convective}}}{L} = h\left(\frac{A}{L}\right)(T_{\text{duct}} - T_{\infty})$$

$$= \left(5.47\frac{\text{W}}{\text{m}^2\cdot\text{K}}\right)\left(0.723\frac{\text{m}^2}{\text{m}}\right)(368\text{K} - 300\text{K})$$

$$= 268.9\text{ W/m}$$

Assuming $\epsilon_{\text{duct}} \approx 0.97$, $F_e = \epsilon_{\text{duct}} = 0.97$. $F_a = 1$ since the duct is enclosed. The radiation losses per unit length are

$$\frac{E_{\text{net}}}{L} = \left(\frac{A}{L}\right)\sigma F_a F_e\left((T_{\text{duct}})^4 - (T_{\text{wall}})^4\right)$$

$$= \left(0.723\frac{\text{m}^2}{\text{m}}\right)\left(5.67\times10^{-8}\frac{\text{W}}{\text{m}^2\cdot\text{K}^4}\right)$$

$$\times (1)(0.97)\left((368\text{K})^4 - (293\text{K})^4\right)$$

$$= 436.2\text{ W/m}$$

The total heat transfer per unit length is

$$\frac{Q_{\text{total}}}{L} = \frac{Q_{\text{convection}}}{L} + \frac{E_{\text{net}}}{L}$$

$$= 268.9\frac{\text{W}}{\text{m}} + 436.2\frac{\text{W}}{\text{m}} = \boxed{705.1\text{ W/m length}}$$

The answer is (A).

3.

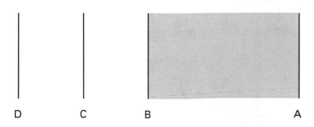

Customary U.S. Solution

(a) The absolute temperatures are

$$T_{\text{D}} = -60°\text{F} + 460 = 400°\text{R}$$

$$T_{\text{A}} = 80°\text{F} + 460 = 540°\text{R}$$

The conductive heat transfer from A to B per unit area is

$$\frac{Q_{\text{A-B}}}{A} = \frac{k(T_{\text{A}} - T_{\text{B}})}{L}$$

$$= \frac{\left(0.025\frac{\text{Btu}}{\text{hr-ft}^2\text{-}°\text{F}}\right)(540°\text{R} - T_{\text{B}})}{(4\text{ in})\left(\frac{1\text{ ft}}{12\text{ in}}\right)}$$

$$= 40.5 - 0.075T_{\text{B}} \quad [\text{Eq. 1}]$$

Since the spaces are evacuated, only radiation should be considered from B to C and from C to D.

The radiation heat transfer per unit area from B to C is

$$\frac{E_{\text{B-C}}}{A} = \sigma F_e F_a(T_{\text{B}}^4 - T_{\text{C}}^4)$$

Since the freezer is assumed to be large, $F_a = 1$.

From Table 34.1 for infinite parallel planes,

$$F_e = \frac{1}{\dfrac{1}{\epsilon_1} + \dfrac{1}{\epsilon_2} - 1} = \frac{1}{\dfrac{1}{0.5} + \dfrac{1}{0.1} - 1}$$

$$= 0.0909$$

$$\frac{E_{\text{B-C}}}{A} = \left(0.1713\times10^{-8}\frac{\text{Btu}}{\text{hr-ft}^2\text{-}°\text{R}^4}\right)$$

$$\times (0.0909)(1)(T_{\text{B}}^4 - T_{\text{C}}^4)$$

$$= (1.56\times10^{-10})T_{\text{B}}^4 - (1.56\times10^{-10})T_{\text{C}}^4 \quad [\text{Eq. 2}]$$

Similarly, radiation heat transfer per unit area from C to D is

$$\frac{E_{\text{C-D}}}{A} = \left(0.1713\times10^{-8}\frac{\text{Btu}}{\text{hr-ft}^2\text{-}°\text{R}^4}\right)$$

$$\times (0.0909)(1)\left(T_{\text{C}}^4 - (400°\text{R})^4\right)$$

$$= (1.56\times10^{-10})T_{\text{C}}^4 - 3.99 \quad [\text{Eq. 3}]$$

The heat transfer from B to C is equal to the heat transfer from C to D.

$$\frac{E_{\text{B-C}}}{A} = \frac{E_{\text{C-D}}}{A}$$

$$(1.56\times10^{-10})T_{\text{B}}^4$$

$$-(1.56\times10^{-10})T_{\text{C}}^4 = (1.56\times10^{-10})T_{\text{C}}^4 - 3.99$$

$$T_{\text{B}}^4 - T_{\text{C}}^4 = T_{\text{C}}^4 - 2.56\times10^{10}$$

$$T_{\text{B}}^4 + 2.56\times10^{10} = 2T_{\text{C}}^4$$

$$T_{\text{C}}^4 = \tfrac{1}{2}T_{\text{B}}^4 + 1.28\times10^{10} \quad [\text{Eq. 4}]$$

The heat transfer from A to B is equal to the heat transfer from B to C.

$$\frac{Q_{A-B}}{A} = \frac{E_{B-C}}{A}$$
$$40.5 - 0.075T_B = (1.56 \times 10^{-10})T_B^4$$
$$- (1.56 \times 10^{-10})T_C^4$$
$$T_C^4 = T_B^4 + (4.81 \times 10^8)T_B$$
$$- 2.60 \times 10^{11} \quad [\text{Eq. 5}]$$

Since Eq. 4 = Eq. 5,

$$\tfrac{1}{2}T_B^4 + 1.28 \times 10^{10} = T_B^4 + (4.81 \times 10^8)T_B$$
$$- 2.60 \times 10^{11}$$
$$T_B^4 + (9.62 \times 10^8)T_B = 5.456 \times 10^{11}$$

By trial and error, $T_B = 501.4°R$.

From Eq. 1,

$$\frac{Q_{A-B}}{A} = 40.5 - (0.075)(501.4°R)$$
$$= \boxed{2.895 \text{ Btu/hr-ft}^2}$$

The answer is (A).

(b) From Eq. 4,

$$T_C = \left(\left(\tfrac{1}{2}\right)(501.4)^4 + 1.28 \times 10^{10} \right)^{\frac{1}{4}}$$
$$= \boxed{459°R}$$

The answer is (B).

SI Solution

(a) The absolute temperatures are

$$T_D = -50°C + 273 = 223K$$
$$T_A = 27°C + 273 = 300K$$

The conductive heat transfer per unit area from A to B is

$$\frac{Q_{A-B}}{A} = \frac{k(T_A - T_B)}{L} = \frac{\left(0.045 \frac{W}{m \cdot K}\right)(300K - T_B)}{(10 \text{ cm})\left(\frac{1 \text{ m}}{100 \text{ cm}}\right)}$$
$$= 135 - 0.45T_B \quad [\text{Eq. 1}]$$

Since spaces are evacuated, only radiation should be considered from B to C and from C to D.

The radiation heat transfer per unit area from B to C is

$$\frac{E_{B-C}}{A} = \sigma F_e F_a(T_B^4 - T_C^4)$$

From the customary U.S. solution, $F_e = 0.0909$ and $F_a = 1$.

$$\frac{E_{B-C}}{A} = \left(5.67 \times 10^{-8} \frac{W}{m^2 \cdot K^4}\right)$$
$$\times (0.0909)(1)(T_B^4 - T_C^4)$$
$$= (5.15 \times 10^{-9})T_B^4 - (5.15 \times 10^9)T_C^4 \quad [\text{Eq. 2}]$$

Similarly, the radiation heat transfer per unit area from C to D is

$$\frac{E_{C-D}}{A} = \left(5.67 \times 10^{-8} \frac{W}{m^2 \cdot K^4}\right)$$
$$\times (0.0909)(1)\left(T_C^4 - (223K)^4\right)$$
$$= (5.15 \times 10^{-9})T_C^4 - 12.746 \quad [\text{Eq. 3}]$$

The heat transfer from B to C is equal to the heat transfer from C to D.

$$\frac{E_{B-C}}{A} = \frac{E_{C-D}}{A}$$
$$(5.15 \times 10^{-9})T_B^4$$
$$- (5.15 \times 10^{-9})T_C^4 = (5.15 \times 10^{-9})T_C^4 - 12.746$$
$$T_B^4 - T_C^4 = T_C^4 - 2.475 \times 10^9$$
$$T_B^4 + 2.475 \times 10^9 = 2T_C^4$$
$$T_C^4 = \tfrac{1}{2}T_B^4 + 1.237 \times 10^9 \quad [\text{Eq. 4}]$$

The heat transfer from A to B is equal to the heat transfer from B to C.

$$\frac{Q_{A-B}}{A} = \frac{E_{B-C}}{A}$$
$$135 - 0.45T_B = (5.15 \times 10^{-9})T_B^4 - (5.15 \times 10^{-9})T_C^4$$
$$T_C^4 = T_B^4 + 8.74 \times 10^7 T_B - 2.62 \times 10^{10}$$
$$[\text{Eq. 5}]$$

Since Eq. 4 = Eq. 5,

$$\tfrac{1}{2}T_B^4 + 1.237 \times 10^9 = T_B^4 + 8.74 \times 10^7 T_B$$
$$- 2.62 \times 10^{10}$$
$$T_B^4 + 17.48 \times 10^7 T_B = 5.49 \times 10^{10}$$

By trial and error, $T_B \approx 278K$.

From Eq. 1,

$$\frac{Q_{A-B}}{A} = 135 - (0.45)(278K) = \boxed{9.9 \text{ W/m}^2}$$

The answer is (A).

(b) From Eq. 4,

$$T_C = \left(\left(\tfrac{1}{2} \right) (278\text{K})^4 + 1.237 \times 10^9 \right)^{\frac{1}{4}} = \boxed{254.9\text{K}}$$

The answer is (B).

4. *Customary U.S. Solution*

The absolute temperature of air entering the duct is

$$45°\text{F} + 460 = 505°\text{R}$$

From the ideal gas law, the density of air entering the duct is

$$\rho = \frac{p}{RT}$$

$$= \frac{\left(14.7 \,\frac{\text{lbf}}{\text{in}^2} \right) \left(144 \,\frac{\text{in}^2}{\text{ft}^2} \right)}{\left(53.35 \,\frac{\text{ft-lbf}}{\text{lbm-°R}} \right) (505°\text{R})} = 0.07857 \text{ lbm/ft}^3$$

The mass flow rate of entering air is

$$\dot{m} = \rho \dot{V} = \left(0.07857 \,\frac{\text{lbm}}{\text{ft}^3} \right) \left(500 \,\frac{\text{ft}^3}{\text{min}} \right) \left(60 \,\frac{\text{min}}{\text{hr}} \right)$$

$$= 2357.1 \text{ lbm/hr}$$

The mass velocity is

$$G = \frac{\dot{m}}{A_{\text{flow}}} = \frac{2357.1 \,\frac{\text{lbm}}{\text{hr}}}{\left(\frac{\pi}{4} \right) (12 \text{ in})^2 \left(\frac{1 \text{ ft}}{12 \text{ in}} \right)^2}$$

$$= 3001.1 \text{ lbm/ft}^2\text{-hr}$$

To calculate the initial film coefficients, estimate the temperature based on the claim. The film coefficients are not highly sensitive to small temperature differences.

$$T_{\text{bulk,air}} = \left(\tfrac{1}{2} \right) (T_{\text{air,in}} + T_{\text{air,out}})$$

$$= \left(\tfrac{1}{2} \right) (45°\text{F} + 50°\text{F}) = 47.5°\text{F}$$

$$T_{\text{surface}} = 70°\text{F} \quad [\text{estimate}]$$

The film coefficient for air flowing inside the duct is given by Eq. 33.39 as

$$h_i \approx (0.00351 + 0.000001583 T_{°\text{F}})$$

$$\times \left(\frac{(G_{\text{lbm/hr-ft}^2})^{0.8}}{(d_{\text{ft}})^{0.2}} \right)$$

$$= \left(0.00351 + (0.000001583)(47.5°\text{F}) \right)$$

$$\times \left(\frac{\left(3001.1 \,\frac{\text{lbm}}{\text{hr-ft}^2} \right)^{0.8}}{\left((12 \text{ in}) \left(\frac{1 \text{ ft}}{12 \text{ in}} \right) \right)^{0.2}} \right)$$

$$h_i = 2.17 \text{ Btu/hr-ft}^2\text{-°F}$$

If Eq. 33.40(b) is used instead, the value of h_i is approximately 2.01 Btu/hr-ft²-°F, indicating that the temperature is not important.

For natural convection on the outside of the duct, estimate the film temperature.

$$T_{\text{film}} = \left(\tfrac{1}{2} \right) (T_{\text{surface}} + T_\infty) = \left(\tfrac{1}{2} \right) (70°\text{F} + 80°\text{F})$$

$$= 75°\text{F}$$

From App. 32.C, the properties of air at 75°F are

$$\text{Pr} \approx 0.72$$

$$\frac{g\beta\rho^2}{\mu^2} = 2.27 \times 10^6 \,\frac{1}{\text{ft}^3\text{-°F}}$$

The characteristic length is the diameter of the duct.

$$L = (12 \text{ in}) \left(\frac{1 \text{ ft}}{12 \text{ in}} \right) = 1 \text{ ft}$$

The Grashof number is

$$\text{Gr} = L^3 \left(\frac{\rho^2 \beta g}{\mu^2} \right) (T_\infty - T_s)$$

$$= (1 \text{ ft})^3 \left(2.27 \times 10^6 \,\frac{1}{\text{ft}^3\text{-°F}} \right) (80°\text{F} - 70°\text{F})$$

$$= 2.27 \times 10^7$$

$$\text{Pr Gr} = (0.72)(2.27 \times 10^7) = 1.63 \times 10^7$$

From Table 32.3, the film coefficient for a horizontal cylinder is

$$h_o = (0.27) \left(\frac{T_\infty - T_s}{d} \right)^{0.25}$$

$$= (0.27) \left(\frac{80°\text{F} - 70°\text{F}}{1 \text{ ft}} \right)^{0.25}$$

$$= 0.48 \text{ Btu/hr-ft}^2\text{-°F}$$

Neglecting the wall resistance, the overall film coefficient from Eq. 33.71 is

$$\frac{1}{U} = \frac{1}{h_o} + \frac{1}{h_i}$$

$$= \frac{1}{2.17 \,\frac{\text{Btu}}{\text{hr-ft}^2\text{-°F}}} + \frac{1}{0.48 \,\frac{\text{Btu}}{\text{hr-ft}^2\text{-°F}}}$$

$$= 2.544 \text{ hr-ft}^2\text{-°F/Btu}$$

$$U = \frac{1}{2.544 \,\frac{\text{hr-ft}^2\text{-°F}}{\text{Btu}}} = 0.393 \text{ Btu/hr-ft}^2\text{-°F}$$

The heat transfer due to convection is

$$Q_{\text{convection}} = U A_{\text{surface}} (T_\infty - T_{\text{bulk,air}})$$

$$= U (\pi d L)(T_\infty - T_{\text{bulk,air}})$$

$$= \left(0.393 \,\frac{\text{Btu}}{\text{hr-ft}^2\text{-°F}} \right)$$

$$\times \pi (1 \text{ ft})(50 \text{ ft})(80°\text{F} - 47.5°\text{F})$$

$$= 2006.3 \text{ Btu/hr}$$

The heat transfer due to radiation is

$$Q_{\text{radiation}} = \sigma F_e F_a A_{\text{surface}}(T_\infty^4 - T_{\text{surface}}^4)$$

Assume the room and duct have an unobstructed view of each other. Then $F_a = 1.0$ and $F_e = \epsilon = 0.28$. The absolute temperatures are

$$T_\infty = 80°\text{F} + 460 = 540°\text{R}$$

$$T_{\text{surface}} = 70°\text{F} + 460 = 530°\text{R}$$

$$Q_{\text{radiation}} = \left(0.1713 \times 10^{-8}\ \frac{\text{Btu}}{\text{hr-ft}^2\text{-}°\text{R}^4}\right)(0.28)(1.0)$$

$$\times\ \pi(1\ \text{ft})(50\ \text{ft})\left((540°\text{R})^4 - (530°\text{R})^4\right)$$

$$= 461.5\ \text{Btu/hr}$$

The total heat transfer to the air is

$$Q_{\text{total}} = Q_{\text{convection}} + Q_{\text{radiation}}$$

$$= 2006.3\ \frac{\text{Btu}}{\text{hr}} + 461.5\ \frac{\text{Btu}}{\text{hr}}$$

$$= 2467.8\ \text{Btu/hr}$$

At 47.5°F, the specific heat of air is approximately 0.240 Btu/lbm-°F. Since the heat transfer is known, the temperature of air leaving the duct can be calculated from

$$Q_{\text{total}} = \dot{m}c_p(T_{\text{air,out}} - T_{\text{air,in}})$$

$$T_{\text{air,out}} = T_{\text{air,in}} + \frac{Q_{\text{total}}}{\dot{m}c_p}$$

$$= 45°\text{F} + \frac{2467.8\ \dfrac{\text{Btu}}{\text{hr}}}{\left(2357.1\ \dfrac{\text{lbm}}{\text{hr}}\right)\left(0.24\ \dfrac{\text{Btu}}{\text{lbm-}°\text{F}}\right)}$$

$$= \boxed{49.4°\text{F}}$$

This agrees with the engineer's estimate.

SI Solution

The absolute temperature of air entering the duct is $7°\text{C} + 273 = 280\text{K}$. From the ideal gas law, the density of air entering the duct is

$$\rho = \frac{p}{RT}$$

$$= \frac{(101.3\ \text{kPa})\left(1000\ \dfrac{\text{Pa}}{\text{kPa}}\right)}{\left(287.03\ \dfrac{\text{J}}{\text{kg·K}}\right)(280\text{K})}$$

$$= 1.2604\ \text{kg/m}^3$$

The mass flow rate of air entering the duct is

$$\dot{m} = \rho\dot{V} = \left(1.2604\ \frac{\text{kg}}{\text{m}^3}\right)\left(0.25\ \frac{\text{m}^3}{\text{s}}\right) = 0.3151\ \text{kg/s}$$

The diameter of the duct is

$$d = (30\ \text{cm})\left(\frac{1\ \text{m}}{100\ \text{cm}}\right) = 0.30\ \text{m}$$

The velocity of air entering the duct is

$$\text{v} = \frac{\dot{V}}{A_{\text{flow}}} = \frac{0.25\ \dfrac{\text{m}^3}{\text{s}}}{\left(\dfrac{\pi}{4}\right)(0.30\ \text{m})^2} = 3.537\ \text{m/s}$$

To calculate the initial film coefficients, estimate the temperatures based on the claim. The film coefficients are not highly sensitive to small temperature differences.

$$T_{\text{bulk,air}} = \left(\tfrac{1}{2}\right)(T_{\text{air,in}} + T_{\text{air,out}})$$

$$= \left(\tfrac{1}{2}\right)(7°\text{C} + 10°\text{C}) = 8.5°\text{C}$$

$$T_{\text{surface}} = 20°\text{C}\quad\text{[estimate]}$$

The film coefficient for air flowing inside the duct is given from Eq. 33.40(a) (independent of temperature).

$$h_i \approx \frac{(3.52)(\text{v}_{\text{m/s}})^{0.8}}{(d_{\text{m}})^{0.2}}$$

$$= \frac{(3.52)\left(3.537\ \dfrac{\text{m}}{\text{s}}\right)^{0.8}}{(0.30\ \text{m})^{0.2}} = 12.3\ \text{W/m}^2\text{·}°\text{C}$$

For natural convection on the outside of the duct, estimate the film coefficient.

$$T_{\text{film}} = \left(\tfrac{1}{2}\right)(T_{\text{surface}} + T_\infty)$$

$$= \left(\tfrac{1}{2}\right)(20°\text{C} + 27°\text{C}) = 23.5°\text{C}$$

From App. 32.D, the properties of air at 23.5°C are

$$\text{Pr} = 0.709$$

$$\frac{g\beta\rho^2}{\mu^2} = 1.43 \times 10^8\ \frac{1}{\text{K·m}^3}$$

The characteristic length, L, is the diameter of the duct, which is 0.30 m. The Grashof number is

$$\text{Gr} = L^3\left(\frac{\rho^2\beta g}{\mu^2}\right)(T_\infty - T_{\text{surface}})$$

$$= (0.30\ \text{m})^3\left(1.43 \times 10^8\ \frac{1}{\text{K·m}^3}\right)(27°\text{C} - 20°\text{C})$$

$$= 2.70 \times 10^7$$

$$\text{Pr}\,\text{Gr} = (0.709)(2.70 \times 10^7) = 1.9 \times 10^7$$

From Table 32.3, the film coefficient for a horizontal cylinder is

$$h_o \approx (1.32)\left(\frac{T_\infty - T_{\text{surface}}}{d}\right)^{0.25}$$

$$= (1.32)\left(\frac{27°\text{C} - 20°\text{C}}{0.30\ \text{m}}\right)^{0.25} = 2.90\ \text{W/m}^2\text{·K}$$

Neglecting the wall resistance, the overall film coefficient from Eq. 34.63 is

$$\frac{1}{U} = \frac{1}{h_o} + \frac{1}{h_i}$$

$$= \frac{1}{12.3 \; \frac{W}{m^2 \cdot K}} + \frac{1}{2.90 \; \frac{W}{m^2 \cdot K}} = 0.426 \; m^2 \cdot K/W$$

$$U = \frac{1}{0.426 \; \frac{m^2 \cdot K}{W}} = 2.35 \; W/m^2 \cdot K$$

The heat transfer due to convection is

$$\begin{aligned}
Q_{\text{convection}} &= U A_{\text{surface}}(T_\infty - T_{\text{bulk,air}}) \\
&= U \pi d L (T_\infty - T_{\text{bulk,air}}) \\
&= \left(2.35 \; \frac{W}{m^2 \cdot K} \right) \pi (0.30 \; m)(15 \; m) \\
&\quad \times (27°C - 8.5°C) \\
&= 614.6 \; W
\end{aligned}$$

The heat transfer due to radiation is

$$Q_{\text{radiation}} = \sigma F_e F_a A_{\text{surface}} \left(T_\infty^4 - T_{\text{surface}}^4 \right)$$

Assume the room and duct have an unobstructed view of each other. Then $F_a = 1.0$ and $F_e = \epsilon = 0.28$. The absolute temperatures are

$$T_\infty = 27°C + 273 = 300K$$

$$T_{\text{surface}} = 20°C + 273 = 293K$$

$$\begin{aligned}
Q_{\text{radiation}} &= \left(5.67 \times 10^{-8} \; \frac{W}{m^2 \cdot K^4} \right)(0.28)(1)\pi \\
&\quad \times (0.30 \; m)(15 \; m) \left((300K)^4 - (293K)^4 \right) \\
&= 163.8 \; W
\end{aligned}$$

The total heat transfer to the air is

$$\begin{aligned}
Q_{\text{total}} &= Q_{\text{convection}} + Q_{\text{radiation}} \\
&= 614.6 \; W + 163.8 \; W = 778.4 \; W
\end{aligned}$$

At 8.5°C, the specific heat of air is

$$\left(1.0048 \; \frac{kJ}{kg \cdot K} \right) \left(1000 \; \frac{J}{kJ} \right) = 1004.8 \; J/kg \cdot K$$

Since the heat transfer is known, the temperature of air leaving the duct can be calculated by

$$Q_{\text{total}} = \dot{m} c_p (T_{\text{air,out}} - T_{\text{air,in}})$$

$$\begin{aligned}
T_{\text{air,out}} &= T_{\text{air,in}} + \frac{Q_{\text{total}}}{\dot{m} c_p} \\
&= 7°C + \frac{778.4 \; W}{\left(0.3151 \; \frac{kg}{s} \right) \left(1004.8 \; \frac{J}{kg \cdot K} \right)} \\
&= \boxed{9.5°C}
\end{aligned}$$

This agrees with the engineer's estimate.

5. *Customary U.S. Solution*

(a) The absolute temperature of entering air is

$$500°F + 460 = 960°R$$

The absolute pressure of entering air is

$$25 \; \text{psig} + 14.7 \; \text{psi} = 39.7 \; \text{psia}$$

The density of air entering, from the ideal gas law, is

$$\begin{aligned}
\rho &= \frac{p}{RT} \\
&= \frac{\left(39.7 \; \frac{lbf}{in^2} \right) \left(144 \; \frac{in^2}{ft^2} \right)}{\left(53.35 \; \frac{ft\text{-}lbf}{lbm\text{-}°R} \right)(960°R)} = 0.1116 \; lbm/ft^3
\end{aligned}$$

The mass flow rate of entering air is

$$\begin{aligned}
\dot{m} &= \rho \dot{V} = \left(0.1116 \; \frac{lbm}{ft^3} \right) \left(200 \; \frac{ft^3}{min} \right) \left(60 \; \frac{min}{hr} \right) \\
&= 1339.2 \; lbm/hr
\end{aligned}$$

At low pressures, the air enthalpy is found from air tables (App. 23.F).

The absolute temperature of leaving air is

$$350°F + 460 = 810°R$$

From App. 23.F,

$$h_1 = 231.06 \; \text{Btu/lbm at } 960°R$$

$$h_2 = 194.25 \; \text{Btu/lbm at } 810°R$$

The heat loss is

$$\begin{aligned}
Q &= \dot{m}(h_1 - h_2) \\
&= \left(1339.2 \; \frac{lbm}{hr} \right) \left(231.06 \; \frac{Btu}{lbm} - 194.25 \; \frac{Btu}{lbm} \right) \\
&= 49{,}296 \; \text{Btu/hr}
\end{aligned}$$

Assuming midpoint pipe surface temperature,

$$\left(\tfrac{1}{2} \right)(T_{\text{in}} + T_{\text{out}}) = \left(\tfrac{1}{2} \right)(500°F + 350°F) = 425°F$$

Since the heat loss is known, the overall heat transfer coefficient can be determined from

$$Q = U A \Delta T = U(\pi d L) \Delta T$$

$$\begin{aligned}
U &= \frac{Q}{(\pi d L)\Delta T} \\
&= \frac{49{,}296 \; \frac{Btu}{hr}}{\pi(4.25 \; in)\left(\frac{1 \; ft}{12 \; in} \right)(35 \; ft)(425°F - 70°F)} \\
&= \boxed{3.57 \; \text{Btu/hr-ft}^2\text{-}°F}
\end{aligned}$$

The answer is (C).

(b) To calculate the overall heat transfer coefficient, disregard the pipe thermal resistance and the inside film coefficient (small compared with outside film and radiation).

Work with the midpoint pipe temperature of 425°F.

The absolute temperatures are

$$T_1 = 425°F + 460 = 885°R$$
$$T_2 = 70°F + 460 = 530°R$$

For radiation heat loss, assume $F_a = 1$. For 500°F enamel paint of any color,

$$F_e = \epsilon \approx 0.9$$

$$\frac{Q_{net}}{A} = E_{net} = \sigma F_e F_a \left(T_1^4 - T_s^4\right)$$

$$= \left(0.1713 \times 10^{-8} \frac{Btu}{hr\text{-}ft^2\text{-}°R^4}\right)$$

$$\times (0.9)(1.0)\left((885°R)^4 - (530°R)^4\right)$$

$$= 824.1 \ Btu/hr\text{-}ft^2$$

From Eq. 34.21, the radiant heat transfer coefficient is

$$h_{radiation} = \frac{E_{net}}{T_1 - T_2} = \frac{824.1 \frac{Btu}{hr}}{885°R - 530°R}$$

$$= 2.32 \ Btu/hr\text{-}ft^2\text{-}°F$$

For the outside film coefficient, evaluate the film at the pipe midpoint. The film temperature is

$$T_f = \left(\tfrac{1}{2}\right)(425°F + 70°F) = 247.5°F$$

From App. 32.C,

$$Pr = 0.72$$

$$\frac{g\beta\rho^2}{\mu^2} = 0.657 \times 10^6 \ \frac{1}{ft^3\text{-}°F}$$

The characteristic length, L, is $d_o = 4.25$ in. The Grashof number is

$$Gr = L^3 \left(\frac{\rho^2 \beta g}{\mu^2}\right) \Delta T$$

$$= (4.25 \ in)^3 \left(\frac{1 \ ft}{12 \ in}\right)^3 \left(0.657 \times 10^6 \ \frac{1}{ft^3\text{-}°F}\right)$$

$$\times (425°F - 70°F)$$

$$= 1.04 \times 10^7$$

$$Pr \, Gr = (0.72)(1.04 \times 10^7) = 7.5 \times 10^6$$

From Table 35.3, the film coefficient for a horizontal cylinder is

$$h_o = (0.27)\left(\frac{\Delta T}{d}\right)^{0.25}$$

$$= (0.27)\left(\frac{425°F - 70°F}{(4.25 \ in)\left(\frac{1 \ ft}{12 \ in}\right)}\right)^{0.25}$$

$$= 1.52 \ Btu/hr\text{-}ft^2\text{-}°F$$

The overall film coefficient is

$$U = h_{total} = h_{radiation} + h_o$$

$$= 2.32 \frac{Btu}{hr\text{-}ft^2\text{-}°F} + 1.52 \frac{Btu}{hr\text{-}ft^2\text{-}°F}$$

$$= \boxed{3.84 \ Btu/hr\text{-}ft^2\text{-}°F}$$

The answer is (C).

This is not too far from the actual value.

(c) • The internal film coefficient was disregarded.

• The emissivity could be lower due to dirt on the outside of the duct.

• h_r and h_o are not really additive.

• Pipe thermal resistance was disregarded.

• The midpoint calculations should be replaced with integration along the length.

SI Solution

(a) The absolute temperature of entering air is

$$260°C + 273 = 533K$$

The absolute pressure of entering air is

$$170 \ kPa + 101.3 \ kPa = 271.3 \ kPa$$

The density of air entering, from the ideal gas law, is

$$\rho = \frac{p}{RT} = \frac{(271.3 \ kPa)\left(1000 \ \frac{Pa}{kPa}\right)}{\left(287.03 \ \frac{J}{kg\cdot K}\right)(533K)}$$

$$= 1.773 \ kg/m^3$$

The mass flow rate of entering air is

$$\dot{m} = \rho \dot{V} = \left(1.773 \ \frac{kg}{m^3}\right)\left(0.1 \ \frac{m^3}{s}\right) = 0.1773 \ kg/s$$

At low pressure, the air enthalpy is found from air tables. The absolute temperature of leaving air is

$$180°C + 273 = 453K$$

From App. 23.S,

$$h_1 = 537.09 \ kJ/kg \ at \ 533K$$
$$h_2 = 454.87 \ kJ/kg \ at \ 453K$$

The heat loss is

$$Q = \dot{m}(h_1 - h_2)$$

$$= \left(0.1773 \ \frac{\text{kg}}{\text{s}}\right)\left(537.09 \ \frac{\text{kJ}}{\text{kg}} - 454.87 \ \frac{\text{kJ}}{\text{kg}}\right)$$

$$\times \left(1000 \ \frac{\text{W}}{\text{kW}}\right)$$

$$= 14\,578 \ \text{W}$$

Assume the midpoint pipe surface temperature.

$$\left(\tfrac{1}{2}\right)(T_{\text{in}} - T_{\text{out}}) = \left(\tfrac{1}{2}\right)(260°\text{C} + 180°\text{C}) = 220°\text{C}$$

Since the heat loss is known, the overall heat transfer coefficient can be determined from

$$Q = UA\Delta T = U(\pi dL)\Delta T$$

$$U = \frac{Q}{(\pi dL)\Delta T}$$

$$= \frac{14\,578 \ \text{W}}{\pi(10.8 \ \text{cm})\left(\dfrac{1 \ \text{m}}{100 \ \text{cm}}\right)(10 \ \text{m})(220°\text{C} - 20°\text{C})}$$

$$= \boxed{21.5 \ \text{W/m}^2\text{·K}}$$

The answer is (C).

(b) From the customary U.S. solution, work with the midpoint pipe temperature of 220°C.

The absolute temperatures are

$$T_1 = 220°\text{C} + 273 = 493\text{K}$$
$$T_2 = 20°\text{C} + 273 = 293\text{K}$$

For radiation heat loss, assume $F_a = 1$. For 260°C enamel paint of any color,

$$F_e = \epsilon \approx 0.9$$

From Eq. 34.21, the radiant heat transfer coefficient is

$$h_r = \frac{\sigma F_a F_e \left(T_1^4 - T_2^4\right)}{T_1 - T_2}$$

$$= \frac{\left(5.67 \times 10^{-8} \ \frac{\text{W}}{\text{m}^2\text{·K}^4}\right)(1)(0.9)}{493\text{K} - 293\text{K}}$$

$$= 13.2 \ \text{W/m}^2\text{·K}$$

For the outside film coefficient, evaluate the film at the pipe midpoint. The film temperature is

$$T_f = \left(\tfrac{1}{2}\right)(220°\text{C} + 20°\text{C}) = 120°\text{C}$$

From App. 32.D, at 120°C,

$$\text{Pr} \approx 0.692$$

$$\frac{g\beta\rho^2}{\mu^2} = 0.528 \times 10^8 \ \frac{1}{\text{K·m}^3}$$

The characteristic length is

$$L = \text{outside diameter} = (10.8 \ \text{cm})\left(\frac{1 \ \text{m}}{100 \ \text{cm}}\right)$$

$$= 0.108 \ \text{m}$$

The Grashof number is

$$\text{Gr} = L^3 \left(\frac{\rho^2 g\beta}{\mu^2}\right)\Delta T$$

$$= (0.108 \ \text{m})^3 \left(0.528 \times 10^8 \ \frac{1}{\text{K·m}^3}\right)$$

$$\times (220°\text{C} - 20°\text{C})$$

$$= 1.33 \times 10^7$$

$$\text{Pr Gr} = (0.692)(1.33 \times 10^7) = 9.20 \times 10^6$$

From Table 32.3, the film coefficient for a horizontal cylinder is

$$h_o = (1.32)\left(\frac{\Delta T}{d}\right)^{0.25}$$

$$= (1.32)\left(\frac{220°\text{C} - 20°\text{C}}{0.108 \ \text{m}}\right)^{0.25}$$

$$= 8.66 \ \text{W/m}^2\text{·K}$$

The overall film coefficient is

$$U = h_{\text{total}} = h_r + h_o = 13.2 \ \frac{\text{W}}{\text{m}^2\text{·K}} + 8.66 \ \frac{\text{W}}{\text{m}^2\text{·K}}$$

$$= \boxed{21.86 \ \text{W/m}^2\text{·K}}$$

The answer is (C).

This solution is almost the same as the actual value.

(c) See the customary U.S. solution.

6. *Customary U.S. Solution*

(a) Heat is lost from the top and sides by radiation and convection. The absolute temperature of the surroundings is

$$75°\text{F} + 460 = 535°\text{R}$$

$$A_{\text{sides}} = \pi dL = \pi(0.75 \ \text{in})(1.5 \ \text{in})\left(\frac{1 \ \text{ft}}{12 \ \text{in}}\right)^2$$

$$= 0.0245 \ \text{ft}^2$$

$$A_{\text{top}} = \frac{\pi}{4}d^2 = \left(\frac{\pi}{4}\right)(0.75 \ \text{in})^2\left(\frac{1 \ \text{ft}}{12 \ \text{in}}\right)^2$$

$$= 0.003068 \ \text{ft}^2$$

$$Q_{\text{total}} = Q_{\text{convection}} + Q_{\text{radiation}}$$

$$= h_{\text{sides}}A_{\text{sides}}(T_s - T_\infty) + h_{\text{top}}A_{\text{top}}(T_s - T_\infty)$$

$$+ \sigma F_e F_a(A_{\text{sides}} + A_{\text{top}})(T_s^4 - T_\infty^4) \quad \text{[Eq. 1]}$$

For the first approximation of T_s, assume $h_{sides} = h_{top}$ = 1.65 Btu/hr-ft²-°F, $F_a = 1$, and $F_e = \epsilon = 0.65$.

$$(5.0 \text{ W})\left(\frac{3.412\ \frac{\text{Btu}}{\text{hr}}}{1\ \text{W}}\right) = \left(1.65\ \frac{\text{Btu}}{\text{hr-ft}^2\text{-}°\text{F}}\right)$$
$$\times (0.0245\ \text{ft}^2)(T_s - 535°\text{R})$$
$$+ \left(1.65\ \frac{\text{Btu}}{\text{hr-ft}^2\text{-}°\text{F}}\right)$$
$$\times (0.003068\ \text{ft}^2)(T_s - 535°\text{R})$$
$$+ \left(0.1713 \times 10^{-8}\ \frac{\text{Btu}}{\text{hr-ft}^2\text{-}°\text{R}^4}\right)$$
$$\times (0.65)(1)(0.0245\ \text{ft}^2$$
$$+ 0.003068\ \text{ft}^2)$$
$$\times \left(T_s^4 - (535°\text{R})^4\right)$$

By trial and error, $T_s \approx 750°$R.

For natural convection on the outside, estimate the film temperature.

$$T_{\text{film}} = \left(\tfrac{1}{2}\right)(T_s + T_\infty) = \left(\tfrac{1}{2}\right)(750°\text{R} + 535°\text{R})$$
$$= 642.5°\text{R}\ \ (182.5°\text{F})$$

From App. 32.C, the properties of air at 182.5°F are

$$\text{Pr} \approx 0.72$$
$$\frac{g\beta\rho^2}{\mu^2} = 1.01 \times 10^6\ \frac{1}{\text{ft}^3\text{-}°\text{F}}$$

For the sides, the characteristic length is 1.5 in. The Grashof number is

$$\text{Gr} = L^3\left(\frac{\rho^2\beta g}{\mu^2}\right)(T_s - T_\infty)$$
$$= (1.5\ \text{in})^3\left(\frac{1\ \text{ft}}{12\ \text{in}}\right)^3\left(1.01 \times 10^6\ \frac{1}{\text{ft}^3\text{-}°\text{F}}\right)$$
$$\times (750°\text{R} - 535°\text{R})$$
$$= 4.24 \times 10^5$$
$$\text{Pr Gr} = (0.72)(4.24 \times 10^5) = 3.05 \times 10^5$$

From Table 32.3, the film coefficient for a vertical surface is

$$h_{\text{sides}} = (0.29)\left(\frac{T_s - T_\infty}{L}\right)^{0.25}$$
$$= (0.29)\left(\frac{750°\text{R} - 535°\text{R}}{(1.5\ \text{in})\left(\frac{1\ \text{ft}}{12\ \text{in}}\right)}\right)^{0.25}$$
$$= 1.87\ \text{Btu/hr-ft}^2\text{-}°\text{F}$$

(This application is slightly outside the useful range of this correlation.)

For the top, the characteristic length is 0.75 in, so

$$\text{Gr} = (0.75\ \text{in})^3\left(\frac{1\ \text{ft}}{12\ \text{in}}\right)^3\left(1.01 \times 10^6\ \frac{1}{\text{ft}^3\text{-}°\text{F}}\right)$$
$$\times (750°\text{R} - 535°\text{R})$$
$$= 5.30 \times 10^4$$
$$\text{Pr Gr} = (0.72)(5.30 \times 10^4) = 3.8 \times 10^4$$

From Table 32.3, the film coefficient for a horizontal surface is

$$h_{\text{top}} = (0.27)\left(\frac{T_s - T_\infty}{L}\right)^{0.25}$$
$$= (0.27)\left(\frac{750°\text{R} - 535°\text{R}}{(0.9)(0.75\ \text{in})\left(\frac{1\ \text{ft}}{12\ \text{in}}\right)}\right)^{0.25}$$
$$= 2.12\ \text{Btu/hr-ft}^2\text{-}°\text{F}$$

Substituting the calculated values of h_{top} and h_{sides} into Eq. 1 gives

$$Q_{\text{total}} = (5.0\ \text{W})\left(\frac{3.412\ \frac{\text{Btu}}{\text{hr}}}{1\ \text{W}}\right)$$
$$= \left(1.87\ \frac{\text{Btu}}{\text{hr-ft}^2\text{-}°\text{F}}\right)(0.0245\ \text{ft}^2)(T_s - 535°\text{R})$$
$$+ \left(2.12\ \frac{\text{Btu}}{\text{hr-ft}^2\text{-}°\text{F}}\right)(0.003068\ \text{ft}^2)(T_s - 535°\text{R})$$
$$+ \left(0.1713 \times 10^{-8}\ \frac{\text{Btu}}{\text{hr-ft}^2\text{-}°\text{R}^4}\right)(0.65)(1)$$
$$\times (0.0245\ \text{ft}^2 + 0.003068\ \text{ft}^2)\left(T_s^4 - (535°\text{R})^4\right)$$

By trial and error, $T_s \approx \boxed{736°\text{R}.}$

(b) Substituting $T_s = 736°$R in the preceding equation,

$$Q_{\text{total}} = 9.209\ \frac{\text{Btu}}{\text{hr}} + 1.277\ \frac{\text{Btu}}{\text{hr}} + 6.492\ \frac{\text{Btu}}{\text{hr}}$$
$$= 16.98\ \text{Btu/hr}$$

$$\frac{Q_{\text{convection}}}{Q_{\text{total}}} = \frac{9.209\ \frac{\text{Btu}}{\text{hr}} + 1.277\ \frac{\text{Btu}}{\text{hr}}}{16.98\ \frac{\text{Btu}}{\text{hr}}}$$
$$= \boxed{0.618\ \ (61.8\%)}$$

$$\frac{Q_{\text{radiation}}}{Q_{\text{total}}} = \frac{6.492\ \frac{\text{Btu}}{\text{hr}}}{16.98\ \frac{\text{Btu}}{\text{hr}}} = \boxed{0.382\ \ (38.2\%)}$$

Heat Transfer

SI Solution

(a) Heat is lost from the top and sides by radiation and convection.

The absolute temperature of the surrounding is

$$24°C + 273 = 297K$$

$$A_{\text{sides}} = \pi dL = \pi(19 \text{ mm})(38 \text{ mm})\left(\frac{1 \text{ m}}{1000 \text{ mm}}\right)^2$$

$$= 0.00227 \text{ m}^2$$

$$A_{\text{top}} = \frac{\pi}{4}d^2 = \left(\frac{\pi}{4}\right)(19 \text{ mm})^2\left(\frac{1 \text{ m}}{1000 \text{ mm}}\right)^2$$

$$= 0.000284 \text{ m}^2$$

$$Q_{\text{total}} = Q_{\text{convection}} + Q_{\text{radiation}}$$
$$= h_{\text{sides}}A_{\text{sides}}(T_s - T_\infty) + h_{\text{top}}A_{\text{top}}(T_s - T_\infty)$$
$$+ \sigma F_e F_a(A_{\text{sides}} + A_{\text{top}})(T_s^4 - T_\infty^4) \quad [\text{Eq. 1}]$$

For a first approximation of T_s, assume $h_{\text{sides}} = h_{\text{top}} = 9.4 \text{ W/m}^2\cdot\text{K}$, $F_a = 1$, and $F_e = \epsilon = 0.65$.

$$5.0 \text{ W} = \left(9.4 \frac{\text{W}}{\text{m}^2\cdot\text{K}}\right)(0.00227 \text{ m}^2)(T_s - 297K)$$

$$+ \left(9.4 \frac{\text{W}}{\text{m}^2\cdot\text{K}}\right)(0.000284 \text{ m}^2)(T_s - 297K)$$

$$+ \left(5.67 \times 10^{-8} \frac{\text{W}}{\text{m}^2\cdot\text{K}^4}\right)(0.65)(1)$$

$$\times (0.00227 \text{ m}^2 + 0.000284 \text{ m}^2)$$

$$\times \left(T_s^4 - (297K)^4\right)$$

By trial and error, $T_s = 416.75K$.

For natural convection on the outside, estimate the film temperature.

$$T_{\text{film}} = \left(\tfrac{1}{2}\right)(T_s + T_\infty) = \left(\tfrac{1}{2}\right)(416.75K + 297K)$$
$$= 356.9K \quad (83.9°C)$$

From App. 32.D, the properties of air at 83.9°C are

$$\text{Pr} \approx 0.697$$

$$\frac{g\beta\rho^2}{\mu^2} \approx 0.616 \times 10^8 \frac{1}{\text{K}\cdot\text{m}^3}$$

For the sides, the characteristic length is 38 mm.

$$\text{Gr} = (38 \text{ mm})^3\left(\frac{1 \text{ m}}{1000 \text{ mm}}\right)^3\left(0.616 \times 10^8 \frac{1}{\text{K}\cdot\text{m}^3}\right)$$

$$\times (416.75K - 297K)$$

$$= 4.048 \times 10^5$$

$$\text{Pr Gr} = (0.697)(4.048 \times 10^5) = 2.82 \times 10^5$$

For the top, the characteristic length is 19 mm.

$$\text{Gr} = (19 \text{ mm})^3\left(\frac{1 \text{ m}}{1000 \text{ mm}}\right)^3\left(0.616 \times 10^8 \frac{1}{\text{K}\cdot\text{m}^3}\right)$$

$$\times (416.75K - 297K)$$

$$= 5.06 \times 10^4$$

$$\text{Pr Gr} = (0.697)(5.06 \times 10^4) = 3.53 \times 10^4$$

From Table 32.3, the film coefficient for a horizontal surface is

$$h_{\text{top}} = (1.32)\left(\frac{T_s - T_\infty}{L}\right)^{0.25}$$

$$= (1.32)\left(\frac{416.75K - 297K}{(0.9)(19 \text{ mm})\left(\frac{1 \text{ m}}{1000 \text{ mm}}\right)}\right)^{0.25}$$

$$= 12.08 \text{ W/m}^2\cdot\text{K}$$

From Table 32.3, the film coefficient for a vertical surface is

$$h_{\text{sides}} = (1.37)\left(\frac{416.75K - 297K}{(38 \text{ mm})\left(\frac{1 \text{ m}}{1000 \text{ mm}}\right)}\right)^{0.25}$$

$$= 10.26 \text{ W/m}^2\cdot\text{K}$$

(This application is slightly outside the useful range of this correlation.)

Using the calculated values of h_{top} and h_{sides} into Eq. 1,

$$Q_{\text{total}} = 5 \text{ W}$$

$$= \left(10.26 \frac{\text{W}}{\text{m}^2\cdot\text{K}}\right)(0.00227 \text{ m}^2)(T_s - 297K)$$

$$+ \left(12.08 \frac{\text{W}}{\text{m}^2\cdot\text{K}}\right)(0.000284 \text{ m}^2)(T_s - 297K)$$

$$+ \left(5.67 \times 10^{-8} \frac{\text{W}}{\text{m}^2\cdot\text{K}^4}\right)(0.65)(1)$$

$$\times (0.00227 \text{ m}^2 + 0.000284 \text{ m}^2)$$

$$\times \left(T_s^4 - (297K)^4\right)$$

By trial and error, $T_s = \boxed{411K.}$

(b) Substitute $T_s = 411K$ into the preceding equation.

$$Q_{\text{total}} = 2.655 \text{ W} + 0.381 \text{ W} + 1.953 \text{ W}$$

$$= 4.989 \text{ W}$$

$$\frac{Q_{\text{convection}}}{Q_{\text{total}}} = \frac{2.655 \text{ W} + 0.381 \text{ W}}{4.989 \text{ W}} = \boxed{0.609 \ (60.9\%)}$$

$$\frac{Q_{\text{radiation}}}{Q_{\text{total}}} = \frac{1.953 \text{ W}}{4.989 \text{ W}} = \boxed{0.391 \ (39.1\%)}$$

7. *Customary U.S. Solution*

The velocity is relatively low, so incompressible flow can be assumed.

The film coefficient on the probe is

$$h = \frac{0.024G^{0.8}}{D^{0.4}}$$

$$= \frac{(0.024)\left(3480 \ \frac{\text{lbm}}{\text{hr-ft}^2}\right)^{0.8}}{\left((0.5 \text{ in})\left(\frac{1 \text{ ft}}{12 \text{ in}}\right)\right)^{0.4}} = 58.3 \text{ Btu/hr-ft}^2\text{-}°\text{F}$$

The absolute temperature of the walls is

$$T_{\text{walls}} = 600°\text{F} + 460 = 1060°\text{R}$$

Neglect conduction and the insignificant kinetic energy loss. The thermocouple gains heat through radiation from the walls and loses heat through convection to the gas.

$$Q_{\text{convection}} = AE_{\text{radiation}}$$
$$hA(T_{\text{probe}} - T_{\text{gas}}) = A\sigma\epsilon(T^4_{\text{walls}} - T^4_{\text{probe}})$$
$$h(T_{\text{probe}} - T_{\text{gas}}) = \sigma\epsilon(T^4_{\text{walls}} - T^4_{\text{probe}})$$

(a) If the actual gas temperature is $300°\text{F} + 460 = 760°\text{R}$,

$$\left(58.3 \ \frac{\text{Btu}}{\text{hr-ft}^2\text{-}°\text{F}}\right)$$
$$\times (T_{\text{probe}} - 760°\text{R}) = \left(0.1713 \times 10^{-8} \ \frac{\text{Btu}}{\text{hr-ft}^2\text{-}°\text{R}^4}\right)$$
$$\times (0.8)\left((1060°\text{R})^4 - T^4_{\text{probe}}\right)$$
$$(1.37 \times 10^{-9})T^4_{\text{probe}}$$
$$+ (58.3)T_{\text{probe}} = 46{,}038$$

By trial and error, $T_{\text{probe}} = \boxed{781°\text{R}.}$

The answer is (D).

(b) If $T_{\text{probe}} = 300°\text{F} + 460 = 760°\text{R}$,

$$\left(58.3 \ \frac{\text{Btu}}{\text{hr-ft}^2\text{-}°\text{F}}\right)$$
$$\times (760°\text{R} - T_{\text{gas}}) = \left(0.1713 \times 10^{-8} \ \frac{\text{Btu}}{\text{hr-ft}^2\text{-}°\text{R}^4}\right)$$
$$\times (0.8)\left((1060°\text{R})^4 - (760°\text{R})^4\right)$$
$$T_{\text{gas}} = \boxed{738.2°\text{R}}$$

The answer is (B).

SI Solution

The velocity is relatively low, so incompressible flow can be assumed.

The film coefficient on the probe is

$$h = \frac{2.9G^{0.8}}{D^{0.4}}$$

$$= \frac{(2.9)\left(4.7 \ \frac{\text{kg}}{\text{s}\cdot\text{m}^2}\right)^{0.8}}{(13 \text{ mm})^{0.4}\left(\frac{1 \text{ m}}{1000 \text{ mm}}\right)^{0.4}} = 56.82 \text{ W/m}^2\cdot\text{K}$$

The absolute temperature of the walls is

$$T_{\text{walls}} = 315°\text{C} + 273 = 588\text{K}$$

Neglecting conduction and the insignificant kinetic energy loss, the thermocouple gains heat through radiation from the walls and loses heat through convection to the gas.

$$\frac{Q_{\text{convection}}}{A} = E_{\text{radiation}}$$
$$h(T_{\text{probe}} - T_{\text{gas}}) = \sigma\epsilon(T^4_{\text{walls}} - T^4_{\text{probe}})$$

(a) If the actual gas temperature is $150°\text{C} + 273 = 423\text{K}$,

$$\left(56.82 \ \frac{\text{W}}{\text{m}^2\cdot\text{K}}\right)$$
$$\times (T_{\text{probe}} - 423\text{K}) = \left(5.67 \times 10^{-8} \ \frac{\text{W}}{\text{m}^2\cdot\text{K}^4}\right)(0.8)$$
$$\times \left((588\text{K})^4 - T^4_{\text{probe}}\right)$$
$$(4.536 \times 10^{-8})T^4_{\text{probe}}$$
$$+ (56.82)T_{\text{probe}} = 29\,457$$

By trial and error, $T_{\text{probe}} = \boxed{477\text{K}.}$

The answer is (D).

(b) If $T_{\text{probe}} = 150°\text{C} + 273 = 423\text{K}$,

$$\left(56.82 \ \frac{\text{W}}{\text{m}^2\cdot\text{K}}\right)$$
$$\times (423\text{K} - T_{\text{gas}}) = \left(5.67 \times 10^{-8} \ \frac{\text{W}}{\text{m}^2\cdot\text{K}^4}\right)(0.8)$$
$$\times \left((588\text{K})^4 - (423\text{K})^4\right)$$
$$T_{\text{gas}} = \boxed{353.1\text{K}}$$

The answer is (B).

Heat Transfer

8. Since the furnace is large compared to the tube, the arrangement factor is taken as 1.0. The radiation from the furnace gas will have a wavelength related to the gas temperature, which is 1000K. The emissivity of the tube at that wavelength is 0.60. The radiant heat transfer rate can be calculated from Eq. 34.24.

$$Q = A\epsilon\sigma \left(T_1^4 - T_2^4\right) = \pi DL\epsilon\sigma(T_1^4 - T_2^4)$$
$$= \pi(0.025 \text{ m})(0.065 \text{ m})(0.60)$$
$$\times \left(5.676 \times 10^{-8} \, \frac{\text{W}}{\text{m}^2 \cdot \text{K}^4}\right)$$
$$\times \left((1000\text{K})^4 - (550\text{K})^4\right)$$
$$= \boxed{1579 \text{ W}}$$

9. (a) Use Eq. 34.26 to calculate the radiant heat transfer coefficient.

$$h_{\text{radiation}} = \epsilon\sigma \left(\frac{T_1^4 - T_2^4}{T_1 - T_2}\right)$$
$$= (0.60) \left(5.676 \times 10^{-8} \, \frac{\text{W}}{\text{m}^2 \cdot \text{K}^4}\right)$$
$$\times \left(\frac{(1000\text{K})^4 - (550\text{K})^4}{1000\text{K} - 550\text{K}}\right)$$
$$= \boxed{68.8 \text{ W/m}^2 \cdot \text{K}}$$

(b) The convective heat transfer coefficient is determined from the given equation.

$$h_{\text{convection}} = 1.32 \left(\frac{T_1 - T_2}{D}\right)^{0.25}$$
$$= (1.32) \left(\frac{1000\text{K} - 550\text{K}}{0.025 \text{ m}}\right)^{0.25}$$
$$= 15.3 \text{ W/m}^2 \cdot \text{K}$$

The combined heat transfer rate is given by Eq. 34.27.

$$Q = (h_{\text{radiation}} + h_{\text{conduction}} + h_{\text{convection}})A(T_1 - T_2)$$
$$= \left(68.8 \, \frac{\text{W}}{\text{m}^2 \cdot \text{K}} + 0 + 15.3 \, \frac{\text{W}}{\text{m}^2 \cdot \text{K}}\right)$$
$$\times \left(\pi(0.025 \text{ m})(0.65 \text{ m})\right)(1000\text{K} - 550\text{K})$$
$$= \boxed{1932 \text{ W}}$$

10. (a) The effective emissivity of the surface is obtained using Eq. 34.29.

$$\epsilon_{\text{eff}} = \frac{\epsilon + 1.0}{2} = \frac{0.9 + 1.0}{2}$$
$$= 0.95$$

Use Eq. 34.30 to calculate the absorptivity from the gas emissivity.

$$\alpha_g = \epsilon_g \left(\frac{T_g}{T_1}\right)^{0.65} = (0.075) \left(\frac{1200\text{K}}{1100\text{K}}\right)^{0.65}$$
$$= 0.079$$

Arrange Eq. 34.28 to solve for radiant heat transfer per unit area.

$$\frac{Q}{A} = \epsilon_{\text{eff}}\sigma(\epsilon_g T_g^4 - \alpha_g T_1^4)$$
$$= (0.95) \left(5.676 \times 10^{-8} \, \frac{\text{W}}{\text{m}^2 \cdot \text{K}^4}\right)$$
$$\times \left((0.075)(1200\text{K})^4 - (0.079)(1100\text{K})^4\right)$$
$$= \boxed{2149 \text{ W/m}^2}$$

(b) The convective heat flux is calculated using the given convective heat transfer coefficient.

$$\frac{Q}{A} = h_{\text{convection}} (T_g - T_1)$$
$$= \left(7.5 \, \frac{\text{W}}{\text{m}^2 \cdot \text{K}}\right)(1200\text{K} - 1100\text{K})$$
$$= \boxed{750 \text{ W/m}^2}$$

PRACTICE PROBLEMS

Phosphorus

1. (*Time limit: one hour*) In a study of a small pond to determine phosphorus impact, the following information was collected.

> pond size: 12 ac-ft (14.8×10^6 L)
> watershed area: 4 ac (grassland)
> average annual rainfall: 5 in
> bioavailable P in runoff: 0.01 mg/L
> runoff coefficient: 0.1

Biological processes in the pond biota convert phosphorus to a nonbioavailable form at the rate of 22% per year. Recycling of sediment phosphorus by rooted plants and by anaerobic conditions in the hypolimnion converts 12% per year of the nonbioavailable phosphorus back to bioavailable forms.

Runoff into the pond evaporates during the year, so no change in pond volume occurs.

(a) Starting from an initial condition of 0.1 mg/L bioavailable P, what is the P concentration after five annual cycles?

(A) 0.06 mg/L
(B) 0.11 mg/L
(C) 0.13 mg/L
(D) 0.18 mg/L

(b) What can be done to reduce the P accumulation?

(A) Add chemicals to precipitate phosphorus in the pond.
(B) Add chemicals to combine with phosphorus in the pond.
(C) Reduce use of phosphorus-based fertilizers in the surrounding fields.
(D) Use natural-based soaps and detergents in the home.

Alkalinity

2. (*Time limit: one hour*) Groundwater is used for a water supply. It is taken from the ground at 25°C. The initial properties of the water are as follows.

CO_2	60 mg/L (as $CaCO_3$)
alkalinity	200 mg/L (as $CaCO_3$)
pH	7.1

The water is treated for CO_2 removal by spraying into the atmosphere through a nozzle. The final CO_2 concentration is 5.6 mg/L (as $CaCO_3$) at 25°C. The first ionization constant of carbonic acid is 4.45×10^{-7}. What is the final pH of the water after spraying and recovery, assuming the alkalinity is unchanged in the process?

(A) 7.3
(B) 7.9
(C) 8.4
(D) 8.8

Solids

3. (*Time limit: one hour*) The solids content of a stream water sample is to be determined. The total solids content is determined by placing a portion of the sample into a porcelain evaporating dish, drying the sample at 105°C, and igniting the residue by placing the dried sample in a muffle furnace at 550°C. The following masses are recorded.

> mass of empty dish: 50.326 g
> mass of dish and sample: 118.400 g
> mass of dish and dry solids: 50.437 g
> mass of dish and ignited solids: 50.383 g

(a) The total solids content is

(A) 900 mg/L
(B) 1000 mg/L
(C) 1100 mg/L
(D) 1600 mg/L

(b) The total volatile solids content is

(A) 630 mg/L
(B) 710 mg/L
(C) 790 mg/L
(D) 830 mg/L

(c) The total fixed solids content is

(A) 270 mg/L
(B) 300 mg/L
(C) 420 mg/L
(D) 840 mg/L

The suspended solids content is determined by filtering a portion of the sample through a glass-fiber filter disk, drying the disk at 105°C, and igniting the residue by placing the dried sample in a muffle furnace at 550°C. The follow masses are recorded.

> volume of sample: 30 mL
> mass of filter disk: 0.1170 g
> mass of disk and dry solids: 0.1278 g
> mass of disk and ignited solids: 0.1248 g

(d) The total suspended solids content is
- (A) 240 mg/L
- (B) 360 mg/L
- (C) 370 mg/L
- (D) 820 mg/L

(e) The volatile suspended solids content is
- (A) 100 mg/L
- (B) 120 mg/L
- (C) 230 mg/L
- (D) 640 mg/L

(f) The fixed suspended solids content is
- (A) 120 mg/L
- (B) 180 mg/L
- (C) 240 mg/L
- (D) 260 mg/L

The dissolved solids content is determined by filtering a portion of the sample through a glass-fiber filter disk into a porcelain evaporating dish, drying the sample at 105°C, and igniting the residue by placing the dried sample in a muffle furnace at 550°C. The following masses are recorded.

> volume of sample: 25 mL
> mass of empty dish: 51.494 g
> mass of dish and dry solids: 51.524 g
> mass of dish and ignited solids: 51.506 g

(g) The total dissolved solids content is
- (A) 1000 mg/L
- (B) 1100 mg/L
- (C) 1200 mg/L
- (D) 1400 mg/L

(h) The volatile dissolved solids content is
- (A) 680 mg/L
- (B) 720 mg/L
- (C) 810 mg/L
- (D) 900 mg/L

(i) The fixed dissolved solids content is
- (A) 230 mg/L
- (B) 300 mg/L
- (C) 410 mg/L
- (D) 480 mg/L

(j) We would expect the total solids determined in part (a) to be equal to the sum of the total suspended solids determined in (d) plus the total dissolved solids determined in (g). Does the total solids content equal the sum of the suspended and dissolved solids?
- (A) yes
- (B) no, probably due to poor laboratory technique
- (C) no, probably due to rounding of measured values
- (D) no, probably because the samples were representative but not identical

Hardness

4. (*Time limit: one hour*) The laboratory analysis of a water sample is as follows. All concentrations are "as substance."

Ca^{++}	74.0 mg/L
Mg^{++}	18.3 mg/L
Na^{+}	27.6 mg/L
K^{+}	39.1 mg/L
pH	7.8
HCO_3^-	274.5 mg/L
SO_4^{--}	72.0 mg/L
Cl^-	49.7 mg/L

(a) The hardness of the water in terms of mg/L of calcium carbonate equivalent is
- (A) 1.3 mg/L as $CaCO_3$
- (B) 4.7 mg/L as $CaCO_3$
- (C) 66 mg/L as $CaCO_3$
- (D) 260 mg/L as $CaCO_3$

(b) Based on the laboratory analysis, which of the following can be said?
- (A) It is surprising that no carbonates were found in the sample.
- (B) The cations in solution, when converted to milliequivalents, will equal the anions, when converted to milliequivalents.
- (C) The large amount of bicarbonate in the solution tends to make the water acidic.
- (D) None of the above can be said.

(c) Assuming that hypothetical compounds are formed proportionally to the relative concentrations of ions, the hypothetical concentration of calcium bicarbonate in the sample is
- (A) 3.7 meq/L
- (B) 4.5 meq/L
- (C) 4.7 meq/L
- (D) 74 meq/L

(d) The hypothetical concentration of magnesium bicarbonate is
- (A) 0.8 meq/L
- (B) 1.5 meq/L
- (C) 1.6 meq/L
- (D) 5.2 meq/L

(e) The hypothetical concentration of magnesium sulfate is
- (A) 0.7 meq/L
- (B) 1.5 meq/L
- (C) 4.5 meq/L
- (D) 5.2 meq/L

(f) The hypothetical concentration of sodium sulfate is

(A) 0.8 meq/L

(B) 1.2 meq/L

(C) 1.5 meq/L

(D) 6.4 meq/L

(g) The hypothetical concentration of sodium chloride is

(A) 0.2 meq/L

(B) 0.4 meq/L

(C) 1.2 meq/L

(D) 1.4 meq/L

(h) The amount of lime (CaO) necessary in a lime softening process to remove the hardness caused by calcium bicarbonate is

(A) 17 mg/L

(B) 28 mg/L

(C) 95 mg/L

(D) 100 mg/L

(i) To remove hardness caused by magnesium bicarbonate, it is necessary to raise the pH by adding 35 mg/L of CaO in excess of the stoichiometric requirements. The amount of lime (CaO) necessary to remove the carbonate hardness caused by magnesium bicarbonate is

(A) 12 mg/L

(B) 45 mg/L

(C) 56 mg/L

(D) 80 mg/L

(j) The water is subsequently recarbonated to reduce the pH. Assume that by this softening process calcium hardness can be reduced to 30 mg/L and magnesium hardness can be reduced to 10 mg/L, both measured in terms of equivalent calcium carbonate. The amount of hardness that remains in the water is

(A) 10 mg/L as $CaCO_3$

(B) 30 mg/L as $CaCO_3$

(C) 45 mg/L as $CaCO_3$

(D) 75 mg/L as $CaCO_3$

SOLUTIONS

1. (a)   annual rainfall $= (4 \text{ ac})(5 \text{ in}) = 20 \text{ ac-in}$

runoff to pond $= (0.1)(20 \text{ ac-in})$

$$\times \left(102{,}790 \; \frac{\text{L}}{\text{ac-in}} \right)$$

$$= 205{,}600 \text{ L}$$

$$\begin{aligned}\text{bioavailable P} \\ \text{reaching pond}\end{aligned} = (205{,}600 \text{ L}) \left(0.01 \; \frac{\text{mg}}{\text{L}} \right)$$

$$= 2056 \text{ mg}$$

$$\begin{aligned}\text{initial bioavailable} \\ \text{P in pond}\end{aligned} = (14.8 \times 10^6 \text{ L}) \left(0.1 \; \frac{\text{mg}}{\text{L}} \right)$$

$$= 14.8 \times 10^5 \text{ mg}$$

The total bioavailable P in the pond is

$$2056 \text{ mg} + 14.8 \times 10^5 \text{ mg} = 14.82058 \times 10^5 \text{ mg}$$

Of this, a net percentage of $100\% - 22\% + ((12\%)(22\%)/100\%) = 80.64\%$ remains after biological processes and recycling.

$$(0.8064)(14.82058 \times 10^5 \text{ mg}) = 11.95132 \times 10^5 \text{ mg}$$

The following table is prepared in a similar manner.

year	P at start of year	P from rainfall	total P available for activity	P at end of year
1	14.8×10^5	2056	14.82058×10^5	11.95132×10^5
2	11.95132×10^5	2056	11.97188×10^5	9.65412×10^5
3	9.65412×10^5	2056	9.67468×10^5	7.80166×10^5
4	7.80166×10^5	2056	7.82222×10^5	6.30784×10^5
5	6.30784×10^5	2056	6.32840×10^5	5.10322×10^5

The concentration after five years will be

$$\frac{5.10322 \times 10^5 \text{ mg}}{14.8 \times 10^6 \text{ L}} = \boxed{0.0345 \text{ mg/L}}$$

The answer is (A).

(b) To reduce the phosphorus accumulation in the pond, the arrival of additional bioavailable phosphorous must be reduced. This entails watershed management to reduce the amount of phosphorus applied as fertilizer and released through other sources. As eutrophication is a natural process accelerated by the availability of plant nutrients (nitrogen and phosphorus especially), reducing phosphorus can slow the process.

While there are chemical means to alter the forms of phosphorus to nonbioavailable states, this usually is not practical on a large scale and is potentially harmful in itself.

It is unlikely that the pond receives untreated discharge from local homes. Use of phosphate-rich detergents in the home will not affect the pond.

The answer is (C).

Environmental

2. With CO_2 present, all alkalinity is in the bicarbonate (HCO_3^-) form. Therefore, the equilibrium expression for the first ionization of carbonic acid may be used.

$$CO_2 + H_2O \rightarrow H_2CO_3 \leftrightarrow H^+ + HCO_3^-$$

$$K_1 = \frac{[H^+][HCO_3^-]}{[H_2CO_3]} = 4.45 \times 10^{-7}$$

$$[H_2CO_3] = [CO_2] = \frac{5.6 \ \frac{mg}{L}}{\left(100 \ \frac{g}{mol}\right)\left(1000 \ \frac{mg}{g}\right)}$$

$$= 5.6 \times 10^{-5}$$

$$[HCO_3^-] = \frac{200 \ \frac{mg}{L}}{\left(100 \ \frac{g}{mol}\right)\left(1000 \ \frac{mg}{g}\right)}$$

$$= 2 \times 10^{-3}$$

Solve for the hydrogen ion concentration.

$$[H^+] = \frac{K_1[H_2CO_3]}{[CO_2]} = \frac{\left(4.45 \times 10^{-7}\right)\left(2 \times 10^{-3}\right)}{5.6 \times 10^{-5}}$$

$$= 1.246 \times 10^{-8}$$

$$pH = -\log[H^+] = -\log(1.246 \times 10^{-8}) = \boxed{7.9}$$

The answer is (B).

3. (a) The density of water is 1 g/mL. The volume of the tested sample is

$$\frac{118.4 \ g - 50.326 \ g}{1 \ \frac{g}{mL}} = 68.1 \ mL$$

$$TS = \frac{(50.437 \ g - 50.326 \ g)\left(1000 \ \frac{mg}{g}\right)\left(1000 \ \frac{mL}{L}\right)}{68.1 \ mL}$$

$$= \boxed{1630 \ mg/L}$$

The answer is (D).

(b) $$TVS = \frac{(50.437 \ g - 50.383 \ g) \times \left(1000 \ \frac{mg}{g}\right)\left(1000 \ \frac{mL}{L}\right)}{68.1 \ mL}$$

$$= \boxed{793 \ mg/L}$$

The answer is (C).

(c) $$TFS = 1630 \ \frac{mg}{L} - 793 \ \frac{mg}{L} = \boxed{837 \ mg/L}$$

The answer is (D).

(d) $$TSS = \frac{(0.1278 \ g - 0.1170 \ g) \times \left(1000 \ \frac{mg}{g}\right)\left(1000 \ \frac{mL}{L}\right)}{30 \ mL}$$

$$= \boxed{360 \ mg/L}$$

The answer is (B).

(e) $$VSS = \frac{(0.1278 \ g - 0.1248 \ g) \times \left(1000 \ \frac{mg}{g}\right)\left(1000 \ \frac{mL}{L}\right)}{30 \ mL}$$

$$= \boxed{100 \ mg/L}$$

The answer is (A).

(f) $$FSS = 360 \ \frac{mg}{L} - 100 \ \frac{mg}{L} = \boxed{260 \ mg/L}$$

The answer is (D).

(g) $$TDS = \frac{(51.524 \ g - 51.494 \ g) \times \left(1000 \ \frac{mg}{g}\right)\left(1000 \ \frac{mL}{L}\right)}{25 \ mL}$$

$$= \boxed{1200 \ mg/L}$$

The answer is (C).

(h) $$VDS = \frac{(51.524 \ g - 51.506 \ g) \times \left(1000 \ \frac{mg}{g}\right)\left(1000 \ \frac{mL}{L}\right)}{25 \ mL}$$

$$= \boxed{720 \ mg/L}$$

The answer is (B).

(i) $$FDS = 1200 \ \frac{mg}{L} - 720 \ \frac{mg}{L} = \boxed{480 \ mg/L}$$

The answer is (D).

(j) If only the first two procedures were performed, the dissolved solids would be calculated as

$$TDS = TS - TSS = 1630 \; \frac{mg}{L} - 360 \; \frac{mg}{L}$$

$$= 1270 \; mg/L \quad (vs. \; 1200 \; mg/L)$$

$$VDS = TVS - VSS = 793 \; \frac{mg}{L} - 100 \; \frac{mg}{L}$$

$$= 693 \; mg/L \quad (vs. \; 720 \; mg/L)$$

$$FDS = TFS - FSS = 837 \; \frac{mg}{L} - 260 \; \frac{mg}{L}$$

$$= 577 \; mg/L \quad (vs. \; 480 \; mg/L)$$

The results of the three sets of tests are not the same. These differences are too great to be caused by improper rounding or faulty laboratory technique. This is probably the result of using samples that are not truly identical. For this reason, the suspended solids values are mathematically determined from the total solids and dissolved solids tests.

The answer is (D).

4. (a) Hardness is caused by multivalent cations: Ca^{++} and Mg^{++} in this example. Calculate the milliequivalents by dividing the measured concentration by the milliequivalent weight.

$$Ca^{++}: \quad \frac{74.0 \; \frac{mg}{L}}{20 \; \frac{mg}{meq}} = 3.7 \; meq/L$$

$$Mg^{++}: \quad \frac{18.3 \; \frac{mg}{L}}{12.2 \; \frac{mg}{meq}} = 1.5 \; meq/L$$

$$\text{total hardness} = 3.7 \; \frac{meq}{L} + 1.5 \; \frac{meq}{L} = 5.2 \; meq/L$$

Determine the calcium carbonate equivalent by multiplying by the equivalent weight of calcium carbonate.

$$\text{hardness} = \left(5.2 \; \frac{meq}{L}\right)\left(50 \; \frac{mg}{meq}\right)$$

$$= \boxed{260 \; mg/L \; as \; CaCO_3}$$

The answer is (D).

(b) Alkalinity in the form of carbonate radical does not exist at a pH below 8.3. Bicarbonate is a form of alkalinity, and it neutralizes, not creates, acidity. To determine the cation/anion relationship, convert all of the concentrations to milliequivalents by dividing the measured concentrations by the milliequivalent weights.

$$Ca^{++}: \quad \frac{74.0 \; \frac{mg}{L}}{20 \; \frac{mg}{meq}} = 3.7 \; meq/L$$

$$Mg^{++}: \quad \frac{18.3 \; \frac{mg}{L}}{12.2 \; \frac{mg}{meq}} = 1.5 \; meq/L$$

$$Na^{+}: \quad \frac{27.6 \; \frac{mg}{L}}{23 \; \frac{mg}{meq}} = 1.2 \; meq/L$$

$$K^{+}: \quad \frac{39.1 \; \frac{mg}{L}}{39.1 \; \frac{mg}{meq}} = 1.0 \; meq/L$$

$$\text{total cations} = 3.7 \; \frac{meq}{L} + 1.5 \; \frac{meq}{L} + 1.2 \; \frac{meq}{L}$$
$$+ 1.0 \; \frac{meq}{L}$$
$$= \boxed{7.4 \; meq/L}$$

$$HCO_3^{-}: \quad \frac{274.5 \; \frac{mg}{L}}{61 \; \frac{mg}{meq}} = 4.5 \; meq/L$$

$$SO_4^{--}: \quad \frac{72.0 \; \frac{mg}{L}}{48 \; \frac{mg}{meq}} = 1.5 \; meq/L$$

$$Cl^{-}: \quad \frac{49.7 \; \frac{mg}{L}}{35.5 \; \frac{mg}{meq}} = 1.4 \; meq/L$$

$$\text{total anions} = 4.5 \; \frac{meq}{L} + 1.5 \; \frac{meq}{L} + 1.4 \; \frac{meq}{L}$$
$$= \boxed{7.4 \; meq/L}$$

The answer is (B).

(c) To determine the hypothetical compounds, construct a milliequivalent per liter bar graph for cations and anions, as shown.

0		3.7	5.2	6.4	7.4
Ca++ 3.7			Mg++ 1.5	Na+ 1.2	K+ 1.0
HCO3− 4.5				SO4−− 1.5	Cl− 1.4
0			4.5	6.0	7.4

The hypothetical compounds are determined by moving from left to right from ions of greater concentration to ions of lesser concentration.

Environmental

hypothetical	compound concentration (meq/L)	remarks
$Ca(HCO_3)_2$	3.7	Ca^{++} exhausted
$Mg(HCO_3)_2$	0.8	HCO_3^- exhausted
$MgSO_4$	0.7	Mg^{++} exhausted
Na_2SO_4	0.8	SO_4^{--} exhausted
$NaCl$	0.4	Na^+ exhausted
KCl	1.0	K^+ and Cl^- exhausted

The answer is (A).

(d) From part (c), the hypothetical concentration of magnesium bicarbonate is $\boxed{0.8 \text{ meq/L.}}$

The answer is (A).

(e) From part (c), the hypothetical concentration of magnesium sulfate is $\boxed{0.7 \text{ meq/L.}}$

The answer is (A).

(f) From part (c), the hypothetical concentration of sodium sulfate is $\boxed{0.8 \text{ meq/L.}}$

The answer is (A).

(g) From part (c), the hypothetical concentration of sodium chloride is $\boxed{0.4 \text{ meq/L.}}$

The answer is (B).

(h) The reactions are

$$CaO + H_2O \rightarrow Ca(OH)_2$$
$$Ca(HCO_3)_2 + Ca(OH)_2 \rightarrow CaCO_3 \downarrow + 2H_2O$$

One molecule of CaO forms one molecule of $Ca(OH)_2$, which in turn reacts with one molecule of $Ca(HCO_3)_2$. There are 3.7 meq/L of $Ca(HCO_3)_2$ in solution. The equivalent weight of CaO is $(40+16)/2 = 28$. Therefore, 3.7 meq/L of CaO are needed.

$$\left(3.7 \frac{\text{meq}}{\text{L}} \text{ CaO}\right)\left(28 \frac{\text{mg}}{\text{meq}}\right) = \boxed{103.6 \text{ mg/L CaO}}$$

The answer is (D).

(i) The reactions are

$$CaO + H_2O \rightarrow Ca(OH)_2$$
$$Mg(HCO_3)_2 + 2Ca(OH)_2 \rightarrow$$
$$2CaCO_3 \downarrow + Mg(OH)_2 \downarrow + 2H_2O$$

Two molecules of $Ca(OH)_2$ are required for each molecule of $Mg(HCO_3)_2$.

$$\left(0.8 \frac{\text{meq}}{\text{L}} \text{ Mg(HCO}_3)_2\right)(2)\left(28 \frac{\text{mg}}{\text{meq}}\right)$$
$$+ 35 \frac{\text{mg}}{\text{L}} \text{ excess} = \boxed{79.8 \text{ mg/L}}$$

The answer is (D).

(j) The original hardness was in the hypothetical forms of calcium bicarbonate, magnesium bicarbonate, and magnesium sulfate. The bicarbonates have been removed in parts (h) and (i), leaving the residuals of 30 mg/L calcium hardness and 10 mg/L magnesium hardness. However, no attempt was made to remove the noncarbonate magnesium hardness represented by magnesium sulfate (this would have required the addition of soda ash and additional lime).

The residual hardness is as follows.

$$\text{calcium hardness:} \quad \frac{30 \dfrac{\text{mg}}{\text{L}}}{50 \dfrac{\text{mg}}{\text{meq}}} = 0.6 \text{ meq/L}$$

$$\text{magnesium carbonate hardness:} \quad \frac{10 \dfrac{\text{mg}}{\text{L}}}{50 \dfrac{\text{mg}}{\text{meq}}} = 0.2 \text{ meq/L}$$

From part (c), the magnesium noncarbonate hardness is 0.7 meq/L.

$$\text{residual hardness} = 0.6 \frac{\text{meq}}{\text{L}} + 0.2 \frac{\text{meq}}{\text{L}}$$
$$+ 0.7 \frac{\text{meq}}{\text{L}}$$
$$= 1.5 \text{ meq/L}$$

$$\left(1.5 \frac{\text{meq}}{\text{L}}\right)\left(50 \frac{\text{mg}}{\text{meq}}\right) = \boxed{75 \text{ mg/L as CaCO}_3}$$

The answer is (D).

36 Water Supply Treatment and Distribution

PRACTICE PROBLEMS

Plain Sedimentation

1. A settling tank has an overflow rate of 100,000 gal/ft²-day. Water carrying sediment of various sizes enters the tank. The sediment has the following distribution of settling velocities.

settling velocity (ft/min)	mass fraction remaining
10.0	0.54
5.0	0.45
2.0	0.35
1.0	0.20
0.75	0.10
0.50	0.03

(a) What is the gravimetric percentage of sediment particles completely removed?

- (A) 32%
- (B) 39%
- (C) 47%
- (D) 55%

(b) What is the total gravimetric percentage of all sediment particles removed?

- (A) 35%
- (B) 40%
- (C) 50%
- (D) 60%

2. A spherical sand particle has a specific gravity of 2.6 and a diameter of 1 mm. What is the settling velocity?

- (A) 0.2 ft/sec
- (B) 0.7 ft/sec
- (C) 1.1 ft/sec
- (D) 1.6 ft/sec

3. A mechanically cleaned circular clarifier is to be designed with the following characteristics.

flow rate	2.8 MGD
detention period	2 hr
surface loading	700 gal/ft²-day

(a) What is the approximate diameter?

- (A) 45 ft
- (B) 60 ft
- (C) 70 ft
- (D) 90 ft

(b) What is the approximate depth?

- (A) 6 ft
- (B) 8 ft
- (C) 12 ft
- (D) 15 ft

(c) If the initial flow rate is reduced to 1.1 MGD, what is the surface loading?

- (A) 190 gal/day-ft²
- (B) 230 gal/day-ft²
- (C) 250 gal/day-ft²
- (D) 280 gal/day-ft²

(d) If the initial flow rate is reduced to 1.1 MGD, what is the average detention period?

- (A) 4 hr
- (B) 5 hr
- (C) 6 hr
- (D) 8 hr

4. (*Time limit: one hour*) A water treatment plant is designed to handle a total flow rate of 1.5 MGD. The current design includes two identical sedimentation basins with the following characteristics.

plan area	90 ft × 16 ft
depth	12 ft
total weir length	48 ft per basin
three-month sustained average low	70% of the design average daily flow
three-month sustained average high	200% of the design average daily flow

The basins must meet the following government standards.

minimum retention time	4.0 hrs
maximum weir load	20,000 gpd/ft
maximum velocity	0.5 ft/min

Do the basins meet the standards?

- (A) Yes, specifications are met at both peak and low flows.
- (B) No, specifications are not met at low flow.
- (C) No, specifications are not met at high flow.
- (D) No, specifications are not met at either high or low flow.

Mixing Physics

5. (*Time limit: one hour*) A flocculator tank with a volume of 200,000 ft^3 uses a paddle wheel to mix the coagulant in 60°F water. The operating characteristics are as follows.

mean velocity gradient	45 sec^{-1}
paddle drag coefficient	1.75
paddle tip velocity	2 ft/sec
relative water/paddle velocity	1.5 ft/sec

(a) What is the theoretical power required to drive the paddle?
- (A) 12 hp
- (B) 17 hp
- (C) 60 hp
- (D) 120 hp

(b) What is the drag force on the paddle?
- (A) 450 lbf
- (B) 910 lbf
- (C) 5500 lbf
- (D) 6400 lbf

(c) What is the required paddle area?
- (A) 900 ft^2
- (B) 1200 ft^2
- (C) 1500 ft^2
- (D) 1700 ft^2

Filtration

6. A water treatment plant has four square rapid sand filters. The flow rate is 4 gal/min-ft^2. Each filter has a treatment capacity of 600,000 gal/day. Each filter is backwashed once a day for 8 min. The rate of rise during washing is 24 in/min.

(a) What are the inside dimensions of each filter?
- (A) 6 ft × 6 ft
- (B) 8 ft × 8 ft
- (C) 10 ft × 10 ft
- (D) 12 ft × 12 ft

(b) What percentage of the filtered water is used for backwashing?
- (A) 2%
- (B) 4%
- (C) 8%
- (D) 10%

7. A water treatment plant has five identical rapid sand filters. Each is square in plan and has a capacity of 1.0 MGD. The application rate is 4 gal/min-ft^2, and the total wetted depth is 10 ft.

(a) What should be the cross-section dimensions of each filter?
- (A) 7 ft × 7 ft
- (B) 9 ft × 9 ft
- (C) 10 ft × 10 ft
- (D) 13 ft × 13 ft

(b) Each filter is backwashed every day for 5 min. The rate of rise of the backwash water is 2 ft/min. What percentage of the plant's filtered water will be used for backwashing?
- (A) 1.3%
- (B) 1.9%
- (C) 2.4%
- (D) 3.1%

Precipitation Softening

8. A town's water supply has the following ionic concentrations.

Al^{+++}	0.5 mg/L
Ca^{++}	80.2 mg/L
Cl$^-$	85.9 mg/L
CO$_2$	19 mg/L
CO$_3^{--}$	0
Fe^{++}	1.0 mg/L
Fl$^-$	0
HCO$_3^-$	185 mg/L
Mg^{++}	24.3 mg/L
Na$^+$	46.0 mg/L
NO$_3^-$	0
SO$_4^{--}$	125 mg/L

(a) What is the total hardness?
- (A) 160 mg/L as CaCO$_3$
- (B) 200 mg/L as CaCO$_3$
- (C) 260 mg/L as CaCO$_3$
- (D) 300 mg/L as CaCO$_3$

(b) How much slaked lime is required to combine with the carbonate hardness?
- (A) 45 mg/L as substance
- (B) 90 mg/L as substance
- (C) 130 mg/L as substance
- (D) 150 mg/L as substance

(c) How much soda ash is required to react with the carbonate hardness?
- (A) none
- (B) 15 mg/L as substance
- (C) 35 mg/L as substance
- (D) 60 mg/L as substance

9. A city's water supply contains the following ionic concentrations.

$Ca(HCO_3)_2$	137 mg/L as $CaCO_3$
$MgSO_4$	72 mg/L as $CaCO_3$
CO_2	0

(a) How much slaked lime is required to soften 1,000,000 gal of this water to a hardness of 100 mg/L if 30 mg/L of excess lime is used?

(A) 930 lbm
(B) 1200 lbm
(C) 1300 lbm
(D) 1700 lbm

(b) How much soda ash is required to soften 1,000,000 gal of this water to a hardness of 100 mg/L if 30 mg/L of excess lime is used?

(A) none
(B) 300 lbm
(C) 500 lbm
(D) 700 lbm

Zeolite Softening

10. The water described in Prob. 9 is to be softened using a zeolite process with the following characteristics: exchange capacity, 10,000 grains/ft³; and salt requirement, 0.5 lbm per 1000 grains hardness removed. How much salt is required to soften the water to 100 mg/L hardness?

(A) 700 lbm/MG
(B) 1800 lbm/MG
(C) 2500 lbm/MG
(D) 3200 lbm/MG

11. The hardness of water from an underground aquifer is to be reduced from 245 mg/L to 80 mg/L using a zeolite process. The volumetric flow rate is 20,000 gal/day. The process has the following characteristics: resin exchange capacity, 20,000 grains/ft³; and zeolite volume, 2 ft³.

(a) What fraction of the water is bypassed around the process?

(A) 0.15
(B) 0.33
(C) 0.67
(D) 0.85

(b) What is the time between regenerations of the softener?

(A) 5 hr
(B) 16 hr
(C) 24 hr
(D) 30 hr

Demand

12. (*Time limit: one hour*) An area is expected to attain a population of 40,000 in 20 years. The cumulative per capita water demand for a peak day in the area is shown in the following diagram.

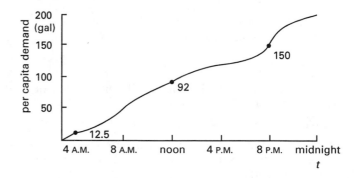

(a) What is the total per capita demand for a peak day?

(A) 160 gal
(B) 200 gal
(C) 240 gal
(D) 290 gal

(b) Assuming uniform operation over 24 hours, what storage volume is required in the treatment plant for all uses, including fire fighting demand?

(A) 1.1 MG
(B) 2.7 MG
(C) 3.2 MG
(D) 5.5 MG

(c) Assume that the pumping station runs uniformly from 4 A.M. until 8 A.M. to fill the storage tanks for the day. What storage is required to meet all uses, including fire fighting?

(A) 3.2 MG
(B) 3.8 MG
(C) 5.5 MG
(D) 8.7 MG

13. The water supply for a town of 15,000 people is taken from a river. The average consumption is 110 gpcd. The water has the following characteristics.

turbidity	20 to 100 NTU (varies)
total hardness	less than 60 mg/L as $CaCO_3$
coliform count	200 to 1000 per 100 mL (varies)

(a) What capacity should the distribution and treatment system have?

(A) 6000 gpm
(B) 7500 gpm
(C) 9500 gpm
(D) 12,000 gpm

Environmental

(b) If the application rate is 4 gal/min-ft^2, what total filter area is required?

(A)   290 ft^2
(B)   580 ft^2
(C)   910 ft^2
(D)   1100 ft^2

(c) Is softening required?

(A)   No, 60 mg/L is soft water.
(B)   No, softening would interfere with turbidity removal.
(C)   Yes, the turbidity and coliform counts would also benefit.
(D)   Yes, all municipal water should be softened.

(d) If a chlorine dose of 2 mg/L is required to obtain the desired chlorine residual, how much chlorine is required every 24 hours?

(A)   15 lbm
(B)   22 lbm
(C)   28 lbm
(D)   32 lbm

SOLUTIONS

1. (a) $$v^* = \dfrac{\left(100{,}000\ \dfrac{\text{gal}}{\text{ft}^2\text{-day}}\right)\left(0.1337\ \dfrac{\text{ft}^3}{\text{gal}}\right)}{\left(24\ \dfrac{\text{hr}}{\text{day}}\right)\left(60\ \dfrac{\text{min}}{\text{hr}}\right)}$$

$$= 9.28\ \text{ft/min}$$

Using interpolation, the mass fraction remaining is

$$0.45 + \left(\dfrac{9.28\ \dfrac{\text{ft}}{\text{min}} - 5.0\ \dfrac{\text{ft}}{\text{min}}}{10.0\ \dfrac{\text{ft}}{\text{min}} - 5.0\ \dfrac{\text{ft}}{\text{min}}}\right)(0.54 - 0.45)$$

$$= 0.527\ \text{remains in flow}$$

$$1 - 0.527 = 0.473$$

$$\boxed{47.3\%\ \text{completely removed}}$$

The answer is (C).

(b) Plot the mass fraction remaining versus the settling velocity.

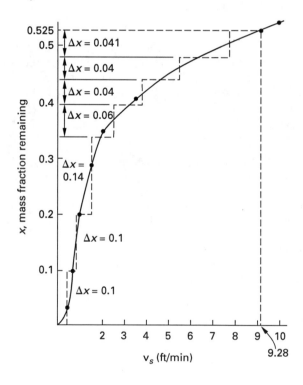

$$v_o = \left(\dfrac{100{,}000\ \text{gal}}{\text{ft}^2\text{-day}}\right)\left(\dfrac{1\ \text{ft}^3}{7.48\ \text{gal}}\right)\left(\dfrac{1\ \text{day}}{24\ \text{hr}}\right)\left(\dfrac{1\ \text{hr}}{60\ \text{min}}\right)$$

$$= 9.28\ \text{ft/min}$$

$$X_o \approx 0.525\quad (52.5\%)$$

Determine $\Delta x v_t$ by graphical integration.

Δx	v_t	$(\Delta x)(v_t)$
0.1	0.5	0.05
0.1	0.8	0.08
0.14	1.4	0.196
0.06	2.5	0.15
0.04	3.75	0.15
0.04	5.5	0.22
0.041	7.75	0.318
	total	1.164

The overall removal efficiency is

$$X = 1 - X_o + \Sigma \left(\frac{\Delta x v_t}{v_o} \right)$$
$$= 1 - 0.525 + \frac{1.164}{9.28}$$
$$= 0.60$$

The answer is (D).

2. From Fig. 36.2, $v_s = \boxed{0.7 \text{ ft/sec}}$.

The answer is (B).

3. (a) The surface area is

$$A_{\text{surface}} = \frac{2.8 \times 10^6 \frac{\text{gal}}{\text{day}}}{700 \frac{\text{gal}}{\text{ft}^2\text{-day}}} = 4000 \text{ ft}^2$$

Since $A = \left(\frac{\pi}{4} \right) D^2$,

$$D = \sqrt{\frac{(4)(4000 \text{ ft}^2)}{\pi}} = \boxed{71.4 \text{ ft}}$$

The answer is (C).

(b) The volume is

$$V = \frac{\left(2.8 \times 10^6 \frac{\text{gal}}{\text{day}} \right)(2 \text{ hr})}{24 \frac{\text{hr}}{\text{day}}} = 2.333 \times 10^5 \text{ gal}$$

The depth is

$$d = \frac{V}{A_{\text{surface}}} = \frac{(2.333 \times 10^5 \text{ gal}) \left(0.1337 \frac{\text{ft}^3}{\text{gal}} \right)}{4000 \text{ ft}^2}$$
$$= \boxed{7.8 \text{ ft}}$$

The answer is (B).

(c) $$v^* = \frac{1.1 \times 10^6 \frac{\text{gal}}{\text{day}}}{4000 \text{ ft}^2} = \boxed{275 \text{ gal/day-ft}^2}$$

The answer is (D).

(d) Use Eq. 36.6.

$$t = \frac{(2.333 \times 10^5 \text{ gal}) \left(24 \frac{\text{hr}}{\text{day}} \right)}{1.1 \times 10^6 \frac{\text{gal}}{\text{day}}} = \boxed{5.09 \text{ hr}}$$

The answer is (B).

4. Assume there are two basins in parallel.

volume per basin $= (90 \text{ ft})(16 \text{ ft})(12 \text{ ft}) = 17{,}280 \text{ ft}^3$

(The freeboard is not given.)

The area per basin is

$$A = (90 \text{ ft})(16 \text{ ft}) = 1440 \text{ ft}^2$$

The three-month peak flow per basin is

$$(2) \left(\frac{1.5 \text{ MGD}}{2} \right) = 1.5 \text{ MGD}$$

$$(1.5 \text{ MGD}) \left(1.547 \frac{\frac{\text{ft}^3}{\text{sec}}}{\text{MGD}} \right) = 2.32 \text{ ft}^3/\text{sec (cfs)}$$

The detention time at peak flow is given by Eq. 36.6.

$$t = \frac{V}{Q} = \frac{17{,}280 \text{ ft}^3}{\left(2.32 \frac{\text{ft}^3}{\text{sec}} \right) \left(60 \frac{\text{min}}{\text{hr}} \right) \left(60 \frac{\text{ft}^3}{\text{hr}} \right)} = 2.07 \text{ hr}$$

Since 2.07 hr < 4 hr, this is not acceptable.

The weir loading is

$$\frac{1.5 \times 10^6 \frac{\text{gal}}{\text{day}}}{48 \text{ ft}} = 31{,}250 \text{ gal/day-ft}$$

Since 31,250 > 20,000, this is not acceptable either.

The overflow rate is

$$v^* = \frac{Q}{A} = \frac{\left(2.32 \frac{\text{ft}^3}{\text{sec}} \right) \left(60 \frac{\text{sec}}{\text{min}} \right)}{1440 \text{ ft}^2} = 0.0967 \text{ ft/min}$$

Since 0.0967 < 0.5, this is acceptable.

At low flow,

$$\text{flow} = \frac{0.7}{2.0} = 0.35 \quad [35\% \text{ of high flow}]$$

$$t = \frac{2.07 \text{ hr}}{0.35} = 5.91 \text{ hr} \quad [\text{acceptable}]$$

$$\text{weir loading} = (0.35)\left(31{,}250 \ \frac{\text{gal}}{\text{day-ft}}\right)$$

$$= 10{,}938 \ \frac{\text{gal}}{\text{day-ft}} \quad [\text{acceptable}]$$

$$v^* = (0.35)\left(0.1 \ \frac{\text{ft}}{\text{min}}\right) = 0.035 \text{ ft/min}$$

$$[\text{acceptable}]$$

> The basins have been correctly designed for low flow but not for peak flow. One or more basins should be used.

The answer is (C).

5. For 60°F water, $\mu = 2.359 \times 10^{-5}$ lbf- sec /ft^2.
From Eq. 36.21,

$$P = \left(2.359 \times 10^{-5} \ \frac{\text{lbf- sec}}{\text{ft}^2}\right)\left(45 \ \frac{1}{\text{sec}}\right)^2 (200{,}000 \text{ ft}^3)$$

$$= 9554 \text{ ft-lbf/ sec}$$

(a) $$\text{water horsepower} = \frac{9554 \ \dfrac{\text{ft-lbf}}{\text{sec}}}{550 \ \dfrac{\text{ft-lbf}}{\text{hp-sec}}}$$

$$= \boxed{17.4 \text{ hp}}$$

The answer is (B).

(b) Since work = force × distance, then power = force × velocity.

$$D = \frac{P}{v} = \frac{9554 \ \dfrac{\text{ft-lbf}}{\text{sec}}}{1.5 \ \dfrac{\text{ft}}{\text{sec}}} = \boxed{6369 \text{ lbf}}$$

The answer is (D).

(c) Use Eq. 36.15.

$$A = \frac{(2)\left(32.2 \ \dfrac{\text{ft}}{\text{sec}^2}\right)(6369 \text{ lbf})}{(1.75)\left(62.4 \ \dfrac{\text{lbf}}{\text{ft}^3}\right)\left(1.5 \ \dfrac{\text{ft}}{\text{sec}}\right)^2}$$

$$= \boxed{1669 \text{ ft}^2}$$

The answer is (D).

6. (a) The area is

$$A = \frac{600{,}000 \ \dfrac{\text{gal}}{\text{day}}}{\left(4 \ \dfrac{\text{gal}}{\text{min-ft}^2}\right)\left(24 \ \dfrac{\text{hr}}{\text{day}}\right)\left(60 \ \dfrac{\text{min}}{\text{hr}}\right)} = 104.2 \text{ ft}^2$$

$$\boxed{\text{use 10 ft} \times \text{10 ft}}$$

The answer is (C).

(b) The volume is

$$V = \left(8 \ \frac{\text{min}}{\text{day}}\right)(100 \text{ ft}^2)\left(2 \ \frac{\text{ft}}{\text{min}}\right) = 1600 \text{ ft}^3/\text{day}$$

$$\left(1600 \ \frac{\text{ft}^3}{\text{day}}\right)\left(7.481 \ \frac{\text{gal}}{\text{ft}^3}\right) = 11{,}970 \text{ gal/day}$$

$$\frac{11{,}970 \ \dfrac{\text{gal}}{\text{day}}}{600{,}000 \ \dfrac{\text{gal}}{\text{day}}} = \boxed{0.01 \ (2\%)}$$

The answer is (A).

7. (a) $$A = \frac{1 \times 10^6 \ \dfrac{\text{gal}}{\text{day}}}{\left(24 \ \dfrac{\text{hr}}{\text{day}}\right)\left(60 \ \dfrac{\text{min}}{\text{hr}}\right)\left(4 \ \dfrac{\text{gal}}{\text{min-ft}^2}\right)}$$

$$= 173.6 \text{ ft}^2$$

$$\text{width} = \sqrt{173.6 \text{ ft}^2} = 13.2 \text{ ft}$$

$$\boxed{13.2 \text{ ft} \times 13.2 \text{ ft}}$$

The answer is (D).

(b) The water volume is

$$V = \left(5 \ \frac{\text{min}}{\text{day}}\right)(173.6 \text{ ft}^2)\left(2 \ \frac{\text{ft}}{\text{min}}\right)(5 \text{ filters})$$

$$= 8680 \text{ ft}^3/\text{day}$$

Convert to gallons.

$$\left(8680 \ \frac{\text{ft}^3}{\text{day}}\right)\left(7.481 \ \frac{\text{gal}}{\text{ft}^3}\right) = 64{,}935 \text{ gal/day}$$

$$\frac{64{,}935 \ \dfrac{\text{gal}}{\text{day}}}{5 \times 10^6 \ \dfrac{\text{gal}}{\text{day}}} = \boxed{0.013 \ (1.3\%)}$$

The answer is (A).

8. (a)

	mg/L as substance		factor from App. 20.C		
Ca^{++}:	80.2	×	2.5	=	200.5 mg/L
Mg^{++}:	24.3	×	4.1	=	99.63 mg/L
Fe^{++}:	1	×	1.79	=	1.79 mg/L
Al^{+++}:	0.5	×	5.56	=	2.78 mg/L

$$\text{hardness} = \boxed{304.7 \text{ mg/L}}$$

The answer is (D).

(b) To remove the carbonate hardness,

$$CO_2: \quad 19 \frac{mg}{L} \times 2.27 = 43.13 \text{ mg/L as } CaCO_3$$

Add lime to remove the carbonate hardness. It does not matter whether the HCO_3^- comes from Mg^{++}, Ca^{++}, or Fe^{++}; adding lime will remove it.

There may be Mg^{++}, Ca^{++}, or Fe^{++} ions left over in the form of noncarbonate hardness, but the problem asked for carbonate hardness. Converting from mg/L of substance to mg/L as $CaCO_3$,

$$HCO_3^-: \quad 185 \frac{mg}{L} \times 0.82 = 151.7 \text{ mg/L}$$

The total equivalents to be neutralized are

$$43.13 \frac{mg}{L} + 151.7 \frac{mg}{L} = 194.83 \text{ mg/L}$$

Convert $Ca(OH)_2$ using App. 20.C.

$$\frac{mg}{L} \text{ of } Ca(OH)_2 = \frac{194.83 \frac{mg}{L}}{1.35}$$

$$= \boxed{144.3 \text{ mg/L as substance}}$$

The answer is (D).

(c) | No soda ash is required since it is used to remove noncarbonate hardness. |
|---|

The answer is (A).

9. (a) $Ca(HCO_3)_2$ and $MgSO_4$ both contribute to hardness. Since 100 mg/L of hardness is the goal, leave all $MgSO_4$ in the water. Take out 137 mg/L + 72 mg/L − 100 mg/L = 109 mg/L of $Ca(HCO_3)_2$. From App. 20.C (including the excess even though the reaction is not complete),

$$\text{pure } Ca(OH)_2 = 30 \frac{mg}{L} + \frac{109 \frac{mg}{L}}{1.35}$$

$$= 110.74 \text{ MG/L}$$

$$\left(110.74 \frac{mg}{L}\right)\left(8.345 \frac{lbm\text{-}L}{mg\text{-}MG}\right) = \boxed{924 \text{ lbm/MG}}$$

The answer is (A).

(b) No soda ash is required.

The answer is (A).

10. The hardness removed is

$$137 \frac{mg}{L} + 72 \frac{mg}{L} - 100 \frac{mg}{L} = 109 \text{ MG/L}$$

$$\left(109 \frac{MG}{L}\right)\left(8.345 \frac{lbm\text{-}L}{mg\text{-}MG}\right) = 909.6 \text{ lbm hardness/MG}$$

$$\left(\frac{0.5 \text{ lbm}}{1000 \text{ gr}}\right)\left(909.6 \frac{lbm}{MG}\right)\left(7000 \frac{gr}{lbm}\right)$$

$$= \boxed{3.18 \times 10^3 \text{ lbm/MG } (3200 \text{ lbm/MG})}$$

The answer is (D).

11. (a) A bypass process is required.

$$\text{fraction bypassed: } \frac{80 \frac{mg}{L}}{245 \frac{mg}{L}} = \boxed{0.326}$$

fraction processed: $1 - 0.326 = 0.674$

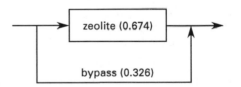

The answer is (B).

(b) The maximum hardness reduction is

$$\frac{(2 \text{ ft}^3)\left(20,000 \frac{gr}{ft^3}\right)}{7000 \frac{gr}{lbm}} = 5.71 \text{ lbm}$$

The hardness removal rate is

$$\frac{(0.674)\left(245 \frac{MG}{L}\right)\left(8.345 \times 10^{-6} \frac{lbm}{gal}\right)\left(20,000 \frac{gal}{day}\right)}{24 \frac{hr}{day}}$$

$$= 1.15 \text{ lbm/hr}$$

$$t = \frac{5.71 \text{ lbm}}{1.15 \frac{lbm}{hr}} = \boxed{4.97 \text{ hr}}$$

The answer is (A).

12. (a) The daily demand is $\boxed{200 \text{ gal.}}$

The answer is (B).

(b)

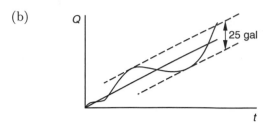

From the cumulative flow quantity, the storage requirement per capita-day is 25 gal.

The population use is

$$\left(25 \ \frac{\text{gal}}{\text{day-person}}\right)(40{,}000 \text{ people}) = 1{,}000{,}000 \text{ gal/day}$$

From Eq. 36.53, the fire fighting requirement is

$$Q = 1020\sqrt{40}\left(1 - 0.01\sqrt{40}\right) = 6043 \text{ gal/min}$$

The flow rate is $6043/1000 = 6$ thousands of gallons, so maintain the flow for 6 hr (approximate ISO specifications).

$$\text{capacity} = 1{,}000{,}000 \text{ gal}$$
$$+ \left(6043 \ \frac{\text{gal}}{\text{min}}\right)\left(60 \ \frac{\text{min}}{\text{hr}}\right)(6 \text{ hr})$$
$$= \boxed{3{,}175{,}000 \text{ gal } (3.2 \text{ MG})}$$

The answer is (C).

(c) The pump supplies demand from 4 A.M. until 8 A.M. The storage supplies demand from 8 A.M. until 4 A.M.

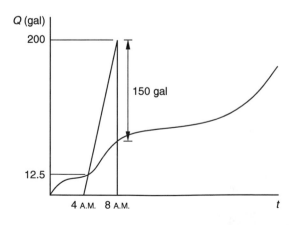

The population use is

$$(150 \text{ gal} + 12.5 \text{ gal})(40{,}000) = 6{,}500{,}000 \text{ gal}$$

Now add fire fighting.

$$6{,}500{,}000 \text{ gal} + \left(6043 \ \frac{\text{gal}}{\text{min}}\right)\left(60 \ \frac{\text{min}}{\text{hr}}\right)(6 \text{ hr})$$
$$= \boxed{8{,}675{,}500 \text{ gal } (8.7 \text{ MG})}$$

The answer is (D).

13. (a) From Table 36.5, use 3 as the peak multiplier.

$$\frac{\left(110 \ \frac{\text{gal}}{\text{day}}\right)(15{,}000)(3)}{\left(24 \ \frac{\text{hr}}{\text{day}}\right)\left(60 \ \frac{\text{min}}{\text{hr}}\right)} = 3437 \text{ gal/min}$$

The fire fighting requirements are given by Eq. 36.53.

$$Q = 1020\sqrt{15}\left(1 - 0.01\sqrt{15}\right) = 3797 \text{ gal/min}$$

The total maximum demand for which the distribution system should be designed is

$$3437 \ \frac{\text{gal}}{\text{min}} + 3797 \ \frac{\text{gal}}{\text{min}} = \boxed{7234 \text{ gal/min}}$$

The answer is (B).

(b) The filter area should not be based on the maximum hourly rate since some of the demand during the peak hours can come from storage (clearwell or tanks). Also, fire requirements can bypass the filters if necessary.

$$\left(110 \ \frac{\text{gal}}{\text{day-person}}\right)(15{,}000 \text{ people}) = 1.65 \times 10^6 \text{ gal/day}$$

Using a flow rate of 4 gpm/ft^2, the required filter area is

$$A = \frac{1.65 \times 10^6 \ \frac{\text{gal}}{\text{day}}}{\left(4 \ \frac{\text{gal}}{\text{min-ft}^2}\right)\left(24 \ \frac{\text{hr}}{\text{day}}\right)\left(60 \ \frac{\text{min}}{\text{hr}}\right)} = \boxed{286.5 \text{ ft}^2}$$

The answer is (A).

(c) $\boxed{\text{No, 60 mg/L is soft water.}}$

The answer is (A).

(d) Disregarding the fire flow, the required average (not peak) daily chlorine mass is given by Eq. 36.11.

$$\left(110 \ \frac{\text{gal}}{\text{day-person}}\right)(15{,}000 \text{ people})$$
$$\frac{\times \left(2 \ \frac{\text{MG}}{\text{L}}\right)\left(8.345 \ \frac{\text{lbm-L}}{\text{MG-mg}}\right)}{10^6 \ \frac{\text{gal}}{\text{MG}}}$$
$$= \boxed{27.54 \text{ lbm/day}}$$

The answer is (C).

37 Biology and Bacteriology

PRACTICE PROBLEMS

1. (*Time limit: one hour*) A fresh wastewater sample is taken containing nitrate ions, sulfate ions, and dissolved oxygen, and it is placed in a sealed jar absent of air.

(a) What is the sequence of oxidation of the compounds?
- (A) nitrate, dissolved oxygen, and then sulfate
- (B) sulfate, nitrate, and then dissolved oxygen
- (C) dissolved oxygen, nitrate, and then sulfate
- (D) none of the above

(b) Obnoxious odors will
- (A) appear in the sample when the dissolved oxygen is exhausted.
- (B) appear in the sample when the dissolved oxygen and nitrates are exhausted.
- (C) appear in the sample when the dissolved oxygen, nitrate, and sulfate are exhausted.
- (D) not appear.

(c) Bacteria will convert ammonia to nitrate if the bacteria are
- (A) phototropic.
- (B) autotrophic.
- (C) thermophilic.
- (D) obligate anaerobes.

(d) Bacteria will generally reduce nitrate to nitrogen gas only if the bacteria are
- (A) photosynthetic.
- (B) obligate aerobic.
- (C) facultative heterotrophic.
- (D) aerobic phototropic.

(e) A bacteriophage is a
- (A) bacterial enzyme.
- (B) virus that infects bacteria.
- (C) mesophilic organism.
- (D) virus that stimulates bacterial growth.

(f) Algal growth in the wastewater
- (A) will be inhibited when the dissolved oxygen is exhausted.
- (B) will be inhibited when toxins are also found in the solution.
- (C) will be inhibited when toxins are found in the solution or when the dissolved oxygen is exhausted.
- (D) cannot be inhibited by chemical means.

(g) In the presence of nitrifying bacteria, nontoxic inorganic compounds, and sunlight, algal growth in the wastewater sample
- (A) will be prevented.
- (B) will be inhibited.
- (C) will be unaffected.
- (D) will be enhanced.

(h) The addition of protozoa to the wastewater will
- (A) not change the wastewater's biochemical composition.
- (B) increase the growth of algae in wastewater.
- (C) increase the growth of bacteria in wastewater.
- (D) decrease the growth of algae and bacteria in the wastewater.

(i) Coliform bacteria in wastewater from human, animal, or soil sources
- (A) can be distinguished in the multiple-tube fermentation test.
- (B) can be categorized into only two groups: human/animal and soil.
- (C) can be distinguished if multiple-tube fermentation and Eschericheiae coli (EC) tests are both used.
- (D) cannot be distinguished.

(j) Under what specific condition would a presence-absence (P-A) coliform test be used on this sample?
- (A) if the sample contains known pathogens
- (B) when multiple-tube fermentation indicates positive dilution in a presumptive test
- (C) when multiple-tube fermentation indicates negative dilution in a presumptive test
- (D) if the sample will be used as drinking water

2. A waste stabilization pond will be used in a municipal wastewater treatment system.

(a) Explain the function of bacteria and algae in the stabilization pond's application.

(b) Explain why algae would be a problem if it were present in the discharge from the pond.

(c) Define mechanisms that could be utilized to control algae and that would prove helpful in the development and use of this wastewater treatment system.

(d) Explain what is meant by *facultative pond.*

SOLUTIONS

1. (a) The sequence of oxygen usage reduced by bacteria is dissolved oxygen, nitrate, and then sulfate.

The answer is (C).

(b) Following the sequence of oxidation, obnoxious odors will occur when dissolved oxygen and nitrate are exhausted.

The answer is (B).

(c) Nitrification is performed by autotrophic bacteria to gain energy for growth by synthesis of carbon dioxide in an aerobic environment.

The answer is (B).

(d) Facultative heterotropic bacteria can decompose organic matter to gain energy under anaerobic conditions by removing the oxygen from nitrate, releasing nitrogen gas.

The answer is (C).

(e) A bacteriophage is a virus that infects bacteria.

The answer is (B).

(f) Algae are photosynthetic (gaining energy from light), releasing oxygen during metabolism. They thrive in aerobic environments. The presence of nitrifying bacteria and toxins and the depletion of dissolved oxygen would inhibit the growth process.

The answer is (C).

(g) Nitrifying bacteria are nonphotosynthetic, gaining energy by taking in oxygen to oxidize reduced inorganic nitrogen. The presence of inorganic nutrients would maintain the nitrification process, thereby limiting algal growth.

The answer is (B).

(h) Protozoa consume bacteria and algae in wastewater treatment and in the aquatic food chain.

The answer is (D).

(i) Fecal coliforms from humans and other warm-blooded animals are the same bacterial species. Coliforms originating from the soil can be separated by a confirmatory procedure using EC medium broth incubated at the elevated temperature of 44.5°C (112°F).

The answer is (B).

(j) The presence-absence technique is used for the testing of drinking water.

The answer is (D).

2. (a) A waste stabilization pond's operation is dependent on the reaction of bacteria and algae. Organic matter is metabolized by bacteria to produce the principle products of carbon dioxide, water, and a small amount of ammonia nitrogen. Algae convert sunlight into energy through photosynthesis. They utilize the end products of cell synthesis and other nutrients to synthesize new cells and produce oxygen. The most important role of the algae is in the production of oxygen in the pond for use by aerobic bacteria. In the absence of sunlight, the algae will consume oxygen in the same manner as bacteria. Algae removal is important in producing a high-quality effluent from the pond.

(b) The discharge of algae increases suspended solids in the discharge and may present a problem in meeting water quality criteria. The algae exert an oxygen demand when they settle to the bottom of the stream and undergo respiration.

(c) The following methods have been suggested for control of algae: (a) multiple ponds in series, (2) drawing off of effluent from below the surface by use of a good baffling arrangement to avoid algae concentrations, (3) sand filter or rock filter for algae removal, (4) alum addition and flocculation, (5) microscreening, and (6) chlorination to kill algae. Chlorination may increase BOD loading due to dead algae cells releasing stored organic material.

(d) Facultative ponds have two zones of treatment: an aerobic surface layer in which oxygen is used by aerobic bacteria for waste stabilization and an anaerobic bottom zone in which sludge decomposition occurs. No artificially induced aeration is used.

Environmental

38 Wastewater Quantity and Quality

PRACTICE PROBLEMS

Sewer Velocities and Sizing

1. (*Time limit: one hour*) A town of 10,000 people (125 gpcd) has its own primary treatment plant.

(a) What mass of total solids should the treatment plant expect?
- (A) 57 lbm/day
- (B) 740 lbm/day
- (C) 3800 lbm/day
- (D) 8300 lbm/day

(b) If the town is 4 mi from the treatment plant and 400 ft above it in elevation, what minimum size pipe should be used between the town and the plant, assuming that the pipe flows full? Disregard infiltration.
- (A) 8 in
- (B) 12 in
- (C) 14 in
- (D) 18 in

2. (*Time limit: one hour*) Your client has just completed a subdivision, and his sewage lines hook up to a collector that goes into a trunk. A problem has developed in the first manhole up from the trunk, and the collector pipe overflows periodically. An industrial plant is hooked directly into the trunk, and it is this plant's flow that is making your client's line back up. The subdivision is in a flood plain, and the sewer lines are very flat and cannot be steepened. (a) Describe two possible solutions to this problem. (b) Sketch plan views of your solutions.

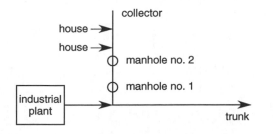

Dilution

3. Town A wants to discharge wastewater into a river 5.79 km upstream of town B. The state standard level for dissolved oxygen is 5 mg/L. Other data pertinent to this problem are given in the following table.

parameter	town A	river
flow (m^3/s)	0.280	0.877
ultimate BOD at 28°C (mg/L)	6.44	7.00
DO (mg/L)	1.00	6.00
K_d at 28°C, day^{-1} (base-10)	N/A	0.199
K_r at 28°C, day^{-1} (base-10)	N/A	0.370
velocity (m/s)	N/A	0.650
temperature (°C)	28	28

(a) Will the dissolved oxygen level at town A be reduced below the state standard level?
- (A) yes; DO$_{min}$ = 4.1 mg/L
- (B) yes; DO$_{min}$ = 4.8 mg/L
- (C) no; DO$_{min}$ = 5.3 mg/L
- (D) no; DO$_{min}$ = 5.9 mg/L

(b) Will the dissolved oxygen level be below the state standard level at town B?
- (A) yes; DO$_{min}$ = 4.1 mg/L
- (B) yes; DO$_{min}$ = 4.8 mg/L
- (C) no; DO$_{min}$ = 5.3 mg/L
- (D) no; DO$_{min}$ = 5.9 mg/L

(c) At any point downstream, will the dissolved oxygen level be below the state standard level?
- (A) yes; DO$_{min}$ = 4.1 mg/L
- (B) yes; DO$_{min}$ = 4.7 mg/L
- (C) no; DO$_{min}$ = 5.1 mg/L
- (D) no; DO$_{min}$ = 5.5 mg/L

(d) Determine if the discharge will reduce the dissolved oxygen in the river at town B below the state standard level if the winter river temperature drops to 12°C and all other characteristics remain the same.
- (A) yes; DO$_{min}$ = 4.5 mg/L
- (B) yes; DO$_{min}$ = 4.9 mg/L
- (C) no; DO$_{min}$ = 5.1 mg/L
- (D) no; DO$_{min}$ = 5.8 mg/L

4. To protect aquatic life, the limit on increase in temperature in a certain stream is 2°C at seasonal low flow. In addition, the stream must not contain more than 0.002 mg/L of un-ionized ammonia.

A manufacturing facility proposes to draw cooling water from the stream and return it to the stream at a higher temperature. They will also be discharging some un-ionized ammonia in the cooling water.

The critical low flow in the stream is 1200 m³/s at a temperature of 20°C. The stream has no measurable un-ionized ammonia in its natural state. The facility intends to withdraw 60 m³/s of cooling water and return it to the stream without loss of mass.

(a) What is the maximum temperature for the discharge water if the stream is not to be damaged?
- (A) 40°C
- (B) 60°C
- (C) 80°C
- (D) 90°C

(b) What is the maximum ammonia concentration for the discharge water if the stream is not to be damaged?
- (A) 0.04 mg/L
- (B) 0.06 mg/L
- (C) 0.09 mg/L
- (D) 0.13 mg/L

5. A rural town of 10,000 people discharges partially treated wastewater into a stream. The stream has the following characteristics.

minimum flow rate	120 ft³/sec
velocity	3 mi/hr
minimum dissolved oxygen	7.5 mg/L at 15°C
temperature	15°C
BOD	0

The reoxygenation and deoxygenation coefficients (base-10) are

reoxygenation coefficient of stream and effluent mixture	0.2 day⁻¹ at 20°C
deoxygenation coefficient	0.15 day⁻¹ at 20°C

reoxygenation coefficient
  of stream and effluent mixture 0.2 day^{-1} at 20°C
deoxygenation coefficient   0.15 day^{-1} at 20°C

The town's effluent has the following characteristics.

source	volume	BOD at 20°C	temp.
domestic	122 gpcd	0.191 lbm/cd	64°F
infiltration	116,000 gpd		51°F
industrial no. 1	180,000 gpd	800 mg/L	95°F
industrial no. 2	76,000 gpd	1700 mg/L	84°F

(a) What is the domestic waste BOD in mg/L at 20°C?
- (A) 140 mg/L
- (B) 160 mg/L
- (C) 190 mg/L
- (D) 220 mg/L

(b) What is the approximate total effluent BOD in mg/L just before discharge into the stream?
- (A) 245 mg/L
- (B) 270 mg/L
- (C) 295 mg/L
- (D) 315 mg/L

(c) What is the average temperature of the effluent just before discharge into the stream?
- (A) 20°C
- (B) 30°C
- (C) 50°C
- (D) 70°C

(d) If the wastewater does not contribute to dissolved oxygen at all, how far downstream is the theoretical point of minimum dissolved oxygen concentration?
- (A) 40 mi
- (B) 70 mi
- (C) 80 mi
- (D) 120 mi

(e) What is the theoretical minimum dissolved oxygen concentration in the stream?
- (A) 4.7 mg/L
- (B) 5.5 mg/L
- (C) 7.3 mg/L
- (D) 8.3 mg/L

6. A sewage treatment plant is being designed to handle both domestic and industrial wastewaters. The city population is 20,000, and an excess capacity factor of 15% is to be used for future domestic expansion. The wastewaters have the following characteristics.

source	volume	BOD
domestic	100 gpcd	0.18 lbm/cd
industrial no. 1	1.3 MGD	1100 mg/L
industrial no. 2	1.0 MGD	500 mg/L

(a) What is the design population equivalent for the plant?
- (A) 60,000
- (B) 75,000
- (C) 90,000
- (D) 110,000

(b) What is the plant's organic loading?
- (A) 2300 lbm/day
- (B) 8100 lbm/day
- (C) 14,000 lbm/day
- (D) 20,000 lbm/day

(c) What is the plant's hydraulic loading?
- (A) 4.6 MGD
- (B) 5.8 MGD
- (C) 6.3 MGD
- (D) 7.5 MGD

SOLUTIONS

1. Assume: 125 gpcd for average flow
800 mg/L total solids

(a)
$$\frac{(10{,}000)\left(125\ \frac{\text{gal}}{\text{day}}\right)\left(800\ \frac{\text{mg}}{\text{L}}\right)\left(8.345\ \frac{\text{lbm-L}}{\text{MG-mg}}\right)}{1\times10^{6}\ \frac{\text{gal}}{\text{MG}}}$$

$$=\boxed{8345\ \text{lbm/day}}$$

(Minor variations in the assumptions should not affect the answer choice.)

The answer is (D).

(b)
$$S=\frac{400\ \text{ft}}{(4\ \text{mi})\left(5280\ \frac{\text{ft}}{\text{mi}}\right)}=0.01894$$

Use Eq. 38.1.

$$\frac{Q_{\text{peak}}}{Q_{\text{ave}}}=\frac{18+\sqrt{\text{P}}}{4+\sqrt{\text{P}}}=\frac{18+\sqrt{10}}{4+\sqrt{10}}$$

$$\approx 3.0$$

$$Q_{\text{peak}}=\frac{\left(125\ \frac{\text{gal}}{\text{day}}\right)(3)(10{,}000)\left(0.1337\ \frac{\text{ft}^3}{\text{gal}}\right)}{\left(24\ \frac{\text{hr}}{\text{day}}\right)\left(60\ \frac{\text{min}}{\text{hr}}\right)\left(60\ \frac{\text{sec}}{\text{min}}\right)}$$

$$=5.80\ \text{ft}^3/\text{sec}\quad(\text{cfs})$$

Assume $n = 0.013$. Use the Manning equation for a pipe flowing full to calculate the normal depth.

$$D=1.33\left(\frac{nQ}{\sqrt{S}}\right)^{\frac{3}{8}}$$

$$=(1.33)\left(\frac{(0.013)\left(5.80\ \frac{\text{ft}^3}{\text{sec}}\right)}{\sqrt{0.01894}}\right)^{\frac{3}{8}}$$

$$=1.06\ \text{ft}\quad(12.7\ \text{in})$$

$$\boxed{14\ \text{in}}$$

(A 12 in pipe does not have the capacity of a 12.7 in pipe.)

The answer is (C).

2. (a) If the first manhole overflows, the piezometric head at the rim must be less than the head in the trunk. Further up at manhole no. 2, the increase in elevation is sufficient to raise rim no. 2, so the head in the trunk must be lowered.

Alternate solutions:

- relief storage (surge chambers) between trunk and manhole no. 1
- storage at plant and gradual release
- private trunk for plant
- private treatment and discharge for plant
- larger trunk capacity using larger or parallel pipes
- backflow preventors

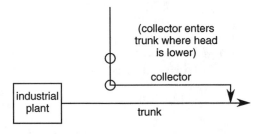

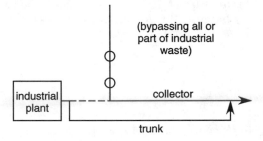

3. (a) Find the dissolved oxygen at town A from Eq. 38.33.

$$\text{DO}=\frac{Q_w\text{DO}_w+Q_r\text{DO}_r}{Q_w+Q_r}$$

$$=\frac{\left(0.280\ \frac{\text{m}^3}{\text{s}}\right)\left(1\ \frac{\text{mg}}{\text{L}}\right)+\left(0.877\ \frac{\text{m}^3}{\text{s}}\right)\left(6\ \frac{\text{mg}}{\text{L}}\right)}{0.280\ \frac{\text{m}^3}{\text{s}}+0.877\ \frac{\text{m}^3}{\text{s}}}$$

$$=\boxed{4.79\ \text{mg/L}}$$

So, immediately after mixing, the water will be reduced below the state standard of 5 mg/L.

The answer is (B).

(b) The composite ultimate BOD is

$$\text{BOD}_{u,28^\circ C}=\frac{Q_w\text{BOD}_w+Q_r\text{BOD}_r}{Q_w+Q_r}$$

$$=\frac{\left(0.280\ \frac{\text{m}^3}{\text{s}}\right)\left(6.44\ \frac{\text{mg}}{\text{L}}\right)+\left(0.877\ \frac{\text{m}^3}{\text{s}}\right)\left(7\ \frac{\text{mg}}{\text{L}}\right)}{0.280\ \frac{\text{m}^3}{\text{s}}+0.877\ \frac{\text{m}^3}{\text{s}}}$$

$$=6.8645\ \text{mg/L}$$

Environmental

From App. 20.D, $DO_{sat} = 7.92$ mg/L at 28°C. The initial deficit is

$$D = DO_{sat} - DO = 7.92 \frac{mg}{L} - 4.79 \frac{mg}{L}$$

$$= 3.13 \text{ mg/L}$$

Calculate travel time from town A to town B.

$$t = \frac{(5.79 \text{ km})\left(1000 \frac{m}{km}\right)}{\left(0.650 \frac{m}{s}\right)\left(86{,}400 \frac{s}{day}\right)}$$

$$= 0.1031 \text{ day}$$

Calculate the deficit 5.79 km downstream. Use base-10 exponents because the K values are in base-10. Use Eq. 38.34.

$$D_t = \left(\frac{K_d BOD_u}{K_r - K_d}\right)(10^{-K_d t} - 10^{-K_r t}) + D_0(10^{-K_r t})$$

$$= \frac{(0.199 \text{ day}^{-1})\left(6.8645 \frac{mg}{L}\right)}{0.370 \frac{mg}{L} - 0.199 \frac{mg}{L}}$$

$$\times \left(10^{-(0.199 \text{ day}^{-1})(0.1031 \text{ day})}\right.$$

$$\left. -10^{-(0.370 \text{ day}^{-1})(0.1031 \text{ day})}\right)$$

$$+ \left(3.13 \frac{mg}{L}\right)\left(10^{-(0.370 \text{ day}^{-1})(0.1031 \text{ day})}\right)$$

$$= 3.17 \text{ mg/L}$$

Calculate the dissolved oxygen downstream.

$$DO_{\text{town B}} = DO_{sat} - D$$

$$= 7.92 \frac{mg}{L} - 3.17 \frac{mg}{L}$$

$$= \boxed{4.75 \text{ mg/L}}$$

This is still below the state standard of 5 mg/L. More distance is required for reoxygenation to increase the oxygen concentration.

The answer is (B).

(c) Determine the critical time. Use Eq. 38.36.

$$t_c = \left(\frac{1}{K_r - K_d}\right)$$

$$\times \log_{10}\left(\left(\frac{K_d BOD_u - K_r D_0 + K_d D_0}{K_d BOD_u}\right)\left(\frac{K_r}{K_d}\right)\right)$$

$$= \left(\frac{1}{0.370 \text{ day}^{-1} - 0.199 \text{ day}^{-1}}\right)$$

$$\times \log_{10}\left(\frac{\begin{array}{l}(0.199 \text{ day}^{-1})\left(6.8645 \frac{mg}{L}\right)\\ \quad - (0.370 \text{ day}^{-1})\left(3.13 \frac{mg}{L}\right)\\ \quad + (0.199 \text{ day}^{-1})\left(3.13 \frac{mg}{L}\right)\end{array}}{(0.199 \text{ day}^{-1})\left(6.8645 \frac{mg}{L}\right)}\right.$$

$$\left. \times \left(\frac{0.370 \text{ day}^{-1}}{0.199 \text{ day}^{-1}}\right)\right)$$

$$= 0.3122 \text{ day}$$

Calculate critical deficit and dissolved oxygen. Use Eq. 38.37.

$$D_c = \left(\frac{K_d BOD_u}{K_r}\right)10^{-K_d t}$$

$$= \left(\frac{(0.199 \text{ day}^{-1})\left(6.8645 \frac{mg}{L}\right)}{0.370 \text{ day}^{-1}}\right)$$

$$\times \left(10^{-(0.199 \text{ day}^{-1})(0.3122 \text{ day})}\right)$$

$$= 3.20 \text{ mg/L}$$

$$DO = DO_{sat} - D_c$$

$$= 7.92 \frac{mg}{L} - 3.20 \frac{mg}{L}$$

$$= \boxed{4.72 \text{ mg/L}}$$

This is below the state standard of 5 mg/L.

The answer is (B).

(d) From Eq. 38.33, the composite dissolved oxygen is

$$DO = \frac{Q_w DO_w + Q_r DO_r}{Q_w + Q_r}$$

$$= \frac{\left(0.280 \ \frac{m^3}{s}\right)\left(1 \ \frac{mg}{L}\right) + \left(0.877 \ \frac{m^3}{s}\right)\left(6 \ \frac{mg}{L}\right)}{0.280 \ \frac{m^3}{s} + 0.877 \ \frac{m^3}{s}}$$

$$= 4.79 \ mg/L \quad [\text{no change}]$$

At 12°C, the saturation dissolved oxygen is found from App. 20.D. $DO_{sat,12°C} = 10.83 \ mg/L$.

Calculate the initial deficit from Eq. 38.33.

$$D_o = 10.83 \ \frac{mg}{L} - 4.79 \ \frac{mg}{L}$$

$$= 6.04 \ mg/L$$

Calculate the temperature of the river water and wastewater mixture.

$$T = \frac{Q_w T_w + Q_r T_r}{Q_w + Q_r}$$

$$= \frac{\left(0.280 \ \frac{m^3}{s}\right)(28°C) + \left(0.877 \ \frac{m^3}{s}\right)(12°C)}{0.280 \ \frac{m^3}{s} + 0.877 \ \frac{m^3}{s}}$$

$$= 15.87°C$$

Calculate K_d at this temperature. Two different values of θ_d are used for the two temperature ranges. Use Eq. 38.27.

$$K_{d,T_1} = K_{d,T_2}\theta^{T_1-T_2}$$

$$K_{d,20°C} = (0.199 \ day^{-1})(1.056)^{20°C-\ 28°C}$$

$$= 0.1287 \ day^{-1}$$

$$K_{d,T_1} = K_{d,T_2}\theta^{T_1-T_2}$$

$$K_{d,15.87°C} = (0.1287 \ day^{-1})(1.135)^{15.87°C-\ 20°C}$$

$$= 0.07628 \ day^{-1}$$

Similarly, use Eq. 38.22 to correct K_r.

$$K_{r,T_1} = K_{r,T_2}(1.024)^{T_1-T_2}$$

$$K_{r,15.87°C} = K_{r,28°C}(1.024)^{15.87°C-\ 28°C}$$

$$= (0.370 \ day^{-1})(1.024)^{15.87°C-\ 28°C}$$

$$= 0.2775 \ day^{-1}$$

Correct BOD_u to the new temperature. Use Eq. 38.31 twice.

$$BOD_{u,20°C} = \frac{BOD_{u,28°C}}{(0.02)(28°C)+0.6}$$

$$BOD_{u,15.87°C} = (BOD_{u,20°C})\big((0.02)(15.87°C)+0.6\big)$$

$$= \frac{\left(6.8645 \ \frac{mg}{L}\right)\big((0.02)(15.87°C)+0.06\big)}{(0.02)(28°C)+0.6}$$

$$= 5.43 \ mg/L$$

Calculate the deficit at town B. Use Eq. 38.34.

$$D_B = \left(\frac{K_d BOD_u}{K_r - K_d}\right)(10^{-K_d t} - 10^{-K_r t}) + D_0(10^{-K_r t})$$

$$= \left(\frac{(0.07628 \ day^{-1})\left(5.43 \ \frac{mg}{L}\right)}{0.2775 \ day^{-1} - 0.07628 \ day^{-1}}\right)$$

$$\times \Big(10^{-(0.07628 \ day^{-1})(0.1031 \ day)}$$

$$-10^{-(0.2775 \ day^{-1})(0.1031 \ day)}\Big)$$

$$+ \left(6.04 \ \frac{mg}{L}\right)(10^{-(0.2775 \ day^{-1})(0.1031 \ day)})$$

$$= 5.75 \ mg/L$$

The dissolved oxygen at town B is

$$DO_B = DO_{sat} - D_B$$

$$= 10.83 \ \frac{mg}{L} - 5.75 \ \frac{mg}{L}$$

$$= \boxed{5.08 \ mg/L}$$

This meets the state standard of 5 mg/L.

The answer is (C).

4. (a)
$$T = 20°C + 2°C = 22°C$$
$$T_1 = 20°C$$
$$Q_1 + Q_2 = 1200 \ m^3/s$$
$$Q_2 = 60 \ m^3/s$$
$$Q_1 = 1200 \ \frac{m^3}{s} - 60 \ \frac{m^3}{s} = 1140 \ m^3/s$$
$$T = \frac{T_1 Q_1 + T_2 Q_2}{Q_1 + Q_2}$$
$$22°C = \frac{(20°C)\left(1140 \ \frac{m^3}{s}\right) + C_2\left(60 \ \frac{m^3}{s}\right)}{1200 \ \frac{m^3}{s}}$$
$$= \boxed{60°C}$$

The answer is (B).

(b)
$$C = 0 + 0.002 \ \frac{mg}{L}$$
$$C_1 = 0$$
$$0.002 \ \frac{mg}{L} = \frac{(0)\left(1140 \ \frac{m^3}{s}\right) + C_2\left(60 \ \frac{m^3}{s}\right)}{1200 \ \frac{m^3}{s}}$$
$$= \boxed{0.04 \ mg/L}$$

The answer is (A).

Environmental

5. (a) The domestic BOD concentration is

$$\frac{\left(0.191 \dfrac{\text{lbm}}{\text{capita-day}}\right)\left(10^6 \dfrac{\text{gal}}{\text{MG}}\right)}{122 \dfrac{\text{gal}}{\text{capita-day}}} \times \left(0.1198 \dfrac{\text{mg-MG}}{\text{L-lbm}}\right)$$

$$= \boxed{187.6 \text{ mg/L}}$$

The answer is (C).

(b) Use Eq. 38.33.

$$\frac{\begin{array}{l}\left(122 \dfrac{\text{gal}}{\text{day}}\right)(10{,}000)\left(187.6 \dfrac{\text{mg}}{\text{L}}\right) \\[4pt] + \left(116{,}000 \dfrac{\text{gal}}{\text{day}}\right)(0) \\[4pt] + \left(180{,}000 \dfrac{\text{gal}}{\text{day}}\right)\left(800 \dfrac{\text{mg}}{\text{L}}\right) \\[4pt] + \left(76{,}000 \dfrac{\text{gal}}{\text{day}}\right)\left(1700 \dfrac{\text{mg}}{\text{L}}\right)\end{array}}{\begin{array}{l}\left(122 \dfrac{\text{gal}}{\text{day}}\right)(10{,}000) + 116{,}000 \dfrac{\text{gal}}{\text{day}} \\[4pt] + 180{,}000 \dfrac{\text{gal}}{\text{day}} + 76{,}000 \dfrac{\text{gal}}{\text{day}}\end{array}}$$

$$= \boxed{315.4 \text{ mg/L}}$$

The answer is (D).

(c) Use Eq. 38.33.

$$\frac{\begin{array}{l}\left(122 \dfrac{\text{gal}}{\text{day}}\right)(10{,}000)(64°\text{F}) \\[4pt] + \left(116{,}000 \dfrac{\text{gal}}{\text{day}}\right)(51°\text{F}) \\[4pt] + \left(180{,}000 \dfrac{\text{gal}}{\text{day}}\right)(95°\text{F}) \\[4pt] + \left(76{,}000 \dfrac{\text{gal}}{\text{day}}\right)(84°\text{F})\end{array}}{\begin{array}{l}\left(122 \dfrac{\text{gal}}{\text{day}}\right)(10{,}000) + 116{,}000 \dfrac{\text{gal}}{\text{day}} \\[4pt] + 180{,}000 \dfrac{\text{gal}}{\text{day}} + 76{,}000 \dfrac{\text{gal}}{\text{day}}\end{array}}$$

$$= \boxed{67.5°\text{F} \ (19.7°\text{C})}$$

The answer is (A).

(d) The total discharge into the river is

$$\left(\left(122 \dfrac{\text{gal}}{\text{day}}\right)(10{,}000) + 116{,}000 \dfrac{\text{gal}}{\text{day}} + 180{,}000 \dfrac{\text{gal}}{\text{day}}\right.$$
$$\left. + 76{,}000 \dfrac{\text{gal}}{\text{day}}\right)\left(1.547 \times 10^{-6} \dfrac{\text{ft}^3\text{-day}}{\text{sec-gal}}\right)$$
$$= 2.46 \text{ ft}^3/\text{sec} \ \ (\text{cfs})$$

step 1: Find the stream conditions immediately after mixing.

$$\text{BOD}_{5,20°\text{C}} = \frac{\left(2.46 \dfrac{\text{ft}^3}{\text{sec}}\right)\left(315.4 \dfrac{\text{mg}}{\text{L}}\right) + \left(120 \dfrac{\text{ft}^3}{\text{sec}}\right)(0)}{2.46 \dfrac{\text{ft}^3}{\text{sec}} + 120 \dfrac{\text{ft}^3}{\text{sec}}}$$

$$= 6.34 \text{ mg/L}$$

$$\text{DO} = \frac{\left(2.46 \dfrac{\text{ft}^3}{\text{sec}}\right)(0) + \left(120 \dfrac{\text{ft}^3}{\text{sec}}\right)\left(7.5 \dfrac{\text{mg}}{\text{L}}\right)}{2.46 \dfrac{\text{ft}^3}{\text{sec}} + 120 \dfrac{\text{ft}^3}{\text{sec}}}$$

$$= 7.35 \text{ mg/L}$$

$$T = \frac{\left(2.46 \dfrac{\text{ft}^3}{\text{sec}}\right)(19.7°\text{C}) + \left(120 \dfrac{\text{ft}^3}{\text{sec}}\right)(15°\text{C})}{2.46 \dfrac{\text{ft}^3}{\text{sec}} + 120 \dfrac{\text{ft}^3}{\text{sec}}}$$

$$= 15.1°\text{C}$$

step 2: Calculate the rate constants at 15.1°C. Use Eq. 38.27.

$$K_{d,15.1°\text{C}} = (0.15 \text{ day}^{-1})(1.135)^{15.1°\text{C}-20°\text{C}}$$
$$= 0.0807 \text{ day}^{-1}$$
$$K_{r,15.1°\text{C}} = (0.2 \text{ day}^{-1})(1.024)^{15.1°\text{C}-20°\text{C}}$$
$$= 0.178 \text{ day}^{-1}$$

step 3: Estimate BOD_u. Use Eq. 38.29.

$$\text{BOD}_{u,20°\text{C}} = \frac{6.34 \dfrac{\text{mg}}{\text{L}}}{1 - 10^{-(0.15 \text{ day}^{-1})(5 \text{ days})}}$$
$$= 7.71 \text{ mg/L}$$

Use Eq. 38.31 to convert BOD_u to 15.1°C.

$$BOD_{u,15.1°C} = BOD_{u,20°C}(0.02T_C + 0.6)$$

$$= \left(7.71 \frac{mg}{L}\right)((0.02)(15.1°C) + 0.6)$$

$$= 6.95 \text{ mg/L}$$

step 4: From App. 20.D at 15°C, saturated DO = 10.15 mg/L. Since the actual is 7.35 mg/L, the deficit is

$$D_0 = 10.15 \frac{mg}{L} - 7.35 \frac{mg}{L} = 2.8 \text{ mg/L}$$

step 5: Calculate t_c. Use Eq. 38.36.

$$t_c = \left(\frac{1}{0.178 \text{ day}^{-1} - 0.0807 \text{ day}^{-1}}\right)$$

$$\times \log_{10}\left(\frac{\begin{array}{c}(0.0807 \text{ day}^{-1})\left(6.95 \frac{mg}{L}\right)\\ - (0.178 \text{ day}^{-1})\left(2.8 \frac{mg}{L}\right)\\ + (0.0807 \text{ day}^{-1})\left(2.8 \frac{mg}{L}\right)\end{array}}{(0.0807 \text{ day}^{-1})\left(6.95 \frac{mg}{L}\right)}\right.$$

$$\left.\times \left(\frac{0.178 \text{ day}^{-1}}{0.0807 \text{ day}^{-1}}\right)\right) = 0.562 \text{ days}$$

step 6: The distance downstream is

$$(0.562 \text{ days})\left(3 \frac{mi}{hr}\right)\left(24 \frac{hr}{day}\right) = \boxed{40.5 \text{ mi}}$$

The answer is (A).

(e) *step 7:* Use Eq. 38.37.

$$D_c = \left(\frac{(0.0807 \text{ day}^{-1})\left(6.95 \frac{mg}{L}\right)}{0.178 \text{ day}^{-1}}\right)$$

$$\times 10^{-(0.0807 \text{ day}^{-1})(0.566 \text{ day})}$$

$$= 2.84 \text{ mg/L}$$

step 8: $$DO_{min} = 10.15 \frac{mg}{L} - 2.84 \frac{mg}{L}$$

$$= \boxed{7.31 \text{ mg/L}}$$

The answer is (C).

6. (a) Do not apply the population expansion factor to the industrial effluents. Use Eq. 38.2, modified for the given population equivalent for the domestic flow contribution.

$$P_e = P_{\text{domestic flow}} + P_{\text{industrial source 1}}$$
$$+ P_{\text{industrial source 2}}$$
$$= (20)(1.15)$$
$$+ \frac{\left(1100 \frac{mg}{L}\right)\left(1.3 \times 10^6 \frac{gal}{day}\right)\times \left(8.345 \times 10^{-9} \frac{lbm\text{-}L}{MG\text{-}mg}\right)}{0.18 \frac{lbm}{day}}$$
$$+ \frac{(500)\left(1.0 \times 10^6 \frac{gal}{day}\right)\times \left(8.345 \times 10^{-9} \frac{lbm\text{-}L}{MG\text{-}mg}\right)}{0.18 \frac{lbm}{day}}$$
$$= \boxed{112.5 \text{ [thousands of people]}}$$

The answer is (D).

(b) Since the plant loading is requested, the organic loading can be given in lbm/day. The population from part (a) is 112,500.

$$L_{BOD} = (112,500)\left(0.18 \frac{lbm}{day}\right) = \boxed{20,250 \text{ lbm/day}}$$

The answer is (D).

(c) $$L_H = (20,000)(1.15)\left(100 \frac{gal}{day}\right)$$
$$+ 1.3 \times 10^6 \frac{gal}{day} + 1.0 \times 10^6 \frac{gal}{day}$$
$$= \boxed{4.6 \times 10^6 \text{ gal/day (4.6 MGD)}}$$

The answer is (A).

Environmental

PRACTICE PROBLEMS

Lagoons

1. A cheese factory located in a normally warm state has liquid waste with the following characteristics.

> *waste no. 1*
> | | |
> |---|---|
> | volume | 10,000 gal/day |
> | BOD | 1000 mg/L |
>
> *waste no. 2*
> | | |
> |---|---|
> | volume | 25,000 gal/day |
> | BOD | 250 mg/L |

The factory will use a 4 ft deep on-site lagoon to stabilize the waste.

(a) What is the total BOD loading?

(A) 60 lbm/day
(B) 85 lbm/day
(C) 110 lbm/day
(D) 140 lbm/day

(b) What lagoon size is required?

(A) 2.7 ac
(B) 6.8 ac
(C) 8.3 ac
(D) 10.9 ac

(c) What is the detention time?

(A) 5 wk
(B) 14 wk
(C) 36 wk
(D) 52 wk

Trickling Filters

2. It is estimated that the BOD of raw sewage received at a treatment plant serving a population of 20,000 will be 300 mg/L. It is estimated that the per capita BOD loading is 0.17 lbm/day. 30% of the influent BOD is removed by settling. One single-stage high-rate trickling filter is to be used to reduce the plant effluent to 50 mg/L. Recirculation is from the filter effluent to the primary settling influent. *Ten States' Standards* is in effect.

(a) What is the design flow rate?

(A) 0.9 MGD
(B) 1.1 MGD
(C) 1.4 MGD
(D) 2.0 MGD

(b) Assuming a design flow rate of 1.35 MGD, what is the total organic load on the filter?

(A) 1700 lbm/day
(B) 2100 lbm/day
(C) 2400 lbm/day
(D) 3400 lbm/day

(c) Assuming an incoming volume of 1.35 MGD and an organic loading of 60 lbm/day-1000 ft^3, what flow should be recirculated?

(A) 0.6 MGD
(B) 0.9 MGD
(C) 1.2 MGD
(D) 2.2 MGD

(d) What is the overall plant efficiency?

(A) 45%
(B) 76%
(C) 83%
(D) 91%

3. The average wastewater flow from a community of 20,000 is 125 gpcd. The 5 day, 20°C BOD is 250 mg/L. The suspended solids content is 300 mg/L. A final plant effluent of 50 mg/L of BOD is to be achieved through the use of two sets of identical settling tanks and trickling filters operating in parallel. The settling tanks are to be designed to a standard of 1000 gpd/ft^2. The trickling filters are to be 6 ft deep. There is no recirculation.

(a) What settling tank surface area is required?

(A) 2500 ft^2
(B) 3000 ft^2
(C) 3500 ft^2
(D) 4500 ft^2

(b) What settling tank diameter is required?

(A) 40 ft
(B) 50 ft
(C) 65 ft
(D) 80 ft

(c) Estimate the BOD removal in the settling tanks.

(A) 15%
(B) 30%
(C) 45%
(D) 60%

(d) What is the trickling filter diameter?
- (A) 60 ft
- (B) 70 ft
- (C) 80 ft
- (D) 90 ft

4. Wastewater from a city with a population of 40,000 has an average daily flow of 4.4 MGD. The sewage has the following characteristics.

BOD_5 at 20°C	160 mg/L
COD	800 mg/L
total solids	900 mg/L
suspended solids	180 mg/L
volatile solids	320 mg/L
settleable solids	8 mg/L
pH	7.8

The wastewater is to be treated with primary settling and secondary trickling filtration. The settling basins are to be circular, 8 ft deep, and designed to a standard overflow rate of 1000 gal/day-ft^2.

(a) Assuming two identical basins operating in parallel, what should be the diameter of each sedimentation basin in order to remove about 30% of the BOD?
- (A) 45 ft
- (B) 53 ft
- (C) 77 ft
- (D) 86 ft

(b) What is the detention time?
- (A) 1.4 hr
- (B) 1.8 hr
- (C) 2.3 hr
- (D) 3.1 hr

(c) What is the weir loading?
- (A) 8200 gal/day-ft
- (B) 11,000 gal/day-ft
- (C) 13,000 gal/day-ft
- (D) 15,000 gal/day-ft

5. (*Time limit: one hour*) A small community has a projected average flow of 1 MGD. Incoming wastewater has the following properties: BOD, 250 mg/L; grit specific gravity, 2.65; and total suspended solids, 400 mg/L. The community wants to have a wastewater treatment plant consisting of a single aerated grit chamber, a single primary clarifier, two identical circular trickling filters, and a single secondary clarifier. There is no equalization basin. Recirculation from the second clarifier to the entrance of the trickling filters will be 100% of the average flow. The final effluent is to have a BOD of 30 mg/L.

(a) What is the peak design flow at the grit chamber?
- (A) 1.0 MGD
- (B) 1.5 MGD
- (C) 2.0 MGD
- (D) 2.5 MGD

(b) Determine the aerated grit chamber width assuming a 3 min detention time, 20 ft length, and a width:depth ratio of 1.25.
- (A) 4 ft
- (B) 6 ft
- (C) 8 ft
- (D) 10 ft

(c) Determine the approximate air requirements for the grit chamber in order to capture approximately 95% of the grit.
- (A) 60 cfm
- (B) 140 cfm
- (C) 280 cfm
- (D) 420 cfm

(d) If the clarifier is 12 ft deep, what should be its diameter?
- (A) 50 ft
- (B) 56 ft
- (C) 63 ft
- (D) 81 ft

(e) What is the peak flow of the primary clarifier effluent?
- (A) 1.0 MGD
- (B) 2.0 MGD
- (C) 3.0 MGD
- (D) 4.0 MGD

(f) Assume the primary clarifier removes 30% of the incoming BOD, and the filters see the average flow only. Determine the diameter of the trickling filters assuming a 6 ft deep rock bed.
- (A) 45 ft
- (B) 55 ft
- (C) 65 ft
- (D) 85 ft

(g) Determine the diameter of the final clarifier.
- (A) 35 ft
- (B) 50 ft
- (C) 65 ft
- (D) 80 ft

Recirculating Biological Contactors

6. (*Time limit: one hour*) 1.5 MGD of wastewater with a BOD of 250 mg/L is processed by a high-rate rock trickling filter followed by recirculating biological contactor (RBC) processing. The trickling filter is 75 ft in diameter and 6 ft deep and was designed to NRC standards. The BOD removal efficiency of the RBC process is given by the following equation. (k has a value of 2.45 gal/day-ft^2, and Q (in units of gal/day) does not include recirculation. A is the immersed area of the RBC in ft^2.)

$$\eta_{BOD} = \frac{1}{\left(1 + \dfrac{kA}{Q}\right)^3}$$

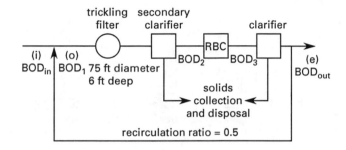

(a) What total exposed RBC surface area is required to achieve an effluent BOD$_{out}$ of 30 mg/L?

  (A) 4.2×10^4 ft^2

  (B) 8.4×10^4 ft^2

  (C) 1.3×10^5 ft^2

  (D) 3.3×10^5 ft^2

(b) The recirculation pick-up point is relocated from after the final clarifier to after the trickling filter. The efficiency of the RBC process is 50%. Determine the recirculation ratio such that BOD$_{out}$ is 30 mg/L.

  (A) 25%

  (B) 50%

  (C) 75%

  (D) 100%

(c) If BOD$_2$ = 85 mg/L, BOD$_{out}$ = 30 mg/L, the yield is 0.4 lbm/lbm BOD removed, and the sludge specific gravity is essentially 1.0, what is the approximate sludge volume produced from the clarifiers?

  (A) 4 ft^3/day

  (B) 12 ft^3/day

  (C) 21 ft^3/day

  (D) 35 ft^3/day

SOLUTIONS

1. (a) The total BOD is

$$\left(1000 \ \frac{mg}{L}\right)\left(8.345 \ \frac{lbm\text{-}L}{mg\text{-}MG}\right)\left(\frac{10{,}000 \ \frac{gal}{day}}{1{,}000{,}000 \ \frac{gal}{MG}}\right)$$

$$+ \left(250 \ \frac{mg}{L}\right)\left(8.345 \ \frac{lbm\text{-}L}{mg\text{-}MG}\right)\left(\frac{25{,}000 \ \frac{gal}{day}}{1{,}000{,}000 \ \frac{gal}{MG}}\right)$$

$$= \boxed{135.6 \text{ lbm/day } (1.4 \text{ MGD})}$$

The answer is (D).

(b) The warm weather and depth contribute to a decrease in pond effectiveness. Assume 20 lbm BOD/ac-day for a nonaerated stabilization pond. From Eq. 39.4, the required area is

$$A = \frac{Q}{v^*} = \frac{135.6 \ \dfrac{lbm}{day}}{20 \ \dfrac{lbm}{ac\text{-}day}} = \boxed{6.8 \text{ ac}}$$

The answer is (B).

(c) Use Eq. 39.5.

$$t_d = \frac{v}{Q}$$

$$= \frac{(6.8 \text{ ac})\left(43{,}560 \ \dfrac{ft^2}{ac}\right)(4 \text{ ft})}{\left(35{,}000 \ \dfrac{gal}{day}\right)\left(1.547 \times 10^{-6} \ \dfrac{ft^3\text{-}day}{sec\text{-}gal}\right)\left(3600 \ \dfrac{sec}{hr}\right)}$$

$$= \boxed{6078 \text{ hr } (36.2 \text{ weeks})}$$

The answer is (C).

2. (a) $Q = \dfrac{\dot{m}}{C} = \dfrac{\left(0.17 \ \dfrac{lbm}{capita\text{-}day}\right)(20{,}000 \text{ people})}{\left(8.345 \times 10^{-6} \ \dfrac{lbm\text{-}L}{gal\text{-}mg}\right)\left(300 \ \dfrac{mg}{L}\right)}$

$$= \boxed{1.358 \times 10^6 \text{ gal/day}}$$

The answer is (C).

(*Ten States' Standards* specifies 100 gpcd in the absence of other information. In that case, Q = (100 gal/capita-day)(20,000 people) = 2×10^6 gal/day.)

(b) The total BOD load leaving the primary clarifier and entering the filter is

$$BOD_i = (1 - 0.30)\left(300 \ \frac{mg}{L}\right) = 210 \ mg/L$$

$$L_{BOD} = (1.35 \ MGD)\left(210 \ \frac{mg}{L}\right)\left(8.345 \ \frac{lbm\text{-}L}{mg\text{-}MG}\right)$$

$$= \boxed{2366 \ lbm/day}$$

The answer is (C).

(c) The efficiency of the filter and secondary clarifier is found from Eq. 39.9.

$$\eta = \frac{210 \ \frac{mg}{L} - 50 \ \frac{mg}{L}}{210 \ \frac{mg}{L}} = 0.762 \ (76.2\%)$$

Use Eq. 39.13.

$$0.762 = \frac{1}{1 + 0.0561\sqrt{\dfrac{60 \ \dfrac{lbm}{day\text{-}1000 \ ft^3}}{F}}}$$

$$F = 1.936$$

Use Eq. 39.15.

$$1.936 = \frac{1 + R}{\left(1 + (0.1)(R)\right)^2}$$

$$R = 1.6$$

Use Eq. 39.10.

$$Q_r = (1.6)(1.35 \ MGD) = \boxed{2.16 \ MGD}$$

The answer is (D).

(d) Use Eq. 39.9.

$$\eta = \frac{300 \ \frac{mg}{L} - 50 \ \frac{mg}{L}}{300 \ \frac{mg}{L}} = \boxed{0.833 \ (83.3\%)}$$

The answer is (C).

3. Use Eq. 39.4. Disregarding variations in peak flow, the average design volume is

$$\frac{\left(125 \ \dfrac{gal}{capita\text{-}day}\right)(20{,}000 \ people)}{1{,}000{,}000 \ \dfrac{gal}{MG}} = 2.5 \ MGD$$

(a) The settling tank surface area is

$$\frac{2.5 \times 10^6 \ \dfrac{gal}{day}}{1000 \ \dfrac{gal}{day\text{-}ft^2}} = \boxed{2500 \ ft^2}$$

The answer is (A).

(b) The required diameter when using two tanks in parallel is

$$D = \sqrt{\frac{(4)(2500 \ ft^2)}{2\pi}} = \boxed{39.9 \ ft \ (use \ 40 \ ft) \ each}$$

The answer is (A).

(c) A 30% removal is typical.

The answer is (B).

(d) BOD entering the filter is

$$(1 - 0.30)\left(250 \ \frac{mg}{L}\right) = 175 \ mg/L$$

The filter efficiency is

$$\eta = \frac{175 \ \frac{mg}{L} - 50 \ \frac{mg}{L}}{175 \ \frac{mg}{L}} = 0.71$$

From Fig. 39.4 with 71% efficiency and $R = 0$, $L_{BOD} = 55 \ lbm/day\text{-}1000 \ ft^3$. (Notice that the abscissa gets larger to the left.)

The total load is found from Eq. 39.12 rearranged in terms of V_1.

$$\frac{(2.5 \ MGD)\left(175 \ \frac{mg}{L}\right)}{\times \left(8.345 \ \frac{lbm\text{-}L}{mg\text{-}MG}\right)\left(1000 \ \frac{ft^3}{1000 \ ft^3}\right)}{\left(55 \ \frac{lbm}{day\text{-}1000 \ ft^3}\right)(2 \ filters)}$$

$$= 33{,}190 \ ft^3/filter$$

With a depth of 6 ft, the total required surface area is

$$A = \frac{V}{depth} = \frac{33{,}190 \ ft^3}{6 \ ft} = 5532 \ ft^2$$

If two filters are used in parallel, the required diameter per filter is

$$D = \sqrt{\frac{4A}{\pi}} = \sqrt{\frac{(4)\left(\dfrac{5532 \ ft^2}{2}\right)}{\pi}} = \boxed{59.3 \ ft}$$

The answer is (A).

4. (a) There is nothing particularly special about a basin that removes 30% BOD. Choose two basins in parallel, each working with half of the total flow. Choose an overflow rate of 1000 gpd/ft². The area per basin is given by Eq. 39.4.

$$A = \frac{4.4 \times 10^6 \ \frac{\text{gal}}{\text{day}}}{(2)\left(1000 \ \frac{\text{gal}}{\text{day-ft}^2}\right)} = 2200 \ \text{ft}^2$$

$$d = \sqrt{\left(\frac{4}{\pi}\right)(2200 \ \text{ft}^2)} = \boxed{52.9 \ \text{ft}}$$

The answer is (B).

(b) The detention time is given by Eq. 39.6.

$$t = \frac{(2200 \ \text{ft}^2)(8 \ \text{ft})}{\left(2.2 \ \frac{\text{MGD}}{\text{tank}}\right)\left(1.547 \ \frac{\frac{\text{ft}^3}{\text{sec}}}{\text{MGD}}\right)\left(3600 \ \frac{\text{sec}}{\text{hr}}\right)}$$

$$= \boxed{1.436 \ \text{hr}}$$

The answer is (A).

(c) circumference $= \pi(52.9 \ \text{ft})$
$\qquad\qquad\qquad = 166.2 \ \text{ft}$

$$\text{weir loading} = \frac{2.2 \times 10^6 \ \frac{\text{gal}}{\text{day}}}{166.2 \ \text{ft}}$$

$$= \boxed{13{,}240 \ \text{gal/day-ft (gpd/ft)}}$$

The answer is (C).

5.

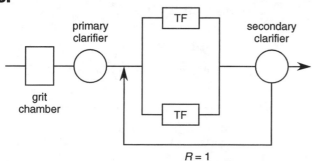

$R = 1$

(a) From Table 28.1, assume a peak flow multiplier of 2 since the population size is unknown. (Alternatively, a population "equivalent" can be used to estimate the population, and then Eq. 38.1 can be used.) The peak flow is

$$(2)(1 \ \text{MGD}) = \boxed{2 \ \text{MGD}}$$

The answer is (C).

(b) The peak flow rate per second is

$$Q_{\text{peak}} = \frac{(2)(1 \ \text{MGD})\left(10^6 \ \frac{\text{gal}}{\text{MG}}\right)\left(0.1337 \ \frac{\text{ft}^3}{\text{gal}}\right)}{\left(24 \ \frac{\text{hr}}{\text{day}}\right)\left(60 \ \frac{\text{min}}{\text{hr}}\right)\left(60 \ \frac{\text{sec}}{\text{min}}\right)}$$

$$= 3.095 \ \text{ft}^3/\text{sec}$$

With a detention time of 3 min, the volume of the grit chamber would be

$$V = Qt = \left(3.095 \ \frac{\text{ft}^3}{\text{sec}}\right)\left(60 \ \frac{\text{sec}}{\text{min}}\right)(3 \ \text{min}) = 557.1 \ \text{ft}^3$$

$$557.1 \ \text{ft}^3 = \text{length} \times \text{width} \times \text{depth}$$
$$= (20 \ \text{ft})(1.25)(\text{depth})^2$$
$$\text{water depth} = 4.72 \ \text{ft} \quad [\text{round to } 4\tfrac{3}{4} \ \text{ft}]$$
$$\text{chamber width} = (1.25)(4.72 \ \text{ft})$$

$$= \boxed{5.9 \ \text{ft} \quad [\text{round to 6 ft}]}$$

The answer is (B).

(c) This is a shallow chamber. Use 3 cfm/ft.

$$\left(3 \ \frac{\text{ft}^3}{\text{min-ft}}\right)(20 \ \text{ft}) = \boxed{60 \ \text{ft}^3/\text{min (cfm) of air}}$$

The answer is (A).

(d) (Note that the *Ten States' Standards* requires two basins.) Surface loading is the primary design parameter. Choose a surface loading of 1000 gal/day-ft². The volume is

$$A = \frac{Q}{v^*} = \frac{(2 \ \text{MGD})\left(10^6 \ \frac{\text{gal}}{\text{day-MGD}}\right)}{1000 \ \frac{\text{gal}}{\text{day-ft}^2}}$$

$$= 2000 \ \text{ft}^2$$

$$D = \sqrt{\frac{4A}{\pi}} = \sqrt{\frac{(4)(2000 \ \text{ft}^2)}{\pi}}$$

$$= \boxed{50.5 \ \text{ft}}$$

The answer is (A).

(e) The volume of the primary clarifier is fixed by the weir height. Therefore, the clarifier does not provide any storage (i.e., no damping of the flow rates). The fluctuations in flow will be passed on to the trickling filters.

$$Q_p + R = 2 \ \text{MGD} + 1 \ \text{MGD} = 3 \ \text{MGD}$$

The answer is (C).

Environmental

(f) Assume the primary sedimentation basin removes 30% of the BOD. BOD incoming to the trickle filter process is given by Eq. 39.12.

$$(1-0.30)\left(250\ \frac{mg}{L}\right) = 175\ mg/L$$

$$\left(175\ \frac{mg}{L}\right)\left(8.345\ \frac{lbm\text{-}L}{mg\text{-}MG}\right)$$
$$\times (1\ MGD) = 1460\ lbm/day$$

The required process efficiency is given by Eq. 39.9.

$$\eta = \frac{BOD_{in} - BOD_{out}}{BOD_{in}} = \frac{175\ \frac{mg}{L} - 30\ \frac{mg}{L}}{175\ \frac{mg}{L}}$$

$$= 0.829\ (82.9\%)$$

From Eq. 39.15 with $w = 0.1$ and $R = 1$,

$$F = \frac{1+R}{(1+wR)^2} = \frac{1+1}{\left(1+(0.1)(1)\right)^2} = 1.65$$

From Eq. 39.13,

$$\eta = \frac{1}{1+0.0561\sqrt{\frac{L_{BOD}}{F}}}$$

$$0.83 = \frac{1}{1+0.0561\sqrt{\frac{L_{BOD}}{1.65}}}$$

$$L_{BOD} \approx 22\ lbm/day\text{-}1000\ ft^3$$

$$\text{filter volume} = \frac{\left(1460\ \frac{lbm}{day}\right)\left(1000\ \frac{ft^3}{1000\ ft^3}\right)}{\left(22\ \frac{lbm}{day\text{-}1000\ ft^3}\right)(2\ filters)}$$

$$= 33{,}180\ ft^3$$

$$D = \sqrt{\frac{(4)(33{,}180\ ft^3)}{\pi(6\ ft)}} = \boxed{83.9\ ft}$$

[round to 85 ft]

Use two 85 ft diameter, 6 ft deep filters.

The answer is (D).

(g) For the final clarifier, the maximum overflow rate is 1100 gpd/ft^2, and the minimum depth is 10 ft. Assume volumetric flow fluctuations will be damped out by previous processes.

$$A = \frac{1\times10^6\ \frac{gal}{day}}{1100\ \frac{gal}{day\text{-}ft^2}} = 909\ ft^2$$

$$D = \sqrt{\frac{(4)(909\ ft^2)}{\pi}} = \boxed{34.0\ ft}\quad \text{[round to 35 ft]}$$

Use a 35 ft diameter, 10 ft deep basin.

The answer is (A).

6. (a) Assume the last clarifier removes 30% of the remaining BOD.

$$BOD_3 = \frac{30\ \frac{mg}{L}}{1-0.3} = 42.857\ mg/L$$

There is no recirculation that matches the NRC model. The NRC model "recirculation" is from the trickling filter discharge directly back to the entrance to the filter. This problem's recirculation is from several processes beyond the trickling filter. The recirculation increases the BOD loading and dilutes the influent.

The BOD loading to the trickling filter must include recirculation and is given by Eq. 39.12.

$$L_{BOD} = \frac{(1.5\ MGD)\left(250\ \frac{mg}{L}\right)}{\pi\left(\frac{75\ ft}{2}\right)^2(6\ ft)}$$
$$\times\left(8.345\ \frac{lbm\text{-}L}{MG\text{-}mg}\right)\left(1000\ \frac{ft^3}{1000\ ft^3}\right)$$

$$+\frac{(0.5)(1.5\ MGD)\left(30\ \frac{mg}{L}\right)}{\pi\left(\frac{75\ ft}{2}\right)^2(6\ ft)}$$
$$\times\left(8.345\ \frac{lbm\text{-}L}{MG\text{-}mg}\right)\left(1000\ \frac{ft^3}{1000\ ft^3}\right)$$

$$= 118.058\ \frac{lbm}{day\text{-}1000\ ft^3} + 7.083\ \frac{lbm}{day\text{-}1000\ ft^3}$$

$$= 125.141\ lbm/day\text{-}1000\ ft^3$$

Since $R = 0$, $F = 1$ [Eq. 39.15].

The filter/clarifier efficiency is given by Eq. 39.13.

$$\eta = \frac{1}{1+0.0561\sqrt{\frac{125.141\ \frac{lbm}{day\text{-}1000\ ft^3}}{1.00}}}$$

$$= 0.6144\ (61.44\%)$$

$$BOD_2 = (1-0.6144)\left(250\ \frac{mg}{L}\right) = 96.4\ mg/L$$

The removal fraction in the RBC must be

$$\frac{96.4\ \frac{mg}{L} - 42.857\ \frac{mg}{L}}{96.4\ \frac{mg}{L}} = 0.5554$$

Solving the given performance equation, the immersed area is

$$0.5554 = \frac{1}{\left(1 + \dfrac{2.45A}{1.5 \times 10^6}\right)^3}$$

$$A = 1.32 \times 10^5 \ \text{ft}^2$$

From Table 39.11, only 40% of the total RBC area is immersed at one time. The total RBC area is

$$A_{\text{total}} = \frac{1.32 \times 10^5 \ \text{ft}^2}{0.4} = \boxed{330{,}000 \ \text{ft}^2}$$

The answer is (D).

(b) This is one of the recirculation modes to which the NRC model applies. Solve the problem backward to get $\text{BOD}_{\text{out}} = 30$ mg/L.

$$\text{BOD}_3 = \frac{30 \ \dfrac{\text{mg}}{\text{L}}}{1 - 0.3} = 42.857 \ \text{mg/L}$$

$$\text{BOD}_2 = \frac{42.857 \ \dfrac{\text{mg}}{\text{L}}}{1 - 0.50} = 85.714 \ \text{mg/L}$$

The efficiency in the NRC model includes the effect of the clarifier, even though the recirculation occurs before the clarifier.

$$\eta_{\text{trickling filter and clarifier}} = \frac{250 \ \dfrac{\text{mg}}{\text{L}} - 85.714 \ \dfrac{\text{mg}}{\text{L}}}{250 \ \dfrac{\text{mg}}{\text{L}}}$$

$$= 0.6571 \ (65.71\%)$$

In this configuration, the BOD loading does not include the effects of recirculation, as the NRC model places a higher emphasis on organic loading than on hydraulic loading. From part (a), $L_{\text{BOD}} = 118.058$ lbm/day-1000 ft^3.

From Eq. 39.13,

$$F = \frac{L_{\text{BOD}}}{\left(\dfrac{1 - \eta}{0.0561\eta}\right)^2}$$

$$= \frac{118.058 \ \dfrac{\text{lbm}}{\text{day-1000 ft}^2}}{\left(\dfrac{1 - 0.6571}{(0.0561)(0.6571)}\right)^2}$$

$$= 1.3644$$

$$1.3644 = \frac{1 + R}{(1 + 0.1R)^2}$$

$$R = \boxed{0.51 \ (51\%)}$$

The answer is (B).

(c) Approximate the BOD removal of the secondary (first in-line) clarifier.

$$\text{BOD}_{\text{entering}} = \frac{\text{BOD}_2}{1 - \eta} = \frac{85 \ \dfrac{\text{mg}}{\text{L}}}{1 - 0.30} = 121.4 \ \text{mg/L}$$

[assuming 30% removal]

The sludge production is

$$\left(0.4 \ \frac{\text{lbm}}{\text{lbm}}\right)\left(\left(121.4 \ \frac{\text{mg}}{\text{L}} - 85 \ \frac{\text{mg}}{\text{L}}\right)\right.$$

$$\left. + \left(43 \ \frac{\text{mg}}{\text{L}} - 30 \ \frac{\text{mg}}{\text{L}}\right)\right)\left(8.345 \ \frac{\text{lbm-L}}{\text{MG-mg}}\right)(1.5 \ \text{MGD})$$

$$= 247.3 \ \text{lbm/day}$$

The sludge volume is

$$V = \frac{247.3 \ \dfrac{\text{lbm}}{\text{day}}}{62.4 \ \dfrac{\text{lbm}}{\text{ft}^3}} = \boxed{3.96 \ \text{ft}^3/\text{day}}$$

The answer is (A).

40 Activated Sludge and Sludge Processing

PRACTICE PROBLEMS

Sludge Quantities

1. 33 m³/day of thickened sludge with a suspended solids content of 3.8% and 13 m³/day of anaerobic digester sludge with a suspended solids content of 7.8% are produced in a wastewater treatment plant.

(a) What would be the effect of using a filter press to increase the solids content of the thickened sludge to 24%?

    (A) 4000 m³/yr
    (B) 8000 m³/yr
    (C) 10,000 m³/yr
    (D) 15,000 m³/yr

(b) What volume of digester sludge must be disposed of from sand drying beds that increase the solids concentration to 35%?

    (A) 500 m³/yr
    (B) 1000 m³/yr
    (C) 2000 m³/yr
    (D) 4000 m³/yr

2. An activated sludge plant processes 10 MGD of wastewater with 240 mg/L BOD and 225 mg/L suspended solids. 70% of the suspended solids are inorganic. The discharge from the final clarifier contains 15 mg/L BOD (all organic) and 20 mg/L suspended solids (all inorganic). Primary clarification removes 60% of the suspended solids and 35% of the BOD. The BOD reduction in the primary clarifier does not contribute to sludge production. The sludge produced has a specific gravity of 1.02 and a solids content of 6%. The cell yield (conversion of BOD reduction to biological solids) is 60%. The final clarifier does not reduce BOD.

(a) What is the daily mass of dry sludge solids produced?

    (A) 2500 lbm/day
    (B) 5000 lbm/day
    (C) 10,000 lbm/day
    (D) 20,000 lbm/day

(b) Assuming that the sludge is completely dried and compressed solid, what is the daily sludge volume?

    (A) 300 ft³/day
    (B) 500 ft³/day
    (C) 1000 ft³/day
    (D) 1600 ft³/day

(c) Assuming that the final moisture content in the sludge is 70% and air voids increase the volume by 10%, what is the daily sludge volume?

    (A) 200 ft³/day
    (B) 500 ft³/day
    (C) 1000 ft³/day
    (D) 1600 ft³/day

The actual application rate depends on sludge strength (primarily nitrogen content), the condition of the receiving soil, the crop or plants using the applied nutrients, and the spreading technology.

Environmental

SOLUTIONS

1. (a) Sludge volume is inversely proportional to the solids content. The sludge volume from the filter press is

$$V_2 = V_1 \left(\frac{\text{SS}_1}{\text{SS}_2} \right)$$

$$= \left(33 \; \frac{\text{m}^3}{\text{day}} \right) \left(\frac{0.038}{0.24} \right) = 5.23 \; \text{m}^3/\text{day}$$

The yearly decrease in volume is

$$\left(33 \; \frac{\text{m}^3}{\text{day}} - 5.23 \; \frac{\text{m}^3}{\text{day}} \right) \left(365 \; \frac{\text{day}}{\text{yr}} \right) = \boxed{10,136 \; \text{m}^3/\text{yr}}$$

The answer is (C).

(b) The sludge volume from the sand drying bed is

$$V_2 = V_1 \left(\frac{\text{SS}_1}{\text{SS}_2} \right)$$

$$= \left(13 \; \frac{\text{m}^3}{\text{day}} \right) \left(\frac{0.078}{0.35} \right) = 2.90 \; \text{m}^3/\text{day}$$

The yearly disposal volume is

$$\left(2.9 \; \frac{\text{m}^3}{\text{day}} \right) \left(365 \; \frac{\text{day}}{\text{yr}} \right) = \boxed{1059 \; \text{m}^3/\text{yr}}$$

The answer is (B).

2. The BOD and SS are not mutually exclusive. Some of the SS is organic in nature, and this shows up as BOD.

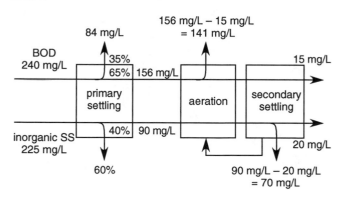

(a) The total dry weight of SS removed in all processes is

$$\left(225 \; \frac{\text{mg}}{\text{L}} - 20 \; \frac{\text{MG}}{\text{L}} \right) \left(8.345 \; \frac{\text{lbm-L}}{\text{MG-MG}} \right) \left(10 \; \frac{\text{MG}}{\text{day}} \right)$$

$$= 17,107 \; \text{lbm/day}$$

The BOD entering the secondary process is

$$(1 - 0.35) \left(240 \; \frac{\text{MG}}{\text{L}} \right) = 156 \; \text{MG/L}$$

Solids from BOD reduction in the secondary process are

$$Y(\Delta \text{BOD})Q = (0.60) \left(156 \; \frac{\text{mg}}{\text{L}} - 15 \; \frac{\text{MG}}{\text{L}} \right)$$

$$\times \left(10 \; \frac{\text{MG}}{\text{day}} \right) \left(8.345 \; \frac{\text{lbm-L}}{\text{MG-MG}} \right)$$

$$= 7060 \; \text{lbm/day}$$

The total dry sludge mass is

$$17,107 \; \frac{\text{lbm}}{\text{day}} + 7060 \; \frac{\text{lbm}}{\text{day}} = \boxed{24,167 \; \text{lbm/day}}$$

The answer is (D).

(b) The specific gravity of the sludge solids can be found from Eq. 40.47.

$$\frac{1}{\text{SG}} = \frac{1 - s}{1} + \frac{s}{\text{SG}_{\text{solids}}}$$

$$\frac{1}{1.02} = \frac{1 - 0.06}{1} + \frac{0.06}{\text{SG}_{\text{solids}}}$$

$$\text{SG}_{\text{solids}} = 1.485$$

The density of the solids is

$$\rho_{\text{solids}} = (1.485) \left(62.4 \; \frac{\text{lbm}}{\text{ft}^3} \right) = 92.66 \; \text{lbm/ft}^3$$

The solid dry sludge volume is

$$V_{\text{solid}} = \frac{m}{\rho} = \frac{24,167 \; \frac{\text{lbm}}{\text{day}}}{92.66 \; \frac{\text{lbm}}{\text{ft}^3}} = \boxed{260.8 \; \text{ft}^3/\text{day}}$$

The answer is (A).

(c) The disposal volume is

$$V_t = V_{\text{solid}} + V_{\text{water}}$$

$$V_{\text{solid}} = \frac{m_{\text{solid}}}{\rho}$$

$$V_{\text{water}} = \frac{m_{\text{water}}}{62.4 \; \frac{\text{lbm}}{\text{ft}^3}} = \frac{0.70 m_t}{62.4 \; \frac{\text{lbm}}{\text{ft}^3}}$$

$$= \frac{0.70 m_{\text{solids}}}{\left(62.4 \; \frac{\text{lbm}}{\text{ft}^3} \right) (1 - 0.70)}$$

$$V \approx (1.10) \left(\frac{24,167 \; \frac{\text{lbm}}{\text{day}}}{92.66 \; \frac{\text{lbm}}{\text{ft}^3}} + \frac{\left(24,167 \; \frac{\text{lbm}}{\text{day}} \right) (0.70)}{\left(62.4 \; \frac{\text{lbm}}{\text{ft}^3} \right) (1 - 0.70)} \right)$$

$$= \boxed{1280.9 \; \text{ft}^3/\text{day}}$$

The answer is (C).

41 Municipal Solid Waste

PRACTICE PROBLEMS

Landfills

1. A town has a current population of 10,000, which is expected to grow linearly, doubling every 15 yr. The town intends to dispose of its municipal solid waste in a 30-acre landfill that will be converted to a park in 20 yr. Solid waste is generated at the rate of 5 lbm/capita-day. The average compacted density in the landfill will be 1000 lbm/yd^3. Disregarding any soil addition for cover and cell construction, how long will it take to fill the landfill to a uniform height of 6 ft?

(A) 2100 days
(B) 4200 days
(C) 6300 days
(D) 9500 days

2. (*Time limit: one hour*) A town of 10,000 people has selected a square landfill site to deposit its solid waste. The minimum unused side borders are 50 ft. The landfill currently consists of a square 50-acre depression with an average depth of 20 ft below the surrounding grade. When the landfill is at final capacity, it will be covered with 10 ft of earth cover. The maximum height of the covered landfill is 20 ft above the surrounding grade. Solid waste is generated at the rate of 5 lbm/capita-day. The average compacted density in the landfill will be 1000 lbm/yd^3.

(a) What is the volumetric capacity of the landfill site?

(A) 1.1×10^6 ft^3
(B) 6.5×10^6 ft^3
(C) 3.1×10^7 ft^3
(D) 5.7×10^7 ft^3

(b) Using a loading factor of 1.25, what is the volume of landfill used each day?

(A) 60 yd^3/day
(B) 120 yd^3/day
(C) 180 yd^3/day
(D) 240 yd^3/day

(c) What is the service life of the landfill site?

(A) 30 yr
(B) 45 yr
(C) 60 yr
(D) 90 yr

SOLUTIONS

1. Find the rate of increase of waste production.

The mass of waste deposited in the first day will be

$$(10{,}000 \text{ people}) \left(5 \, \frac{\text{lbm}}{\text{person-day}} \right) = 50{,}000 \text{ lbm/day}$$

The mass of waste deposited on the last day will be

$$(20{,}000 \text{ people}) \left(5 \, \frac{\text{lbm}}{\text{person-day}} \right) = 100{,}000 \text{ lbm/day}$$

The increase in rate is

$$\frac{\Delta m}{\Delta t} = \frac{100{,}000 \, \frac{\text{lbm}}{\text{day}} - 50{,}000 \, \frac{\text{lbm}}{\text{day}}}{(15 \text{ years}) \left(365 \, \frac{\text{days}}{\text{year}} \right)} = 9.132 \text{ lbm/day}^2$$

The mass deposited on day D is

$$m_D = 50{,}000 + (9.132)(D - 1)$$
$$\approx 50{,}000 + 9.132 D$$

The cumulative mass deposited is

$$m_t = \int_0^t m_D \, dt = 50{,}000 t + \frac{9.132 t^2}{2}$$

With a compacted density of 1000 lbm/yd^3 and a loading factor of 1.00 (no soil cover), the capacity of the site with a 6 ft lift is

$$m_{\text{max}} = \frac{(30 \text{ ac}) \left(43{,}560 \, \frac{\text{ft}^2}{\text{ac}} \right) (6 \text{ ft}) \left(1000 \, \frac{\text{lbm}}{\text{yd}^3} \right)}{27 \, \frac{\text{ft}^3}{\text{yd}^3}}$$
$$= 2.9 \times 10^8 \text{ lbm}$$

The time to fill is found by solving the quadratic equation.

$$2.9 \times 10^8 = 50{,}000 t + \frac{9.132 t^2}{2}$$
$$t^2 + 10{,}951 t = 6.351 \times 10^7$$
$$t = \boxed{4193 \text{ days}}$$

The answer is (B).

Environmental

2. The side length of the square disposal site is

(a) $\text{side} = \sqrt{(50 \text{ ac})\left(43{,}560 \ \dfrac{\text{ft}^2}{\text{ac}}\right)} = 1476 \text{ ft}$

With 50 ft borders, the usable area is

$$A = \left(1476 \text{ ft} - (2)(50 \text{ ft})\right)^2 = 1.893 \times 10^6 \text{ ft}^2$$

If the site is excavated 20 ft, 10 ft of soil is used as cover, and the maximum above-ground height is 20 ft, the service capacity of compacted waste is

$$(1.893 \times 10^6 \text{ ft}^2)(30 \text{ ft}) = \boxed{5.68 \times 10^7 \text{ ft}^3}$$

The answer is (D).

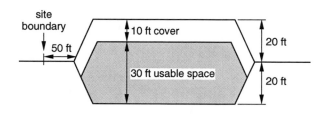

site boundary

10 ft cover

50 ft

20 ft

30 ft usable space

20 ft

(not to scale or representative of actual construction)

(b) The volume of landfill per day is

$$\dfrac{(10{,}000 \text{ people})\left(5 \ \dfrac{\text{lbm}}{\text{day-person}}\right)(1.25)}{1000 \ \dfrac{\text{lbm}}{\text{yd}^3}}$$

$$= \boxed{62.5 \text{ yd}^3/\text{day}}$$

The answer is (A).

(c) The service life is

$$\dfrac{5.68 \times 10^7 \text{ ft}^3}{\left(27 \ \dfrac{\text{ft}^3}{\text{yd}^3}\right)\left(62.5 \ \dfrac{\text{yd}^3}{\text{day}}\right)\left(365 \ \dfrac{\text{days}}{\text{yr}}\right)} = \boxed{92.2 \text{ yr}}$$

The answer is (D).

42 Environmental Engineering

PRACTICE PROBLEMS

There are no practice problems corresponding with Ch. 42 of the *Chemical Engineering Reference Manual*.

PRACTICE PROBLEMS

1. Using the equilibrium data for the ethanol-water system at 1 atm total pressure provided in the Txy diagram, determine the equilibrium composition of both the liquid and the vapor phases at 95.5°C.

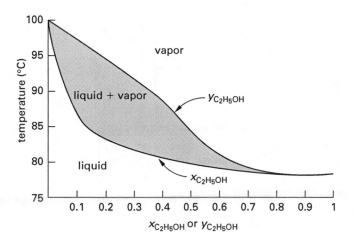

2. An ethanol-water mixture at 1 atm total pressure and 95.5°C contains 10% ethanol (mole basis). Use the Txy diagram provided to determine the total amounts of liquid and of vapor in 1000 mol of the mixture at equilibrium.

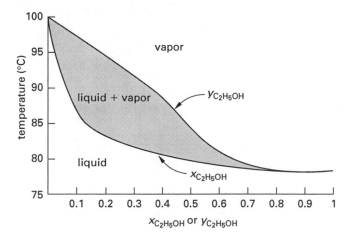

3. The Txy diagram for a methanol-carbon tetrachloride system at 1 atm total pressure is shown. This system forms an azeotrope at $x = y = 0.555$. Determine the equilibrium composition of the liquid and the vapor phases at 60°C.

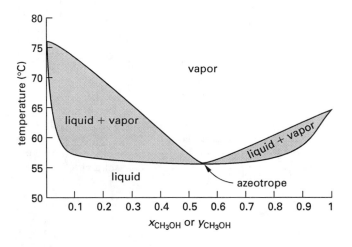

4. A Txy diagram is provided for a methanol-carbon tetrachloride mixture consisting of 845 mol of mixture with 10% methanol (mole basis) at 60°C and 1 atm total pressure. Determine the equilibrium composition and amounts of the liquid and the vapor phases.

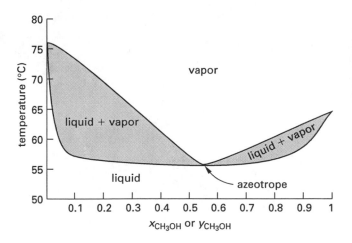

5. The equilibrium data in the xy diagram provided describe a methanol-carbon tetrachloride mixture. What is the relative volatility of both species at the azeotropic composition of $x_{CH_3OH} = 0.55$?

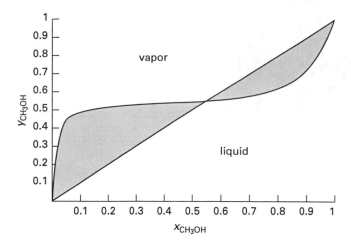

6. An equilibrium methanol-carbon tetrachloride mixture with a liquid mole fraction, $x_{CH_3OH} = 0.40$ is described by the plotted data shown. What is the relative volatility of methanol to carbon tetrachloride?

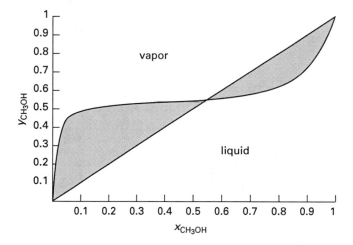

Problems 7 and 8 are based on the following information. 800 mol of a liquid mixture contain 40 mol% acetone $((CH_3)_2CO)$ in water at 62°C and 1 atm total pressure. The following Txy diagram presents the equilibrium data for the acetone-water system at 1 atm total pressure.

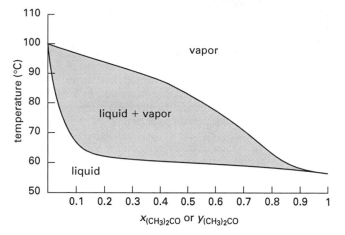

7. At equilibrium, what is the mole fraction of acetone in each phase?

(A) $x_{(CH_3)_2CO} = 0.21$, $y_{(CH_3)_2CO} = 0.40$

(B) $x_{(CH_3)_2CO} = 0.21$, $y_{(CH_3)_2CO} = 0.82$

(C) $x_{(CH_3)_2CO} = 0.40$, $y_{(CH_3)_2CO} = 0.40$

(D) $x_{(CH_3)_2CO} = 0.40$, $y_{(CH_3)_2CO} = 0.84$

8. At equilibrium, how many moles of liquid and of vapor are there?

(A) $L = 530$ mol, $V = 270$ mol

(B) $L = 550$ mol, $V = 250$ mol

(C) $L = 570$ mol, $V = 230$ mol

(D) $L = 590$ mol, $V = 210$ mol

SOLUTIONS

1.

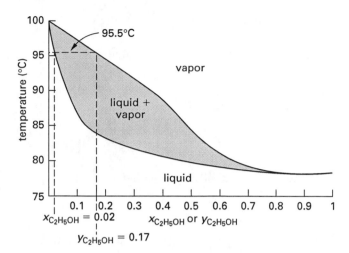

For an equilibrium liquid and vapor mixture at 95.5°C, the isotherm's intersection with the saturated liquid line indicates that the equilibrium composition of the liquid phase, $x_{C_2H_5OH}$, is 0.02. The isotherm's intersection with the saturated vapor line gives the equilibrium composition of the vapor phase, $y_{C_2H_5OH}$, as 0.17. The mole fraction of water in each phase is calculated using Eqs. 43.4 and 43.5.

$$x_{C_2H_5OH} + x_{H_2O} = 1$$
$$x_{H_2O} = 1 - x_{C_2H_5OH} = 1 - 0.02 = 0.98$$

$$y_{C_2H_5OH} + y_{H_2O} = 1$$
$$y_{H_2O} = 1 - y_{C_2H_5OH} = 1 - 0.17 = 0.83$$

(The same result could be obtained directly from the data contained in Table 43.1.)

2. From Prob. 1, the equilibrium composition of each phase at 95.5°C is

$$x_{C_2H_5OH} = 0.02$$
$$x_{H_2O} = 0.98$$
$$y_{C_2H_5OH} = 0.17$$
$$y_{H_2O} = 0.83$$

From the problem statement,

$$F = 1000 \text{ mol}$$
$$z_{C_2H_5OH} = 0.10$$

Use the lever rule to determine the ratio of L/V.

$$\frac{L}{V} = \frac{\overline{yz}}{\overline{zx}} = \frac{y_{C_2H_5OH} - z_{C_2H_5OH}}{z_{C_2H_5OH} - x_{C_2H_5OH}} = \frac{0.17 - 0.1}{0.1 - 0.02} = 0.875$$

$$L = 0.875V$$

Substituting the expression for L and the given value of $F = 1000$ mol into the overall mass-balance equation, Eq. 43.16,

$$F = L + V$$
$$1000 \text{ mol} = 0.875V + V$$
$$V = 533 \text{ mol}$$
$$L = 467 \text{ mol}$$

3.

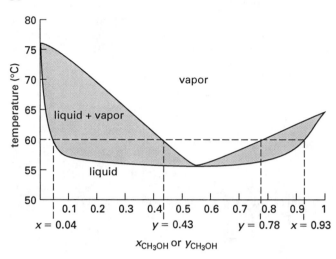

Draw the isotherm at 60°C and read the equilibrium composition of the liquid and the vapor phases. There are two possible equilibrium compositions. These are

$$x_{CH_3OH} = 0.04$$
$$y_{CH_3OH} = 0.43$$

and

$$x_{CH_3OH} = 0.93$$
$$y_{CH_3OH} = 0.78$$

Additional information, such as initial mixture composition, is required to determine which of these two possibilities applies.

4.

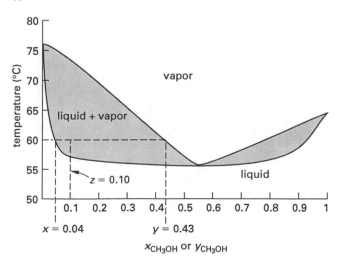

Mass Transfer

From Prob. 3, the equilibrium composition of the liquid and the vapor at 60°C is either

$$x_{CH_3OH} = 0.04$$

$$y_{CH_3OH} = 0.43$$

or

$$x_{CH_3OH} = 0.93$$

$$y_{CH_3OH} = 0.78$$

The feed composition is given as $z_{CH_3OH} = 0.10$. Plotting the feed composition on the equilibrium diagram reveals that the equilibrium composition of the liquid and the vapor phases lies on the left-hand side of the two-phase envelope. So,

$$x_{CH_3OH} = 0.04$$

$$y_{CH_3OH} = 0.43$$

From the problem statement,

$$F = 845 \text{ mol}$$

The lever rule is used to determine the ratio L/V.

$$\frac{L}{V} = \frac{\overline{yz}}{\overline{zx}} = \frac{y_{CH_3OH} - z_{CH_3OH}}{z_{CH_3OH} - x_{CH_3OH}} = \frac{0.43 - 0.1}{0.1 - 0.04} = 5.5$$

$$L = 5.5V$$

Substituting the expression for L and the given value of $F = 845$ mol into Eq. 43.16, the overall mass balance is

$$F = L + V$$

$$845 \text{ mol} = 5.5V + V$$

$$V = 130 \text{ mol}$$

$$L = 715 \text{ mol}$$

5. From the xy diagram, at the azeotrope, $x_{CH_3OH} = 0.55, y_{CH_3OH} = 0.55$. From Eq. 43.7, the equilibrium ratio for methanol is

$$K_{CH_3OH} = \frac{y_{CH_3OH}}{x_{CH_3OH}} = \frac{0.55}{0.55} = 1.0$$

The mole fractions for carbon tetrachloride are calculated by using Eqs. 43.4 and 43.5.

$$x_{CCl_4} + x_{CH_3OH} = 1$$

$$x_{CCl_4} = 1 - x_{CH_3OH} = 1 - 0.55 = 0.45$$

$$y_{CCl_4} + y_{CH_3OH} = 1$$

$$y_{CCl_4} = 1 - y_{CH_3OH} = 1 - 0.55 = 0.45$$

The equilibrium ratio for carbon tetrachloride is

$$K_{CCl_4} = \frac{y_{CCl_4}}{x_{CCl_4}} = \frac{0.45}{0.45} = 1.0$$

The relative volatility of methanol to carbon tetrachloride is

$$\alpha_{CH_3OH-CCl_4} = \frac{K_{CH_3OH}}{K_{CCl_4}} = \frac{1.0}{1.0} = 1$$

The relative volatility of carbon tetrachloride to methanol is also equal to 1.

6. From the xy diagram, at $x_{CH_3OH} = 0.40, y_{CH_3OH} = 0.54$. The equilibrium ratio is

$$K_{CH_3OH} = \frac{y_{CH_3OH}}{x_{CH_3OH}} = \frac{0.54}{0.40} = 1.35$$

Using Eqs. 43.4 and 43.5, the mole fractions of carbon tetrachloride are

$$x_{CCl_4} + x_{CH_3OH} = 1$$

$$x_{CCl_4} = 1 - x_{CH_3OH} = 1 - 0.40 = 0.60$$

$$y_{CCl_4} + y_{CH_3OH} = 1$$

$$y_{CCl_4} = 1 - y_{CH_3OH} = 1 - 0.54 = 0.46$$

The equilibrium ratio for carbon tetrachloride is

$$K_{CCl_4} = \frac{y_{CCl_4}}{x_{CCl_4}} = \frac{0.46}{0.60} = 0.77$$

The relative volatility of methanol to carbon tetrachloride is

$$\alpha_{CH_3OH-CCl_4} = \frac{K_{CH_3OH}}{K_{CCl_4}} = \frac{1.35}{0.77} = 1.75$$

7. The liquid and vapor equilibrium compositions are found from the points where the 62°C isotherm intersects the liquid and the vapor equilibrium lines. The equilibrium mole fractions of acetone are $x_{(CH_3)_2CO} = 0.21$ and $y_{(CH_3)_2CO} = 0.82$.

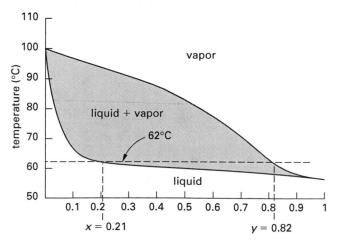

The answer is (B).

8.

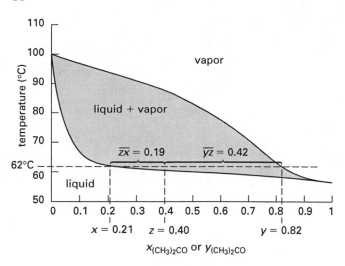

Substitute the expression for L and the given value of $F = 800$ mol into Eq. 43.16.

$$F = L + V$$
$$800 \text{ mol} = 2.21V + V$$
$$V = 249 \text{ mol} \quad (250 \text{ mol})$$
$$L = 551 \text{ mol} \quad (550 \text{ mol})$$

The answer is (B).

For a 40 mol% acetone solution, $z_{(CH_3)_2CO} = 0.40$. From Prob. 7, the equilibrium compositions of acetone are $x_{(CH_3)_2CO} = 0.21$ and $y_{(CH_3)_2CO} = 0.82$. From the lever rule, Eq. 43.17, the mole ratio of liquid, L, to vapor, V, is

$$\frac{L}{V} = \frac{y_{(CH_3)_2CO} - z_{(CH_3)_2CO}}{z_{(CH_3)_2CO} - x_{(CH_3)_2CO}} = \frac{0.82 - 0.40}{0.40 - 0.21} = 2.21$$

$$L = 2.21V$$

Mass Transfer

44 Vapor-Liquid Processes

PRACTICE PROBLEMS

Distillation

1. A distillation column with a total condenser and a partial reboiler is used to separate an acetone-acetic acid mixture. The feed, F, is saturated liquid containing 50 mol% acetone ($z_F = 0.50$). The desired distillate composition is 95 mol% acetone ($x_D = 0.95$), and the desired bottoms composition is 2 mol% acetone ($x_B = 0.02$). The external reflux ratio, L_0/D, is 2, and the reflux is a saturated liquid. The column pressure is 1 atm. The equilibrium data are shown.

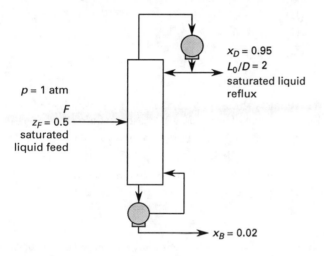

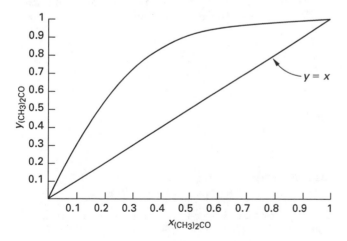

(a) Assuming constant molar overflow, what is the optimum number of stages required to achieve the separation? (b) What is the optimum feed stage location? (c) What is the minimum internal reflux ratio?

(d) What is the minimum number of stages required to achieve the separation?

2. A distillation column with a total condenser and a partial reboiler is used to separate an ethanol-water mixture at 40°C. The feed, F, is saturated liquid containing 30 mol% ethanol ($z_F = 0.3$). The design distillate composition is 85 mol% ethanol ($x_D = 0.85$), and the design bottoms composition 3 mol% ethanol ($x_B = 0.03$). The external reflux ratio, L_0/D, is 3, and the reflux is a saturated liquid. The column pressure is 1 atm. Refer to the illustration for additional data, including enthalpies (h) of the streams.

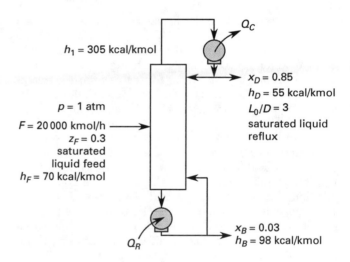

(a) What is the distillate molar flow rate? (b) What is the bottoms molar flow rate? (c) What is the heat duty of the condenser? (d) What is the heat duty of the reboiler?

3. A byproduct stream contains 20 mol% methanol in water. An equilibrium-stage distillation column is needed to remove the methanol from this stream and to recycle the water stream to another process. The column operates at 1 atm, using 50% more than the minimum internal reflux ratio, a total condenser, and a partial reboiler. The other process can tolerate a maximum of 2 mol% methanol in the water stream. The methanol, containing as much as 10 mol% water, can also be sold to a commercial recycler. The byproduct stream flows at 80 000 mol/h and is approximately 30% vapor at 1 atm. The equilibrium data are shown.

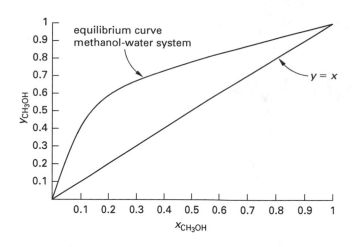

(a) How many equilibrium stages are required? (b) What is the optimum feed location?

Information for Problems 4–8

A new material, $ARBC^3$, leaves the synthesis process mixed with 55 mol% water. The temperature of the $ARBC^3$-water mixture is 80°C, and the molar flow rate is 5000 kmol/h. Distillation is being evaluated as a means of removing the water from this process stream. Empirical equilibrium data and several enthalpies of the feed composition are available. The enthalpy of the feed, H_F, at 80°C and 1 atm is 45 kcal/kmol. The enthalpy of the saturated vapor, H_V, at 1 atm pressure is 900 kcal/kmol. At 1 atm, the saturated liquid enthalpy, H_L, is 175 kcal/kg. The target distillate composition, x_D, is 90 mol% $ARBC^3$, and the desired bottoms composition, x_B, is 3 mol% $ARBC^3$. As a starting point, a column with a total condenser and a partial reboiler is being evaluated. The external reflux ratio, L_0/D, is $^4/_3$, and the liquid reflux is saturated. Constant molar overflow is a valid assumption, as is operation at atmospheric pressure.

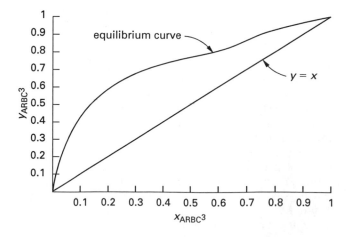

4. What is the quality of the feed at 80°C and 1 atm pressure?

    (A) 0.49
    (B) 0.85
    (C) 1.2
    (D) 1.4

5. Approximately how many equilibrium stages does the separation require?

    (A) 3 equilibrium stages
    (B) 4 equilibrium stages plus a partial reboiler
    (C) 6 equilibrium stages plus a partial reboiler
    (D) 9 equilibrium stages

6. Counting from the top of the column, where is the optimum feed location?

    (A) stage 3
    (B) stage 4
    (C) stage 6
    (D) stage 8

7. What is the approximate minimum number of equilibrium stages that this separation requires?

    (A) 2 equilibrium stages plus a partial reboiler
    (B) 3 equilibrium stages plus a partial reboiler
    (C) 6 equilibrium stages plus a partial reboiler
    (D) 8 equilibrium stages plus a partial reboiler

8. If the feed enthalpy is equal to the enthalpy of the saturated liquid, what minimum internal reflux ratio will achieve the desired separation?

    (A) 0.35
    (B) 0.68
    (C) 0.92
    (D) 1.5

Absorption

9. An absorption tower uses water to remove ammonia from air. The tower operates at 1 atm and 27°C. Its inlet water is pure ($x_0 = 0$), and its inlet gas has 0.045 mole fraction of ammonia ($y_{n+1} = 0.045$). The tower removes 95.75% of the ammonia from the air. The feed rate is 15 000 mol/h and the liquid-to-gas flow-rate ratio, L/G, is 2.25.

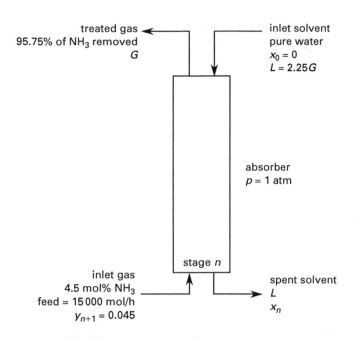

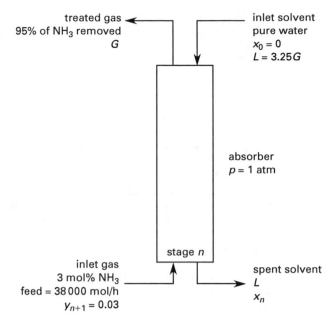

The equilibrium data for the ammonia-water system at 1 atm, in terms of mole ratios of ammonia, X and Y, are given by the following diagram.

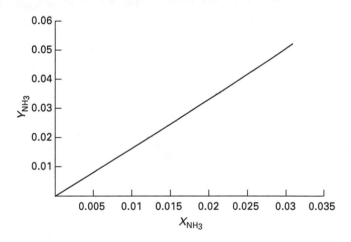

The equilibrium data for the ammonia-water system at 1 atm, in terms of the mole ratios of ammonia, X and Y, are given in the following diagram.

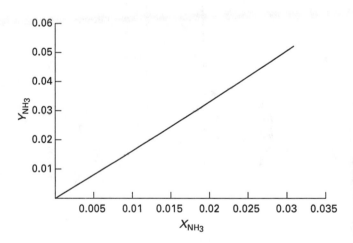

(a) How many equilibrium stages does this separation require? (b) What is the mole fraction of ammonia in the exiting water stream? (c) What is the minimum ratio of liquid-to-gas flow rates?

Description for Problems 10–12

An absorption tower uses water to remove ammonia from air. The tower operates at 1 atm and 27°C. The inlet water is pure ($x_0 = 0$). The inlet gas is 0.03 mole fraction ammonia ($y_{n+1} = 0.03$). The tower is required to remove 95% of the ammonia from the air. The inlet gas feed rate is 38 000 mol/h, and the liquid-to-gas flow rate ratio, L/G, is 3.25.

10. What is the mole ratio of ammonia in the water stream leaving the unit?

    (A) 0.0009

    (B) 0.009

    (C) 0.09

    (D) 0.9

11. Approximately how many equilibrium stages does this separation require?

    (A) $2\frac{1}{4}$

    (B) $3\frac{1}{3}$

    (C) 4

    (D) $4\frac{1}{2}$

12. What is the minimum ratio of liquid-to-gas flow rates?

(A) 0.31

(B) 0.68

(C) 1.6

(D) 2.6

Stripping

13. A gas stripper uses air to remove component A from a water stream. The tower operates at 1 atm and 30°C. The water feed rate, L_F, is 1000 mol/h. The inlet water has 9.1 mol% of component A. The stripper must reduce the mole fraction of component A in the water to 0.008. The mole fraction of component A in the air may be increased to 0.074. The constant molar overflow assumption is valid.

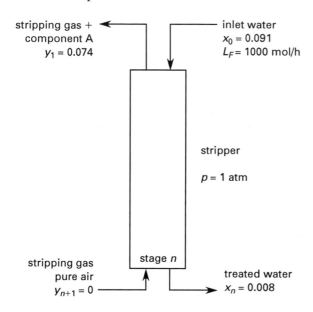

The equilibrium data in terms of mole ratios of component A, X and Y, are given in the following diagram.

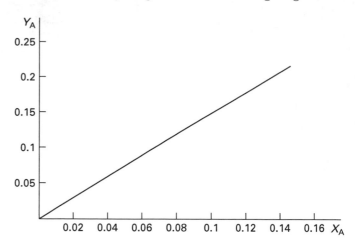

(a) How many equilibrium stages does this separation require? (b) What is the ratio of the solvent to the carrier gas molar flow rates? (c) What is the solvent molar flow rate? (d) What is the carrier gas molar flow rate?

SOLUTIONS

Distillation

1. (a) Feed Line

Since the feed is a saturated liquid, the feed quality, q, is 1. Therefore, the feed line is vertical at the feed composition, $z_F = x_F = 0.5$.

Top (Rectifying Section) Operating Line

From Eq. 44.35, the operating equation for the rectifying section is

$$y = \left(\frac{L}{V}\right) x + \left(1 - \frac{L}{V}\right) x_D$$

$$\frac{L}{V} \quad \text{[slope]}$$

$$\left(1 - \frac{L}{V}\right) x_D \quad \text{[y-intercept]}$$

Using Eq. 44.36 to calculate the internal reflux ratio (L/V) from the external reflux ratio (L_0/D),

$$\frac{L}{V} = \frac{\dfrac{L_0}{D}}{1 + \dfrac{L_0}{D}} = \frac{2}{1+2} = {}^2/_3 \quad \text{[slope]}$$

The y-intercept is

$$\left(1 - \frac{L}{V}\right) x_D = \left(1 - \frac{2}{3}\right) (0.95)$$

$$= 0.317$$

The top operating line can be plotted on the equilibrium diagram using the slope and y-intercept.

Bottom (Stripping Section) Operating Line

The operating equation for the stripping section is found from Eq. 44.37.

$$y = \left(\frac{\overline{L}}{\overline{V}}\right) x + \left(1 - \frac{\overline{L}}{\overline{V}}\right) x_B$$

$$\frac{\overline{L}}{\overline{V}} \quad \text{[slope]}$$

$$\left(1 - \frac{\overline{L}}{\overline{V}}\right) x_B \quad \text{[y-intercept]}$$

One point on the line is the intersection of the bottom operating line with the top operating line at the feed line. One additional point is needed to define the bottom operating line. Rather than determine the slope of the bottom operating line, $\overline{L}/\overline{V}$, the second point can be found by determining where the bottom operating line intersects the $y = x$ line. Substitute $y = x$ into the operating line equation.

$$y = \left(\frac{\overline{L}}{\overline{V}}\right) x + \left(1 - \frac{\overline{L}}{\overline{V}}\right) x_B$$

$$x = \left(\frac{\overline{L}}{\overline{V}}\right) x + \left(1 - \frac{\overline{L}}{\overline{V}}\right) x_B$$

$$x - \left(\frac{\overline{L}}{\overline{V}}\right) x = \left(1 - \frac{\overline{L}}{\overline{V}}\right) x_B$$

$$\left(1 - \frac{\overline{L}}{\overline{V}}\right) x = \left(1 - \frac{\overline{L}}{\overline{V}}\right) x_B$$

$$x = x_B = 0.02$$

Refer to the accompanying illustration. The separation requires 4 equilibrium stages and 1 partial reboiler.

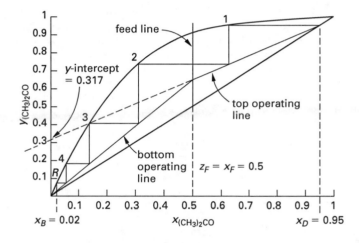

(b) Referring to the illustration in part (a), the optimum feed location is at stage 2, where the horizontal step crosses the feed line.

(c) The minimum internal reflux ratio is determined by pivoting the top operating line so that it intersects the equilibrium data curve at the feed line. (Drawing the top operating line in this way creates a pinch point at the feed composition.) The slope of this line is the minimum internal reflux ratio.

The slope is

$$\left(\frac{L}{V}\right)_{\min} = \frac{\Delta y}{\Delta x} = \frac{0.95 - 0.86}{0.95 - 0} = 0.095$$

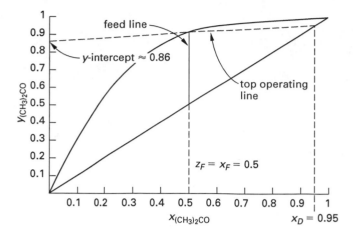

(d) Drawing the operating lines with total reflux and stepping off stages determines the minimum number of stages to achieve the separation. With total reflux, $L = V$ and $L/V = 1$, so $\overline{L} = \overline{V}$ and $\overline{L}/\overline{V} = 1$. The slopes of the top and the bottom operating lines are 1, which corresponds to the $y = x$ line on the McCabe-Thiele diagram. From the following illustration, the minimum number of stages is 3 plus a partial reboiler.

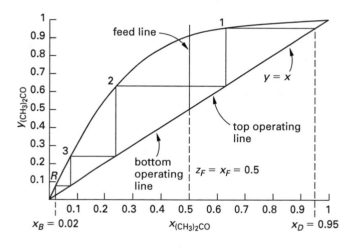

2. (a) The distillate (D) and the bottoms (B) flow rates can be determined from the overall and the ethanol mass balances.

$$F = B + D$$
$$B = F - D$$
$$Fz_F = Bx_B + Dx_D$$
$$Fz_F = (F - D)x_B + Dx_D$$

Rearrange the ethanol mass-balance equation.

$$D = F\left(\frac{z_F - x_B}{x_D - x_B}\right) = \left(20\,000\ \frac{\text{kmol}}{\text{h}}\right)\left(\frac{0.3 - 0.03}{0.85 - 0.03}\right)$$
$$= 6585\ \text{kmol/h}$$

(b) From the overall mass balance, the bottoms flow rate is

$$B = F - D = 20\,000\ \frac{\text{kmol}}{\text{h}} - 6585\ \frac{\text{kmol}}{\text{h}}$$
$$= 13\,415\ \text{kmol/h}$$

(c) The energy balance is used to determine the condenser heat duty. For the condenser, with Eq. 44.17,

$$Q_C = \left(1 + \frac{L_0}{D}\right)\left(\frac{z_F - x_B}{x_D - x_B}\right)F(h_1 - h_D)$$
$$= (1 + 3)\left(\frac{0.3 - 0.03}{0.85 - 0.03}\right)\left(20\,000\ \frac{\text{kmol}}{\text{h}}\right)$$
$$\times \left(305\ \frac{\text{kcal}}{\text{kmol}} - 55\ \frac{\text{kcal}}{\text{kmol}}\right)$$
$$= 6.59 \times 10^6\ \text{kcal/h}$$

The condenser heat duty is positive, indicating that heat is removed from the overhead stream.

(d) For the reboiler, from Eq. 44.13, the heat duty is found from the energy balance.

$$Q_R = Dh_D + Bh_B - Fh_F + Q_C$$
$$= \left(6585\ \frac{\text{kmol}}{\text{h}}\right)\left(55\ \frac{\text{kcal}}{\text{kmol}}\right)$$
$$+ \left(13\,415\ \frac{\text{kmol}}{\text{h}}\right)\left(98\ \frac{\text{kcal}}{\text{kmol}}\right)$$
$$- \left(20\,000\ \frac{\text{kmol}}{\text{h}}\right)\left(70\ \frac{\text{kcal}}{\text{kmol}}\right) + 6.59 \times 10^6\ \frac{\text{kcal}}{\text{h}}$$
$$= 6.87 \times 10^6\ \text{kcal/h}$$

The reboiler duty is positive, indicating that heat is added by the reboiler.

3. (a) The bottoms constitute the "water stream." With a maximum of 2 mol% methanol, $x_B = 0.02$. The overhead stream (the distillate) will contain the more volatile, concentrated methanol. With a maximum of 10 mol% water allowed in this stream, the target mole fraction of methanol is $x_D = 0.90$.

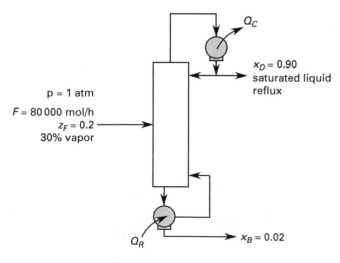

Feed Line

One point on the feed line is the intersection of the liquid feed composition with the $y = x$ line, or $y = x = z_F = 0.2$. Since the feed is 30% vapor, the feed is 70% liquid. The feed quality, q, is 0.7. From Eq. 44.44, the feed equation is

$$y = \left(\frac{q}{q-1}\right) x + \left(\frac{1}{1-q}\right) z_F$$

$$\frac{q}{q-1} \quad \text{[slope]}$$

$$\left(\frac{1}{1-q}\right) z_F = \left(\frac{1}{1-0.7}\right)(0.2) = 0.67 \quad \text{[y-intercept]}$$

Using the points $(0.2, 0.2)$ and $(0, 0.67)$, the feed line can be plotted on the equilibrium diagram.

Top (Rectifying Section) Operating Line

The intersection of the top operating line with the feed line at the equilibrium data creates a pinch point, which corresponds to the minimum internal reflux. The top operating line for the rectifying section at the minimum reflux ratio can be plotted on the equilibrium diagram by connecting the distillate composition ($y = x = x_D = 0.9$) to the intersection of the feed line at the equilibrium data curve (at $(0.1, 0.42)$). The slope of this line is 0.6.

$$\left(\frac{L}{V}\right)_{min} = \frac{\Delta y}{\Delta x} = \frac{0.9 - 0.42}{0.9 - 0.1} = 0.6$$

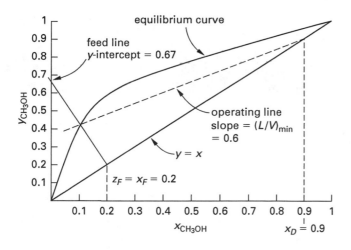

The problem statement specifies an internal reflux ratio 50% greater than the minimum internal reflux ratio.

$$\frac{L}{V} = 1.5 \left(\frac{L}{V}\right)_{min} = (1.5)(0.6) = 0.9$$

With one point ($y = x = x_D = 0.9$) and the operating line slope ($L/V = 0.9$), the top operating line can be

plotted on the equilibrium diagram. To check the operating line, calculate the y-intercept. From Eq. 44.35, the top operating equation is

$$y = \left(\frac{L}{V}\right) x + \left(1 - \frac{L}{V}\right) x_D$$

$$\frac{L}{V} = 0.9 \quad \text{[slope]}$$

$$\left(1 - \frac{L}{V}\right) x_D = (1 - 0.9)(0.9) = 0.09 \quad \text{[y-intercept]}$$

Bottom (Stripping Section) Operating Line

The bottom operating line intersects the top operating line at the feed line. The bottom operating line also intersects the $y = x$ line at $x_B = 0.02$. Using these two points, the bottom operating line can be plotted on the equilibrium diagram.

Starting at the distillate and stepping off stages shows that this separation requires 4 equilibrium stages and a partial reboiler to achieve a bottoms composition of just under 2 mol%.

(b) From the illustration, the optimum feed location is at stage 3, since the stages were stepped off to minimize the total number of stages.

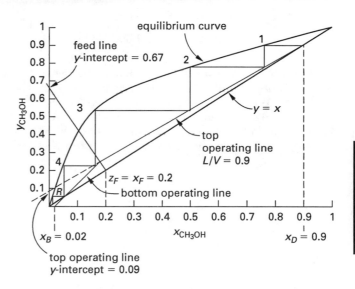

4. Calculate the feed quality, q, from Eq. 44.38.

$$q = \frac{H_V - H_F}{H_V - H_L} = \frac{900 \ \dfrac{\text{kcal}}{\text{kmol}} - 45 \ \dfrac{\text{kcal}}{\text{kmol}}}{900 \ \dfrac{\text{kcal}}{\text{kmol}} - 175 \ \dfrac{\text{kcal}}{\text{kmol}}} = 1.18 \quad (1.20)$$

The feed is subcooled, which is evident from $H_F < H_L$.

The answer is (C).

5.

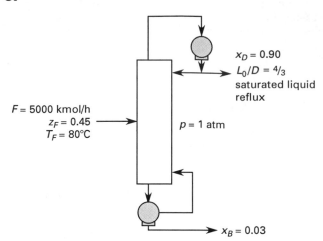

F = 5000 kmol/h
$z_F = 0.45$
$T_F = 80°C$

$x_D = 0.90$
$L_0/D = {}^4/_3$
saturated liquid reflux

$p = 1$ atm

$x_B = 0.03$

Feed Line

The slope of feed line can be calculated using Eq. 44.44 and the feed quality, q, calculated in Prob. 4.

$$\frac{q}{q-1} = \frac{1.18}{1.18-1} = 6.56$$

Since the process stream is 55 mol% water, the mole fraction of water in the feed, z_{H_2O}, is 0.55. The mole fraction of $ARBC^3$ in the feed, z_F, is

$$1 = z_{H_2O} + z_F$$
$$z_F = 1 - z_{H_2O} = 1 - 0.55 = 0.45$$

The feed line intersects the $y = x$ line at the feed composition, z_F, providing $(0.45, 0.45)$ as one point on the feed line. With this point and the slope, the feed line can be plotted on the equilibrium diagram. Alternatively, a second feed line point can be obtained by calculating the y-intercept of the feed line from the feed equation, Eq. 44.44.

$$\left(\frac{1}{1-q}\right) z_F = \left(\frac{1}{1-1.18}\right)(0.45) = -2.5$$

These two points, $x = y = z_F$ $(0.45, 0.45)$ and the y-intercept $(0, -2.5)$ can also be used to plot the feed line.

Top (Rectifying Section) Operating Line

The slope of the top operating line is calculated from Eq. 44.36, using the given external reflux ratio.

$$\frac{L}{V} = \frac{\dfrac{L_0}{D}}{1 + \dfrac{L_0}{D}} = \frac{\dfrac{4}{3}}{1 + \dfrac{4}{3}} = 0.57$$

The y-intercept can be obtained from Eq. 44.35.

$$\left(1 - \frac{L}{V}\right) x_D = (1 - 0.57)(0.9) = 0.39$$

Alternatively, the top operating line can be plotted using two points, the intersection of the operating line with the $y = x$ line at x_D $(0.9, 0.9)$ and the y-intercept $(0, 0.39)$.

Bottom (Stripping Section) Operating Line

Knowing that the bottom operating line intersects the top operating line at the feed line provides one point on the line. A second point needed to plot the line can be identified as the intersection of the bottom operating line with the $y = x$ line, at x_B $(0.03, 0.03)$. The feed line and the operating lines are plotted on the equilibrium diagram as shown.

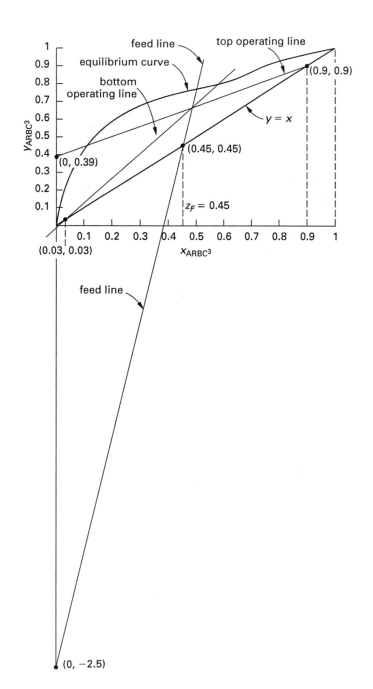

Stepping off stages shows that the total number of equilibrium stages required is approximately 4 plus a reboiler.

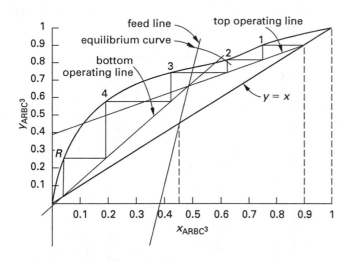

The answer is (B).

6. Use the graphical solution from Prob. 5. The optimum feed location is where the operating lines intersect the feed line, at stage 3.

The answer is (A).

7. A minimum number of equilibrium stages for separation represents a condition of total reflux. This occurs where the slopes of both the top and the bottom operating lines are equal to 1, making them coincidental with the $y = x$ line. Using the $y = x$ operating line, step off stages starting at the target distillate mole fraction, $x_D = 0.9$. The minimum number of equilibrium stages that this separation requires is 3 plus a partial reboiler. The minimum number of stages (while operating at total reflux) theoretically gives a better separation than is required.

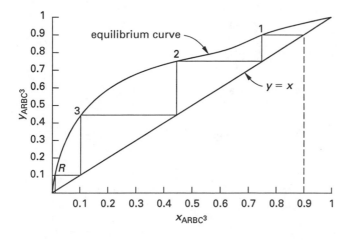

The answer is (B).

8. If the enthalpy of the feed is equal to the enthalpy of saturated liquid ($h_F = h_L$), then the slope of the feed line is calculated using Eq. 44.38.

$$q = \frac{h_V - h_F}{h_V - h_L} = \frac{h_V - h_L}{h_V - h_L} = 1$$

When the feed quality, q, is 1, the feed is saturated liquid and the feed line is vertical at the feed mole fraction, $z_F = 0.45$. The minimum internal reflux ratio is determined graphically by drawing the top operating line from $y = x = x_B$ (0.9, 0.9) to the feed line, so that it touches the equilibrium curve. This creates a pinch point at $x_{ARBC^3} \approx 0.52$. This line intersects the points (0.9, 0.9) and has its y-intercept at (0, 0.285). The slope of the top operating line is the minimum internal reflux ratio.

$$\frac{L}{V} = \frac{\Delta y}{\Delta x} = \frac{0.9 - 0.285}{0.9 - 0} = 0.68$$

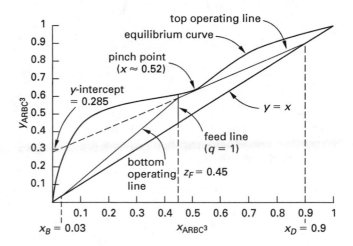

The answer is (B).

Absorption

9. (a) The inlet gas and exit (treated) gas compositions in terms of mole ratios, Y_i, and the air flow rate, G, are needed to plot the operating line.

The inlet gas stream has 4.5 mol% ammonia. From Eq. 44.53, the mole ratio of the inlet gas, Y_{n+1}, is

$$Y_{n+1} = \frac{y_{NH_3}}{1 - y_{NH_3}} = \frac{0.045}{1 - 0.045} = 0.047$$

The amount of ammonia in the feed is

$$G_{NH_3,in} = y_{n+1}G_F = (0.045)\left(15\,000\,\frac{mol}{h}\right)$$

$$= 675\,mol/h$$

The absorber removes 95.75% of the ammonia from the inlet air stream, leaving 4.25% of the inlet ammonia remaining in the gas.

$$G_{NH_3,out} = (4.25\%)G_{NH_3,in} = (0.0425)\left(675\,\frac{mol}{h}\right)$$

$$= 28.7\,mol/h$$

Mass Transfer

The airflow rate is

$$G_{\text{air}} = G_F - G_{\text{NH}_3,\text{in}} = 15\,000\ \frac{\text{mol}}{\text{h}} - 675\ \frac{\text{mol}}{\text{h}}$$
$$= 14\,325\ \text{mol/h}$$

From Eq. 44.53, the mole ratio of ammonia in the treated gas stream is

$$Y_1 = \frac{G_{\text{NH}_3,\text{out}}}{G} = \frac{28.7\ \dfrac{\text{mol NH}_3}{\text{h}}}{14\,325\ \dfrac{\text{mol air}}{\text{h}}}$$
$$= 0.002\ \text{mol NH}_3/\text{mol air}$$

The operating equation, from Eq. 44.57, is

$$Y_{j+1} = \left(\frac{L}{G}\right)X_j + \left(Y_1 - \left(\frac{L}{G}\right)X_0\right)$$

From the problem statement, $L/G = 2.25$. The y-intercept is

$$Y_1 - \left(\frac{L}{G}\right)X_0 = 0.002 - (2.25)(0) = 0.002$$

Plot the operating line on the equilibrium diagram and step off stages. This separation requires approximately 7 equilibrium stages.

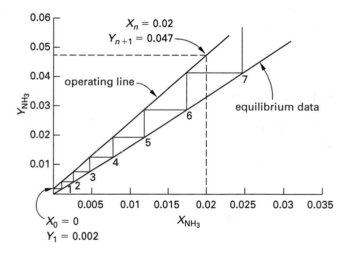

(b) The mole ratio of ammonia in the exit water stream (spent solvent) is the x-coordinate on the operating line corresponding to the mole ratio of ammonia in the inlet gas, $Y_{n+1} = 0.047$. From the diagram, the exiting water composition is $X_n = 0.02$. The mole fraction is

$$x_n = \frac{X_n}{1 + X_n} = \frac{0.02}{1 + 0.02} = 0.0196$$

(c) Since X_0, Y_1, and Y_{n+1} are fixed quantities, the minimum liquid-to-gas flow-rate ratio for the absorber, $(L/G)_{\text{min}}$, is found by pivoting the operating line at the point (X_0, Y_1) so that it intersects the equilibrium curve at the inlet gas solute composition found in part (a), $Y_{n+1} = 0.047$. The corresponding equilibrium exit solvent solute mole ratio, X_n, as read from the diagram is 0.028. From the new operating line, the slope, $(L/G)_{\text{min}}$, is calculated using Eq. 44.59.

$$\left(\frac{L}{G}\right)_{\text{min}} = \frac{Y_{n+1} - Y_1}{X_n - X_0} = \frac{0.047 - 0.002}{0.028 - 0} = 1.61$$

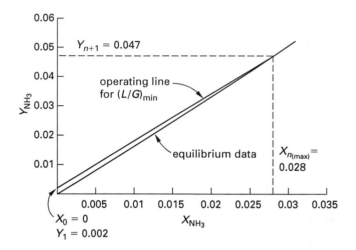

10. The inlet gas stream has a mole fraction of ammonia of 0.03. The molar flow rate of ammonia in the inlet (feed) gas is

$$G_{\text{NH}_3,\text{in}} = y_{n+1}G_F = (0.03)\left(38\,000\ \frac{\text{mol}}{\text{h}}\right)$$
$$= 1140\ \text{mol/h}$$

The absorber removes 95% of the ammonia from the inlet air stream. The molar flow rate of ammonia in the exit water is

$$L_{\text{NH}_3,\text{out}} = (95\%)G_{\text{NH}_3,\text{in}} = (0.95)\left(1140\ \frac{\text{mol}}{\text{h}}\right)$$
$$= 1083\ \text{mol/h}$$

The air (carrier gas) flow rate, G, is

$$G_{\text{air}} = G_F - G_{\text{NH}_3,\text{in}} = 38\,000\ \frac{\text{mol gas}}{\text{h}} - 1140\ \frac{\text{mol NH}_3}{\text{h}}$$
$$= 36\,860\ \text{mol/h}$$

Given the ratio $L/G = 3.25$, the solvent flow rate, L, is

$$L_{\text{water}} = 3.25G = (3.25)\left(36\,860\ \frac{\text{mol air}}{\text{h}}\right)$$
$$= 119\,800\ \text{mol/h}$$

The mole ratio of ammonia in the exiting water stream, using Eq. 44.52, is

$$X_n = \frac{L_{\text{NH}_3,\text{out}}}{L} = \frac{1083 \ \frac{\text{mol NH}_3}{\text{h}}}{119\,800 \ \frac{\text{mol water}}{\text{h}}}$$

$$= 0.009 \ \text{mol NH}_3/\text{mol air}$$

The answer is (B).

11. The operating line can be plotted by identifying the two points $(X_0, \ Y_1)$ and $(X_n, \ Y_{n+1})$. The mole ratio of ammonia in the exit water stream, X_n, was calculated in Prob. 10 as 0.009. From the problem statement, the pure water feed means that $X_0 = 0$. Also from the problem statement, the inlet gas stream is 0.03 mole fraction ammonia or

$$y_{n+1} = 0.03$$

The corresponding mole ratio of ammonia in the inlet gas stream, Y_{n+1}, using Eq. 44.53, is

$$Y_{n+1} = \frac{y_{\text{NH}_3}}{1 - y_{\text{NH}_3}} = \frac{0.03}{1 - 0.03} = 0.031$$

One point on the operating line, $(X_n, \ Y_{n+1})$, is $(0, 0.031)$.

As calculated in Prob. 10, the molar flow rate of ammonia in the inlet gas is

$$G_{\text{NH}_3,\text{in}} = y_{n+1}G_F = (0.03)\left(38\,000 \ \frac{\text{mol}}{\text{h}}\right)$$

$$= 1140 \ \text{mol/h}$$

The absorber removes 95% of the ammonia from the inlet air stream, leaving $100\% - 95\% = 5\%$ of the ammonia in the treated gas stream.

$$G_{\text{NH}_3,\text{out}} = (5\%)G_{\text{NH}_3,\text{in}} = (0.05)\left(1140 \ \frac{\text{mol}}{\text{h}}\right)$$

$$= 57 \ \text{mol/h}$$

The air (carrier gas) flow rate, G, from Prob. 10, is

$$G_{\text{air}} = G_F - G_{\text{NH}_3,\text{in}} = 38\,000 \ \frac{\text{mol gas}}{\text{h}} - 1140 \ \frac{\text{mol NH}_3}{\text{h}}$$

$$= 36\,860 \ \text{mol/h}$$

The mole ratio of ammonia in the treated gas stream, using Eq. 44.53, is

$$Y_1 = \frac{G_{\text{NH}_3,\text{out}}}{G} = \frac{57 \ \frac{\text{mol NH}_3}{\text{h}}}{36\,860 \ \frac{\text{mol air}}{\text{h}}} = 0.0015$$

From Eq. 44.57, the operating equation is

$$Y_{j+1} = \left(\frac{L}{G}\right)X_j + \left(Y_1 - \left(\frac{L}{G}\right)X_0\right)$$

The problem statement specifies that the liquid-to-gas ratio, L/G, is 3.25, which is the slope of the operating line. The y-intercept of this line is

$$Y_1 - \left(\frac{L}{G}\right)X_0 = 0.0015 - (3.25)(0)$$

$$= 0.0015$$

This separation requires approximately $3\tfrac{3}{5}$ equilibrium stages.

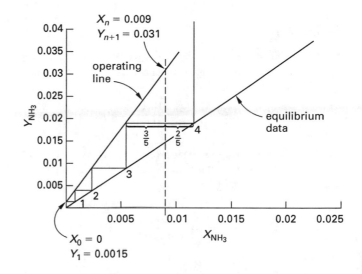

The answer is (B).

12. The inlet and exit gas compositions, Y_{n+1} and Y_1, as well as the composition of the solvent stream, X_0, are fixed quantities. Therefore, the minimum liquid-to-gas flow-rate ratio for the absorber, $(L/G)_{\text{min}}$, can be found graphically by pivoting the operating line at the point $(X_0, \ Y_1)$ so that it touches the equilibrium curve at the inlet gas solute mole ratio found in Prob. 11, $Y_{n+1} = 0.031$. The corresponding equilibrium exit solvent solute mole ratio, X_n, as read from the diagram, is 0.0188. The slope of this new operating line is $(L/G)_{\text{min}}$.

$$\left(\frac{L}{G}\right)_{\text{min}} = \frac{Y_{n+1} - Y_1}{X_n - X_0} = \frac{0.031 - 0.0015}{0.0188 - 0}$$

$$= 1.57 \quad (1.6)$$

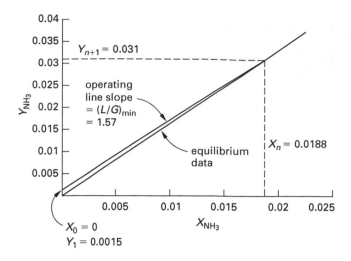

$X_0 = 0$
$Y_1 = 0.0015$

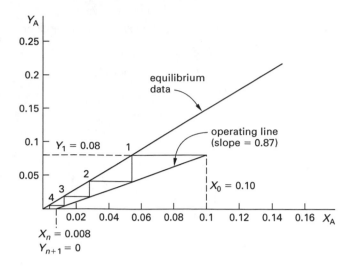

$X_n = 0.008$
$Y_{n+1} = 0$

The answer is (C).

Stripping

13. (a) The inlet water stream has 9.1 mol% A. The mole ratio of component A in the inlet solvent stream is

$$X_0 = \frac{x_0}{1 - x_0} = \frac{0.091}{1 - 0.091} = 0.10$$

Given the desired mole fraction of component A of 0.008 in the treated water, the mole ratio of component A needed in this stream is

$$X_n = \frac{x_n}{1 - x_n} = \frac{0.008}{1 - 0.008} = 0.008$$

The inlet gas stream is pure air. The mole ratio of solute in the inlet air is

$$Y_{n+1} = 0$$

With a target mole fraction of A in the exit gas stream of 0.074, the corresponding mole ratio of component A in this stream is

$$Y_1 = \frac{y_1}{1 - y_1} = \frac{0.074}{1 - 0.074} = 0.08$$

In contrast to absorption processes, the operating line lies below the equilibrium data. The operating line can be plotted using the conditions at the entrance and the exit of the stripping tower; one point is (X_n, Y_{n+1}) and another point is (X_0, Y_1). Plotting the operating line with the equilibrium data and stepping off stages shows that this separation requires approximately 3.5 stages.

(b) The ratio (L/G) is the slope of the operating line. The slope can be found using the operating line coordinates determined in part (a) and Eq. 44.59.

$$\frac{L}{G} = \frac{Y_{n+1} - Y_1}{X_n - X_0} = \frac{0 - 0.08}{0.008 - 0.1} = 0.87$$

(c) The solvent (water) molar flow rate, L, can be obtained from the inlet feed rate. The feed is comprised of water and solute. The mole fraction of water in the feed is given by

$$x_0 + x_{\text{water}} = 1$$
$$x_{\text{water}} = 1 - x_0 = 1 - 0.091 = 0.909$$
$$L = L_F x_{\text{water}} = \left(1000 \; \frac{\text{mol}}{\text{h}}\right)(0.909)$$
$$= 909 \; \text{mol/h water}$$

(d) The carrier gas molar flow rate, G, is found from the slope of the operating line found in part (b).

$$\frac{L}{G} = 0.87$$

$$G = \frac{L}{0.87} = \frac{909 \; \frac{\text{mol}}{\text{h}}}{0.87} = 1045 \; \text{mol/h air}$$

45 Liquid-Liquid Extraction

PRACTICE PROBLEMS

Partially Miscible Phases

1. 100 g of 45 wt% acetic acid in water and 180 g of isopropyl ether are combined in a beaker. Given the following phase diagram, what are the equilibrium compositions of the two phases?

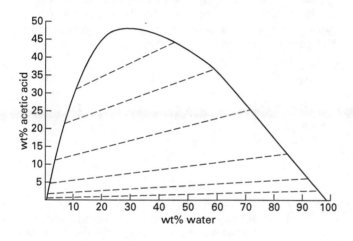

2. Isopropyl ether is to be used to remove acetic acid from water in a countercurrent extractor. The extractor reduces the acetic acid concentration in the water from 45 wt% to 3 wt%. The ratio of feeds and the equilibrium data are the same as in Prob. 1. How many equilibrium stages are needed?

Immiscible Phases

3. A byproduct stream flowing at the rate of 2323 kg/h is water containing 1 wt% nicotine. The nicotine content needs to be reduced to 0.1 wt% using pure kerosene. The equilibrium data is provided. (a) What is the minimum solvent flow rate needed to achieve the separation? (b) How many equilibrium stages in a countercurrent extractor are required to achieve this separation if 52% more than the minimum solvent flow rate is used?

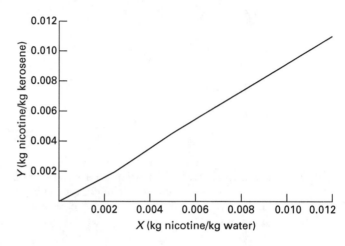

4. Equal amounts of a solution of 50 wt% 1,1,2-trichloroethane in water and pure acetone are mixed. What will be the composition of the resulting phases?

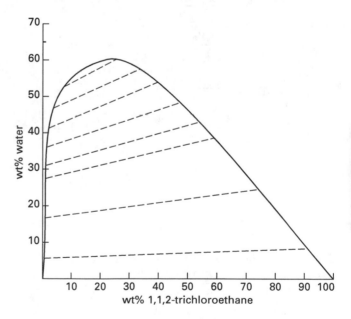

SOLUTIONS

1. This is a mixing problem. The beginning liquid compositions are plotted on the equilibrium diagram: acetic acid-water at (45, 55) and pure isopropyl ether at (0, 0). The position of the mixing point is determined by using the lever rule, Eq. 45.10, where F_1 is the weight of pure isopropyl ether, and F_2 is the weight of the acetic acid-water solution.

$$\frac{F_1}{F_2} = \frac{\overline{MF_2}}{\overline{F_1M}} = \frac{180 \text{ g}}{100 \text{ g}} = 1.8$$

$$\overline{MF_2} = 1.8\overline{F_1M}$$

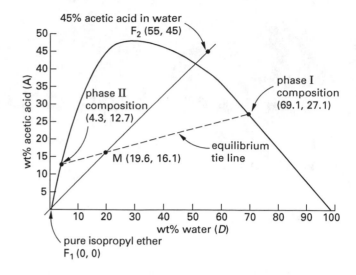

The mixing point, M, can be located from the ratio of line segment lengths. Alternatively, the composition of the diluent, D (water), and of the solute, A (acetic acid), for the mixing point can be calculated using Eqs. 45.9 and 45.8.

$$x_{D,M} = \frac{F_1 x_{D,F_1} + F_2 x_{D,F_2}}{F_1 + F_2}$$

$$= \frac{(180 \text{ g})(0) + (100 \text{ g})(0.55)}{180 \text{ g} + 100 \text{ g}} = 0.196$$

$$x_{A,M} = \frac{F_1 x_{A,F_1} + F_2 x_{A,F_2}}{F_1 + F_2}$$

$$= \frac{(180 \text{ g})(0) + (100 \text{ g})(0.45)}{180 \text{ g} + 100 \text{ g}} = 0.161$$

In terms of wt%, the mixing point is at (19.6, 16.1).

By drawing the tie line that passes through the mixing point, the equilibrium compositions can be read. For each phase,

Phase I at (69.1, 27.1):   69.1 wt% water,
                           27.1 wt% acetic acid,
                           3.8 wt% isopropyl ether

Phase II at (4.3, 12.7):   4.3 wt% water,
                           12.7 wt% acetic acid,
                           83 wt% isopropyl ether

Very little separation was achieved by using one equilibrium stage.

2. Using data from Prob. 1, the two phases are the same, and the position of the mixing point does not change. To maintain the feed ratios of Prob. 1, a basis of 100 g/h of feed (acetic acid-water) and 180 g/h of pure isopropyl ether is selected. The illustration shows the known compositions and flow rates.

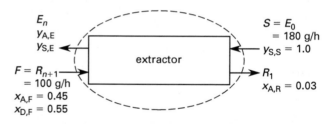

The raffinate stream exits with 3 wt% acetic acid, and the exiting raffinate stream is in equilibrium with the extract phase leaving stage 1 (E_1). This places R_1 on the equilibrium diagram on the two-phase envelope at 3 wt% acetic acid.

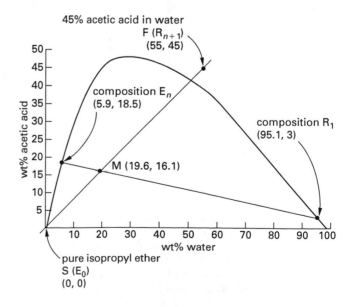

The composition of the extract leaving the extractor, E_n, is found by constructing a line from the raffinate composition, R_1, through the mixing point, M, to the two-phase envelope. From the equilibrium diagram, E_n is 5.9 wt% water and 18.5 wt% acetic acid. Since the difference point, Δ, is at the intersection of lines $\overline{R_1 E_0}$ and $\overline{E_n F}$, the difference point can be located graphically. Alternatively, the position of the difference point can be calculated. The mass balance equations for water and acetic acid are

$$55 \text{ g/h water} = 0.951 R_1 + 0.059 E_n$$
$$45 \text{ g/h acetic acid} = 0.03 R_1 + 0.185 E_n$$

Solving the two equations for the two unknowns, R_1 and E_n,

$$R_1 = 43 \text{ g/h}$$
$$E_n = 236 \text{ g/h}$$

Substitute these values into Eqs. 45.12 and 45.13 to determine the coordinates of the difference point.

$$x_{A,\Delta} = \frac{E_0 y_{A,0} - R_1 x_{A,1}}{E_0 - R_1}$$

$$= \frac{\left(180 \frac{\text{g}}{\text{h}}\right)(0) - \left(43 \frac{\text{g}}{\text{h}}\right)(0.03)}{180 \frac{\text{g}}{\text{h}} - 43 \frac{\text{g}}{\text{h}}} = -0.0094$$

$$x_{D,\Delta} = \frac{E_0 y_{D,0} - R_1 x_{D,1}}{E_0 - R_1}$$

$$= \frac{\left(180 \frac{\text{g}}{\text{h}}\right)(0) - \left(43 \frac{\text{g}}{\text{h}}\right)(0.951)}{180 \frac{\text{g}}{\text{h}} - 43 \frac{\text{g}}{\text{h}}} = -0.298$$

In terms of wt%, the difference point plots at $(-29.8, -0.94)$. (See *Illustrations for Sol. 2*).

Stepping off stages using the difference point indicates that the total number of equilibrium stages required is approximately 8. (See *Illustrations for Sol. 2*).

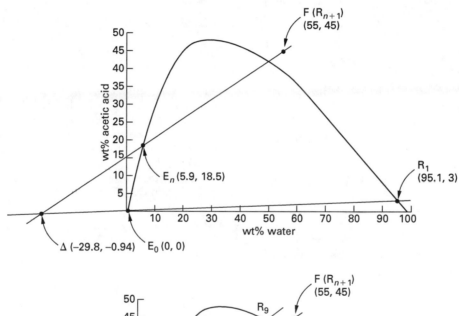

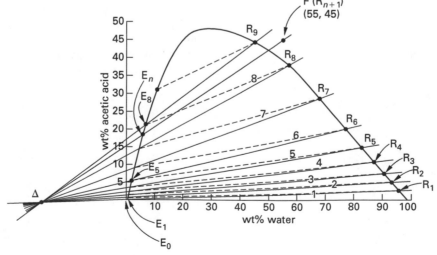

Illustrations for Sol. 2

3. (a) Since both streams are dilute, x is approximately the same as X, and y is approximately the same as Y.

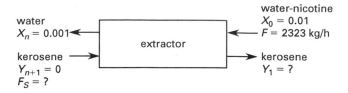

The minimum solvent flow rate can be determined graphically or from the operating equation, Eq. 45.4. The slope of the operating line is the ratio of the diluent feed to the solvent feed, F_D/F_S. The minimum solvent flow rate will be at the maximum slope of this line. One point on the operating line is at (X_n, Y_{n+1}) or (0.001, 0) which is plotted on the equilibrium diagram. The operating line is drawn from this point to intersect the equilibrium curve at $X_0 = 0.01$. From the illustration, $X_0 = 0.01$ at $Y_{1,\max} = 0.0092$. The slope of this line is

$$\frac{F_D}{F_{S,\min}} = \frac{\Delta Y}{\Delta X} = \frac{Y_1 - Y_{n+1}}{X_0 - X_n} = \frac{0.0092 - 0}{0.01 - 0.001} = 1$$

$$F_D = F_{S,\min}$$

Since the byproduct stream feed is 99% water (diluent),

$$F_D = F(0.99) = \left(2323 \; \frac{\text{kg}}{\text{h}}\right)(0.99) = 2300 \; \text{kg/h}$$

$$F_D = F_{S,\min} = 2300 \; \text{kg/h}$$

This separation requires a minimum solvent flow rate, F_S, of 2300 kg/h kerosene.

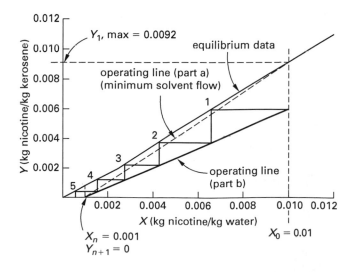

(b) To determine the number of equilibrium stages if 52% more than the minimum solvent flow rate is used, the new operating line must be plotted. The new solvent flow rate is

$$F_S = (1.52)F_{S,\min} = (1.52)\left(2300 \; \frac{\text{kg}}{\text{h}}\right) = 3496 \; \text{kg/h}$$

From Eq. 45.4, the operating equation is

$$Y_{j+1} = \left(\frac{F_D}{F_S}\right)X_j + \left(Y_1 - \left(\frac{F_D}{F_S}\right)X_0\right)$$

The new slope of the operating line is

$$\frac{F_D}{F_S} = \frac{2300 \; \dfrac{\text{kg}}{\text{h}}}{3496 \; \dfrac{\text{kg}}{\text{h}}} = 0.66$$

The new operating line can be plotted using (X_n, Y_{n+1}) or (0.001, 0) and the new slope. Stepping off stages indicates that this separation requires approximately 4.3 stages, which would typically be rounded up to 5 stages.

4. This is a mixing problem. The beginning liquid compositions are plotted on the equilibrium diagram: 1,1,2-trichloroethane-water at (50, 50) and pure acetone at (0, 0). The position of the mixing point is calculated using the lever rule, Eq. 45.10, using F_1 to denote the weight of pure acetone, and F_2 as the weight of the 1,1,2-trichloroethane-water solution. A basis of 100 g of each liquid is chosen.

$$\frac{F_1}{F_2} = \frac{\overline{MF_2}}{\overline{MF_1}} = \frac{100 \; \text{g}}{100 \; \text{g}} = 1$$

From the ratio of line-segment lengths, the location of the mixing point, M, can be identified. Alternatively, the composition of the diluent, D (1,1,2-trichloroethane) and of the solute, A (water), at the mixing point can be calculated using Eqs. 45.9 and 45.8.

$$\begin{aligned}
x_{D,M} &= \frac{F_1 x_{D,F_1} + F_2 x_{D,F_2}}{F_1 + F_2} \\
&= \frac{(100 \; \text{g})(0) + (100 \; \text{g})(0.50)}{100 \; \text{g} + 100 \; \text{g}} = 0.25
\end{aligned}$$

$$\begin{aligned}
x_{A,M} &= \frac{F_1 x_{A,F_1} + F_2 x_{A,F_2}}{F_1 + F_2} \\
&= \frac{(100 \; \text{g})(0) + (100 \; \text{g})(0.50)}{100 \; \text{g} + 100 \; \text{g}} = 0.25
\end{aligned}$$

In terms of wt%, the mixing point is at (25, 25).

The equilibrium compositions can be read by drawing the tie line that passes through the mixing point. For each phase,

Phase I at (67, 32): 67 wt% 1,1,2-trichloroethane,
                       32 wt% water,
                       1 wt% acetone

Phase II at (1, 22): 1 wt% 1,1,2-trichloroethane,
                       22 wt% water,
                       77 wt% acetone

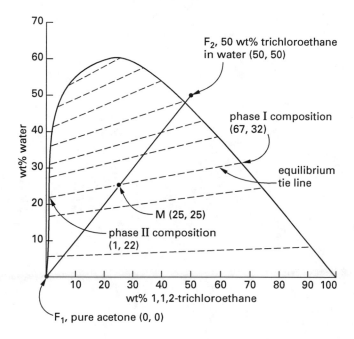

46 Solid-Liquid Processes

PRACTICE PROBLEMS

Drying

1. A batch of material with a 20% moisture content must be dried to a 2.5% moisture content. The drying curve is shown. The initial solids weight is 700 lbm, and the drying surface is 0.2 ft^2/lbm dry weight. Estimate the total drying time given that the drying rate in falling rate period II is e^X.

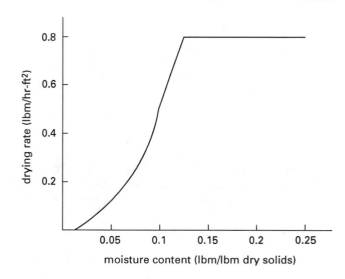

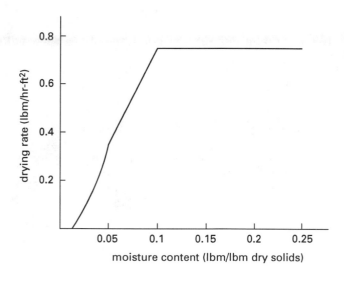

2. A batch of material that has a 22.5% moisture content must be dried to a 2.5% moisture content. The drying curve is shown. The initial solids weight is 1000 lbm, and the drying surface is 0.4 ft^2/lbm dry weight. Estimate the total drying time given that the drying rate in falling rate period II is e^{5X}.

Leaching

3. A proprietary solvent leaches caffeine from coffee beans. Analyses indicate that nothing else dissolves in this solvent except caffeine. The equilibrium relationship is known to be $Y = 0.8X$ for caffeine in the solvent, where

$$Y = \frac{\text{g caffeine}}{\text{kg solvent}}$$

$$X = \frac{\text{g caffeine}}{\text{kg insoluble solids}}$$

Starting with pure solvent and coffee beans that have 20 g caffeine/kg beans, the goal is to reduce the caffeine content in the beans to 5% of the original content (i.e., 1 g caffeine/kg beans).

(a) Determine the minimum amount of solvent required to achieve this separation using a countercurrent, equilibrium-staged system.

(b) Determine how many equilibrium stages this separation would require using twice the minimum solvent flow rate found in part (a).

Mass Transfer

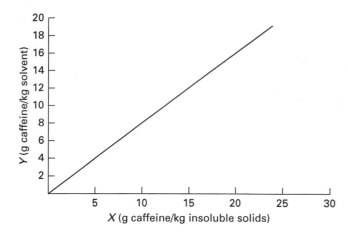

Washing

4. Sand mined from a beach contains 0.035 wt% salt. The sand also contains 40% seawater by volume. The salt content of the sand must be reduced to 0.002 wt% by washing with 1.0 kg of pure water per 1000 cm³ of wet sand. The following density data is known.

$$\rho_{water} = 1.0 \text{ g/cm}^3$$

$$\rho_{dry\ sand} = 1.8 \text{ g/cm}^3 \quad [\text{including air voids}]$$

$$\rho_{dry\ sand} = 3.0 \text{ g/cm}^3 \quad [\text{excluding air voids}]$$

(a) How many equilibrium washing stages would this separation require?

(b) Starting at the wet sand inlet to the second stage, what are the compositions of the streams leaving the second equilibrium stage?

SOLUTIONS

1. *Constant Rate Period:*

The moisture content of the material is reduced from the initial moisture content of 20% to the critical moisture content, X_c, in the constant rate period. The initial material has 20% moisture, so the initial moisture content, X_1, is

$$X_1 = \frac{X_T}{L_s}$$

$$= \frac{\left(0.2 \dfrac{\text{lbm moisture}}{\text{lbm total}}\right)(700 \text{ lbm total})}{\left((1-0.2) \dfrac{\text{lbm dry solids}}{\text{lbm total}}\right)(700 \text{ lbm total})}$$

$$= 0.25 \text{ lbm moisture/lbm dry solids}$$

From the given drying curve, the critical moisture content, X_c, is

$$X_c = 0.1 \text{ lbm moisture/lbm dry solids}$$

Reading from the drying curve, in the constant rate period, the drying rate, N_c, is

$$N_c = 0.75 \text{ lbm moisture/hr-ft}^2$$

In the constant rate drying period, the drying time, t_c, is

$$t_c = \left(\frac{L_s}{A}\right)\left(\frac{X_1 - X_c}{N_c}\right)$$

$$= \left(\frac{\text{lbm dry solids}}{0.2 \text{ ft}^2}\right)$$

$$\times \left(\frac{0.25 \dfrac{\text{lbm moisture}}{\text{lbm dry solids}} - 0.1 \dfrac{\text{lbm moisture}}{\text{lbm dry solids}}}{0.75 \dfrac{\text{lbm moisture}}{\text{hr-ft}^2}}\right)$$

$$= 1 \text{ hr}$$

Falling Rate Period I:

In falling rate period I, the initial drying rate, N_1, is N_c.

$$N_1 = N_c = 0.75 \text{ lbm moisture/hr-ft}^2$$

From the drying curve, the final drying rate, N_2, is

$$N_2 = 0.35 \text{ lbm moisture/hr-ft}^2$$

The initial moisture content, X_1, is the critical moisture content, X_c.

$$X_1 = X_c = 0.1 \text{ lbm moisture/lbm dry solids}$$

From the drying curve, the final moisture content, X_2, is

$$X_2 = 0.05 \text{ lbm moisture/lbm dry solids}$$

Since the drying curve in falling rate period I is linear, the total drying time, t_{I}, is

$$t_{\text{I}} = \left(\frac{L_s}{A}\right)\left(\frac{X_1 - X_2}{N_1 - N_2}\right)\ln\frac{N_1}{N_2}$$

$$= \left(\frac{\text{lbm dry solids}}{0.2 \text{ ft}^2}\right)$$

$$\times \left(\frac{0.1 \dfrac{\text{lbm moisture}}{\text{lbm dry solids}} - 0.05 \dfrac{\text{lbm moisture}}{\text{lbm dry solids}}}{0.75 \dfrac{\text{lbm moisture}}{\text{hr-ft}^2} - 0.35 \dfrac{\text{lbm moisture}}{\text{hr-ft}^2}}\right)$$

$$\times \left(\ln\frac{0.75 \dfrac{\text{lbm moisture}}{\text{hr-ft}^2}}{0.35 \dfrac{\text{lbm moisture}}{\text{hr-ft}^2}}\right)$$

$$= 0.48 \text{ hr}$$

Falling Rate Period II:

In falling rate period II, the drying rate, N_{II}, is given as

$$N_{\text{II}} = e^X$$

From the drying curve, the initial moisture content, X_1, is

$$X_1 = 0.05 \text{ lbm moisture/lbm dry solids}$$

From the problem statement, the final material has 2.5% moisture. The final moisture content, X_2, is

$$X_2 = \frac{X_T}{L_s} = \frac{0.025}{1 - 0.025} = \frac{0.025 \text{ lbm moisture}}{0.975 \text{ lbm dry solids}}$$

$$= 0.026 \text{ lbm moisture/lbm dry solids}$$

In falling rate period II, the drying time, t_{II}, is

$$t_{\text{II}} = \left(\frac{L_s}{A}\right)\int_{X_2}^{X_1}\frac{dX}{N}$$

$$= \left(\frac{1}{0.2}\right)\int_{0.026}^{0.05} e^{-X}dX$$

$$= (5)\left(-e^{-X}\Big|_{0.026}^{0.05}\right)$$

$$= (5)(-e^{-0.05} - (-e^{-0.026}))$$

$$= 0.12 \text{ hr}$$

The total drying time, t_D, is the sum of the drying times for each period.

$$t_D = t_c + t_{\text{I}} + t_{\text{II}} = 1 \text{ hr} + 0.48 \text{ hr} + 0.12 \text{ hr} = \boxed{1.6 \text{ hr}}$$

2. *Constant Rate Period:*

The moisture content of the material is reduced from the initial moisture content of 22.5% to the critical moisture content, X_c, in the constant rate period. The initial material has 22.5% moisture, so by definition the initial moisture content, X_1, is

$$X_1 = \frac{X_T}{L_s}$$

$$= \frac{\left(0.225 \dfrac{\text{lbm moisture}}{\text{lbm total}}\right)(1000 \text{ lbm total})}{\left((1 - 0.225) \dfrac{\text{lbm dry solids}}{\text{lbm total}}\right)(1000 \text{ lbm total})}$$

$$= 0.29 \text{ lbm moisture/lbm dry solids}$$

From the given drying curve, the critical moisture content, X_c, is

$$X_c = 0.125 \text{ lbm moisture/lbm dry solids}$$

Reading from the drying curve, in the constant rate period, the drying rate, N_c, is

$$N_c = 0.8 \text{ lbm moisture/hr-ft}^2$$

In the constant rate drying period, the drying time, t_c, is

$$t_c = \left(\frac{L_s}{A}\right)\left(\frac{X_1 - X_c}{N_c}\right)$$

$$= \left(\frac{\text{lbm dry solids}}{0.4 \text{ ft}^2}\right)$$

$$\times \left(\frac{0.29 \dfrac{\text{lbm moisture}}{\text{lbm dry solids}} - 0.125 \dfrac{\text{lbm moisture}}{\text{lbm dry solids}}}{0.8 \dfrac{\text{lbm moisture}}{\text{hr-ft}^2}}\right)$$

$$= 0.52 \text{ hr}$$

Falling Rate Period I:

In falling rate period I, the initial drying rate, N_1, is N_c.

$$N_1 = N_c = 0.8 \text{ lbm moisture/hr-ft}^2$$

From the drying curve, the final drying rate, N_2, is

$$N_2 = 0.5 \text{ lbm moisture/hr-ft}^2$$

The initial moisture content, X_1, is the critical moisture content, X_c.

$$X_1 = X_c = 0.125 \text{ lbm moisture/lbm dry solids}$$

From the drying curve, the final moisture content, X_2, is

$$X_2 = 0.075 \text{ lbm moisture/lbm dry solids}$$

Since the drying curve in falling rate period I is linear, the total drying time, t_I, is

$$t_I = \left(\frac{L_s}{A}\right)\left(\frac{X_1 - X_2}{N_1 - N_2}\right)\ln\frac{N_1}{N_2}$$

$$= \left(\frac{\text{lbm dry solids}}{0.4 \text{ ft}^2}\right)$$

$$\times \left(\frac{0.125\,\dfrac{\text{lbm moisture}}{\text{lbm dry solids}} - 0.075\,\dfrac{\text{lbm moisture}}{\text{lbm dry solids}}}{0.8\,\dfrac{\text{lbm moisture}}{\text{hr-ft}^2} - 0.5\,\dfrac{\text{lbm moisture}}{\text{hr-ft}^2}}\right)$$

$$\times \left(\ln\frac{0.8\,\dfrac{\text{lbm moisture}}{\text{hr-ft}^2}}{0.5\,\dfrac{\text{lbm moisture}}{\text{hr-ft}^2}}\right)$$

$$= 0.20 \text{ hr}$$

Falling Rate Period II:

In falling rate period II, the drying rate N_{II}, is given as

$$N_{II} = e^{5X}$$

From the drying curve, the initial moisture content, X_1, is

$$X_1 = 0.075 \text{ lbm moisture/lbm dry solids}$$

From the problem statement, the final material has 2.5% moisture. The final moisture content, X_2, is

$$X_2 = \frac{X_T}{L_s} = \frac{0.025}{1 - 0.025} = \frac{0.025 \text{ lbm moisture}}{0.975 \text{ lbm dry solids}}$$

$$= 0.026 \text{ lbm moisture/lbm dry solids}$$

In falling rate period II, the drying time, t_{II}, is

$$t_{II} = \left(\frac{L_s}{A}\right)\int_{X_2}^{X_1}\frac{dX}{N}$$

$$= \left(\frac{1}{0.4}\right)\int_{0.026}^{0.075}e^{-5X}dX$$

$$= \left(\frac{5}{2}\right)\left(-\frac{1}{5}e^{-5X}\Big|_{0.026}^{0.075}\right)$$

$$= \left(\frac{5}{2}\right)\left(-\frac{1}{5}e^{-(5)(0.075)} - \left(-\frac{1}{5}e^{-(5)(0.026)}\right)\right)$$

$$= 0.095 \text{ hr}$$

The total drying time, t_D, is the sum of the drying times for each period.

$$t_D = t_c + t_I + t_{II} = 0.52 \text{ hr} + 0.20 \text{ hr} + 0.095 \text{ hr}$$

$$= \boxed{0.82 \text{ hr}}$$

3. (a) To find the minimum solvent required, use the McCabe-Thiele method to draw the operating curve.

The data given in the problem are shown in the following illustration.

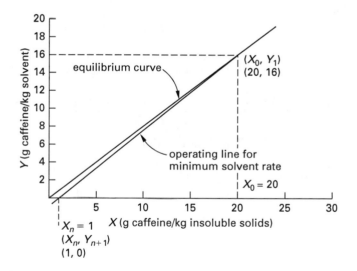

The operating equation for leaching is

$$Y_{j+1} = \left(\frac{F_{\text{solid}}}{F_{\text{solvent}}}\right)X_j + \left(Y_1 - \left(\frac{F_{\text{solid}}}{F_{\text{solvent}}}\right)X_0\right)$$

The slope of the operating line represents the ratio of insoluble solids to solvent flow rates. For the operating line, only enough information is available to plot one point at $(X_1, Y_{n+1}) = (1, 0)$. However, the minimum solvent flow rate is related to the slope of the operating line that intersects the equilibrium curve (to create a pinch point) at the point where $X_0 = 20$ g caffeine/kg insoluble solids. Using these points and drawing this operating line on the equilibrium diagram reveals the corresponding value of $Y_1 = 16$ g caffeine/kg solvent.

The slope is

$$\text{slope} = \frac{F_{\text{solid}}}{F_{\text{solvent}}} = \frac{\Delta Y}{\Delta X} = \frac{Y_1 - Y_{n+1}}{X_0 - X_n}$$

$$= \frac{16\,\dfrac{\text{g caffeine}}{\text{kg solvent}} - 0\,\dfrac{\text{g caffeine}}{\text{kg solvent}}}{\left(20\,\dfrac{\text{g caffeine}}{\text{kg insoluble solids}} - 1\,\dfrac{\text{g caffeine}}{\text{kg insoluble solids}}\right)}$$

$$= 0.84 \text{ kg insoluble solids/kg solvent}$$

The minimum amount of solvent needed is the reciprocal of the minimum slope.

$$\frac{F_{\text{solvent}}}{F_{\text{solid}}} = \frac{1}{\text{slope}} = \frac{1}{0.84 \; \dfrac{\text{kg insoluble solids}}{\text{kg solvent}}}$$

$$= \boxed{1.2 \text{ kg solvent/kg insoluble solids}}$$

Therefore, the separation requires a minimum of 1.2 kg of solvent per kg of coffee beans.

(b) At a solvent flow rate, F_{solvent}, that is twice the minimum solvent flow rate, the new operating line slope becomes

$$\text{slope} = \frac{F_{\text{solid}}}{2F_{\text{solvent}}} = \frac{0.84 \; \dfrac{\text{kg insoluble solids}}{\text{kg solvent}}}{2}$$

$$= 0.42 \text{ kg insoluble solids/kg solvent}$$

Using this slope and the specified conditions at one end of the countercurrent leaching system, $(X_n, Y_{n+1}) = (1, 0)$, the new operating line can be plotted on the equilibrium diagram and the stages stepped off. The separation requires approximately $3\frac{1}{2}$ equilibrium stages, which rounds to $\boxed{4 \text{ stages.}}$

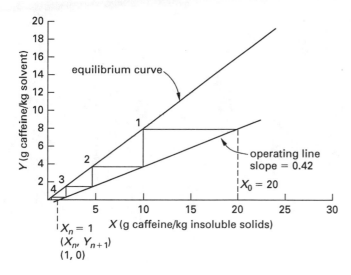

4. (a) The equilibrium relationship for washing processes is $y = x$. The following illustration shows the specified parameters.

Assuming that the porosity of solids in the underflow, ϵ, and the densities, ρ_i, are constant throughout the washer, all the underflow and overflow liquid mass flow rates, U_i and O_i, are also constant between stages. The applicable operating line equation for washing is

$$y_{j+1} = \left(\frac{U}{O}\right) x_j + \left(y_{n+1} - \left(\frac{U}{O}\right) x_n \right)$$

From the problem statement, 1.0 kg of pure water is used per 1000 cm^3 of wet sand. Using the basis of 1000 cm^3/h of wet sand processed,

$$O = \left(\frac{1.0 \text{ kg water}}{1000 \text{ cm}^3 \text{ wet sand}} \right) \left(1000 \; \frac{\text{cm}^3 \text{ wet sand}}{\text{h}} \right)$$

$$= 1.0 \text{ kg water/h}$$

Using the basis, the volumetric flow rate of wet solids is 1000 cm^3/h. The sand contains 40% water by volume, making the porosity

$$\epsilon = 0.4$$

The fluid density, ρ_f, is the density of water, ρ_w.

Since the problem statement specifies the amount of wet solids processed, determine the underflow rate, U, with

$$U = \dot{V}_{\text{wet solids}} \epsilon \rho_f$$

$$= \left(1000 \; \frac{\text{cm}^3}{\text{h}} \right) (0.4) \left(1.0 \; \frac{\text{g}}{\text{cm}^3} \right) \left(\frac{1 \text{ kg}}{1000 \text{ g}} \right)$$

$$= 0.4 \text{ kg/h}$$

The slope of the operating line is

$$\frac{U}{O} = \frac{0.4 \; \dfrac{\text{kg}}{\text{h}}}{1.0 \; \dfrac{\text{kg}}{\text{h}}} = 0.4$$

The y-intercept for the operating line is

$$\text{y-intercept} = y_{n+1} - \left(\frac{U}{O}\right) x_n = 0 - (0.4)(0.002)$$

$$= -0.0008$$

Plotting the operating line on a $y = x$ equilibrium diagram shows that this separation requires 2.7 equilibrium stages, which rounds up to $\boxed{3 \text{ stages.}}$

Mass Transfer

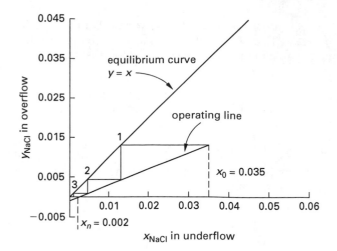

(b) From the equilibrium diagram where the second step intersects the equilibrium curve ($y = x$ for washing), the equilibrium composition of both the underflow and the overflow streams leaving the second stage is approximately 0.0044 wt% salt. Since $y = x$, both compositions are equal ($x_2 = y_2 = 0.0044$).

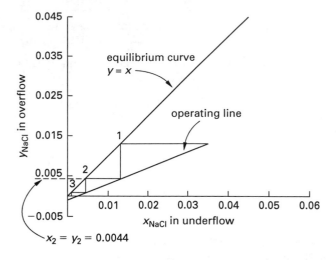

47 Kinetics

PRACTICE PROBLEMS

1. The rate constant for a reaction doubles when the temperature is raised from 37°C to 57°C. (a) What is the activation energy of this reaction? (b) At what temperature will the rate constant double again?

2. Compound A reacts to form compound P in a constant-volume, isothermal batch reactor. Determine the reaction order from the tabulated data.

time (h)	concentration of reactant A (mol/L)
0	5.00
0.5	0.69
1.0	0.37
1.5	0.25
2.0	0.19
2.5	0.16

3. Consider the reversible reaction

$$A + B \leftrightarrow C + D$$

The feed stream contains 3 mol of A, 2 mol of B, and 1 mol of C for each mole of D present. The equilibrium constant, K_{eq}, is 1.45. Calculate the component mole fractions and extent of reaction at equilibrium.

4. An endothermic first-order reaction, A→P, is carried out in a batch reactor.

heat of reaction	97 000 kJ/kmol
reactor volume	1 m³
heat capacity of mixture	2.5 kJ/kg·K
total mass	975 kg
initial number of moles	10.5 kmol
rate constant	$4750e^{-8500/T}\text{s}^{-1}$

After the reaction mixture is heated to 300°C, it is allowed to proceed adiabatically. 15 mol% of A is converted during the heat-up. What is the temperature in the reactor at 75% conversion?

5. The reaction $A \xrightarrow{k_1} P \xrightarrow{k_2} S$ is carried out in a tubular plug-flow reactor. Both reactions are first order.

reactor diameter	10 cm
feed rate	0.15 m³/h pure A
feed concentration	0.10 kmol A/m³
rate constants	$k_1 = 175$ h^{-1}
	$k_2 = 12.5$ h^{-1}

(a) Determine the reactor length that will maximize the yield of P. (b) What are the exit concentrations of A, P, and S?

6. The reaction $A \xrightarrow{k_1} P \xrightarrow{k_2} S$ is carried out in a perfectly mixed flow reactor. Both reactions are first order.

reactor volume	2.5 L
feed rate	0.15 m³/h pure A
feed concentration	0.10 kmol A/m³
rate constants	$k_1 = 175$ h^{-1}
	$k_2 = 12.5$ h^{-1}

Assuming constant density, determine the effluent concentrations of A, P, and S.

7. A CSTR is to be used to produce B at a rate of 55 kmol/h by the first-order reaction, A→B, with a rate constant of 7.0 h^{-1}. The feed concentration of reactant A is 1 kmol/m³. (a) What reactor volume is required to achieve 70% conversion in a single reactor? (b) What total reactor volume is required to achieve 70% conversion in a series of two CSTRs?

SOLUTIONS

1. (a) From the problem statement, $k_2 = 2k_1$. The activation energy, E, can be found by taking a ratio of rate constants as defined by the Arrhenius law equation.

$$T_1 = 37°C + 273 = 310K$$

$$T_2 = 57°C + 273 = 330K$$

$$\frac{k_1}{k_2} = \frac{A_0 e^{-\frac{E}{R^* T_1}}}{A_0 e^{-\frac{E}{R^* T_2}}} = e^{-\left(\frac{E}{R^*}\right)\left(\frac{1}{T_1} - \frac{1}{T_2}\right)}$$

$$\ln\frac{k_1}{k_2} = -\left(\frac{E}{R^*}\right)\left(\frac{1}{T_1} - \frac{1}{T_2}\right)$$

$$\ln\frac{1}{2} = -\left(\frac{E}{8.314 \ \frac{J}{mol\cdot K}}\right)\left(\frac{1}{310K} - \frac{1}{330K}\right)$$

$$E = \boxed{29\,477 \ J/mol \quad (29\,500)}$$

(b) The same approach used in part (a) is used to determine the temperature. Doubling the rate constant again implies that

$$k_2' = 2k_2 = (2)(2k_1) = 4k_1$$

$$\ln\frac{k_1}{k_2'} = -\left(\frac{E}{R^*}\right)\left(\frac{1}{T_1} - \frac{1}{T_2'}\right)$$

$$\ln\frac{1}{4} = -\left(\frac{29\,500 \ \frac{J}{mol}}{8.314 \ \frac{J}{mol\cdot K}}\right)\left(\frac{1}{310K} - \frac{1}{T_2'}\right)$$

$$T_2' = \boxed{353K \quad (80°C)}$$

2. The integrated form of the rate equation can be plotted to determine reaction order. The rate data is reorganized as in the following table.

Time (h)	C_A 0 order	$\ln\frac{C_{A_0}}{C_A}$ 1st order	$\frac{1}{C_A}$ 2nd order	$\frac{0.1}{C_A^2}$ 3rd order
0	5.00	0.00	0.20	0.00
0.5	0.69	1.98	1.45	0.21
1.0	0.37	2.60	2.70	0.73
1.5	0.25	3.00	4.00	1.60
2.0	0.19	3.27	5.26	2.77
2.5	0.16	3.44	6.25	3.91

Plotting the pertinent concentration expression versus time will give a straight line when the correct reaction order has been assumed.

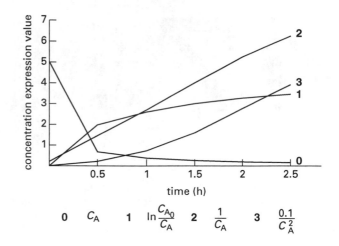

$$0 \quad C_A \qquad 1 \quad \ln\frac{C_{A_0}}{C_A} \qquad 2 \quad \frac{1}{C_A} \qquad 3 \quad \frac{0.1}{C_A^2}$$

In this case, the reaction is $\boxed{\text{second order.}}$

3. The total number of moles is conserved. A total of 7 mol will be taken as the basis for calculations, since $N_t = N_A + N_B + N_C + N_D = 3 + 2 + 1 + 1 = 7$.

The extent of reaction is expressed as

$$-\frac{dN_A}{a} = -\frac{dN_B}{b} = \frac{dN_C}{c} = \frac{dN_D}{d} = d\varepsilon$$

Integrating yields

$$-\int_3^{N_A} dN_A = -\int_2^{N_B} dN_B = \int_1^{N_C} dN_C$$

$$= \int_1^{N_D} dN_D = \int_0^\varepsilon d\varepsilon$$

For component A,

$$-\int_3^{N_A} dN_A = \int_0^\varepsilon d\varepsilon$$

$$-N_A\Big|_3^{N_A} = \varepsilon\Big|_0^\varepsilon$$

$$3 - N_A = \varepsilon$$

$$N_A = 3 - \varepsilon$$

The mole fraction of A is then

$$x_A = \frac{N_A}{N_t} = \frac{3 - \varepsilon}{7}$$

Following the same procedure for each component, at any given extent of reaction the number of moles of each component and the mole fraction are as given in the table shown.

component	number of moles	mole fraction
A	$3 - \varepsilon$	$\dfrac{3 - \varepsilon}{7}$
B	$2 - \varepsilon$	$\dfrac{2 - \varepsilon}{7}$
C	$1 + \varepsilon$	$\dfrac{1 + \varepsilon}{7}$
D	$1 + \varepsilon$	$\dfrac{1 + \varepsilon}{7}$
total	7	1

The extent of reaction is calculated from the equilibrium constant definition.

$$K_{eq} = \frac{x_C x_D}{x_A x_B}$$

$$1.45 = \frac{\left(\dfrac{1 + \varepsilon}{7}\right)\left(\dfrac{1 + \varepsilon}{7}\right)}{\left(\dfrac{3 - \varepsilon}{7}\right)\left(\dfrac{2 - \varepsilon}{7}\right)}$$

$$\varepsilon = \boxed{0.869}$$

The mole fractions at equilibrium are obtained by using this value for ε in the expressions tabulated above.

$$x_A = \frac{3 - \varepsilon}{7} = \frac{3 - 0.869}{7} = \boxed{0.304}$$

$$x_B = \frac{2 - \varepsilon}{7} = \frac{2 - 0.869}{7} = \boxed{0.162}$$

$$x_C = x_D = \frac{1 + \varepsilon}{7} = \frac{1 + 0.869}{7} = \boxed{0.267}$$

4. The mass balance and the energy balance around the batch reactor are solved simultaneously to determine the final temperature in the reactor. For a batch reactor,

$$\frac{dX_A}{d\theta} = \left(\frac{V}{N_{A_0}}\right)(-r_A)$$

$$m_t c_p \frac{dT}{d\theta} = V(-\Delta H_r)(-r_A) + qA_k$$

In an adiabatic reactor, the heat flux, q, is 0. Since the reaction is endothermic, ΔH_r is positive. The negative term, $-\Delta H_r$, indicates that the temperature will decrease during the reaction. Combining the two equations results in the following expression.

$$\frac{dT}{dX} = \frac{(-\Delta H_r)N_{A_0}}{m_t c_p}$$

Upon integration,

$$T = T_0 + \left(\frac{-\Delta H_r N_{A_0}}{m_t c_p}\right)(X_A - X_{A_0})$$

$$= 300°\text{C} + 273$$

$$+ \left(\frac{\left(-97\,000\ \dfrac{\text{kJ}}{\text{mol}}\right)(10.5\ \text{kmol})}{(975\ \text{kg})\left(2.5\ \dfrac{\text{kJ}}{\text{kg·K}}\right)}\right)(0.75 - 0.15)$$

$$= \boxed{322\text{K} \quad (49°\text{C})}$$

5. (a) The rate equation for the first-order consumption of A is

$$-\frac{dC_A}{d\tau} = k_1 C_A$$

In the integrated form,

$$k_1 \tau = \ln \frac{C_{A_0}}{C_A}$$

Solving for concentration,

$$C_A = C_{A_0} e^{-k_1 \tau}$$

Likewise, the rate expression for the first-order production of P is related to the amount of P produced from the reaction of A to form P, and the amount of P consumed to produce S.

$$\frac{dC_P}{d\tau} = k_1 C_A - k_2 C_P$$

$$= k_1 C_{A_0} e^{-k_1 \tau} - k_2 C_P$$

Integrating the above expression and recalling that $C_P = 0$ at $\tau = 0$,

$$\frac{C_P}{C_{A_0}} = \left(\frac{k_1}{k_2 - k_1}\right)(e^{-k_1 \tau} - e^{-k_2 \tau})$$

To obtain the reactor length that maximizes the production of product P, differentiate the above equation with respect to space time and set it equal to zero. Solving for space time yields

$$\tau_{\max} = \left(\frac{C_{A_0} V}{F_{A_0}}\right)_{\max} = \frac{\ln \dfrac{k_2}{k_1}}{k_2 - k_1} = \frac{\ln\left(\dfrac{12.5\ \text{h}^{-1}}{175\ \text{h}^{-1}}\right)}{12.5\ \text{h}^{-1} - 175\ \text{h}^{-1}}$$

$$= 0.0162\ \text{h}$$

Substituting $V = \pi(D^2/4)L$ into the previous equation and rearranging to solve for reactor length,

$$L = \tau_{\max}\left(\frac{F_{A_0}}{C_{A_0}}\right)\left(\frac{4}{\pi D^2}\right) = \tau_{\max} \dot{V}_{A_0}\left(\frac{4}{\pi D^2}\right)$$

$$= (0.0162\ \text{h})\left(0.15\ \frac{\text{m}^3}{\text{h}}\right)$$

$$\times \left(\frac{4}{\pi\left((10\ \text{cm})\left(\dfrac{1\ \text{m}}{100\ \text{cm}}\right)\right)^2}\right)$$

$$= \boxed{0.31\ \text{m}}$$

(b) The final concentrations of reactant A and of product P can be calculated directly from the integrated rate expressions derived in part (a).

$$C_A = C_{A_0} e^{-k_1 \tau} = \left(0.10 \ \frac{\text{kmol}}{\text{m}^3}\right) e^{-(175 \ \text{h}^{-1})(0.0162 \ \text{h})}$$

$$= \boxed{0.00587 \ \text{kmol/m}^3}$$

$$C_P = C_{A_0} \left(\frac{k_1}{k_2 - k_1}\right) \left(e^{-k_1 \tau} - e^{-k_2 \tau}\right)$$

$$= \left(0.10 \ \frac{\text{kmol}}{\text{m}^3}\right) \left(\frac{175 \ \text{h}^{-1}}{12.5 \ \text{h}^{-1} - 175 \ \text{h}^{-1}}\right)$$

$$\times \left(e^{-(175 \ \text{h}^{-1})(0.0162 \ \text{h})} - e^{-(12.5 \ \text{h}^{-1})(0.0162 \ \text{h})}\right)$$

$$= \boxed{0.0816 \ \text{kmol/m}^3}$$

The final concentration of side-product S is calculated from the material balance.

$$C_S = C_{A_0} - C_A - C_P$$

$$= 0.10 \ \frac{\text{kmol}}{\text{m}^3} - 0.00587 \ \frac{\text{kmol}}{\text{m}^3} - 0.0816 \ \frac{\text{kmol}}{\text{m}^3}$$

$$= \boxed{0.0125 \ \text{kmol/m}^3}$$

6. For a perfectly mixed flow reactor (CSTR), the differential mass balance for a constant-density system can be expressed as

$$C_{A_0} - C_A = \tau(-r_A)$$

Space time can be calculated from

$$\tau = \frac{V}{\dot{V}_{A_0}} = \frac{(2.5 \ \text{L})\left(\dfrac{1 \ \text{m}^3}{1000 \ \text{L}}\right)}{0.15 \ \dfrac{\text{m}^3}{\text{h}}} = 0.0167 \ \text{h}$$

The first-order rate equation for the consumption of A is

$$-r_A = -\frac{dC_A}{dt} = k_1 C_A$$

Substituting the rate expression into the mass balance,

$$C_{A_0} - C_A = \tau(k_1 C_A)$$

$$C_A = \frac{C_{A_0}}{1 + k_1 \tau}$$

$$= \frac{0.10 \ \dfrac{\text{kmol}}{\text{m}^3}}{1 + (175 \ \text{h}^{-1})(0.0167 \ \text{h})}$$

$$= \boxed{0.0255 \ \text{kmol/m}^3}$$

Similarly, the rate expression for the production of P is

$$r_P = \frac{dC_P}{dt} = k_1 C_A - k_2 C_P$$

Realizing that the feed concentration of product P is zero, a material balance yields

$$C_{P_0} - C_P = \tau(-r_P)$$

$$0 - C_P = \tau(-k_1 C_A + k_2 C_P)$$

$$C_P = \frac{\tau k_1 C_A}{1 + k_2 \tau}$$

$$= \frac{\tau k_1 C_{A_0}}{(1 + k_1 \tau)(1 + k_2 \tau)}$$

$$= \frac{(0.0167 \ \text{h})(175 \ \text{h}^{-1})\left(0.10 \ \dfrac{\text{kmol}}{\text{m}^3}\right)}{\left(1 + (175 \ \text{h}^{-1})(0.0167 \ \text{h})\right)}$$

$$\times \left(1 + (12.5 \ \text{h}^{-1})(0.0167 \ \text{h})\right)$$

$$= \boxed{0.0616 \ \text{kmol/m}^3}$$

The final concentration of side-product S is calculated from a material balance.

$$C_S = C_{A_0} - C_A - C_P$$

$$= 0.10 \ \frac{\text{kmol}}{\text{m}^3} - 0.0255 \ \frac{\text{kmol}}{\text{m}^3} - 0.0616 \ \frac{\text{kmol}}{\text{m}^3}$$

$$= \boxed{0.0129 \ \text{kmol/m}^3}$$

Comparison with the solution for Prob. 5(b) reveals that a plug-flow reactor has a higher yield of desired product P.

7. (a) From the problem statement, B is produced at 55 kmol/h with a 70% conversion ($X_A = 0.7$) of A to B. Therefore, the feed concentration of A is

$$F_{A_0} X_A = F_B$$

$$F_{A_0} = \frac{F_B}{X_A} = \frac{55 \ \dfrac{\text{kmol}}{\text{h}}}{0.7}$$

$$= 78.6 \ \text{kmol/h}$$

Conversion is a function of rate constant and space time for a single CSTR.

$$X_A = 1 - \frac{1}{1 + k\tau}$$

$$= 1 - \frac{1}{1 + k\left(\dfrac{C_{A_0} V}{F_{A_0}}\right)}$$

Rearranging to solve for reactor volume,

$$V = \left(\frac{1}{1-X_A} - 1\right)\left(\frac{F_{A_0}}{kC_{A_0}}\right)$$

$$= \left(\frac{1}{1-0.7} - 1\right)\left(\frac{78.6\ \frac{kmol}{h}}{(7.0\ h^{-1})\left(1\ \frac{kmol}{m^3}\right)}\right)$$

$$= \boxed{26.2\ m^3}$$

(b) For two CSTRs placed in series, the function for conversion is

$$X_{A_2} = 1 - \frac{1}{(1+k\tau)^2}$$

$$= 1 - \frac{1}{\left(1 + k\left(\frac{C_{A_0}V}{F_{A_0}}\right)\right)^2}$$

Rearranging to solve for the volume of one reactor,

$$V = \left(\frac{F_{A_0}}{kC_{A_0}}\right)\left(\sqrt{\frac{1}{1-X_{A_2}}} - 1\right)$$

$$= \left(\frac{78.6\ \frac{kmol}{h}}{(7.0\ h^{-1})\left(1\ \frac{kmol}{m^3}\right)}\right)\left(\sqrt{\frac{1}{1-0.7}} - 1\right)$$

$$= \boxed{9.27\ m^3}$$

The total reactor volume is twice that of a single reactor, or $(2)(9.27\ m^3) = 18.5\ m^3$.

Comparing this to the solution to part (a) indicates that two reactors in series are more efficient for this system than a single reactor.

48 Basic Chemical Plant Design

PRACTICE PROBLEMS

1. Component A and component B, both liquids, are mixed at feed conditions of 30°C and 1 bar. The reaction process follows the reaction A + B → C and requires two reaction vessels operating at the feed conditions.

The second reactor's effluent, also a liquid, must be cooled before entering a distillation column. The distillate from the column is 98 mol% component A and 2 mol% component B. The bottoms consists of 99 mol% C and a combined 1 mol% of A and B. The second reactor's effluent temperature is an excellent indicator of the desired conversion of components A and B to C. Increasing the concentration of component A in the second reactor enables the desired conversion of reactants to component C to occur. The distillate of the column is stored in a tank that is the source of component A to the second reactor, to achieve the desired conversion.

Construct a process flow sheet for this process. Indicate a temperature measuring device that would control the flow of component A from the storage tank.

2. A gaseous stream of components A and B is available at 25°C and 2 bars. Create a process flow sheet for the catalytic reaction of components A and B at 100°C and 4 bars, to produce component C.

Include in the diagram, a separation unit (represented as a column with distillate and bottoms) that operates at reactor temperature and pressure to separate unused reactants (distillate) and product C (bottoms). By design of the separation unit, product C leaves the column at 75°C and atmospheric pressure while the distillate leaves the column at 50°C and 2 bars and is recycled.

3. Design and show the flow sheet for a three-cycle cascade cooling tower system that uses turbine expanders and in which refrigeration cycle A cools refrigeration cycle B, which subsequently cools refrigeration cycle C.

4. Use the laboratory pilot-study data provided to design a reactor with a 35 000 L capacity. The fluid has a viscosity of 0.01 kg/m·s and specific gravity of 1.1.

laboratory data

vessel diameter	25 cm
impeller diameter	7.5 cm
impeller type	4-bladed, 45° pitched-blade turbine
liquid level	25 cm
baffle width	2 cm
number of baffles	4
impeller speed	690 rpm
vessel volume	13.5 L
power	7.0 W

5. Heat transfer in clean, round pipes follows the empirical correlation,

$$Nu = 0.0225 Re^{0.8} Pr^{0.4}$$

In a pilot-scale unit, a heat-transfer coefficient of 300 W/m²·K is measured for fluid flowing at 5 m/s through a 3 cm diameter pipe. Calculate the heat-transfer coefficient when the process is scaled up to a geometrically similar 15 cm diameter pipe, using the same fluid.

SOLUTIONS

1. The process flow sheet shown incorporates all the process requirements.

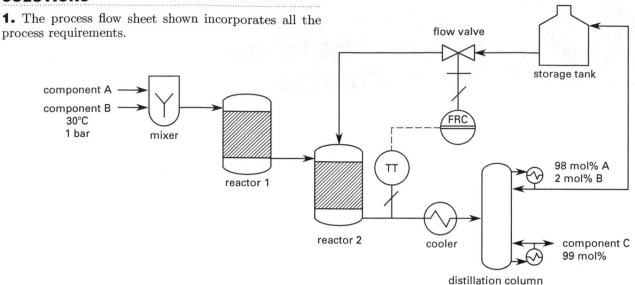

2.

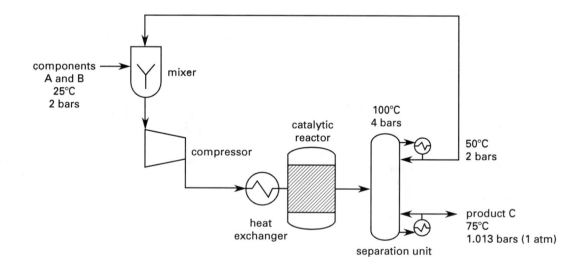

3. Cascade cooling towers are sometimes employed to obtain better cooling capacities and colder temperatures. A cascade system is a series of refrigeration cycles, one cooling the next.

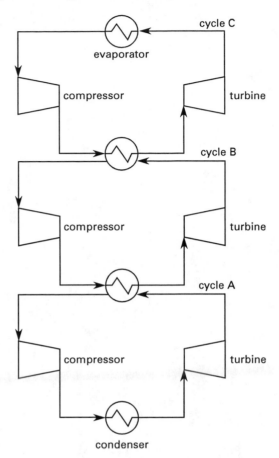

4. For geometric similarity, the number of baffles must be the same as the laboratory unit (4) and the impeller type must be the same (4-bladed, 45° pitched-blade turbine). The other geometric quantities are calculated with a scale factor (R) that is estimated from the reactor volumes.

$$R = \frac{D_f}{D_p} \approx \left(\frac{V_f}{V_p}\right)^{\frac{1}{3}} = \left(\frac{35\,000 \text{ L}}{13.5 \text{ L}}\right)^{\frac{1}{3}} = 13.7$$

The full-scale vessel diameter is

$$D_f = RD_p = (13.7)(25 \text{ cm})\left(\frac{1 \text{ m}}{100 \text{ cm}}\right) = 3.43 \text{ m}$$

The liquid level in the scaled-up reactor is

$$z_f = Rz_p = (13.7)(25 \text{ cm})\left(\frac{1 \text{ m}}{100 \text{ cm}}\right) = 3.43 \text{ m}$$

The new reactor baffle diameter is

$$B_f = RB_p = (13.7)(2 \text{ cm}) = 27.4 \text{ cm}$$

The scaled-up impeller diameter is

$$D_f = RD_p = (13.7)(7.5 \text{ cm})\left(\frac{1 \text{ m}}{100 \text{ cm}}\right) = 1.03 \text{ m}$$

The required impeller speed for the full-scale reactor is calculated by maintaining kinematic similarity.

$$n_f = n_p\left(\frac{D_p}{D_f}\right) = \frac{n_p}{R} = \frac{690 \frac{\text{rev}}{\text{min}}}{13.7} = 50.4 \text{ rpm}$$

The power number is held constant for dynamic similarity.

$$\text{Po}_f = \text{Po}_p$$

$$\frac{P_f}{\rho n_f^3 D_f^5} = \frac{P_p}{\rho n_p^3 D_p^5}$$

$$P_f = P_p\left(\frac{n_f}{n_p}\right)^3\left(\frac{D_f}{D_p}\right)^5 = P_p R^8$$

$$= (7.0 \text{ W})(13.7)^8 = 8.7 \times 10^9 \text{ W}$$

5. Maintaining a constant Reynolds number provides kinematic similarity.

$$\text{Re}_p = \text{Re}_f$$

$$\left(\frac{Dv\rho}{\mu}\right)_p = \left(\frac{Dv\rho}{\mu}\right)_f$$

$$v_f = v_p\left(\frac{D_p}{D_f}\right) = \left(5 \frac{\text{m}}{\text{s}}\right)\left(\frac{3 \text{ cm}}{15 \text{ cm}}\right) = 1 \text{ m/s}$$

The heat-transfer coefficient in the full-scale unit can be obtained by scaling the given design equation. The constant term and fluid properties will cancel.

$$\frac{\text{Nu}_p}{\text{Nu}_f} = \frac{0.0225\text{Re}_p^{0.8}\text{Pr}_p^{0.4}}{0.0225\text{Re}_f^{0.8}\text{Pr}_f^{0.4}} = \left(\frac{\text{Re}_p}{\text{Re}_f}\right)^{0.8}\left(\frac{\text{Pr}_p}{\text{Pr}_f}\right)^{0.4}$$

$$\frac{h_p D_p}{h_f D_f} = \left(\frac{v_p D_p}{v_f D_f}\right)^{0.8}$$

$$h_f = h_p\left(\frac{D_p}{D_f}\right)^{0.2}\left(\frac{v_f}{v_p}\right)^{0.8}$$

$$= \left(300 \frac{\text{W}}{\text{m}^2\cdot\text{K}}\right)\left(\frac{3 \text{ cm}}{15 \text{ cm}}\right)^{0.2}\left(\frac{1 \frac{\text{m}}{\text{s}}}{5 \frac{\text{m}}{\text{s}}}\right)^{0.8}$$

$$= \boxed{60 \text{ W/m}^2\cdot\text{K}}$$

49 Psychrometrics

PRACTICE PROBLEMS

1. A room contains air at 80°F (27°C) dry-bulb and 67°F (19°C) wet-bulb. The total pressure is 1 atm.

(a) What is the humidity?
- (A) 0.009 lbm/lbm (0.009 kg/kg)
- (B) 0.011 lbm/lbm (0.011 kg/kg)
- (C) 0.014 lbm/lbm (0.014 kg/kg)
- (D) 0.018 lbm/lbm (0.018 kg/kg)

(b) What is the enthalpy?
- (A) 30.2 Btu/lbm (51.3 kJ/kg)
- (B) 30.8 Btu/lbm (52.4 kJ/kg)
- (C) 31.5 Btu/lbm (53.9 kJ/kg)
- (D) 31.9 Btu/lbm (54.2 kJ/kg)

(c) What is the specific heat?
- (A) 0.234 Btu/lbm-°F (0.979 kJ/kg·K)
- (B) 0.237 Btu/lbm-°F (0.991 kJ/kg·K)
- (C) 0.239 Btu/lbm-°F (0.999 kJ/kg·K)
- (D) 0.242 Btu/lbm-°F (1.012 kJ/kg·K)

2. If one row of cooling coils bypasses one-third of the air passing through it, what is the theoretical bypass factor for four rows of identical cooling coils in series?
- (A) 0.01
- (B) 0.09
- (C) 0.33
- (D) 0.67

3. 1000 ft^3/min (0.5 m^3/s) of air at 50°F (10°C) dry-bulb and 95% relative humidity are mixed with 1500 ft^3/min (0.75 m^3/s) of air at 76°F (24°C) and 45% relative humidity.

(a) What is the dry-bulb temperature of the mixture?
- (A) 55°F (13°C)
- (B) 65°F (16°C)
- (C) 68°F (18°C)
- (D) 70°F (21°C)

(b) What is the specific humidity of the mixture?
- (A) 0.008 lbm/lbm (0.008 kg/kg)
- (B) 0.009 lbm/lbm (0.009 kg/kg)
- (C) 0.010 lbm/lbm (0.010 kg/kg)
- (D) 0.011 lbm/lbm (0.011 kg/kg)

(c) What is the dew point of the mixture?
- (A) 45°F (7.2°C)
- (B) 48°F (8.9°C)
- (C) 51°F (11°C)
- (D) 54°F (12°C)

4. Air at 60°F (16°C) dry-bulb and 45°F (7°C) wet-bulb passes through an air washer with a humidifying efficiency of 70%.

(a) What is the effective bypass factor of the system?
- (A) 0.30
- (B) 0.50
- (C) 0.67
- (D) 0.70

(b) What is the dry-bulb temperature of the air leaving the washer?
- (A) 45°F (7.2°C)
- (B) 50°F (9.7°C)
- (C) 54°F (12°C)
- (D) 57°F (14°C)

5. 95°F (35°C) dry-bulb, 75°F (24°C) wet-bulb air passes through a cooling tower and leaves at 85°F (29°C) dry-bulb and 90% relative humidity.

(a) What is the enthalpy change per cubic foot (meter) of air?
- (A) 0.57 Btu/ft^3 (18 kJ/m^3)
- (B) 1.3 Btu/ft^3 (40 kJ/m^3)
- (C) 3.2 Btu/ft^3 (99 kJ/m^3)
- (D) 7.6 Btu/ft^3 (240 kJ/m^3)

(b) What is the change in moisture content per cubic foot (meter) of air?
- (A) 1.8×10^{-4} lbm/ft^3 (2.7×10^{-3} kg/m^3)
- (B) 3.3×10^{-4} lbm/ft^3 (4.5×10^{-3} kg/m^3)
- (C) 6.7×10^{-4} lbm/ft^3 (9.9×10^{-3} kg/m^3)
- (D) 9.2×10^{-4} lbm/ft^3 (14×10^{-3} kg/m^3)

6. An air washer receives 1800 ft^3/min (0.85 m^3/s) of air at 70°F (21°C) and 40% relative humidity and discharges the air at 75% relative humidity. A recirculating water spray with a constant temperature of 50°F (10°C) is used.

(a) What will be the condition of the discharged air?

(b) What mass of makeup water is required per minute?

7. Repeat Probs. 6(a) and 6(b) using saturated steam at atmospheric pressure in place of the 50°F (10°C) water spray.

8. During performances, a theater experiences a sensible heat load of 500,000 Btu/hr (150 kW) and a moisture load of 175 lbm/hr (80 kg/h). Air enters the theater at 65°F (18°C) and 55% relative humidity and is removed when it reaches 75°F (24°C) or 60% relative humidity, whichever comes first.

(a) What is the ventilation rate in mass of air per hour?

(b) What are the conditions of the air leaving the theater?

9. 500 ft^3/min (0.25 m^3/s) of air at 80°F (27°C) dry-bulb and 70% relative humidity are removed from a room. 150 ft^3/min (0.075 m^3/s) pass through an air conditioner and leave saturated at 50°F (10°C). The remaining 350 ft^3/min (0.175 m^3/s) bypass the air conditioner and mix with the conditioned air at 1 atm.

(a) What is the mixture's temperature?
  (A) 55°F (12.8°C)
  (B) 66°F (18.8°C)
  (C) 71°F (21.9°C)
  (D) 74°F (23.3°C)

(b) What is the mixture's humidity ratio?
  (A) 0.010 lbm/lbm (1.0 g/kg)
  (B) 0.013 lbm/lbm (1.3 g/kg)
  (C) 0.017 lbm/lbm (1.7 g/kg)
  (D) 0.021 lbm/lbm (2.1 g/kg)

(c) What is the mixture's relative humidity?
  (A) 45%
  (B) 57%
  (C) 73%
  (D) 81%

(d) What is the heat load (in tons) of the air conditioner?
  (A) 0.9 ton
  (B) 1.3 ton
  (C) 2.4 tons
  (D) 2.9 tons

10. (*Time limit: one hour*) A dehumidifier takes 5000 ft^3/min (2.36 m^3/s) of air at 95°F (35°C) dry-bulb and 70% relative humidity and discharges it at 60°F (16°C) dry-bulb and 95% relative humidity. The dehumidifier uses a wet R-12 refrigeration cycle operating between 100°F (saturated) (38°C) and 50°F (10°C).

(a) Locate the air entering and leaving points on the psychrometric chart.

(b) Find the quantity of water removed from the air.

(c) Find the quantity of heat removed from the air.

(d) Draw the temperature-entropy and enthalpy-entropy diagrams for the refrigeration cycle.

(e) Find the temperature, pressure, enthalpy, entropy, and specific volume for each endpoint of the refrigeration cycle.

11. (*Time limit: one hour*) 1500 ft^3/min (0.71 m^3/s) of saturated 25 psia (170 kPa) air is heated from 200°F to 400°F (93°C to 204°C) in a constant pressure, constant moisture drying process.

(a) What is the final relative humidity?
  (A) 4.7%
  (B) 9.2%
  (C) 13%
  (D) 23%

(b) What is the final specific humidity?
  (A) 0.41 lbm/lbm (0.41 kg/kg)
  (B) 0.53 lbm/lbm (0.53 kg/kg)
  (C) 0.66 lbm/lbm (0.66 kg/kg)
  (D) 0.79 lbm/lbm (0.79 kg/kg)

(c) How much heat is required per unit mass of dry air?
  (A) 31 Btu/lbm (71 kJ/kg)
  (B) 57 Btu/lbm (130 kJ/kg)
  (C) 99 Btu/lbm (230 kJ/kg)
  (D) 120 Btu/lbm (280 kJ/kg)

(d) What is the final dew point?
  (A) 180°F (82°C)
  (B) 200°F (93°C)
  (C) 220°F (104°C)
  (D) 240°F (115°C)

12. (*Time limit: one hour*) 410 lbm/hr (0.052 kg/s) of dry 800°F (427°C) air pass through a scrubber to reduce particulate emissions. To protect the elastomeric seals in the scrubber, the air temperature is reduced to 350°F (177°C) by passing the air through a spray of 80°F (27°C) water. The pressure in the spray chamber is 20 psia (140 kPa).

(a) How much water is evaporated per hour?
  (A) 18 lbm/hr (0.0023 kg/s)
  (B) 27 lbm/hr (0.0035 kg/s)
  (C) 31 lbm/hr (0.0040 kg/s)
  (D) 39 lbm/hr (0.0050 kg/s)

(b) What will be the relative humidity of the air leaving the spray chamber?
  (A) 1.3%
  (B) 2.0%
  (C) 3.1%
  (D) 4.4%

13. (*Time limit: one hour*) An evaporative counter-flow air cooling tower removes 1×10^6 Btu/hr (290 kW) from a water flow. The temperature of the water is reduced from 120°F to 110°F (49°C to 43°C). Air enters the cooling tower at 91°F (33°C) and 60% relative humidity, and air leaves at 100°F (38°C) and 82% relative humidity.

(a) Calculate the air flow rate.

(b) Calculate the quantity of makeup water.

SOLUTIONS

1. *Customary U.S. Solution*

((a) and (b)) Locate the intersection of 80°F dry bulb and 67°F wet bulb on the psychrometric chart (App. 49.A). Read the value of humidity and enthalpy.

$$\omega = 0.0112 \text{ lbm moisture/lbm dry air}$$
$$h = 31.5 \text{ Btu/lbm dry air}$$

The answer is (B) for Prob. 1(a).

The answer is (C) for Prob. 1(b).

(c) c_p is gravimetrically weighted. c_p for air is 0.240 Btu/lbm-°F, and c_p for steam is approximately 0.40 Btu/lbm-°F.

$$G_{\text{air}} = \frac{1}{1 + 0.0112} = 0.989$$

$$G_{\text{steam}} = \frac{0.0112}{1 + 0.0112} = 0.011$$

$$c_{p,\text{mixture}} = G_{\text{air}}c_{p,\text{air}} + G_{\text{steam}}c_{p,\text{steam}}$$

$$= (0.989)\left(0.240 \; \frac{\text{Btu}}{\text{lbm-}°\text{F}}\right)$$

$$+ (0.011)\left(0.40 \; \frac{\text{Btu}}{\text{lbm-}°\text{F}}\right)$$

$$= \boxed{0.242 \text{ Btu/lbm-}°\text{F}}$$

The answer is (D).

SI Solution

((a) and (b)) Locate the intersection of 27°C dry bulb and 19°C wet bulb on the psychrometric chart (App. 49.B). Read the value of humidity and enthalpy.

$$\omega = \left(10.5 \; \frac{\text{g}}{\text{kg dry air}}\right)\left(\frac{1 \text{ kg}}{1000 \text{ g}}\right)$$

$$\boxed{\begin{array}{l} = 0.0105 \text{ kg/kg dry air} \\ h = 53.9 \text{ kJ/kg dry air} \end{array}}$$

The answer is (B) for Prob. 1(a).

The answer is (C) for Prob. 1(b).

(c) c_p is gravimetrically weighted. c_p for air is 1.0048 kJ/kg·K, and c_p for steam is approximately 1.675 kJ/kg·K.

$$G_{\text{air}} = \frac{1}{1 + 0.0105} = 0.990$$

$$G_{\text{steam}} = \frac{0.0105}{1 + 0.0105} = 0.010$$

$$c_{p,\text{mixture}} = G_{\text{air}}c_{p,\text{air}} + G_{\text{air}}c_{p,\text{steam}}$$

$$= (0.990)\left(1.0048 \ \frac{\text{kJ}}{\text{kg·K}}\right)$$

$$+ (0.010)\left(1.675 \ \frac{\text{kJ}}{\text{kg·K}}\right)$$

$$= \boxed{1.0115 \ \text{kJ/kg·K}}$$

The answer is (D).

2.
$$\text{BF}_{n \text{ rows}} = (\text{BF}_{1 \text{ row}})^n$$

$$= \left(\tfrac{1}{3}\right)^4 = \boxed{0.0123}$$

The answer is (A).

3.

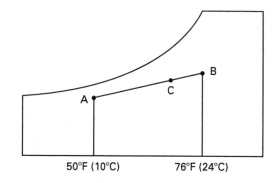

A   C   B

50°F (10°C)   76°F (24°C)

Customary U.S. Solution

(a) Locate the two points on the psychrometric chart and draw a line between them.

Reading from the chart (specific volumes),

$$v_A = 13.0 \ \text{ft}^3/\text{lbm}$$
$$v_B = 13.7 \ \text{ft}^3/\text{lbm}$$

The density at each point is

$$\rho_A = \frac{1}{v_A} = \frac{1}{13.0 \ \dfrac{\text{ft}^3}{\text{lbm}}} = 0.0769 \ \text{lbm/ft}^3$$

$$\rho_B = \frac{1}{v_B} = \frac{1}{13.7 \ \dfrac{\text{ft}^3}{\text{lbm}}} = 0.0730 \ \text{lbm/ft}^3$$

The mass flow at each point is

$$\dot{m}_A = \rho_A \dot{V}_A = \left(0.0769 \ \frac{\text{lbm}}{\text{ft}^3}\right)\left(1000 \ \frac{\text{ft}^3}{\text{min}}\right)$$

$$= 76.9 \ \text{lbm/min}$$

$$\dot{m}_B = \rho_B \dot{V}_B = \left(0.0730 \ \frac{\text{lbm}}{\text{ft}^3}\right)\left(1500 \ \frac{\text{ft}^3}{\text{min}}\right)$$

$$= 109.5 \ \text{lbm/min}$$

The gravimetric fraction of flow A is

$$\frac{76.9 \ \dfrac{\text{lbm}}{\text{min}}}{76.9 \ \dfrac{\text{lbm}}{\text{min}} + 109.5 \ \dfrac{\text{lbm}}{\text{min}}} = 0.413$$

Since the scales are all linear,

$$0.413 = \frac{T_B - T_C}{T_B - T_A}$$

$$T_C = T_B - (0.413)(T_B - T_A)$$

$$= 76°F - (0.413)(76°F - 50°F)$$

$$= \boxed{65.3°F}$$

The answer is (B).

(b) $\omega = \boxed{0.0082 \ \text{lbm moisture/lbm dry air}}$

The answer is (A).

(c) $T_{\text{dp}} = \boxed{51°F}$

The answer is (C).

SI Solution

(a) Locate the two points on the psychrometric chart and draw a line between them.

Reading from the chart (specific volumes),

$$v_A = 0.813 \ \text{m}^3/\text{kg dry air}$$
$$v_B = 0.856 \ \text{m}^3/\text{kg dry air}$$

The density at each point is

$$\rho_A = \frac{1}{v_A} = \frac{1}{0.813 \ \dfrac{\text{m}^3}{\text{kg}}} = 1.23 \ \text{kg/m}^3$$

$$\rho_B = \frac{1}{v_B} = \frac{1}{0.856 \ \dfrac{\text{m}^3}{\text{kg}}} = 1.17 \ \text{kg/m}^3$$

The mass flow at each point is

$$\dot{m}_A = \rho_A \dot{V}_A = \left(1.23 \ \frac{\text{kg}}{\text{m}^3}\right)\left(0.5 \ \frac{\text{m}^3}{\text{s}}\right) = 0.615 \ \text{kg/s}$$

$$\dot{m}_B = \rho_B \dot{V}_B = \left(1.17 \ \frac{\text{kg}}{\text{m}^3}\right)\left(0.75 \ \frac{\text{m}^3}{\text{s}}\right) = 0.878 \ \text{kg/s}$$

The gravimetric fraction of flow A is

$$\frac{0.615 \ \frac{kg}{s}}{0.615 \ \frac{kg}{s} + 0.878 \ \frac{kg}{s}} = 0.412$$

The answer is (B).

Since the scales are linear,

$$0.412 = \frac{T_B - T_C}{T_B - T_A}$$
$$T_C = T_B - (0.412)(T_B - T_A)$$
$$= 24°C - (0.412)(24°C - 10°C)$$
$$= \boxed{18.2°C}$$

(b) $\qquad \omega = \left(8.0 \ \frac{g}{kg \ dry \ air}\right)\left(\frac{1 \ kg}{1000 \ g}\right)$

$$= \boxed{0.008 \ kg/kg \ dry \ air}$$

The answer is (A).

(c) $\qquad\qquad T_{dp} = \boxed{10.6°C}$

The answer is (C).

4. *Customary U.S. Solution*

(a) From Eq. 49.24, the bypass factor is

$$BF = 1 - \eta_{sat}$$

$$= 1 - 0.70 = \boxed{0.30}$$

The answer is (D).

(b) From Eq. 49.34, the dry-bulb temperature of air leaving the washer can be determined.

$$\eta_{sat} = \frac{T_{db,in} - T_{db,out}}{T_{db,in} - T_{wb,in}}$$
$$0.70 = \frac{60°F - T_{db,out}}{60°F - 45°F}$$
$$T_{db,out} = \boxed{49.5°F}$$

The answer is (B).

SI Solution

(a) See the customary U.S. solution.

(b) From Eq. 49.34, the dry-bulb temperature of air leaving the washer can be determined.

$$\eta_{sat} = \frac{T_{db,in} - T_{db,out}}{T_{db,in} - T_{wb,in}}$$
$$0.70 = \frac{16°C - T_{db,out}}{16°C - 7°C}$$
$$T_{db,out} = \boxed{9.7°C}$$

The answer is (B).

5. *Customary U.S. Solution*

(a) Refer to the psychrometric chart (App. 49.A).

At point 1, properties of air at $T_{db} = 95°F$ and $T_{wb} = 75°F$ are

$$\omega_1 = 0.0141 \ lbm \ moisture/lbm \ air$$
$$h_1 = 38.4 \ lbm \ air$$
$$v_1 = 14.3 \ ft^3/lbm \ air$$

At point 2, properties of air at $T_{db} = 85°F$ and 90% relative humidity are

$$\omega_2 = 0.0237 \ lbm \ moisture/lbm \ air$$
$$h_2 = 46.6 \ Btu/lbm \ air$$

The enthalpy change is

$$\frac{h_2 - h_1}{v_1} = \frac{46.6 \ \frac{Btu}{lbm \ air} - 38.4 \ \frac{Btu}{lbm \ air}}{14.3 \ \frac{ft^3}{lbm \ air}}$$

$$= \boxed{0.573 \ Btu/ft^3 \ air}$$

The answer is (A).

(b) The moisture added is

$$\frac{\omega_2 - \omega_1}{v_1} = \frac{0.0237 \ \frac{lbm \ moisture}{lbm \ air} - 0.0141 \ \frac{lbm \ moisture}{lbm \ air}}{14.3 \ \frac{ft^3}{lbm \ air}}$$

$$= \boxed{6.71 \times 10^{-4} \ lbm/ft^3 \ air}$$

The answer is (C).

Plant Design

SI Solution

(a) Refer to the psychrometric chart (App. 49.B).

At point 1, properties of air at $T_{db} = 35°C$ and $T_{wb} = 24°C$ are

$$\omega_1 = 14.3 \text{ g/kg air}$$
$$h_1 = 71.8 \text{ kJ/kg air}$$
$$v_1 = 0.8893 \text{ m}^3/\text{kg air}$$

At point 2, properties of air at $T_{db} = 29°C$ and 90% relative humidity are

$$\omega_2 = 23.1 \text{ g/kg air}$$
$$h_2 = 88 \text{ kJ/kg air}$$

The enthalpy change is

$$\frac{h_2 - h_1}{v_1} = \frac{88 \frac{\text{kJ}}{\text{kg air}} - 71.8 \frac{\text{kJ}}{\text{kg air}}}{0.8893 \frac{\text{m}^3}{\text{kg air}}}$$

$$= \boxed{18.2 \text{ kJ/m}^3 \text{ air}}$$

The answer is (A).

(b) The moisture added is

$$\frac{\omega_2 - \omega_1}{v_1} = \frac{\left(\left(23.1 \frac{\text{g}}{\text{kg air}}\right) - \left(14.3 \frac{\text{kg}}{\text{kg air}}\right)\right) \times \left(\frac{1 \text{ kg}}{1000 \text{ g}}\right)}{0.8893 \frac{\text{m}^3}{\text{kg air}}}$$

$$= \boxed{9.90 \times 10^{-3} \text{ kg/m}^3 \text{ air}}$$

The answer is (C).

6.

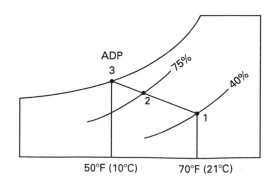

Customary U.S. Solution

(a) Refer to the psychrometric chart (App. 49.A).

At point 1, properties of air at $T_{db} = 70°F$ and $\phi = 40\%$ are

$$h_1 = 23.6 \text{ Btu/lbm air}$$
$$\omega_1 = 0.00623 \text{ lbm moisture/lbm air}$$
$$v_1 = 13.48 \text{ ft}^3/\text{lbm air}$$

The mass flow rate of incoming air is

$$\dot{m}_{a,1} = \frac{\dot{V}_1}{v_1} = \frac{1800 \frac{\text{ft}^3}{\text{min}}}{13.48 \frac{\text{ft}^3}{\text{lbm air}}} = 133.53 \text{ lbm air/min}$$

Locate point 1 on the psychrometric chart.

Notice that the temperature of the recirculating water is constant but not equal to the air's entering wet-bulb temperature. Therefore, this is not an adiabatic process.

Locate point 3 as 50°F saturated condition (water being sprayed) on the psychrometric chart.

Draw a line from point 1 to point 3. The intersection of this line with 75% relative humidity defines point 2 as

$$\boxed{\begin{array}{l} h_2 = 21.4 \text{ Btu/lbm air} \\ \omega_2 = 0.0072 \text{ lbm moisture/lbm air} \\ T_{db,2} = 56°F \\ T_{wb,2} = 51.8°F \end{array}}$$

(b) The moisture (water) added is

$$\dot{m}_w = \dot{m}_{a,1}(\omega_2 - \omega_1)$$
$$= \left(133.53 \frac{\text{lbm air}}{\text{min}}\right) \left(0.0072 \frac{\text{lbm moisture}}{\text{lbm air}}\right.$$
$$\left. - 0.00625 \frac{\text{lbm moisture}}{\text{lbm air}}\right)$$
$$= \boxed{0.127 \text{ lbm/min}}$$

SI Solution

(a) Refer to the psychrometric chart (App. 49.B).

At point 1, properties of air at $T_{db} = 21°C$ and $\phi = 40\%$ are

$$h_1 = 36.75 \text{ kJ/kg air}$$
$$\omega_1 = 6.2 \text{ g moisture/kg air}$$
$$v_1 = 0.842 \text{ m}^3/\text{kg air}$$

The mass flow rate of incoming air is

$$\dot{m}_{a,1} = \frac{\dot{V}_1}{v_1} = \frac{0.85 \frac{\text{m}^3}{\text{s}}}{0.842 \frac{\text{m}^3}{\text{kg air}}} = 1.010 \text{ kg air/s}$$

Locate point 1 on the psychrometric chart.

Notice that the temperature of the recirculating water is constant but not equal to the air's entering wet-bulb temperature. Therefore, this is not an adiabatic process.

Locate point 3 as 10°C saturated condition (water being sprayed) on the psychrometric chart.

Draw a line from point 1 to point 3. The intersection of this line with 75% relative humidity defines point 2 as

$$\boxed{\begin{aligned} T_{2,\text{db}} &= 14.7°C \\ h_2 &= 34.1 \text{ kJ/kg air} \\ \omega_2 &= 7.6 \text{ g moisture/kg air} \end{aligned}}$$

(b) The water added is

$$\dot{m}_w = \dot{m}_{a,1}(\omega_2 - \omega_1)$$
$$= \left(1.010 \frac{\text{kg air}}{\text{s}}\right)\left(7.6 \frac{\text{g moisture}}{\text{kg air}} - 6.2 \frac{\text{g moisture}}{\text{kg air}}\right)$$
$$\times \left(\frac{1 \text{ kg}}{1000 \text{ g}}\right)$$
$$= \boxed{0.00141 \text{ kg/s}}$$

7. *Customary U.S. Solution*

((a) and (b)) From Prob. 6,

$$\omega_1 = 0.00623 \text{ lbm moisture/lbm air}$$
$$h_1 = 23.6 \text{ Btu/lbm air}$$
$$\dot{m}_{a,1} = 133.53 \text{ lbm air/min}$$

From the steam table (App. 23.B) for 1 atm steam, $h_{\text{steam}} = 1150.5$ Btu/lbm.

From the conservation of energy equation (Eq. 49.32),

$$\dot{m}_{a,1}h_1 + \dot{m}_{\text{steam}}h_{\text{steam}} = \dot{m}_{a,1}h_2$$
$$\left(133.53 \frac{\text{lbm air}}{\text{min}}\right)\left(23.6 \frac{\text{Btu}}{\text{lbm air}}\right)$$
$$+ \dot{m}_{\text{steam}}\left(1150.5 \frac{\text{Btu}}{\text{lbm}}\right) = \left(133.53 \frac{\text{lbm air}}{\text{min}}\right)h_2$$
$$\text{[Eq. 1]}$$

From conservation of mass for the water (Eq. 49.33),

$$\dot{m}_{a,1}\omega_1 + \dot{m}_{\text{steam}} = \dot{m}_{a,2}\omega_2$$
$$\left(133.53 \frac{\text{lbm air}}{\text{min}}\right)\left(0.00623 \frac{\text{lbm moisture}}{\text{lbm air}}\right)$$
$$+ \dot{m}_{\text{steam}} = \left(133.53 \frac{\text{lbm air}}{\text{min}}\right)\omega_2 \quad \text{[Eq. 2]}$$

Since no single relationship exists between ω_2, $\dot{m}_{\text{steam}}$, and h_2, a trial-and-error solution is required. Once $\dot{m}_{\text{steam}}$ is selected, ω_2 and h_2 can be found from Eq. 1 and Eq. 2 as

$$h_2 = 8.616\dot{m}_{\text{steam}} + 23.6$$
$$\omega_2 = 0.00623 + 0.00749\dot{m}_{\text{steam}}$$

Once h_2 and ω_2 are known, the relative humidity can be determined from the psychrometric chart. Continue the process until a relative humidity of 75% is achieved.

$\dot{m}_{\text{steam}}$ $\left(\dfrac{\text{lbm}}{\text{min}}\right)$	ω_2 $\left(\dfrac{\text{lbm moisture}}{\text{lbm air}}\right)$	h_2 $\left(\dfrac{\text{Btu}}{\text{lbm air}}\right)$	ϕ_2 (%)
0.3	0.00848	26.18	53
0.4	0.00923	27.05	58
0.5	0.00998	27.91	62
0.6	0.01072	28.77	66
0.7	0.01147	29.63	70
0.8	0.01222	30.49	74
0.82	0.01237	30.67	74.8
0.83	0.01245	30.75	75.3
0.825	0.01241	30.71	75.0

$$\boxed{\begin{aligned} \dot{m}_{\text{steam}} &= 0.825 \text{ lbm/min} \\ \omega_2 &= 0.01241 \text{ lbm moisture/lbm air} \\ h_2 &= 30.71 \text{ Btu/lbm air} \\ T_{\text{db}} &= 71.5°F \\ T_{\text{wb}} &= 65.9°F \end{aligned}}$$

SI Solution

((a) and (b)) From Prob. 6,

$$\omega_1 = 6.2 \text{ g moisture/kg air}$$
$$h_1 = 36.75 \text{ kJ/kg air}$$
$$\dot{m}_{a,1} = 1.010 \text{ kg air/s}$$

From the steam table (App. 23.O), for 1 atm steam, $h_{\text{steam}} = 2675.5$ kJ/kg.

From the conservation of energy equation (Eq. 49.32),

$$\dot{m}_{a,1}h_1 + \dot{m}_{\text{steam}}h_{\text{steam}} = \dot{m}_{a,1}h_2$$
$$\left(1.010 \frac{\text{kg air}}{\text{s}}\right)\left(36.75 \frac{\text{kJ}}{\text{kg air}}\right)$$
$$+ \dot{m}_{\text{steam}}\left(2675.5 \frac{\text{kJ}}{\text{kg}}\right) = \left(1.010 \frac{\text{kg air}}{\text{s}}\right)h_2$$
$$\text{[Eq. 1]}$$

Plant Design

From conservation of mass of water (Eq. 49.33),

$$\dot{m}_{a,1}\omega_1 + \dot{m}_{steam} = \dot{m}_{a,2}\omega_2$$

$$\left(1.010 \ \frac{kg \ air}{s}\right)\left(6.2 \ \frac{g \ moisture}{kg \ air}\right)\left(\frac{1 \ kg}{1000 \ kg}\right)$$

$$+\dot{m}_{steam} = \left(1.010 \ \frac{kg \ air}{s}\right)\omega_2$$

[Eq. 2]

Since no single relationship exists between ω_2, $\dot{m}_{steam}$, and h_2, a trial-and-error solution is required. Once $\dot{m}_{steam}$ is selected, ω_2 and h_2 can be found from Eq. 1 and Eq. 2 as

$$h_2 = 36.75 + 2649.0\dot{m}_{steam}$$

$$\omega_2 = 0.0062 + 0.99\dot{m}_{steam}$$

Once h_2 and ω_2 are known, the relative humidity can be determined from the psychrometric chart. Continue the process until a relative humidity of 75% is achieved.

$\dot{m}_{steam}$ $\left(\dfrac{kg}{s}\right)$	ω_2 $\left(\dfrac{kg \ moisture}{kg \ air}\right)$	h_2 $\left(\dfrac{kJ}{kg \ air}\right)$	ϕ_2 (%)
0.005	0.0112	50.00	69.5
0.0055	0.0116	51.32	74.5
0.0056	0.0117	51.58	75.0

$\dot{m}_{steam} = 0.0056 \ kg/s$

$$\omega_2 = \left(0.0117 \ \frac{kg \ moisture}{kg \ air}\right)\left(1000 \ \frac{g}{kg}\right)$$

$$= 11.7 \ g \ moisture/kg \ air$$

$$T_{db} = 21.3°C$$

$$T_{wb} = 18.2°C$$

8. *Customary U.S. Solution*

(a) From the psychrometric chart (App. 49.A), for incoming air at 65°F and 55% relative humidity, $\omega_1 = 0.0072$ lbm moisture/lbm air.

With sensible heating as a limiting factor, calculate the mass flow rate of air entering the theater from Eq. 49.26 (ventilation rate).

$$\dot{q} = \dot{m}_a(c_{p,air} + \omega c_{p,moisture})(T_2 - T_1)$$

$$500,000 \ \frac{Btu}{hr} = \dot{m}_a \left(0.240 \ \frac{Btu}{lbm\text{-}°F}\right.$$

$$+ \left(0.0072 \ \frac{lbm \ moisture}{lbm \ air}\right)$$

$$\times \left. \left(0.444 \ \frac{Btu}{lbm\text{-}°F}\right)\right)(75°F - 65°F)$$

$$\dot{m}_a = \boxed{2.056 \times 10^5 \ lbm \ air/hr}$$

(b) Assume that this air absorbs all the moisture. Then, the final humidity ratio is given by

$$\dot{m}_w = \dot{m}_a(\omega_2 - \omega_1)$$

$$\omega_2 = \left(\frac{\dot{m}_w}{\dot{m}_a}\right) + \omega_1$$

$$= \frac{175 \ \dfrac{lbm \ moisture}{hr}}{2.056 \times 10^5 \ \dfrac{lbm \ air}{hr}} + 0.0072 \ \frac{lbm \ moisture}{lbm \ air}$$

$$= 0.00805 \ lbm \ moisture/lbm \ air$$

The final conditions are

$$\boxed{\begin{array}{l} T_{db} = 75°F \quad [given] \\ \omega_2 = 0.00805 \ lbm \ moisture/lbm \ air \end{array}}$$

From the psychrometric chart (App. 49.A), the relative humidity is 44%. This is below 60%.

SI Solution

(a) From the psychrometric chart (App. 49.B), for incoming air at 18°C and 55% relative humidity, $\omega_1 = 7.1$ g moisture/kg air.

With sensible heating as a limiting factor, calculate the mass flow rate of air entering the theater from Eq. 49.26 (ventilation rate).

$$\dot{q} = \dot{m}_a(c_{p,air} + \omega c_{p,moisture})$$

$$\times (T_2 - T_1)$$

$$(150 \ kW)\left(1000 \ \frac{W}{kW}\right) = \dot{m}_a\left(\left(1.005 \ \frac{kJ}{kg\cdot°C}\right)\left(1000 \ \frac{J}{kJ}\right)\right.$$

$$+ \left(1.805 \ \frac{kJ}{kg\cdot°C}\right)\left(1000 \ \frac{J}{kJ}\right)$$

$$\times \left. \left(7.1 \ \frac{g \ moisture}{kg \ air}\right)\left(\frac{1 \ kg}{1000 \ g}\right)\right)$$

$$\times (24°C - 18°C)$$

$$\dot{m}_a = \boxed{24.56 \ kg/s}$$

(b) Assume that this air absorbs all the moisture. Then, the final humidity ratio is given by

$$\dot{m}_w = \dot{m}_a(\omega_2 - \omega_1)$$

$$\omega_2 = \frac{\dot{m}_w}{\dot{m}_a} + \omega_1$$

$$= \frac{\left(80 \ \dfrac{\text{kg}}{\text{h}}\right)\left(\dfrac{1 \ \text{h}}{3600 \ \text{s}}\right)}{24.56 \ \dfrac{\text{kg}}{\text{s}}}$$

$$+ \left(7.0 \ \frac{\text{g moisture}}{\text{kg air}}\right)\left(\frac{1 \ \text{kg}}{1000 \ \text{g}}\right)$$

$$= 0.00790 \ \text{kg moisture/kg air}$$

The final conditions are

$$\boxed{\begin{aligned} T_{db} &= 24°C \\ \omega_2 &= \left(0.00790 \ \frac{\text{kg moisture}}{\text{kg air}}\right)\left(1000 \ \frac{\text{g}}{\text{kg}}\right) \\ &= 7.9 \ \text{g moisture/kg air} \end{aligned}}$$

From the psychrometric chart (App. 49.B), the relative humidity is 44%. This is below 60%.

9.

$T_{db} = 80°F \ (27°C)$
$\phi = 70\%$
$\dot{V} = 500 \ \text{ft}^3/\text{min} \ (0.25 \ \text{m}^3/\text{s})$

$\dot{V}_1 = 150 \ \text{ft}^3/\text{min}$
$(0.075 \ \text{m}^3/\text{s})$

$\dot{V}_2 = 350 \ \text{ft}^3/\text{min}$
$(0.475 \ \text{m}^3/\text{s})$

(out)

room

$50°F$
$(10°C)$

air conditioner

(in)

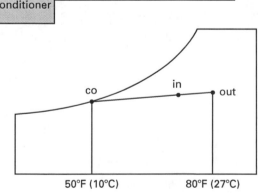

50°F (10°C)   80°F (27°C)

Customary U.S. Solution

Locate point "out" ($T_{db} = 80°F$ and $\phi = 70\%$) and point "co" (saturated at 50°F) on the psychrometric chart. At point "out" from App. 49.A,

$$v_{out} = 13.95 \ \text{ft}^3/\text{lbm air}$$

$$h_{out} = 36.2 \ \text{Btu/lbm air}$$

At point "co," $h_{co} = 20.3$ Btu/lbm air.

The air mass flow rate through the air conditioner is

$$\dot{m}_1 = \frac{\dot{V}_1}{v_1} = \frac{\dot{V}_1}{v_{out}} = \frac{150 \ \dfrac{\text{ft}^3}{\text{min}}}{13.95 \ \dfrac{\text{ft}^3}{\text{lbm air}}}$$

$$= 10.75 \ \text{lbm air/min}$$

The mass flow rate of the bypass air is

$$\dot{m}_2 = \frac{\dot{V}_2}{v} = \frac{350 \ \dfrac{\text{ft}^3}{\text{min}}}{13.95 \ \dfrac{\text{ft}^3}{\text{lbm air}}} = 25.09 \ \text{lbm air/min}$$

The percentage of bypass air is

$$x = \frac{25.09 \ \dfrac{\text{lbm air}}{\text{min}}}{10.75 \ \dfrac{\text{lbm air}}{\text{min}} + 25.09 \ \dfrac{\text{lbm air}}{\text{min}}} = 0.70 \quad (70\%)$$

Using the lever rule and the fact that all of the temperature scales are linear,

$$\begin{aligned} T_{db,in} &= T_{co} + (0.70)(T_{out} - T_{co}) \\ &= 50°F + (0.70)(80°F - 50°F) \\ &= 71°F \end{aligned}$$

At that point,

(a) $$\boxed{T_{db,in} = 71°F}$$

The answer is (C) for Prob. 9(a).

(b) $$\boxed{\omega_{in} = 0.0132 \ \text{lbm moisture/lbm air}}$$

The answer is (B) for Prob. 9(b).

(c) $$\boxed{\phi_{in} = 81\%}$$

The answer is (D) for Prob. 9(c).

(d) The air conditioner capacity is given by

$$\dot{Q} = \dot{m}_{air}(h_{t,2} - h_{t,1}) = \dot{m}_1(h_{out} - h_{co})$$

$$= \left(10.75 \ \frac{\text{lbm air}}{\text{min}}\right)\left(36.2 \ \frac{\text{Btu}}{\text{lbm air}} - 20.3 \ \frac{\text{Btu}}{\text{lbm air}}\right)$$

$$\times \left(\frac{1 \ \text{ton}}{200 \ \dfrac{\text{Btu}}{\text{min}}}\right)$$

$$= \boxed{0.85 \ \text{ton}}$$

The answer is (A).

SI Solution

Locate point "out" ($T_{db} = 27°C$, $\phi = 70\%$) and point "co" (saturated at $10°C$) on the psychrometric chart. At point "out" from App. 49.B,

$$v_{out} = 0.872 \text{ m}^3/\text{kg air}$$
$$h_{out} = 67.3 \text{ kJ/kg air}$$

At point "co" from App. 49.B, $h_{co} = 29.26$ kJ/kg air.

At mass flow rate through the air conditioner,

$$\dot{m}_1 = \frac{\dot{V}_1}{v} = \frac{0.075 \dfrac{\text{m}^3}{\text{s}}}{0.872 \dfrac{\text{m}^3}{\text{kg air}}} = 0.0860 \text{ kg air/s}$$

The flow rate of bypass air is

$$\dot{m}_2 = \frac{\dot{V}_2}{v} = \frac{0.175 \dfrac{\text{m}^3}{\text{s}}}{0.872 \dfrac{\text{m}^3}{\text{kg air}}} = 0.2007 \text{ kg air/s}$$

The percentage bypass air is

$$x = \frac{0.2007 \dfrac{\text{kg air}}{\text{s}}}{0.0860 \dfrac{\text{kg air}}{\text{s}} + 0.2007 \dfrac{\text{kg air}}{\text{s}}}$$
$$= 0.70 \ (70\%)$$

Using the lever rule and the fact that all of the temperature scales are linear,

$$T_{db,in} = T_{co} + (0.70)(T_{out} - T_{co})$$
$$= 10°C + (0.70)(27°C - 10°C)$$
$$= 21.9°C$$

At that point,

(a) $$\boxed{T_{db,in} = 21.9°C}$$

The answer is (C) for Prob. 9(a).

(b) $$\boxed{\omega_{in} = 13.4 \text{ g moisture/kg air}}$$

The answer is (B) for Prob. 9(b).

(c) $$\boxed{\phi_{in} = 81\%}$$

The answer is (D) for Prob. 9(c).

(d) The air conditioner capacity is given by

$$\dot{Q} = \dot{m}_{air}(h_{t,2} - h_{t,1}) = \dot{m}_1(h_{out} - h_{co})$$
$$= \left(0.0860 \ \frac{\text{kg air}}{\text{s}}\right)\left(67.3 \ \frac{\text{kJ}}{\text{kg air}} - 29.26 \ \frac{\text{kJ}}{\text{kg air}}\right)$$
$$\times \left(0.2843 \ \frac{\text{ton}}{\text{kW}}\right)$$
$$= \boxed{0.93 \text{ ton}}$$

The answer is (A).

10.

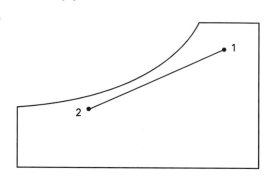

Customary U.S. Solution

(a) At point 1, from the psychrometric chart (App. 49.A) at $T_{db} = 95°F$ and $\phi = 70\%$,

$$\boxed{\begin{aligned} h_1 &= 50.7 \text{ Btu/lbm air} \\ v_1 &= 14.56 \text{ ft}^3/\text{lbm air} \\ \omega_1 &= 0.0253 \text{ lbm water/lbm air} \end{aligned}}$$

At point 2, from the psychrometric chart (App. 49.A) at $T_{db} = 60°F$ and $\phi = 95\%$,

$$\boxed{\begin{aligned} h_2 &= 25.8 \text{ Btu/lbm air} \\ \omega_2 &= 0.0105 \text{ lbm water/lbm air} \end{aligned}}$$

(b) The air mass flow rate is

$$\dot{m}_a = \frac{\dot{V}}{v_1} = \frac{5000 \ \dfrac{\text{ft}^3}{\text{min}}}{14.56 \ \dfrac{\text{ft}^3}{\text{lbm air}}} = 343.4 \text{ lbm air/min}$$

From Eq. 49.27, the water removed is

$$\dot{m}_w = \dot{m}_a(\omega_1 - \omega_2)$$
$$= \left(343.4 \ \frac{\text{lbm air}}{\text{min}}\right)$$
$$\times \left(0.0253 \ \frac{\text{lbm water}}{\text{lbm air}} - 0.0105 \ \frac{\text{lbm water}}{\text{lbm air}}\right)$$
$$= \boxed{5.08 \text{ lbm water/min}}$$

(c) From Eq. 49.28, the quantity of heat removed is

$$\dot{q} = \dot{m}_a(h_1 - h_2)$$
$$= \left(343.4 \; \frac{\text{lbm air}}{\text{min}}\right)\left(50.7 \; \frac{\text{Btu}}{\text{lbm air}} - 25.8 \; \frac{\text{Btu}}{\text{lbm air}}\right)$$
$$= \boxed{8551 \; \text{Btu/min}}$$

(d) Considering an R-12 refrigeration cycle operating at saturated condition at 100°F, the T-s and h-s diagrams are as follows.

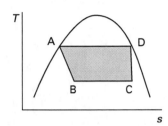
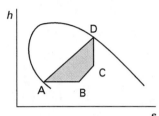

Use App. 23.G for saturated conditions.

(e) At A,

$$T = 100°\text{F} \quad \text{[given]}$$
$$p = p_\text{sat} \text{ at } 100°\text{F} = 131.6 \text{ psia}$$
$$h_\text{A} = h_f \text{ at } 100°\text{F} = 31.16 \text{ Btu/lbm}$$
$$s_\text{A} = s_f \text{ at } 100°\text{F} = 0.06316 \text{ Btu/lbm-}°\text{R}$$
$$v_\text{A} = v_f \text{ at } 100°\text{F} = 0.0127 \text{ ft}^3/\text{lbm}$$

At point B,

$$T = 50°\text{F} \quad \text{[given]}$$
$$p = p_\text{sat} \text{ at } 50°\text{F} = 61.39 \text{ psia}$$
$$h_\text{B} = h_a = 31.16 \text{ Btu/lbm}$$
$$h_{f,\text{B}} = 19.27 \text{ Btu/lbm}$$
$$h_{fg,\text{B}} = 64.51 \text{ Btu/lbm}$$
$$x = \frac{h_\text{B} - h_{f,\text{B}}}{h_{fg,\text{B}}}$$
$$= \frac{31.16 \; \frac{\text{Btu}}{\text{lbm}} - 19.27 \; \frac{\text{Btu}}{\text{lbm}}}{64.51 \; \frac{\text{Btu}}{\text{lbm}}}$$
$$= 0.184$$
$$s_\text{B} = s_{f,\text{B}} + x s_{fg,\text{B}}$$
$$= 0.04126 \; \frac{\text{Btu}}{\text{lbm-}°\text{R}}$$
$$\quad + (0.184)\left(0.12659 \; \frac{\text{Btu}}{\text{lbm-}°\text{R}}\right)$$
$$= 0.06455 \text{ Btu/lbm-}°\text{R}$$

$$v_\text{B} = v_{f,\text{B}} + x(v_g - v_f)$$
$$= 0.0118 \; \frac{\text{ft}^3}{\text{lbm}}$$
$$\quad + (0.184)\left(0.673 \; \frac{\text{ft}^3}{\text{lbm}} - 0.0118 \; \frac{\text{ft}^3}{\text{lbm}}\right)$$
$$= 0.1335 \text{ ft}^3/\text{lbm}$$

At point D,

$$T = 100°\text{F} \quad \text{[given]}$$
$$p = p_\text{sat} \text{ at } 100°\text{F} = 131.6 \text{ psia}$$
$$h_\text{D} = h_g \text{ at } 100°\text{F} = 88.62 \text{ Btu/lbm}$$
$$s_\text{D} = s_g \text{ at } 100°\text{F} = 0.16584 \text{ Btu/lbm-}°\text{R}$$
$$v_\text{D} = v_g \text{ at } 100°\text{F} = 0.319 \text{ ft}^3/\text{lbm}$$

At point C,

$$T = 50°\text{F} \quad \text{[given]}$$
$$p = p_\text{sat} \text{ at } 50°\text{F} = 61.39 \text{ psia}$$
$$s_\text{C} = s_\text{D} = 0.16584 \text{ Btu/lbm-}°\text{R}$$
$$s_{f,\text{C}} = 0.04126 \text{ Btu/lbm-}°\text{R}$$
$$s_{fg,\text{C}} = 0.12659 \text{ Btu/lbm-}°\text{R}$$

$$x = \frac{s_\text{C} - s_{f,\text{C}}}{s_{fg,\text{C}}}$$
$$= \frac{0.16584 \; \frac{\text{Btu}}{\text{lbm-}°\text{R}} - 0.04126 \; \frac{\text{Btu}}{\text{lbm-}°\text{R}}}{0.12659 \; \frac{\text{Btu}}{\text{lbm-}°\text{R}}}$$
$$= 0.984$$
$$h_\text{C} = h_{f,\text{C}} + x h_{fg,\text{C}}$$
$$= 19.27 \; \frac{\text{Btu}}{\text{lbm}} + (0.984)\left(64.51 \; \frac{\text{Btu}}{\text{lbm}}\right)$$
$$= 82.75 \text{ Btu/lbm}$$
$$v_\text{C} = v_{f,\text{C}} + x(v_{g,\text{C}} - v_{f,\text{C}})$$
$$= 0.0118 \; \frac{\text{ft}^3}{\text{lbm}}$$
$$\quad + (0.984)\left(0.673 \; \frac{\text{ft}^3}{\text{lbm}} - 0.0118 \; \frac{\text{ft}^3}{\text{lbm}}\right)$$
$$= 0.662 \text{ ft}^3/\text{lbm}$$

SI Solution

(a) The psychrometric chart is shown in the customary U.S. solution. At point 1, from the psychrometric chart (App. 49.B) at $T_\text{db} = 35°\text{C}$ and $\phi = 70\%$,

$$\boxed{\begin{array}{l} h_1 = 99.9 \text{ kJ/kg air} \\ v_1 = 0.91 \text{ m}^3/\text{kg air} \\ \omega_1 = \left(25.3 \; \frac{\text{g moisture}}{\text{kg air}}\right)\left(\frac{1 \text{ kg}}{1000 \text{ g}}\right) \\ \quad = 0.0253 \text{ kg moisture/kg air} \end{array}}$$

At point 2, from the psychrometric chart (App. 49.B) at $T_{db} = 16°C$ and $\phi = 95\%$,

$$\boxed{\begin{aligned} h_2 &= 43.4 \text{ kJ/kg air} \\ \omega_2 &= \left(10.8 \, \frac{\text{g moisture}}{\text{kg air}}\right)\left(\frac{1 \text{ kg}}{1000 \text{ g}}\right) \\ &= 0.0108 \text{ kg moisture/kg air} \end{aligned}}$$

(b) The air mass flow rate is

$$\dot{m}_a = \frac{\dot{V}_1}{v_1} = \frac{\dot{V}_1}{v_{\text{out}}} = \frac{2.36 \, \frac{\text{m}^3}{\text{s}}}{0.91 \, \frac{\text{m}^3}{\text{kg air}}} = 2.593 \text{ kg air/s}$$

The water removed from Eq. 49.27 is

$$\begin{aligned} \dot{m}_w &= \dot{m}_a(\omega_1 - \omega_2) \\ &= \left(2.593 \, \frac{\text{kg air}}{\text{s}}\right)\left(0.0253 \, \frac{\text{kg moisture}}{\text{kg air}}\right. \\ &\qquad \left. - 0.0108 \, \frac{\text{kg moisture}}{\text{kg air}}\right) \\ &= \boxed{0.0376 \text{ kg moisture/s}} \end{aligned}$$

(c) From Eq. 49.28, the quantity of heat removed is

$$\begin{aligned} \dot{q} &= \dot{m}_a(h_1 - h_2) \\ &= \left(2.593 \, \frac{\text{kg air}}{\text{s}}\right)\left(99.9 \, \frac{\text{kJ}}{\text{kg air}} - 43.4 \, \frac{\text{kJ}}{\text{kg air}}\right) \\ &= \boxed{146.5 \text{ kW}} \end{aligned}$$

(d) The T-s and h-s diagrams are shown in the customary U.S. solution.

(e) Use App. 23.T for saturated conditions.

At point A,

$$\begin{aligned} T &= 38°C \quad \text{[given]} \\ p &= p_{\text{sat}} \text{ at } 38°C = 0.91324 \text{ MPa} \\ h_A &= h_f \text{ at } 38°C = 237.23 \text{ kJ/kg} \\ s_A &= s_f \text{ at } 38°C = 1.1259 \text{ kJ/kg·°C} \\ v_A &= v_f \text{ at } 38°C = \frac{1}{\rho_f} = \frac{1}{1261.9 \, \frac{\text{kg}}{\text{m}^3}} \\ &= 0.0007925 \text{ m}^3/\text{kg} \end{aligned}$$

At point B,

$$\begin{aligned} T &= 10°C \quad \text{[given]} \\ p &= p_{\text{sat}} \text{ at } 10°C = 0.42356 \text{ MPa} \\ h_B &= h_A = 237.23 \text{ kJ/kg} \\ h_{f,B} &= 209.48 \text{ kJ/kg} \\ h_{g,B} &= 356.79 \text{ kJ/kg} \end{aligned}$$

$$x = \frac{h_B - h_{f,B}}{h_{g,B} - h_{f,B}} = \frac{237.23 \, \frac{\text{kJ}}{\text{kg}} - 209.48 \, \frac{\text{kJ}}{\text{kg}}}{356.79 \, \frac{\text{kJ}}{\text{kg}} - 209.48 \, \frac{\text{kJ}}{\text{kg}}}$$

$$= 0.188$$

$$\begin{aligned} s_B &= s_{f,B} + x(s_{g,B} - s_{f,B}) \\ &= 1.0338 \, \frac{\text{kJ}}{\text{kg·K}} \\ &\quad + (0.188)\left(1.5541 \, \frac{\text{kJ}}{\text{kg·K}} - 1.0338 \, \frac{\text{kJ}}{\text{kg·K}}\right) \\ &= 1.1316 \text{ kJ/kg·K} \end{aligned}$$

$$\begin{aligned} v_B &= v_{f,B} + x(v_{g,B} - v_{f,B}) \\ v_{f,B} &= \frac{1}{\rho_{f,B}} = \frac{1}{1363.0 \, \frac{\text{kg}}{\text{m}^3}} = 0.0007337 \text{ m}^3/\text{kg} \\ v_{g,B} &= 0.04119 \text{ m}^3/\text{kg} \\ v_B &= 0.0007337 \, \frac{\text{m}^3}{\text{kg}} \\ &\quad + (0.188)\left(0.04119 \, \frac{\text{m}^3}{\text{kg}} - 0.0007337 \, \frac{\text{m}^3}{\text{kg}}\right) \\ &= 0.008339 \text{ m}^3/\text{kg} \end{aligned}$$

At point D,

$$\begin{aligned} T &= 38°C \quad \text{[given]} \\ p &= p_{\text{sat}} \text{ at } 38°C = 0.42356 \text{ MPa} \\ h_D &= h_g \text{ at } 38°C = 367.95 \text{ kJ/kg} \\ s_D &= s_g \text{ at } 38°C = 1.5461 \text{ kJ/kg·K} \\ v_D &= v_g \text{ at } 38°C = 0.01931 \text{ m}^3/\text{kg} \end{aligned}$$

At point C,

$$\begin{aligned} T &= 10°C \\ p &= p_{\text{sat}} \text{ at } 10°C = 0.42356 \text{ MPa} \\ s_C &= s_D = 1.5461 \text{ kJ/kg·K} \\ s_{f,C} &= 1.0338 \text{ kJ/kg·K} \\ s_{g,C} &= 1.5541 \text{ kJ/kg·K} \end{aligned}$$

$$x = \frac{s_C - s_{f,C}}{s_{g,C} - s_{f,C}} = \frac{1.5461 \, \frac{\text{kJ}}{\text{kg·K}} - 1.0338 \, \frac{\text{kJ}}{\text{kg·K}}}{1.5541 \, \frac{\text{kJ}}{\text{kg·K}} - 1.0338 \, \frac{\text{kJ}}{\text{kg·K}}}$$

$$= 0.985$$

$$h_C = h_{f,C} + x(h_{g,C} - h_{f,C})$$

$$= 209.48 \frac{\text{kJ}}{\text{kg}} + (0.985)\left(356.79 \frac{\text{kJ}}{\text{kg}} - 209.48 \frac{\text{kJ}}{\text{kg}}\right)$$

$$= 354.58 \text{ kJ/kg}$$

$$v_C = v_{f,C} + x(v_{g,C} - v_{f,C})$$

$$v_{g,C} = 0.04119 \text{ m}^3/\text{kg}$$

$$v_C = 0.0007337 \frac{\text{m}^3}{\text{kg}}$$

$$+ (0.985)\left(0.04119 \frac{\text{m}^3}{\text{kg}} - 0.0007337 \frac{\text{m}^3}{\text{kg}}\right)$$

$$= 0.04058 \text{ m}^3/\text{kg}$$

11. *Customary U.S. Solution*

(a) The saturation pressure at $200°\text{F}$ from App. 23.A is $p_{\text{sat},1} = 11.529$ psia.

Since air is saturated (100% relative humidity), the water vapor pressure is equal to the saturation pressure.

$$p_{w,1} = p_{\text{sat},1} = 11.529 \text{ psia}$$

The partial pressure of the air is

$$p_{a,1} = p_1 - p_{w,1} = 25 \text{ psia} - 11.529 \text{ psia}$$

$$= 13.471 \text{ psia}$$

From Table 23.7, the specific gas constants are

$$R_w = 85.78 \text{ ft-lbf/lbm-}°\text{R}$$

$$R_{\text{air}} = 53.35 \text{ ft-lbf/lbm-}°\text{R}$$

The mass of water vapor from the ideal gas law is

$$\dot{m}_{w,1} = \frac{p_{w,1}\dot{V}}{R_w T}$$

$$= \frac{\left(11.529 \frac{\text{lbf}}{\text{in}^2}\right)\left(144 \frac{\text{in}^2}{\text{ft}^2}\right)\left(1500 \frac{\text{ft}^3}{\text{min}}\right)}{\left(85.78 \frac{\text{ft-lbf}}{\text{lbm-}°\text{R}}\right)(200°\text{F} + 460)}$$

$$= 43.99 \text{ lbm/min water}$$

The mass of air from the ideal gas law is

$$\dot{m}_{a,1} = \frac{p_{a,1}\dot{V}}{R_{\text{air}} T}$$

$$= \frac{\left(13.471 \frac{\text{lbf}}{\text{in}^2}\right)\left(144 \frac{\text{in}^2}{\text{ft}^2}\right)\left(1500 \frac{\text{ft}^3}{\text{min}}\right)}{\left(53.35 \frac{\text{ft-lbf}}{\text{lbm-}°\text{R}}\right)(200°\text{F} + 460)}$$

$$= 82.64 \text{ lbm/min air}$$

The humidity ratio is

$$\omega_1 = \frac{\dot{m}_{w,1}}{\dot{m}_{a,1}} = \frac{43.99 \frac{\text{lbm water}}{\text{min}}}{82.64 \frac{\text{lbm air}}{\text{min}}}$$

$$= 0.532 \text{ lbm water/lbm air}$$

Since it is a constant pressure, constant moisture drying process, mole fractions and partial pressures do not change.

$$p_{w,2} = p_{w,1} = 11.529 \text{ psia}$$

The saturation pressure at $400°\text{F}$ from App. 23.A is $p_{\text{sat},2} = 247.1$ psia.

The relative humidity at state 2 is

$$\phi_2 = \frac{p_{w,2}}{p_{\text{sat},2}} = \frac{11.529 \text{ psia}}{247.1 \text{ psia}} = \boxed{0.0467 \quad (4.67\%)}$$

The answer is (A).

(b) The specific humidity remains constant.

$$\omega_2 = \omega_1 = \boxed{0.532 \text{ lbm water/lbm air}}$$

The answer is (B).

(c) The heat required consists of two parts.

Obtain enthalpy for air from App. 23.F.

The absolute temperatures are

$$T_1 = 200°\text{F} + 460 = 660°\text{R}$$

$$T_2 = 400°\text{F} + 460 = 860°\text{R}$$

$$h_1 = 157.92 \text{ Btu/lbm}$$

$$h_2 = 206.46 \text{ Btu/lbm}$$

The heat absorbed by the air is

$$q_1 = h_2 - h_1 = 206.46 \frac{\text{Btu}}{\text{lbm}} - 157.92 \frac{\text{Btu}}{\text{lbm}}$$

$$= 48.54 \text{ Btu/lbm dry air}$$

(There will be a small error if constant specific heat is used instead.)

For water, use the Mollier diagram. From App. 23.E, h_1 at $200°\text{F}$ and 11.529 psia is 1146 Btu/lbm (almost saturated).

Follow a constant 11.529 psia pressure curve up to $400°\text{F}$.

$$h_2 = 1240 \text{ Btu/lbm}$$

(There will be a small error if Eq. 49.19(b) is used instead.)

The heat absorbed by the steam is

$$q_2 = \omega(h_2 - h_1)$$
$$= \left(0.532 \, \frac{\text{lbm water}}{\text{lbm air}}\right)\left(1240 \, \frac{\text{Btu}}{\text{lbm}} - 1146 \, \frac{\text{Btu}}{\text{lbm}}\right)$$
$$= 50.01 \, \text{Btu/lbm air}$$

The total heat absorbed is

$$q_{\text{total}} = q_1 + q_2 = 48.54 \, \frac{\text{Btu}}{\text{lbm air}} + 50.01 \, \frac{\text{Btu}}{\text{lbm air}}$$
$$= \boxed{98.55 \, \text{Btu/lbm air}}$$

The answer is (C).

(d) The dew point is the temperature at which water starts to condense out in a constant pressure process. Following the constant 11.529 psia pressure line back to the saturation line, $\boxed{T_{\text{dp}} = 200°\text{F}.}$

The answer is (B).

SI Solution

(a) From App. 23.N, the saturation pressure at 93°C is

$$p_{\text{sat},1} = (0.7879 \, \text{bar})\left(100 \, \frac{\text{kPa}}{\text{bar}}\right) = 78.79 \, \text{kPa}$$

Since air is saturated (100% relative humidity), water vapor pressure is equal to saturation pressure.

$$p_{w,1} = p_{\text{sat},1} = 78.79 \, \text{kPa}$$

The partial pressure of air is

$$p_{a,1} = p_1 - p_{w,1} = 170 \, \text{kPa} - 78.79 \, \text{kPa} = 91.21 \, \text{kPa}$$

From Table 23.7, the specific gas constants are

$$R_w = 461.50 \, \text{J/kg·K}$$
$$R_{\text{air}} = 287.03 \, \text{J/kg·K}$$

The mass of water vapor from the ideal gas law is

$$\dot{m}_{w,1} = \frac{p_{w,1}\dot{V}}{R_w T}$$
$$= \frac{(78.79 \, \text{kPa})\left(1000 \, \frac{\text{Pa}}{\text{kPa}}\right)\left(0.71 \, \frac{\text{m}^3}{\text{s}}\right)}{\left(461.50 \, \frac{\text{J}}{\text{kg·K}}\right)(93°\text{C} + 273)}$$
$$= 0.3312 \, \text{kg/s water}$$

The mass of air from the ideal gas law is

$$\dot{m}_{a,1} = \frac{p_{a,1}\dot{V}}{R_{\text{air}} T}$$
$$= \frac{(91.21 \, \text{kPa})\left(1000 \, \frac{\text{Pa}}{\text{kPa}}\right)\left(0.71 \, \frac{\text{m}^3}{\text{s}}\right)}{\left(287.03 \, \frac{\text{J}}{\text{kg·K}}\right)(93°\text{C} + 273)}$$
$$= 0.6164 \, \text{kg/s air}$$

The humidity ratio is

$$\omega_1 = \frac{\dot{m}_{w,1}}{\dot{m}_{a,1}} = \frac{0.3312 \, \frac{\text{kg water}}{\text{s}}}{0.6164 \, \frac{\text{kg air}}{\text{s}}}$$
$$= 0.537 \, \text{kg water/kg air}$$

Since it is a constant pressure, constant moisture drying process, mole fractions and partial pressure do not change.

$$p_{w,2} = p_{w,1} = 78.79 \, \text{kPa}$$

The saturation pressure at 204°C from App. 23.N is

$$p_{\text{sat},2} = (16.95 \, \text{bar})\left(100 \, \frac{\text{kPa}}{\text{bar}}\right) = 1695 \, \text{kPa}$$

The relative humidity at state 2 is

$$\phi_2 = \frac{p_{w,2}}{p_{\text{sat},2}} = \frac{78.79 \, \text{kPa}}{1695 \, \text{kPa}} = \boxed{0.0465 \, (4.65\%)}$$

The answer is (A).

(b) The specific humidity remains constant.

$$\omega_2 = \omega_1 = \boxed{0.537 \, \text{kg water/kg air}}$$

The answer is (B).

(c) The heat required consists of two parts.

Obtain enthalpy for air from App. 23.S.

The absolute temperatures are

$$T_1 = 93°\text{C} + 273 = 366\text{K}$$
$$T_2 = 204°\text{C} + 273 = 477\text{K}$$
$$h_1 = 366.63 \, \text{kJ/kg}$$
$$h_2 = 479.42 \, \text{kJ/kg}$$

The heat absorbed by air is

$$q_1 = h_2 - h_1 = 479.42 \, \frac{\text{kJ}}{\text{kg}} - 366.63 \, \frac{\text{kJ}}{\text{kg}}$$
$$= 112.79 \, \text{kJ/kg air}$$

(There will be a small error if constant specific heat is used instead.)

For water use the Mollier diagram. From App. 23.R, h_1 at 93°C and 78.79 kPa is 2670 kJ/kg (almost saturated).

Follow a constant 78.79 kPa pressure curve up to 204°C.

$$h_2 = 2890 \, \text{kJ/kg}$$

(There will be a small error if Eq. 49.19(a) is used instead.)

The heat absorbed by steam is

$$q_2 = \omega(h_2 - h_1)$$
$$= \left(0.537 \frac{\text{kg water}}{\text{kg air}}\right)\left(2890 \frac{\text{kJ}}{\text{kg}} - 2670 \frac{\text{kJ}}{\text{kg}}\right)$$
$$= 118.14 \text{ kJ/kg air}$$

The total heat absorbed is

$$q_{\text{total}} = q_1 + q_2$$
$$= 112.79 \frac{\text{kJ}}{\text{kg air}} + 118.14 \frac{\text{kJ}}{\text{kg air}}$$
$$= \boxed{230.93 \text{ kJ/kg air}}$$

The answer is (C).

(d) The dew point is the temperature at which water starts to condense out in a constant pressure process. Following the constant 78.79 kPa pressure line back to the saturation line, $\boxed{T_{\text{dp}} \approx 93°\text{C}.}$

The answer is (B).

12. *Customary U.S. Solution*

(a) The absolute air temperatures are

$$T_1 = 800°\text{F} + 460 = 1260°\text{R}$$
$$T_2 = 350°\text{F} + 460 = 810°\text{R}$$

At low pressures, use air tables. From App. 23.F,

$$h_1 = 306.65 \text{ Btu/lbm}$$
$$h_2 = 194.25 \text{ Btu/lbm}$$

From App. 23.A, the enthalpy of water at 80°F is $h_{w,1}$ = 48.09 Btu/lbm.

From Eq. 49.19(b), the enthalpy of steam at 350°F is

$$h_{w,2} \approx \left(0.444 \frac{\text{Btu}}{\text{lbm-}°\text{F}}\right)(350°\text{F}) + 1061 \frac{\text{Btu}}{\text{lbm}}$$
$$= 1216.4 \text{ Btu/lbm}$$

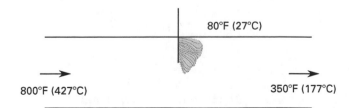

80°F (27°C)

800°F (427°C)　　　　350°F (177°C)

Air temperature is reduced from 800°F to 350°F, and this energy is used to change water at 80°F to steam at 350°F. From the energy balance equation,

$$\dot{m}_w(h_{w,2} - h_{w,1}) = \dot{m}_a(h_1 - h_2)$$
$$\dot{m}_w = \frac{\dot{m}_a(h_1 - h_2)}{h_{w,2} - h_{w,1}}$$
$$= \frac{\left(410 \frac{\text{lbm}}{\text{hr}}\right)\left(306.65 \frac{\text{Btu}}{\text{lbm}} - 194.25 \frac{\text{Btu}}{\text{lbm}}\right)}{1216.4 \frac{\text{Btu}}{\text{lbm}} - 48.09 \frac{\text{Btu}}{\text{lbm}}}$$
$$= \boxed{39.4 \text{ lbm/hr water}}$$

The answer is (D).

(b) The number of moles of water evaporated (in the air mixture) is

$$\dot{n}_w = \frac{39.4 \frac{\text{lbm}}{\text{hr}}}{18.016 \frac{\text{lbm}}{\text{lbmol}}} = 2.19 \text{ lbmol/hr}$$

The number of moles of air at the exit in the mixture is

$$\dot{n}_a = \frac{410 \frac{\text{lbm}}{\text{hr}}}{28.967 \frac{\text{lbm}}{\text{lbmol}}} = 14.15 \text{ lbmol/hr}$$

The mole fraction of water in the mixture is

$$x_w = \frac{\dot{n}_w}{\dot{n}_a + \dot{n}_w}$$
$$= \frac{2.19 \frac{\text{lbmol}}{\text{hr}}}{14.15 \frac{\text{lbmol}}{\text{hr}} + 2.19 \frac{\text{lbmol}}{\text{hr}}} = 0.134$$

Partial pressure of water vapor is

$$p_w = x(p_{\text{chamber}}) = (0.134)(20 \text{ psia}) = 2.68 \text{ psia}$$

From App. 23.A, the saturation pressure at 350 is p_{sat} = 134.53 psia.

The relative humidity is

$$\phi = \frac{p_w}{p_{\text{sat}}} = \frac{2.68 \text{ psia}}{134.53 \text{ psia}} = \boxed{0.020 \ \ (2.0\%)}$$

The answer is (B).

SI Solution

(a) The absolute temperatures are

$$T_1 = 427°C + 273 = 700K$$
$$T_2 = 177°C + 273 = 450K$$

Air tables can be used at low pressures. From App. 23.S,

$$h_1 = 713.27 \text{ kJ/kg}$$
$$h_2 = 451.80 \text{ kJ/kg}$$

From App. 23.N, the enthalpy of water at 27°C is $h_{w,1} = 113.25$ kJ/kg.

From Eq. 49.19(a), the enthalpy of steam at 177°C is

$$h_{w,2} = \left(1.805 \frac{\text{kJ}}{\text{kg·°C}}\right)(177°C) + 2501 \frac{\text{kJ}}{\text{kg}}$$
$$= 2820.5 \text{ kJ/kg}$$

Air temperature is reduced from 427°C to 177°C, and this energy is used to change water at 27°C to steam at 177°C. From the energy balance equation,

$$\dot{m}_w = \frac{\dot{m}_a(h_1 - h_2)}{h_{w,2} - h_{w,1}}$$

$$= \frac{\left(0.052 \frac{\text{kg}}{\text{s}}\right)\left(713.27 \frac{\text{kJ}}{\text{kg}} - 451.80 \frac{\text{kJ}}{\text{kg}}\right)}{2820.5 \frac{\text{kJ}}{\text{kg}} - 113.25 \frac{\text{kJ}}{\text{kg}}}$$

$$= \boxed{0.00502 \text{ kg/s}}$$

The answer is (D).

(b) The number of moles of water evaporated (in the air mixture) is

$$\dot{n}_w = \frac{0.00502 \frac{\text{kg}}{\text{s}}}{18.016 \frac{\text{kg}}{\text{kmol}}} = 2.79 \times 10^{-4} \text{ kmol/s}$$

The number of moles of air in the mixture at the exit is

$$\dot{n}_a = \frac{0.052 \frac{\text{kg}}{\text{s}}}{28.967 \frac{\text{kg}}{\text{kmol}}} = 1.80 \times 10^{-3} \text{ kmol/s}$$

The mole fraction of water in the mixture is

$$x_w = \frac{\dot{n}_w}{\dot{n}_a + \dot{n}_w}$$

$$= \frac{2.79 \times 10^{-4} \frac{\text{kmol}}{\text{s}}}{1.80 \times 10^{-3} \frac{\text{kmol}}{\text{s}} + 2.79 \times 10^{-4} \frac{\text{kmol}}{\text{s}}}$$

$$= 0.134$$

The partial pressure of water vapor is

$$p_w = x p_{\text{chamber}} = (0.134)(140 \text{ kPa})$$
$$= 18.76 \text{ kPa}$$

From App. 23.N, the saturation pressure at 177°C is

$$p_{\text{sat}} = (9.389 \text{ bar})\left(100 \frac{\text{kPa}}{\text{bar}}\right) = 938.9 \text{ kPa}$$

The relative humidity is

$$\phi = \frac{p_w}{p_{\text{sat}}} = \frac{18.76 \text{ kPa}}{938.9 \text{ kPa}} = \boxed{0.020 \ (2.0\%)}$$

The answer is (B).

13. *Customary U.S. Solution*

(a) The cooled water flow rate is given by

$$Q = \dot{m}_w c_p \Delta T$$

$$1 \times 10^6 \text{ Btu} = \dot{m}_w \left(1.0 \frac{\text{Btu}}{\text{lbm-°F}}\right)(120°F - 110°F)$$

$$\dot{m}_w = 1 \times 10^5 \text{ lbm/hr}$$

From the psychrometric chart (App. 49.A), for air in at $T_{\text{db}} = 91°F$ and $\phi = 60\%$,

$$h_{\text{in}} \approx 42.7 \text{ Btu/lbm}$$
$$\omega_{\text{in}} = 0.0190 \text{ lbm moisture/lbm air}$$

For air out, the normal psychrometric chart (offscale) cannot be used, so use Eq. 49.11 to find the humidity ratio and Eqs. 49.17, 49.18(b), and 49.19(b) to calculate the enthalpy of air. (App. 49.D could also be used as a simpler solution.)

From App. 38.A, the saturated steam pressure at 100°F is 0.9503 psia.

$$p_w = \phi p_{\text{sat}}$$
$$= (0.82)(0.9503 \text{ psia})$$
$$= 0.7792 \text{ psia}$$

From Eq. 49.1,

$$p_a = p - p_w = 14.696 \text{ psia} - 0.7792 \text{ psia}$$
$$= 13.9168 \text{ psia}$$

From Eq. 49.11, the humidity ratio for air out is

$$\phi = 1.608 \omega_{\text{out}} \left(\frac{p_a}{p_{\text{sat}}}\right)$$

$$0.82 = 1.608 \omega_{\text{out}} \left(\frac{13.9168 \text{ psia}}{0.9503 \text{ psia}}\right)$$

$$\omega_{\text{out}} = 0.0348 \text{ lbm moisture/lbm air}$$

From Eqs. 49.17, 49.18(b), and 49.19(b), the enthalpy of air out is

$$h_2 = h_a + \omega_2 h_w$$

$$= \left(0.240 \; \frac{\text{Btu}}{\text{lbm-°F}}\right) T_{2,°F} + \omega_2$$

$$\times \left(\left(0.444 \; \frac{\text{Btu}}{\text{lbm-°F}}\right) T_{2,°F} + 1061 \; \frac{\text{Btu}}{\text{lbm}}\right)$$

$$= \left(0.240 \; \frac{\text{Btu}}{\text{lbm-°F}}\right)(100°\text{F})$$

$$+ \left(0.0348 \; \frac{\text{lbm moisture}}{\text{lbm air}}\right)\left(\left(0.444 \; \frac{\text{Btu}}{\text{lbm-°F}}\right)\right.$$

$$\times (100°\text{F}) + 1061 \; \frac{\text{Btu}}{\text{lbm}}\bigg)$$

$$= 62.47 \; \text{Btu/lbm air}$$

The mass flow rate of air can be determined from Eq. 49.26.

$$q = \dot{m}_a(h_2 - h_1)$$

$$\dot{m}_{\text{air}} = \frac{q}{h_2 - h_1} = \frac{1 \times 10^6 \; \frac{\text{Btu}}{\text{hr}}}{62.47 \; \frac{\text{Btu}}{\text{lbm air}} - 42.7 \; \frac{\text{Btu}}{\text{lbm air}}}$$

$$= \boxed{5.058 \times 10^4 \; \text{lbm air/hr}}$$

(b) From conservation of water vapor,

$$\omega_1 \dot{m}_{\text{air}} + \dot{m}_{\text{make-up}} = \omega_2 \dot{m}_{\text{air}}$$

$$\dot{m}_{\text{make-up}} = \dot{m}_{\text{air}}(\omega_{\text{out}} - \omega_{\text{in}})$$

$$= \left(5.058 \times 10^4 \; \frac{\text{lbm air}}{\text{hr}}\right)$$

$$\times \left(\begin{matrix} 0.0348 \; \dfrac{\text{lbm moisture}}{\text{lbm air}} \\[2mm] - \; 0.0191 \; \dfrac{\text{lbm moisture}}{\text{lbm air}} \end{matrix}\right)$$

$$= \boxed{794 \; \text{lbm water/hr}}$$

SI Solution

(a) The cooled water flow rate is given by

$$Q = \dot{m}_w c_p \Delta T$$

$$290 \; \text{kW} = \dot{m}_w \left(4.187 \; \frac{\text{kJ}}{\text{kg-°C}}\right)(49°\text{C} - 43°\text{C})$$

$$\dot{m}_w = 11.54 \; \text{kg/s}$$

From the psychrometric chart (App. 49.B), for air in at $T_{\text{db}} = 33°\text{C}$ and $\phi = 60\%$,

$$h_{\text{in}} = 82.3 \; \text{kJ/kg air}$$

$$\omega_{\text{in}} = \left(19.2 \; \frac{\text{g moisture}}{\text{kg air}}\right)\left(\frac{1 \; \text{kg}}{1000 \; \text{g}}\right)$$

$$= 0.0192 \; \text{kg moisture/kg air}$$

For air out, the psychrometric chart (off scale) cannot be used, so use Eq. 49.11 to find the humidity ratio and Eqs. 49.17, 49.18(a), and 49.19(a) to calculate enthalpy of air.

From App. 23.N, the saturated steam pressure at 38°C is 0.06632 bars.

$$p_w = \phi p_{\text{sat}}$$

$$= (0.82)(0.06632 \; \text{bar})$$

$$= 0.05438 \; \text{bar}$$

From Eq. 49.1,

$$p_a = p - p_w = 1 \; \text{bar} - 0.05438 \; \text{bar}$$

$$= 0.94562 \; \text{bar}$$

From Eq. 49.11, the humidity ratio for air out is

$$\phi = 1.608\omega_{\text{out}}\left(\frac{p_a}{p_{\text{sat}}}\right)$$

$$0.82 = 1.608\omega_{\text{out}}\left(\frac{0.94562 \; \text{bar}}{0.06632 \; \text{bar}}\right)$$

$$\omega_{\text{out}} = 0.0358 \; \text{kg moisture/kg air}$$

From Eqs. 49.17, 49.18(a), and 49.19(a), the enthalpy of air out is

$$h_2 = h_a + \omega_2 h_w$$

$$= \left(1.005 \; \frac{\text{kJ}}{\text{kg-°C}}\right) T_{°C}$$

$$+ \omega_{\text{out}}\left(\left(1.805 \; \frac{\text{kJ}}{\text{kg-°C}}\right) T_{°C} + 2501 \; \frac{\text{kJ}}{\text{kg}}\right)$$

$$= \left(1.005 \; \frac{\text{kJ}}{\text{kg-°C}}\right)(38°\text{C}) + \left(0.0358 \; \frac{\text{kg moisture}}{\text{kg air}}\right)$$

$$\times \left(\left(1.805 \; \frac{\text{kJ}}{\text{kg-°C}}\right)(38°\text{C}) + 2501 \; \frac{\text{kJ}}{\text{kg}}\right)$$

$$= 130.2 \; \text{kJ/kg air}$$

The mass flow rate of air can be determined from Eq. 49.26.

$$q = \dot{m}_a(h_2 - h_1)$$

$$\dot{m}_{\text{air}} = \frac{q}{h_2 - h_1} = \frac{290 \; \text{kW}}{130.2 \; \frac{\text{kJ}}{\text{kg air}} - 82.3 \; \frac{\text{kJ}}{\text{kg air}}}$$

$$= \boxed{6.054 \; \text{kg air/s}}$$

Plant Design

(b) From conservation of water vapor,

$$\omega_1 \dot{m}_{air} + \dot{m}_{make\text{-}up} = \omega_2 \dot{m}_{air}$$

$$\dot{m}_{make\text{-}up} = \dot{m}_{air}(\omega_{out} - \omega_{in})$$

$$= \left(6.054 \; \frac{\text{kg air}}{\text{s}}\right)$$

$$\times \left(0.0358 \; \frac{\text{kg moisture}}{\text{kg air}} - 0.0192 \; \frac{\text{kg moisture}}{\text{kg air}}\right)$$

$$= \boxed{0.100 \text{ kg water/s}}$$

50 Ventilation and Humidification

PRACTICE PROBLEMS

1. An office room has floor dimensions of 60 ft by 95 ft (18 m by 29 m) and a ceiling height of 10 ft (3 m). 45 people occupy the office, and half of them smoke.

(a) Calculate the ventilation rate based on six air changes per hour.

(b) State your assumptions and determine the ventilation rate based on occupancy.

2. 150 ppm of methanol (TLV = 200 ppm; MW = 32.04; SG = 0.792) and 285 ppm of methylene chloride (TLV = 500 ppm; MW = 84.94; SG = 1.336) are found in the air in a plating booth. Two pints of each are evaporated per hour. Use a mixing safety factor (i.e., a K value) of 6. What ventilation rate is required?

    (A) 2500 ft^3/min

    (B) 5200 ft^3/min

    (C) 10,000 ft^3/min

    (D) 13,000 ft^3/min

3. An auditorium is designed to seat 4500 people. The ventilation rate is 60 ft^3/min (1.68 m^3/min) per person of outside air. The outside temperature is 0°F (-18°C) dry-bulb, and the outside pressure is 14.6 psia (100.6 kPa). Air leaves the auditorium at 70°F (21°C) dry-bulb. There is no recirculation. The furnace has been sized assuming an internal heat gain of 1,250,000 Btu/hr (370 kW) and the outside conditions described.

(a) At what temperature should the air enter the auditorium?

    (A) 52°F (11.1°C)

    (B) 55°F (12.8°C)

    (C) 62°F (16.7°C)

    (D) 67°F (19.3°C)

(b) How much heat should be supplied to the ventilation air?

    (A) 1.5×10^7 Btu/hr (4.5 MW)

    (B) 2.2×10^7 Btu/hr (6.5 MW)

    (C) 3.9×10^7 Btu/hr (12 MW)

    (D) 5.1×10^7 Btu/hr (16 MW)

(c) Has the furnace been sized properly?

    (A) The furnace is less than half the required capacity.

    (B) The furnace is more than half the required capacity.

    (C) The furnace is less than twice the required capacity.

    (D) The furnace is more than twice the required capacity.

4. A room is maintained at design conditions of 75°F (23.9°C) dry-bulb and 50% relative humidity. The air outside is at 95°F (35°C) dry-bulb and 75°F (23.9°C) wet-bulb. Conditioned air enters the room and increases 20°F (11.1°C) in temperature before being removed from the room. The sensible and latent loads are 200,000 Btu/hr (60 kW) and 50,000 Btu/hr (15 kW), respectively. 2000 ft^3/min (57 m^3/min) of outside air are used.

(a) What is the apparatus dew point?

    (A) 45°F (7.2°C)

    (B) 51°F (10°C)

    (C) 56°F (13°C)

    (D) 62°F (17°C)

(b) What is the volume of air flowing through the coil?

    (A) 4200 ft^3/min (120 m^3/min)

    (B) 5800 ft^3/min (160 m^3/min)

    (C) 7600 ft^3/min (220 m^3/min)

    (D) 9300 ft^3/min (260 m^3/min)

Plant Design

SOLUTIONS

1. *Customary U.S. Solution*

(a) The office volume is

$$V = (60 \text{ ft})(95 \text{ ft})(10 \text{ ft}) = 57{,}000 \text{ ft}^3$$

Based on six air changes per hour, the flow rate is

$$\dot{V} = \left(57{,}000 \ \frac{\text{ft}^3}{\text{air change}}\right) \left(6 \ \frac{\text{air changes}}{\text{hr}}\right) \left(\frac{1 \text{ hr}}{60 \text{ min}}\right)$$

$$= \boxed{5700 \text{ ft}^3/\text{min}}$$

(b) For preferred ventilation based on Sec. 50-2, "Ventilation Standards," assume the following.

- The ventilation rate for a nonsmoking area is 20 ft^3/min.

- The ventilation rate for a smoking area ranges from 30 to 60 ft^3/min, with an average value of 45 ft^3/min.

The preferred ventilation rate is

$$\dot{V} = \left(\tfrac{1}{2}\right)(45 \text{ persons}) \left(\frac{20 \ \frac{\text{ft}^3}{\text{min}}}{1 \text{ person}}\right)$$

$$+ \left(\tfrac{1}{2}\right)(45 \text{ persons}) \left(\frac{45 \ \frac{\text{ft}^3}{\text{min}}}{1 \text{ person}}\right)$$

$$= \boxed{1463 \text{ ft}^3/\text{min}}$$

SI Solution

(a) The office volume is

$$(18 \text{ m})(29 \text{ m})(3 \text{ m}) = 1566 \text{ m}^3$$

Based on six air changes per hour, the flow rate is

$$\dot{V} = \left(1566 \ \frac{\text{m}^3}{\text{air change}}\right) \left(6 \ \frac{\text{air changes}}{\text{h}}\right) \left(\frac{1 \text{ h}}{60 \text{ min}}\right)$$

$$= \boxed{156.6 \text{ m}^3/\text{min}}$$

(b) For preferred ventilation based on Sec. 50-2, "Ventilation Standards," assume the following.

- The ventilation rate for a nonsmoking area is 0.57 m^3/min.

- The ventilation rate for a smoking area ranges from 0.84 to 1.68 m^3/min, with an average value of 1.26 m^3/min.

The preferred ventilation rate is

$$\dot{V} = \left(\tfrac{1}{2}\right)(45 \text{ persons}) \left(\frac{0.57 \ \frac{\text{m}^3}{\text{min}}}{1 \text{ person}}\right)$$

$$+ \left(\tfrac{1}{2}\right)(45 \text{ persons}) \left(\frac{1.26 \ \frac{\text{m}^3}{\text{min}}}{1 \text{ person}}\right)$$

$$= \boxed{41.2 \text{ m}^3/\text{min}}$$

2. From Eq. 50.12(b), the ventilation rate required is

$$\dot{V}_{\text{cfm}} = \frac{(4.03 \times 10^8) K (\text{SG}) R_{\text{pints/min}}}{(\text{MW}) \text{TLV}_{\text{ppm}}}$$

For the methanol,

$$\dot{V}_{\text{cfm}} = \frac{(4.03 \times 10^8)(6)(0.792) \left(2 \ \frac{\text{pints}}{\text{h}}\right) \left(\frac{1 \text{ h}}{60 \text{ min}}\right)}{(32.04)(200 \text{ ppm})}$$

$$= 9962 \text{ ft}^3/\text{min}$$

For the methylene chloride,

$$\dot{V}_{\text{cfm}} = \frac{(4.03 \times 10^8)(6)(1.336) \left(2 \ \frac{\text{pints}}{\text{hr}}\right) \left(\frac{1 \text{ hr}}{60 \text{ min}}\right)}{(84.94)(500 \text{ ppm})}$$

$$= 2535 \text{ ft}^3/\text{min}$$

The total ventilation rate is

$$9962 \ \frac{\text{ft}^3}{\text{min}} + 2535 \ \frac{\text{ft}^3}{\text{min}} = \boxed{12{,}497 \text{ ft}^3/\text{min}}$$

The answer is (D).

3. *Customary U.S. Solution*

(a) The total ventilation rate is

$$\dot{V} = \left(\frac{60 \ \frac{\text{ft}^3}{\text{min}}}{1 \text{ person}}\right)(4500 \text{ persons}) \left(60 \ \frac{\text{min}}{\text{hr}}\right)$$

$$= 1.62 \times 10^7 \text{ ft}^3/\text{hr}$$

The absolute temperature of outside air is

$$T = 0{}^\circ\text{F} + 460 = 460{}^\circ\text{R}$$

From the ideal gas law, the density of outside air is

$$\rho = \frac{p}{RT} = \frac{\left(14.6 \ \frac{\text{lbf}}{\text{in}^2}\right) \left(144 \ \frac{\text{in}^2}{\text{ft}^2}\right)}{\left(53.35 \ \frac{\text{ft-lbf}}{\text{lbm-}{}^\circ\text{R}}\right)(460{}^\circ\text{R})}$$

$$= 0.08567 \text{ lbm/ft}^3$$

The mass flow rate is

$$\dot{m} = \dot{V}\rho = \left(1.62 \times 10^7 \ \frac{\text{ft}^3}{\text{hr}}\right)\left(0.08567 \ \frac{\text{lbm}}{\text{ft}^3}\right)$$
$$= 1.388 \times 10^6 \ \text{lbm/hr}$$

Assume no latent heat (no moisture) at 0°F.

From Table 50.1, the sensible heat generated by each person seated in the theater is 225 Btu/hr. Therefore,

$$Q_{\text{in from people}} = \left(225 \ \frac{\frac{\text{Btu}}{\text{hr}}}{\text{person}}\right)(4500 \ \text{persons})$$
$$= 1.01 \times 10^6 \ \text{Btu/hr}$$

The air leaves the auditorium at 70°F. From App. 32.C, the specific heat is 0.240 Btu/lbm-°F (remains fairly constant). Since Q is known, the air temperature entering the auditorium can be calculated by

$$Q = \dot{m}c_p(T_{\text{out,air}} - T_{\text{in,air}})$$
$$T_{\text{in,air}} = T_{\text{out,air}} - \frac{Q}{\dot{m}c_p}$$
$$= 70°\text{F} - \frac{1.01 \times 10^6 \ \frac{\text{Btu}}{\text{hr}}}{\left(1.388 \times 10^6 \ \frac{\text{lbm}}{\text{hr}}\right)\left(0.240 \ \frac{\text{Btu}}{\text{lbm-}°\text{F}}\right)}$$
$$= \boxed{67.0°\text{F}}$$

The answer is (D).

(b) The heat needed to heat dry ventilation air from 0 to 67°F is

$$Q = \dot{m}c_p\Delta T$$
$$= \left(1.388 \times 10^6 \ \frac{\text{lbm}}{\text{hr}}\right)\left(0.240 \ \frac{\text{Btu}}{\text{lbm-}°\text{F}}\right)(67°\text{F} - 0°\text{F})$$
$$= \boxed{2.23 \times 10^7 \ \text{Btu/hr}}$$

The answer is (B).

(c) Since 1.01×10^6 Btu/hr $< 1.25 \times 10^6$ Btu/hr, the furnace is too small.

The answer is (A).

SI Solution

(a) The total ventilation rate is

$$\dot{V} = \left(\frac{1.68 \ \frac{\text{m}^3}{\text{min}}}{\text{person}}\right)(4500 \ \text{persons})\left(60 \ \frac{\text{min}}{\text{h}}\right)$$
$$= 4.54 \times 10^5 \ \text{m}^3/\text{h}$$

The absolute temperature of outside air is

$$T = -18°\text{C} + 273 = 255\text{K}$$

From the ideal gas law, the density of outside air is

$$\rho = \frac{p}{RT} = \frac{(100.6 \ \text{kPa})\left(1000 \ \frac{\text{Pa}}{\text{kPa}}\right)}{\left(287.03 \ \frac{\text{J}}{\text{kg·K}}\right)(255\text{K})}$$
$$= 1.374 \ \text{kg/m}^3$$

The mass flow rate is

$$\dot{m} = \rho\dot{V} = \left(1.374 \ \frac{\text{kg}}{\text{m}^3}\right)\left(4.54 \times 10^5 \ \frac{\text{m}^3}{\text{h}}\right)$$
$$= 6.238 \times 10^5 \ \text{kg/h}$$

Assume no latent heat (no moisture) at −18°C.

From Table 50.1 and the table footnote, the sensible heat generated by people seated in the theater is

$$\left(225 \ \frac{\text{Btu}}{\text{h}}\right)\left(0.293 \ \frac{\text{W}}{\frac{\text{Btu}}{\text{h}}}\right) = 65.93 \ \text{W}$$

Therefore,

$$Q_{\text{in from people}} = \left(65.93 \ \frac{\text{W}}{\text{persons}}\right)(4500 \ \text{persons})$$
$$\times \left(\frac{1 \ \text{kW}}{1000 \ \text{W}}\right)$$
$$= 296.7 \ \text{kW}$$

The air leaves the auditorium at 21°C. From App. 32.D, the specific heat ≈ 1.0048 kJ/kg·K (remains fairly constant). The air temperature entering the auditorium can be found from known Q as

$$Q = \dot{m}c_p(T_{\text{out,air}} - T_{\text{in,air}})$$
$$T_{\text{in,air}} = T_{\text{out,air}} - \frac{Q}{\dot{m}c_p}$$
$$= 21°\text{C} - \frac{296.7 \ \text{kW}}{\left(6.238 \times 10^5 \ \frac{\text{kg}}{\text{h}}\right)}$$
$$\times \left(\frac{1 \ \text{h}}{3600 \ \text{s}}\right)\left(1.0048 \ \frac{\text{kJ}}{\text{kg·K}}\right)$$
$$= \boxed{19.3°\text{C}}$$

The answer is (D).

Plant Design

(b) The heat needed to heat dry ventilation air from $-18°C$ to $19.3°C$ is

$$Q = \dot{m}c_p\Delta T$$

$$= \left(6.238 \times 10^5 \; \frac{\text{kg}}{\text{h}}\right)\left(\frac{1 \text{ h}}{3600 \text{ s}}\right)\left(1.0048 \; \frac{\text{kJ}}{\text{kg}\cdot\text{K}}\right)$$

$$\times \left(19.3°C - (-18°C)\right)$$

$$= \boxed{6494 \text{ kW}}$$

The answer is (B).

(c) Since $296.7 \text{ kW} < 370 \text{ kW}$, the furnace is too small.

The answer is (A).

4.

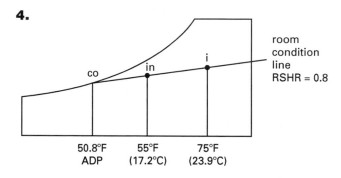

room condition line
RSHR = 0.8

| 50.8°F | 55°F | 75°F |
| ADP | (17.2°C) | (23.9°C) |

Customary U.S. Solution

(a) From Eq. 49.22, the room sensible heat ratio is

$$\text{RSHR} = \frac{q_s}{q_t} = \frac{q_s}{q_s + q_l}$$

$$= \frac{200{,}000 \; \dfrac{\text{Btu}}{\text{hr}}}{200{,}000 \; \dfrac{\text{Btu}}{\text{hr}} + 50{,}000 \; \dfrac{\text{Btu}}{\text{hr}}}$$

$$= 0.8$$

Locate the point i for design conditions of 75°F dry bulb and 50% relative humidity. Draw a condition line with a slope of 0.8 through point i.

$$\text{ADP} = \boxed{50.8°F} \qquad \left[\begin{array}{l}\text{intersection of room condi-}\\\text{tion and saturation line}\end{array}\right]$$

$$T_w = \text{ADP}$$

$$T_{\text{db,in}} = 75°F - 20°F = 55°F$$

The answer is (B).

(b) From Eq. 50.6(b), the volumetric flow rate of air entering the room is

$$\dot{V}_{\text{in,cfm}} = \frac{\dot{q}_{s,\text{Btu/hr}}}{\left(1.08 \; \dfrac{\text{Btu-min}}{\text{ft}^3\text{-hr-°F}}\right)(T_{i,°F} - T_{\text{in},°F})}$$

$$= \frac{200{,}000 \; \dfrac{\text{Btu}}{\text{hr}}}{\left(1.08 \; \dfrac{\text{Btu-min}}{\text{ft}^3\text{-hr-°F}}\right)(75°F - 55°F)}$$

$$= 9259 \text{ ft}^3/\text{min}$$

This is a mixing problem.

(i)

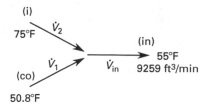

Using the lever rule, and since the temperature scales are all linear, the fraction of air passing through the conditioner is

$$\frac{T_i - T_{\text{in}}}{T_i - T_{\text{co}}} = \frac{75°F - 55°F}{75°F - 50.8°F} = 0.826$$

$$\dot{V}_1 = (0.826)\left(9259 \; \frac{\text{ft}^3}{\text{min}}\right) = \boxed{7648 \text{ ft}^3/\text{min}}$$

The answer is (C).

SI Solution

(a) From Eq. 49.22, the room sensible heat ratio is

$$\text{RSHR} = \frac{q_s}{q_s + q_l} = \frac{60 \text{ kW}}{60 \text{ kW} + 15 \text{ kW}} = 0.8$$

Locate the point i for design conditions of 23.9°C dry bulb and 50% relative humidity. Draw a condition line with a slope of 0.8 through point i.

$$\text{ADP} = \boxed{10°C} \qquad \left[\begin{array}{l}\text{intersection of room condi-}\\\text{tion and saturation line}\end{array}\right]$$

$$T_w = \text{ADP}$$

$$T_{\text{db,in}} = 23.9°C - 11.1°C = 12.8°C$$

The answer is (B).

(b) From Eq. 50.6(a),

$$\dot{V}_{\text{in,m}^3/\text{min}} = \frac{\dot{q}_{s,\text{kW}}}{\left(0.02 \; \dfrac{\text{kJ}\cdot\text{min}}{\text{m}^3\cdot\text{s}\cdot°C}\right)(T_{i,°C} - T_{\text{in},°C})}$$

$$= \frac{60 \text{ kW}}{\left(0.02 \; \dfrac{\text{kJ}\cdot\text{min}}{\text{m}^3\cdot\text{s}\cdot°C}\right)(23.9°C - 12.8°C)}$$

$$= 270.3 \text{ m}^3/\text{min}$$

This is a mixing problem.

(i)

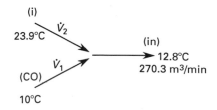

Using the lever rule, and since the temperature scales are all linear, the fraction of air passing through the conditioner is

$$\frac{T_i - T_{\text{in}}}{T_i - T_{\text{co}}} = \frac{23.9°\text{C} - 12.8°\text{C}}{23.9°\text{C} - 10°\text{C}}$$

$$= 0.799$$

$$\dot{V}_1 = (0.799)\left(270.3 \ \frac{\text{m}^3}{\text{min}}\right)$$

$$= \boxed{216.0 \ \text{m}^3/\text{min}}$$

The answer is (C).

Plant Design

51 Properties of Solid Bodies

PRACTICE PROBLEMS

Center of Gravity

1. Locate the center of gravity if the weights of the sphere, rod, and cylinder are 64.4, 32.2, and 64.4 pounds (29.2, 14.6, and 29.2 kg), respectively.

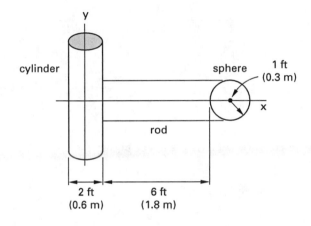

2. Find the center of gravity of the object shown.

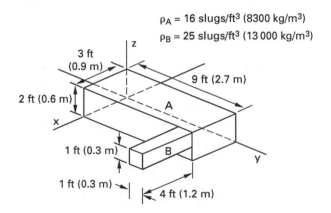

3. Find the center of gravity with respect to the x-axis of the object shown.

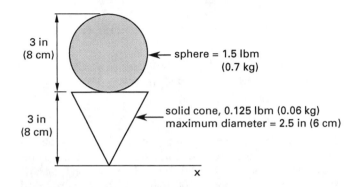

Mass Moments of Inertia

4. What is the mass moment of inertia about the x-axis of the object shown in Prob. 2?

5. What is the mass moment of inertia about the x-axis for the object shown in Prob. 3?

6. Find the moment of inertia about the BB axis if I_{AA} is 90 slug-ft^2 (120 kg·m^2) and the mass is 64.4 pounds (29.2 kg).

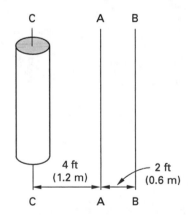

7. A spoked flywheel has an outside diameter of 60 in (1500 mm). The rim thickness is 6 in (150 mm), and the width is 12 in (300 mm). The cylindrical hub has an outside diameter of 12 in (300 mm), a thickness of 3 in (75 mm), and a width of 12 in (300 mm). The rim and hub are connected by six equally spaced cylindrical radial spokes, each with a diameter of 4.25 in (110 mm). All parts of the flywheel are ductile cast iron with a density of 0.256 lbm/in^3 (7080 kg/m^3). What is the rotational mass moment of inertia of the entire flywheel?

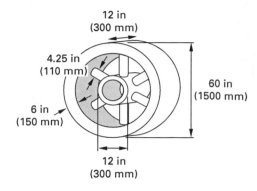

(A) 14,000 lbm-ft^2 (530 kg·m^2)
(B) 15,000 lbm-ft^2 (570 kg·m^2)
(C) 16,000 lbm-ft^2 (600 kg·m^2)
(D) 17,000 lbm-ft^2 (650 kg·m^2)

8. The rim of a spoked cast-iron flywheel in a punch press has a radius of gyration of 15 in (38.1 cm) and a face width of 12 in (30.5 cm). The flywheel rotates at 200 rpm. It supplies 1500 ft-lbf (2 kJ) of energy and slows to 175 rpm each time a punch press is cycled. The density of the cast iron is 0.26 lbm/in^3 (7200 kg/m^3). The hub and arms increase the rotational moment of inertia by 10%. What is the rim thickness?

(A) 1.9 in (4.7 cm)
(B) 2.3 in (5.8 cm)
(C) 2.9 in (7.4 cm)
(D) 3.5 in (8.9 cm)

SOLUTIONS

1. *Customary U.S. Solution*

Since the object is symmetrical about the x-axis, $y_c = 0$ ft.

$$m_S = 64.4\,\text{lbm}$$
$$x_{cS} = 1\,\text{ft} + 6\,\text{ft} + 1\,\text{ft} = 8\,\text{ft}$$
$$m_R = 32.2\,\text{lbm}$$
$$x_{cR} = 1\,\text{ft} + 3\,\text{ft} = 4\,\text{ft}$$
$$m_C = 64.4\,\text{lbm}$$
$$x_{cC} = 0\,\text{ft}$$
$$x_c = \frac{\Sigma m_i x_{ci}}{\Sigma m_i}$$

$$= \frac{\begin{matrix}(64.4\,\text{lbm})(8\,\text{ft}) + (32.2\,\text{lbm})(4\,\text{ft}) \\ + (64.4\,\text{lbm})(0\,\text{ft})\end{matrix}}{(64.4\,\text{lbm} + 32.2\,\text{lbm} + 64.4\,\text{lbm})}$$

$$= \boxed{4\,\text{ft}}$$

SI Solution

Since the object is symmetrical about the x-axis, $y_c = 0$ m.

$$m_S = 29.2\,\text{kg}$$
$$x_{cS} = 0.3\,\text{m} + 1.8\,\text{m} + 0.3\,\text{m} = 2.4\,\text{m}$$
$$m_R = 14.6\,\text{kg}$$
$$x_{cR} = 0.3\,\text{m} + 0.9\,\text{m} = 1.2\,\text{m}$$
$$m_C = 29.2\,\text{kg}$$
$$x_{cC} = 0\,\text{m}$$
$$x_c = \frac{\Sigma m_i x_{ci}}{\Sigma m_i}$$

$$= \frac{\begin{matrix}(29.2\,\text{kg})(2.4\,\text{m}) + (14.6\,\text{kg})(1.2\,\text{m}) \\ + (29.2\,\text{kg})(0\,\text{m})\end{matrix}}{(29.2\,\text{kg} + 14.6\,\text{kg} + 29.2\,\text{kg})}$$

$$= \boxed{1.20\,\text{m}}$$

2. *Customary U.S. Solution*

$$m_A = (2\,\text{ft})\,(3\,\text{ft})\,(9\,\text{ft})\left(16\,\frac{\text{slug}}{\text{ft}^3}\right)$$
$$= 864\,\text{slugs}$$
$$m_B = (1\,\text{ft})\,(1\,\text{ft})\,(4\,\text{ft})\left(25\,\frac{\text{slug}}{\text{ft}^3}\right)$$
$$= 100\,\text{slugs}$$

$x_{cA} = 1.5\,\text{ft}$

$x_{cB} = 3\,\text{ft} + 2\,\text{ft} = 5\,\text{ft}$

$x_c = \dfrac{\Sigma m_i x_{ci}}{\Sigma m_i} = \dfrac{(864\,\text{slug})(1.5\,\text{ft}) + (100\,\text{slug})(5\,\text{ft})}{864\,\text{slug} + 100\,\text{slug}}$

$= \boxed{1.863\,\text{ft}}$

$y_{cA} = 4.5\,\text{ft}$

$y_{cB} = 8\,\text{ft} + 0.5\,\text{ft} = 8.5\,\text{ft}$

$y_c = \dfrac{(864\,\text{slug})(4.5\,\text{ft}) + (100\,\text{slug})(8.5\,\text{ft})}{864\,\text{slug} + 100\,\text{slug}}$

$= \boxed{4.915\,\text{ft}}$

$z_{cA} = 1\,\text{ft}$

$z_{cB} = 1\,\text{ft} + 0.5\,\text{ft} = 1.5\,\text{ft}$

$z_c = \dfrac{(864\,\text{slug})(1\,\text{ft}) + (100\,\text{slug})(1.5\,\text{ft})}{864\,\text{slug} + 100\,\text{slug}}$

$= \boxed{1.052\,\text{ft}}$

SI Solution

$m_A = (0.6\,\text{m})(0.9\,\text{m})(2.7\,\text{m})\left(8300\,\dfrac{\text{kg}}{\text{m}^3}\right)$

$= 12\,101\,\text{kg}$

$m_B = (0.3\,\text{m})(0.3\,\text{m})(1.2\,\text{m})\left(13\,000\,\dfrac{\text{kg}}{\text{m}^3}\right)$

$= 1404\,\text{kg}$

$x_{cA} = 0.45\,\text{m}$

$x_{cB} = 0.9\,\text{m} + 0.6\,\text{m} = 1.5\,\text{m}$

$x_c = \dfrac{\Sigma m_i x_{ci}}{\Sigma m_i}$

$= \dfrac{(12\,101\,\text{kg})(0.45\,\text{m}) + (1404\,\text{kg})(1.5\,\text{m})}{12\,101\,\text{kg} + 1404\,\text{kg}}$

$= \boxed{0.5592\,\text{m}}$

$y_{cA} = 1.35\,\text{m}$

$y_{cB} = 2.4\,\text{m} + 0.15\,\text{m} = 2.55\,\text{m}$

$y_c = \dfrac{(12\,101\,\text{kg})(1.35\,\text{m}) + (1404\,\text{kg})(2.55\,\text{m})}{12\,101\,\text{kg} + 1404\,\text{kg}}$

$= \boxed{1.475\,\text{m}}$

$z_{cA} = 0.3\,\text{m}$

$z_{cB} = 0.3\,\text{m} + 0.15\,\text{m} = 0.45\,\text{m}$

$z_c = \dfrac{(12\,101\,\text{kg})(0.3\,\text{m}) + (1404\,\text{kg})(0.45\,\text{m})}{12\,101\,\text{kg} + 1404\,\text{kg}}$

$= \boxed{0.3156\,\text{m}}$

3. *Customary U.S. Solution*

$m_S = 1.5\,\text{lbm}$

$y_{cS} = 3\,\text{in} + 2.75\,\text{in} - 1.5\,\text{in} = 4.25\,\text{in}$

$m_C = 0.125\,\text{lbm}$

$y_{cC} = \dfrac{3h}{4} = \dfrac{(3)(3\,\text{in})}{4} = 2.25\,\text{in}$

$y_c = \dfrac{\Sigma m_i y_{ci}}{\Sigma m_i}$

$= \dfrac{(1.5\,\text{lbm})(4.25\,\text{in}) + (0.125\,\text{lbm})(2.25\,\text{in})}{1.5\,\text{lbm} + 0.125\,\text{lbm}}$

$= \boxed{4.096\,\text{in from the }x\text{-axis}}$

SI Solution

$m_S = 0.70\,\text{kg}$

$y_{cS} = 8\,\text{cm} + \dfrac{8\,\text{cm}}{2} = 12\,\text{cm}$

$m_C = 0.06\,\text{kg}$

$y_{cC} = \dfrac{3h}{4} = \dfrac{(3)(8\,\text{cm})}{4} = 6.0\,\text{cm}$

$y_c = \dfrac{\Sigma m_i y_{ci}}{\Sigma m_i}$

$= \dfrac{(0.70\,\text{kg})(12\,\text{cm}) + (0.06\,\text{kg})(6.0\,\text{cm})}{0.70\,\text{kg} + 0.06\,\text{kg}}$

$= \boxed{11.53\,\text{cm from the }x\text{-axis}}$

4. *Customary U.S. Solution*

Since A is a rectangular parallelepiped,

$I_{cA} = \dfrac{1}{12}m_A(a^2 + b^2)$

$= \left(\dfrac{1}{12}\right)(864\,\text{slug})\left((2\,\text{ft})^2 + (9\,\text{ft})^2\right)$

$= 6120\,\text{slug-ft}^2$

The distance, d, from the x-centroidal axis to the x-axis is found by

$d = \sqrt{(y_c)^2 + (z_c)^2} = \sqrt{(4.5\,\text{ft})^2 + (1\,\text{ft})^2} = 4.61\,\text{ft}$

$I_{xA} = I_{cA} + m_A d^2$

$= 6120\,\text{slug-ft}^2 + (864\,\text{slug})(4.61\,\text{ft})^2$

$= 24{,}482\,\text{slug-ft}^2$

For B,

$I_{cB} = \dfrac{1}{12}m_B(a^2 + b^2)$

$= \left(\dfrac{1}{12}\right)(100\,\text{slug})\left((1\,\text{ft})^2 + (1\,\text{ft})^2\right)$

$= 16.67\,\text{slug-ft}^2$

$$d = \sqrt{(8.5 \text{ ft})^2 + (1.5 \text{ ft})^2} = 8.63 \text{ ft}$$

$$I_{xB} = I_{cB} + m_B d^2$$
$$= 16.67 \text{ slug-ft}^2 + (100 \text{ slug}) (8.63 \text{ ft})^2$$
$$= 7464 \text{ slug-ft}^2$$

$$I_x = I_{xA} + I_{xB}$$
$$= 24{,}482 \text{ slug-ft}^2 + 7464 \text{ slug-ft}^2$$
$$= \boxed{31{,}946 \text{ slug-ft}^2}$$

SI Solution

Since A is a rectangular parallelepiped,

$$I_{cA} = \frac{1}{12} m_A \left(a^2 + b^2 \right)$$
$$= \left(\frac{1}{12} \right) (12\,101 \text{ kg}) \left((0.6 \text{ m})^2 + (2.7 \text{ m})^2 \right)$$
$$= 7714 \text{ kg·m}^2$$

The distance, d, from the x-centroidal axis to the x-axis is found by

$$d = \sqrt{(y_c)^2 + (z_c)^2} = \sqrt{(1.35 \text{ m})^2 + (0.3 \text{ m})^2} = 1.38 \text{ m}$$

$$I_{xA} = I_{cA} + m_A d^2$$
$$= 7714 \text{ kg·m}^2 + (12\,101 \text{ kg}) (1.38 \text{ m})^2$$
$$= 30\,759 \text{ kg·m}^2$$

For B,

$$I_{cB} = \frac{1}{12} m_B \left(a^2 + b^2 \right)$$
$$= \left(\frac{1}{12} \right) (1404 \text{ kg}) \left((0.3 \text{ m})^2 + (0.3 \text{ m})^2 \right)$$
$$= 21.06 \text{ kg·m}^2$$

$$d = \sqrt{(2.55 \text{ m})^2 + (0.45 \text{ m})^2} = 2.59 \text{ m}$$

$$I_{xB} = I_{cB} + m_B d^2$$
$$= 21.06 \text{ kg·m}^2 + (1404 \text{ kg}) (2.59 \text{ m})^2$$
$$= 9439 \text{ kg·m}^2$$

$$I_x = I_{xA} + I_{xB}$$
$$= 30\,759 \text{ kg·m}^2 + 9439 \text{ kg·m}^2$$
$$= \boxed{40\,198 \text{ kg·m}^2}$$

5. *Customary U.S. Solution*

For the cone,

$$I_{xC} = \frac{3}{5} m_C \left(\frac{r^2}{4} + h^2 \right)$$
$$= \left(\frac{3}{5} \right) (0.125 \text{ lbm}) \left(\frac{(1.25 \text{ in})^2}{4} + (3 \text{ in})^2 \right)$$
$$= 0.704 \text{ lbm-in}^2$$

For the sphere,

$$I_{cS} = \frac{2}{5} m_S r^2 = \left(\frac{2}{5} \right) (1.5 \text{ lbm}) (1.5 \text{ in})^2$$
$$= 1.35 \text{ lbm-in}^2$$
$$I_{xS} = I_{cS} + m_S d^2$$
$$= 1.35 \text{ lbm-in}^2 + (1.5 \text{ lbm})(4.25 \text{ in})^2$$
$$= 28.44 \text{ lbm-in}^2$$

$$I_x = I_{xC} + I_{xS} = 0.704 \text{ lbm-in}^2 + 28.44 \text{ lbm-in}^2$$
$$= \boxed{29.14 \text{ lbm-in}^2}$$

SI Solution

For the cone,

$$I_{xC} = \frac{3}{5} m_C \left(\frac{r^2}{4} + h^2 \right)$$
$$= \left(\frac{3}{5} \right) (0.06 \text{ kg}) \left(\frac{(0.03 \text{ m})^2}{4} + (0.08 \text{ m})^2 \right)$$
$$= 0.000239 \text{ kg·m}^2$$

For the sphere,

$$I_{cS} = \frac{2}{5} m_S r^2 = \left(\frac{2}{5} \right) (0.70 \text{ kg}) (0.04 \text{ m})^2$$
$$= 0.000448 \text{ kg·m}^2$$
$$I_{xS} = I_{cS} + m_S d^2$$
$$= 0.000448 \text{ kg·m}^2 + (0.7 \text{ kg})(0.12 \text{ m})^2$$
$$= 0.010528 \text{ kg·m}^2$$

$$I_x = I_{xC} + I_{xS} = 0.000239 \text{ kg·m}^2 + 0.010528 \text{ kg·m}^2$$
$$= \boxed{0.010767 \text{ kg·m}^2}$$

6. *Customary U.S. Solution*

$$m = (64.4 \text{ lbm}) \left(\frac{\text{slug}}{32.2 \text{ lbm}} \right) = 2 \text{ slug}$$

$$I_{CC} = I_{AA} - m d^2 = 90 \text{ slug-ft}^2 - (2 \text{ slug}) (4 \text{ ft})^2$$
$$= 58 \text{ slug-ft}^2$$

$$I_{BB} = I_{CC} + md'^2 = 58 \text{ slug-ft}^2 + (2 \text{ slug})(6 \text{ ft})^2$$

$$= \boxed{130 \text{ slug-ft}^2}$$

SI Solution

$$I_{CC} = I_{AA} - md^2$$

$$= 120 \text{ kg·m}^2 - (29.2 \text{ kg})(1.2 \text{ m})^2$$

$$= 77.95 \text{ kg·m}^2$$

$$I_{BB} = I_{CC} + md'^2$$

$$= 77.95 \text{ kg·m}^2 + (29.2 \text{ kg})(1.8 \text{ m})^2$$

$$= \boxed{172.56 \text{ kg·m}^2}$$

7. *Customary U.S. Solution*

From the hollow circular cylinder in App. 51.A, the mass moment of inertia of the rim is

$$I_{x,1} = \left(\frac{\pi \rho L}{2}\right)(r_o^4 - r_i^4)$$

$$= \left(\frac{\pi}{2}\right)\left(0.256 \frac{\text{lbm}}{\text{in}^3}\right)(12 \text{ in})$$

$$\times \left(\left(\frac{60 \text{ in}}{2}\right)^4 - \left(\frac{60 \text{ in} - 12 \text{ in}}{2}\right)^4\right)\left(\frac{1 \text{ ft}^2}{144 \text{ in}^2}\right)$$

$$= 16{,}025 \text{ lbm-ft}^2$$

From the hollow circular cylinder in App. 51.A, the mass moment of inertia of the hub is

$$I_{x,2} = \left(\frac{\pi \rho L}{2}\right)(r_o^4 - r_i^4)$$

$$= \left(\frac{\pi}{2}\right)\left(0.256 \frac{\text{lbm}}{\text{in}^3}\right)(12 \text{ in})$$

$$\times \left(\left(\frac{12 \text{ in}}{2}\right)^4 - \left(\frac{12 \text{ in} - 6 \text{ in}}{2}\right)^4\right)\left(\frac{1 \text{ ft}^2}{144 \text{ in}^2}\right)$$

$$= 42 \text{ lbm-ft}^2$$

The length of a cylindrical spoke is

$$L = 24 \text{ in} - 6 \text{ in} = 18 \text{ in}$$

The mass of a spoke is

$$m = \rho A L = \rho \pi r^2 L$$

$$= \left(0.256 \frac{\text{lbm}}{\text{in}^3}\right)\pi \left(\frac{4.25 \text{ in}}{2}\right)^2 (18 \text{ in})$$

$$= 65.37 \text{ lbm}$$

From the solid circular cylinder in App. 51.A, the mass moment of inertia of a spoke about its own centroidal axis is

$$I_z = \frac{m(3r^2 + L^2)}{12}$$

$$= \frac{(65.37 \text{ lbm})\left((3)\left(\frac{4.25 \text{ in}}{2}\right)^2 + (18 \text{ in})^2\right)\left(\frac{1 \text{ ft}^2}{144 \text{ in}^2}\right)}{12}$$

$$= 13 \text{ lbm-ft}^2$$

Use the parallel axis theorem, Eq. 51.12, to find the mass moment of inertia of a spoke about the axis of the flywheel.

$$d = \frac{12 \text{ in}}{2} + \frac{18 \text{ in}}{2} = 15 \text{ in}$$

$$I_{x,3} \text{ per spoke} = I_z + md^2$$

$$= 13 \text{ lbm-ft}^2 + (65.37 \text{ lbm})\left(\frac{15 \text{ in}}{12 \frac{\text{in}}{\text{ft}}}\right)^2$$

$$= 115 \text{ lbm-ft}^2$$

The total for six spokes is

$$I_{x,3} = 6I_{x,3} \text{ per spoke}$$

$$= (6)(115 \text{ lbm-ft}^2) = 690 \text{ lbm-ft}^2$$

Finally, the total rotational mass moment of inertia of the flywheel is

$$I = I_{x,1} + I_{x,2} + I_{x,3}$$

$$= 16{,}025 \text{ lbm-ft}^2 + 42 \text{ lbm-ft}^2 + 690 \text{ lbm-ft}^2$$

$$= \boxed{16{,}757 \text{ lbm-ft}^2}$$

The answer is (D).

SI Solution

From the hollow circular cylinder in App. 51.A, the mass moment of inertia of the rim is

$$I_{x,1} = \left(\frac{\pi \rho L}{2}\right)(r_o^4 - r_i^4)$$

$$= \left(\frac{\pi \left(7080 \frac{\text{kg}}{\text{m}^3}\right)(0.3 \text{ m})}{2}\right)$$

$$\times \left(\left(\frac{1.5 \text{ m}}{2}\right)^4 - \left(\frac{1.5 \text{ m} - 0.30 \text{ m}}{2}\right)^4\right)$$

$$= 623.3 \text{ kg·m}^2$$

Plant Design

From the hollow circular cylinder in App. 51.A, the mass moment of inertia of the hub is

$$I_{x,2} = \left(\frac{\pi \rho L}{2}\right)(r_o^4 - r_i^4)$$

$$= \left(\frac{\pi \left(7080 \frac{\text{kg}}{\text{m}^3}\right)(0.3 \text{ m})}{2}\right)$$

$$\times \left(\left(\frac{0.3 \text{ m}}{2}\right)^4 - \left(\frac{0.3 \text{ m} - 0.15 \text{ m}}{2}\right)^4\right)$$

$$= 1.6 \text{ kg·m}^2$$

The length of a cylindrical spoke is

$$L = 0.6 \text{ m} - 0.15 \text{ m} = 0.45 \text{ m}$$

The mass of a spoke is

$$m = \rho A L = \rho \pi r^2 L$$

$$= \left(7080 \frac{\text{kg}}{\text{m}^3}\right)\pi \left(\frac{0.11 \text{ m}}{2}\right)^2 (0.45 \text{ m})$$

$$= 30.3 \text{ kg}$$

From the solid circular cylinder in App. 51.A, the mass moment of inertia of a spoke about its own centroidal axis is

$$I_z = \frac{m(3r^2 + L^2)}{12}$$

$$= \frac{(30.3 \text{ kg})\left((3)\left(\frac{0.11 \text{ m}}{2}\right)^2 + (0.45 \text{ m})^2\right)}{12}$$

$$= 0.53 \text{ kg·m}^2$$

Use the parallel axis theorem, Eq. 51.12, to find the mass moment of inertia of a spoke about the axis of the flywheel.

$$d = \frac{0.30 \text{ m}}{2} + \frac{0.45 \text{ m}}{2} = 0.375 \text{ m}$$

$$I_{x,3} \text{ per spoke} = I_z + md^2$$

$$= 0.53 \text{ kg·m}^2 + (30.3 \text{ kg})(0.375 \text{ m})^2$$

$$= 4.8 \text{ kg·m}^2$$

The total for six spokes is

$$I_{x,3} = 6 I_{x3} \text{ per spoke}$$

$$= (6)(4.8 \text{ kg·m}^2) = 28.8 \text{ kg·m}^2$$

Finally, the total rotational mass moment of inertia of the flywheel is

$$I = I_{x,1} + I_{x,2} + I_{x,3}$$

$$= 623.3 \text{ kg·m}^2 + 1.6 \text{ kg·m}^2 + 28.8 \text{ kg·m}^2$$

$$= \boxed{653.7 \text{ kg·m}^2}$$

The answer is (D).

8. *Customary U.S. Solution*

The angular velocity of the flywheel at the start of the cycle is

$$\omega_1 = \frac{\left(2\pi \frac{\text{rad}}{\text{rev}}\right)\left(200 \frac{\text{rev}}{\text{min}}\right)}{60 \frac{\text{sec}}{\text{min}}} = 20.94 \text{ rad/sec}$$

The angular velocity of the flywheel at the end of the cycle is

$$\omega_2 = \frac{\left(2\pi \frac{\text{rad}}{\text{rev}}\right)\left(175 \frac{\text{rev}}{\text{min}}\right)}{60 \frac{\text{sec}}{\text{min}}} = 18.33 \text{ rad/sec}$$

By the work-energy principle, the decrease in kinetic energy is equal to the work supplied by the flywheel.

$$\Delta \text{KE} = W$$

$$\left(\tfrac{1}{2}\right)\left(\frac{I}{g_c}\right)\omega_1^2 - \left(\tfrac{1}{2}\right)\left(\frac{I}{g_c}\right)\omega_2^2 = W$$

Solve for the mass moment of inertia, I.

$$I = \frac{2 g_c W}{\omega_1^2 - \omega_2^2}$$

$$= \frac{(2)\left(32.2 \frac{\text{ft-lbm}}{\text{lbf-sec}^2}\right)(1500 \text{ ft-lbf})}{\left(20.94 \frac{\text{rad}}{\text{sec}}\right)^2 - \left(18.33 \frac{\text{rad}}{\text{sec}}\right)^2}$$

$$= 942.5 \text{ lbm-ft}^2$$

Assume all of the rim mass is concentrated at the mean radius. The mean circumference is

$$L_{\text{mean}} = 2\pi r_{\text{mean}}$$

$$= 2\pi \left(\frac{30 \text{ in}}{2}\right) = 94.25 \text{ in}$$

The mass moment of inertia of the rim is

$$I_{\text{rim}} = m r_{\text{mean}}^2 = \rho t w L_{\text{mean}} r_{\text{mean}}^2$$

$$= \left(0.26 \frac{\text{lbm}}{\text{in}^3}\right)t(12 \text{ in})(94.25 \text{ in})$$

$$\times \left(\frac{30 \text{ in}}{2}\right)^2 \left(\frac{1 \text{ ft}^2}{144 \text{ in}^2}\right)$$

$$= 459.5t$$

Adding 10% for the hub and arms,

$$I_{\text{total}} = 1.10 I_{\text{rim}} = (1.10)(459.5t)$$

$$= 505.5t$$

Solve for t.

$$t = \frac{I}{\frac{I_{\text{total}}}{t}} = \frac{942.5 \text{ lbm-ft}^2}{505.5 \frac{\text{lbm-ft}^2}{\text{in}}}$$

$$= \boxed{1.86 \text{ in}}$$

The answer is (A).

SI Solution

The angular velocity of the flywheel at the start of the cycle is

$$\omega_1 = \frac{\left(2\pi \frac{\text{rad}}{\text{rev}}\right)\left(200 \frac{\text{rev}}{\text{min}}\right)}{60 \frac{\text{sec}}{\text{min}}} = 20.94 \text{ rad/sec}$$

The angular velocity of the flywheel at the end of the cycle is

$$\omega_2 = \frac{\left(2\pi \frac{\text{rad}}{\text{rev}}\right)\left(175 \frac{\text{rev}}{\text{min}}\right)}{60 \frac{\text{sec}}{\text{min}}} = 18.33 \text{ rad/sec}$$

By the work-energy principle, the decrease in kinetic energy is equal to the work supplied by the flywheel.

$$\Delta \text{KE} = W$$
$$\tfrac{1}{2}I\omega_1^2 - \tfrac{1}{2}I\omega_2^2 = W$$

Solve for the mass moment of inertia, I.

$$I = \frac{2W}{\omega_1^2 - \omega_2^2}$$

$$= \frac{(2)(2 \text{ kJ})\left(1000 \frac{\text{J}}{\text{kJ}}\right)}{\left(20.94 \frac{\text{rad}}{\text{sec}}\right)^2 - \left(18.33 \frac{\text{rad}}{\text{sec}}\right)^2}$$

$$= 39.03 \text{ kg·m}^2$$

Assume all of the rim mass is concentrated at the mean radius. The mean circumference is

$$L_{\text{mean}} = 2\pi r_{\text{mean}}$$
$$= 2\pi \left(\frac{0.762 \text{ m}}{2}\right) = 2.394 \text{ m}$$

The mass moment of inertia of the rim is

$$I_{\text{rim}} = mr_{\text{mean}}^2 = \rho t w L_{\text{mean}} r_{\text{mean}}^2$$
$$= \left(7200 \frac{\text{kg}}{\text{m}^3}\right) t (0.305 \text{ m})(2.394 \text{ m}) \left(\frac{0.762 \text{ m}}{2}\right)^2$$
$$= 763.1t$$

Adding 10% for the hub and arms,

$$I_{\text{total}} = 1.10 I_{\text{rim}} = (1.10)(763.1t)$$
$$= 839.4t$$

Solve for t.

$$t = \frac{I}{\frac{I_{\text{total}}}{t}} = \frac{39.03 \text{ kg·m}^2}{839.4 \frac{\text{kg·m}^2}{\text{m}}}$$

$$= \boxed{0.0465 \text{ m}}$$

The answer is (A).

Plant Design

52 Engineering Materials

PRACTICE PROBLEMS

1. How much (in molecules of HCl per gram of PVC) HCl should be used as an initiator in PVC if the efficiency is 20% and an average molecular weight of 7000 g/mol is desired? The final polymer has the following structure.

(A) 7.8×10^{18} molecules HCl per gram PVC

(B) 4.3×10^{20} molecules HCl per gram PVC

(C) 9.5×10^{21} molecules HCl per gram PVC

(D) 3.6×10^{23} molecules HCl per gram PVC

2. 10 ml of a 0.2% solution (by weight) of hydrogen peroxide is added to 12 g of ethylene to stabilize the polymer. What is the average degree of polymerization if the hydrogen peroxide is completely utilized? Assume that hydrogen breaks down according to

$$H_2O_2 \rightarrow 2(OH^-) + \cdots$$

Assume that the stabilized polymer has the following structure.

(A) 180

(B) 730

(C) 910

(D) 1200

SOLUTIONS

1. The vinyl chloride mer is

With 20% efficiency, 5 molecules of HCl per PVC molecule are required to supply each end Cl atom. This is the same as 5 mol HCl per mole PVC. Using Avogadro's number of 6.022×10^{23} molecules per mole, the number of molecules of HCl per gram of PVC is

$$\frac{\left(5 \; \dfrac{\text{mol HCl}}{\text{mol PVC}}\right)\left(6.022 \times 10^{23} \; \dfrac{\text{molecules}}{\text{mol}}\right)}{7000 \; \dfrac{\text{g}}{\text{mol}}}$$

$$= \boxed{4.30 \times 10^{20} \text{ molecules HCl/gram PVC}}$$

The answer is (C).

2. The molecular weight of hydrogen peroxide, H_2O_2, is

$$(2)\left(1 \; \frac{\text{g}}{\text{mol}}\right) + (2)\left(16 \; \frac{\text{g}}{\text{mol}}\right) = 34 \text{ g/mol}$$

The weight of H_2O_2 is

$$(10 \text{ mL})\left(1 \; \frac{\text{g}}{\text{mL}}\right) = 10 \text{ g}$$

The number of H_2O_2 molecules in a 0.2% solution is

$$\frac{(10 \text{ g})\left(\dfrac{0.2\%}{100\%}\right)\left(6.022 \times 10^{23} \; \dfrac{\text{molecules}}{\text{mol}}\right)}{34 \; \dfrac{\text{g}}{\text{mol}}}$$

$$= 3.54 \times 10^{20} \text{ molecules}$$

The molecular weight of ethylene, C_2H_4, is

$$(2)\left(12 \; \frac{\text{g}}{\text{mol}}\right) + (4)\left(1 \; \frac{\text{g}}{\text{mol}}\right) = 28 \text{ g/mol}$$

The number of ethylene molecules is

$$\frac{(12 \text{ g}) \left(6.022 \times 10^{23} \, \frac{\text{molecules}}{\text{mol}}\right)}{28 \, \frac{\text{g}}{\text{mol}}}$$

$$= 2.58 \times 10^{23} \text{ molecules}$$

Since it takes one H_2O_2 molecule (i.e., $2OH^-$ radicals) to stabilize a polyethylene molecule, there are 3.54×10^{20} polymers.

The degree of polymerization is

$$\text{DP} = \frac{2.58 \times 10^{23} \text{ C}_2\text{H}_4 \text{ molecules}}{3.54 \times 10^{20} \text{ polymers}} = \boxed{729}$$

The answer is (B).

53 Physical Properties of Construction Materials

PRACTICE PROBLEMS

1. A batch of fine aggregate was sieve graded. The percentages retained on each sieve are shown. What is the sand's fineness modulus?

sieve	percentage retained
4	4
8	11
16	21
30	22
50	24
100	17
dust (pan)	1

(A) 0.99
(B) 1.8
(C) 2.9
(D) 99

2. (*Time limit: one hour*) The 6 in × 6 in concrete beam section shown was tested in a third-point loading apparatus. The failure occurred outside the middle third as shown under a maximum total load of 5000 lbf. The beam was normal-weight concrete with a compressive strength of 5000 psi. Neglect the beam's self-weight.

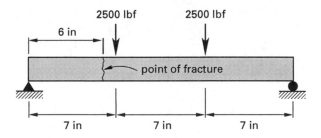

(a) The maximum shear force is most nearly
   (A) 625 lbf
   (B) 1250 lbf
   (C) 2500 lbf
   (D) 5000 lbf

(b) The maximum bending moment is most nearly
   (A) 5000 in-lbf
   (B) 10,000 in-lbf
   (C) 15,000 in-lbf
   (D) 17,500 in-lbf

(c) The shear force within the middle third of the beam is most nearly
   (A) zero
   (B) 625 lbf
   (C) 1250 lbf
   (D) 2500 lbf

(d) The bending moment within the middle third of the beam
   (A) is zero.
   (B) increases linearly from right to left.
   (C) is parabolic in shape.
   (D) is a straight line with zero slope.

(e) The modulus of rupture is most nearly
   (A) 320 psi
   (B) 420 psi
   (C) 520 psi
   (D) 620 psi

(f) If the fracture occurred within the middle third, the modulus of rupture is most nearly
   (A) 420 psi
   (B) 450 psi
   (C) 470 psi
   (D) 490 psi

(g) The modulus of rupture using the ACI empirical equation is most nearly
   (A) 530 psi
   (B) 550 psi
   (C) 570 psi
   (D) 590 psi

(h) If the beam is made of lightweight concrete, the modulus of rupture is most nearly
   (A) 380 psi
   (B) 400 psi
   (C) 420 psi
   (D) 440 psi

(i) The center-point loading test is no longer standard in determining the modulus of rupture for concrete because

    (A) it yields higher values of modulus of rupture than the true values.

    (B) it is difficult to perform accurately in the laboratory.

    (C) it costs more than the third-point loading test.

    (D) of all of the above.

(j) The modulus of rupture test yields a higher value of strength than a direct tensile test and splitting tensile test made on the same specimen because

    (A) the assumed stress block shape does not match the real shape.

    (B) direct tensile tests are sensitive to any accidental eccentricity.

    (C) the concrete is assumed to be perfectly elastic.

    (D) of all of the above.

3. If a concrete has a density of 2300 kg/m^3 and a compressive strength of 15 MPa, what modulus of elasticity is predicted by ACI 318?

    (A) 14 MPa

    (B) 18 MPa

    (C) 28 MPa

    (D) 18 GPa

4. (*Time limit: one hour*) A normal-weight concrete specimen tested at 28 days yielded the following results.

(a) A 6 in × 12 in concrete cylinder failed at an axial compressive force of 105,000 lbf. The ultimate compressive strength is most nearly

    (A) 1200 psi

    (B) 1900 psi

    (C) 3700 psi

    (D) 17,500 psi

(b) A 6 in × 12 in concrete cylinder resisted a transverse force of 45,000 lbf in a split tensile cylinder test. The concrete tensile strength is most nearly

    (A) 400 psi

    (B) 450 psi

    (C) 625 psi

    (D) 666 psi

(c) A 6 in square unreinforced beam section resisted a force of 5400 lbf on a 21 in span length. A third-point loading test was used, and the fracture occurred within the middle third. The modulus of rupture is most nearly

    (A) 150 psi

    (B) 400 psi

    (C) 525 psi

    (D) 900 psi

(d) The ratio of the modulus of rupture to the compressive strength is most nearly

    (A) 4%

    (B) 5%

    (C) 12%

    (D) 14%

(e) The ratio of the split tensile strength to the compressive strength is most nearly

    (A) 8%

    (B) 11%

    (C) 17%

    (D) 20%

(f) The conclusion that could be stated based on the answers of parts (d) and (e) is

    (A) concrete is weak in tension and strong in compression.

    (B) concrete is weak in tension and strong in shear.

    (C) concrete is strong in tension and weak in compression.

    (D) none of the above.

(g) Given a compressive strength of 3700 lbf/in^2, the approximate value of the modulus of rupture is most nearly

    (A) 460 psi

    (B) 500 psi

    (C) 550 psi

    (D) 600 psi

(h) Given a compressive strength of 3700 psi, the approximate value of the split tensile strength is most nearly

    (A) 300 psi

    (B) 410 psi

    (C) 500 psi

    (D) 600 psi

(i) From the compression test, the axial strain was 0.0015 in/in and the lateral strain was 0.00027 in/in. The Poission's ratio is most nearly

    (A) 0.00027

    (B) 0.0015

    (C) 0.10

    (D) 0.18

(j) For the beam in part (c), if the fracture occurred 1 in away from the left support, the value of the modulus of rupture obtained from this test

    (A) is less than the correct value.

    (B) equals the correct value.

    (C) is higher than the correct value.

    (D) becomes indeterminate.

5. Stress-strain curves are shown for grade-60 steel rebar and 4000 psi normal-weight concrete.

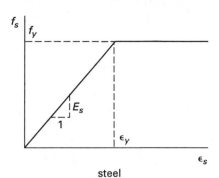

steel

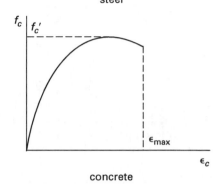

concrete

(a) What is the yield strength of the steel?
 (A) 30 ksi
 (B) 36 ksi
 (C) 50 ksi
 (D) 60 ksi

(b) What is the modulus of elasticity of the steel?
 (A) 10×10^6 psi
 (B) 29×10^6 psi
 (C) 30×10^6 psi
 (D) 36×10^6 psi

(c) What is the steel strain at the yield point?
 (A) 0.002 in/in
 (B) 0.003 in/in
 (C) 0.06 in/in
 (D) 0.09 in/in

(d) What is the concrete's compressive strength?
 (A) 2000 psi
 (B) 4000 psi
 (C) 6000 psi
 (D) 8000 psi

(e) What is the concrete's modulus of elasticity?
 (A) 1.1×10^6 psi
 (B) 1.7×10^6 psi
 (C) 2.9×10^6 psi
 (D) 3.6×10^6 psi

(f) What is the concrete's ultimate compressive strain?
 (A) 0.001 in/in
 (B) 0.003 in/in
 (C) 0.005 in/in
 (D) 0.009 in/in

Plant Design

SOLUTIONS

1. Consider the sieves in sequence from finest to coarsest. If the no. 10 sieve had been used first, only the pan (dust) would have passed through, and the retained percentage would have been $100\% - 1\% = 99\%$.

For the no. 50 sieve, the retained amount would have been $100\% - 1\% - 17\% = 82\%$.

The following table is prepared similarly.

sieve	cumulative retained
100	$100\% - 1\% = 99\%$
50	$99\% - 17\% = 82\%$
30	$82\% - 24\% = 58\%$
16	$58\% - 22\% = 36\%$
8	$36\% - 21\% = 15\%$
4	$15\% - 11\% = 4\%$
	TOTAL: 294

The fineness modulus is $294/100 = \boxed{2.94.}$

The answer is (C).

2. (a) From the shear diagram, the maximum shear force is $\boxed{2500 \text{ lbf.}}$

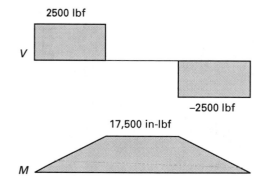

The answer is (C).

(b) From the moment diagram, the maximum moment is $\boxed{17,500 \text{ in-lbf.}}$

The answer is (D).

(c) From the shear diagram, the shear force in the middle third is $\boxed{\text{zero.}}$

The answer is (A).

(d) From the moment diagram, the moment is constant in the middle third.

The answer is (D).

(e) Use Eq. 53.7.

$$f_r = \frac{Mc}{I}$$
$$= \frac{(2500 \text{ lbf})(6 \text{ in})\left(\dfrac{6 \text{ in}}{2}\right)}{\dfrac{(6 \text{ in})(6 \text{ in})^3}{12}}$$
$$= \boxed{417 \text{ lbf/in}^2}$$

The answer is (B).

(f) Use Eq. 53.7.

$$f_r = \frac{Mc}{I}$$
$$= \frac{(17{,}500 \text{ in-lbf})\left(\dfrac{6 \text{ in}}{2}\right)}{\dfrac{(6 \text{ in})(6 \text{ in})^3}{12}}$$
$$= \boxed{486 \text{ lbf/in}^2}$$

The answer is (D).

(g) Use the empirical ACI equation, Eq. 53.8.

$$f_r = 7.5\sqrt{f_c'} = 7.5\sqrt{5000 \ \frac{\text{lbf}}{\text{in}^2}}$$
$$= \boxed{530 \text{ lbf/in}^2}$$

The answer is (A).

(h) Use the empirical ACI equation, Eq. 53.8.

$$f_r = (0.75)(7.5)\sqrt{f_c'} = (0.75)(7.5)\sqrt{5000 \ \frac{\text{lbf}}{\text{in}^2}}$$
$$= \boxed{397 \text{ lbf/in}^2}$$

The answer is (B).

(i) **The answer is (B).**

(j) **The answer is (D).**

3. Use Eq. 53.1. (Equation 53.2 could also be used.)

$$E_c = w^{1.5}(0.043)\sqrt{f_c'}$$
$$= \left(2300 \ \frac{\text{kg}}{\text{m}^3}\right)^{1.5}(0.043)\sqrt{15 \text{ MPa}}$$
$$= \boxed{18{,}369 \text{ MPa} \ \ (18.4 \text{ GPa})}$$

The answer is (D).

4. (a) Use Eq. 53.3.

$$f_c' = \frac{P}{A} = \frac{105{,}000 \text{ lbf}}{\left(\frac{\pi}{4}\right)(6 \text{ in})^2}$$

$$= \boxed{3714 \text{ lbf/in}^2}$$

The answer is (C).

(b) Use Eq. 53.7.

$$f_{ct} = \frac{2P}{\pi DL} = \frac{(2)(45{,}000 \text{ lbf})}{\pi(6 \text{ in})(12 \text{ in})}$$

$$= \boxed{398 \text{ lbf/in}^2}$$

The answer is (A).

(c) $$f_r = \frac{Mc}{I} = \frac{\left(\dfrac{5400 \text{ lbf}}{2}\right)\left(\dfrac{21 \text{ in}}{3}\right)\left(\dfrac{6 \text{ in}}{2}\right)}{\dfrac{(6 \text{ in})(6 \text{ in})^3}{12}}$$

$$= \boxed{525 \text{ lbf/in}^2}$$

The answer is (C).

(d) $$\frac{f_r}{f_c'} = \frac{525 \ \dfrac{\text{lbf}}{\text{in}^2}}{3714 \ \dfrac{\text{lbf}}{\text{in}^2}} = \boxed{0.141 \ (14\%)}$$

The answer is (D).

(e) $$\frac{f_{ct}}{f_c'} = \frac{398 \ \dfrac{\text{lbf}}{\text{in}^2}}{3714 \ \dfrac{\text{lbf}}{\text{in}^2}} = \boxed{0.107 \ (11\%)}$$

The answer is (B).

(f) *The answer is (A).*

(g) Use Eq. 53.8.

$$f_r = 7.5\sqrt{f_c'} = 7.5\sqrt{3700 \ \frac{\text{lbf}}{\text{in}^2}}$$

$$= \boxed{456 \text{ lbf/in}^2}$$

The answer is (A).

(h) Use Eq. 53.4.

$$f_{ct} = 6.7\sqrt{f_c'} = 6.7\sqrt{3700 \ \frac{\text{lbf}}{\text{in}^2}}$$

$$= \boxed{408 \text{ lbf/in}^2}$$

The answer is (B).

(i) $$\nu = \frac{\text{lateral strain}}{\text{axial strain}}$$

$$= \frac{0.00027 \ \dfrac{\text{in}}{\text{in}}}{0.0015 \ \dfrac{\text{in}}{\text{in}}} = \boxed{0.18}$$

The answer is (D).

(j) Refer to ASTM C78.

The answer is (D).

5. (a) The yield strength is 60 ksi for grade-60 steel.

The answer is (D).

(b) The standard modulus of elasticity for reinforcing steel is $\boxed{29 \times 10^6 \text{ psi.}}$

The answer is (B).

(c) The strain at yield is

$$\epsilon_y = \frac{f_y}{E}$$

$$= \frac{60{,}000 \ \dfrac{\text{lbf}}{\text{in}^2}}{29 \times 10^6 \ \dfrac{\text{lbf}}{\text{in}^2}}$$

$$= \boxed{0.00207 \text{ in/in}}$$

The answer is (A).

(d) The compressive strength is the concrete's designation.

The answer is (B).

(e) For normal-weight concrete,

$$E_c = 57{,}000\sqrt{f_c'} = 57{,}000\sqrt{4000 \ \frac{\text{lbf}}{\text{in}^2}}$$

$$= \boxed{3.6 \times 10^6 \text{ lbf/in}^2}$$

The answer is (D).

(f) Concrete is assumed to crack when it reaches a standard strain value of $\boxed{0.003 \text{ in/in.}}$

The answer is (B).

Plant Design

54 Material Testing

PRACTICE PROBLEMS

1. The engineering stress and engineering strain for a copper specimen are 20,000 lbf/in² (140 MPa) and 0.0200 in/in (0.0200 mm/mm), respectively. Poisson's ratio for the specimen is 0.3.

(a) What is the true stress?
- (A) 14,000 lbf/in² (98 MPa)
- (B) 18,000 lbf/in² (130 MPa)
- (C) 20,000 lbf/in² (140 MPa)
- (D) 22,000 lbf/in² (160 MPa)

(b) What is the true strain?
- (A) 0.0182 in/in (0.0182 mm/mm)
- (B) 0.0189 in/in (0.0189 mm/mm)
- (C) 0.0194 in/in (0.0194 mm/mm)
- (D) 0.0198 in/in (0.0198 mm/mm)

2. A graph of engineering stress-strain is shown. Poisson's ratio for the material is 0.3.

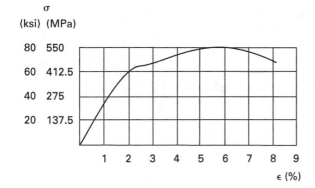

(a) Find the 0.5% yield strength.
- (A) 70,000 lbf/in² (480 MPa)
- (B) 76,000 lbf/in² (530 MPa)
- (C) 84,000 lbf/in² (590 MPa)
- (D) 98,000 lbf/in² (690 MPa)

(b) Find the elastic modulus.
- (A) 2.4×10^6 lbf/in² (17 GPa)
- (B) 2.7×10^6 lbf/in² (19 GPa)
- (C) 2.9×10^6 lbf/in² (20 GPa)
- (D) 3.0×10^6 lbf/in² (21 GPa)

(c) Find the ultimate strength.
- (A) 72,000 lbf/in² (500 MPa)
- (B) 76,000 lbf/in² (530 MPa)
- (C) 80,000 lbf/in² (550 MPa)
- (D) 84,000 lbf/in² (590 MPa)

(d) Find the fracture strength.
- (A) 62,000 lbf/in² (430 MPa)
- (B) 70,000 lbf/in² (480 MPa)
- (C) 76,000 lbf/in² (530 MPa)
- (D) 84,000 lbf/in² (590 MPa)

(e) Find the percentage of elongation at fracture.
- (A) 6%
- (B) 8%
- (C) 10%
- (D) 12%

(f) Find the shear modulus.
- (A) 0.9×10^6 lbf/in² (6.3 GPa)
- (B) 1.1×10^6 lbf/in² (7.7 GPa)
- (C) 1.2×10^6 lbf/in² (7.9 GPa)
- (D) 1.5×10^6 lbf/in² (11 GPa)

(g) Find the toughness.
- (A) 5100 in-lbf/in³ (35 MJ/m³)
- (B) 5700 in-lbf/in³ (38 MJ/m³)
- (C) 6300 in-lbf/in³ (42 MJ/m³)
- (D) 8900 in-lbf/in³ (60 MJ/m³)

3. A specimen with an unstressed cross-sectional area of 4 in² (25 cm²) necks down to 3.42 in² (22 cm²) before breaking in a standard tensile test. What is the reduction in area of the material?
- (A) 0.094 (0.089)
- (B) 0.10 (0.094)
- (C) 0.13 (0.10)
- (D) 0.15 (0.12)

4. A constant 15,000 lbf/in^2 (100 MPa) tensile stress is applied to a specimen. The stress is known to be less than the material's yield strength. The strain is measured at various times. What is the steady-state creep rate for the material?

time (hr)	strain (in/in)
5	0.018
10	0.022
20	0.026
30	0.031
40	0.035
50	0.040
60	0.046
70	0.058

(A) 0.00037 hr^{-1}
(B) 0.00041 hr^{-1}
(C) 0.00046 hr^{-1}
(D) 0.00049 hr^{-1}

SOLUTIONS

1. *Customary U.S. Solution*

The fractional reduction in diameter is

$$\nu e = (0.3)\left(0.020 \; \frac{in}{in}\right) = 0.006$$

(a) The true stress is given by Eq. 54.5.

$$\sigma = \frac{F}{A_o(1 - \nu e)^2} = \left(\frac{F_o}{A_o}\right)\left(\frac{1}{(1 - \nu e)^2}\right)$$

$$= \left(20{,}000 \; \frac{lbf}{in^2}\right)\left(\frac{1}{(1 - 0.006)^2}\right)$$

$$= \boxed{20{,}242 \; lbf/in^2}$$

The answer is (C).

(b) The true strain is given by Eq. 54.6.

$$\epsilon = \ln(1 + e) = \ln(1 + 0.020) = \boxed{0.0198 \; in/in}$$

The answer is (D).

SI Solution

The fractional reduction in diameter is

$$\nu e = (0.3)\left(0.020 \; \frac{mm}{mm}\right) = 0.006$$

(a) The true stress is given by Eq. 54.5.

$$\sigma = \frac{F}{A_o(1 - \nu e)^2} = \left(\frac{F}{A_o}\right)\left(\frac{1}{(1 - \nu e)^2}\right)$$

$$= (140 \; MPa)\left(\frac{1}{(1 - 0.006)^2}\right)$$

$$= \boxed{141.7 \; MPa}$$

The answer is (C).

(b) The true strain is

$$\epsilon = \ln(1 + e) = \ln(1 + 0.020) = \boxed{0.0198 \; mm/mm}$$

The answer is (D).

2. *Customary U.S. Solution*

(a) Extend a line from the 0.5% offset strain value parallel to the linear portion of the curve.

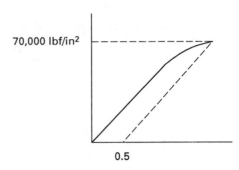

70,000 lbf/in²

0.5

The 0.5% yield strength is $\boxed{70{,}000 \text{ lbf/in}^2.}$

The answer is (A).

(b) At the elastic limit, the stress is 60,000 lbf/in² and the percent strain is 2. The elastic modulus is

$$E = \frac{\text{stress}}{\text{strain}} = \frac{60{,}000 \ \dfrac{\text{lbf}}{\text{in}^2}}{0.02}$$

$$= \boxed{3 \times 10^6 \text{ lbf/in}^2}$$

The answer is (D).

(c) The ultimate strength is the highest point of the curve. This value is $\boxed{80{,}000 \text{ lbf/in}^2.}$

The answer is (C).

(d) The fracture strength is at the end of the curve. This value is $\boxed{70{,}000 \text{ lbf/in}^2.}$

The answer is (B).

(e) The percent elongation just prior to fracture is determined by extending a straight line down from the fracture point. This gives an approximate value of $\boxed{6\%.}$

The answer is (A).

(f) The shear modulus is given by Eq. 54.21.

$$G = \frac{E}{2(1+\nu)} = \frac{3 \times 10^6 \ \dfrac{\text{lbf}}{\text{in}^2}}{(2)(1+0.3)}$$

$$= \boxed{1.15 \times 10^6 \text{ lbf/in}^2}$$

The answer is (C).

(g) The toughness is the area under the stress-strain curve. Divide the area into squares of 20 ksi × 1%. There are about 25.5 squares covered.

$$(25.5)\left(20{,}000 \ \frac{\text{lbf}}{\text{in}^2}\right)\left(0.01 \ \frac{\text{in}}{\text{in}}\right) = \boxed{5100 \text{ in-lbf/in}^3}$$

The answer is (A).

SI Solution

(a) Extend a line from the 0.5% offset strain value parallel to the linear portion of the curve.

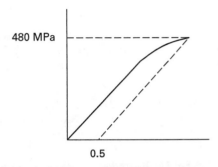

480 MPa

0.5

The 0.5% yield strength is $\boxed{480 \text{ MPa.}}$

The answer is (A).

(b) At the elastic limit, the stress is 410 MPa and the percent strain is 2. The elastic modulus is

$$E = \frac{\text{stress}}{\text{strain}} = \frac{410 \text{ MPa}}{(0.02)\left(1000 \ \dfrac{\text{MPa}}{\text{GPa}}\right)}$$

$$= \boxed{20.5 \text{ GPa}}$$

The answer is (D).

(c) The ultimate strength is the highest point of the curve. This value is $\boxed{550 \text{ MPa.}}$

The answer is (C).

(d) The fracture strength is at the end of the curve. This value is $\boxed{480 \text{ MPa.}}$

The answer is (B).

Plant Design

(e) The percent elongation at fracture is determined by extending a straight line parallel to the initial strain line from the fracture point. This gives an approximate value of $\boxed{6\%.}$

The answer is (A).

(f) The shear modulus is given by Eq. 54.21.

$$G = \frac{E}{2(1+\nu)} = \frac{20.5 \text{ GPa}}{(2)(1+0.3)}$$

$$= \boxed{7.88 \text{ GPa}}$$

The answer is (C).

(g) The toughness is the area under the stress-strain curve. Divide the area into squares of 137.5 MPa × 1%. There are about 25.5 squares covered.

$$(25.5)(137.5 \text{ MPa})\left(0.01 \ \frac{\text{m}}{\text{m}}\right) = \boxed{35 \text{ MJ/m}^3}$$

The answer is (A).

3. *Customary U.S. Solution*

The reduction in area is found from Eq. 54.12.

$$\text{reduction in area} = \frac{A_o - A_f}{A_o}$$

$$= \frac{4.0 \text{ in}^2 - 3.42 \text{ in}^2}{4.0 \text{ in}^2} = \boxed{0.145}$$

The answer is (D).

SI Solution

Use the reduction in area as the measure of ductility. Use Eq. 54.12.

$$\text{ductility} = \text{reduction in area} = \frac{A_o - A_f}{A_o}$$

$$= \frac{25 \text{ cm}^2 - 22 \text{ cm}^2}{25 \text{ cm}^2} = \boxed{0.12}$$

The answer is (D).

4. Plot the data and draw a straight line. Disregard the first and last data points, as these represent primary and tertiary creep, respectively.

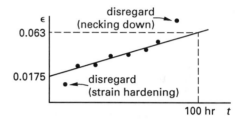

The creep rate is the slope of the line.

$$\text{creep rate} = \frac{\Delta\epsilon}{\Delta t} = \frac{0.063 \ \frac{\text{in}}{\text{in}} - 0.0175 \ \frac{\text{in}}{\text{in}}}{100 \text{ hr}}$$

$$= \boxed{0.000455 \ \frac{1}{\text{hr}}}$$

The answer is (C).

55 Thermal Treatment of Metals

PRACTICE PROBLEMS

1. Refer to the following equilibrium diagram for an alloy of elements A and B.

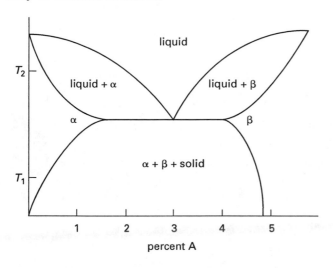

percent A

(a) For a 4% A alloy at temperature T_1, what are the compositions of solids α and β?

(b) For a 1% A alloy at temperature T_2, how much liquid and how much solid are present?

2. Write the procedures used with 2011 aluminum for (a) annealing and (b) precipitation hardening.

SOLUTIONS

1. (a) For temperature T_1, the equilibrium diagram is

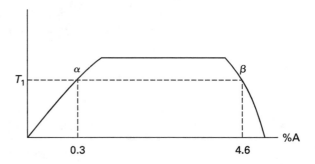

From the diagram, solid α is $\boxed{0.3\% \text{ A.}}$

The %B for solid α is

$$\%B = 100\% - 0.3\% = \boxed{99.7\%}$$

For solid β,

$$\%A = \boxed{4.6\%}$$

$$\%B = 100\% - 4.6\% = \boxed{95.4\%}$$

(b) For temperature T_2, the equilibrium diagram is

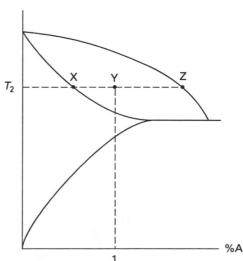

From the diagram, the following distances are measured.

The distance from point X to point Y is 9 mm.

The distance from point X to point Z is 21 mm.

From Eq. 55.2, the percent liquid is

$$\left(\frac{XY}{XZ}\right)(100\%) = \left(\frac{9 \text{ mm}}{21 \text{ mm}}\right)(100\%) = \boxed{42.9\%}$$

From Eq. 55.1, the percent solid is

$$100\% - \text{percent liquid} = 100\% - 42.9\% = \boxed{57.1\%}$$

2. (a) Since 2011 aluminum is a nonferrous substance, it does not readily form allotropes and thus its properties cannot be changed by the controlled cooling in an annealing process.

(b) The procedures for precipitation hardening of 2011 aluminum are

1. precipitation

2. rapid quenching

3. artificial aging

56 Modeling of Engineering Systems

PRACTICE PROBLEMS

For each of the systems of ideal elements shown, (a) draw the system diagram and (b) write the differential equations.

1.

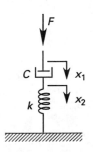

2.

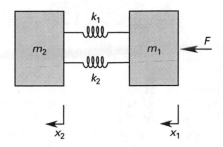

3.

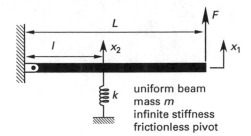

uniform beam
mass m
infinite stiffness
frictionless pivot

4.

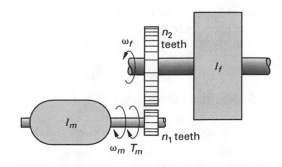

5.

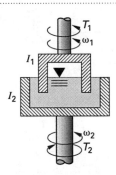

6. The coupling of a railroad car is modeled as the mechanical system shown. Assume all elements are linear. What are the system equations that describe the positions x_1 and x_2 as functions of time?

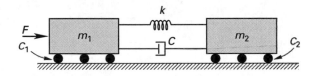

7. Water is discharged freely at a constant rate into an open tank. Water flows out of the tank through a drain with a resistance to flow. (a) Draw the system diagram using idealized elements, and (b) write the differential equations that describe the response of the system.

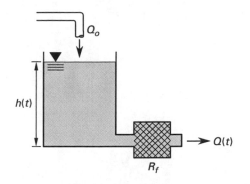

8. Water is pumped into the bottom of an open tank. (a) Draw the system diagram using idealized elements, and (b) write the differential equations that describe the response of the system.

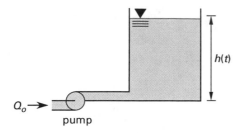

SOLUTIONS

1. (a) The velocity of the plunger is v_1. The velocity of the body of the damper is the same as the upper part of the spring, v_2. By Rule 56.2, the other end of the force and the spring is attached to the stationary wall at $v = 0$. The system diagram is

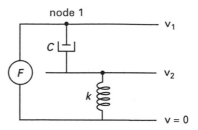

(b) By Rule 56.3, the force from the source is the same force experienced by the dashpot. One of the system equations is based on node 1. Using Rule 56.4 and expanding with Eq. 56.5,

$$F = F_C = C(v_1 - v_2) = C(x_1' - x_2')$$

By Rule 56.3, the force from the source is the same force experienced by the spring. A second system equation is based on node 2. Using Rule 56.4 and expanding with Eq. 56.4,

$$F = F_k = k(x_2 - 0) = kx_2$$

2. (a) The velocity of the ends of the springs connected to m_i is v_1. The velocity of the ends of the springs connected to m_2 is v_2. By Rule 56.1, the other end of each mass connects to $v = 0$. The system diagram is

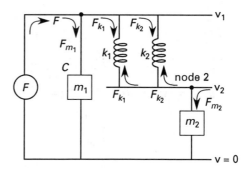

(b) The force leaving the source splits: some of it goes through m_1, some of it goes through k_1, and some of it goes through k_2. One of the system equations is based on node 1.

$$F = F_{m_1} + F_{k_1} + F_{k_2}$$

Using Rule 56.4 and expanding with Eqs. 56.3 and 56.4,

$$F = m_1 a_1 + k_1(x_1 - x_2) + k_2(x_1 - x_2)$$
$$= m_1 x_1'' + (k_1 + k_2)(x_1 - x_2)$$

A second system equation is based on node 2. The conservation law is written to conserve force in the v_2 line.

$$0 = F_{m_2} + F_{k_2} + F_{k_1}$$

Using Rule 56.4 and expanding with Eqs. 56.3 and 56.4,

$$0 = m_2 a_2 + k_2(x_2 - x_1) + k_1(x_2 - x_1)$$
$$= m_2 x_2'' + (k_1 + k_2)(x_2 - x_1)$$

3. (a) Treat this as a rotational system. The applied rotational torque is

$$T = FL$$

The equivalent torsional spring constant is

$$k_r = \frac{M_{\text{resisting}}}{\theta} = \frac{F_k l}{\theta} = \frac{kx_2 l}{\theta}$$

However, $x_2 = l \sin \theta$ and $\theta \approx \sin \theta$ for small angles.

$$k_r = kl^2$$

The moment of inertia of the beam about the hinge point is

$$I = \tfrac{1}{3} mL^2$$

The equivalent rotational system is

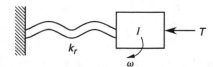

The angular velocity of the end of the spring connected to the inertial element is ω. By Rule 56.2, the other end of the spring is attached to the stationary wall at $\omega = 0$. The system diagram is

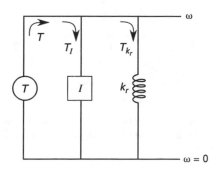

(b) The torque leaving the source splits: some of it goes through I and some of it goes through k_r. The conservation law is written to conserve torque in the ω line.

$$T = T_I + T_{k_r}$$

Using Rule 56.4 and expanding with Eqs. 56.6 and 56.7,

$$T = I\alpha + k_r(\theta - 0)$$
$$FL = \left(\tfrac{1}{3} mL^2\right) \theta'' + kl^2 \theta$$

4. (a) The angular velocity of the small gear is ω_m, and the angular velocity of the large gear is ω_f. By Rule 56.2, the other end of each inertia connects to $\omega = 0$. The gearing transforms the torque and angular displacement from gear 1 to gear 2. The system diagram is

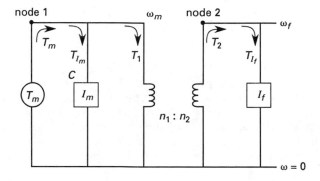

(b) The conservation law based on node 1 is written to conserve torque in the ω_m line.

$$T_m = T_{I_m} + T_1 = I_m \alpha_m + T_1 = I_m \theta_m'' + T_1$$

The same conservation principle based on node 2 is used to conserve torque in the ω_f line.

$$T_2 = T_{I_f} = I_f \alpha_f = I_f \theta_f''$$

The transformer equations are

$$T_2 = \left(\frac{n_2}{n_1}\right) T_1$$

$$\theta_m = \left(\frac{n_2}{n_1}\right) \theta_f$$

5. (a) Consider the fluid to act as a damper with coefficient C_r. The plunger is connected to velocity ω_1, and the body is connected to velocity ω_2. By Rule 56.2, the ends of the inertia elements are connected to $\omega = 0$. The system diagram is

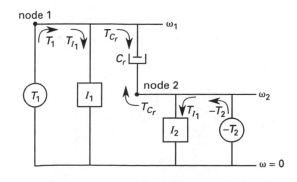

(b) One of the system equations is based on conservation of torque at node 1.

$$T_1 = T_{I_1} + T_{C_r}$$

Using Rule 56.4 and expanding with Eqs. 56.6 and 56.8,

$$T_1 = I_1\alpha_1 + C_r(\omega_1 - \omega_2) = I_1\theta_1'' + C_r(\theta_1' - \theta_2')$$

The second system equation is based on conservation of torque at node 2.

$$-T_2 = T_{I_2} + T_{C_r} = I_2\alpha_2 + C_r(\omega_2 - \omega_1)$$
$$= I_2\theta_2'' + C_r(\theta_2' - \theta_1')$$

6. (a) The velocity of the end of the spring connected to m_1 is v_1. This is also the velocity of the plunger and the velocity of the viscous damper, C_1. The velocity of the end of the spring connected to m_2 is v_2. This is also the velocity of the body of the damper and the velocity of the viscous damper C_2. By Rule 56.2, the other end of each mass connects to $v = 0$. The system diagram is

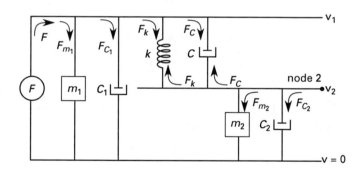

(b) The force leaving the source splits: some of it goes through m_1, C_1, k, and C. The conservation law is written to conserve force in the v_1 line. The equation is based on node 1.

$$F = F_{m_1} + F_{C_1} + F_k + F_C$$

Using Rule 56.4 and expanding with Eqs. 56.3, 56.4, and 56.5,

$$F = m_1 a_1 + C_1(v_1 - 0) + C(v_1 - v_2) + k(x_1 - x_2)$$
$$= m_1 x_1'' + C_1 x_1' + C(x_1' + x_2') + k(x_1 - x_2)$$

The same conservation principle based on node 2 is used to conserve force in the v_2 line. Using Rule 56.4 and expanding with Eqs. 56.3, 56.4, and 56.5,

$$0 = F_{C_2} + F_{m_2} + F_C + F_k$$
$$= C_2(v_2 - 0) + m_2 a_2 + C(v_2 - v_1) + k(x_2 - x_1)$$
$$= C_2 x_2' + m_2 x_2'' + C(x_2' - x_1') + k(x_2 - x_1)$$

7. (a) The fluid capacitance of the water in the tank is C_f. From Rule 56.2, one end of each of the two energy sources Q_1 and Q_2 connects to $p = 0$. The system diagram is

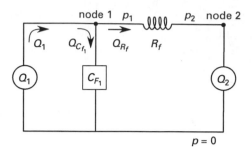

(b) From Eq. 56.10, the flow through the capacitor is

$$Q_{C_{f_1}} = C_{f_1}\left(\frac{dp_1}{dt}\right)$$

From Eq. 56.12, the flow through the resistor is

$$Q_{R_f} = \frac{p_1 - p_2}{R_f}$$

One of the system equations is based on conservation of flow at node 1.

$$Q_1 = Q_{f_1} + Q_{R_f} = C_{f_1}\left(\frac{dp_1}{dt}\right) + \left(\frac{1}{R_f}\right)(p_1 - p_2)$$

The second system equation is based on conservation of flow at node 2.

$$Q_2 = \left(\frac{1}{R_f}\right)(p_1 - p_2)$$

8. (a) The fluid resistance in the entrance pipe is R_f. The pressure at the entrance is p_1, and the pressure in the tank is p_2. The fluid capacitance of the water is C_f. The pressure at the open top of the tank is $p = 0$. The system diagram is

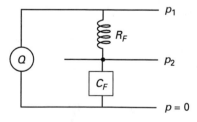

(b) By Rule 56.3, the source flow Q is the same flow through the resistor and the capacitor. Use Eq. 56.12 for the resistor.

$$Q = \frac{p_1 - p_2}{R_f}$$

Use Eq. 56.10 for the capacitor.

$$Q = C_f\left(\frac{dp_2}{dt}\right)$$

57 Analysis of Engineering Systems

PRACTICE PROBLEMS

1. Simplify the following block diagrams and determine the overall system gain.

(a)

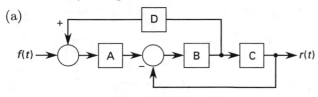

(b)

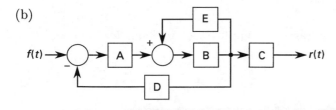

2. *(Time limit: one hour)* A mass of 100 lbm (45 kg) is supported uniformly by a spring system. The spring system has a combined stiffness of 1200 lbf/ft (17.5 kN/m). A dashpot with a damping coefficient of 60 lbf-sec/ft (880 N·s/m) has been installed.

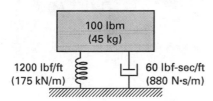

(a) What is the undamped natural frequency?

(b) What is the damping ratio?

(c) Sketch the magnitude and phase characteristics of the frequency response.

(d) Sketch the response to a unit step input.

3. *(Time limit: one hour)* A constant-speed motor/ magnetic clutch drive train is monitored and controlled by a speed-sensing tachometer. The entire system is modeled as a control system block diagram, as shown on the following page. (The lowercase letters represent small-signal increments from the reference values.) When the control system is operating, the desired motor speed, n (in rpm), is set with a speed-setting potentiometer. The setting is compared to the tachometer output. The comparator output error (in volts), controls the clutch. A current, i (in amps), passes through

the clutch coil. The external load torque, $t_L m$ (in in-lbf), is seen by the clutch and is countered by the clutch output torque, t (in in-lbf).

(a) Plot the open-loop frequency response.

(b) What is the open-loop steady-state gain?

(c) Plot the unity feedback closed-loop frequency response.

(d) What is the closed-loop steady-state gain?

(e) Plot the system sensitivity.

(f) Describe the closed-loop response to a step change in the desired output angular velocity. Is it damped or oscillatory? Is there a steady-state error? Why or why not?

(g) Describe the closed-loop response to a step change in the load torque. Is the response damped or oscillatory? Is there a steady-state error? Why or why not?

(h) Assume that you have to select the comparator gain and that it doesn't have to be 0.1. Using the root-locus method or either the Routh or Nyquist stability criterion, find the limits of the comparator gain that cause the closed-loop system to be unstable.

(i) How can you improve the steady-state response of the closed-loop system to constant disturbances in the load torque, t_L?

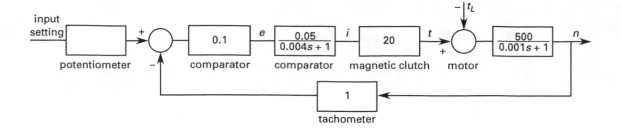

SOLUTIONS

1. (a) Draw the first block diagram.

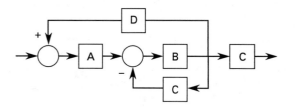

From the rules of simplifying block diagrams, use case 7 to move the extreme right pick-off point to the left of C.

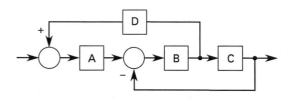

Use case 6 to combine the two summing points on the left.

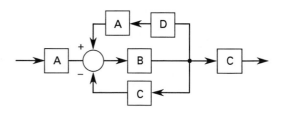

Use case 1 to combine boxes in series in the upper feedback loop.

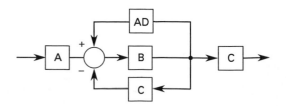

Use case 2 to combine the two feedback loops.

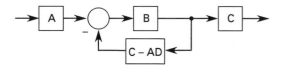

Use case 3 to simplify the remaining feedback loop.

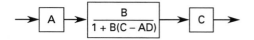

Use case 1 to combine boxes in series to determine the system gain.

$$G_{\text{loop}} = \frac{ABC}{1 + BC - ABD}$$

(b) Draw the second block diagram.

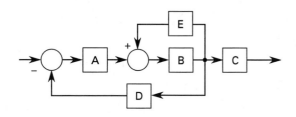

From the rules of simplifying block diagrams, use case 6 to combine the two summing points on the left.

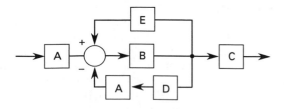

Use case 1 to combine boxes in series in the lower feedback loop.

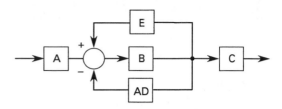

Use case 2 to combine the two feedback loops.

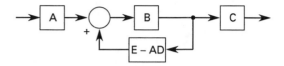

Use case 3 to simplify the remaining feedback loop.

$$\rightarrow \boxed{A} \rightarrow \boxed{\frac{B}{1 - B(E - AD)}} \rightarrow \boxed{C} \rightarrow$$

Use case 1 to combine boxes in series to determine the system gain.

$$G_{\text{loop}} = \frac{ABC}{1 - BE + ABD}$$

2. Customary U.S. Solution

The system differential equation for a force input f is

$$mx'' + Bx' + kx = f$$

Divide by m to write the equation in terms of natural frequency and damping factor.

$$x'' + \left(\frac{B}{m}\right) x' + \left(\frac{k}{m}\right) x = \frac{f}{m}$$

$$x'' + 2\zeta\omega_n x' + \omega_n^2 x = \frac{f}{\frac{k}{\omega_n^2}} = \omega_n^2 \left(\frac{f}{k}\right)$$

Define the forcing function as $h = f/k$.

The equation is the same as Eq. 57.3.

$$x'' + 2\zeta\omega_n x' + \omega_n^2 x = \omega_n^2 h$$

(a) The undamped natural frequency is

$$\omega_n = \sqrt{\frac{kg_c}{m}} = \sqrt{\frac{\left(1200 \ \frac{\text{lbf}}{\text{ft}}\right)\left(32.2 \ \frac{\text{lbm-ft}}{\text{lbf-sec}^2}\right)}{100 \ \text{lbm}}}$$

$$= \boxed{19.66 \ \text{rad/sec}}$$

(b) The damping ratio is

$$\zeta = \frac{\frac{B}{m}}{2\omega_n} = \frac{B}{2\omega_n m} = \frac{\left(60 \ \frac{\text{lbf-sec}}{\text{ft}}\right)\left(32.2 \ \frac{\text{lbm-ft}}{\text{lbf-sec}^2}\right)}{(2)\left(19.66 \ \frac{\text{rad}}{\text{sec}}\right)(100 \ \text{lbm})}$$

$$= \boxed{0.491}$$

(c) Take the Laplace transform of Eq. 57.3. Consider zero initial conditions.

$$s^2 x(s) + 2\zeta\omega_n s x(s) + \omega_n^2 x(s) = \omega_n^2 H(s)$$

Determine the transfer function.

$$T(s) = \frac{x(s)}{H(s)} = \frac{\omega_n^2}{s^2 + 2\zeta\omega_n s + \omega_n^2}$$

The frequency response is obtained by letting $s = j\omega$.

$$T(j\omega) = \frac{\omega_n^2}{(\omega_n^2 - \omega^2) + j2\zeta\omega_n\omega}$$

Write the equation in polar form.

$$T(j\omega) = |T(j\omega)|e^{j\phi(\omega)}$$

The magnitude is

$$|T(j\omega)| = \frac{\omega_n^2}{\sqrt{(\omega_n^2 - \omega^2)^2 + (2\zeta\omega_n\omega)^2}}$$

At $\omega = 0$, $|T(j\omega)| = 1$.

At $\omega = \omega_n$,

$$|T(j\omega)| = \frac{1}{2\zeta} = \frac{1}{(2)(0.491)} \approx 1.0$$

At $\omega \to \infty$, $|T(j\omega)| \to 0$.

A peak occurs near $\omega = \omega_n$. To obtain the location ω_p, set the derivative of $|T(j\omega)|$ equal to zero.

$$\frac{d|T(j\omega)|}{d\omega} = -\left(\frac{3}{2}\right)\omega_n^2\left((\omega_n^2 - \omega^2) + (2\zeta\omega_n\omega)^2\right)^{-\frac{3}{2}}$$

$$\times \left((2)(\omega_n^2 - \omega^2)(-2\omega)\right.$$

$$\left. + (2)(2\zeta\omega_n\omega)(2\zeta\omega_n)\right)$$

$$= 0$$

$$\omega_p = \omega_n\sqrt{1 - 2\zeta^2}$$

$$= \left(19.66 \ \frac{\text{rad}}{\text{sec}}\right)\sqrt{1 - (2)(0.491)^2}$$

$$= 14.15 \ \text{rad/sec}$$

Plant Design

At $\omega = \omega_p$,

$$|T(j\omega)| = \frac{\omega_n^2}{\sqrt{(\omega_n^2 - \omega_p^2)^2 + (2\zeta\omega_n\omega_p)^2}}$$

$$= \frac{\left(19.66 \ \frac{\text{rad}}{\text{sec}}\right)^2}{\sqrt{\left(\left(19.66 \ \frac{\text{rad}}{\text{sec}}\right)^2 - \left(14.15 \ \frac{\text{rad}}{\text{sec}}\right)^2\right)^2 + \left(\begin{array}{c}(2)(0.491)\left(19.66 \ \frac{\text{rad}}{\text{sec}}\right) \\ \times \left(14.15 \ \frac{\text{rad}}{\text{sec}}\right)\end{array}\right)^2}}$$

$$= 1.17$$

A sketch of the frequency response magnitude is

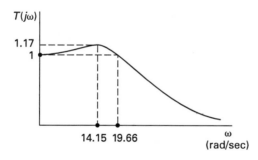

The phase is

$$\phi(\omega) = \tan^{-1}\left(\frac{-2\zeta\omega_n\omega}{\omega_n^2 - \omega^2}\right)$$

At $\omega = 0$, $\phi(\omega) = 0$.
At $\omega = \omega_n$, $\phi(\omega) = -\pi/2$.
At $\omega \to \infty$, $\phi(\omega) = -\pi$.
A sketch of the phase is

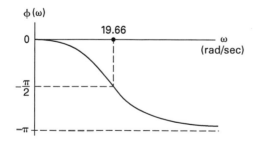

(d) The final value of the step response is obtained from the final value theorem.

$$x_{\text{final}} = \lim_{s \to 0} sx(s) = \lim_{s \to 0} sT(s)H(s)$$

$$= \lim_{s \to 0} sT(s)\left(\frac{1 \ \text{ft}}{s}\right) = \left(\frac{\omega_n^2}{\omega_n^2}\right)(1 \ \text{ft})$$

$$= 1.0 \ \text{ft}$$

From Eq. 57.7, the fraction of overshoot is

$$M_p = \exp\left(\frac{-\pi\zeta}{\sqrt{1 - \zeta^2}}\right) = \exp\left(\frac{-\pi(0.491)}{\sqrt{1 - (0.491)^2}}\right)$$

$$= 0.17$$

The peak is $x_{\text{max}} = 1.17$ ft.

The damped natural frequency is

$$\omega_d = \omega_n\sqrt{1 - \zeta^2} = \left(19.66 \ \frac{\text{rad}}{\text{sec}}\right)\sqrt{1 - (0.491)^2}$$

$$= 17.1 \ \text{rad/sec}$$

The peak time from Eq. 57.6 is

$$t_p = \frac{\pi}{\omega_d} = \frac{\pi}{17.1 \ \frac{\text{rad}}{\text{sec}}} = 0.18 \ \text{sec}$$

The 5% criterion settling time from Eq. 57.9 is

$$t_s = \frac{3.00}{\zeta\omega_n} = \frac{3.00}{(0.491)\left(19.66 \ \frac{\text{rad}}{\text{sec}}\right)} = 0.31 \ \text{sec}$$

The sketch of the response to a unit step input is obtained from Fig. 57.2 and the preceding calculations.

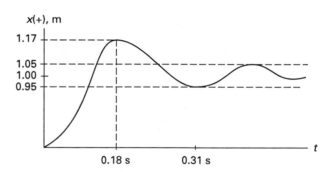

SI Solution

The system differential equation for a force input f is

$$mx'' + Bx' + kx = f$$

Divide by m to write the equation in terms of natural frequency and damping factor.

$$x'' + \left(\frac{B}{m}\right)x' + \left(\frac{k}{m}\right)x = \frac{f}{m}$$

$$x'' + 2\zeta\omega_n x' + \omega_n^2 x = \frac{f}{\frac{k}{\omega_n^2}} = \omega_n^2\left(\frac{f}{k}\right)$$

Define the forcing function as $h = f/k$.

The equation is the same as Eq. 57.3.

$$x'' + 2\zeta\omega_n x' + \omega_n^2 x = \omega_n^2 h$$

(a) The undamped natural frequency is

$$\omega_n = \sqrt{\frac{k}{m}} = \sqrt{\frac{17\,500\ \dfrac{\text{N}}{\text{m}}}{45\ \text{kg}}} = \boxed{19.7\ \text{rad/s}}$$

(b) The damping ratio is

$$\zeta = \frac{\dfrac{B}{m}}{2\omega_n} = \frac{\dfrac{880\ \dfrac{\text{N}\cdot\text{s}}{\text{m}}}{45\ \text{kg}}}{(2)\left(19.7\ \dfrac{\text{rad}}{\text{s}}\right)} = \boxed{0.50}$$

(c) Take the Laplace transform of Eq. 57.3. Consider zero initial conditions.

$$s^2 x(s) + 2\zeta\omega_n s x(s) + \omega_n^2 x(s) = \omega_n^2 H(s)$$

Determine the transfer function.

$$T(s) = \frac{x(s)}{H(s)} = \frac{\omega_n^2}{s^2 + 2\zeta\omega_n s + \omega_n^2}$$

The frequency response is obtained by letting $s = j\omega$.

$$T(j\omega) = \frac{\omega_n^2}{(\omega_n^2 - \omega^2) + j2\zeta\omega_n\omega}$$

Write the equation in polar form.

$$T(j\omega) = |T(j\omega)|e^{j\phi(\omega)}$$

The magnitude is

$$|T(j\omega)| = \frac{\omega_n^2}{\sqrt{(\omega_n^2 - \omega^2)^2 + (2\zeta\omega_n\omega)^2}}$$

At $\omega = 0$, $|T(j\omega)| = 1$.

At $\omega = \omega_n$,

$$|T(j\omega)| = \frac{1}{2s} = \frac{1}{(2)(0.50)} = 1.0$$

At $\omega \to \infty$, $|T(j\omega)| \to 0$.

A peak occurs near $\omega = \omega_n$. To obtain the location ω_p, set the derivative of $|T(j\omega)|$ equal to zero.

$$\frac{d|T(j\omega)|}{d\omega} = -\left(\frac{3}{2}\right)\omega_n^2\left((\omega_n^2 - \omega^2) + (2\zeta\omega_n\omega)^2\right)^{-\frac{3}{2}}$$
$$\times\left((2)(\omega_n^2 - \omega^2)(-2\omega) + (2)(2\zeta\omega_n\omega)2\zeta\omega_n\right)$$
$$= 0$$

$$\omega_p = \omega_n\sqrt{1 - 2\zeta^2}$$
$$= \left(19.7\ \frac{\text{rad}}{\text{s}}\right)\sqrt{1 - (2)(0.50)^2}$$
$$= 13.9\ \text{rad/s}$$

At $\omega = \omega_p$,

$$|T(j\omega)| = \frac{\omega_n^2}{\sqrt{(\omega_n^2 - \omega_p^2)^2 + (2\zeta\omega_n\omega_p)^2}}$$

$$= \frac{\left(19.7\ \dfrac{\text{rad}}{\text{s}}\right)^2}{\sqrt{\left(\left(19.7\ \dfrac{\text{rad}}{\text{s}}\right)^2 - \left(13.9\ \dfrac{\text{rad}}{\text{s}}\right)^2\right)^2 + \left(\begin{array}{c}(2)(0.50)\left(19.7\ \dfrac{\text{rad}}{\text{s}}\right) \\ \times\left(13.9\ \dfrac{\text{rad}}{\text{s}}\right)\end{array}\right)^2}}$$

$$= 1.16$$

A sketch of the frequency response magnitude is

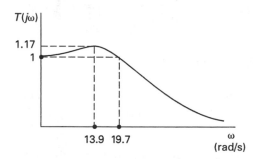

The phase is

$$\phi(\omega) = \tan^{-1}\left(\frac{-2\zeta\omega_n\omega}{\omega_n^2 - \omega^2}\right)$$

At $\omega = 0$, $\phi(\omega) = 0$.

At $\omega = \omega_n$, $\phi(\omega) = -\pi/2$.

At $\omega \to \infty$, $\phi(\omega) = -\pi$.

A sketch of the phase is

Plant Design

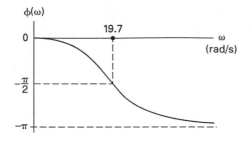

(d) The final value of the step response is obtained from the final value theorem.

$$x_{\text{final}} = \lim_{s \to 0} sx(s) = \lim_{s \to 0} sT(s)H(s)$$

$$= \lim_{s \to 0} sT(s)\left(\frac{1 \text{ m}}{s}\right) = \left(\frac{\omega_n^2}{\omega_n^2}\right)(1 \text{ m})$$

$$= 1.0 \text{ m}$$

From Eq. 57.7, the fraction of overshoot is

$$M_p = \exp\left(\frac{-\pi\zeta}{\sqrt{1-\zeta^2}}\right) = \exp\left(\frac{-\pi(0.50)}{\sqrt{1-(0.50)^2}}\right)$$

$$= 0.16$$

The peak is $x_{\text{max}} = 1.16$ m.

The damped natural frequency is

$$\omega_d = \omega_n\sqrt{1-\zeta^2} = \left(19.7 \; \frac{\text{rad}}{\text{s}}\right)\sqrt{1-(0.50)^2}$$

$$= 17.1 \text{ rad/s}$$

The peak time from Eq. 57.6 is

$$t_p = \frac{\pi}{\omega_d} = \frac{\pi}{17.1 \; \frac{\text{rad}}{\text{s}}} = 0.18 \text{ s}$$

The 5% criterion settling time from Eq. 57.9 is

$$t_s = \frac{3.00}{\zeta\omega_n} = \frac{3.00}{(0.50)\left(19.7 \; \frac{\text{rad}}{\text{s}}\right)} = 0.30 \text{ s}$$

The sketch of the response to a unit step input is obtained from Fig. 57.2 and the preceding calculations.

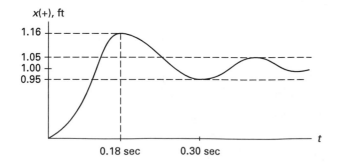

3. First, redraw the system in more traditional form.

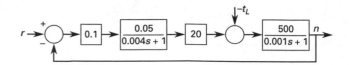

From Fig. 57.5, use case 1 to combine boxes in series.

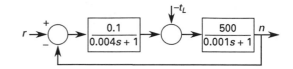

Use case 5 to combine the two summing points.

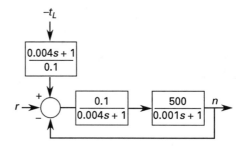

Use case 1 to combine boxes in series.

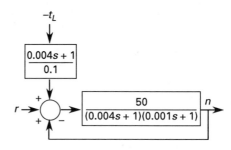

(a) Ignoring the feedback loop, the open-loop transfer function from r to n is

$$T_1(s) = \frac{N(s)}{R(s)}$$

$$= \frac{50}{(0.004s+1)(0.001s+1)}$$

The open-loop transfer function from $-t_L$ to n is

$$T_2(s) = \frac{N(s)}{-T_L(s)}$$

$$= \left(\frac{0.004s+1}{0.1}\right)\left(\frac{50}{(0.004s+1)(0.001s+1)}\right)$$

$$= \frac{500}{0.001s+1}$$

The open-loop frequency response for $T_1(s)$ is

$$T_1(j\omega) = |T_1(j\omega)|e^{j\phi_1(\omega)}$$

From Eq. 57.34, the gain is $20 \log |T_1(j\omega)|$ (in dB).

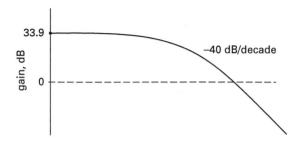

The phase is $\phi_1(\omega)$.

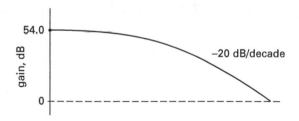

The open-loop frequency response for $T_2(s)$ is

$$T_2(j\omega) = |T_2(j\omega)|e^{j\phi_2(\omega)}$$

From Eq. 63.34, the gain is $20 \log |T_2(j\omega)|$ (in dB).

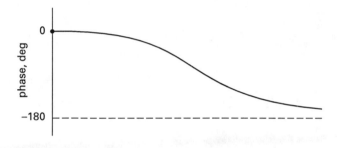

The phase is $\phi_2(\omega)$.

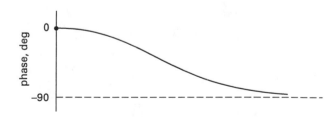

(b) The open-loop steady-state gain for $T_1(s)$ is

$$T_1(0) = \frac{50}{\big((0.004)(0) + 1\big)\big((0.001)(0) + 1\big)} = \boxed{50}$$

The open-loop steady-state gain for $T_2(s)$ is

$$T_2(0) = \frac{500}{(0.001)(0) + 1} = \boxed{500}$$

(c) From Fig. 57.5, use case 3 to simplify the feedback loop.

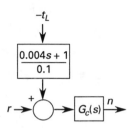

$$
\begin{aligned}
G_c(s) &= \frac{\dfrac{50}{(0.004s + 1)(0.001s + 1)}}{1 + \dfrac{50}{(0.004s + 1)(0.001s + 1)}} \\[2mm]
&= \frac{50}{(0.004s + 1)(0.001s + 1) + 50} \\[2mm]
&= \frac{50}{(4 \times 10^{-6})s^2 + 0.005s + 51}
\end{aligned}
$$

The closed-loop transfer function from r to n is

$$
\begin{aligned}
T_1(s) &= \frac{N(s)}{R(s)} = G_c(s) \\[2mm]
&= \frac{50}{(4 \times 10^{-6})s^2 + 0.005s + 51}
\end{aligned}
$$

The closed-loop transfer function from $-t_L$ to n is

$$
\begin{aligned}
T_2(s) &= \frac{N(s)}{-T_L(s)} = \left(\frac{0.004s + 1}{0.1}\right)G_c(s) \\[2mm]
&= \frac{(500)(0.004s + 1)}{(4 \times 10^{-6})s^2 + 0.005s + 51}
\end{aligned}
$$

The gain for the closed-loop frequency response of $T_1(s)$ is

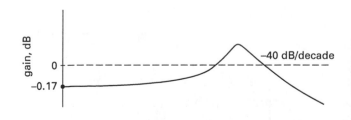

Plant Design

The phase for the closed-loop frequency response of $T_1(s)$ is

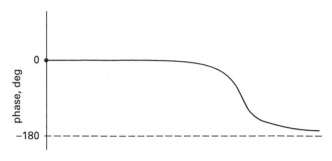

The gain for the closed-loop frequency response of $T_2(s)$ is

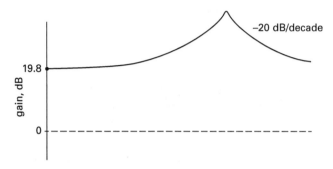

The phase for the closed-loop frequency response of $T_2(s)$ is

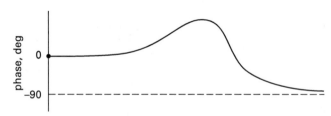

(d) The closed-loop steady-state gain for $T_1(s)$ is

$$T_1(0) = \frac{50}{(4 \times 10^{-6})(0)^2 + (0.005)(0) + 51} = \boxed{0.98}$$

The closed-loop steady-state gain for $T_2(s)$ is

$$T_2(0) = \frac{(500)\big((0.004)(0) + 1\big)}{(4 \times 10^{-6})(0)^2 + (0.005)(0) + 51} = \boxed{9.80}$$

(e) From Fig. 57.4, use Eq. 57.22 to find the system sensitivity.

$$S = \frac{1}{1 + GH} = \frac{1}{1 + \left(\dfrac{50}{(0.004s + 1)(0.001s + 1)}\right)} \quad (1)$$

$$= \frac{(0.004s + 1)(0.001s + 1)}{50 + (0.004s + 1)(0.001s + 1)}$$

$$= \frac{(0.004s + 1)(0.001s + 1)}{(4 \times 10^{-6})s^2 + 0.005s + 51}$$

The frequency response for S is

$$S(j\omega) = |S(j\omega)|e^{j\phi(\omega)}$$

From Eq. 57.34, the gain is $20 \log |S(j\omega)|$ (in dB).

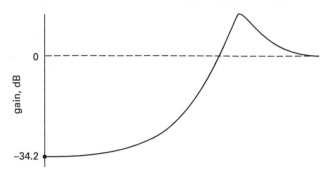

The phase is $\phi(\omega)$.

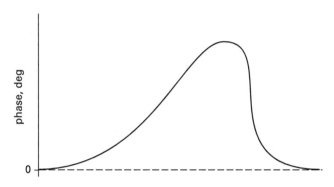

(f) From part (c), the response to a step change in input r is

$$N(s) = R(s)G_c(s)$$

$$= \left(\frac{R}{s}\right)\left(\frac{50}{4 \times 10^{-6}s^2 + 0.005s + 51}\right)$$

$$= \frac{(1.25 \times 10^7)\left(\dfrac{R}{s}\right)}{s^2 + 1250s + 1.275 \times 10^7}$$

Equate the denominator of $N(s)$ to the standard second-order form $s^2 + 2\zeta\omega_n s + \omega_n^2$.

$$\omega_n = \sqrt{1.275 \times 10^7 \frac{\text{rad}^2}{\text{sec}^2}} = 3570.7 \text{ rad/sec}$$

Set $2\zeta\omega_n$ equal to 1250 and solve for ζ.

$$\zeta = \frac{1250 \frac{\text{rad}}{\text{sec}}}{2\omega_n} = \frac{1250 \frac{\text{rad}}{\text{sec}}}{(2)\left(3570.7 \frac{\text{rad}}{\text{sec}}\right)} = 0.175$$

Since ζ is less than one, the response is oscillatory.

Use Eq. 57.7 to find the fraction overshoot.

$$M_p = \exp\left(\frac{-\pi\zeta}{\sqrt{1-\zeta^2}}\right) = \exp\left(\frac{-\pi(0.175)}{\sqrt{1-(0.175)^2}}\right)$$
$$= 0.57$$

Use Eq. 57.5 to find the 90% rise time.

$$t_r = \frac{\pi - \arccos\zeta}{\omega_d} = \frac{\pi - \arccos\zeta}{\omega_n\sqrt{1-\zeta^2}}$$
$$= \frac{\pi - \arccos(0.175)}{\left(3570.7 \ \dfrac{\text{rad}}{\text{sec}}\right)\sqrt{1-(0.175)^2}}$$
$$= 4.97 \times 10^{-4} \ \text{sec}$$

The response is very fast but highly oscillatory.

Use the final value theorem, Eq. 57.27, to find the steady-state value.

$$n = \lim_{s\to 0} sN(s) = \lim_{s\to 0}\left(\frac{s(1.25\times 10^7)\left(\dfrac{R}{s}\right)}{s^2 + 1250s + 1.275\times 10^7}\right)$$
$$= 0.98R$$

The steady-state error is

$$e = R - n = R - 0.98R = 0.02R$$

There is a steady-state error proportional to the desired output angular velocity R.

The plot for $n(t)$ is

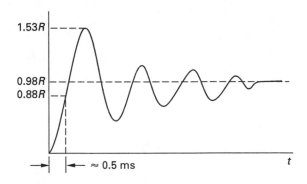

(g) From part (c), the response to a step change in input t_L is

$$N(s) = -T_L(s)\left(\frac{0.004s+1}{0.1}\right)G_c(s)$$
$$= \left(\frac{-T_L}{s}\right)\left(\frac{(500)(0.004s+1)}{(4\times 10^{-6})s^2 + 0.005s + 51}\right)$$
$$= \frac{-(1.25\times 10^8)\left(\dfrac{T_L}{s}\right)(0.004s+1)}{s^2 + 1250s + 1.275\times 10^7}$$

Since the denominator is the same as that for $N(s)$ in Sol. 3.6, the response to a step change in t_L will be similar to a step change in r. However, the numerator, $0.004s + 1$, will cause the response to deviate from second order. The response will still be oscillatory and very fast.

Use the final value theorem, Eq. 57.27, to find the steady-state error.

$$n = \lim_{s\to 0} sN(s)$$
$$= \lim_{s\to 0}\left(\frac{-s(1.25\times 10^8)\left(\dfrac{T_L}{s}\right)(0.004s+1)}{s^2 + 1250s + 1.27\times 10^7}\right)$$
$$= -9.84T_L$$

The new steady-state error for both r and t_L inputs is

$$e = R - n = R - (0.98R - 9.84T_L) = 0.02R + 9.84T_L$$

Thus the load torque, $-t_L$, contributes to the steady-state error.

The response is oscillatory. The plot of the response due to a step change in $-t_L$ is

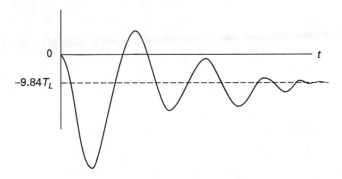

(h) Replace the comparator gain of 0.1 with K. The reduced system is

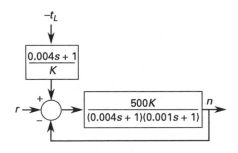

The closed-loop transfer function from r to n is

$$T_1(s) = \frac{N(s)}{R(s)} = \frac{\dfrac{500K}{(0.004s+1)(0.001s+1)}}{1 + \dfrac{500K}{(0.004s+1)(0.001s+1)}}$$
$$= \frac{500K}{(4\times 10^{-6})s^2 + 0.005s + 1 + 500K}$$

The Routh-Hurwitz table is

$$\begin{bmatrix} a_0 & a_2 \\ a_1 & a_3 \\ b_1 & b_2 \end{bmatrix} = \begin{bmatrix} 4 \times 10^{-6} & 1 + 500K \\ 0.005 & 0 \\ b_1 & b_2 \end{bmatrix}$$

Use Eq. 57.40.

$$\begin{aligned} b_1 &= \frac{a_1 a_2 - a_0 a_3}{a_1} \\ &= \frac{(0.005)(1 + 500K) - (4 \times 10^{-6})(0)}{0.005} \\ &= 1 + 500K \end{aligned}$$

For a stable system, there can be no sign changes in the first column of the table.

Thus,

$$1 + 500K > 0$$

$$K > -0.002$$

(i) The closed-loop system steady-state response can be improved by adding integral control. This will effectively compensate for any steady-state disturbances due to t_L and will provide a zero steady-state error for a step input for r. This addition has a side effect of reducing the stability margin of the system. However, if properly designed, the system will still be stable.

58 Process Monitoring and Instrumentation

Plant Design

59 Electrical Systems and Equipment

PRACTICE PROBLEMS

Induction Motors

1. Calculate the full-load phase current drawn by a 440 V (rms) 20-hp (per phase) induction motor having a full-load efficiency of 86% and a full-load power factor of 76%.

2. A 200-hp, three-phase, four-pole, 60 Hz, 440 V (rms) squirrel-cage induction motor operates at full load with an efficiency of 85%, power factor of 91%, and 3% slip.

(a) Find the speed in rpm.

(b) Find the torque developed.

(c) Find the line current.

3. A factory's induction motor load draws 550 kW at 82% power factor. What size synchronous motor is required to carry 250 hp and raise the power factor to 95%? The line voltage is 220 V (rms).

4. The nameplate of an induction motor lists 960 rpm as the full-load speed. For what frequency was the motor designed?

SOLUTIONS

1. $I_{\text{phase}} = \dfrac{P_{\text{phase}}}{\eta V \cos \phi}$

$$= \frac{(20 \text{ hp}) \left(0.7457 \times 10^3 \, \dfrac{\text{W}}{\text{hp}} \right)}{(0.86)(440 \text{ V})(0.76)} = \boxed{51.86 \text{ A}}$$

2. (a) $n_r = \left(\dfrac{(2) \left(60 \, \dfrac{\text{sec}}{\text{min}} \right)}{p} \right) (1 - s)$

$$= \left(\frac{(2) \left(60 \, \dfrac{\text{sec}}{\text{min}} \right) (60 \text{ Hz})}{4} \right) (1 - 0.03)$$

$$= \boxed{1746 \text{ rpm}}$$

(b) $T = \dfrac{P}{\omega} = \dfrac{(200 \text{ hp}) \left(550 \, \dfrac{\text{ft-lbf}}{\text{hp-sec}} \right)}{2\pi \left(1746 \, \dfrac{\text{rev}}{\text{min}} \right) \left(\dfrac{\text{min}}{60 \text{ sec}} \right)}$

$$= \boxed{602 \text{ ft-lbf}}$$

(c) $I_l = \left(\dfrac{1}{\sqrt{3}} \right) \left(\dfrac{P}{\eta V_l \cos \phi} \right)$

$$= \left(\frac{1}{\sqrt{3}} \right) \left(\frac{(200 \text{ hp}) \left(0.7457 \times 10^3 \, \dfrac{\text{W}}{\text{hp}} \right)}{(0.85)(440 \text{ V})(0.91)} \right)$$

$$= \boxed{253 \text{ A}}$$

3.

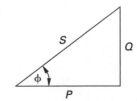

The original power angle is

$$\phi_i = \arccos(0.82) = 34.92°$$
$$P_1 = 550 \text{ kW}$$
$$Q_i = P_1 \tan \phi_i = (550 \text{ kW})(\tan 34.92°) = 384.0 \text{ kVAR}$$

The new conditions are

$$P_2 = (250 \text{ hp}) \left(0.7457 \frac{\text{kW}}{\text{hp}} \right) = 186.4 \text{ kW}$$
$$\phi_f = \arccos(0.95) = 18.19°$$

Since both motors perform real work,

$$P_f = P_1 + P_2 = 550 \text{ kW} + 186.4 \text{ kW} = 736.4 \text{ kW}$$

The new reactive power is

$$Q_f = P_f \tan \phi_f = (736.4 \text{ kW})(\tan 18.19°)$$
$$= 242.0 \text{ kVAR}$$

The change in reactive power is

$$\Delta Q = 384.0 \text{ kVAR} - 242.0 \text{ kVAR} = 142 \text{ kVAR}$$

Synchronous motors used for power factor correction are rated by apparent power.

$$S = \sqrt{(\Delta P)^2 + (\Delta Q)^2}$$
$$= \sqrt{(186.4 \text{ kW})^2 + (142 \text{ kVAR})^2} = \boxed{234.3 \text{ kVA}}$$

4. $f = \dfrac{pn_s}{(20) \left(60 \frac{\text{sec}}{\text{min}} \right)} = \dfrac{pn}{(2) \left(60 \frac{\text{sec}}{\text{min}} \right)(1 - s)}$

The slip and number of poles are unknown. Assume $s = 0$ and $p = 4$.

$$f = \frac{(4) \left(960 \frac{\text{rev}}{\text{min}} \right)}{(2) \left(60 \frac{\text{sec}}{\text{min}} \right)} = 32 \text{ Hz}$$

This is not close to anything in commercial use. Try $p = 6$.

$$f = \frac{(6) \left(960 \frac{\text{rev}}{\text{min}} \right)}{(2) \left(60 \frac{\text{sec}}{\text{min}} \right)} = 48 \text{ Hz}$$

With a 4% slip, $f = 50$ Hz.

$$\boxed{50 \text{ Hz (European)}}$$

60 Illumination and Sound

PRACTICE PROBLEMS

1. What is the total sound pressure level when a 40 dB source is placed adjacent to a 35 dB source?

(A) 17 dB
(B) 28 dB
(C) 37 dB
(D) 41 dB

2. With no machinery operating, the background noise in a room has a sound pressure level of 43 dB. With the machinery operating, the sound pressure level is 45 dB. What is the sound pressure level due to the machinery alone?

(A) 31 dB
(B) 41 dB
(C) 47 dB
(D) 49 dB

3. An unenclosed source produces a sound pressure level of 100 dB. An enclosure is constructed from a material having a transmission loss of 30 dB. The sound pressure level inside the enclosure from the enclosed source increases to 110 dB. What is the reduction in sound pressure level outside the enclosure?

(A) 20 dB
(B) 30 dB
(C) 80 dB
(D) 90 dB

4. 4 ft (1.2 m) from an isotropic sound source, the sound pressure level is 92 dB. What is the sound pressure level 12 ft (3.6 m) from the source?

(A) 66 dB
(B) 74 dB
(C) 90 dB
(D) 110 dB

5. Octave band measurements of a noise source were made. The measurements were 85, 90, 92, 87, 82, 78, 65, and 54 dB at frequencies of 63, 125, 250, 500, 1000, 2000, 4000, and 8000 Hz, respectively. What is the overall A-weighted sound pressure?

(A) 6.1 dB
(B) 13 dB
(C) 18 dB
(D) 25 dB

6. What is the maximum possible reduction in sound pressure level if the number of sabins is 50% of the total room area?

(A) 1.2 dB decrease
(B) 3.0 dB decrease
(C) 4.6 dB decrease
(D) 6.1 dB decrease

7. A room has dimensions of 100 ft × 400 ft × 20 ft (30 m × 120 m × 6 m). All surfaces are precast concrete. 40% of the walls are treated acoustically with a material having a sound absorption coefficient of 0.8. What is the reduction in sound pressure level?

(A) 1.2 dB decrease
(B) 3.0 dB decrease
(C) 4.6 dB decrease
(D) 6.1 dB decrease

8. An office has dimensions of 20 ft × 50 ft × 10 ft (6 m × 15 m × 3 m). The floor is covered with roll vinyl. The walls and ceiling are sheetrock. 20% of the walls are glass windows. There are 15 desks, 15 occupants, and 5 miscellaneous sabins. After complaints from the occupants, the ceiling is treated with a sound absorbing material with a sound absorption coefficient of 0.7. What is the reduction in sound level?

(A) 3.5 dB decrease
(B) 5.7 dB decrease
(C) 6.3 dB decrease
(D) 12 dB decrease

9. A room has dimensions of 15 ft × 20 ft × 10 ft (4.5 m × 6 m × 3 m). The sound absorption coefficients are 0.03, 0.5, and 0.06 for the floor, ceiling, and walls, respectively. A machine with a sound power level of 65 dB is located at the intersection of the floor and wall, 7.5 ft (2.25 m) from the perpendicular walls. The ambient sound pressure level is 50 dB everywhere in the room. What is the sound pressure level 5 ft (1.5 m) from the machine?

(A) 45 dB
(B) 61 dB
(C) 73 dB
(D) 81 dB

PROFESSIONAL PUBLICATIONS, INC.

SOLUTIONS

1. Use Eq. 60.16.

$$L = 10\log\sum 10^{\frac{W_i}{10}} = 10\log\left(10^{\frac{40}{10}} + 10^{\frac{35}{10}}\right)$$

$$= \boxed{41.2 \text{ dB}}$$

The answer is (D).

2. Use Eq. 60.16.

$$L = 10\log\sum 10^{\frac{W_i}{10}} = 10\log\left(10^{\frac{43}{10}} + 10^{\frac{W}{10}}\right) = 45 \text{ dB}$$

Solve for the unknown machinery sound pressure level, W.

$$10^{\frac{W}{10}} = 10^{\frac{45}{10}} - 10^{\frac{43}{10}}$$

$$W = 10\log\left(10^{\frac{45}{10}} - 10^{\frac{43}{10}}\right) = \boxed{40.7 \text{ dB}}$$

The answer is (B).

3. Define $L_{W,1}$ as the sound pressure level inside the enclosure, and define $L_{W,2}$ as the sound pressure level outside the enclosure. Use Eq. 60.23.

$$L_{W,2} = L_{W,1} - \text{TL} = 110 \text{ dB} - 30 \text{ dB} = 80 \text{ dB}$$

Define $L_{W,1}$ as the sound pressure level for the unenclosed source, and define $L_{W,2}$ as the sound pressure level for the enclosed source.

Use Eq. 60.22 to solve for the insertion loss.

$$\text{IL} = 100 \text{ dB} - 80 \text{ dB} = \boxed{20 \text{ dB}}$$

The answer is (A).

4. From Eq. 60.15, the free-field sound pressure is inversely proportional to the square of the distance from the source.

$$\frac{p_2}{p_1} = \left(\frac{r_1}{r_2}\right)^2$$

From Eq. 60.13,

$$L_{p,2} = L_{p,1} + 10\log\left(\frac{r_1}{r_2}\right)^2$$

$$= L_{p,1} + 20\log\left(\frac{r_1}{r_2}\right)$$

Customary U.S. Solution

$$L_{p,2} = 92 \text{ dB} + 20\log\left(\frac{4 \text{ ft}}{12 \text{ ft}}\right)$$

$$= 92 \text{ dB} - 9.5 \text{ dB}$$

$$= \boxed{82.5 \text{ dB}}$$

The answer is (C).

SI Solution

$$L_{p,2} = 92 \text{ dB} + 20\log\left(\frac{1.2 \text{ m}}{3.6 \text{ m}}\right)$$

$$= 92 \text{ dB} - 9.5 \text{ dB}$$

$$= \boxed{82.5 \text{ dB}}$$

The answer is (C).

5. Add the corrections from Table 60.5 to the measurements.

frequency	measurement	correction	corrected value
63	85	−26.2	58.8
125	90	−16.1	73.9
250	92	−8.6	83.4
500	87	−3.2	83.8
1000	82	0	82.0
2000	78	+1.2	79.2
4000	65	+1.0	66.0
8000	54	−1.1	52.9

Use Eq. 60.16.

$$L = 10\log\sum 10^{\frac{W_i}{10}}$$

$$= 10\log\left(10^{5.88} + 10^{7.39} + 10^{8.34} + 10^{8.38} + 10^{8.2}\right.$$
$$\left. + 10^{7.92} + 10^{6.6} + 10^{5.29}\right)$$

$$= \boxed{88.6 \text{ dBA}}$$

The answer is (A).

6. Define A as the total room area.

$$\sum S_1 = 0.50A$$

The maximum number of sabins is equal to the room area.

$$\sum S_2 = A$$

Use Eq. 60.21.

$$\text{NR} = 10\log\left(\frac{\sum S_1}{\sum S_2}\right) = 10\log\left(\frac{0.50A}{A}\right)$$
$$= 10\log(0.50)$$
$$= \boxed{-3.0\text{ dB}\quad[\text{decrease}]}$$

The answer is (B).

7. *Customary U.S. Solution*

The surface area of the room walls is

$$A_1 = \big((2)(100\text{ ft}) + (2)(400\text{ ft})\big)(20\text{ ft}) = 20{,}000\text{ ft}^2$$

The surface area of the room floor and ceiling is

$$A_2 = (2)(100\text{ ft})(400\text{ ft}) = 80{,}000\text{ ft}^2$$

The sound absorption coefficient of precast concrete is the NRC value of 0.02 from App. 60.A.

Define the sound absorption coefficient of precast concrete as α_{concrete}.

The sabin area of the room with all precast concrete is

$$\sum S_2 = \alpha_{\text{concrete}}(A_1 + A_2)$$
$$= (0.02)(20{,}000\text{ ft}^2 + 80{,}000\text{ ft}^2)$$
$$= 2000\text{ ft}^2$$

Define the sound absorption coefficient of the wall acoustical treatment as α_{wall}.

The sabin area of the room with 40% of the walls treated with $\alpha_{\text{wall}} = 0.8$ is

$$\sum S_1 = \alpha_{\text{concrete}}A_2 + \alpha_{\text{concrete}}(0.6A_1) + \alpha_{\text{wall}}(0.4A_1)$$
$$= (0.02)(80{,}000\text{ ft}^2) + (0.02)(0.6)(20{,}000\text{ ft}^2)$$
$$\quad + (0.8)(0.4)(20{,}000\text{ ft}^2)$$
$$= 8240\text{ ft}^2$$

Use Eq. 60.21.

$$\text{NR} = 10\log\left(\frac{\sum S_1}{\sum S_2}\right)$$
$$= 10\log\left(\frac{8240\text{ ft}^2}{2000\text{ ft}^2}\right) = \boxed{6.1\text{ dB}}$$

The answer is (D).

SI Solution

The surface area of the room walls is

$$A_1 = \big((2)(30\text{ m}) + (2)(120\text{ m})\big)(6\text{ m}) = 1800\text{ m}^2$$

The surface area of the room floor and ceiling is

$$A_2 = (2)(30\text{ m})(120\text{ m}) = 7200\text{ m}^2$$

The sound absorption coefficient of precast concrete is the NRC value of 0.02 from App. 60.A.

Define the sound absorption coefficient of precast concrete as α_{concrete}.

The sabin area of the room with all precast concrete is

$$\sum S_2 = \alpha_{\text{concrete}}(A_1 + A_2)$$
$$= (0.02)(1800\text{ m}^2 + 7200\text{ m}^2)$$
$$= 180\text{ m}^2$$

Define the sound absorption coefficient of the wall acoustical treatment as α_{wall}.

The sabin area of the room with 40% of the walls treated with $\alpha_{\text{wall}} = 0.8$ is

$$\sum S_1 = \alpha_{\text{concrete}}A_2 + \alpha_{\text{concrete}}(0.6A_1) + \alpha_{\text{wall}}(0.4A_1)$$
$$= (0.02)(7200\text{ m}^2) + (0.02)(0.6)(1800\text{ m}^2)$$
$$\quad + (0.8)(0.4)(1800\text{ m}^2)$$
$$= 741.6\text{ m}^2$$

Use Eq. 60.21.

$$\text{NR} = 10\log\left(\frac{\sum S_1}{\sum S_2}\right)$$
$$= 10\log\left(\frac{741.6\text{ m}^2}{180\text{ m}^2}\right) = \boxed{6.1\text{ dB}}$$

The answer is (D).

8. *Customary U.S. Solution*

The area of the walls is

$$A_1 = \big((2)(20\text{ ft}) + (2)(50\text{ ft})\big)(10\text{ ft}) = 1400\text{ ft}^2$$

From App. 60.A, the sound absorption coefficient of sheetrock and glass is $\alpha_1 = 0.03$.

The area of the floor is

$$A_2 = (20\text{ ft})(50\text{ ft}) = 1000\text{ ft}^2$$

From App. 60.A, the sound absorption coefficient of roll vinyl is $\alpha_2 = 0.03$.

The area of the ceiling is

$$A_3 = (20\text{ ft})(50\text{ ft}) = 1000\text{ ft}^2$$

The sound absorption coefficient of the sheetrock is $\alpha_3 = 0.03$.

From App. 60.A, the desks have approximately 1.5 sabins each and the occupants have approximately 5.0 sabins each.

The total sabin area of the untreated room is

$$\sum S_2 = \alpha_1 S_1 + \alpha_2 S_2 + \alpha_3 S_3 + \text{desks} + \text{occupants}$$
$$+ \text{miscellaneous}$$
$$= (0.03)(1400 \text{ ft}^2) + (0.03)(1000 \text{ ft}^2)$$
$$+ (0.03)(1000 \text{ ft}^2) + (15)(1.5 \text{ ft}^2)$$
$$+ (15)(5.0 \text{ ft}^2) + 5.0 \text{ ft}^2$$
$$= 204.5 \text{ ft}^2$$

The total sabin area excluding the ceiling is

$$204.5 \text{ ft}^2 - \alpha_3 S_3 = 204.5 \text{ ft}^2 - (0.03)(1000 \text{ ft}^2)$$
$$= 174.5 \text{ ft}^2$$

The total sabin area of the room with the ceiling treated with $\alpha_3 = 0.7$ sound absorption is

$$\sum S_1 = 174.5 \text{ ft}^2 + \alpha_3 S_3$$
$$= 174.5 \text{ ft}^2 + (0.7)(1000 \text{ ft}^2)$$
$$= 874.5 \text{ ft}^2$$

Use Eq. 60.21.

$$\text{NR} = 10 \log \left(\frac{\sum S_1}{\sum S_2} \right)$$
$$= 10 \log \left(\frac{874.5 \text{ ft}^2}{204.5 \text{ ft}^2} \right) = \boxed{6.3 \text{ dB}}$$

The answer is (C).

SI Solution

The area of the walls is

$$A_1 = \big((2)(6 \text{ m}) + (2)(15 \text{ m})\big)(3 \text{ m}) = 126 \text{ m}^2$$

From App. 60.A, the sound absorption coefficient of sheetrock and glass is $\alpha_1 = 0.03$.

The area of the floor is

$$A_2 = (6 \text{ m})(15 \text{ m}) = 90 \text{ m}^2$$

From App. 60.A, the sound absorption coefficient of roll vinyl is $\alpha_2 = 0.03$.

The area of the ceiling is

$$A_3 = (6 \text{ m})(15 \text{ m}) = 90 \text{ m}^2$$

The sound absorption coefficient of the sheetrock is $\alpha_3 = 0.03$.

From App. 60.A, the desks have approximately 1.5 sabins each and the occupants have approximately 5.0 sabins each.

The total sabin area (m²) of the untreated room is

$$\sum S_2 = \alpha_1 S_1 + \alpha_2 S_2 + \alpha_3 S_3 + \text{desks} + \text{occupants}$$
$$+ \text{miscellaneous}$$
$$= (0.03)(126 \text{ m}^2) + (0.03)(90 \text{ m}^2)$$
$$+ (0.03)(90 \text{ m}^2) + (15)(1.5 \text{ ft}^2)$$
$$\times \left(\frac{1 \text{ m}}{3.28 \text{ ft}} \right)^2 + (15)(5.0 \text{ ft}^2) \left(\frac{1 \text{ m}}{3.28 \text{ ft}} \right)^2$$
$$+ (5.0 \text{ ft}^2) \left(\frac{1 \text{ m}}{3.28 \text{ ft}} \right)^2$$
$$= 18.7 \text{ m}^2$$

The total sabin area excluding the ceiling is

$$18.7 \text{ m}^2 - \alpha_3 S_3 = 18.7 \text{ m}^2 - (0.03)(90 \text{ m}^2)$$
$$= 16.0 \text{ m}^2$$

The total sabin area of the room with the ceiling treated with $\alpha_3 = 0.7$ sound absorption is

$$\sum S_1 = 16.0 \text{ m}^2 + \alpha_3 S_3$$
$$= 16.0 \text{ m}^2 + (0.7)(90 \text{ m}^2)$$
$$= 79.0 \text{ m}^2$$

Use Eq. 60.21.

$$\text{NR} = 10 \log \left(\frac{\sum S_1}{\sum S_2} \right) = 10 \log \left(\frac{79.0 \text{ m}^2}{18.7 \text{ m}^2} \right)$$
$$= \boxed{6.3 \text{ dB}}$$

The answer is (C).

9. *Customary U.S. Solution*

Define the sound absorption coefficients of the floor, ceiling, and walls as α_1, α_2, and α_3, respectively.

The floor area is

$$A_1 = (15 \text{ ft})(20 \text{ ft}) = 300 \text{ ft}^2$$

The ceiling area is

$$A_2 = (15 \text{ ft})(20 \text{ ft}) = 300 \text{ ft}^2$$

The area of the walls is

$$A_3 = \big((2)(15 \text{ ft}) + (2)(20 \text{ ft})\big)(10 \text{ ft}) = 700 \text{ ft}^2$$

The total surface area of the room is

$$A = \sum A_i = A_1 + A_2 + A_3$$
$$= 300 \text{ ft}^2 + 300 \text{ ft}^2 + 700 \text{ ft}^2$$
$$= 1300 \text{ ft}^2$$

From Eq. 60.18, the average sound absorption coefficient of the room is

$$\bar{\alpha} = \frac{\sum S_i}{\sum A_i} = \frac{\alpha_1 A_1 + \alpha_2 A_2 + \alpha_3 A_3}{A}$$
$$= \frac{(0.03)(300 \text{ ft}^2) + (0.5)(300 \text{ ft}^2) + (0.06)(700 \text{ ft}^2)}{1300 \text{ ft}^2}$$
$$= 0.155$$

From Eq. 60.19, the room constant is

$$R = \frac{\bar{\alpha} A}{1 - \bar{\alpha}} = \frac{(0.155)(1300 \text{ ft}^2)}{1 - 0.155} = 238.5 \text{ ft}^2$$

From Eq. 60.15, the sound pressure level due to the machine is

$$L_{p,\text{dB}} = 10.5 + L_W + 10 \log \left(\frac{Q}{4\pi r^2} + \frac{4}{R} \right)$$
$$= 10.5 + 65 + 10 \log \left(\frac{4}{4\pi(5 \text{ ft})^2} + \frac{4}{238.5 \text{ ft}^2} \right)$$
$$= 60.2 \text{ dB}$$

Use Eq. 60.16 to combine the machine sound pressure level with the ambient sound pressure level.

$$L = 10 \log \sum 10^{\frac{W_i}{10}} = 10 \log \left(10^{\frac{60.2}{10}} + 10^{\frac{50}{10}} \right)$$
$$= \boxed{60.6 \text{ dB}}$$

The answer is (B).

SI Solution

Define the sound absorption coefficients of the floor, ceiling, and walls as α_1, α_2, and α_3, respectively.

The floor area is

$$A_1 = (4.5 \text{ m})(6 \text{ m}) = 27 \text{ m}^2$$

The ceiling area is

$$A_2 = (4.5 \text{ m})(6 \text{ m}) = 27 \text{ m}^2$$

The area of the walls is

$$A_3 = ((2)(4.5 \text{ m}) + (2)(6 \text{ m}))(3 \text{ m}) = 63 \text{ m}^2$$

The total surface area of the room is

$$A = \sum A_i = A_1 + A_2 + A_3$$
$$= 27 \text{ m}^2 + 27 \text{ m}^2 + 63 \text{ m}^2$$
$$= 117 \text{ m}^2$$

From Eq. 60.18, the average sound absorption coefficient of the room is

$$\bar{\alpha} = \frac{\sum S_i}{\sum A_i} = \frac{\alpha_1 A_1 + \alpha_2 A_2 + \alpha_3 A_3}{A}$$
$$= \frac{(0.03)(27 \text{ m}^2) + (0.5)(27 \text{ m}^2) + (0.06)(63 \text{ m}^2)}{117 \text{ m}^2}$$
$$= 0.155$$

From Eq. 60.19, the room constant is

$$R = \frac{\bar{\alpha} A}{1 - \bar{\alpha}} = \frac{(0.155)(117 \text{ m}^2)}{1 - 0.155} = 21.5 \text{ m}^2$$

From Eq. 60.15, the sound pressure level due to the machine is

$$L_{p,\text{dB}} = L_W + 10 \log \left(\frac{Q}{4\pi r^2} + \frac{4}{R} \right)$$
$$= 65 + 10 \log \left(\frac{4}{4\pi(1.5 \text{ m})^2} + \frac{4}{21.5 \text{ m}^2} \right)$$
$$= 60.2 \text{ dB}$$

Use Eq. 60.16 to combine the machine sound pressure level with the ambient sound pressure level.

$$L = 10 \log \sum_{10} 10^{\frac{W_i}{10}} = 10 \log \left(10^{\frac{60.2}{10}} + 10^{\frac{50}{10}} \right) = \boxed{60.6 \text{ dB}}$$

The answer is (B).

61 Workplace Safety

PRACTICE PROBLEMS

There are no practice problems corresponding with Ch. 61 of the *Chemical Engineering Reference Manual*.

Plant Design

62 Process and Production Optimization

PRACTICE PROBLEMS

1. *(Time limit: one hour)* Printed circuit boards are manufactured in four consecutive departmental operations. Each operation occurs on a different machine. Employees in all departments work from 8:00 a.m. to 5:00 p.m. and have one hour total for lunch and personal breaks. The target rate is 900,000 completed units per year. Units found to be defective are discarded. No units are produced during set-up, downtime, maintenance, or record-keeping periods.

| | department | | | |
	1	2	3	4
production time (sec/unit)	6	10	11	45
set-up time (min/day)	16	8	20	5
downtime (min/day)	12	10	15	0
maintenance time (min/day)	8	12	8	0
record-keeping (min/day)	6	6	6	30
percentage defects	4%	6%	3%	2%

(a) What is the maximum number of completed circuit boards that can be produced in one year if there is one machine per department?

(b) If additional machines can be added for any or all operations, what is the most efficient method of meeting the target production rate? (There are no changes in the defect rates.)

(c) What is the efficiency of each department if the capacity is increased per part (b)?

2. *(Time limit: one hour)* Four workers perform operations 1, 2, 3, and 4 in sequence on a manual assembly line. Each station performs its operation only once on the product before sending the product on to the next operation. Operation times at the stations are as given. (Travel times are included in the operation times.)

station	time (min)
1	0.6
2	0.6
3	0.9
4	0.8

A fifth "floating" station has the ability to assist any of the four stations. The fifth station works with the same efficiencies and times as the four stations. There is no fixed assignment for this fifth station. The fifth station is allowed to help any station that needs it.

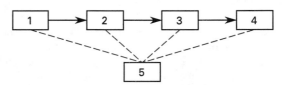

The operators of all five stations are permitted a 10 minute break each hour.

(a) What is the maximum number of products that can be produced assuming that the fifth station is assigned to work optimally? Neglect the initial (transient) performance.

(b) What is the fraction of station 5's time allocated to each operation?

3. The activities that constitute a project are listed below. The project starts at $t = 0$.

activity	predecessors	successors	duration
start	–	A	0
A	start	B,C,D	7
B	A	G	6
C	A	E,F	5
D	A	G	2
E	C	H	13
F	C	H,I	4
G	D,B	I	18
H	E,F	finish	7
I	F,G	finish	5
finish	H,I	–	0

(a) Draw the critical path network.

(b) Indicate the critical path.

(c) What is the earliest finish?

(d) What is the latest finish?

(e) What is the slack along the critical path?

(f) What is the float along the critical path?

4. PERT activities constituting a short project are listed with their characteristic completion times. If the project starts at $t = 15$, what is the probability that the project will be completed on or before $t = 42$?

activity	predecessors	successors	t_{min}	t_{likely}	t_{max}
start	–	A	0	0	0
A	start	B,D	1	2	5
B	A	C	7	9	20
C	B	D	5	12	18
D	A,C	finish	2	4	7
finish	D	–	0	0	0

   (A) 18%
   (B) 29%
   (C) 56%
   (D) 71%

5. *(Time limit: one hour)* Your manufacturing facility produces two models of municipal transit buses, designated B-1 and B-2. You can sell as many of either model as you produce. The per-bus profits are $800,000 for model B-1 and $650,000 for model B-2. You would like to maximize your company's profit by determining the number of each model buses to produce. However, it is not a a simple matter of producing only B-1 models because certain common parts are in limited supply.

part	number available
gage sending units	2000
wheel housing flares	1800
intake grilles	3600

Due to differences in design, the quantity of each common part varies between the two models.

part	number in model	
	B-1	B-2
gage sending units	8	10
wheel housing flares	6	4
intake grilles	3	2

What quantity of each bus should your facility produce in order to maximize profit?

6. A small company makes two chemicals, designated C-1 and C-2. The process for both includes fermentation and purification. Labor limits all fermentation operations to 300 hr per month, and purification is limited to 120 hr per month. Each unit of product C-1 requires 10 hr for fermentation and 8 hr for purification. Each unit of product C-2 requires 20 hr for fermentation and 3 hr for purification. The profit per unit of product C-1 is $3000; the profit per unit of product C-2 is $5000.

(a) How much should the company make of each chemical per month if partial units are permitted?

(b) How much should the company make of each chemical per month if only whole units are permitted?

7. A linear wage incentive program provides for 50% participation and a bonus that begins at a productivity level of 66.7% of standard. A worker produces 1900 units in 8 hr. The standard time for each unit is 0.004 hr. What is the worker's relative earnings?

   (A) 67%
   (B) 95%
   (C) 130%
   (D) 160%

SOLUTIONS

1. Assume the following.

$$5 \; \frac{\text{work days}}{\text{week}}; \; 52 \; \frac{\text{weeks}}{\text{year}}$$

$$\left(5 \; \frac{\text{work days}}{\text{week}}\right)\left(52 \; \frac{\text{weeks}}{\text{year}}\right)$$

$$= 260 \; \text{work days/year}$$

Let t equal production time, in minutes/unit, for 1 unit.

Let m equal the minutes/day available for production.

Note:

$$\left(8 \; \frac{\text{hours}}{\text{day}}\right)\left(60 \; \frac{\text{minutes}}{\text{hour}}\right) = 480 \; \text{minutes/day}$$

$$m = 480 \; \frac{\text{minutes}}{\text{day}} - \left(\frac{\text{setup time}}{\text{day}} + \frac{\text{downtime}}{\text{day}}\right.$$
$$\left. + \frac{\text{maintenance time}}{\text{day}} + \frac{\text{recordkeeping time}}{\text{day}}\right)$$

$$\left(900{,}000 \; \frac{\text{units}}{\text{year}}\right)\left(\frac{1 \; \text{year}}{260 \; \text{work days}}\right)$$

$$= 3462 \; \text{units/work day}$$

n = number of machines needed to reach a production rate of 3462 units/day

e = efficiency of each department

$$= \frac{\begin{array}{c}\text{number of units}\\\text{produced in each department}\end{array}}{\begin{array}{c}\text{maximum possible number of units}\\\text{produced in each department}\end{array}}$$

p = percentage of defective units

$\left(\frac{m}{t}\right)(1-p)$ = maximum production in each department using one machine

Based on the preceding definitions,

$$n = \frac{3462}{\left(\dfrac{m}{t}\right)(1-p)} + 1 \quad \text{[greatest integer function]}$$

$$e = \frac{3462}{n\left(\dfrac{m}{t}\right)(1-p)}$$

dept	t	m	$\left(\dfrac{m}{t}\right)(1-p)$	n	e
1	6/60	438	4204	1	82.4%
2	10/60	444	2504	2	69.1%
3	11/60	431	2280	2	75.9%
4	45/60	445	581	6	99.3%

(a) Since each circuit board must go through all four departments, maximum production is 581 units/day.

$$\left(581 \; \frac{\text{units}}{\text{day}}\right)\left(260 \; \frac{\text{work days}}{\text{year}}\right)$$

$$= \boxed{151{,}060 \; \text{units/year}}$$

(b) The values of n are computed in the chart.

(c) The values of e are computed in the chart.

2. (b) For the time being, disregard the 10 min per hour shift break since this break reduces the capacity of all stations by the same percentage.

Determine the maximum output per hour for each station.

station	output
1	$\dfrac{60 \; \frac{\text{min}}{\text{hr}}}{0.6 \; \frac{\text{min}}{\text{unit}}} = 100 \; \text{units/hr}$
2	$\dfrac{60 \; \frac{\text{min}}{\text{hr}}}{0.6 \; \frac{\text{min}}{\text{unit}}} = 100 \; \text{units/hr}$
3	$\dfrac{60 \; \frac{\text{min}}{\text{hr}}}{0.9 \; \frac{\text{min}}{\text{unit}}} = 66.67 \; \text{units/hr}$
4	$\dfrac{60 \; \frac{\text{min}}{\text{hr}}}{0.8 \; \frac{\text{min}}{\text{unit}}} = 75 \; \text{units/hr}$

Stations 3 and 4 are the bottleneck operations. Intuitively, we would want to help operation 3 the most, followed by helping operation 4.

Start by allocating station 5 capacity to the slowest operation, operation 3. Try to bring station 3 up to the same capacity as stations 1 and 2. To do so requires station 5 to produce $100 - 66.67 = 33.33$ units per hour. Since station 5 works at the same speed as station 3, the fraction of time station 5 needs to assist station 3 is $33.33/66.67 = 0.5$ (50%). This leaves 50% of station 5's time available to assist other stations.

Plant Design

Next, allocate the remaining station 5 time to station 4. To bring station 4 up to 100 units per hour will require station 5 to produce $100 - 75 = 25$ units per hour. Since station 5 works at the same rate as station 4, the fraction of time station 5 needs to assist station 4 is $25/75 = 0.3333$ (33.33%).

So, we have brought all of the stations up to 100 units per hour. Station 5 still has some remaining capacity: $100\% - 50\% - 33.33\% = 16.67\%$. This remaining capacity needs to be allocated to all of the remaining stations to bring them all up to the same output rate.

Suppose we want to raise the output of the assembly line by 1 unit (i.e., from 100 to 101 units/hr). How much time would this take? It would take 0.6 min for operation 1, 0.6 min for operation 2, 0.9 min for operation 3, and 0.8 min for operation 4. Suppose we want to raise the output by 2 units/hr. That would take 1.2 min for operation 1, 1.2 min for operation 2, 1.8 min for operation 3, and 1.6 min for operation 4. Notice that the ratios of times between stations remain the same. The additional time for operation 2, for example, is always the same as for operation 1.

All of the extra time must come from the remaining capacity of station 5, since all other stations are working at their individual capacities. Station 5 has 17% of its time left, and it must allocate its time in the same fraction (ratio) as the assembly times.

operation	time	fraction of total	ratio × 17%
1	0.6	0.2069	3.52%
2	0.6	0.2069	3.52%
3	0.9	0.3103	5.27%
4	0.8	0.2759	4.69%
totals	2.9	1.0000	17.00%

Therefore, 3.52% of station 5's time will be given to operations 1 and 2. Operation 3 will receive $50\% + 5.27\% = 55.27\%$ of station 5's time. Operation 4 will receive $33.33\% + 4.69\% = 38.02\%$.

(a) The production rates of each of the operations are as follows.

operations 1 and 2:
$$\frac{(1.0352)\left(60\ \frac{min}{hr}\right)}{0.6\ \frac{min}{unit}} = 103.5 \text{ units/hr}$$

operation 3:
$$\frac{(1.5527)\left(60\ \frac{min}{hr}\right)}{0.9\ \frac{min}{unit}} = 103.5 \text{ units/hr}$$

operation 4:
$$\frac{(1.3802)\left(60\ \frac{min}{hr}\right)}{0.8\ \frac{min}{unit}} = 103.5 \text{ units/hr}$$

So, the capacity of the assembly line is 103 units per hour.

(Check: The total time required per product is 2.9 minutes. With all five 5 stations working optimally, the total available time per hour is (5 stations)(60 min/station) = 300 min. The optimal production rate would be (300 min/hr)/(2.9 min/unit) = 103.5 units/hr. This checks.)

However, everybody takes a 10 min break per hour, and the stations only work 50 min per hour. So, the overall capacity is reduced proportionally.

$$\text{capacity} = \left(103.5\ \frac{\text{units}}{\text{hr}}\right)\left(\frac{50\ \text{min}}{60\ \text{min}}\right)$$

$$= \boxed{86.3 \text{ units/hr}}$$

3. (a) The critical path network diagram is as follows.

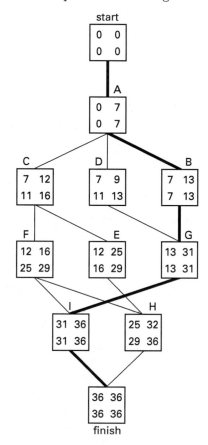

ES (earliest start) Rule: The earliest start time for an activity leaving a particular node is equal to the largest of the earliest finish times for all activities entering the node.

LF (latest finish) Rule: The latest finish time for an activity entering a particular node is equal to the smallest of the latest start times for all activities leaving the node.

The activity is critical if the earliest start equals the latest start.

(b) The critical path is | A-B-G-I. |

(c) The earliest finish is | 36. |

(d) The latest finish is | 36. |

(e) The slack along the critical path is | 0. |

(f) The float along the critical path is | 0. |

4. From Eq. 62.1,

$$\mu = \frac{t_{\text{minimum}} + 4t_{\text{likely}} + t_{\text{maximum}}}{6}$$

From Eq. 62.2,

$$\sigma^2 = \left(\frac{t_{\text{maximum}} - t_{\text{minimum}}}{6}\right)^2$$

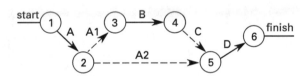

The critical path is A-A1-B-C-D.

The following probability calculations assume that all activities are independent. Use the following theorems for the sum of independent random variables and use the normal distribution for T (project time).

$$\mu_{\text{total}} = t_A + t_B + t_C + t_D$$
$$\sigma^2_{\text{total}} = \sigma^2_A + \sigma^2_B + \sigma^2_C + \sigma^2_D$$

The variance is 10.52778 and the standard deviation is 3.244654.

$$\mu_{\text{total}} = 43.83333$$
$$\sigma^2_{\text{total}} = 10.52778$$
$$\sigma_{\text{total}} = 3.244654$$
$$z = \frac{t - \mu_{\text{total}}}{\sigma} = \left|\frac{42 - 43.83333}{3.244654}\right| = 0.565$$

The probability of finishing for $T \le 42$ is 0.286037 | (28.6%). |

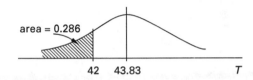

The answer is (B).

5. This is a two-dimensional linear programming problem.

$$x_1 = \text{no. of B-1 buses produced}$$
$$x_2 = \text{no. of B-2 buses produced}$$
$$Z = \text{total profit}$$

PERT Analysis for Prob. 4

no.	name	activity exp. time	variance	earliest start	latest start	earliest finish	latest finish	slack LS-ES
1	A	+2.33333	+0.44444	15	15	+17.3333	+17.3333	0
2	A1	0	0	+17.3333	+17.3333	+17.3333	+17.3333	0
3	A2	0	0	+17.3333	+39.6667	+17.3333	+39.6667	+22.3333
4	B	+10.5000	+4.69444	+17.3333	+17.3333	+27.8333	+27.8333	0
5	C	+11.8333	+4.69444	+27.8333	+27.8333	+39.6667	+39.6667	0
6	D	+4.16667	+0.69444	+39.6667	+39.6667	+43.8333	+43.8333	0

expected completion time = 43.83333

Plant Design

The objective function is

$$\text{maximize } Z = 800{,}000x_1 + 650{,}000x_2$$

The constraints are

$$8x_1 + 10x_2 \le 2000$$
$$6x_1 + 4x_2 \le 1800$$
$$3x_1 + 2x_2 \le 3600$$
$$x_1 \ge 0, x_2 \ge 0$$

The simplex theory states that the optimal solution will be a feasible corner point.

feasible corner points (x_1, x_2)	Z
$(0,0)$	0
$(0, 200)$	$130{,}000{,}000$
optimal solution → $(250, 0)$	$200{,}000{,}000$

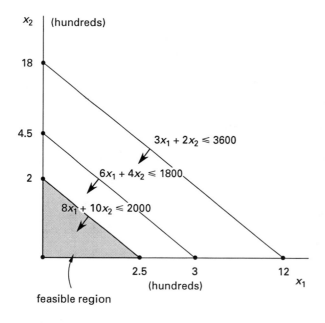

feasible region

6. (a) Solve as a linear programming problem, since partial units are permitted.

$$x_1 = \text{no. of units of C-1 produced/month}$$
$$x_2 = \text{no. of units of C-2 produced/month}$$
$$Z = \text{total profit}$$

The objective function is

$$\text{maximize } Z = 3000x_1 + 5000x_2$$

The constraints are

$$10x_1 + 20x_2 \le 300$$
$$8x_1 + 3x_2 \le 120$$
$$x_1 \ge 0, x_2 \ge 0$$

The simplex theory states that the optimal solution will be a feasible corner point.

feasible corner points (x_1, x_2)	Z
$(0,0)$	0
$(0, 15)$	$75{,}000$
optimal solution → $\left(\dfrac{150}{13}, \dfrac{120}{13}\right)$	$80{,}769.23$
$(15, 0)$	$45{,}000$

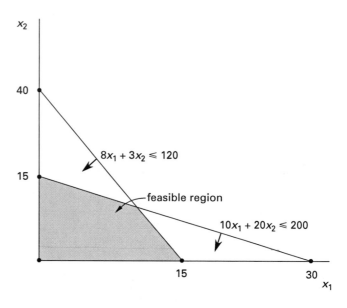

At the intersection point of the two constraints,

$$65x_2 = 600$$
$$x_2 = \frac{600}{65} = 120/13$$

Substituting x_2 into the first constraint yields

$$10x_1 + (20)\left(\frac{120}{13}\right) \le 300$$
$$x_1 = 150/13$$
$$(x_1, x_2) = \left(\frac{150}{13}, \frac{120}{13}\right)$$

(b) Integer programming methods are required when variables are constrained to integer values. Merely deleting the fractional parts in the solution doesn't always work. In this case, deleting the fractional parts yields

$$x_1 = \text{INT}\left(\frac{150}{13}\right) = 11$$
$$x_2 = \text{INT}\left(\frac{120}{13}\right) = 9$$
$$Z = 3000x_1 + 5000x_2 = 78{,}000$$

However, $x_1 = x_2 = 10$ also satisfies the constraints and yields the optimum $Z = 80,000$. In the absence of integer programming tools, trial and error in the vicinity of the feasible corners is required.

7. RE = relative earnings

From Eq. 62.70,

$$\text{bonus factor} = \frac{\text{standard bonus}}{\text{participation}}$$

$$\text{productivity} = \frac{\text{standard time}}{\text{actual time}}$$

From Eq. 62.71,

$$\text{RE} = 1 + (\text{productivity})(\text{participation})$$
$$\times (1 + \text{bonus factor}) - (\text{participation})$$

$$\text{bonus factor} = \frac{0.667}{0.500} = 1.334$$

$$\text{productivity} = \frac{0.004 \text{ hour}}{\dfrac{8}{1900} \text{ hour}} = 0.95$$

$$\text{RE} = 1 + (0.95)(0.50)(1 + 1.334) - 0.5$$

$$= 1.60865 \quad \boxed{(161\%)}$$

The answer is (D).

63 Engineering Economic Analysis

PRACTICE PROBLEMS

1. At 6% effective annual interest, how much will be accumulated if $1000 is invested for ten years?

2. At 6% effective annual interest, what is the present worth of $2000 that becomes available in four years?

3. At 6% effective annual interest, how much should be invested to accumulate $2000 in 20 years?

4. At 6% effective annual interest, what year-end annual amount deposited over seven years is equivalent to $500 invested now?

5. At 6% effective annual interest, what will be the accumulated amount at the end of ten years if $50 is invested at the end of each year for ten years?

6. At 6% effective annual interest, how much should be deposited at the start of each year for ten years (a total of ten deposits) in order to empty the fund by drawing out $200 at the end of each year for ten years (a total of ten withdrawals)?

7. At 6% effective annual interest, how much should be deposited at the start of each year for five years to accumulate $2000 on the date of the last deposit?

8. At 6% effective annual interest, how much will be accumulated in ten years if three payments of $100 are deposited every other year for four years, with the first payment occurring at $t = 0$?

9. $500 is compounded monthly at a 6% nominal annual interest rate. How much will have accumulated in five years?

10. What is the effective annual rate of return on an $80 investment that pays back $120 in seven years?

11. A new machine will cost $17,000 and will have a resale value of $14,000 after five years. Special tooling will cost $5000. The tooling will have a resale value of $2500 after five years. Maintenance will be $2000 per year. The effective annual interest rate is 6%. What will be the average annual cost of ownership during the next five years?

12. An old covered wooden bridge can be strengthened at a cost of $9000, or it can be replaced for $40,000. The present salvage value of the old bridge is $13,000. It is estimated that the reinforced bridge will last for 20 years, will have an annual cost of $500, and will have a salvage value of $10,000 at the end of 20 years. The estimated salvage value of the new bridge after 25 years is $15,000. Maintenance for the new bridge would cost $100 annually. The effective annual interest rate is 8%. Which is the best alternative?

13. A firm expects to receive $32,000 each year for 15 years from sales of a product. An initial investment of $150,000 will be required to manufacture the product. Expenses will run $7530 per year. Salvage value is zero, and straight-line depreciation is used. The income tax rate is 48%. What is the after-tax rate of return?

14. A public works project has initial costs of $1,000,000, benefits of $1,500,000, and disbenefits of $300,000.

(a) What is the Winfrey benefit/cost ratio?

(b) What is the excess of benefits over costs?

15. A speculator in land pays $14,000 for property that he expects to hold for ten years. $1000 is spent in renovation, and a monthly rent of $75 is collected from the tenants. (Use the year-end convention.) Taxes are $150 per year, and maintenance costs are $250 per year. What must be the sale price in ten years to realize a 10% rate of return?

16. What is the effective annual interest rate for a payment plan of 30 equal payments of $89.30 per month when a lump sum payment of $2000 would have been an outright purchase?

17. A depreciable item is purchased for $500,000. The salvage value at the end of 25 years is estimated at $100,000.

(a) What is the depreciation in each of the first three years using the straight line method?

(b) What is the depreciation in each of the first three years using the sum-of-the-years' digits method?

(c) What is the depreciation in each of the first three years using the double declining balance method?

18. Equipment that is purchased for $12,000 now is expected to be sold after ten years for $2000. The estimated maintenance is $1000 for the first year, but it is expected to increase $200 each year thereafter. The effective annual interest rate is 10%.

(a) What is the present worth?

(b) What is the annual cost?

19. A new grain combine with a twenty year life can remove seven pounds of rocks from its harvest per hour. Any rocks left in its output hopper will cause $25,000 damage in subsequent processes. Several investments are available to increase the rock-removal capacity, as listed in the table. The effective annual interest rate is 10%. What should be done?

rock removal rate	annual probability of exceeding rock removal rate	required investment to achieve removal rate
7	0.15	0
8	0.10	$15,000
9	0.07	$20,000
10	0.03	$30,000

20. (*Time limit: one hour*) A mechanism that costs $10,000 has operating costs and salvage values as given. An effective annual interest rate of 20% is to be used.

year	operating cost	salvage value
1	$2000	$8000
2	$3000	$7000
3	$4000	$6000
4	$5000	$5000
5	$6000	$4000

(a) What is the economic life of the mechanism?

(b) Assuming that the mechanism has been owned and operated for four years already, what is the cost of owning and operating the mechanism for one more year?

21. (*Time limit: one hour*) A salesperson intends to purchase a car for $50,000 for personal use, driving 15,000 miles per year. Insurance for personal use costs $2000 per year, and maintenance costs $1500 per year. The car gets 15 miles per gallon, and gasoline costs $1.50 per gallon. The resale value after five years will be $10,000. The salesperson's employer has asked that the car be used for business driving of 50,000 miles per year and has offered a reimbursement of $0.30 per mile. Using the car for business would increase the insurance cost to $3000 per year and maintenance to $2000 per year. The salvage value after five years would be reduced to $5000. If the employer purchased a car for the salesperson to use, the initial cost would be the same, but insurance, maintenance, and salvage would be $2500, $2000, and $8000, respectively. Use 10% for the employer's and the salesperson's effective annual interest rates.

(a) Is the reimbursement offer adequate?

(b) With a reimbursement of $0.30 per mile, how many miles must the car be driven per year to justify the employer buying the car for the salesperson to use?

22. (*Time limit: one hour*) Alternatives A and B are being evaluated. The effective annual interest rate is 10%. What alternative is economically superior?

	alternative A	alternative B
first cost	$80,000	$35,000
life	20 years	10 years
salvage value	$7000	0
annual costs		
years 1–5	$1000	$3000
years 6–10	$1500	$4000
years 11–20	$2000	0
additional cost		
year 10	$5000	0

23. (*Time limit: one hour*) A car is needed for three years. Plans A and B for acquiring the car are being evaluated. An effective annual interest rate of 10% is to be used. Which plan is economically superior?

plan A: lease the car for $0.25/mile (all inclusive)

plan B: purchase the car for $30,000
keep the car for three years
sell the car after three years for $7200
pay $0.14 per mile for oil and gas
pay other costs of $500 per year

24. (*Time limit: one hour*) Two methods are being considered to meet strict air pollution control requirements over the next ten years. Method A uses equipment with a life of ten years. Method B uses equipment with a life of five years that will be replaced with new equipment with an additional life of five years. Capacities of the two methods are different, but operating costs do not depend on the throughput. Operation is 24 hours per day, 365 days per year. The effective annual interest rate for this evaluation is 7%.

	method A	method B	
	years 1–10	years 1–5	years 6–10
installation cost	$13,000	$6000	$7000
equipment cost	$10,000	$2000	$2200
operating cost			
per hour	$10.50	$8.00	$8.00
salvage value	$5000	$2000	$2000
capacity (tons/yr)	50	20	20
life	10 years	5 years	5 years

(a) What is the uniform annual cost per ton for each method?

(b) Over what range of throughput (in tons/yr) does each method have the minimum cost?

25. (*Time limit: one hour*) A transit district has asked for your assistance in determining the proper fare for its bus system. An effective annual interest rate of 7% is to be used. The following additional information was compiled for your study.

cost per bus	$60,000
bus life	20 years
salvage value	$10,000
miles driven per year	37,440
number of passengers per year	80,000
operating cost	$1.00 per mile in the first year, increasing $0.10 per mile each year thereafter

(a) If the fare is to remain constant for the next 20 years, what is the break-even fare per passenger?

(b) If the transit district decides to set the per-passenger fare at $0.35 for the first year, by what amount should the per-passenger fare go up each year thereafter such that the district can break even in 20 years?

(c) If the transit district decides to set the per-passenger fare at $0.35 for the first year and the per-passenger fare goes up $0.05 each year thereafter, what additional governmental subsidy (per passenger) is needed for the district to break even in 20 years?

26. Make a recommendation to your client to accept one of the following alternatives. Use the present worth comparison method. (Initial costs are the same.)

A. a 25 year annuity paying $4800 at the end of each year, where the interest rate is a nominal 12% per annum

B. a 25 year annuity paying $1200 every quarter at 12% nominal annual interest

27. A firm has two alternatives for improvement of its existing production line. The data are as follows.

	alternative A	alternative B
initial installment cost	$1500	$2500
annual operating cost	$800	$650
service life	5 years	8 years
salvage value	0	0

Determine the best alternative using an interest rate of 15%.

28. Two mutually exclusive alternatives requiring different investments are being considered. The life of both alternatives is estimated at 20 years with no salvage values. The minimum rate of return that is considered acceptable is 4%. Which alternative is best?

	alternative A	alternative B
investment required	$70,000	$40,000
net income per year	$5620	$4075
rate of return on total investment	5%	8%

29. Compare the costs of two plant renovation schemes, A and B. Assume equal lives of 25 years, no salvage values, and interest at 25%.

	alternative A	alternative B
first cost	$20,000	$25,000
annual expenditure	$3000	$2500

(a) Make the comparison on the basis of present worth. Which alternative is best?

(b) Make the comparison on the basis of capitalized cost. Which alternative is best?

(c) Make the comparison on the basis of annual cost. Which alternative is best?

30. With interest at 8%, obtain the solutions to the following to the nearest dollar.

(a) A machine costs $18,000 and has a salvage value of $2000. It has a useful life of eight years. What is its book value at the end of five years using straight line depreciation?

(b) Using data from part (a), find the depreciation in the first three years using the sinking fund method.

(c) Repeat part (a) using double declining balance depreciation to find the first five years' depreciation.

31. A chemical pump motor unit is purchased for $14,000. The estimated life is eight years, after which it will be sold for $1800. Find the depreciation in the first two years by the sum-of-the-years' digits method. Calculate the after-tax depreciation recovery using 15% interest with 52% income tax.

32. A soda ash plant has the water effluent from processing equipment treated in a large settling basin. The settling basin eventually discharges into a river that runs alongside the basin. Recently enacted environmental regulations require all rainfall on the plant to be diverted and treated in the settling basin. A heavy rainfall will cause the entire basin to overflow. An uncontrolled overflow will cause environmental damage and heavy fines. The construction of additional height on the existing basic walls is under consideration.

Data on the costs of construction and expected costs for environmental cleanup and fines are shown. Data on 50 typical winters have been collected. The soda ash plant management considers 12% to be their minimum rate of return, and it is felt that after 15 years the plant will be closed. Which alternative minimizes the company's total expected costs?

Plant Design

additional basin height (ft)	number of winters with basin overflow	expense for environmental clean up per year	construction cost
0	24		0
5	14	$600,000	$600,000
10	8	$650,000	$710,000
15	3	$700,000	$900,000
20	1	$800,000	$1,000,000
	50		

33. A wood processing plant installed a waste gas scrubber at a cost of $30,000 to remove pollutants from the exhaust discharged into the atmosphere. The scrubber has no salvage value and will cost $18,700 to operate next year, with operating costs expected to increase at the rate of $1200 per year thereafter. When should the company consider replacing the scrubber? Money can be borrowed at 12%.

34. Two alternative piping schemes are being considered by a water treatment facility. On the basis of a ten year life and an interest rate of 12%, determine the number of hours of operation for which the two installations will break even.

	alternative A	alternative B
pipe diameter	4 in	6 in
head loss for required flow	48 ft	26 ft
size motor required	20 hp	7 hp
energy cost per hour of operation	$ 0.30	$ 0.10
cost of motor installed	$3600	$2800
cost of pipes and fittings	$3050	$5010
salvage value at end of 10 years	$200	$280

35. An 88% learning curve is used with an item whose first production time was six weeks.

(a) How long will it take to produce the fourth item?

(b) How long will it take to produce the 6th through 14th items?

36. *(Time limit: one hour)* A company is considering two alternatives, only one of which can be selected.

alternative	initial investment	salvage value	annual net profit	life
A	$120,000	$15,000	$57,000	5 yr
B	$170,000	$20,000	$67,000	5 yr

The net profit is after operating and maintenance costs, but before taxes. The company pays 45% of its year-end profit as income taxes. Use straight line depreciation. Do not use investment tax credit. Find the best alternative if the company's minimum attractive rate of return is 15%.

37. *(Time limit: one hour)* A company is considering the purchase of equipment to expand its capacity. The equipment cost is $300,000. The equipment is needed for five years, after which it will be sold for $50,000.

The company's before-tax cash flow will be improved $90,000 annually by the purchase of the asset.

The corporate tax rate is 48%, and straight line depreciation will be used. The company will take an investment tax credit of 6.67%. What is the after-tax rate of return associated with this equipment purchase?

38. *(Time limit: one hour)* A 120-room hotel is purchased for $2,500,000. A 25 year loan is available for 12%. A study was conducted to determine the various occupancy rates.

occupancy	probability
65% full	0.40
70%	0.30
75%	0.20
80%	0.10

The operating costs of the hotel are as follows.

taxes and insurance	$20,000 annually
maintenance	$50,000 annually
operating	$200,000 annually

The life of the hotel is figured to be 25 years when operating 365 days per year. The salvage value after 25 years is $500,000.

Neglect tax credit and income taxes. Determine the average rate that should be charged per room per night to return 15% of the initial cost each year.

39. *(Time limit: one hour)* A company is insured for $3,500,000 against fire. The insurance rate is $0.69/1000. The insurance company will decrease the rate to $0.47/1000 if fire sprinklers are installed. The initial cost of the sprinklers is $7500. Annual costs are $200; additional taxes are $100 annually. The system life is 25 years. What is the rate of return on this investment?

40. *(Time limit: one hour)* Heat losses through the walls in an existing building cost a company $1,300,000 per year. This amount is considered excessive, and two alternatives are being evaluated. Neither of the alternatives will increase the life of the existing building beyond the current expected life of six years, and neither of the alternatives will produce a salvage value.

Alternative A: Do nothing. Continue with current losses.

Alternative B: Spend $2,000,000 immediately to upgrade the building and reduce the loss by 80%. This alternative will require annual maintenance of $150,000.

Alternative C: Spend $1,200,000 immediately. Repeat the $1,200,000 expenditure three years from now. Heat loss the first year will be reduced 80%. Due to deterioration, the reduction will be 55% and 20% in the second and third years. (The pattern is repeated starting after the second expenditure.) There are no maintenance costs.

All energy and maintenance costs are considered expenses for tax purposes. The company's tax rate is 48%, and straight line depreciation is used. 15% is regarded as the effective annual interest rate. Evaluate each alternative on an after-tax basis, and recommend the best alternative.

41. *(Time limit: one hour)* You have been asked to determine if a seven year old machine should be replaced. Give a full explanation for your recommendation. Base your decision on a before-tax interest rate of 15%.

The existing machine is presumed to have a ten year life. It has been depreciated on a straight line basis from its original value of $1,250,000 to a current book value of $620,000. Its ultimate salvage value was assumed to be $350,000 for purposes of depreciation. Its present salvage value is estimated at $400,000, and this is not expected to change over the next three years. The current operating costs are not expected to change from $200,000 per year.

A new machine costs $800,000, with operating costs of $40,000 the first year, and increasing by $30,000 each year thereafter. The new machine has an expected life of ten years. The salvage value depends on the year the new machine is retired.

year retired	salvage
1	$600,000
2	$500,000
3	$450,000
4	$400,000
5	$350,000
6	$300,000
7	$250,000
8	$200,000
9	$150,000
10	$100,000

SOLUTIONS

1.

$i = 6\%$ a year

By the formula from Table 63.1,

$$F = P(1+i)^n = (\$1000)(1+0.06)^{10} = \boxed{\$1790.85}$$

By the factor converting P to F, $(F/P, i, n) = 1.7908$ for $i = 6\%$ a year and $n = 10$ years.

$$F = P(F/P, 6\%, 10) = (\$1000)(1.7908) = \boxed{\$1790.80}$$

2.

$F = \$2000$

$t = 0 \qquad t = 10$

$t = 0 \qquad t = 4$

P

$i = 6\%$ a year

By the formula from Table 63.1,

$$P = \frac{F}{(1+i)^n} = \frac{\$2000}{(1+0.06)^4} = \boxed{\$1584.19}$$

From the factor converting F to P, $(P/F, i, n) = 0.7921$ for $i = 6\%$ a year and $n = 4$ years.

$$P = F(P/F, 6\%, 4) = (\$2000)(0.7921) = \boxed{\$1584.20}$$

Plant Design

3.

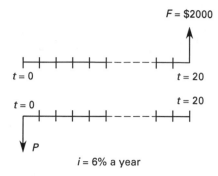

$F = \$2000$

$i = 6\%$ a year

By the formula from Table 63.1,

$$P = \frac{F}{(1+i)^n} = \frac{\$2000}{(1+0.06)^{20}} = \boxed{\$623.61}$$

From the factor converting F to P, $(P/F, i, n) = 0.3118$ for $i = 6\%$ a year and $n = 20$ years.

$$P = F(P/F, 6\%, 20) = (\$2000)(0.3118) = \boxed{\$623.60}$$

4.

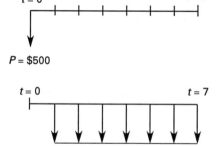

$P = \$500$

$i = 6\%$ a year

By the formula from Table 63.1,

$$A = P\left(\frac{i(1+i)^n}{(1+i)^n - 1}\right) = (\$500)\left(\frac{(0.06)(1+0.06)^7}{(1+0.06)^7 - 1}\right)$$

$$= \boxed{\$89.57}$$

By the factor converting P to A, $(A/P, i, n) = 0.17914$ for $i = 6\%$ a year and $n = 7$ years.

$$A = P(A/P, 6\%, 7) = (\$500)(0.17914) = \boxed{\$89.57}$$

5.

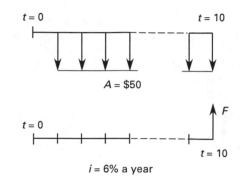

$A = \$50$

F

$i = 6\%$ a year

By the formula from Table 63.1,

$$F = A\left(\frac{(1+i)^n - 1}{i}\right) = (\$50)\left(\frac{(1+0.06)^{10} - 1}{0.06}\right)$$

$$= \boxed{\$659.04}$$

By the factor converting A to F, $(F/A, i, n) = 13.181$ for $i = 6\%$ a year and $n = 10$ years.

$$F = A(F/A, 6\%, 10) = (\$50)(13.181) = \boxed{\$659.05}$$

6.

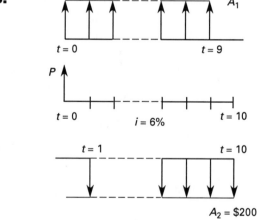

A_1

$i = 6\%$

$A_2 = \$200$

By the formula from Table 63.1, for each cash flow diagram,

$$P = A_1 + A_1\left(\frac{(1+0.06)^9 - 1}{(0.06)(1+0.06)^9}\right)$$

$$= A_2\left(\frac{(1+0.06)^{10} - 1}{(0.06)(1+0.06)^{10}}\right)$$

Therefore for $A_2 = \$200$,

$$A_1 + A_1\left(\frac{(1+0.06)^9 - 1}{(0.06)(1+0.06)^9}\right)$$

$$= (\$200)\left(\frac{(1+0.06)^{10} - 1}{(0.06)(1+0.06)^{10}}\right)$$

$$7.80A_1 = \$1472.02$$

$$A_1 = \boxed{\$188.72}$$

By the factor converting A to P,

$$(P/A, 6\%, 9) = 6.8017$$
$$(P/A, 6\%, 10) = 7.3601$$
$$A_1 + A_1(6.8017) = (\$200)(7.3601)$$
$$7.8017 A_1 = \$1472.02$$
$$A_1 = \frac{\$1472.02}{7.8017} = \boxed{\$188.68}$$

7.

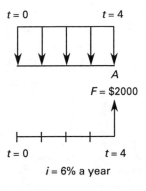

$i = 6\%$ a year

By the formula from Table 63.1,

$$F = A\left(\frac{(1+i)^n - 1}{i}\right)$$

Since the deposits start at the start of each year, $n = 4 + 1$, for a total of five deposits.

$$F = A\left(\frac{(1+i)^{n+1} - 1}{i}\right) = \$2000$$
$$= A\left(\frac{(1+0.06)^5 - 1}{0.06}\right)$$
$$\$2000 = 5.6371 A$$
$$A = \frac{\$2000}{5.6371} = \boxed{\$354.79}$$

By the factor converting P and A to F,

$$F = A\big((F/P, 6\%, 4) + (F/A, 6\%, 4)\big)$$
$$\$2000 = A(1.2625 + 4.3746)$$
$$A = \boxed{\$354.79}$$

8.

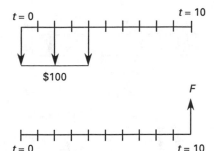

$i = 6\%$ a year

By the formula from Table 63.1, $F = P(1+i)^n$. If each deposit is considered as P, each will accumulate interest for periods of 10, 8, and 6 years.

Therefore,

$$F = (\$100)(1 + 0.06)^{10} + (\$100)(1 + 0.06)^8$$
$$+ (\$100)(1 + 0.06)^6$$
$$= (\$100)(1.7908 + 1.5938 + 1.4185)$$
$$= (\$100)(4.8031)$$
$$= \boxed{\$480.31}$$

By the factor converting P to F,

$$(F/P, i, n) = 1.7908 \text{ for } i = 6\% \text{ and } n = 10$$
$$= 1.5938 \text{ for } i = 6\% \text{ and } n = 8$$
$$= 1.4185 \text{ for } i = 6\% \text{ and } n = 6$$

By summation,

$$F = (\$100)(1.7908 + 1.5938 + 1.4185)$$
$$= (\$100)(4.8031)$$
$$= \boxed{\$480.31}$$

9.

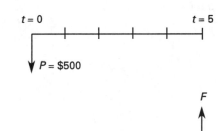

$r = 6\%$ a year

Since the deposit is compounded monthly, the effective interest rate should be calculated as shown by Eq. 63.54.

$$i = \left(1 + \frac{r}{k}\right)^k - 1 = \left(1 + \frac{0.06}{12}\right)^{12} - 1$$
$$= \boxed{0.061677 \quad (6.1677\%)}$$

By the formula from Table 63.1,

$$F = P(1+i)^n = (\$500)(1 + 0.061677)^5 = \$674.42$$

Plant Design

To use a table of factors, interpolation is required.

$i\%$	factor F/P
6	1.3382
6.1677	desired
7	1.4026

$$i = \left(\frac{6.1677 - 6}{7 - 6}\right)(1.4026 - 1.3382)$$

$$= 0.0108$$

Therefore,

$$F/P = 1.3382 + 0.0108 = 1.3490$$

$$F = P(F/P, 6.1677\%, 5) = (\$500)(1.3490)$$

$$= \boxed{\$674.50}$$

10.

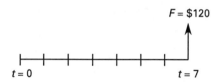

By the formula from Table 63.1,

$$F = P(1 + i)^n$$

Therefore,

$$(1 + i)^n = F/P$$

$$i = (F/P)^{\frac{1}{n}} - 1 = \left(\frac{\$120}{\$80}\right)^{\frac{1}{7}} - 1$$

$$= 0.059 \approx \boxed{6\%}$$

By the factor coverting P to F,

$$F = P(F/P, i\%, 7)$$

$$(F/P, i\%, 7) = F/P = \frac{\$120}{\$80} = 1.5$$

By checking the interest tables,

$$(F/P, i\%, 7) = 1.4071 \text{ for } i = 5\%$$

$$= 1.5036 \text{ for } i = 6\%$$

$$= 1.6058 \text{ for } i = 7\%$$

Therefore, $i = \boxed{6\%}$.

11.

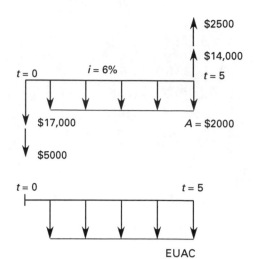

Annual cost of ownership, EUAC, can be obtained by the factors converting P to A and F to A.

$$P = \$17,000 + \$5000$$

$$= \$22,000$$

$$F = \$14,000 + \$2500$$

$$= \$16,500$$

$$\text{EUAC} = A + P(A/P, 6\%, 5) - F(A/F, 6\%, 5)$$

$$(A/P, 6\%, 5) = 0.23740$$

$$(A/F, 6\%, 5) = 0.17740$$

$$\text{EUAC} = \$2000 + (\$22,000)(0.23740)$$

$$- (\$16,500)(0.17740)$$

$$= \boxed{\$4295.70}$$

12.

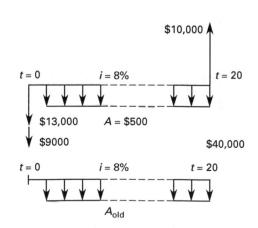

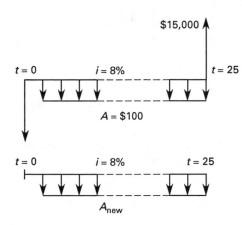

Consider the salvage value as a benefit lost (cost).

$$EUAC_{old} = \$500 + (\$22,000)(A/P, 8\%, 20)$$
$$- (\$10,000)(A/F, 8\%, 20)$$
$$(A/P, 8\%, 20) = 0.10185$$
$$(A/F, 8\%, 20) = 0.02185$$
$$EUAC_{old} = \$500 + (\$22,000)(0.10185)$$
$$- (\$10,000)(0.02185)$$
$$= \$2522.20$$

Similarly,

$$EUAC_{new} = \$100 + (\$40,000)(A/P, 8\%, 25)$$
$$- (\$15,000)(A/F, 8\%, 25)$$
$$(A/P, 8\%, 25) = 0.09368$$
$$(A/F, 8\%, 25) = 0.01368$$
$$EUAC_{new} = \$100 + (\$40,000)(0.09368)$$
$$- (\$15,000)(0.01368)$$
$$= \$3642$$

Therefore, the new bridge is going to be more costly.

The best alternative is to strengthen the old bridge.

13.

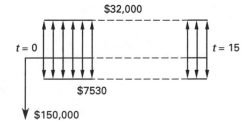

The annual depreciation is

$$D = \frac{C - S_n}{n} = \frac{\$150,000}{15}$$
$$= \$10,000/\text{year}$$

The taxable income is

$$\$32,000 - \$7530 - \$10,000 = \$14,470/\text{year}$$

Taxes paid are

$$(\$14,470)(0.48) = \$6945.60/\text{year}$$

The after-tax cash flow is

$$\$24,470 - \$6945.60 = \$17,524.40$$

The present worth of the alternate is zero when evaluated at its ROR.

$$0 = -\$150,000 + (\$17,524.40)(P/A, i\%, 15)$$

Therefore,

$$(P/A, i\%, 15) = \frac{\$150,000}{\$17,524.40} = 8.55949$$

By checking the tables, this factor matches $i = 8\%$.

ROR = 8%

14. The Winfrey benefit/cost ratio is

$$B/C = \frac{B - D}{C}$$

(a) The benefit/cost ratio will be

$$B/C = \frac{\$1,500,000 - \$300,000}{\$1,000,000} = \boxed{1.2}$$

(b) The excess of benefits over cost are $200,000.

15.

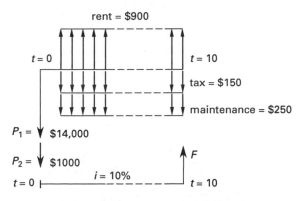

The annual rent is

$$(\$75)\left(12\ \frac{\text{months}}{\text{year}}\right) = \$900$$

$$P = P_1 + P_2 = \$15,000$$
$$A_1 = -\$900$$
$$A_2 = \$250 + \$150 = \$400$$

By the factors converting P to F and A to F,

$$F = (\$15,000)(F/P, 10\%, 10)$$
$$+ (\$400)(F/A, 10\%, 10)$$
$$- (\$900)(F/A, 10\%, 10)$$
$$(F/P, 10\%, 10) = 2.5937$$
$$(F/A, 10\%, 10) = 15.937$$
$$F = (\$15,000)(2.5937) + (\$400)(15.937)$$
$$- (\$900)(15.937)$$
$$= \boxed{\$30,937}$$

16.

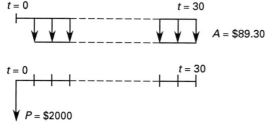

By the formula relating P to A,

$$P = A\left(\frac{(1+i)^n - 1}{i(1+i)^n}\right)$$

$$\frac{(1+i)^{30} - 1}{i(1+i)^{30}} = \frac{\$2000}{\$89.30} = 22.30$$

By trial and error,

i	$(1+i)^{30}$	$\dfrac{(1+i)^{30}-1}{i(1+i)^{30}}$
10	17.45	9.42
6	5.74	13.76
4	3.24	17.28
2	1.81	22.37

2% per month is close.

$$i = (1+0.02)^{12} - 1 = \boxed{0.2682\ \ (26.82\%)}$$

17. (a) Use the straight line method, Eq. 63.25.

$$D = \frac{C - S_n}{n}$$

Each year depreciation will remain the same.

$$D = \frac{\$500,000 - \$100,000}{25} = \boxed{\$16,000}$$

(b) Sum-of-the years digits (SOYD) can be calculated as shown by Eq. 63.28,

$$D_j = \frac{(C - S_n)(n - j + 1)}{T}$$

Use Eq. 63.27.

$$T = \tfrac{1}{2}n(n+1) = \left(\tfrac{1}{2}\right)(25)(25+1) = 325$$
$$D_1 = \frac{(\$500,000 - \$100,000)(25 - 1 + 1)}{325}$$
$$= \boxed{\$30,769}$$
$$D_2 = \frac{(\$500,000 - \$100,000)(25 - 2 + 1)}{325}$$
$$= \boxed{\$29,538}$$
$$D_3 = \frac{(\$500,000 - \$100,000)(25 - 3 + 1)}{325}$$
$$= \boxed{\$28,308}$$

(c) The double-declining balance (DDB) method can be used. By Eq. 63.32,

$$D_j = dC(1-d)^{j-1}$$

Use Eq. 63.31.

$$d = \frac{2}{n}$$
$$= \frac{2}{25}$$
$$D_1 = \left(\frac{2}{25}\right)(\$500,000)\left(1 - \frac{2}{25}\right)^0 = \boxed{\$40,000}$$
$$D_2 = \left(\frac{2}{25}\right)(\$500,000)\left(1 - \frac{2}{25}\right)^1 = \boxed{\$36,800}$$
$$D_3 = \left(\frac{2}{25}\right)(\$500,000)\left(1 - \frac{2}{25}\right)^2 = \boxed{\$33,856}$$

18.

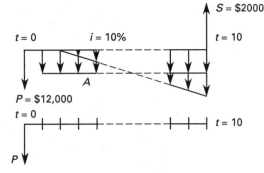

(a) $A = \$1000$ and $G = \$200$ for $t = n - 1 = 9$ years.

$F = S = \$2000$

$P = \$12,000 + A(P/A, 10\%, 10) + G(P/G, 10\%, 10)$
$\quad - F(P/F, 10\%, 10)$

$\quad = \$12,000 + (\$1000)(6.1446) + (\$200)(22.8913)$
$\quad - (\$2000)(0.3855)$

$\quad = \boxed{\$21,952}$

(b) $A = (\$12,000)(A/P, 10\%, 10) + \$1000$
$\quad + (\$200)(A/G, 10\%, 10)$
$\quad - (\$2000)(A/F, 10\%, 10)$

$\quad = (\$12,000)(0.16275) + \$1000 + (\$200)(3.7255)$
$\quad - (\$2000)(0.06275)$

$\quad = \boxed{\$3572.60}$

19. An increase in rock removal capacity can be achieved by a 20-year loan (investment). Different cases available can be compared by equivalent uniform annual cost (EUAC).

$$EUAC = \text{annual loan cost}$$
$$+ \text{expected annual damage}$$
$$= \text{cost } (A/P, 10\%, 20)$$
$$+ (\$25,000)(\text{probability})$$

$(A/P, 10\%, 20) = 0.11746$

A table can be prepared for different cases.

rock removal rate	cost ($)	annual loan cost ($)	expected annual damage ($)	EUAC ($)
7	0	0	3750	3750.00
8	15,000	1761.90	2500	4261.90
9	20,000	2349.20	1750	4099.20
10	30,000	3523.80	750	4273.80

It is cheapest to do nothing.

20. Calculate the cost of owning and operating for each year.

$A_1 = (\$10,000)(A/P, 20\%, 1) + \$2000$
$\quad - (\$8000)(A/F, 20\%, 1)$

$(A/P, 20\%, 1) = 1.2$
$(A/F, 20\%, 1) = 1.0$

$A_1 = (\$10,000)(1.2) + \$2000 - (\$8000)(1.0)$
$\quad = \$6000$

$A_2 = (\$10,000)(A/P, 20\%, 2) + \$2000$
$\quad + (\$1000)(A/G, 20\%, 2)$
$\quad - (\$7000)(A/F, 20\%, 2)$

$(A/P, 20\%, 2) = 0.6545$
$(A/G, 20\%, 2) = 0.4545$
$(A/F, 20\%, 2) = 0.4545$

$A_2 = (\$10,000)(0.6545) + \$2000$
$\quad + (\$1000)(0.4545) - (\$7000)(0.4545)$
$\quad = \$5818$

$A_3 = (\$10,000)(A/P, 20\%, 3) + \$2000$
$\quad + (\$1000)(A/G, 20\%, 3)$
$\quad - (\$6000)(A/F, 20\%, 3)$

$(A/P, 20\%, 3) = 0.4747$
$(A/G, 20\%, 3) = 0.8791$
$(A/F, 20\%, 3) = 0.2747$

$A_3 = (\$10,000)(0.4747) + \$2000$
$\quad + (\$1000)(0.8791)$
$\quad - (\$6000)(0.2747)$
$\quad = \$5977.90$

$A_4 = (\$10,000)(A/P, 20\%, 4)$
$\quad + \$2000 + (\$1000)(A/G, 20\%, 4)$
$\quad - (\$5000)(A/F, 20\%, 4)$

$(A/P, 20\%, 4) = 0.3863$
$(A/G, 20\%, 4) = 1.2762$
$(A/F, 20\%, 4) = 0.1863$

$A_4 = (\$10,000)(0.3863) + \$2000$
$\quad + (\$1000)(1.2762) - (\$5000)(0.1863)$
$\quad = \$6207.70$

$A_5 = (\$10,000)(A/P, 20\%, 5) + \$2000$
$\quad + (\$1000)(A/G, 20\%, 5)$
$\quad - (\$4000)(A/F, 20\%, 5)$

Plant Design

$(A/P, 20\%, 5) = 0.3344$

$(A/G, 20\%, 5) = 1.6405$

$(A/F, 20\%, 5) = 0.1344$

$$A_5 = (\$10,000)(0.3344) + \$2000$$
$$+ (\$1000)(1.6405) - (\$4000)(0.1344)$$
$$= \$6446.90$$

(a) Since the annual owning and operating cost is smallest after two years of operation, it is advantageous to sell the mechanism after the second year.

The economic life is two years.

(b) After four years of operation, the owning and operating cost of the mechanism for one more year will be

$$A = \$6000 + (\$5000)(1 + i) - \$4000$$
$$i = 0.2 \quad (20\%)$$
$$A = \$6000 + (\$5000)(1.2) - \$4000$$
$$= \boxed{\$8000}$$

21. (a) To find out if the reimbursement is adequate, calculate the business-related expense.

Charge the company for business travel.

$$\text{insurance: } \$3000 - \$2000 = \$1000$$
$$\text{maintenance: } \$2000 - \$1500 = \$500$$
$$\text{drop in salvage value: } \$10,000 - \$5000 = \$5000$$

The annual portion of the drop in salvage value is

$$A = (\$5000)(A/F, 10\%, 5)$$
$$(A/F, 10\%, 5) = 0.1638$$
$$A = (\$5000)(0.1638) = \$819/\text{year}$$

The cost of gas is

$$\left(\frac{50,000 \text{ mi}}{15 \frac{\text{mi}}{\text{gal}}}\right)\left(\frac{\$1.50}{\text{gal}}\right) = \$5000/\text{yr}$$

$$\text{EUAC per mile} = \frac{\$1000 + \$500 + \$819 + \$5000}{50,000 \text{ mi}}$$
$$= \boxed{\$0.14638/\text{mi}}$$

Since the reimbursement per mile was \$0.30 and since \$0.30 > \$0.14638, the reimbursement is adequate.

(b) Next, determine (with reimbursement) how many miles must the car be driven to break even.

If the car is driven M miles per year,

$$\left(\frac{\$0.30}{1 \text{ mi}}\right) M = (\$50,000)(A/P, 10\%, 5) + \$2500$$
$$+ \$2000 - (\$8000)(A/F, 10\%, 5)$$
$$+ \left(\frac{M}{15 \frac{\text{mi}}{\text{gal}}}\right)(\$1.50)$$

$$(A/P, 10\%, 5) = 0.2638$$
$$(A/F, 10\%, 5) = 0.1638$$
$$0.3M = (\$50,000)(0.2638) + \$2500 + \$2000$$
$$- (\$8000)(0.1638) + 0.1M$$
$$0.2M = \$16,379.60$$
$$M = \frac{\$16,379.60}{0.2 \frac{\$}{\text{mi}}} = \boxed{81,898 \text{ mi}}$$

22.

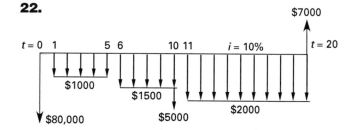

$$P_A = \$80,000 + (\$1000)(P/A, 10\%, 5)$$
$$+ (\$1500)(P/A, 10\%, 5)(P/F, 10\%, 5)$$
$$+ (\$2000)(P/A, 10\%, 10)(P/F, 10\%, 10)$$
$$+ (\$5000)(P/F, 10\%, 10)$$
$$- (\$7000)(P/F, 10\%, 20)$$

$(P/A, 10\%, 5) = 3.7908$

$(P/F, 10\%, 5) = 0.6209$

$(P/A, 10\%, 10) = 6.1446$

$(P/F, 10\%, 10) = 0.3855$

$(P/F, 10\%, 10) = 0.3855$

$(P/F, 10\%, 20) = 0.1486$

$$P_A = \$80,000 + (\$1000)(3.7908)$$
$$+ (\$1500)(3.7908)(0.6209)$$
$$+ (\$2000)(6.1446)(0.3855)$$
$$+ (\$5000)(0.3855) - (\$7000)(0.1486)$$
$$= \$92,946.15$$

Since the lives are different, compare by EUAC.

$$\text{EUAC}(A) = (\$92,946.14)(A/P, 10\%, 20)$$
$$= (\$92,946.14)(0.1175) = \$10,921$$

Similarly, evaluate alternative B.

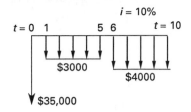

$$P_B = \$35,000 + (\$3000)(P/A, 10\%, 5)$$
$$\quad + (\$4000)(P/A, 10\%, 5)(P/F, 10\%, 5)$$
$$(P/A, 10\%, 5) = 3.7908$$
$$(P/F, 10\%, 5) = 0.6209$$
$$P_B = \$35,000 + (\$3000)(3.7908)$$
$$\quad + (\$4000)(3.7908)(0.6209)$$
$$= \$55,787.23$$
$$\text{EUAC}(B) = (\$55,787.23)(A/P, 10\%, 10)$$
$$= (\$55,787.23)(0.1627) = \$9077$$

Since EUAC(B) < EUAC(A),

Alternative B is economically superior.

23. For both cases, if the annual cost is compared with a total annual mileage of M,

$$A_A = \$0.25M$$
$$A_B = (\$30,000)(A/P, 10\%, 3) + \$0.14M$$
$$\quad + \$500 - (\$7200)(A/F, 10\%, 3)$$
$$(A/P, 10\%, 3) = 0.40211$$
$$(A/F, 10\%, 3) = 0.30211$$
$$A_B = (\$30,000)(0.40211) + \$0.14M + \$500$$
$$\quad - (\$7200)(0.30211)$$
$$= \$12,063.30 + \$0.14M$$
$$\quad + \$500 - \$2175.19$$
$$= \$10,388.11 + \$0.14M$$

For an equal annual cost $A_A = A_B$,

$$\$0.25M = \$10,388.11 + \$0.14M$$

An annual mileage would be $M = 94,437$ mi.

For an annual mileage less than that, $A_A < A_B$.

Plan A is economically superior until that mileage is exceeded.

24. ((a) and (b))

Method A:

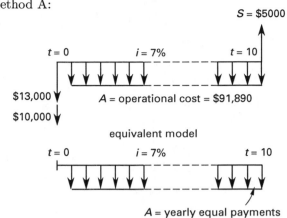

$$\begin{aligned} &24 \text{ hours/day} \\ &365 \text{ days/year} \\ &\text{total of } (24)(365) = 8760 \text{ hours/year} \\ &\$10.50 \text{ operational cost/hour} \\ &\text{total of } (8760)(\$10.50) = \$91,890 \text{ operational cost/year} \end{aligned}$$

$$A = \$91,980 + (\$23,000)(A/P, 7\%, 10)$$
$$\quad - (\$5000)(A/F, 7\%, 10)$$
$$(A/P, 7\%, 10) = 0.14238$$
$$(A/F, 7\%, 10) = 0.07238$$
$$A = \$91,980 + (\$23,000)(0.14238)$$
$$\quad - (\$5000)(0.07238)$$
$$= \$94,892.84/\text{yr}$$

Therefore, the uniform annual cost per ton each year will be

$$\frac{\$94,892.84}{50 \text{ ton}} = \boxed{\$1897.86}$$

Method B:

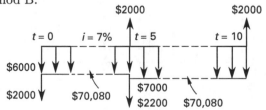

$$\begin{aligned} &8760 \text{ hours/year} \\ &\$8 \text{ operational cost/hour} \\ &\text{total of } \$70,080 \text{ operational cost/year} \end{aligned}$$

$$A = \$70{,}080 + (\$6000 + \$2000)(A/P, 7\%, 10)$$
$$+ (\$7000 + \$2200 - \$2000)(P/F, 7\%, 5)$$
$$\times (A/P, 7\%, 10)$$
$$- (\$2000)(A/F, 7\%, 10)$$

$$(A/P, 7\%, 10) = 0.1424$$
$$(A/F, 7\%, 10) = 0.07238$$
$$(P/F, 7\%, 5) = 0.7130$$

$$A = \$70{,}080 + (\$8000)(0.1424)$$
$$+ (\$7200)(0.7130)(0.1424)$$
$$- (\$2000)(0.07238)$$
$$= \$71{,}805.46/\text{yr}$$

Therefore, the uniform annual cost per ton each year will be

$$\frac{\$71{,}805.46}{20 \text{ ton}} = \boxed{\$3590.27}$$

tons/hr	cost of using A		cost of using B		cheapest
0–20	\$94,893	(1x)	\$71,805	(1x)	B
20–40	\$94,893	(1x)	\$143,610	(2x)	A
40–50	\$94,893	(1x)	\$215,415	(3x)	A
50–60	\$189,786	(2x)	\$215,415	(3x)	A
60–80	\$189,786	(2x)	\$287,220	(4x)	A

25.

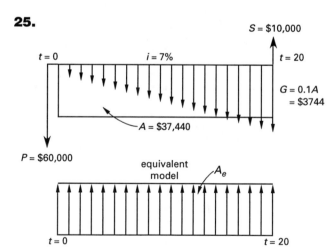

$$A_e = (\$60{,}000)(A/P, 7\%, 20) + A$$
$$+ G(P/G, 7\%, 20)(A/P, 7\%, 20)$$
$$- (\$10{,}000)(A/F, 7\%, 20)$$
$$(A/P, 7\%, 20) = 0.09439$$

$$A = (37{,}440 \text{ mi})\left(\frac{\$1.0}{1 \text{ mi}}\right) = \$37{,}440$$
$$G = 0.1A = (0.1)(\$37{,}440) = \$3744$$
$$(P/G, 7\%, 20) = 77.509$$
$$(A/F, 7\%, 20) = 0.02439$$

$$A_e = (\$60{,}000)(0.09439) + \$37{,}440$$
$$+ (\$3744)(77.509)(0.09439)$$
$$- (\$10{,}000)(0.02439)$$
$$= \$70{,}250.88$$

(a) With 80,000 passengers a year, the break-even fare per passenger would be

$$\text{fare} = \frac{A_e}{80{,}000} = \frac{\$70{,}250.88}{80{,}000} = \boxed{\$0.878/\text{passenger}}$$

(b)
$$\$0.878 = \$0.35 + G(A/G, 7\%, 20)$$
$$G = \frac{\$0.878 - \$0.35}{7.3163}$$
$$= \boxed{\$0.072 \text{ increase per year}}$$

(c) As in part (b), the subsidy should be

$$\text{subsidy} = \text{cost} - \text{revenue}$$
$$P = \$0.878 - \big(\$0.35 + (\$0.05)(A/G, 7\%, 20)\big)$$
$$= \$0.878 - \big(\$0.35 + (\$0.05)(7.3163)\big)$$
$$= \boxed{\$0.162}$$

26.
$$P(A) = (\$4800)(P/A, 12\%, 25)$$
$$= (\$4800)(7.8431)$$
$$= \$37{,}646.88$$

$$(4 \text{ quarters})(25 \text{ years}) = 100 \text{ compounding periods}$$
$$P(B) = (\$1200)(P/A, 3\%, 100)$$
$$= (\$1200)(31.5989)$$
$$= \$37{,}918.68$$

Alternative B is economically superior.

27. $$\text{EUAC(A)} = (\$1500)(A/P, 15\%, 5) + \$800$$
$$= (\$1500)(0.2983) + \$800$$
$$= \$1247.45$$
$$\text{EUAC(B)} = (\$2500)(A/P, 15\%, 8) + \$650$$
$$= (\$2500)(0.2229) + \$650$$
$$= \$1207.25$$

Alternative B is economically superior.

28. The data given imply that both investments return 4% or more. However, the increased investment of \$30,000 may not be cost effective. Do an incremental analysis.

incremental cost = \$70,000 − \$40,000 = \$30,000

$$\text{incremental} \atop \text{income} = \$5620 - \$4075 = \$1545$$

$$0 = -\$30,000 + (\$1545)(P/A, i\%, 20)$$

$$(P/A, i\%, 20) = 19.417$$

$$i \approx 0.25\% < 4\%$$

Alternative B is economically superior.

(The same conclusion could be reached by taking the present worths of both alternatives at 4%.)

29. (a) $P(A) = (-\$3000)(P/A, 25\%, 25) - \$20,000$

$$= (-\$3000)(3.9849) - \$20,000$$

$$= -\$31,954.70$$

$$P(B) = (-\$2500)(3.9849) - \$25,000$$

$$= -\$34,962.25$$

A is better.

(b) $\quad CC(A) = \$20,000 + \dfrac{\$3000}{0.25} = \$32,000$

$$CC(B) = \$25,000 + \dfrac{\$2500}{0.25} = \$35,000$$

A is better.

(c) $EUAC(A) = (\$20,000)(A/P, 25\%, 25) + \$3000$

$$= (\$20,000)(0.2509) + \$3000$$

$$= \$8018.00$$

$$EUAC(B) = (\$25,000)(0.2509) + \$2500$$

$$= \$8772.50$$

A is better.

30. (a) $BV = \$18,000 - (5)\left(\dfrac{\$18,000 - \$2000}{8}\right)$

$$= \$8000$$

(b) With the sinking fund method, the basis is

$$(\$18,000 - \$2000)(A/F, 8\%, 8)$$

$$= (\$18,000 - \$2000)(0.0940)$$

$$= \$1504$$

$$D_1 = (\$1504)(1.000) = \$1504$$

$$D_2 = (\$1504)(1.0800) = \$1624$$

$$D_3 = (\$1504)(1.0800)^2 = (\$1504)(1.1664)$$

$$= \$1754$$

(c) $D_1 = \left(\frac{2}{8}\right)(\$18,000) = \$4500$

$$D_2 = \left(\frac{2}{8}\right)(\$18,000 - \$4500) = \$3375$$

$$D_3 = \left(\frac{2}{8}\right)(\$18,000 - \$4500 - \$3375) = \$2531$$

$$D_4 = \left(\frac{2}{8}\right)(\$18,000 - \$4500 - \$3375 - \$2531)$$

$$= \$1898$$

$$D_5 = \left(\frac{2}{8}\right)(\$18,000 - \$4500 - \$3375$$
$$- \$2531 - \$1898)$$

$$= \$1424$$

$$BV = \$18,000 - \$4500 - \$3375 - \$2531$$
$$- \$1898 - \$1424$$

$$= \$4272$$

31. $\quad T = \left(\frac{1}{2}\right)(8)(9) = 36$

$$D_1 = \left(\frac{8}{36}\right)(\$14,000 - \$1800) = \$2711$$

$$\Delta D = \left(\frac{1}{36}\right)(\$14,000 - \$1800) = \$339$$

$$D_2 = \$2711 - \$339 = \$2372$$

$$DR = (0.52)(\$2711)(P/A, 15\%, 8)$$

$$- (0.52)(\$339)(P/G, 15\%, 8)$$

$$= (0.52)(\$2711)(4.4873)$$

$$- (0.52)(\$339)(12.4807)$$

$$= \$4125.74$$

32. $EUAC_{5\,\text{ft}} = (\$600,000)(A/P, 12\%, 15)$

$$+ \left(\frac{14}{50}\right)(\$600,000) + \left(\frac{8}{50}\right)(\$650,000)$$

$$+ \left(\frac{3}{50}\right)(\$700,000) + \left(\frac{1}{50}\right)(\$800,000)$$

$$= \$418,080$$

$$EUAC_{10\,\text{ft}} = (\$710,000)(A/P, 12\%, 15)$$

$$+ \left(\frac{8}{50}\right)(\$650,000) + \left(\frac{3}{50}\right)(\$700,000)$$

$$+ \left(\frac{1}{50}\right)(\$800,000)$$

$$= \$266,228$$

$$EUAC_{15\,\text{ft}} = (\$900,000)(A/P, 12\%, 15)$$

$$+ \left(\frac{3}{50}\right)(\$700,000) + \left(\frac{1}{50}\right)(\$800,000)$$

$$= \$190,120$$

$$EUAC_{20\,\text{ft}} = (\$1,000,000)(A/P, 12\%, 15)$$

$$+ \left(\frac{1}{50}\right)(\$800,000)$$

$$= \$162,800$$

Build to 20 ft.

Plant Design

33. Assume replacement after 1 year.

$$\text{EUAC}(1) = (\$30{,}000)(A/P, 12\%, 1) + \$18{,}700$$
$$= (\$30{,}000)(1.12) + \$18{,}700 = \$52{,}300$$

Assume replacement after 2 years.

$$\text{EUAC}(2) = (\$30{,}000)(A/P, 12\%, 2)$$
$$+ \$18{,}700 + (\$1200)(A/G, 12\%, 2)$$
$$= (\$30{,}000)(0.5917) + \$18{,}700$$
$$+ (\$1200)(0.4717) = \$37{,}017$$

Assume replacement after 3 years.

$$\text{EUAC}(3) = (\$30{,}000)(A/P, 12\%, 3)$$
$$+ \$18{,}700 + (\$1200)(A/G, 12\%, 3)$$
$$= (\$30{,}000)(0.4163) + \$18{,}700$$
$$+ (\$1200)(0.9246) = \$32{,}299$$

Similarly, calculate to obtain the numbers in the following table.

years in service	EUAC
1	$52,300
2	$37,017
3	$32,299
4	$30,207
5	$29,152
6	$28,602
7	$28,335
8	$28,234
9	$28,240
10	$28,312

Replace after 8 yr.

34. Assume the head and horsepower data are already reflected in the hourly operating costs.

Let N = no. of hours operated each year.

$$\text{EUAC(A)} = (\$3600 + \$3050)(A/P, 12\%, 10)$$
$$- (\$200)(A/F, 12\%, 10) + 0.30N$$
$$= (\$6650)(0.1770) - (\$200)(0.0570) + 0.30N$$
$$= 1165.65 + 0.30N$$
$$\text{EUAC(B)} = (\$2800 + \$5010)(A/P, 12\%, 10)$$
$$- (\$280)(A/F, 12\%, 10) + 0.10N$$
$$= (\$7810)(0.1770) - (\$280)(0.0570) + 0.10N$$
$$= 1366.41 + 0.10N$$

$$\text{EUAC(A)} = \text{EUAC(B)}$$
$$1165.65 + 0.30N = 1366.41 + 0.10N$$

$$N = \boxed{1003.8 \text{ hr}}$$

35. (a) From Eq. 63.78,

$$\frac{T_2}{T_1} = 0.88 = 2^{-b}$$
$$\log 0.88 = -b \log 2$$
$$-0.0555 = -(0.3010)b$$
$$b = 0.1843$$
$$T_4 = (6)(4)^{-0.1843} = \boxed{4.65 \text{ wk}}$$

(b) From Eq. 63.79,

$$T_{6-14} = \left(\frac{6}{1 - 0.1843}\right)$$
$$\times \left(\left(14 + \tfrac{1}{2}\right)^{1-0.1843} - \left(6 - \tfrac{1}{2}\right)^{1-0.1843}\right)$$
$$= \left(\frac{6}{0.8157}\right)(8.857 - 4.017)$$
$$= \boxed{35.6 \text{ wk}}$$

36. First check that both alternatives have an ROR greater than the MARR. Work in thousands of dollars.

Evaluate alternative A.

$$P(\text{A}) = -\$120 + (\$15)(P/F, i\%, 5)$$
$$+ (\$57)(P/A, i\%, 5)(1 - 0.45)$$
$$+ \left(\frac{\$120 - \$15}{5}\right)(P/A, i\%, 5)(0.45)$$
$$= -\$120 + (\$15)(P/F, i\%, 5)$$
$$+ (\$40.8)(P/A, i\%, 5)$$

Try 15%.

$$P(\text{A}) = -\$120 + (\$15)(0.4972) + (\$40.8)(3.3522)$$
$$= \$24.23$$

Try 25%.

$$P(\text{A}) = -\$120 + (\$15)(0.3277) + (\$40.8)(2.6893)$$
$$= -\$5.36$$

Since $P(\text{A})$ goes through 0,

$$(\text{ROR})_\text{A} > \text{MARR} = 15\%$$

Next, evaluate alternative B.

$$P(\text{B}) = -\$170 + (\$20)(P/F, i\%, 5)$$
$$+ (\$67)(P/A, i\%, 5)(1 - 0.45)$$
$$+ \left(\frac{\$170 - \$20}{5}\right)(P/A, i\%, 5)(0.45)$$
$$= -\$170 + (\$20)(P/F, i\%, 5)$$
$$+ (\$50.35)(P/A, i\%, 5)$$

Try 15%.

$$P(\text{B}) = -\$170 + (\$20)(0.4972) + (\$50.35)(3.352)$$
$$= \$8.72$$

Since $P(\text{B}) > 0$ and will decrease as i increases,

$$(\text{ROR})_\text{B} > 15\%$$

ROR > MARR for both alternatives.

Do an incremental analysis to see if it is worthwhile to invest the extra $\$170 - \$120 = \$50$.

$$P(\text{B} - \text{A}) = -\$50 + (\$20 - \$15)(P/F, i\%, 5)$$
$$+ (\$50.35 - \$40.8)(P/A, i\%, 5)$$

Try 15%.

$$P(\text{B} - \text{A}) = -\$50 + (\$5)(0.4972)$$
$$+ (\$9.55)(3.3522)$$
$$= -\$15.50$$

Since $P(\text{B}-\text{A}) < 0$ and would become more negative as i increases, the ROR of the added investment is $< 15\%$.

Alternative A is superior.

37. Use the year-end convention with the tax credit. The purchase is made at $t = 0$. However, the tax credit is received at $t = 1$ and must be multiplied by $(P/F, i\%, 1)$.

(Note that 0.0667 is actually ²/₃ of 10%.)

$$P = -\$300,000 + (0.0667)(\$300,000)(P/F, i\%, 1)$$
$$+ (\$90,000)(P/A, i\%, 5)(1 - 0.48)$$
$$+ \left(\frac{\$300,000 - \$50,000}{5}\right)(P/A, i\%, 5)(0.48)$$
$$+ (\$50,000)(P/F, i\%, 5)$$
$$= -\$300,000 + (\$20,000)(P/F, i\%, 1)$$
$$+ (\$46,800)(P/A, i\%, 5)$$
$$+ (\$24,000)(P/A, i\%, 5)$$
$$+ (\$50,000)(P/F, i\%, 5)$$

By trial and error,

i	P
10%	$17,616
15%	-$20,412
12%	$1448
13%	-$6142
12¼%	-$479

i is between 12% and 12¹/₄%.

38. Assume loan payments are made at the end of each year. Find the annual payment.

$$\text{payment} = (\$2,500,000)(A/P, 12\%, 25)$$
$$= (\$2,500,000)(0.1275)$$
$$= \$318,750$$
$$\text{distributed profit} = (0.15)(\$2,500,000)$$
$$= \$375,000$$

After paying all expenses and distributing the 15% profit, the remainder should be 0.

$$0 = \text{EUAC} = \$20,000 + \$50,000 + \$200,000$$
$$+ \$375,000 + \$318,750 - \text{annual receipts}$$
$$- (\$500,000)(A/F, 15\%, 25)$$
$$= \$963,750 - \text{annual receipts}$$
$$- (\$500,000)(0.0047)$$

This calculation assumes $i = 15\%$, which equals the desired return. However, this assumption only affects the salvage calculation, and since the number is so small, the analysis is not sensitive to the assumption.

$$\text{annual receipts} = \$961,400$$

The average daily receipts are

$$\frac{\$961,400}{365} = \$2634$$

Use the expected value approach. The average occupancy is

$$(0.40)(0.65) + (0.30)(0.70) + (0.20)(0.75)$$
$$+ (0.10)(0.80) = 0.70$$

The average number of rooms occupied each night is

$$(0.70)(120 \text{ rooms}) = 84 \text{ rooms}$$

The minimum required average daily rate per room is

$$\frac{\$2634}{84} = \boxed{\$31.36}$$

39. $$\text{annual savings} = \left(\frac{0.69 - 0.47}{1000}\right)(\$3,500,000) = \$770$$
$$P = -\$7500 + (\$770 - \$200 - \$100)$$
$$\times (P/A, i\%, 25) = 0$$
$$(P/A, i\%, 25) = 15.957$$

Searching the tables and interpolating,

$$i \approx \boxed{3.75\%}$$

Plant Design

40. Work in millions of dollars.

$$P(A) = -(\$1.3)(1 - 0.48)(P/A, 15\%, 6)$$
$$= (\$1.3)(0.52)(3.7845)$$
$$= -\$2.56 \quad [\text{millions}]$$

Since this is an after-tax analysis and since the salvage value was mentioned, assume that the improvements can be depreciated.

Use straight line depreciation.

Evaluate alternative B.

$$D_j = \tfrac{2}{6} = 0.333$$
$$P(B) = -\$2 - (\$0.20)(\$1.3)(1 - 0.48)(P/A, 15\%, 6)$$
$$- (\$0.15)(1 - 0.48)(P/A, 15\%, 6)$$
$$+ (\$0.333)(0.48)(P/A, 15\%, 6)$$
$$= -\$2 - (\$0.20)(\$1.3)(0.52)(3.7845)$$
$$- (\$0.15)(0.52)(3.7845)$$
$$+ (\$0.333)(0.48)(3.7845)$$
$$= -\$2.206 \quad [\text{millions}]$$

Next, evaluate alternative C.

$$D_j = \frac{1.2}{3} = 0.4$$
$$P(C) = -(\$1.2)\big(1 + (P/F, 15\%, 3)\big)$$
$$- (\$0.20)(\$1.3)(1 - 0.48)$$
$$\times \big((P/F, 15\%, 1) + (P/F, 15\%, 4)\big)$$
$$- (\$0.45)(\$1.3)(1 - 0.48)$$
$$\times \big((P/F, 15\%, 2) + (P/F, 15\%, 5)\big)$$
$$- (\$0.80)(\$1.3)(1 - 0.48)$$
$$\times \big((P/F, 15\%, 3) + (P/F, 15\%, 6)\big)$$
$$+ (\$0.4)(0.48)(P/A, 15\%, 6)$$
$$= -(\$1.2)(1.6575)$$
$$- (\$0.20)(\$1.3)(0.52)(0.8696 + 0.5718)$$
$$- (\$0.45)(\$1.3)(0.52)(0.7561 + 0.4972)$$
$$- (\$0.80)(\$1.3)(0.52)(0.6575 + 0.4323)$$
$$+ (\$0.4)(0.48)(3.7845)$$
$$= -\$2.436 \quad [\text{millions}]$$

> Alternative B is superior.

41. This is a replacement study. Since production capacity and efficiency are not a problem with the defender, the only question is when to bring in the challenger.

Since this is a before-tax problem, depreciation is not a factor, nor is book value.

The cost of keeping the defender one more year is

$$\text{EUAC(defender)} = \$200,000 + (0.15)(\$400,000)$$
$$= \$260,000$$

For the challenger,

$$\text{EUAC(challenger)}$$
$$= (\$800,000)(A/P, 15\%, 10) + \$40,000$$
$$+ (\$30,000)(A/G, 15\%, 10)$$
$$- (\$100,000)(A/F, 15\%, 10)$$
$$= (\$800,000)(0.1993) + \$40,000$$
$$+ (\$30,000)(3.3832)$$
$$- (\$100,000)(0.0493)$$
$$= \$296,006$$

Since the defender is cheaper, keep it. The same analysis next year will give identical answers. Therefore, keep the defender for the next 3 years, at which time the decision to buy the challenger will be automatic.

Having determined that it is less expensive to keep the defender than to maintain the challenger for 10 years, determine whether the challenger is less expensive if retired before 10 years.

If retired in 9 years,

$$\text{EUAC(challenger)} = (\$800,000)(A/P, 15\%, 9) + \$40,000$$
$$+ (\$30,000)(A/G, 15\%, 9)$$
$$- (\$150,000)(A/F, 15\%, 9)$$
$$= (\$800,000)(0.2096)$$
$$+ \$40,000 + (\$30,000)(3.0922)$$
$$- (\$150,000)(0.0596)$$
$$= \$291,506$$

Similar calculations yield the following results for all the retirement dates.

n	EUAC
10	$296,000
9	$291,506
8	$287,179
7	$283,214
6	$280,016
5	$278,419
4	$279,909
3	$288,013
2	$313,483
1	$360,000

Since none of these equivalent uniform annual costs is less than that of the defender, it is not economical to buy and keep the challenger for any length of time.

> Keep the defender.

64 Engineering Law

PRACTICE PROBLEMS

Company Ownership

1. List the different forms of company ownership. What are the advantages and disadvantages of each?

General Contracts

2. Define the requirements for a contract to be enforceable.

3. What standard features should a written contract include?

Consulting Fee Structure

4. Describe the ways a consulting fee can be structured.

5. What is a retainer fee?

SOLUTIONS

1. The three different forms of company ownership are the (1) sole proprietorship, (2) partnership, and (3) corporation.

A *sole proprietor* is his or her own boss. This satisfies the proprietor's ego and facilitates quick decisions, but unless the proprietor is trained in business, the company will usually operate without the benefit of expert or mitigating advice. The sole proprietor also personally assumes all the debts and liabilities of the company. A sole proprietorship is terminated upon the death of the proprietor.

A *partnership* increases the capitalization and the knowledge base beyond that of a proprietorship, but offers little else in the way of improvement. In fact, the partnership creates an additional disadvantage of one partner's possible irresponsible actions creating debts and liabilities for the remaining partners.

A *corporation* has sizable capitalization (provided by the stockholders) and a vast knowledge base (provided by the board of directors). It keeps the company and owner liability separate. It also survives the death of any employee, officer, or director. Its major disadvantage is the administrative work required to establish and maintain the corporate structure.

2. To be legal, a contract must contain an *offer*, some form of *consideration* (which does not have to be equitable), and an *acceptance* by both parties. To be enforceable, the contract must be voluntarily entered into, both parties must be competent and of legal age, and the contract cannot be for illegal activities.

3. A written contract will identify both parties, state the purpose of the contract and the obligations of the parties, give specific details of the obligations (including relevant dates and deadlines), specify the consideration, state the boilerplate clauses to clarify the contract terms, and leave places for signatures.

4. A consultant will either charge a fixed fee, a variable fee, or some combination of the two. A one-time fixed fee is known as a *lump-sum fee*. In a *cost plus fixed fee* contract, the consultant will also pass on certain costs to the client. Some charges to the client may depend on other factors, such as the salary of the consultant's staff, the number of days the consultant works, or the eventual cost or value of an item being designed by the consultant.

5. A *retainer* is a (usually) nonreturnable advance paid by the client to the consultant. While the retainer may be intended to cover the consultant's initial expenses until the first big billing is sent out, there does not need to be any rational basis for the retainer. Often, a small retainer is used by the consultant to qualify the client (i.e., to make sure the client is not just shopping around and getting free initial consultations) and as a security deposit (to make sure the client does not change consultants after work begins).

65 Engineering Ethics

PRACTICE PROBLEMS

(Note: Each problem has two parts. Determine whether the situation is (or can be) permitted legally. Then, determine whether the situation is permitted ethically.)

1. (a) Was it legal and/or ethical for an engineer to sign and seal plans that were not prepared by him or prepared under his responsible direction, supervision, or control?

(b) Was it legal and/or ethical for an engineer to sign and seal plans that were not prepared by him but were prepared under his responsible direction, supervision, and control?

2. Under what conditions would it be legal and/or ethical for an engineer to rely on the information (e.g., elevations and amounts of cuts and fills) furnished by a grading contractor?

3. Was it legal and/or ethical for an engineer to alter the soils report prepared by another engineer for his client?

4. Under what conditions would it be legal and/or ethical for an engineer to assign work called for in his contract to another engineer?

5. A licensed professional engineer was convicted of a felony totally unrelated to his consulting engineering practice.

(a) What actions would you recommend be taken by the state registration board?

(b) What actions would you recommend be taken by the professional or technical society (e.g., ASCE, ASME, IEEE, NSPE, etc.)?

6. An engineer came across some work of a predecessor. After verifying the validity and correctness of all assumptions and calculations, the engineer used the work. Under what conditions would such use be legal and/or ethical?

7. A building contractor made it a policy to provide cellular car telephones to the engineers of projects he was working on. Under what conditions could the engineers accept the telephones?

8. An engineer designed a tilt-up slab building for a client. The design engineer sent the design out to another engineer for checking. The checking engineer himself sent the plans to a concrete contractor for review. The concrete contractor made suggestions that were incorporated into the recommendations of the checking contractor. These recommendations were subsequently incorporated into the plans by the original design engineer. What steps must be taken to keep the design process legal and/or ethical?

9. A consulting engineer registered his corporation as "John Williams, P.E. and Associates, Inc." even though he had no associates. Under what conditions would this name be legal and/or ethical?

10. When it became known that a chemical plant was planning on producing a toxic product, an engineer wrote to the local newspaper condemning the action. Under what conditions would this action be legal and/or ethical?

11. An engineer signed a contract with a client. The fee the client agreed to pay was based on the engineer's estimate of time required. The engineer was able to complete the contract satisfactorily in half the time he expected. Under what conditions would it be legal and/or ethical for the engineer to keep the full fee?

12. After working on a project for a client, the engineer was asked by a competitor of the client to perform design services. Under what conditions would it be legal and/or ethical for the engineer to work for the competitor?

13. Two engineers submitted bids to a prospective client for a design project. The client told engineer A how much engineer B had bid and invited engineer A to beat the amount. Under what conditions could engineer A legally/ethically submit a lower bid?

14. A registered civil engineer specializing in well-drilling, irrigation pipelines, and farmhouse sanitary systems took a booth at a county fair located in a farming town. By a random drawing, the engineer's booth was located next to a hog-breeder's booth, complete with live (prize) hogs. The engineer gave away helium balloons with his name and phone number to all visitors to the booth. Did the engineer violate any laws/ethical guidelines?

PROFESSIONAL PUBLICATIONS, INC.

15. While in a developing country supervising construction of a project he designed, an engineer discovered his client's project manager was treating local workers in an unsafe and inhuman (but for that country, legal) manner. When he objected, the client told the engineer to mind his own business. Later, the local workers asked the engineer to participate in a walkout and strike with them.

(a) What legal/ethical positions should the engineer take?

(b) Should it have made any difference if the engineer had or had not yet accepted any money from the client?

16. While working for a client, an engineer learns confidential knowledge of a proprietary production process being used by the client's chemical plant. The process is clearly destructive to the environment, but the client will not listen to the objections of the engineer. To inform the proper authorities will require the engineer to release information that he gained in confidence. Is it legal and/or ethical for the engineer to expose the client?

17. While working for an engineering design firm, an engineer was moonlighting as a soils engineer. At night, the engineer used the facilities of his employer to analyze and plot the results of soils tests. He then used his employer's computers and word processors to write his reports. The equipment, computers, and word processors would otherwise be unused. Under what conditions could the engineer's actions be considered legal and/or ethical?

SOLUTIONS

Introduction to the Answers

Case studies in law and ethics can be interpreted in many ways. The problems presented are simple thumbnail outlines. In most real cases, there will be more facts to influence a determination than are presented in the case scenarios. In some cases, a state may have specific laws affecting the determination; in other cases, prior case law will have been established.

The determination of whether an action is legal can be made in two ways. The obvious interpretation of an illegal action is one that violates a specific law or statute. An action can also be *found to be illegal* if it is judged in court to be a breach of a written, verbal, or implied contract. Both of these approaches are used in the following solutions.

These answers have been developed to teach legal and ethical principals. While being realistic, they are not necessarily based on actual incidents or prior case law.

1. (a) Stamping plans for someone else is illegal. The registration laws of all states permit a registered engineer to stamp/sign/seal only plans that were prepared by him personally or were prepared under his direction, supervision, or control. This is sometimes called being in *responsible charge*. The stamping/signing/sealing, for a fee or gratis, of plans produced by another person, whether that person is registered or not and whether that person is an engineer or not, is illegal.

(b) The act is unethical. An illegal act, being a concealed act, is intrinsically unethical. In addition, stamping/signing/sealing plans that have not been checked violates the rule contained in all ethical codes that requires an engineer to protect the public.

2. Unless the engineer and contractor worked together such that the engineer had personal knowledge that the information was correct, accepting the contractor's information is illegal. Not only would using unverified data violate the state's registration law (for the same reason that stamping/signing/sealing unverified plans in problem 1 was illegal), but the engineer's contract clause dealing with assignment of work to others would probably be violated.

The act is unethical. An illegal act, being a concealed act, is intrinsically unethical. In addition, using unverified data violates the rule contained in all ethical codes that requires an engineer to protect the client.

3. It is illegal to alter a report to bring it "more into line" with what the client wants unless the alterations represent actual, verified changed conditions. Even when the alterations are warranted, however, use of the unverified remainder of the report is a violation of the state registration law requiring an engineer only to stamp/sign/seal plans developed by or under him. Furthermore, this would be a case of fraudulent misrepresentation unless the originating engineer's name was removed from the report.

Unless the engineer who wrote the original report has given permission for the modification, altering the report would be unethical.

4. Assignment of engineering work is legal (1) if the engineer's contract permitted assignment, (2) all prerequisites (i.e., notifying the client) were met, and (3) the work was performed under the direction of another licensed engineer.

Assignment of work is ethical (1) if it is not illegal, (2) if it is done with the awareness of the client, and (3) if the assignor has determined that the assignee is competent in the area of the assignment.

5. (a) The registration laws of many states require a hearing to be held when a licensee is found guilty of unrelated, but nevertheless unforgivable, felonies (e.g., moral turpitude). The specific action (e.g., suspension, revocation of license, public censure, etc.) taken depends on the customs of the state's registration board.

(b) By convention, it is not the responsibility of technical and professional organizations to monitor or judge the personal actions of their members. Such organizations do not have the authority to discipline members (other than to revoke membership), nor are they immune from possible retaliatory libel/slander lawsuits if they publicly censure a member.

6. The action is legal because, by verifying all the assumptions and checking all the calculations, the engineer effectively does the work himself. Very few engineering procedures are truly original; the fact that someone else's effort guided the analysis does not make the action illegal.

The action is probably ethical, particularly if the client and the predecessor are aware of what has happened (although it is not necessary for the predecessor to be told). It is unclear to what extent (if at all) the predecessor should be credited. There could be other extenuating circumstances that would make referring to the original work unethical.

7. Gifts, per se, are not illegal. Unless accepting the phones violates some public policy or other law, or is in some way an illegal bribe to induce the engineer to favor the contractor, it is probably legal to accept the phones.

Ethical acceptance of the phones requires (among other considerations) that (1) the phones be required for the job, (2) the phones be used for business only, (3) the phones are returned to the contractor at the end of the job, and (4) the contractor's and engineer's clients know and approve of the transaction.

8. There are two issues here: (1) the assignment and (2) the incorporation of work done by another. To avoid a breach, the contracts of both the design and checking engineers must permit the assignments. To avoid a violation of the state registration law requiring engineers to be in responsible charge of the work

they stamp/sign/seal, both the design and checking engineers must verify the validity of the changes.

To be ethical, the actions must be legal and all parties (including the design engineer's client) must be aware that the assignments have occurred and that the changes have been made.

9. The name is probably legal. If the name was accepted by the state's corporation registrar, it is a legally formatted name. However, some states have engineering registration laws that restrict what an engineering corporation may be named. For example, all individuals listed in the name (e.g., "Cooper, Williams, and Somerset—Consulting Engineers") may need to be registered. Whether having "Associates" in the name is legal depends on the state.

Using the name is unethical. It misleads the public and represents unfair competition with other engineers running one-person offices.

10. Unless the engineeer's accusation is known to be false or exaggerated, or the engineer has signed an agreement (e.g., confidentiality, non-disclosure, etc.) with his employer forbidding the disclosure, the letter to the newspaper is probably not illegal.

The action is probably unethical. (If the letter to the newspaper is unsigned it is a concealed action and is definitely unethical.) While whistle-blowing to protect the public is implicitly an ethical procedure, unless the engineer is reasonably certain that manufacture of the toxic product represents a hazard to the public, he has a responsibility to the employer. Even then, the engineer should exhaust all possible remedies to render the manufacture nonhazardous before blowing the whistle. Of course, the engineer may quit working for the chemical plant and be as critical as the law allows without violating engineer-employer ethical considerations.

11. Unless the engineer's payment was explicitly linked in the contract to the amount of time spent on the job, taking the full fee would not be illegal or a breach of the contract.

An engineer has an obligation to be fair in estimates of cost, particularly when the engineer knows no one else is providing a competitive bid. Taking the full fee would be ethical if the original estimate was arrived at logically and was not meant to deceive or take advantage of the client. An engineer is permitted to take advantage of economies of scale, state-of-the-art techniques, and break-through methods. (Similarly, when a job costs more than the estimate, the engineer may be ethically bound to stick with the original estimate.)

12. In the absence of a nondisclosure or noncompetition agreement or similar contract clause, working for the competitor is probably legal.

Working for both clients is unethical. Even if both clients know and approve, it is difficult for the engineer not to "cross-pollinate" his work and improve one client's position with knowledge and insights gained at the expense of the other client. Furthermore, the mere appearance of a conflict of interest of this type is a violation of most ethical codes.

13. In the absence of a sealed-bid provision mandated by a public agency and requiring all bids to be opened at once (and the award going to the lowest bidder), the action is probably legal.

It is unethical for an engineer to undercut the price of another engineer. Not only does this violate a standard of behavior expected of professionals, it unfairly benefits one engineer because a similar chance is not given to the other engineer. Even if both engineers are bidding openly against each other (in an auction format), the client must understand that a lower price means reduced service. Each reduction in price is an incentive to the engineer to reduce the quality or quantity of service.

14. It is generally legal for an engineer to advertise his services. Unless the state has relevant laws, the engineer probably did not engage in illegal actions.

Most ethical codes prohibit unprofessional advertising. The unfortunate location due to a random drawing might be excusable, but the engineer should probably refuse to participate. In any case, the balloons are a form of unprofessional advertising, and as such, are unethical.

15. (a) As stated in the scenario statement, the client's actions are legal for that country. The fact that the actions might be illegal in another country is irrelevant. Whether or not the strike is legal depends on the industry and the laws of the land. Some or all occupations (e.g., police and medical personnel) may be forbidden to strike. Assuming the engineer's contract does not prohibit his own participation, the engineer should determine the legality of the strike before making a decision to participate.

If the client's actions are inhuman, the engineer has an ethical obligation to withdraw from the project. Not doing so associates the profession of engineering with human misery.

(b) The engineer has a contract to complete the project for the client. (It is assumed that the contract between the engineer and client was negotiated in good faith, that the engineer had no knowledge of the work conditions prior to signing, and that the client did not falsely induce the engineer to sign.) Regardless of the reason for withdrawing, the engineer is breaching his contract. In the absence of proof of illegal actions by the client, withdrawal by the engineer requires a return of all fees received. Even if no fees have been received, withdrawal exposes the engineer to other delay-related claims by the client.

16. A contract for an illegal action cannot be enforced. Therefore, any confidentiality or nondisclosure agreement that the engineer has signed is unenforceable if the production process is illegal, uses illegal chemicals, or violates laws protecting the environment. If the production process is not illegal, it is not legal for the engineer to expose the client.

Society and the public are at the top of the hierarchy of an engineer's responsibilities. Obligations to the public take precedence over the client. If the production process is illegal, it would be ethical to expose the client.

17. It is probably legal for the engineer to use the facilities, particularly if the employer is aware of the use. (The question of whether the engineer is trespassing or violating a company policy cannot be answered without additional information.)

Moonlighting, in general, is not ethical. Most ethical codes prohibit running an engineering consulting business while receiving a salary from another employer. The rationale is that the moonlighting engineer is able to offer services at a much lower price, placing other consulting engineers at a competitive disadvantage. The use of someone else's equipment only compounds the problem. Since the engineer does not have to pay for using the equipment, he does not have to charge his clients for it. This places him at an unfair competitive advantage compared to other consultants who have invested heavily in equipment.

66 Engineering Licensing in the U.S.

PRACTICE PROBLEMS

There are no practice problems corresponding with Ch. 66 of the *Chemical Engineering Reference Manual*.